Urban Dynamics, Environment and Health

Braj Raj Kumar Sinha
Editor

Urban Dynamics, Environment and Health

An International Perspective

Springer

Editor
Braj Raj Kumar Sinha
Former Professor
Department of Geography
Institute of Science
Banaras Hindu University
Varanasi, Uttar Pradesh, India

ISBN 978-981-99-5743-9 ISBN 978-981-99-5744-6 (eBook)
https://doi.org/10.1007/978-981-99-5744-6

This Springer imprint is published by the registered company Springer Nature Singapore Pte Ltd.
The registered company address is: 152 Beach Road, #21-01/04 Gateway East, Singapore 189721, Singapore

Foreword by Michael E. Meadows

Our Anthropocene world is a rapidly urbanizing one and there is intense concern as to the impact of this accelerating process-on society, the economy, and on the environment. We have arguably already pushed many of the so-called planetary boundaries beyond the zone of uncertainty and, in some cases, we are perhaps even edging towards the point of no return. Geographers must be in the vanguard of research, teaching, and outreach that addresses not only the nature of human impact in our increasingly urban context but also explores the range of options that exist towards resolving the numerous challenges that arise from that impact. Studies of urbanization, and its diverse manifestations, must play a key role in helping to address those challenges towards the goal of a sustainable future.

Geography and geographers have a long and distinguished history of contributions to understanding the spatial drivers, processes, and effects of urbanization. The present volume, "Urban Dynamics, Environment and Health: An International Perspective" in this respect under the thorough and careful guidance of Prof. Braj Raj Kumar Sinha, gathers together a high calibre international team of researchers to further enlighten us as the nature of this crucial issue. The challenges of "environment" and "health", especially in the urban context, have been brought into sharp focus as a consequence of the COVID-19 pandemic, as well as the ongoing global climate and biodiversity crises. Studies of urbanization by geographers, with their uniquely spatial perspective, are important components of the engine of scientific understanding and, when the economic, environmental, and health challenges associated with rapid urbanization are considered, the need for such studies becomes all the more urgent.

In compiling research studies from a range of cities that span the global north and south, this volume offers a timely and authoritative set of perspectives. The chapters in this volume testify to the role that geographers can play in addressing the multiple challenges of urbanization and I congratulate Prof. Braj Raj Kumar Sinha and the team on such a comprehensive set of studies in the quest for a more sustainable urban future to ensure the well-being of all people.

November 2022

Prof. Michael E. Meadows
B.Sc. (Hons) Sussex, Ph.D. Cantab.
FSSAG, FRGS, FRSSAf, FAAS,
MEA, FGSG;
President: International Geographical
Union, 2020-2024.
Department of Environmental
& Geographical Science
University of Cape Town
Cape Town, South Africa

School of Geographical
and Oceanographic Sciences
Nanjing University
Nanjing, PR China

Foreword by George Pomeroy

Born of the same intellectual godfather, the late Prof. Dr. Ashok K. Dutt (Professor Emeritus, Department of Geography & Planning, the University of Akron, Ohio, USA) and given his recent passing, it is an incredibly heartfelt honour to write the foreword of this valuable book edited by Prof. (Dr.) Braj Raj Kumar Sinha. Indeed, Dr. Dutt not only introduced us to each other, but he was also the common thread in nearly all of our early academic interactions. Our earliest professional collaborative bonds were tied in the spaces of the Asian Urbanization Conferences and the American Association of Geographers (AAG) annual meetings. Our friendship also developed in those same spaces and provided opportunities for the exchange of novel ideas in the field of urban geography.

In addition to sharing Dr. Dutt as a professional mentor, I've had the pleasure of knowing and working with Prof. Sinha for two decades. A productive and engaging scholar, with a truly global network of intellectual connections, he is at every turn conscientious to his professional colleagues and his students. An additional reward in writing this foreword comes from having met many of the contributors on several occasions. Furthermore, the contributors, coming from 18 different countries, bring a very international and multidisciplinary perspective.

Broadly speaking, this timely volume synthesizes the three important topics of urban dynamics, environment, and health. The four parts, encompassing 32 chapters, span urban dynamics, urban environments, concepts related to urban health and urban development, planning, and policies. The authors including Dr. Sinha have worked hard in developing new and fresh insights into their chapters. The geographic focus and scale varies, some from the local (small towns, the urban fringe), to the city, to metropolitan region, and then to the urban system. While there is a decidedly South and Southeast Asian focus to the book, perspectives are diverse including European (France, Spain, Belgium, Poland, Romania, and the United Kingdom), North American (USA), African (Nigeria & Ghana), Australian, and Chinese case studies. A wide range of methods are employed across the 32 chapters and showcase the expertise brought to bear by these authors. Each chapter offers deep insight into its respective topic and geographic setting. The book offers rich lessons for those wanting to explore urban dynamics, the environment, and health, as well as the linkages between these topics. Indeed, this book is especially timely. We are at the intersection of food insecurity, the need for improved and equitable quality of life, and environmental threats, with climate change looming most of all. We arrive at this intersection at a time where urbanization is still inexorably and haphazardly spreading across the landscape. The scholarship contained in this volume speaks clearly to addressing these important issues.

It will prove useful and rewarding to urbanists, public health professionals, and planning practitioners. The authors and editor have contributed to a commendable edited book and one that honours the legacy of Prof. (Dr.) Ashok K. Dutt.

December 2022

Prof. George Pomeroy
Professor of Geography
Shippensburg University
Shippensburg, Pennsylvania, USA

Preface

Urban dynamics, environment, and health are the three major global challenges for the twenty-first century as the interaction of various elements of them on one hand provides the means of progress, prosperity, peace, and wellness, but on the other hand it leads to worsen the urban environmental conditions in several forms which in turn harms the health and quality of life of the urban people. This scenario is expected to further continue due to the uninterrupted rise in urban population through ongoing urbanization and the consequences emerging out of this are to be faced by the present and future urban generations at local, regional, and global levels. These three themes comprise a complex and multidimensional concept and hence become difficult to study from spatial and temporal angles and remain the focal areas of research in the field of urban geography and other allied subjects from the perspective of social well-being.

The available literature related to urban dynamics, environment, and health shows that several scholars from different disciplines have worked on these themes but their works are in piecemeal form and with certain disciplinary limitations. This volume is an attempt to cover the above three major challenging global urban issues together comprehensively in one volume to provide the first systematic in-depth study with a view to open up new opportunities to benefit the present and future global scholars for further research in the field of urban studies. For this the editor collected and compiled well crafted scholarly contributions of distinguished scholars from different countries of the world to fill the gap in literature on urban dynamics, environment, and health. In fact, this book takes a comprehensive view of different dimensions of urban problems and becomes unique in respect of international perspective.

This book comprises thirty-two well prepared important chapters of eminent authors from across different countries of the world and these are nicely organized into four thematic parts: urban dynamics; urban environmental issues; urban health problems; and urban development, planning, and policies as other related issues to enable readers to easily understand the interconnections of various key elements of each urban-related topical issue locally, regionally and globally.

Part I includes ten well prepared papers and presents an extensive global picture of urban dynamics.

"Introduction: A Broad Perspective on the Concepts of Urban Dynamics, Environment, and Health" as the first chapter by **the Editor** is of special interest to the scholarly readers engaged in teaching, research, writing books and research papers for publication, students pursuing degree-oriented courses at graduation and post graduation levels regardless of their disciplinary affinities as it addresses in greater detail the urban dynamics-related concepts, fundamental aspects of urban environment, concepts related to urban health, and urban dynamics-environment-health linkages including the way forward and concluding remarks as a whole. Such a comprehensive picture of the three major challenging global urban issues in a holistic form is scarcely available in the academic world.

The work of **Yves Boquet (France)** on "Sustainable Urbanization in Southeast Asian Megacities: The Contrasting Cases of Singapore and Manila" is an outstanding contribution to this volume and is of great academic interest. This paper attempts to feature the major characteristics of Singapore and Manila as the examples of Asian megacities. This paper first examines the dominant concepts of sustainability in general and urban sustainability in particular, then it proceeds to highlight some of the achievements that have built Singapore into an international model for urban sustainability policies, especially in respect of urban mobilities and the experiences in city greening. It also takes a critical look at the multiple failures seen in metropolitan Manila, which may be considered as almost the complete opposite of Singapore, with multiple problems such as heavy road congestion, housing woes, air and water pollution, or flooding. In the case of Manila megacity, he stressed on the fulfilment of the needs of the urban poor as sustainable urbanization provides economic and environmental opportunities for better life of the common urban population in general and the disadvantaged subgroups of urban population in particular.

Krishna Prasad Poudel and Upendra Paudel (Nepal) took into consideration the increasing trend of urbanization in Nepal as a challenging issue from the point of view of sustainable urban development. According to them urbanization in Nepal is a new phenomenon because prior to 1951 there were only few towns and they were confined only in the Kathmandu Valley. The development of such towns was mainly due to trade and commercial activities. The actual urbanization in Nepal started after the democratic movement of 1951 and by 2020 the urban population of Nepal reached 30 million and such an increase in urban population has transformed the whole picture of the country as the urban centres have to face several urban problems. The trend of urbanization has been very fast particularly after 2001 and has caused a severe lack of social amenities. Their work is totally based on the published sources facts and figures and they have tried to critically analyse the trend of urbanization, urbanization processes, and population living within the constituent urban centres and finally they have suggested to follow objective-based planning for proper development of the urban centres.

The study on changes in urban residential space occurring due to several factors of urban dynamics in the era of fast urbanization is vital from the point of view of searching the ways to solve the problems of extremely high demand for affordable housing especially in a metropolis. The chapter of **Mihaela-Daniela Preda (Romania)** in this direction is a very significant contribution to this volume. Her study is mainly based on the respondent's opinion and this has enriched the contents of the chapter. Her study introduces to feature the characteristics of the residential space in Bucharest City during the pre-and post-communist period, the main dysfunctions related to the quality of housing which appeared in the new residential areas especially in the peri-urban area, and also to feature the prospective ways to solve the emerging issues of urban housing. The study found local authorities and political decision-makers as the chief actors who could find solutions regarding the reported problems concerning the changes and challenges of urban residential space in Bucharest City. The approaches adopted in this study can be applicable to the similar research in case of other big cities of the world.

The contribution of **Philippe Grall, Martin Duplantier (France), and Camille Kamga (USA)** on "Rational Use of Urban Land for Sustainable Urban Growth: A Case Study of Metropolization in France" is of great importance to this part of the book. The origin of this chapter is very much related to the consequences of the early 2020 COVID-19 pandemic which affected almost the whole world and exposed worldwide sanitary crisis. Rational utilization of urban land is understood to play an immense role in sustainable urban growth determining better environment, economy, health, and better urban society. They have tried to demonstrate the real picture of Metropolization in France as the Metropolization provides the opportunity of economy that becomes crucial in transforming the overall situation of the Metropolitan areas. They chose Bordeaux a metropolitan city from France as a case study. They stressed on the policy recommendation for rational utilization of urban land and introduced 29 points as directions and recommendations to launch the debate with policy makers. They are of the opinion that sustainable urban growth highly depends on proper use of urban land as well as on better coordination and organization in different urban sectors. All this will lead to better environment, quality of life, and better health of the metropolitan inhabitants.

"Accessibility, Land Use, and Land Economics: A Factor Analysis Approach in Butwal Urban Area of Nepal" is also an important contribution to this volume. Urban land use and land use planning are generally the important means of reaching physical, economic, and social community goals. In this paper, **Chhabi Lal Chidi (Nepal)** has applied factor analysis approach to assess the dimension of relationship between accessibility, urban land use, and urban land economics. For this, he selected the historically important Butwal Urban Area located at the break of bulk point between Hill and Tarai region of Nepal as case study. His study is exclusively based on the topographic maps of the Survey Department of the Government of Nepal (scale 1:25000), Google Earth images, aerial photos, and city maps. To find the detailed picture of land use and land economics he randomly selected fifty sampled plots from the urban area of Butwal Municipality. These sampled areas were the distinct units surrounded by streets, trails, fences, streams, etc. After correlation matrix he found that the urban land use pattern such as urban functions, industrial intensity, and institutional intensity is related to radial and travel distance to CBD while the land economics and land value is mostly related to travel distance to CBD rather than the radial distance to CBD.

Land is very vital natural resource and its use in fringe area of a city is a matter of concern for the planners, academicians, local authorities, and the governments because land in fringe area is used in a haphazard manner which creates several problems. Since most of the open land of the fringe area always remains under the process of conversion into built-up area due to the effect of expansion of the main city, land use pattern of a fringe area becomes an indicator of urbanization. Looking into the ongoing transformation in the fringe area land use the authors- **Chandan Kumar kharwar and Braj Raj Kumar Sinha (India)** selected urban fringe area of Varanasi City to study the spatial and temporal changes of land use pattern with a view to develop judicious planning for better fringe area development. For this, the authors applied appropriate cartographic tools and techniques for mapping the changes in the pattern of land use over the period of twenty years between 1991 and 2011. Results show that there has been 33.78 % decline in the area under agriculture. The waste land and water bodies have also declined, while fallow land and vegetation cover have slightly increased. In such conditions careful planning and management of land use pattern of the urban fringe is essential.

"An Evidence-Based Study on Locational Attributes, Space Use, and Operational Characteristics of Online Shopping in Dhaka City Corporation Areas" as carried out by **Mamata Jafrin Mouli, Ishrat Islam, and Marufa Khan (Bangladesh)** is quite distinct from other urban dynamics-related papers as it focuses on a new and an emerging subject of research in geography. Online shopping has a significant impact on social, economic, and environmental aspects of a city people. Taking this into consideration they selected Dhaka (the capital of Bangladesh) one of the fastest growing cities of the world to explore the locational characteristics and operation system of the online shops as well as their potentials and challenges in the two city corporation areas of Dhaka. Their research is the result of the information collected through a questionnaire survey conducted on the selected 81 online shops between 2017 and 2018. The findings related to locational attributes suggest that the size of office, storage space, and their locations differ from that of the traditional commercial activities. Motorcycles were found as the most convenient vehicle in providing services and delivering the products within the city area, while couriers and trucks were found as the major means of delivering services and products outside the city area. Finally, they suggested undertaking the innovative and integrated policies, rules, and national guidelines for better e-commerce-related business and to ensure sustainable growth and efficiency of the city in future.

The study of **Daniela Zamfir and Ilinca Valentina Stoica (Romania)** on "Demographic Changes and Challenges of Small Towns in Romania" is another important addition to this edited book because small towns in Romania are considered to be the most vulnerable to structural transformations and ongoing demographic changes. They are also the basis of the urban network and are very heterogeneous in terms of economic and demographic characteristics. Several small towns receive significant inflows of younger groups and enjoy favourable situations due to their location in the surroundings of cities, whereas others developed new economic activities to attract people. Looking into these characteristics of the Romanian small towns the authors aimed to analyse the demographic changes which took place and the challenges that emerged during the post-communism. For this, they applied population dynamics, vitality index, net migration rate, old-age dependency ratio, and the composite index of demographic vulnerability. After analysis they found a general tendency of demographic decline and ageing among the small towns of Romania owing to various causes of urban dynamics and hence the demographic mosaic confirms the divergent trajectories of small towns in post-communist Romania. In view of the magnitude of population decline and ageing in recent years, they emphasized on the need for designing a growth-oriented future planning, policies, and strategies to deal with the challenges of demographic change from the perspective of balanced growth of the Romanian small towns.

The contribution of **Daniela Dumbraveanu (Romania)** on the Romanian "urban eventscape" is new and interesting to this volume and for the academic scholars because urban eventscape nowadays is gaining importance as one of the important commercial activities worldwide. The work of Dumbraveanu in this direction is appreciable as only few academicians including geographers have made an attempt to carry out research on urban eventscape. According to her "urban eventscape" is one of the fastest growing and most dynamic forms of post-communist tourism in Romania and plays an important role in local development. In light of such facts, she made an attempt to analyse how dynamic event tourism has its location in selected urban units of the country; assess the extent to which Romania has managed to develop a particular "eventscape"; discuss and analyse ways in which events and event tourism have contributed to shaping selected "festivalscapes". Her research is in fact centred on Sighisoara and Cluj-Napoca nationally relevant festivalscapes as post-communist tourist destinations. Other "eventful cities", such as Sibiu and Alba Iulia follow to develop "festivalscapes" on the pattern of the Sighisoara and Cluj-Napoca. Romania currently experiences a positive trend in the evolution of the event industry and event tourism by indicating a sign for further development.

Part II incorporates the description and analysis of seven chapters relating to different urban environmental issues and their brief picture is as follows:

The study of **Rini Rachmawati, Ulfah Choirunnisa, Qorih Imami, Lely A. Nasution (Indonesia)** on "Social, Cultural, and Information and Communication Technology Transformation towards Smart Environment: Lessons Learned from the City of Magelang, Indonesia" is of great academic interest and a valuable contribution to this part of the volume. The scope of their study is very vast as it encompasses social, cultural, and information and communication technology transformation as four major dimensions of smart environment. Their study focuses first on the prevailing condition of smart city and smart environment of Singapore, Hong Kong, and India, then their detailed study concentrates on the City of Magelang, Indonesia. For this, they gathered data undergoing in-depth interviews with several local officials in charge of environmental management and communities. The study mainly incorporates the smart environment-related programmes and activities as planned and implemented by both the local government and communities. The result of their research shows that the implementation of the smart environment-related programmes, plans, and policies in the city of Magelang work well as indicated by the roles of both the local government and communities. Secondly, the transformation in social and cultural aspects, information and communication technology favours smart environment in the city of Magelang. This is evident from the availability of better service quality related to environmental management, the establishment of communities that provide support for environmental management, the local people who become aware of the importance of conserving environment, and the use of information and communication technology in facilitating the activities of environmental management.

A study on "A Bus Rapid Transit System to Help Alleviate Air Pollution in Kathmandu, Nepal" conducted by **Keshav Bhattarai and Ambika P. Adhikari (USA)** is very informative and important contribution to this book from the perspective of better urban environment. In this paper they have made a sincere attempt to address the average vehicular speeds on various roads, the increasing population densities from 1990 to 2022, the increasing numbers of types of vehicles registered in different years, the number of people travelling by various fleet modes in the Kathmandu Valley, pollutants generated by the total number of buses, vans (minibuses and microbuses), taxis, motorcycles, etc., per kilometre in kilogramme while ferrying passengers. All the vehicles running at either optimal or at suboptimal speeds are the source of gaseous emissions. Their study is based on the review of related literature as well as on the statistical information gathered from various published sources. Since the city leaders in the Kathmandu Valley have aspiration to make their cities smart, bus rapid transit running on electricity would be one of the options to help the Kathmandu transportation in alleviating air pollution. Bus Rapid Transit (BRT) system is expected to help reduce daily vehicular emission by approximately 7,500 tonnes. Thus, an electricity-run BRT system will support increased mobility and transportation efficiency and help reduce drastically the transportation-related air pollution in the Kathmandu Valley.

Urban poor settlements across the world are one of the important urban environmental issues of the twenty-first century as poor settlements are related to several socio-economic, environmental, and health problems and need proper attention to upgrade them. The work of Nurul **Islam Nazem and Md. Anwar Hossain (Bangladesh)** in this direction is quite important and interesting to this edited book because the identification and mapping of poor settlements in urban areas are difficult tasks but it is important from the point of view of poverty alleviation programmes. In this paper, they have focused on the methods of mapping and collection of some basic data on poor settlements in Chattagram City. They have prepared a good map of the distribution of poor settlements and applied certain criteria such as land ownership, land tenure, and housing condition in analysing the overall picture of poor settlements. Further they took into account the five major areas of concern to arrive at conclusions of the study. These were land and housing conditions, environmental services, infrastructure, and economic and social conditions. Their findings suggest that the poor settlements are scattered all over the city with more concentration within the old, central part and industrial zones and they are comparatively new. Housing and environmental services were found in most deplorable situation and land ownership and holding rights are limited and the management of the environment is poor. All these are noticed because of the very dynamic nature of Chattagram City in terms of education, tourism, commerce, industry, and port facility that attract a wide range of poor and low-income people from the country's vast rural areas.

Mumita Tanjeela's (Bangladesh) study on gender-based landslide vulnerability and role of women in landslide preparedness is another informative contribution to this edited volume. Her study is important and distinctive as it exclusively focuses on the impact of vulnerability of landslides on women as well as on the role of young women at the time of occurrence of landslides. Landslide nowadays has become a global challenge as it adversely affects every aspect of life of the people prone to landslide hazards. The participation of women during the period of landslides is no doubt commendable as they have to face several adverse conditions. In light of this, her paper addresses the nexus among climate-induced migration, poverty, and gender vulnerabilities during the occurrence of landslide hazards in Cox's Bazar district of Chottogram division, Bangladesh. The paper also throws light on formal preparedness practices of the community at risk and importance of women's involvement in this certain type of disaster preparedness activities. This study is based on the information collected through focus group discussions, in-depth interviews of women living in the community under study, key informant interviews of project officials, and participant observation of their community activities. The findings clearly indicate that gender is a critical factor in shaping women's vulnerability to any hazard. Their study also reveals that despite gender differentiation women's agency and their involvement in disaster preparedness activities have positive impacts on community resiliency. Thus women's role in emergency preparedness and responses to disaster risk reduction is not less significant than that of their male counterpart.

The topic of study of **Sufia Khanom (Bangladesh), Fiona Miller and Sandie Sauchet-Pearson (Australia)** is attractive, informative, and quite fit this part of the book because this deals with one of the important urban environmental issues. For this they chose Dhaka, the fastest growing city of Bangladesh, and considered this city as a place of development and deprivation for environmentally induced migrants. This city is characterized by the presence of poverty, informal settlements, high-rise buildings, shopping malls, restaurants, insecurity, inequity, and inequality in terms of social, cultural, economic, political, and environmental aspects, yet the city has gravitational force and pulls a large number of environmentally induced migrants. In this study they have tried to analyse how Dhaka City attracts rural environmental migrants as cheap labour due to employment opportunities especially in the textile and construction industries. Dhaka being the centre of economic opportunities compels larger section of in-migrants to live on the streets and in a place which is environmentally hazardous to health. The people generally migrate towards Dhaka with a hope of development to find relief from the stress of their precarious rural lives due to environmental calamities but their dreams remain unfulfilled. The study concludes that the uneven geography of urban growth in Dhaka is one where development and (in) security co-evolve and indeed reflect the inequalities as embedded in the development of the City.

These days affordable housing in urban areas across the world has become a matter of great concern as millions of urban people especially urban poor do not have housing affordability capacity. The rapid urbanization is the key factor in creating the overcrowding and scarcity of housing in urban areas. In this context the research of **Urmi Sengupta (U.K.)** on "Housing Quality, Affordability, and the Middle Class in Urban India" is a significant contribution to this edited volume. In her study she has tried to highlight the contradictions and paradoxes and the manner in which the concept of affordability, quality, and lifestyles are embedded in both discourse and practice in India. The study mainly concentrates on housing and macro-economy: trends and new frontiers, affordable housing development antecedents, rising number of middle-class people, and their effects on urban housing consumption and on affordability, quality, and lifestyle. Lastly, she recommends checking the unbridled market-defining of housing quality and lifestyles and for making complete provisions of affordable housing to meet the needs of housing in urban India.

Environmental pollution is a big global challenging issue for the twenty-first century as it has a very serious adverse impact on the living system of the earth. It affects every aspect of human life at local, regional, and global levels. The research paper of **Anuradha Sahay and Rashmi Ranjana (India)** in accordance with the current worldwide environmental pollution scenario is a valuable contribution to this volume from the perspective of better urban life. They have chosen Patna City, one of the most polluted cities of India, as case study to fulfil the objectives of their study. They have mainly focused on the types of environmental pollution, informal workers, informal economic activities, and impact of pollution on health of the informal workers engaged in different informal activities located at different commercial places in Patna City. Informal workers generally face health hazards like asthma, diarrhoea, hypertension, headache, respiratory diseases, vector borne diseases, hearing problem, etc. At the end of their paper they have put forth some suggestions to mitigate the effect of pollution on health of the urban people in general and informal workers in particular.

Part III includes eight chapters drawn from certain developed and developing countries and these are well written and connected with the problems of urban health.

The hospital being one of the important medical institutions plays a great role in the treatment of people who suffer from diseases, ill health, and injuries. Its contribution towards well-being of the people of a society is of paramount importance. In this direction the work of **Liliana Dumitrache and Mariana Nae (Romania)** on the changing landscape of hospitals in urban Romania is very informative. It is a rather new addition to the field of medical geography as it provides some new ideas which are quite useful for the scholars interested in further research, planners and administrators who are in the capacity of developing plans, and in executing policies from the perspective of better urban health. Their paper is an attempt to identify the current issues and driving forces of the geographical distribution and reorganization of the public and private hospitals network during the past three decades and to understand different dimensions of the changing landscape of hospitals in urban Romania. They have basically focused on hospital landscape infrastructure, privatization of Romanian hospitals, clinical activity of hospitals, and the volume-outcome relationship. Their paper demonstrates that the geographical distribution of hospitals in Romania is consistent with those of the urban areas though hospitals vary widely across the country depending on the size, specialization, and medical services they provide. Hospitalization rates in Romania are much higher compared to that of other European countries. The planning of hospitals at the spatial level is not in accordance with the choice of types of services, hence there is need for a viable planning for proper structural reforms in hospitals with a view to provide better healthcare facilities in urban Romania.

The study of **Onipede Wusu and Soniyi, R. A. (Nigeria)** is new of its kind and empirically examined a noteworthy contribution to this part of the book because few scholars have worked on sexual health behaviour of young people and hence there is scanty literature in the field of medical health. Young people's sexual health behaviour is a global challenge from the perspective of health of younger generations. The authors of this paper selected Lagos metropolis, a commercial hub of Nigeria as a case study. Their study on "Influence of Social Environment and Unconventional Substance Consumption on Young People's Sexual Health Behaviour in Lagos Metropolis, Nigeria" is based on the statistical data collected through a survey conducted among young people aged between 10 and 24 years in Lagos metropolis in 2019. They have used univariate, bivariate, multivariate, and chi-square statistical techniques to fulfil the framed objectives of the study. Their study reveals that use of modern contraceptives works as an effective tool in preventing sexually transmitted infections (STIs) and unwanted or untimely pregnancy during school environment. Sexually active males in monogamous family environment are less likely to report modern contraceptive use for STIs prevention. Similarly, those who ever used unconventional substance were more likely to use modern contraceptives to prevent STIs and unwanted pregnancy compared to those who never used. As concluding remarks they emphasized the execution of effectively proper policies that can increase the sexual subjectivity of young people to discourage use of unconventional substance as well as to check STIs in school and family environments of the Lagos City.

Mental health condition especially of the young people is a very serious global concern as many young people out of the economically more active human resources so-called demographic dividend suffer from poor mental health which keeps them away from their productive contribution to development of the society. From this point of view the work of **Alina Mareci and AncaTudoricu (Romania)** as twentieth chapter is an educative, interesting, and useful addition to this volume as this focuses on the mental health issues of teenagers and young adults ranging from 15 to 35 years of age. This age group population is deemed to be the backbone and possesses a great human wealth upon which any type of development of the concerned region or nation highly depends. To conduct a research on mental health issues is a very difficult task nevertheless the authors of this chapter have made a sincere attempt to intensively deal with the youth's perception of mental health issues in urban Romania. To accomplish the objectives of their research they employed multi stage quantitative and qualitative methodological approaches and gathered detailed information concerning mental health-related depression or sadness from more than 1000 urban respondents. Findings of their research reveal that the target groups have a solid understanding of the importance of the issue owing to their own experiences or experiences gained from their family or friends circle. However, the target group mostly lacks the mechanisms to address depression and mental illness. Lastly they suggested further research to explore the actual reasons of poor mental health and to execute concrete action plans towards improving the accessibility to both information about depression and means of specialized care.

Continuously increasing proportion of the elderly population through the process of ageing is a worldwide challenging issue from the point of view of planning for making all those health-related requirements available which can promote and maintain the health of such elderly population. This issue particularly for the developing countries is of great concern and much more challenging because of probability of concentration of more old people in future owing to the effect of expected faster ageing process. In light of this the work of **Ana-Maria Talos (Romania)** on "Active Ageing and Health Status of the Aged Population of Bucharest City, Romania" is academically very valuable and adds a new dimension to this part of the book. Her study tries to explore the underlying facts related to status of health of the aged people of Bucharest City. She first focused on the proportion of elderly population of the European countries and European Union and their expectancy of life at birth. The study is based on the primary data relating to satisfaction of health, needs, and lifestyles of the elderly collected from 400 respondents selected from different socio-economic backgrounds and spatial locations. This study reveals that gender and socio-economic variables such as education and income have greater influence on lifestyle and elderly needs and adverse health conditions have a strong effect on family activity, rate of dependency, social life, and daily routine. A futuristic planning for the wellness of the elderly population has been proposed by the author.

The paper of **Zhu Qian (Canada)** on the life satisfaction of displaced people in resettlement community in Nanjing City is another important academic contribution to this volume. Displacement of people and their adjustment to the level of their satisfaction in a new society of their destination is a challenging one. Displacement of people due to several reasons has always been a topic of academic debate as displaced people have to face several socio-economic consequences. Displacement always transforms the life of displaced people owing to conversion of their land for certain purposes. The author of this paper has taken into consideration Nanjing City of China for the assessment of displaced people's life satisfaction in government-designated urban resettlement communities after acquisition of their land. This study is based on the primary information of 98 resettled households from two urbanized towns of Nanjing however, this involves theoretical aspects also. A Likert Scale was employed in the assessment of the life satisfaction of the displaced. The study exhibits that while compensation and resettlement policy reforms have improved resettled people's material living conditions but they still struggle with urban life transformation and adaptation from the social and attitudinal perspectives. The author has suggested an integrated resettlement approach to facilitate better accessibility to social security programmes, non-agricultural employment opportunities and to address issues in identity adaptation, lifestyle transformation, and social activity participation.

"Socio-Economic Status of Mothers and Stunting among their Children of Under-Five Years Age in Urban Areas of DKI Jakarta" of **Titik Harsanti and Azka Ubaidillah (Indonesia)** is an important study from the perspective of health of the urban children. Stunting, mainly caused by malnutrition, is one of the indicators of health and still remains a global issue, although it is particularly of greater concern to poor developing countries. From this viewpoint, the present chapter on socio-economic condition of mothers and their stunting children is a lone contribution to this volume. The authors of this paper have tried to focus on the maternal factors that influence under-five years stunting children in the urban areas of Jakarta. According to them, mother's education, working status of the mother, share of family income, and purchase decision of durable goods are the major factors of stunting. This research is based on the sample statistical data collected from Jakarta Utara, Jakarta Pusat, Jakarta Timur, Jakarta Barat, and Jakarta Selatan areas. They have employed suitable statistical techniques in analysing the facts and arriving at the findings.

The study of **Prabhakar Nishad and Braj Raj Kumar Sinha (India)** is another important contribution in this volume with respect to the health of urban children. Health of children is of paramount importance to every country in the world as it is a potentially precious wealth from the point of view of progress of a nation. Child delivery, post natal check-up, and child immunization are very much related to health of the children and in this context the role of educational attainment of mothers becomes very important. Better educated mothers are generally more aware of maintaining their own health and the health of their children. They remain more conscious at the time of delivery, post natal check-up, and immunization of their children. The timing of the delivery of the baby is very important for both the mother and the baby as most of the neonatal deaths occur during this period. In light of the above the authors chose urban areas of India as a whole for demonstrating the relationship between educational attainment of mothers and place of child delivery in urban India; assessing the impact of educational attainment of mothers on immunization of children aged 12-23 months; and investigating the effect of mother's educational attainment on the post natal check-up of mother and newborn baby. This study is based on the secondary data obtained from the National Family Health Survey-4 (NFHS-4), 2015-2016. Findings of the study show that educational attainment of mothers is highly correlated with the place of delivery. The institutional child delivery is more in the case of mothers having more education. Improving maternal education is important for increasing child vaccination uptake and coverage. The role of educational attainment of mothers in postnatal check-up is also important.

A study on migration and lifestyle as carried out by **Priyanka Thakur and Braj Raj Kumar Sinha (India)** is academically another important and interesting contribution to this volume from the point of view of health. Health of migrants is always associated with their lifestyles which change through the process of migration. In fact, most of the chronic (non-infectious) diseases of migrants are the result of their way of living. From this viewpoint the authors of this study selected a city (Asansol, West Bengal) from eastern India as a case study to examine the facts of lifestyles that are related to the migrants who arrived in Asansol City from different surrounding rural and urban areas. The study is basically based on the statistical information collected through personal field survey from three hundred urban households during 2019-2020. The study is supported by the theoretical concepts also. Therefore, this study involves both theoretical and empirical approaches. The study shows that particular lifestyle varies through demographic, social, cultural, and economic conditions of the surveyed in-migrants. From health perspective it was found that the majority of in-migrants in the city generally suffer more from lifestyle-related diseases such as gastric, blood pressure, obesity, arthritis, and diabetes.

Part IV Comprises seven well written chapters by eminent scholars and exhibits the detailed picture of urban development, planning, and policies.

The concept of green city is very useful as green spaces improve the quality of the environment by reducing air pollution and enhancing the sustainability of urban ecosystem for the well-being of urban people and other urban lives. Making the city green is a dire need of today globally from the perspective of sustainable urban development, planning, and policies. In this respect, the chapter of **Upuli Perera (Sri Lanka)** jointly with **Collins Adjei Mensah (Ghana)** and **Lakshi Karunarathne** (**Sri Lanka**) on "Going Green in Urban Development: Conceptual Advancements and Applications in Selected Cities" is a very important contribution to this part of the book as they have tried to introduce some new ideas to be applied while planning for urban development. Their study is entirely based on the review of 324 academic publications related to the going green concept from the early nineteenth century to the present day twenty-first century. To fulfil their objectives they selected four cities-Milton Keynes New Town from England, climate resilient Canberra City from Australia, Accra Metropolis (the capital city of Ghana), and The Kottawa-Makumbura multi modal transport hub city from Sri Lanka as urban development case studies. Findings derived from these four case studies exhibit that going green projects have provided multiple benefits in terms of providing better housing, transport, amenities, employment, minimizing risks from climatic changes, attracting funding for cities, and increasing quality of life in urban areas.

Migration being an element of human behaviour or a part of usual human activity plays a significant role in influencing various aspects of human life and the environment of an area over certain period of time. Internal migration is the most prominent type of migration across the world which involves a large number of people who migrate from different areas within the territory of a nation. Rural-urban migration is very prominent in this type of migration and contributes a lot to the progress and development of a town or city, although it also has some adverse effects on the urban society. The contribution of **Xingna Nina Zhang (China)** in this respect is academically very important for the researchers involved in research on internal migration and development. She opined that internal migration has played an important part in an unprecedented urbanization and industrialization as well as in shaping the overall urban development across the country. Her study is basically theoretical in nature and is based on the review of major researches of several experts on internal migration around the world. In this study she has examined the major trends and characteristics of internal migration in mainland of China, the role of drivers of internal migration, and investigated links between internal migration and socio-economic development with a focus on the critical role of the state policies. She put forth suggestions that China's experiences of internal migration and urban development can be employed in policy making regarding migration management and regional development governance and can be shared by other developing countries too.

The study of **Bartosz Bartosiewicz (Poland)** and **Christopher Cusack (USA)** is an example of the unique and innovative contribution of scholars to this part of the book because this deals with a challenging global issue of shrinking cities. Their study is closely related to the cases of declining population of cities. The declining or shrinking cities (due to a decline in urban population caused by deindustrialization, decline in economic and social resource base, poor urban infrastructure leading to out-migration), nowadays especially in the United States and European countries have become a focal theme of research in geography and in other allied subjects. From this perspective, they have tried to develop new strategic approach of resiliency and redevelopment for determining solutions to check the process of counter-urbanization and to enhance the quality of life of urban population of such cities. According to them the Great Lakes manufacturing region of the United States and the central industrial region of Poland comprising multiple cities face the plight of depopulation. Taking this situation into account they have selected the two postindustrial cities-Detroit and Lodz, respectively-as case study to accomplish the objectives of their research. The results show that both Detroit and Lodz have implemented redevelopment strategies focusing on resiliency and innovation for a bright future.

The polycentric megacity region, being a bunch of interrelated multiple large urban centres, is an emerging global issue of the twenty-first century as it poses a complex problem for the urban planners, policy makers, administrators, and social scientists from the perspective of spatial planning for regional development. Despite this, the polycentric megacity region offers various social, cultural, economic, and political opportunities in the life of the urban people. In this context, the chapter of Linda **McCarthy (USA)** is of great academic importance to this section of the book. Her work is an outcome of the review of the academic- and public policy-related literatures on polycentric megacity regions, regional planning, and governance to establish the state of the theoretical and empirical research for the United States and Europe and also to develop a conceptual framework to examine polycentric megacity regions, regional planning, and governance from a comparative perspective. To deeply understand the interconnections of different elements of polycentric megacity regions she chose New York from the United States and the Rhine-Ruhr from Europe as case studies to open up the ongoing urban scenario and to establish some new ideas for further regional planning and governance towards better urban life. Hence, the results of the study of New York and the Rhine-Ruhr polycentric megacities may be of interest for policy and planning in other large polycentric megacity regions like Delhi, Beijing, and other large cities of global status.

The contribution of **Carmen Vazquez-Verela and José M. Martartínez-Navarro (Spain)** on "Medium-Sized Spanish Cities: Urban planning and Urban Policies" is academically very significant to this part of the book because the medium-sized cities generally receive less attention of scholars in comparison to the bigger cities while the contribution of medium-sized cities in the development of the nation is not less. Nowadays the life of urban people in bigger cities across the world is becoming uneasy and more challenging and as a result many people are moving to the medium-sized cities. The medium-sized cities are comparatively affordable, less expensive, less crowded, and less problematic and offer better social amenities and opportunities for the urban dwellers. Probably these are the reasons that prompted the authors to select the medium-sized Spanish cities for their study to address the dynamics of urbanization over the period of forty years. The findings of their study reveal that the medium-sized Spanish cities have undergone a dramatic change and show the trend and pattern of rapid expansion, hence there is requirement of a concrete urban planning and urban policies to enhance the sustainability of such cities for better urban life.

The research work of **Nguyen Viet Long (Vietnam) and Hana Nguyen (USA)** on the topic, "Unfolding Binh Duong's Formula to Success: A Breakthrough Strategy towards Industry 4.0, Knowledge Economy, and Sustainable City" is a literature review-based contribution in this volume. It is a skillfully crafted important chapter from the perspective of industrialization, capability building, and sustainable urban development. Their study is the first of its type and is based on documentary information and a qualitative approach, with participant observation and semi-structured interviews of sixty-four experts and leaders in Vietnam. The goal of this study is to build solid foundations, creating an innovative community and favourable business environment, increasing economic competitiveness, improving living standards, and gravitating towards the knowledge and digital economy. Binh Duong succeeded in achieving an exponential growth over the past 4 years. The success was so influential that the province aspired to re-implement and reinforce the strategy for the period of 2021-2025. This research provides the means of valuable documents and insights imperative to fulfil the process of documentary research in the fields of social sciences.

The chapter of **Ria Liza Centeno-Canlas, Joseph D. Retumban, and Custer C. Deocaris (Philippines)** on "A Future Design for the Sustainable Urban Renewal of Manila Megacity of the Philippines" is the last but academically not the least important addition to this edited volume. This is an interesting study on a megacity of global status characterized by the unique geographical location, combination of seventeen cities, historical features, and the complex social, cultural, economic, political, demographic, and environmental challenges. A continuous increase in urban population in metro Manila mainly due to rural-urban migration has created complex socio-economic, health, and environmental problems. Such problems prompted the authors to undertake this study to explore the way for sustainable urban renewal of Manila megacity of the Philippines. In this study, they have employed the modified Delphi approach to prioritize specific measures or indices that can assist in guiding for the future urban renewal of the Manila Mega City. In this approach the authors employed the 12 Key Elements and their sub-indicators to serve as basis of utilizing technologies for "future design framework" to enhance the assessment and decision-making process for urban planning. The study, however, uses limited quantitative facts in the case study approaches confining to specific areas of Manila. The authors, however, are confident that the proposed future design framework of this study can be empirically tested and applied to urban renewal assessment of Metro Manila as well as other megacities in the world.

Varanasi, India

Braj Raj Kumar Sinha

Acknowledgement

I express my sincere thanks to each of the contributors to this edited volume for producing his/her scholarly chapter on time, tremendous support, and patience in coping with the delays on my part. I would like to express my deep sense of appreciation to late Prof. P. P. Karan, an iconic social and cultural geographer of Indian origin, who worked for a very long time in the Department of Geography/Japanese Studies, University of Kentucky, Lexington, USA, for his ever encouragement with full of inspiration. I also wish to keep on record my deep sense of gratitude to late Prof. Ashok K. Dutt, an internationally reputed legendry geographer in the field of urban geography and planning, and an ambassador of Indian geography to the United States of America; a Professor Emeritus, Geography and Planning; and Professor Emeritus of Urban Studies, the University of Akron, OH 44325-5005 (USA), for his great help, affection, benevolence, good suggestions, and academic motivation. I remain ever grateful to my wife Mrs. Meera Rani Sinha and my daughters Ekta Sinha and Chaitanya Sinha for their love, constant inspiration, assistance, sacrifice, perseverance, and endurance in achieving the goal of this publication. Thanks are also due to Ms. Satvinder Kaur, Publishing Editor, Springer Nature Social Sciences and Humanities, New Delhi for accepting the manuscript, meticulously handling it, and finally publishing this book.

Braj Raj Kumar Sinha

Dr. Braj Raj Kumar Sinha
Former Professor, Department of
Geography, Institute of Science
Banaras Hindu University
Varanasi, Uttar Pradesh, India

About this Book

This comprehensive book on "**Urban Dynamics, Environment and Health: An International Perspective**" is the first of its type and provides a breadth of information on conceptual and empirical studies of various elements of these three vital urban issues. This volume is a combination of a striking collection of thirty-two well written contributions of established academicians and emerging scholars from across the globe-North America, Europe, Africa, Asia (South and South East Asia), and Australia. Involving spatial, temporal, conceptual, empirical, and multidisciplinary approaches and techniques this book provides a veritable wealth of knowledge in the field of urban, environmental, human and medical geography, and other allied subjects and thus has a tremendous scope. It presents a multidisciplinary account of the major dimensions of urbanization and urban life. All the chapters in this book are organized into four thematic parts: Part I-Urban Dynamics; Part II-Urban Environmental Issues; Part III-Urban Health Problems; and Part IV-Urban Development, Planning, and Policies to enable readers to easily understand the interconnections of various key elements of each urban-related topical issue locally, regionally, and globally.

There have been a variety of works on the urban dynamics, environment, and health but with certain disciplinary limitations. Geography, being a branch of earth and human sciences, maintains the trans-disciplinary approach in studying the whole physical (natural) and human elements and their relationship in spatial and temporal contexts towards social well-being in general, and here it takes into account the different dimensions of urban dynamics, environment, and health (the source of wealth/resources for the quality of life and happiness) in particular. The chapters have been appropriately arranged to make the book most comprehensive from theoretical, empirical, and methodological points of view. This book is quite different from other similar books because this demonstrates the numerous experiences and the cutting-edge findings of the empirical studies of the national and international contributors and provides the first systematic in-depth study of the three major and challenging urban issues of the twenty-first century together. Extensive maps, charts, diagrams, and tables as cartographic representations accompany the analysis and will facilitate the present and future global readers to understand the theme at regional, national, and international levels. Indeed, this book takes a holistic view of different dimensions of urban dynamics, urban environment, and urban health and provides fresh insights into the thought and a broad understanding of the current issues. The book will add a new facet and fill the gap of existing literature in the field of urban studies.

The entire chapters, incorporated in this book, are informative, convincing, and quite useful to understand the complex interconnectedness of different aspects of urban dynamics, environment, and health from the broader perspective of social relevance. Thus, this volume is an invaluable reference and serves the primary interest of students and scholars of urban geography, human geography, environmental geography, medical geography, sociology, economics, demography, anthropology, philosophy, biological science, architecture, and applied technology. It is particularly useful for students, researchers, and those engaged in further research on different dimensions and issues of the urban dynamics, environment, and health regardless of their disciplinary affinities.

Dr. Braj Raj Kumar Sinha
Former Professor
Department of Geography
Institute of Science
Banaras Hindu University
Varanasi, Uttar Pradesh, India

Contents

Editor and Contributors

About the Editor

Dr. Braj Raj Kumar Sinha has been Professor of Geography at Banaras Hindu University, Varanasi, India. Presently, he is Pro Vice-Chancellor of Magadh University, Bodh-Gaya (Bihar). He obtained a B.A. (Hons.) and M.A. from Patna University, India and a Ph. D. from the Maharaja Sayajirao University of Baroda, India. He has previously worked at Visva-Bharati (a Central University), Santiniketan, West Bengal from 25 Sept. 1988 to 3 June 2007 (last appointment as professor). Professor Sinha's areas of academic interest are population geography, cartography, resource geography, and development studies. He has more than 80 research papers/articles and three books to his credit with national and international publishers. Professor Sinha has been the Vice-President and member of the Executive Council of the National Association of Geographers of India. He is currently member of the Editorial Board of *Human Geographies-Journal of Studies and Research in Human Geography,* the Scientific Annals of the Alexandru Ioan Cuza (Romania) and a member of the International Board of Directors of the Asian Urban Research Association (AURA, Akron, USA). He is the recipient of national and international level prizes/awards. He has visited the Czech Republic, South Korea, Romania, UK, USA, Canada, Australia, Thailand, Indonesia, and Vietnam for academic purposes.

Contributors

Ambika P. Adhikari City of Tempe, Tempe, USA

Bartosz Bartosiewicz Institute of the Built Environment and Spatial Policy, Faculty of Geographical Sciences, University of Lodz, Lodz, Poland

Keshav Bhattarai Department of Physical Sciences, University of Central Missouri, Warrensburg, USA

Yves Boquet Department of Geography, Université de Bourgogne-Franche-Comté, Dijon, France

Ria Liza Centeno-Canlas Center for Innovation, National University, Manila, Philippines

Chhabi Lal Chidi Central Department of Geography, Tribhuvan University, Kathmandu, Nepal

Ulfah Choirunnisa Graduate School on Geography, Universitas Gadjah Mada, Yogyakarta, Indonesia

Christopher Cusack Department of Geography, Keene State College, Keene, NH, USA

Custer C. Deocaris Department of Science and Technology, Philippine Nuclear Research Institute, Diliman, Quezon City, Philippines

Daniela Dumbrăveanu University of Bucharest, Bucharest, Romania

Liliana Dumitrache Department of Human and Economic Geography, Faculty of Geography, University of Bucharest, Bucharest, Romania

Martin Duplantier French Association of Architects and Contractors (AMO), Paris, France

Philippe Grall Policy Institute for Local Officials (IFER), Paris, France

Titik Harsanti Politeknik Statistika STIS, Jakarta, Indonesia

Md. Anwar Hossain Department of Geography and Environment, University of Dhaka, Dhaka, Bangladesh

Qorih Imami Graduate School on Environmental Science, Universitas Gadjah Mada, Yogyakarta, Indonesia

Ishrat Islam Department of Urban and Regional Planning, Bangladesh University of Engineering and Technology (BUET), Dhaka, Bangladesh

Camille Kamga Urban Transportation Research Center (UTRC) at City College of New-York (CUNY), New Yok, NY, USA

Lakshi Karunarathne Department of Estate Management and Valuation, University of Sri Jayewardenepura, Nugegoda, Sri Lanka

Marufa Khan Department of Urban and Regional Planning, Bangladesh University of Engineering and Technology (BUET), Dhaka, Bangladesh

Sufia Khanom Bangladesh Institute of International and Strategic Studies (BIISS), Dhaka, Bangladesh

Chandan Kumar Kharwar Department of Geography, Institute of Science, Banaras Hindu University, Varanasi, U.P., India

Nguyen Viet Long International University, National University of Ho Chi Minh City, Ho Chi Minh City, Vietnam

Alina Mareci Faculty of Geography, University of Bucharest, Bucharest, Romania

José M. Martínez-Navarro Department of Geography, Universidad Autónoma de Madrid, Madrid, Spain

Linda McCarthy Department of Geography, University of Wisconsin-Milwaukee, Milwaukee, USA

Collins Adjei Mensah Department of Geography and Regional Planning, University of Cape Coast, Cape Coast, Ghana

Fiona Miller Department of Geography and Planning, Macquarie University, Syndey, Australia

Mamata Jafrin Mouli Department of Urban and Regional Planning, Bangladesh University of Engineering and Technology (BUET), Dhaka, Bangladesh

Mariana Nae Department of Human and Economic Geography, Faculty of Geography, University of Bucharest, Bucharest, Romania

Lely A. Nasution Graduate School on Environmental Science, Universitas Gadjah Mada, Yogyakarta, Indonesia

Nurul Islam Nazem Department of Geography and Environment, University of Dhaka, Dhaka, Bangladesh

Hana Nguyen Georgia State University, Atlanta, USA

Prabhakar Nishad Department of Geography, D. A. V. Post Graduate College, Siwan, J. P. University, Chapra, Bihar, India

Upendra Paudel Department of Geography, Prithvi Narayan Campus, Tribhuvan University, Pokhara, Nepal

Upuli Perera Department of Estate Management and Valuation, University of Sri Jayewardenepura, Nugegoda, Sri Lanka

Krishna Prasad Poudel Department of Geography Education, Tribhuvan University, Kirtipur, Kathmandu, Nepal

Mihaela-Daniela Preda Faculty of Geography, University of Bucharest, Bucharest, Romania

Zhu Qian School of Planning, University of Waterloo, Waterloo, ON, Canada

Rini Rachmawati Department of Development Geography, Faculty of Geography, Universitas Gadjah Mada, Yogyakarta, Indonesia

Rashmi Ranjana Department of Geography, College of Commerce, Arts and Science, Patliputra University, Patna, India

Joseph D. Retumban Center for Resilient Philippines, National University, Manila, Philippines

Anuradha Sahay Department of Geography, Patna University, Patna, India

Urmi Sengupta School of Natural and Built Environment, Queen's University, Belfast, Belfast, U.K.

Braj Raj Kumar Sinha Former Professor, Department of Geography, Institute of Science, Banaras Hindu University, Varanasi, UP, India

R. A. Soniyi Department of Sociology, Lagos State University, Lagos, Nigeria

Ilinca-Valentina Stoica Faculty of Geography and Interdisciplinary Centre for Advanced Research on Territorial Dynamics (CICADIT), University of Bucharest, Bucharest, Romania

Sandie Suchet-Pearson Department of Geography and Planning, Macquarie University, Syndey, Australia

Ana-Maria Talos Department of Human and Economic Geography, Faculty of Geography, University of Bucharest, Bucharest, Romania

Mumita Tanjeela Department of Sociology, East West University, Dhaka, Bangladesh

Priyanka Thakur Department of Geography, Banaras Hindu University, Varanasi, India

Anca Tudoricu Faculty of Geography, University of Bucharest, Bucharest, Romania

Azka Ubaidillah Politeknik Statistika STIS, Jakarta, Indonesia

Carmen Vázquez-Varela Department of Geography and Spatial Planning, Universidad de Castilla-La Mancha, Cuenca, Spain

Onipede Wusu Department of Sociology, Lagos State University, Lagos, Nigeria

Daniela Zamfir Faculty of Geography and Interdisciplinary Centre for Advanced Research on Territorial Dynamics (CICADIT), University of Bucharest, Bucharest, Romania

Xingna Nina Zhang Department Public Health, Policy and Systems, Institute of Population Health, University of Liverpool, Liverpool, UK

Part I
Urban Dynamics

Chapter 1
Introduction: A Broad Perspective on the Concepts of Urban Dynamics, Environment, and Health

Braj Raj Kumar Sinha

Abstract Urban dynamics, environment, and health are the three major interlinked global challenges for the twenty-first century and continue to be the focal areas of research in the field of urban geography and other allied subjects. These three themes comprise a complex and multidimensional concept and accordingly become difficult to identify, define, classify, and analyse from a spatial and temporal angle. Urban dynamics is completely linked with the environment and health aspects of urban people. The fast growing urban population and rapid expansion of the world's cities in haphazard way owing to incessant urbanization are the major causes of deterioration in the quality of the environment, health, and quality of life of the people. Such deterioration has created a big threat to the human society and civilization. In fact, urban areas in both the developed and developing countries have grown merely in population and environmental problems, not in prosperity. This scenario will further continue owing to continuous rise in urban population through different urban dynamics. Even despite this, the cities on the one hand provide the opportunity of better educational and health facilities, better human capital formation, innovation of science and technology, employment and economy for the workers, better standard of living, manufacturing and industrial growth, trade and commerce, etc., but on the other hand, unplanned and haphazard expansion of the cities and unfriendly relation between population and environment due to increasing trend of urbanization have created a big threat in the form of challenges like environmental pollution; rural–urban migration, growing slum population; loss of agricultural land, natural vegetation; traffic congestion, diseases and health, shortage of housing and potable water, sanitation and hygiene, headforemost use of energy, extreme economic inequalities, etc. Therefore, in light of the above, this chapter aims to address the urban dynamics-related concepts, fundamental aspects of urban environment, concepts related to urban health, and urban dynamics–environment–health linkages including the way forward and concluding remarks as a whole.

B. R. K. Sinha (✉)
Former Professor, Department of Geography, Institute of Science, Banaras Hindu University, Varanasi, UP 221005, India
e-mail: sinha_brk@yahoo.co.in

B. R. K. Sinha (ed.), *Urban Dynamics, Environment and Health*,
https://doi.org/10.1007/978-981-99-5744-6_1

Keyword Urban dynamics · Incessant urbanization · Ruralization · Quality of life · Haphazard expansion · Environmental problems · Urban health · Filth theory · Sustainable urban development · Globalization

1.1 A Prologue

Urban dynamics, environment, and health are the three major interlinked areas in the field of urban geography and interdisciplinary subjects and portray the wider urban issue-related concepts. The interconnection of various elements of these three areas has great bearing on the quality of life of people in space and time. The sequential arrangement of these three vital aspects in this chapter is an expression of priority action of the process of change in urban space, environment, and human health along with time.

Urban dynamics, environment, and human health are the global challenges for the twenty-first century. Cities have long been the world's economic dynamos, but today the speed and scale of their expansion in the world and especially in the developing countries are astonishing. The interaction of these three components on the one hand provides the means of progress, prosperity, peace, and wellness of the people (in terms of opportunity) but on the other hand irrational and unplanned growth of urban areas, careless and ruthless behaviour of human beings with environment to fulfil their greed have led to worsening conditions in several forms which in turn harms the health and quality of life of human population. Unplanned and haphazard expansion of urban centres and imbalanced and unfriendly relation between population and environment in the human–environment interactive arena has created a big threat to peace, prosperity, health and quality of life and successively to the human society and civilization. In fact, urban areas in both the developed and developing countries have grown merely in population and environmental problems but not much in prosperity. This scenario will further continue owing to continuous rise in urban population through different urban dynamics including rapid expansion in urban area, particularly in Asian and African countries. Currently, the world is inhabited by 55% urban population and in the next three decades the world is expected to be inhabited by a huge majority (exceeding more than two-thirds) of the world's urban population. Asia and Africa together will contribute above 74% urban population. The world's total future urban population (estimated to be more than two-thirds of the world's total population) will mount maximum pressure on all types of resources and services to meet the needs from within urban as well as from rural areas and in turn it will lead to further challenging environmental and health problems.

In this way, urban dynamics, environment, and health are the major issues and challenges to be faced by the present and future generation at the local, regional, national, and international level in the twenty-first century. Another fact is that small proportion of urban population lives in abundance and greater proportion in inadequate resources and services. In urban areas, extreme social disorder, severe class conflicts, crimes, widespread violence, urban civil war, extreme economic inequalities, much disparity in the level of urbanization, environmental and health problems do exist in different countries. Therefore, this chapter attempts to address the underlying concepts of urban dynamics, environment, and health; elucidate the synergetic relation of urban dynamics, urban environment, and urban human health with different physical and human aspects from local, regional, national, and global perspective; and throw light on the way forward concerning concrete action plan for policy programme change in both the more developed and less developed countries toward sustainable urban development and environment for better health, prosperity and quality of life of the present and future urban population. Finally, this chapter ends with the concluding remarks as a whole.

1.1.1 Urban Dynamics-Related Concepts

This section sheds light on the urbanization in space and time, concepts and meaning of dynamics, urban dynamics-cum-urban growth, challenges and opportunities, and policy measures for urban dynamics.

1.1.1.1 Geo-Historical Perspective of Urbanization

When we look into the places and time of beginning of urbanization, it is found that there is a dearth of precise statistics concerning the exact place and time of origin of urbanization and that is why the urban scholars generally take into consideration the archaeological evidences concerning technological innovations and advancement in Neolithic culture of the period between 6000 and 4000 BC as an indicator of urbanization. On this basis (Davis, 1955, pp. 429–430) stated that many towns and a few cities arose prior to the Christian era in regions around the Mediterranean ocean and in areas of Southern and Western Asia. By 3000 BC several science and technological inventions seem to have influenced Egypt, Mesopotamia, and India where true cities did exist. Second source of information regarding history of urbanization is Ritchie and Roser (2020). This source provides information that before 5000 BC there was no urban population. In 5000 BC the share of urban population in the world was only 0.10%. At Zero Point of time (i.e. between BC and AD) the world's urban population was only 1%. Prior to 1600, the share of the world population living in urban areas was less than 5% and by 1800 this share reached 7.30%. By 1900, it increased to 16.40%. United Nations (2019) data shows that the percentage of urban population of the world in 1950 was 29.61%, which became 36.60% in 1970. In 1990 it rose to 43.00%, and in 2000 it reached 46.70%. The year 2007 is considered as rural–urban divide as it marked 50.16% urban population. The year 2015 showed 53.90% urban population, whereas the year 2018 marked 55% urban population in the world. The share of population living at present in urban areas is expected to increase to 60.40% in 2030; 64.50% in 2040 and 68.40% in 2050, i.e., by the mid of the twenty-first century. It is quite apparent from the above picture that the world's urban population grew very slowly between 5000 BC and 1800 AD. The speed of growth of urban population from 1800 AD onward and till the end of 1950 (i.e., over the period of 150 years) was moderate; but the tempo of urbanization beyond 1950 and till 2018 (i.e., over the period of 68 years) remained very fast. Although the rate of urbanization (in percent) generally continued to decline except during a few particular periods of time. The interesting picture here is that the percentage growth in urban population in both the more developed and less developed regions of the world from 1950 to the present as well as to the projected period till 2050 continued and will continue to a significant increase. But such increase has always been more in the case of the more developed regions as against to that of the less developed regions. Similarly, the rate of urbanization in per cent in both the more developed and less developed regions of the world from 1950 to the present as well as to the projected period till 2050 continued to a gradual decline but such decline has always been more in case of the more developed regions compared to that of the less developed regions. It is indicative of appreciably more, or in other words, faster increase (over the period of 100 years) in the less developed regions compared to that of the more developed regions of the world. Owing to this, urbanization in the less developed or developing regions is expected to remain faster than that of the more developed regions of the world. Another interesting feature is related to the spatial variation with regard to the 2018 year level of urbanization. In this year, 82% population of Northern America; 81% population of Latin America and the Caribbean; 74% population of Europe; and 68% population of Oceania was living in different urban areas, whereas the percentage of urban population in Asia and Africa at the same time was 50 and 43, respectively. Currently Africa and Asia are urbanizing faster than other regions of the world and are projected to have 59 and 66% urban population, respectively, as a proportion to their respective total population by 2050. It is also interesting to note that in 2050, Asia and Africa together (52.10 and 22.30%, respectively) will constitute 74.40% of the total urban population of the world and the remaining geographical regions of the world will stand with only 25.60% urban population of the world by observing a decline in their shares of urban population. The above picture of urban growth is a result of the interplay of natural and migratory forces, expansion in urban area's size and conversion of rural land into urban by land reclassification scheme. In fact, future increase in the proportion of urban population (even beyond 2050) is likely to continue in all countries, though more in developing countries, in the same way unless some unexpected happenings (like Coronavirus Disease (COVID-19), fourteenth-century Plague, famine, war, and the like) take place. Therefore, the current trends, patterns, processes, and implications of rapid urban growth would further affect the environment, health, and well-being of the present and future population.

1.1.1.2 Literary Meaning of the Dynamics

Before anything else, it is imperative to understand the concept and meaning of the term "Dynamics". According to Webster's Encyclopedic Unabridged Dictionary of the English language (1996), Oxford Advanced Learner's Dictionary (2000), and The Compact Oxford Reference Dictionary (2001) the word "dynamics" is the plural form of dynamic, the meaning of which is related to the motivating or driving forces that stimulate or produce change, action or effects within a process of growth and development in any field of study. While dealing with the history of geography (Richard, 2000, pp. 176–182) used the word dynamic to denote development or change in the phenomena a geographer studies. The term "dynamic" in his view is generally used as opposite of the word "static" in regional context. The changes introduced in an area are the result of the natural and human forces which may be from within or outside the area. Kormondy (1991, pp. 168–169) used "dynamics" in the form of controlling or causal factors that make changes in ecological communities. The word "dynamics" in case of population and demographic studies connote the same meaning. Population dynamics, also known as population processes, are by far the most important among population traits of an area, so-called rural or urban. Day (1983, pp. 2–3), Newman and Matzke (1984, p. 5); and Ross (1982, p. 325) took into consideration fertility, mortality, and migration as the dynamics of population growth. These three determine the demographic character of population of a region in terms of size, rate of population increase, geographic distribution, age and sex structure, life expectancy at birth, and family composition. Similarly, Newell (1988, p. 82) stated that births and deaths determine natural change and in-migration and out-migration determine net migration (unnatural change) and all these in turn determine total population change. Zelinsky (1966, p. 10) and Weeks (1989, p. 85) regarded births, deaths, and migration as major demographic events or vital events or dynamic elements of population that change the gross number of people and affect the population size of a given area. Population growth, either in rural or urban areas, occurs as a result of the combination of these vital events. Moore and Mc Guinness (1999, p. 144) and Blanchet (1998, p. 21) noted birth (fertility), death (mortality), and migration as the three components of population change and are related to population ageing and economic growth of an area. Out of these three vital components, births and deaths cum infant mortality rate are very important and play a major role in changing the composition of all population characteristics. These are the elements of natural population growth. Birth, death, and infant mortality rates also play a crucial role in population transition. Different regions or countries experienced variations in the birth rate and death rate of their population. Such variation in natural components of population growth is generally due to variation in their condition caused by positive or negative effect of persistent surrounding environment. Lutz (2002, pp. 132–133) also emphasized the role of birth and death rates in population transition of a region. In the words of Ferris and Wilson (1987, p. 372) population dynamics conveys quantitative change in the size of population determined by number of persons, change in population of different age groups, in the ratio of sex, as well as in population behaviour of an area along with the passage of time. They also focused on fertility (reproduction), mortality, and migration in the concept of population dynamics. Some others opined that births, deaths, immigration, and emigration (the product of natality, mortality, and migration, respectively) are the vital ecological events that bring change in size of population, age structure, and density of population over a specific period of time. According to Sinha (2005, p. 413) migration plays a significant role in both the spatial and non-spatial context. In spatial context, in-migration and out-migration make changes at both the origin and destination places of migrants and in non-spatial context, in health, education, earnings, lifestyles, psychological motives, socio-economic profile, and demographic as well as environmental aspects of them. In-migration and out-migration are also important agents in diffusion of culture and in changing pattern of opportunities in space. In this way, the concept of population dynamics includes birth, death, and migration as the factors or determinants or forces or elements or components of population growth and all these bring change, through their interplay, in population of both the rural and urban areas of the world along with time. In addition to this, other dynamics related to social, cultural, economic, political, and physical or natural aspects are also important in producing change in any spatial phenomenon of physical or human group over time. This concept of dynamics is applicable as a whole with urban dynamics. Thus, the word dynamics is used in two senses. First, it is used to denote "change", in any geographical phenomenon under study, as a result introduced by human activities with add and advice of the nature in a certain region over given points of time. Second, it is used to mean as a motivating or driving factor or force that produces change or effects within a process of growth and development in any field of study in spatial and temporal context.

1.1.1.3 Concepts of Urban Dynamics

Urban Dynamics are those forces that determine the shape of urban areas through time. It is also known as a gradual change in urban areas over time. In fact, urban dynamics do not only mean changes in urban structure. In a more appropriate and broader sense, urban dynamics are the components or the driving forces of urban growth. They help turn land from rural to urban and create conditions of modernization in lifestyle, diversification in social, cultural, economic and political activities of people, spatial coverage by clustering of settlements at both the horizontal and vertical scales. The major dynamics or components of urban growth are spatial dynamics (spatial spread or spatial expansion of urban occupancy over a period of time). It is the manifestation of conversion of rural or agricultural land of suburban or immediate urban hinterland; change in the pattern of urban land use and urban land cover and classification and reclassification of land for urban use leading towards the physical or morphological changes in urban centre. Coast, river bank, river delta, lake, hill, valley, and mining are also the drivers of urban growth. Population or demographic dynamics (the birth, death, and migration) work as a driving force in changing urban population and urban growth. Development dynamics (the creation and distribution of cities hardware in the form of social infrastructure such as transportation system, water supply and sewerage system, electricity supply system and waste material treatment system, educational and medical facility system; and cities software in the form of legal system, market economy mechanism, plans, policies, and programmes relating to above social infrastructure and services) help grow urban areas. Knowledge economy (information technology and knowledge); business composition (industries, commerce; and other economic functions including employment) establish the pathway of urban development. In other words, urban dynamics relate to change in urban society, urban infrastructure, urban economy, urban governance, and urban environment. Urban Dynamics are therefore meant for change in the size of urban population, change or expansion in urban areas, change in urban land use, occupation, commerce, transportation, technology, etc. (i.e., change in physical, social, cultural, economic, and political aspects of urban areas).

Manuel Garcıa Docampo (2014, pp. 1–11) focused on territorial dynamics in which he identified three types of urban spaces such as compact or urban core or city, periphery or ring area, and rural. The growth or change in such urban spaces takes place through four phases of the process of dynamics. These are urbanization (the process of concentration of population in the city as well as in periphery by creating suburban ring to the loss of rural territory), de-concentration or ex-urbanization (the process of growing urban ring at the cost of core city), de-urbanization or counter-urbanization (the process of loss of population in both the core city and periphery or ring area), and re-urbanization (the process of urbanizing the core area again and loss of population in ring area). Infrastructure, transport facilities, productive organizations, management, and governance also play a role of urban dynamics in increasing or decreasing population in such urban spaces. Ross (1982, pp. 656, 662–663) noted that expansion or decline in urban areas is decided by economic, technological, demographic, and plan-policy-related factors. Further, he considered the natural change (the result of birth and death process), the migration effect (the result of in- and out-migration), reclassification of population and areas or localities from rural to urban or vice versa that are annexed to other urban areas as the dynamics of urban growth.

Urbanization initially follows a systematic growth step. First of all a small settlement occurs where supportive conditions exist, and gradually the small settlement changes into a town and then into a city due to positive growth of population and diversification in life supporting socio-cultural, economic, and political activities. Such changes enable the concerned settlement or settlements to attract the people initially from the surrounding areas to invest capital, build institutions, and develop other required infrastructures. All this further triggers migration to the town or city as a process of urban growth. Bloom et al. (2008, pp. 772–274) considered three means of urbanization and urban growth. Urbanization and urban growth occur first by rural-to-urban migration. The study of urban growth and migration experts provides an idea that the contribution of migration to total urban growth population ranges between 40 and 50%. In this way, migration plays a great role in urbanization and growth of urban areas. Secondly, urban growth takes place through natural population growth, the result of births and deaths. However, the rate of natural population growth varies due to variation in births and deaths across different regions. Thirdly, urban growth occurs due to reclassification of rural settlements as urban. Therefore, urbanization or urban growth takes place owing to migration, births, deaths, and reclassification of rural settlements as urban or rural population or rural land as urban population or urban land.

While studying transformation of settlements in Presevo Valley in Serbia, Ejupi (2014, p. 655) explained that growth of population either in the low plains or hills is a result of the components of natural movement (natural population growth) and population migration, but the role of mechanical population growth induced by in-migration or immigration process is more dominant. The major reasons of a permanent increase in the size of population in the Presevo Valley settlements are favourable relief and natural resources, water resources, fertile agricultural land, availability of pastures and forests, better transport and infrastructures, etc. Almost all these conditions are applicable as the dynamics of urban population growth. Similar study is of Zhitin (2011, p. 19) who considered positive natural population growth and more inbound migration rate as the factors of fast urban growth in the Northern Caucasus. This process causes the same results in other urban areas too. Rural-to-urban migration plays a dominant role in urban growth. Demeny and McNicoll (1998, pp. 1–3) gave due consideration to population growth, fertility and mortality as dynamics of transition. According to them, "urbanization is a demographic consequence or concomitant of development". Heimlich and Anderson (2001, p. 15) studied population growth (the result of fertility, mortality, and mobility), redistribution of the metropolitan population, household formation (caused by social, cultural, and economic aspects as well as marriage, divorce, movement to a new place, etc.), in-migration, i.e. internal (from within country), and immigration, so-called external (from another country) as the major drivers that spur or drive urban growth. In the words of Heimlich and Anderson (2001, p. 15) the ultimate driver of urban growth is population growth and household formation. Once the new housings are developed or built and occupied by the new residents then, other requirements such as new schools, roads, sewers, and water supply services, desired infrastructures are realized for use. After this, shopping centres and businesses start growing to serve the needs of the residents. All these human activities require consumption of land which is possible from the non-built-up rural tracts. All these become an ongoing process leading to continuous rise in the urban built-up area and urban growth.

Forrester (1994, pp. 6–10) used the term urban dynamics with reference to growth, stagnation, and decline of cities in the United States. According to him, when the city ages, industrial and commercial activities start growing older and ceases employment opportunities. Similarly, housing also starts ageing along with the ageing city and this in turn causes low rent and attracts more people towards higher density of population. In the same way, construction of less expensive houses on city land snatches the scope of job-oriented commercial and economic activities and pulls more people of lower socio-economic status to concentrate with poverty. Therefore, there is need for visionary and humanitarian government policies. Alfeld and Edward (1995, pp. 200–207) pointed out that urban dynamics involves the ageing of housing, business structure, migration, and linkages among them as the driving forces of growth and decline of a city. Ageing of houses attracts people of lower social and economic group because lower income group persons generally do not afford costly houses. Older houses become less expensive and as a result more people tend to live in the areas of such houses where density of population becomes high. In the same way, industrial and commercial buildings also age and lower the rent and consequent upon that the entrepreneurs requiring more space are attracted to start business by employing less workers and getting lesser profit. This also increases population density. Similarly, migration is always due to attractiveness of city life and people in-migrate to city as long as its attractiveness continues, but over a longer period, when attractiveness of the city fades, the rate of migration starts ceasing and causes decline in city. In fact, ageing of houses and business structure as well as decaying feedback connections also start working against the expectation and choice of people and lead to decline in city. Cities, therefore, work both in positive and negative ways in providing opportunities for better life of its residents and also in trapping many persons in poverty and in causing poor health behaving against the expectation and desire of its inhabitants.

Davis (1955, pp. 429–430) also stated that technological innovations, favourable "climate, soil, water and topography" as well as river valleys played an important role in urban growth. Hence, all these are considered as important factors (dynamics) in urban growth. Climatic conditions, knowledge-based educational and health-related facilities/services, job opportunities, commercial and transportation-related services, administration, and governance also play an important role as motivating factors in growth and expansion of a town or city. In fact, urban dynamics play a great role in our urban life. According to Nong and Du (2011, pp. 63–66) large-scale change in the environment in general and urban environment in particular is largely the result of human activities. In the same way, urban growth or urban expansion through urban land use and land cover changes is determined by the biophysical and socio-economic determinants. Similarly, urban development depends on better accessibility of roads, centre of economy, and high density of population. Such factors are termed as the dynamics of urban growth.

1.1.1.4 Urban Dynamics and Behaviour of Urban Growth

To study or statistically examine the changing pattern and future trend of urban growth (urban dynamics) caused by human induced land transformations or land use-land cover changes and its impacts on the climate of San Francisco Bay area, Clarke et al. (1997, pp. 250–253), Clarke and Gaydos (1998, p. 705) went through a practical exercise for preparing a "Cellular Automation Model". For this, they collected and employed multi-sources statistical data concerning "land cover, topography, climate, population, settlement, transportation, aerial photography, satellite imageries and other data source elements". After synthesis, digitization, and automation of data related to these elements the Cellular Automation Model was developed that provided a strong reference-base to describe and understand the future scenario of land transformation and urban growth in the urbanizing Bay area of San Francisco. Finally, they arrived at the conclusion that "Diffusion, Breed, Spread, Slope resistance and Road gravity" are the five factors or dynamics that control the behaviour of urban growth system. Here, *Diffusion* decides the outward side distribution of dispersion of new settlements through the roads; *Breed* ascertains the probability of generating growth cycle by a newly developed detached settlement on its own; *Spread* determines that to what extent a normal outward organic expansion takes place within the growth system; *Slope Resistance* determines the possibility of a settlement that extends up steeper slopes; and *Road Gravity* works as a pulling force and attracts new settlements towards and along the existing roads from within a certain distance of the roads. Further, they also found four types of urban growth in their Model. These are: "Spontaneous, Diffusive, Organic, and Road-influenced" where, *Spontaneous* urban growth takes place due to the influence of an urban centre in its surrounding or neighbourhood areas. *Diffusive* growth occurs in the desirable locations of flat areas even though such areas may not be close to an already established urban centre. *Organic* growth expands outward side of an existing urban centre showing the trend of city to spread in future. *Road-influenced* growth occurs along the networks of transportation, so-called transport links, by indicating enlarged accessibility. The study of Silvaa and Clarke (2002, pp. 525–530) is exclusively based on the principle of "SLEUTH" model of Clarke et al. (1997) to test whether the result of the application of this model will remain similar in case of other cities (European or non Western) located in different local environments of urban growth. SLEUTH model is a phrase of "slope, land use, exclusion, urban extent, transportation and hill-shade". In their study here also, Silvaa and Clarke, considered "diffusion, breed, spread, slope and road coefficients" as the dynamics or factors or parameters that control the behaviour of urban growth system, and "spontaneous, diffusive, organic and road-influenced" as the four types of urban growth. In other words, spontaneous growth takes place in neighbourhood areas of an urban centre with suitable slope under the influence of the diffusion factor; diffusive growth occurs in the form of new growth centres under the effect of breed factor at the suitable places that may be even away from the established urban centre; *organic* growth is a type of expansion of cities into their vacant neighbouring areas; *road-influenced* growth happens near and along the roads due to the importance of roads in terms of easier movement of people and transportation.

Feng et al. (2012, pp. 206–207) considered social and economic development a cause of urban growth through the expansion of urban land and urban land use changes. They also put emphasis on protecting the land resources that is lost due to urbanization. Further, they used SLEUTH Model in their study to predict urban dynamics with the help of four urban growth rules—spontaneous growth, new spreading centre growth, edge growth, and road-influenced growth controlled by the diffusion, breed, spread, slope and road gravity factors based on Clarke et al. (1997). Wilson et al. (1987, p. 277) used urban growth model and classified urban growth into "infill, expansion and outlying" type. Outlying urban growth was further subdivided into "isolated, linear branch and clustered branch growth". This classification was made on the basis of distance between the existing or already developed urban areas and the type of urban growth that takes place. *Infill growth*, according to them, occurs in that neighbourhood non-built-up open urban spaces (vacant areas) which are surrounded by at least forty percent of the already developed (built-up) urban areas. Such non-built-up urban areas lie close to the existing urban area where basic social amenities are available and is being converted into urban area by gap-filling process.

Expansion/Extension type of growth takes place in the non-urban or non-built-up area lying at the edge or periphery as expansion of the existing developed urban area. Such expanding areas are surrounded by not "more than 40% existing developed pixels" (developed urban areas). This is also known as the edge, periphery, border, or boundary development. Heimlich and Anderson (2001, pp. 2–18) called this type of development as metropolitan fringe or urban fringe development. *Outlying* growth occurs away from the existing urban centres on account of change in the status of land from non-urban to urban. Heimlich and Anderson (2001, pp. 9–12) parsed this type of growth as a "development beyond the urban fringe". *"Isolated growth"* develops in the form of more isolated large lot housing normally at some distance from an existing developed area being developed, or in other words, beyond the urban fringe. Such development is a distinctive feature of a new house or new construction surrounded by little or non-urban space. *"Linear growth"* takes place along physical features such as river, mountain, hill, sea coast, and cultural features such as road, railway line, canal, etc., in narrow and elongated form. *"Clustered branch"* is defined as a new urban growth. It is "neither linear nor isolated, rather, a cluster or a group". It is considered as a typical form of large, compact, and dense growth of housing (Wilson et al., 2003, pp. 275–277). Harvey and Clark (1965, p. 1) defined this type of growth as a leap-frog development which is the "settlement of discontinuous, although possibly compact, patches of urban uses".

Nong et al. (2018, pp. 2–7) also noted and explained the three urban growth types or urban landscape expansion modes with illustrations. These are "Infilling or gap-filling growth" that takes place on the patches of open space lying vacant within the heavily urbanized areas and by such filling process open spaces merge with the already existing built-up area. "Edge-expansion" occurs on the non-infill or non-built-up areas situated on the periphery of the highly urbanized centre. Through this development process edge expansion joins or merges with the existing urban footprint (previously developed urban area) expanding outward side from the city. "Spontaneous growth" happens at the gap of some distance from the existing urban footprint (existing urban centre). It is also known as the "outlying growth" and this neither intersects the existing urban footprint nor becomes contiguous with previously developed urban areas. This plays a great role in fragmenting the open lands. Viana et al. (2019, p. 623) considered urban growth as an urban land pattern in terms of a static phenomenon as well as a process in terms of a dynamic phenomenon. While studying different aspects or dimensions of urban sprawl land use pattern, Galster et al. (2001, pp. 683–686, 709) noted sprawl as a type of urban growth and as a process of urban development. It also denotes the condition of urban land use pattern over a given time. Further, they also noted the effect of different types of sprawl on the distribution of income and externalities of the pattern of urban growth.

According to Viana et al. (2019, pp. 621–622) and Foley et al. (2005, pp. 570–574) urban growth, a result of the expansion of urban land, is normally influenced by both the natural and human induced factors and leads to transform the land use and land cover towards the satisfaction of needs of rising population. The twenty-first century is expected to witness further increase in the size of urban population and cities leading to diminish the productive agricultural lands and life supporting ecosystem's capacity particularly in the developing countries. In fact, future scenario of urban growth will be a great challenge for the world in the coming time of this century. According to Xiaoqing et al. (2010, pp. 33–38) the fast urban growth and land use change, the result of rapid social and economic progress, have profound effect on worldwide change and have become a focal theme of research in the field of urban dynamics. Erdogan et al. (2014, pp. 11–19, 259) noted that change in land use and land cover is a result of the modification introduced by the action of human population on the land surface of a given spatial unit or a region. Change in land use and land cover directly or indirectly affects the principal ingredients of climate change, biodiversity; soil, water, and air pollutions; population and urban growth as an important driving force at local, regional and global levels. Likewise, urban growth also works as an important factor in determining land use and land cover changes. In this way, change in land use and land cover is complex, and dynamic process of growth and development in space and time and becomes a vital issue for those engaged in academic research and in formulating plans and policies for development. The spatial data related to land use and land cover change helps to understand the dynamics of present and future urban growth and development system.

Atkinson and Wulff (2009, pp. 6–19) took into account "gentrification" as one of the factors or dynamics of transformation in urban areas. It is also considered as a process of converting the central areas or important suburb areas of a city into higher housing status where the lower socio-economically status people used to live. Through this process the people of socio-economically lower status households are displaced or dislocated from the areas undergoing considerable rise in the rent and cost of housing to other areas where the housing cost, rent, and price of land is low. This creates housing, psychological, and health stress on the lower income household's people, especially on the aged, the lone-parents, and the low waged people. In this way, gentrification causes both positive and negative impacts on a city's population. Positive impacts of gentrification are stabilization of declining urban areas, increase in real estate value, increase in local income, decrease in social crimes, concentration of higher status people, etc. Negative impacts of gentrification encompass loss of options of housing for poor and vulnerable section of the urban society, loss of social mixing and the demographic balance, psychological stress on health, education, income, lack of support of social network, loss of better work opportunities, loss of housing, reduction in opportunities of commercial and industrial activities for the displaced and dislocated people, increase in crime in the areas of displaced people and the like.

The author considered urban growth as a process of adding more land from surrounding open spaces of a town or city to urban built-up area and this is possible on account of an increase in population, through both the natural as well as mechanical population growth processes, and extra needs for housing, infrastructures, and services. In another sense, urban growth is a process of an outward extension or expansion of urban areas over a certain period of time owing to multiple demographic, social, cultural, political, and natural or physical factors operating in and around a town or city.

1.1.1.5 Challenges and Opportunities Caused by Urban Dynamics

According to the UN-HABITAT (2012, pp. 85, 108) uninterrupted urban expansion, headforemost use of energy, climate change, sustainability of environment, unemployment, poverty, shortage of housing, lack of adequate water and sanitation, inequality and social exclusion, demographic imbalance, lack of infrastructures and services, and the like are the major issues and challenges for the cities especially in the developing countries. Such issues are importantly related to urban growth and economic prosperity which vary from one urban centre to another depending on a variety of factors operating in every urban area regardless of its size. To meet all these challenges, the cities should grow with a visionary and reinvigorated planning and good governance for rational use of urban space, better technological innovations, social harmony and integration, better cultural and economic opportunities leading to better livelihood, better urban living and quality of life.

Open Learn Create (2017, pp. 9–16) stated that positive population growth driven by birth, death, and migration as well as infrastructure development are the major reasons of expansion of urban areas at the cost of rural land. Such a process of change in geographical areas of towns or cities put pressure on environment, sanitation, and health which becomes a challenge for the local government. Growing towns and cities are the centres of a better economy, human capital, and development in commerce, industry, science, and technology. Educational, health, and social services are better available in urban areas than that of rural areas. Similarly, infant and general mortality rates are lower in urban areas because of the availability of better healthcare facilities. Likewise, other basic amenities such as supply of fresh and potable water and electricity are also better in urban than that of rural areas. People of different social classes and religions live and work together in cities by ignoring social and cultural barriers. This is a sign of better understanding and harmony among the people living in urban areas. Despite these, there are some negative consequences on environment and human health owing to haphazard urban growth. Also, all urban families do not avail equal opportunities of basic urban facilities and services of housing, schooling, health care, water supply, disposal of solid waste, sewerage, and the like.

Lerch (2017, p. 1) also highlighted the world population growth and population growth rate and mentioned that urban population grows much faster than rural population. The rapid increase in urban population, particularly in less developed geographical regions becomes a great challenge in the way of achieving the sustainable development goals of the United Nations by 2030, when the slum population in most cities of the world continues to grow faster. Similarly, Bloom and Fink (2008, p. 772) accentuated that urban areas are the source of economy, earning, employment, education, human capital, prosperity, better quality of life and better market facilities. Workers in urban areas are more productive than rural workers. Though, rapid urban growth and urban population are found associated with crowding, poor population coming from rural areas, urban unemployment, poverty, impediments to productivity, urban violence, traffic congestion and environmental degradation, etc.

Kapluhan (2014, p. 732) explained that, currently, the world population faces considerable challenges. These are related to environment, migration, poverty, haphazard urbanization, social clashes and disputes in many areas, concentration of youth population, elderly population and their health, sanitary requirements, etc. These are challenging issues for the world.

Urban expansion determined by continuous population growth negatively affects agricultural lands. This process causes loss of prime cultivated lands in which the food crops are grown for the consumption by both the urban and rural population. The gradual loss of croplands due to urban expansion in the world and especially in the developing countries will remain a challenging issue in the twenty-first century. The study of Bren et al. (2017, pp. 8939–8940) revealed that about "80% of global cropland loss from urban expansion will take place in Asia and Africa". This cropland loss is expected to threaten environmental sustainability, food systems, and livelihoods in the coming urbanized world.

Nantapo et al. (2015, pp. 896–905) emphasized that current rise in human population owing to natural population growth in combination with rural-to-urban migration has led to several food- and health-related challenges in both the food efficient and food deficient regions. The growing global population especially in the developing regions generally suffers from calorie-related health conditions and diseases such as obesity and coronary heart diseases. The food availability is also related to social and economic condition of the people. According to Davis (1955, pp. 429–430) cities are the centres of education, intellect, arts and science, social mixing, economic activities especially trade and commerce, industries, recreational facilities, political power, etc. Cities work as a magnetic force in pulling the talents or brilliance and even the persons of moderate quality often born in rural areas and develop or transform them into a better productive human resource to be utilized in different economic activities to produce a use value for satisfying social needs. In fact, city has better educational and health facilities for human resource development. According to Mccall (1955, p. 151) investment is one of the factors of urban growth and considered "towns are a symbol and an index of economic development". Urban growth works towards human and social development and economic growth for poverty reduction. It provides a roadmap for policy makers interested in implementing high-quality programmes. Ross (1982, p. 649) also considered urbanization as an essential and desirable step for economic development and rational utilization of resources with the purpose of social well-being. Yet, it has many ill effects resulting in the creation of slum, unhygienic condition, pollution, ecological imbalance, housing problem, traffic congestion, social vice or social pathology, and scarcity of medical and educational facilities and civil supplies. In the words of Liddle (2017, p. 13) cities are the source of both wealth and poverty. An increase in gross domestic product evidently lowers poverty and narrows gaps between rural and urban areas. Conversely, higher levels of urbanization increase poverty and gaps between rural and urban areas. Therefore, cities have both positive and negative impacts on society, culture, economy, polity, and environment in space and time.

According to Demiroğlu et al. (2014, pp. 579–580, 595) cities are those living spatial units which swiftly grow and attract more people and provide them shelter, safety, opportunities of education, services, technological development, income, employment, industries, trade and commerce, and the like activities. Conversely, the urbanization speed brings negative impact on the environment, economy, social, cultural, and political aspects, quality of life as well as natural and other resources leading to the collapse of the urban system. Seto et al. (2010, pp. 168–179) focused on urbanization that will decide the social, cultural, economic, political, population, and environmental traits and resources of this century. Scholars of the second half of the last century considered urban growth and urban centres as environmental problems, but recent researches reveal complicated rapport between urbanization and environment. There are two opinions concerning urbanization and its effects. Some scholars have opinions about the positive impact in terms of increasing awareness towards environment, safety, security, and economic growth, but the opinion of some others is about an increase in the rate of use of resources and energy, degradation in environment through increasing environmental pollutions as a negative influence. It is also noticed that the rate of urban land expansion is faster than the rate of growth of urban population. In other words, the dynamics of expansion of urban land is faster than the dynamics of urban population growth. This is due to footloose international capital, governance and institutional structures, and agglomeration of economies. Therefore, the twenty-first century is meant for the world of urban population and for drawing attention of the government, planners, architectures, academicians, and researchers to make sustainable urban development for the well-being and better quality of life of the present and future population.

Nong et al. (2018, pp. 1–4) also considered varied socio-economic, demographic, and political forces as the drivers for the conversion of non-urban land into urban (Urbanization). This process has several positive and negative consequences. Urban growth provides an opportunity to boost the economy and productivity of the people, enhance social status, cultural values, and build the centre of education. In common with this, urban growth becomes the cause of loss of cropped or farm lands leading to future food insecurity and livelihoods, environmental degradation. It creates problems concerning disposal of urban waste, housing, supply of water and energy, increases incidence of diseases, and decreases social assistance infrastructure and services. Bounoua et al. (2009, pp. 217–229) assessed the impact of urban growth, the result of land use and land cover changes, on surface energy, water, and carbon balances. They considered cities as the centre of a rapid change in social, cultural, and economic as well as environmental and population aspects. In other words, cities work as a centre of economic development and advantages, good working and living conditions, but at the same time drastically reduce the per capita use of land, cause severe damage to health of the people, environment, air quality, vegetative cover, and water resources. Hence, there is a need for balanced urbanization through a comprehensive programme of land use and land cover.

Xiaoqing et al. (2010, pp. 37–38) pointed out that the conversion of open urban space or land especially adjacent to or within highly urbanized urban centres occurs by infill or gap-fill urban growth process. This type of expansion of urban growth is the result of human induced driving forces or activities and causes loss of resources such as farm land, forest land, water bodies, and unutilized land leading finally to urban landscape and environmental damage. Tan et al. (2005, pp. 187–196) stated that expansion of urban land provides economic and social development opportunities; side by side it mounts pressure on environment and various types of resources. The process of urban growth through urban land expansion causes loss of prime arable land. Therefore, an approach of sustainable urban development is prerequisite to save agricultural land. Grimm et al. (2008, p. 756) considered urban areas as the hot spots and the expansion of them affects the rate of material consumption, land use and land cover, sustainability, environment (particularly biodiversity, hydro systems, biogeochemical cyclic pattern and climate, balance of surface energy), human behaviour, habits and at local, regional, and global levels. Urban population growth particularly in more densely populated countries driven by economic growth and demographic changes is expected to ever increase the demands for various types of services from the nearby and distant ecosystem areas.

While studying dynamics of real estate and struggle by the poor people for housing in Sao Paulo (Facius, 2015, p. 163) pointed out that the process of abandonment and gentrification causes displacement of people and inflict different parts of the city and definite neighbourhoods. In fact, gentrification is a big problem in the urban areas particularly for the poor and middle-class society. Sandal and Adiguzel (2014, p. 570) noted that social and economic insufficiencies in rural areas work as motivating factors for migration. The cities are considered as a centre of enough resources and attract people and owing to this people migrate from rural-to-urban areas. Urban areas are also considered as the centres of industries, attractive city life, technological progress, transport and other infrastructures, political power, and social status, cultural values, and other better opportunities. All this attracts the rural people to the urban areas. According to Curebal et al. (2014, p. 48), urbanization, being a gradual process of conversion of non-built-up land into built-up, on any type of landform negatively impacts the environment and natural resources.

1.1.1.6 Policy Measures for Urban Dynamics

Urban dynamics build urban areas as the places of different opportunities and threats for urban population. The worldwide rapid growth of urban population is a matter of grave concern for the sustainable urban growth. Urban dynamics have a vast scope for the geographers, scientists from science and social science subjects, technical experts, medical practitioners, researchers, administrators, and policy makers to work in order to explore the way for looking ahead to mitigate urban problems and to create new prospects of urban opportunities and resources in order to sustain the growing urban population and urban sustainability.

Since the maximum future urbanization is to take place in the developing countries, the prospects of emergence of urban threats would be far greater there than that of the developed countries. Owing to this, the pressure of achieving the goal of urban sustainability by 2030 would be much more on the part of the developing countries in comparison to that of the developed countries. For this, there is need for a sincere, well planned, and viable urban planning, policy, and management to check the faster growth of population in urban areas with special focus on the growing cities of the developing countries for sustainable urban dynamics. The effective execution of such measures at local, regional, and global levels would control the ongoing negative effects of urban dynamics and would result into the well-managed cities of the world in general and of the developing countries in particular to offer better social, cultural, economic, and political opportunities to benefit all the urban people particularly the urban poor and most vulnerable groups of population.

In addition to the above, there is also a necessity of framing a well thought out population policy at both the national and international levels to look ahead to the role of urban population dynamics in increasing the unplanned and haphazard expansion and growth of urban areas. In fact, the condition of urban population dynamics (fertility, mortality, and migration) considerably varies across the world and has varying impacts on urbanization of different nations. As a result, some particular countries intend to lower the fertility, whereas some other particular countries intend either to raise or to maintain the same level of fertility depending upon their current demographic conditions determining the level of urbanization. From this standpoint the fast urbanizing developing world needs to lower the fertility rate by promoting family planning measures to curb the speed of urbanization for the better urban life. The developed countries with declining fertility may plan to increase the fertility and those having lower fertility may plan to maintain the same level of fertility to make their cities sustainable.

To check the higher natural population growth and massive rural–urban migration contributing to urban growth in the developing countries is quite necessary. The check on emigration of people from developing to the developed countries will result in contracting the speed of urbanization in the developed countries and increase the prospects of maximizing the stock of human resources for use in different economic activities to strengthen the speed of national progress of the developing countries. From this viewpoint, a check on rural–urban migration to the cities of a developed nation from within its national boundary and immigration from other developed nations as well as from the developing nations would be an effective migration policy intervention in the direction of sustainable urban dynamics. This pattern of national and international migration can be universally applicable in case of any developed or developing country of the world. Therefore, a strong national and international level migration policy would be probably an effective approach in making the urban dynamics and urban areas sustainable.

Also, a long-term visionary policy measure or policy intervention at national and international levels in terms of rational use of urban land for the purpose of industrialization, commercialization, transportation, innovation, etc., would also be a step forward in making social, cultural, economic, political, and population urban dynamics sustainable for better environment and better urban life.

Therefore, the need of today is to create a viable condition of urban dynamics for rational use of urban space, better technological innovations, social harmony and integration, better social, cultural, economic, and political opportunities. All this would be possible only through a well planned urban policy and good governance worldwide. Such policy interventions would go a long way to play a key role in making the worldwide cities to thrive well for better livelihood, better urban living, and better quality of life.

1.1.2 Urban Environment

This Section aims to focus on different dimensions of urban environment. However, it becomes necessary to first highlight the fundamental concepts and meaning, contextual aspects, significance, dynamics, multidimensional picture of problems and challenges and opportunities of urban environment. Urban environment is one of the challenging global issues for the twenty-first century because it is very complicated and dynamic in nature and hence it remains a focal theme of study in geography and in other allied subjects. The word "environment" was first used by Thomas Carlyle in the South of Scotland in 1828 (Jessop, 2012, p. 708). This term is universally applied all over the world whether it is urban or rural, though, it varies in terms of its magnitude, quality, and its impact or effect on living and non-living components across different spatial units/regions at different stages of time.

1.1.2.1 Concepts and Meaning of Environment

Understanding the underlying facts or ideas concerning environment is very important for any scholar, researcher, and academician working in the field of urbanization, environment, and health. Concept provides a general idea or notion about a given subject required in its analysis and explanation. In other words, concepts stand for abstract ideas or underlying facts as primary attributes of a concerned subject. Concepts can also be considered as the basic components of thought that allow someone to create or develop a knowledge base with the help of past experience. Meaning, on the other hand, is often considered as an idea that represents something in terms of words or language to communicate information about something or a case under question by eliminating vagueness. Meaning represents the fundamental ideas or facts of the concerned topic. Here, environment is an integral part of this chapter and reveals its nature and scope. Earth's environment is commonly known as a thin life bearing layer (biosphere) for all forms of life that lies closely interconnected with the lithosphere, atmosphere, and hydrosphere. In other words, this life supporting layer is found surrounding the land, gaseous, and water realms. According to Strahler and Strahler (1977, pp. 2–3) environment can generally be defined as that which surrounds all life forms between the realm of atmosphere, hydrosphere, and lithosphere. Richard (1968, pp. 42–47) considered the physical body or physical world as natural environment which environs human beings. Normally, natural environment is considered as environment (in solid, liquid, and gaseous form called lithospheric, hydrospheric, and atmospheric environment). Therefore, environment can be treated as all those phenomena on this living earth which surround both the human and living world. Park (2009, p. 616) regarded environment as external factor and comprises soil, water, air, climate, and social and economic conditions under which we live. Simply, it can be defined as the sum of all external conditions that affect the life and development of an organism, human behaviour, or society. Turk (1985, p. 1) stated that the earth on which we live is environment. It encompasses all the natural elements of the earth such as soil, minerals, rocks, air, water, and all living organisms, such as animals and plants. These are all very important and surround human population. Environment is considered by Daubenmire (1974, p. 2) as the set of surroundings that influences an organism. Actually, any internal force, substance, or condition that affects in any way becomes a factor of their environment and the sum total of these factors constitutes an environment. According to Bhattacharya (1991, p. 9) environment is a living space in which human and other living species are likely to change due to the activities of the species. De and Bose (1991, p. 135) were of the opinion that environment is the sum total of surrounding conditions within which an organism or group of organisms and an object exists. In the words of Mitra (1991, p. 199) environment is the aggregate of all external conditions and influences affecting the life and development of an organism, human behaviour, or society. According to Swarup et al. (1992, pp. 4–7), environment pervades right from indoor to work place and throughout the socio-economic activities related to industry, agriculture, urbanization, transportation, and so on. Michael (2000, p. 94) took environment into consideration as conditions of work or the type of work an individual carries out, the culture, climate, and management style in the organization, and the social group within which individuals work. Jeanne and Roger (2001, p. 309) included all types of services, products, and resources in the concept of environment. They stated that environment is a result of several interactive processes of different components of a system. Saxena (1999, p. 1) treated the term environment as surroundings which is a composite form of several components. This surrounds human population and other living organisms. It is the sum total of all conditions and influences which affect the growth, development, life, and death of all organisms, species, or races. Singh (2019, pp. 20–21) also considered environment as a circumstance which influences development and growth of people, animals and plants, living and working conditions. This according to him is an inseparable whole and is constituted by the interacting systems of physical, biological, and cultural elements. In fact, the environment is equated with nature wherein physical components such as land, water, air, etc., support and affect life in the biosphere. In the opinion of Dikshit (1984, pp. 68–83) environment should be taken into consideration in the context of our concern and priorities because our immediate concern is the quality of space we live in, the air we breathe, the food we eat, the water we drink and the resources we draw from the environment to support our economy to sustain life.

After having gone through the views or ideas of the above scholars the author is of the opinion that the environment should be considered to include everything or conditions around and about a geographic space (urban or rural) at a certain point of time in which human beings and other living organisms live and function. The above concepts can be easily and fully applied in the context of urban space or urban areas or city regions. Therefore, Urban Environment can be defined as those conditions or states that exist or happen within or around an urban centre in which all living and non-living things have their existence. In fact, all living components (such as human beings, animals, different organisms, plants, etc.) and non-living things (such as land, air, water, soils, rock, landforms, minerals, climate, etc.) are the integral parts of the surrounding environment in spatial and temporal context. From a living system's perspective urban environment includes all those surrounding conditions within which human and other beings live, survive, and function. The role of non-living components in the life of living things is very important as non-living elements in solid, liquid, and gaseous forms help create the life support layer and provide several opportunities in biosphere in which living things live, grow, survive, function, and die. Urban environment cannot be solely considered in terms of natural/physical state or condition in which all living objects live but also in terms of social, cultural, economic, political, and demographic conditions (from human perspective) that affect human and other lives. In this way, urban environment can be defined as the sum total of all circumstances or conditions of natural/physical, biological, and cultural components existing or prevailing within and around a town, city, or city region at a specific point of time. Such composite conditions always affect the life of human beings and other living things on this planet earth.

1.1.2.2 Contextual Aspects of Urban Environment

The environment (urban or rural) is viewed differently by different experts of different subjects from different angles. A geographer's angle of viewing environment is different than that of others. Geographers view environment in context of the earth on which the natural or physical (abiotic or non-living) and biological (biotic or living) things, components, elements, or features exist. In this context, environment is termed as both physical and biological where the physical or natural environment includes land, water, air, etc., as natural components and the biological environment includes plants, animals including humans, and other organisms as biological components. On the basis of land, water, and air components physical environment is further divided into lithospheric, hydrospheric, and atmospheric conditions or environment. Likewise, on the basis of physical features such as mountain, plateau, plain, ocean, coast, river, lake, etc., environment is also considered as mountain, plateau, and plain environment; desert environment, oceanic environment, coastal environment, river and lake environment, and the like. Similarly, biological environment can also be divided into floral, faunal, and other organism-based environment. Among all the biological components, human population is the most important environmental factor that creates and develops cultural or built environment through interaction with natural living and non-living components or elements in geographical space along with time. In this way, the resultant part of the interaction between human beings and nature becomes a cultural environment which comprises social, economic, political, etc., environment. When one looks into the importance of the four components of geography—nature, human beings, space, and time—environment can also be treated as natural and human environment in space and time. Human population among all the components of the environment is solely responsible for the creation of congenial and uncongenial urban environment through industrialization, urbanization, and other human activities. Urban environment in this respect should be considered as a set of intertwined social, cultural, economic, political, and demographic facts and the natural state of physical environment. Therefore, urban environment should be taken into consideration from the viewpoint of solving urban problems resulting from human activities and other urban facts. *In a relational context,* environment is the surroundings in which an organization operates, including air, water, land, natural resources, flora, fauna, humans, and their interrelations. *In an organizational context,* environment can be considered as a working condition and relation in which both human beings and machines work in producing a use value of any description to satisfy the human needs. *Sociologists* consider environment in social context and give emphasis on social environment that refers to the social conditions or circumstances in which both the living and non-living systems of a society coexist. *Anthropologists* look at the environment in a cultural context and give importance to cultural environment that includes housing, diet, culture, lifestyle, tourism, politics, legislation, technology, science, religion, language, mass media and communications, organizations, educational and health institutions, and infrastructure which influence individuals of urban society at a particular point of time in space. *Economists* see the environment in an economic context and lay emphasis on economic condition or state of an area that affects both the human and natural phenomena at a given point of time. It also comprises an investment climate (a condition in which someone has to invest money, talents, energies, etc., in an economic activity), land, labour, capital, and other resource conditions. In *an occupational context* environment can also be considered as agricultural, industrial, mining, commercial, organizational, and technological environment. *Psychologists* regard environment in a psychological context and stress on psychological environment that is meant in terms of physical or interpersonal conditions or states that affect mood, attitude, aptitude, interest, feelings, and behaviour of a person or persons of an urban area at any point of time. Thus, the urban environment is a result of the combination of interconnected social, cultural, economic, political, demographic, and natural elements. In other words, the urban environment from human perspective can be viewed as an urban condition surrounding us and is the consequence of human activities performed in and around both the nature and human made urban space over a given period of time. Environment may be stable or unstable, balanced or imbalanced which has its bearings on health and prosperity of the living organisms of that space. When the balance of the environment is disturbed either due to natural forces or calamity or by the action of human beings, it affects the health, mind, and life of urban and other population locally, regionally, and globally. For instance, outbreak of COVID-19 pandemic caught the whole world, especially the urbanized areas, under its influence and damaged natural, social, cultural, economic, demographic, psychological, and political environment.

1.1.2.3 Significance of Urban Environment

Significance of environment is another important aspect of urban area. Decidedly, environment is an important, integral, complex, dynamic, and continuously developing system. It has always been considered as a vital factor for the survival of life on earth. Richard (1968, pp. 42–47) considered the earth as physical or natural body which is the home of humans. According to him, it provides the means and nurtures our lives. Here, he viewed physical body (the earth) as physical environment which is a source of life supporting means. Ryabchikov (1975, p. Preface) has rightly focused on the importance of natural environment. According to him, the natural environment is essential for the existence of human beings and is the primary source of their life and social production. The study of environmental dimension in the world of human society becomes very significant as environment or each environmental region is endowed with varying quality of life support resources on the living earth. Environment is the source of resources that support the life of human population, animals, plants, and other organisms of this planet earth. It directs and determines the existence, growth, and development of humans and other life forms and their activities. Environment provides various types of conditions in terms of terrain, soil, water, air, weather, etc., for the plants and animals and affects their life cycle, lifestyle, and activities. The successful survival of all lives on the earth in general, and human beings in particular highly depends on the prevailing environment conditions of the places along with time. Actually, every aspect of life (social, cultural, economic, and political) of the people of a place is completely related to the environment of their places in space and time. Purohit and Agrawal (2004, pp. 1–4) highlighted the importance of environment and stated that the environment, particularly physical environment, provides both habitation and raw materials for the synthesis of organic food. All life processes are dependent on this environment. Environment (the circumstances or conditions) affects an individual or group of people or community located at a particular place at a particular point of time. Human population inhabits the natural world as well as the built or technological, social, and cultural world and is influenced by the environment related to such aspects. In this way, environment is the key factor and plays a great role in the growth and development of all forms of lives. In the context of human population, it becomes very vital and influences all kinds of activities. The role and effect of environment is apparently seen in agriculture, industry, commerce, construction, transport, and communications, in all types of institutions of learning, production, and of health as well as in all other kinds of human activities relating to social, cultural, economic, and political aspects.

1.1.2.4 Dynamics of Urban Environment

Urban environment is not static rather it is always changing. The meaning of "urban environmental dynamics" is perceived in two forms—first, in terms of change (positive or negative as well as increase or decrease both in qualitative and quantitative sense) and second, in terms of factors or determinants or driving forces that cause a change in a phenomenon or object under study. Here, environmental dynamics are meant for changes in the state of urban surroundings. Changes may be in terms of improvement in the quality of urban environment or deterioration in the quality of urban milieu. The latter one leads to unhealthy or unsuitable conditions from the point of view of urban life. Such changes take place due to several reasons. In other words, environmental dynamics (environmental changes) may be within an urban space or may be beyond a city. In terms of forces also the determinants of environmental changes may be from within a city (internal urban environmental factors) as well as from outside the city (external factors influencing urban environment).

Factors influencing urban environment can simply be divided into two groups-physical/natural factors and anthropogenic or human induced factors. Natural factors can further be divided into internal or endogenic/endogenetic and external or exogenic/exogenetic natural factors. The endogenetic factors are those forces which originate from the interior of the earth and are of diastrophic and sudden nature and in turn directly or indirectly affect urban environment. However, effect of the diastrophic forces on the environment is observed in larger geographical areas after a very long span of time owing to its slow movements whereas, the consequences of the sudden forces on the environment is observed at local and regional levels within a short period of time due to abrupt movements. Diastrophic (tectonic) force includes plate tectonics and sudden forces include earthquakes and volcanoes as major one.

Internal Natural Factors and Urban Environment: The major internal natural factors are plate tectonics, earthquake, and volcanic eruption. These are also known as endogenic or endogenetic forces and are very much related to urban environment.

Plate Tectonics: This is a widely acknowledged scientific theory and is related to the earth's crust and mechanism of movement of its plates (parts of the crust) which leads to the environmental change. The word "tectonic" refers to the geologic dynamic forces or conditions inside the earth that cause motions or movements of the earth's crust—the continental and the oceanic crust. It is also related to the structure of the earth's crust. "Lithosphere" is the combination of both the crust (continental + oceanic) and upper portion of the mantle and moves as a single rigid layer over the soft asthenosphere (situated just below the lithosphere). The rigid lithosphere (around 80 km thick) is broken into six major plates: the North and South American plate (continental crust and all of the oceanic crust up to the Mid-Atlantic Ridge), the Pacific plate (exclusively oceanic crust), the Eurasian plate (mostly continental crust), the African plate (continental and neighbouring oceanic crust), the Indian-Australian plate (continental and oceanic crust of the Indian and Pacific ocean), and the Antarctic plate (continental and oceanic crust). All these are known as the lithospheric crust or lithospheric plates or lithospheric tectonic plates (Strahler and Strahler, 1977, pp. 153–157). "Plate tectonics" is a scientific study of the entire mechanism or process of large-scale motion of the plates leading to cause the deformation and changes in the existing features of the lithosphere (the outermost layer of the earth).

Plate Tectonics and environment: Geologically and geographically, plate tectonic being an important endogenic force has wide ranging effects on the environment and in support of this Foley (2015, pp. 1–2) asserted the role of plate tectonics in the operation of long-term cycle of the atmospheric carbon dioxide (CO2) throughout the habitable zone of the earth. Nwankwoala (2015, p. 227) also focused on the relationship of the plate tectonics and the environment. The drifting or shifting apart of different plates leads to repositioning of the continents and in turn affects the terrestrial land forms, large-scale carbon storage, and increased glaciations and gradually this process causes climate and environmental changes. Ashokmuthu (2008, pp. 1 and 11) also pointed out that plate tectonics play a considerable role in climate change over a long span of time owing to both the vertical and horizontal displacements. The solid outermost layer of the earth is always in direct contact with the atmospheric and oceanic life-oriented environment which is disturbed at and after the time of occurrence of the movement of the plates. A tectonic force also greatly affects the areas of human agglomeration. Strahler and Strahler (1977, p. 157) were of the opinion that there is a deep effect of plate tectonics on the terrestrial environment as the movement of tectonic plates drastically disturbs and changes the continents and ocean basins (the basic structure of human and other life support environment). However, the consequence of plate tectonics on the earth appears after thousands and millions of years. For instance, the extensive coverage of Pleistocene period glaciations in major parts of North America and Eurasia was the result of displacement of major continental blocks caused by plate movements during the Canozoic era. Similarly, the Tertiary period mountain ranges and high plateaus formed owing to continental collision work as an effective relief barrier in controlling the global atmospheric circulation as well as in developing new types of climates such as the South Asia monsoon climate. In fact, the changes in global pattern in plate motions cause changes in climatic conditions on a global scale (Singh, 2019, pp. 257–258) and largely affect humans, animals, plants, and other living things on the earth. The urban area's environment is always more affected.

Earthquake and Urban Environment: Earthquake is one of the terrestrial extreme natural events or hazards or disasters that occur due to sudden tectonic movement of the continental and oceanic plates caused by the thermal conditions of the interior of the earth. In other words, earthquake is also known as an extreme tectonic event that takes place due to abrupt endogenetic forces working within the earth. Strahler and Strahler (1977, p. 159) define earthquake as a movement, motion, tremor, or vibration of the surface of the ground. According to them, "earthquake is a form of energy of wave motion transmitted through the surface layer of the earth in widening circles from a point of sudden energy release-the focus". Earthquake widely affects population (particularly urban) and changes demographic structure through the dynamics of population change. It badly affects urban cultural features such as residential, commercial, institutional, and other buildings, parks, roads, rails, airports, ports and harbours, supply of electricity, telecommunication networks, dams, bridges, factories, water supply pipe lines, sewerage, drainage, etc.; economic aspects such as land and its use, employment, economic activities, and economic condition. It affects human health, medical assistance, and helps originate epidemics and several health-related problems during and after the earthquake. It causes deformation of ground surface, disturbance in ground water conditions, damages vegetation cover, and deteriorates water and air quality and overall urban environment. Nola (2018, pp. 19–20) quite clearly explains the direct and indirect impacts of earthquake on environment, medical, and public health. Earthquake terribly changes the environment and badly affects the population of the affected area. The consequences may be fatal and can remain in the environment for a longer time. The result of direct impacts of earthquake is seen in terms of damage of land, buildings, structures, and harms to inhabitants which took place just at the time of occurrence of the quake. Earthquake also directly causes ground failure, tsunami, landslide, volcanic eruption, disappearance, and formation of lakes or ponds. The consequences of indirect impacts of earthquake are observed with regard to drinking water pollution due to leakage and breaking of water pipes, gas and fuel pipes; air pollution due to leakage of gas, fuel, and presence of dust particles and other substances originated from damaged factories/industries and buildings, etc. Hazardous waste materials such as sewage, biomedical waste, radioactive, heavy metals, and other poisonous material released becomes the source of pollution. For example, 11 March 2011 Fukushima (Japan) earthquake, extremely polluted air, water, and land for long time and caused several health problems. The 1737 and 1985 earthquakes severely damaged Kolkata City and Mexico City, respectively, and disturbed the urban environment to a great extent. Particularly, health and public health-related aspects during and after the earthquake become more challenging in urban areas.

Nowadays, earthquake is considered as one of the challenging and complicated global urban environmental issues. Rapid urbanization, irrational use of land and other resources, unplanned construction, inadequate arrangement of infrastructure, and availability of up-to-date information and communication technologies have increased the risk of an earthquake in major cities of the world.

Earthquake is more disastrous than any other type of natural hazard in terms of its adverse effects on human population; human, cultural, and natural resources, natural habitat, etc. Globally, earthquake kills more people in comparison to other nature induced hazards. Related literature shows that on an average one person out of three in the world is under earthquake exposure and majority of this is found in urban areas.

According to United Nations (2018, p. 9) the world's total number of cities with at least 500,000 population in 2018 were 1146 and out of them 679 constituting 59% were found highly vulnerable "to at least one of six types of natural disaster, namely cyclones, floods, droughts, earthquakes, landslides and volcanic eruptions", while the cities with 500,000 population and above together were found highly susceptible to at least one type of natural hazard. 189 cities, mostly located along coastlines were found associated with greater risk of vulnerability to two or more types of natural disasters. Twenty-six cities including megacities namely Manila, Osaka, and Tokyo have faced high risk of exposure to three or more types of disasters.

The distribution pattern of past earthquakes evidently shows three major earthquake zones or belts in the world and these are Circum-Pacific belt (the Ring of Fire), Mid-Continental belt, and Mid-Atlantic belt. Together the first and second belts account for about 80% of the total earthquakes of the world, and those cities of the world which are located along these belts always remain at high seismic risk and there have been many examples of disastrous earthquakes which took place in the past and accordingly disturbed the urban environment at large.

Volcanic Eruption and Urban Environment: Volcanic eruption is another complex endogenic force like an earthquake and is known as a natural terrestrial hazard or disaster. Volcanic hazard is associated with lava and mud flows, volcanic gas and ash, dust, cinders, fragmented rock materials, etc. Volcanic eruption comprising of aforementioned hazardous materials has a wide-ranging disastrous impact on the life and properties of human beings, human settlements, animal life, and on overall development process of the areas affected by volcano. However, sometimes volcano works as a boon to human beings by providing some natural resources. The fertile black soil (suitable especially for cotton cultivation) of the Deccan Traps of India formed by the spread of bulk of volcanic lava erupted at the end of the Cretaceous period is one of the exemplary examples of volcanic benefit. Leaving aside such instance, most of the volcanic eruptions have serious impacts on physical, social, cultural, economic, demographic, psychological, and political environment of the concerned towns and cities. Just for instance, the Mt. Pelee volcanic eruption of 1902 snuffed out 30,000 people except only two alive persons and totally destroyed the St. Pierre city, Martinique (West Indes) at once. This fully describes the impact of a volcanic eruption on various aspects of urban environment. According to Strahler and Strahler (1977, p. 148) history has been the witness of serious loss of life and destruction of towns and cities that existed in the neighbourhood of active volcanoes. The Laki volcanic eruption of 1783 caused enormous lava flows that covered a large area of Iceland and killed a large number of people, animals, and damaged several human properties. The Laki eruption also caused weather and climatic change in Europe in the years after its eruption resulting into crop failure owing to famine. Volcanic eruption generally causes tsunamis, weather, and climatic change. Huge quantity of fine volcanic dust and ashes rise high into the troposphere and stratosphere and remain suspended for years and build layers of dust and in turn cause change in weather and climate at regional and global levels. This is because of the fact that accumulation of dust alters the transparency of the atmosphere and decreases the amount of short wave solar radiation reaching the surface of the earth by scattering and reflecting some amount of incoming solar radiation, but at the same time the dust layer does not restrict the outgoing long wave radiation from the earth's surface. For example, enormous quantity of volcanic dusts, ashes, and fragmented rock materials erupted from Krakatoa volcano (Indonesia) in 1883 reached the atmosphere and formed a thick layer of dust in the stratosphere leading to a global decline of solar radiation between 10 and 20%. Emission of lava from erupting volcanoes also act as an environmental hazard and often destroys plants, vegetation, species of animals, human activities, etc. and thus causes an imbalance in urban environment.

Most of the active volcanoes of the world are found lying in the Circum-Pacific belt, Mid-Continental belt (Alpine Mt. chains and the Mediterranean Sea, also known as the Alpine-Mediterranean Trans-Asiatic belt), and in the Basaltic plateau areas. Some volcanoes are also found along the Mid-Atlantic oceanic ridge. There are twenty-five major cities around the world that lie in the vicinity of active volcanoes and are likely to be severely affected in terms of overall urban environment at any time. Such cities and their nearby active volcanoes are: Hilo, (Hawaii Volcanoes), USA; Naples, (Vesuvius), Italy; Goma, (Mount Nyiragongo), Congo; Legazpi City, (Mount Mayon), Philippines; Puebla, (Popocatépetl), Mexico; Anchorage, (Mt. Spur and Mt. Redoubt), Alaska, USA; Kagoshima, (Sakurajima), Japan; St. Pierre, (Mount Pelée), Martinique; Managua, (Masaya), Nicaragua; Batangas City, (Mount Taal), Philippines; Puerto De La Cruz, (Mount Teide), Spain; Petropavlovsk-Kamchatsky, (Mount K lyuchevskaya Sopka), Russia; Pasto, (Galeras), Colombia; Vancouver, (Mt. Baker), Canada; Santorini, (Santorini and Kolumbo), Greece; Shimabara, (Mount Unzen), Japan; Arequipa, (El Misti), Peru; Yogyakarta, (Mount Merrapi), Indonesia; Stromboli, (Stromboli), Italy; Seattle, (Mount Rainier), Washington, USA; Reykjavik, (Eyjafjallajokull and Katla), Iceland; Catania, (Mount Etna), Sicily; Quito, (Cotopaxi and Pichincha), Ecuador; Jackson Hole, (Yellowstone caldera), Wyoming, USA; and Auckland, (Rangitoto and Mount Wellington), New Zealand (https://www.thetravel.com/cities-that-could-be-buried-by-lava-at-any-moment/).

External Natural Factors and Urban Environment: The main external natural factors are landslides, geomorphic landforms (mountain, plateau, and plain), rivers/lakes, and natural vegetation.

Landslide: Landslides may be nature as well as human induced. Nature induced landslide is caused by the natural forces such as feeble geological formation, tectonism, volcanic activities, and heavy rainfall whereas human induced landslide is caused by human activities such as deforestation, construction activities, unscientific agricultural practices, excavation, etc. From the point of view of landslides, the Himalayan regions in India and similar regions in other parts of the world are vulnerable. The 1980 and 1984 landslides of Nainital city, the 1998 landslide of Ukhimath, and the 2003 landslide of Uttarkashi (Vernavat landslide) caused heavy loss of human lives and property. These are the best examples of the disastrous landslides that occurred in the Himalayas. Landslides being associated with different types of debris or sediment act as an agent in disturbing or blocking the roads, streams or rivers, water bodies and pollute the quality of water and air, and disturb the natural resources, the living wildlife habitats by destroying forest and natural vegetation. In fact, the physical, socio-economic, and demographic urban environment is completely disturbed by the occurrence of landslide in towns or cities. Therefore, in order to check the impacts of landslides there is a need to prepare maps of landslide vulnerable areas; not allowing human settlements and other development activities such as construction works, excavation, mining and quarrying, deforestation, etc., especially along geologically fragile hill slopes. Reforestation, increasing and enlarging people's awareness towards safety and security from landslides would also play a fundamental role in landslide disaster management.

Mountain: This plays a significant role in maintaining the living environment (composed of humans, animals, plants, and other organisms) by providing water, forest, mineral, and other natural resources. Such natural resources become the life support system for mountain people and other living things. Mountain being high in elevation always has a cold climate due to lower temperature and affects health of people. Sometimes it positively acts in maintaining favourable climate. For instance, the Himalayas prevent cold winds moving from Central Asia to enter into the Indian subcontinent. Similarly, it also works in causing rainfall on its southern slope by drifting the South-West Monsoon wind to the west coming northwards from the Bay of Bengal. Mountain also helps in forming fertile plain in its lower valleys by transporting silts through river water from higher elevation. Mountain also offers a healthy climate and picturesque scene to hill stations that attracts a large number of tourists for recreational purpose during summer season and also during other times of the year. Srinagar, Shimla, Manali, Mussoorie, Nainital, Darjeeling, Gangtok, and Shillong are some examples of hill stations located in the Himalayas. Mountain covers around 26% area of the surface of the earth and about 30% of the mountain people live in varying sizes of mountain towns or cities. Examples of varying sizes of mountain towns and cities from around the world are Mexico City, Quito, Caracas, Denver, Vancouver, Geneva, Nairobi, Tehran, Kathmandu, and Dehradun (Wikipedia). However, the mountain urban and non-urban people are always under the influence of harsh mountain environment. They have to face the rugged terrains, harsh climatic conditions, heavy rains, cloudburst, landslides, floods, and tremors. Side by side, deforestation, overgrazing, irrational use of land, increasing urbanization, emergence of slums, increasing number of tourists, problems of liquid and solid waste treatments become the reasons of air, water, and noise pollution in the mountainous areas which the mountain urban people have to endure.

Plateau: This covers around 33% area of the surface of the earth. Plateaus are generally sparsely populated. Plateau environment also has some significant contribution to the life of humans and other living things. It is the major source of environmental resources such as minerals, fertile soils, river and underground water, hydro electricity, pastures for grazing, and rearing animals. The plateau particularly at greater elevation maintains favourable climatic condition that attracts people to settle and run their lives there. Most of the mineral resources of the world are found in plateau areas, for example, gold is found in the western Australian plateau, diamonds, copper and gold are available in African plateau and similarly, coal, iron ore, manganese, mica, and other environmental resources are extracted from the Chota Nagpur Plateau of Jharkhand, India. In most of the plateau areas, the ground water storage capacity is low and confined to limited extent. The average amount of rainfall remains low but temperature considerably varies between winter and summer. This situation largely affects the life of both the urban and rural plateau people. There are several cities which are located in plateau areas and can be called as plateau cities. Bengaluru, Nagpur, Aurangabad, Pune, and Hyderabad from the peninsular plateau and Ranchi from the Chota Nagpur plateau of India; Lhasa from Tibetan plateau, Moses Lake and Ellensburg from Columbia plateau, and several other cities of different plateau areas are examples of plateau cities. The natural and human induced environments of such cities vary from one plateau to another plateau area.

Plain: This is a major natural geomorphic landform and covers about 41% of the earth's surface. It is characterized by almost flat expanse of fertile land. Plains have a very gentle slope and low elevation. The Great Plains provide a congenial environment for better agriculture, easier human life and settlement, growth of industries, towns and cities, trade and commercial activities, transport and communication and for other human activities. The plains have a continental climate with cold winter and hot summer. Most of the plain areas in the world are densely populated and more urbanized and often suffer from different types of atmospheric hazards such as floods, water logging, famines, diseases, and cyclonic storms. Specifically, the towns or cities located in low-lying plains and in coastal areas round the world suffer more from such natural environmental hazards. In other words, those towns or cities which are located either on the bank of the main river or on the tributaries or surrounded by the tributaries of the main river are always at risk of inundation. For instance, several alluvial urban centres of the Ganga plain, such as Allahabad, Varanasi, Ghazipur, Patna, Muzaffarpur, Samastipur, Khagaria, Kithar Bhagalpur, Malda, and Kolkata always suffer from floods and water logging caused by either overflow of the river's water or due to occurrence of torrential rain. The cities most vulnerable to atmospheric hazards, for example, from different countries of the world are Guangzhou and Zhanjiang, China; Mumbai, Kolkata and Chennai, India; Saint Petersburg, Miami and New Orleans, USA; Guayaquil, Ecuador; Ho Chi Minh City, Vietnam; Jakarta, Indonesia; Nagoya, Japan; Bangkok, and Thailand.

In fact, the towns and cities located in fragile environmental zones like volatile hill slopes, alluvial riverine belts, coastal areas, tectonically and seismically susceptible zones, weak ground conditions, arid and desertic condition areas, are always at high risks of geomorphic land forms related environmental hazards.

Rivers/lakes and natural vegetation: These also play a great role in the urban environment. Rivers and lakes provide several natural benefits to the human society. Rivers and lakes are the means of navigation and water resource for irrigation, and industries and for other purposes both in urban and rural areas. These have been the attractive natural features and helped increase the urban civilization. However, rivers, lakes and other water bodies if get polluted with harmful materials, become the source of water pollution and sometimes create urban flooding. Similarly, natural vegetation cover significantly affects weather and climatic conditions and improves urban air quality. It also works effectively in mitigating the impact of frequent sandstorms arising and moving from the arid deserts to the urban areas.

Anthropogenic or Human Factors: The word "anthropogenic" is equated with the word "human" and anthropogenic factors refer to those forces or causes which are related to social, cultural, economic, and political human activities. Human beings are always in the process of changing or modifying the environment through their economic and other developmental activities looking to fulfil their needs. Anthropogenic activities are related to the primary, secondary, tertiary, and several other development-related human activities. Humans continue to interfere in the physical (land, water, air, river, lakes, vegetation, and other features), social, cultural, economic, and political environment to achieve their goals but at the same time the consequences of those activities have adverse effects that impact the urban people. The role of anthropogenic activities in affecting urban environment is greater than that of the natural activities or factors.

Anthropogenic activities in urban environment include urban agriculture, urban forestry, mining and quarry works, industrial activities, construction activities, trade and commercial activities, transport and communication-related activities, and other anthropogenic activities.

Urban agriculture: Urban agriculture-related literature shows that different scholars have worked differently and in their own ways, however, the study of Makaya and Todzwo (2019, pp. 1–3), Hallett et al. (2017, pp. 65–109) and Stewart et al. (2013, pp. 1–4) is extensive and provides a comprehensive idea about the urban agriculture. The common elements that emerged from their studies are as follows. Urban agriculture is one of the types of human economic activities and the resulting effects of that on urban environment are termed as anthropogenic results. Urban farming, being a part of anthropogenic activities in both the developed and developing worlds, is considered as an important source of livelihood and practised by urban residents generally in the backyards of the residential places, open urban spaces, wetlands, marshy lands, and in the lands adjacent to the waterways and streams suitable for agriculture with a view to produce any form of food to mitigate food crisis, unemployment and to promote farmer's economy, community building, recreation, social, economic, and environmental sustainability within the towns or cities as well as in their fringe or suburban areas. Some scholars advocate that urban agriculture promotes urban environmental sustainability and use of agricultural wastes and by-products as manure in the agricultural fields and reduces waste disposal costs in urban centres. Despite such potential benefits, urban agriculture causes huge diversion of lake's and river's limited water and pollute water, land resources, and disturb and destroy ecosystem and biodiversity in urban towns and cities especially in the developing countries. Urban farming also causes loss of some important indigenous plant species and disturbs the pristine urban environment through use of waste water, organic materials, chemical fertilizers, pesticides, animal wastes, etc. Urban agriculture also increases the health risk by polluting the urban environment. The use of waste water can contaminate food, standing rain and irrigated water in the cropping fields may increase malaria and different types of waterborne diseases, close contact with livestock increases the risk of zoonotic diseases, more use of chemical fertilizers and pesticides are likely to further harm human health and urban environment.

Even non-urban agricultural activities carried out far away from the cities also pollute urban environment. For instance, large-scale burning of crop stubble or residue and biomass in Punjab and Haryana produces carbon dioxide, nitrogen oxide, and particulate matter which reaches Delhi through the movement of air from west to east and increases air pollution in Delhi ever year particularly during winter season and causes serious health problems. So, some appropriate measures should be taken into consideration to check the urban agriculture-related problems for better urban environment and human health.

Urban Forest and Forestry: Urban forest (tree/plant covered urban area) and urban forestry (mechanism of planning, planting, caring, and managing forest) are very much related to anthropogenic activities and play a significant role in the overall urban environment. Urban forest provides social, cultural, economic, political, and environmental benefits to people towards their well-being in the towns, cities, and suburbs. The main domain of urban forestry is the urban forest and comprises green areas covered by trees in an urban setting. Urban forestry includes street and road side trees, trees in parks, private yards, cemeteries, and in other places; fruit trees and orchards, forests-natural and planted associated with vegetation (Konijnendijk & Randrup, 2004, pp. 471–472).

Urban forest reduces particulate matter and air pollution, energy use, makes the city cool by reducing urban temperature, benefits human health and social well-being, reduces stress, negative thoughts and symptoms of depression, promotes peace, concentration, physical activity, social ties, moods and life satisfaction, provides economy, forest resources, habitat and foods for animals, works in mitigating the effects of urban heat island, stores and sequesters carbon, checks obstruct runoff and storm water, support tourism, combats desertification, halts degradation of land and loss of biodiversity, protects and promotes terrestrial ecosystem's sustainability and overall makes the urban environment sustainable.

Despite all these benefits, lack of urban forest or urban forestry can increase the risks of runoff and can increase pollution in the air causing health problems and damage to property (Turner-Skoff & Cavender, 2019, pp. 323–339). Similarly, according to Referowska-Chodak (2019, pp. 1–3) human activities, due to rapid urbanization and pressure of increasing demands on urban green spaces, lead to deforestation, fragmentation of forest, unsatisfactory management of forest, alteration in habitat, deterioration in environment, translocation of plants, animals, and microorganisms, over use of urban and suburban forests. The effect of all this increases the concentration of carbon dioxide, intensifies greenhouse effects in the atmosphere, and raises urban temperature by creating urban heat islands.

Therefore, the local governments, non-governmental organizations, public-based associations, and urban residents should focus on planting, maintaining, and managing right types of trees/plants (as some particular tree species are likely to increase the risks of pests and diseases) along urban streets and road sides, in vacant areas, forests, parks, and gardens with a view to increase green urban spaces as urban trees are more effective in mitigating greenhouse gas emissions and improving urban environmental quality.

Mining and Quarrying: Mining and quarrying are the important anthropogenic activities for extracting useful mineral and other geological resources from the earth. Despite several economic benefits mining and quarrying harm people and surrounding environment. If mining and quarrying activities are practised within or near the urban areas, then this affects the urban people's health and quality of urban environment in varied forms. An important representative example of adverse effects of mining activity is the open pit mine in Cerro de Pasco city in Peru where the city residents badly suffer with their health problems caused by mine induced water and air pollution. Another important representative example of adverse impacts of coal mines and coal-based industrial pollution is Dhanbad (the coal capital city of India). Such anthropogenic activities significantly affect human health, quality of water, air, and other environmental resources of other mining towns or cities by mixing up with sulphides exposed to the soil, air, and water. Once the sulphide minerals are dissolved in rain or stream water then water becomes acidic (known as acid mine drainage). This acidic water further dissolves other harmful surface materials and badly affects nearby urban streams, rivers and aquatic life. Dust and other particulates originated during mining operation pollute air and when such particles are transported by air to the urban centres located even away from the mines affect health of the urban people and quality of urban environment. Toxic mining substances and other mining wastes contaminate both the surface and ground water which become harmful for living organisms. Noise created by blasting, transporting, grinding, crushing, etc., also becomes harmful for urban people. The impact of mining activities is also observed on loss of habitat, wildlife existing in the town's nearby mining areas (Guidebook for Evaluating Mining Project EIAs, 2010, pp. 1–17). The impact of mining and quarrying activities on urban environment depends on the use of mining and quarry methods such as the air blasting, extraction of mining materials, and transportation. Such activities generate noise, dust, fume, and gases and these are transported by air to the urban residents leading to cause human health problems (Bacci & Diniz, 2013, p. 371).

Industrial Activities: Many large- and small-scale industries are located in and in nearby areas of the cities and play a vital role in the life of human beings because every aspect of human life is directly or indirectly related to industrial activities. Industry being a part of human economic functions is the major source of human prosperity and industrialization is an indicator of modernization of human society. In fact, industries provide opportunities for social, cultural, economic, and political development. Despite all this, industrial activities are considered as a vital source of pollution of air, water, soil, and also cause global warming, natural resource depletion, and destruction of natural habitat in and around towns or cities.

Industrial activities release carbon dioxides, nitrogen oxides, organic solvents, and other pollutants into the atmosphere. Most of these pollutants are produced by burning coal, natural gases, mineral oils, and other industrial minerals in generating and supplying energy and power in the industries, transportation and domestic works and badly affect urban people's health and the quality of urban environment in a number of ways. Emission of carbon dioxides, nitrogen oxides, organic solvents, and other pollutants in the atmosphere leads to increase in the greenhouse effects and ozone hole which in turn raises the temperature of the surface of the earth and creates urban heat islands. Industries functioning within or near towns or cities sometimes cause hazardous environmental conditions by emitting huge quantities of toxic and poisonous gases that very badly affect the urban population and even the non-urban population living in the vicinity of industrial centres. For instance, the leakage of poisonous gas (2–4 December 1984) from the Union Carbide India Limited plant located in densely populated old Bhopal city was the world's worst industrial disastrous incident which killed thousands of urban people and damaged many people's eyes. Similar examples are the 1986 Chernobyl nuclear disaster in Ukraine and the 2011 Fukushima nuclear plant disaster in Japan. These are the representative examples of the disastrous environmental effects of urban industries (Singh, 2019, p. 438). Similarly, several urban area-based industries and factories discharge poisonous dust, smoke, and toxic gases and adversely affect population at both the local and regional levels. Industrialization continues to put pressure on the exploitation of natural resources to meet the increasing demands of industries and leads to reduce urban forest's cover by the process of encroachment of forest land into industrial activities, excavation of land for mineral and rock material resources, collapse of ground surface owing to withdrawal of minerals, mineral oil, and ground water, reduce urban and non-urban agricultural land by conversion of land into urban industrial activities, pollute the ground water, and lower the ground water table. Apart from these, urban industrial wastes, polluted industrial water, toxic gases, chemicals, aerosols, ashes, smokes, and concentration of other industrial pollutants are becoming a big threat to the existence of global human society. Therefore, the government, factory workers, NGOs, and local people together must adopt such measures that may promote positive effects of industries on human health and sustainable urban environment.

Construction activities: Construction activities are one of the important anthropogenic activities and are very much related to the development process in space and time. Construction activities are considered as a major part of human economic activity particularly in urban areas in order to meet the needs of residential houses, commercial complexes, educational and medical institutions, administrative buildings and offices, factories, roads, railways, airports, bridges, and other social, cultural, and economic development infrastructures. Construction is of course a means of social, cultural, and economic development of an urban society and contributes to increase in standard of living and quality of urban people's lives. Despite important contributions to the human society, construction activities become the source of noise, dust, emission of greenhouse gas, and other harmful gases, temperature increase, air pollution, waste materials (solid and liquid), generation of waste water, and contamination of surface and ground water; destruction of natural vegetation, forest cover, wildlife and natural landscape; exploitation of minerals, fossil fuels, and energy; traffic and population growth, construction induced accidental hazards, odour and foul materials, and other hazardous materials and in turn negatively affect overall urban environment and human health. Ametepey and Ansah (2015, pp.21–23) identified several impacts of construction activities on the environment such as the exploitation and consumption of natural and human made resource materials, disturbance in biodiversity; generation of noise, vibration, and dust as local issues; transport-related road traffic problems; generation of different types of waste materials; land and soil pollution; emissions of toxic substances like nitrogen and sulphur oxides (volatile organic compounds) and chlorofluorocarbons in the atmosphere through the use of air conditioning, refrigeration and fire fighting activities; construction-related incidental hazards; consumption and pollution of water; consumption of electricity and other sources of energy. In fact, construction, so-called, construction industry uses both the natural and cultural resources such as forest and natural vegetation-based materials, sand, clay, land, water, limestone, cements, rock and iron materials, human resources, etc., and generates and expands the particulate matters in the air of the surrounding of construction sites right from initial stage of construction to operational and final stage of construction and even during demolition of old buildings. Recent researchers show that construction activities are responsible for 23% of air pollution, 50% of change in the climate, 40% of drinking water pollution, 50% of landfill-related wastes, and 40% of global energy use (https://gocontractor.com/blog/how-does-construction-impact-the-environment/). Therefore, all the stakeholders relating to construction activities should honestly follow the policies, rules, and regulations developed by the concerned governments from the point of view of sustainable urban environment and urban development. In addition to this, the government should from time to time strictly review the performance of construction companies, contractors, and other owners engaged in construction activities.

Trade and Commercial Activities: Among different economic functions, trade and commercial activities are also important human economic activities and contribute a lot to the welfare of society, economic growth, and other developmental aspects at local, regional, and global levels. Commercial activities according to Kim and Park (2018, p. 2) are fundamental and necessary activities in the daily life of people and their nature, types and characteristics vary depending on the place of their location and surrounding environment. Commercial activities are generally found located along the roads or streets of residential areas of small, medium, and big towns as well as in malls in particular buildings of a city. The intensity of trade and commercial activities is higher in central business district of big towns or cities compared to that of small or medium towns. Central business district is the hub of business, commercial complex, buildings of commercial functions, hotels, apartments, institutions of culture, services and finance and other supporting facilities such as appropriate and convenient traffic, smooth communications and other favourable places and environment suitable for business activities. Despite several social, cultural, and economic advantages, commercial activities have an obvious negative impact on the urban market area's environment and in its surroundings. The multi-storey commercial buildings prevent the air circulation and exacerbate acid rain and greenhouse effect. Apart from this, large and dense population as well as commercial and other human economic activities discharge artificial heat that directly affect the thermal condition and quality of the city air (Yaguang, 2011, pp. 258–260). In fact, the trade and commercial activity areas in the city increases the crowd of population and traffic during the business hours between the opening and the closing of the business activities especially in the day time and as a result the crowded population and traffic increase air and noise pollution and in turn harms the health of both the local residents and those visiting commercial areas. The trade and commercial activities are also responsible for dumping of both the liquid and solid waste unused/useless materials along the roads or at certain places in commercial areas. Such waste materials create unhygienic conditions and become the cause of spread of some diseases. Sometimes, such waste materials block the local drainages and increase pollution in the commercial areas. Therefore, there is need for increasing awareness among the people engaged in trade and commercial activities for proper management of waste materials with a view to keep the commercial areas clean and hygienic for maintaining better physical and built environment towards social well-being.

Transport and Communication-related Activities: Transport and communication are one of the important human economic functions everywhere in the world and play a vital role in the overall development of both the urban and rural areas. Transport provides essential services in moving humans, goods, and animals from one place to another place and communication becomes a source of conveying messages or information from one corner to another corner of the world. In fact, transport and communication fulfil in the economy of the area concerned the same function as arteries and veins do in a human body (Sinha, 1987, pp. 372–373) and provide opportunities of employment for human resources ranging from unskilled to semi-skilled and highly skilled. Transport and communication activities are not less important than other human activities in the development of an urban area. The cutting-edge technological development in the transport and communication systems has made the world more accessible by reducing transportation or travel distance and by reducing time to seconds in receiving information about any happening from across the world through information and communication technologies such as the internet, telephone, mobile phone/cell phone and the mass media such as radio, television, newspapers, etc. Satellites have made the communication even much faster. Reduction in transportation or travel distance and communication receiving time reflects an idea to denote the term "the world is shrinking". In other words, the influence of distance on travel and on information receiving time from one part to another part of the world has obviously declined owing to development and use of advanced level information and communication technologies (ICT) in transport and communication sectors or activities and that is why, the world is treated as becoming small or contracted. However, here the word "shrinking" does not practically mean that the actual size of the world is "contracting". It is just a hypothetical idea for the purpose of nomenclature in the changing scenario of the world. Ioannides et al. (2008, pp. 202 and 222) in their study also provided similar idea of reduction in the effect of length of spatial distance on travelling time and cost of transmitting and receiving communication from one place to another place in the world. Hence, they mentioned that the distance and physical location of economic activity will no longer matter in the era of ICTs in urban economic environment.

Along with all these positive impacts, transport and communication activities also have negative impacts on urban the environment. Transport and communication are the major source of urban environmental pollution because they emit atmospheric pollutants in different forms. Enormous growth in transport and communication sectors in recent years has further created several environmental problems in urban areas. Urban areas are the major user of automobiles (cars, buses, trucks, lorries, tractors, auto rickshaws, scooters, motorcycles), trains, ships, aircrafts, etc. These are the major means of transportation and use maximum amount of diesel, petrol, and other fossil fuels and emit carbon dioxides (CO_2), carbon monoxides (CO), nitrogen oxides (N_2O), and other nitrogenous compounds such as nitrogen dioxide (NO_2), nitric oxide (NO), nitrogen trioxide (NO_3) as atmospheric pollutants. Emission of such atmospheric pollutants from transportation in gaseous form becomes highly responsible for large-scale air pollution in urban areas. Worldwide increasing rate of the use of petroleum by different modes of transport particularly in urban areas is a main factor of increasing concentration of carbon dioxide at an alarming rate in the atmosphere. Such increasing concentration of CO_2 in the atmosphere continues to intensify the greenhouse effect which in turn raises the earth's temperature and causes global warming and change in climate. Fuglestvedt et al. (2008, p. 554) also stated that transport sector consumes and burns the largest amount of petroleum of the world and becomes a major contributor to global warming by emitting carbon dioxide, nitrogen oxides, carbon monoxides, and other atmospheric pollutants. They expected further increase in global greenhouse gas emissions from transport. Rodrigue et al. (2013, pp. 213–260) considered traffic congestion, scarcity of parking space, and transport-related accidents as serious problems particularly in larger cities of the world. Transport-related activities create several other environmental problems in urban areas. Since transport vehicles consume large quantity of energy and oil, discharge numerous pollutants like carbon dioxide, nitrogen oxide, etc., in the atmosphere and degrade air, water, and soil quality at local, regional, and global levels. Different modes of transportation produce differently the dust, smog, noise, and odours that become harmful to human health and adversely affect the quality of life of urban people. Transport activities also cause loss of habitat and biodiversity and disturb urban landscape. The worldwide human generated urban transportation activities and emissions of pollutants from transport vehicles are considered responsible for causing global warming and climate change.

ICTs and especially the internet play a vital role in social, cultural, economic, and political development and in protecting environment from pollution in the interest of the present and future generations. ICTs being an important source of such multi-dimensional benefits to urban people have negative impacts also on the urban environment through both the electronic waste and emission of greenhouse gases. The internet and an unprecedented growth in the number of users of the internet including ICT devices across the world are currently contributing to the emission of greenhouse gases. Similarly, the worldwide consumption of energy during the course of production, distribution, and use of information and communication technology hardware (equipment and devices), recycling and disposal of ICT-related discarded materials (e waste) become the source of air, water, soil, and other types of pollution in urban areas. These facts are based on the information and communication technology-related studies of Souter (2012, pp. 1–14), Vickery (2012, pp. 2–3), Cerf (2012, p. 5) and Houghton (https://www.oecd.org/sti/ieconomy/40833025.pdf).

Therefore, to check the phenomenal rise in greenhouse gas emissions from transport and communication there is need for major cut in the use of energy and fossil fuels in both the developed and the developing countries. The need is to search and find alternative source of energy and fuels such as hydrogen and methanol to be used to operate different modes of transport. Development of solar energy and biogas may be another alternative. The need is to find new ways to make ICT carbon free. Emphasis should also be on developing greener electronics. Worldwide large-scale afforestation and green vegetation in urban areas would reduce the greenhouse effect.

Other Anthropogenic Activities: This focuses on population growth and urbanization as population growth is held only responsible for urban environmental deterioration while other reasons (other human activities) as discussed in the above paragraphs are indeed the offshoots of overpopulation growth. Ever increasing world population puts pressure on the use of limited resources of the earth. Obviously, over exploitation of natural resources by fast increasing population is expected to cause depletion and exhaustion of valuable natural resources especially of non-renewable resources of the earth. Urbanization consequent upon rapid population growth is another anthropogenic activity that extensively affects the urban environment. In fact, continuous rise in world's human population (due to more natural population increase and fast rural–urban migration) is mainly responsible for the world urbanization, growing concentration of urban population, and for intensifying human activities in urban areas leading to cause the urban environmental problems in different forms. Hence, urban population itself through urbanization, industrialization, and other anthropogenic activities is the root cause of the urban environmental degradation and ecological imbalance at local, regional, and global levels.

Rapidly increasing concentration of population in urban areas of the world continues to mount pressure on the extra resources and services which may be actually or potentially available within or outside the urban centres to meet the continuously increasing demands for satisfying the needs of the growing urban population. All this successively leads to the generation and expansion of further human activities relating to construction, industry, trade and commerce, transport and communication, automobiles, ICTs, etc., and all these happen at the cost of valuable agricultural, forest, barren, and water body's lands available primarily beyond the outer limit of the main cities through the origin, expansion, and development of new urban areas in the form of urban sprawl or suburban sprawl. In this way, the ongoing rapid and haphazard geographic expansion of the urban centres completely changes the physical form of land use and land cover and causes deforestation, de-ruralization, loss of productive agricultural land, and degradation in the quality of the urban environment within and around the urban centres. To be very specific, urbanization increases the residential, commercial, cultural, and administrative buildings, roads/streets, sewage, drains, concrete surface area, automobiles, industrial and commercial activities, solid and liquid waste materials, smoke, dusts, aerosols, sewage water. These in turn cause air, water, noise, and other pollutions in urban centres. Urbanization also causes depletion of surface and ground water resources and pollutes sea, river, and lake water. Urbanization is the source of creating urban smog, urban heat islands, dust pollution domes, and other environmental problems like global warming and climatic change as a result of cumulative effect of the urban environmental degradation at global level. In a broader sense, fast urbanization, sprawling urban growth, and expansion of industrial activities change the thermal properties of the surface of the ground, decrease the albedo, alter the aerodynamic character of the surface, change the pattern of air circulation, reduce the speed of wind and total insolation, increase temperature, create urban heat islands and pollution dome in the lower atmosphere above the major urban centres. Urban industries and transportation create huge quantity of dust and poisonous gases in urban areas. Urbanization is also held responsible for nuclear catastrophe, chemical wars as anthropogenic hazards; crimes, wars, communal riots as social pollution, and formation of slums and poverty as cultural and economic problems. Some religious group people and their religions-related activities from across different parts of the world also contribute to some extent the environmental pollution to urban areas. Urban tourism is also a source of creating environmental problems in urban areas (Singh, 2019, pp. 2020–2021 and 422–440).

Therefore, there is a need for an urgency of checking urban population growth, unplanned and haphazard urban growth, and expansion of urban centres to make the cities sustainable and environmental friendly to enhance the well-being of the present and future urban population. A check on rural–urban migration and priority focus on ruralization and on the development of growth poles as counter-urbanization will be a step forward in making the world's cities sustainable and environmental friendly towards saving the planet earth.

1.1.2.5 Multidimensional Picture of Urban Environmental Problems

Urban environmental problems are complicated and multidimensional and are the result of a wide range of natural and human factors as stated above. Following are the detailed pictures of urban environmental problems which need to be addressed and intensively studied with a view to tackle and solve them in order to maintain and improve the quality of the urban environment to better the well-being and quality of life of urban people at the regional, national, and international levels. Since urban space and its environment are composed of several natural and human components, it would be pertinent to focus on each type of urban environmental problems.

Natural environment-related problems: These are urban environmental degradation, urban pollution (air, water, noise, dust, toxic wastes, and electromagnetic radiations), natural catastrophes (floods, droughts, cyclones, storms, earthquakes, tsunamis, and landslides), global warmin,g and climate change; *Social environmental problems* include ethnic-religious conflicts, surge of anti-social elements, crimes, corruptions, violence, social insecurity, terrorism, consumption of alcohol and drugs, human and drug trafficking, and the like. Such social crises frequently occur in urban areas; C*ultural environment*al problems are solid and liquid waste material disposal and its recycling, sanitation and unhygienic conditions, transport-related problems and risks (traffic congestion, accidents, and injuries), industrial and other occupational hazards and working conditions, poor housing condition and shortage of affordable housing, marginalization of local communities in urban areas (exclusion and inclusion forces), scarcity of drinking water, depletion of ground water, power supply cut, etc.; *Urban problems related to economic environment* are economic crises (unemployment, urban poverty, wage structure, economic inequality, utilization of child labour and food insecurity, cost of housing, price of real estate, etc.); Population environment-related problems are high urban population growth, large-scale rural–urban migration, concentration of overpopulation, slums population, population and resources, human health problems and diseases, declining fertility rate, family size and ageing population, etc.; and Problems related to Political environment consists of urban environmental governance and people's participation, integrated approaches and initiatives to the valuation of the urban environment, urban environmental and urban disasters planning, management, perception, and awareness.

1.1.2.6 *Challenges and Opportunities of Urban Environment*

Urban environment locally, regionally, and globally is viewed as a challenge and an opportunity for the present and future generations. This is because the urban environment has a major impact on global life irrespective of territorial or administrative boundaries, caste, creed, party, politics, and religion. Impact of the urban environment is generally noticed in productivity, human health, safety, amenity, and ecology. Coastal cities and towns located particularly at low elevation are expected to suffer more frequently by storms and floods which in turn will increase risk to livelihoods, health, and lives. The governments and other administrative bodies all across the world will finally be put under much stress. Environmental challenges include all natural or physical, social, cultural, economic, demographic, and political environment-related urban problems as stated above. Environmental issues such as *pollution* (air, water, noise, dust and toxic wastes); *degradation of natural support systems* (forests, land, minerals, ground water, and aquatic ecosystems, state of coastal areas, global warming, climate change, loss of agricultural land); *social crises* (social dispute, social deprivation, social disintegration, racial conflict, terrorism, etc.); *cultural* problems (traffic congestion, solid waste management system, scarcity of water, scarcity of open space, housing problem and inadequate infrastructure, etc.); *economic problems* (unemployment, poverty, rising real estate pricing, and other economic crises); *demographic problems* (migration, urban population growth, slum population and its development, etc.), and political inconsistency (weak governance of local governmental bodies, inappropriate urban planning, political instability, political conflicts, etc.) are the major challenges of deteriorating urban environment. All these are the result of rapid population growth, diversification in economic activities, and other factors operating at the local, regional, national, and international levels. The low and medium income countries are affected most severely by deteriorating environmental quality.

Notwithstanding several urban environment-related challenges the urban environment also offers various opportunities. Different types of urban environments are the source of different kinds of opportunities in terms of tangible and intangible resources. Each urban setting requires necessary resources for its growth and survival. Natural environment in this respect becomes of paramount importance for the development of an urban area. In fact, each urban centre is endowed with certain natural resources which become the source of its growth and for the survival of its people, animals, plants, and other organisms. In this context, *the urban natural environment* becomes a major source of several natural resources such as land, water, air, mineral resources, lake and river water, forest, plant and animal resources, marine and atmospheric resources, fossil fuels, biomass, solar and wind energy, etc. This is also true that most of such resources are exhaustible and limited in quantity and that is why there is a need for judicious use of such natural resources for the longtime benefit of the urban people. Other urban environments such as social, cultural, economic, and political are also very important for the development of entire urban areas however, these are closely related to surrounding natural environment and needs to be separately addressed. *Urban social environment* plays a great role in urban growth and in the prosperity and well-being of the urban people. Urban society in urban areas is the source of indispensable tangible or intangible social resources in the form of cash or kinds, services, information, love and affection, and hope through social interactions and interrelationships. Urban social environment provides opportunities of forming social organizations, social solidarity, supportive social networks, social relationships, social support systems, social service camps, etc., to help the elderly and their family, poor and the needy urban people through community participation, human welfare-based social services, social integration, and social interactions. Different social and religious groups have their social organizations and help the needy urban people particularly during disasters, accidents, and during any unexpected harmful events within or in nearby urban areas. *Urban cultural environment* is another important part of urban areas and is a result of the relationship of human population and nature. Urban population through different anthropogenic activities shapes the cultural environment (cultural system or landscape) by using physical space and physical environment for the benefit of the urban people. Urban cultural environment always remains the source of opportunities of several cultural resources which become beneficial for the urban inhabitants. Cultural resources in terms of wealth or property in urban areas may include human induced resources, cultural organizations, language, customs, beliefs and taboos, education and health-related services, educational, medical, and other institutions, tourism, certain archaeological structures, buildings and housing, roads, railways, air and waterways, bridges, power lines and electricity, gas lines and gases, petroleum, industries, trade and commerce, transportation, science and communication technologies and other cultural property and resources. All these are essential means and are used by urban people in everyday life to fulfil and satisfy their needs. *Urban economic environment or system* is one of the important dimensions of urban environment and has a vast scope of study from the point of view fast growth and expansion of urban areas and heavy concentration of population in cities at global level. Globally, urban areas are considered as the dynamos of economic growth. Urban economic environment is linked with different elements of physical, social, cultural, and political urban environment and creates provide opportunities of utilization of human capital, earnings, wage labour, income generation, standard of living, employment, informal and formal economy, work participation, job opportunity in urban agriculture, dairying, urban mining, industries, trade and commerce, transport and communication activities and in other types of governmental and non-governmental urban services; generation of tax and revenue collection, gross domestic products and the like. Urban economic environment in urban areas is the source of livelihood, human well-being, and the source of money sent by the persons working in city to support their families in rural areas. Urban areas due to their economic environment become the centre of attraction of in-migrants, consumers, businesses, manufacturing, economic properties and infrastructure or assets, economic management and entrepreneurship, and other economic functions. *Human (population) environment* in urban areas refers to population condition and focuses on human development, human resources, demographic dividends, occupational structure, urban population growth, and urbanization. All these are linked with urban prosperity and urban spatial development along with time. Similarly, urban political environment includes good political governance, development, and territorial security-based political strategies, visionary political organizations, good political plans and policies, political stability, and good international relations and all these can be considered as urban political opportunities in building environmental friendly sustainable urban development for peace, prosperity, and well-being of the present and future urban generations.

1.1.2.7 Prospective Measures for Improving Urban Environment

Here, the prospective measures stand for focusing on the possible ways to check the ongoing urban environmental problems. In this direction, the notion of environmental planning and management (relating to local municipal policies and programmes), environmental sustainability (in support of environmentally sustainable urban growth at local, regional, national, and international levels), environmental pollution, environmental hazards and health-related strategies (for combating environmental degradation and pollution hazards, including innovative technical, social and community-based responses to water, sanitation and waste disposal, industrial and urban energy pollution, together with public health protection and hazard reduction measures); and some local agenda such as innovative schemes for community natural resource conservation, community partnerships and cooperatives for greening local development programmes, recycling projects and other essentialities would be important prospective measures. All these prospective courses of action are expected to serve effectively in controlling urban environmental problems in order to fulfil the dream of eco-friendly urban environment and successively to promote the prosperity and well-being of the present and future urban population at local, regional, and global levels.

1.1.3 Urban Health

Urban health as a field of study did not get an official recognition till it became a sub-discipline of international public health during the mid-1980s, though the relation between health of the urban people and condition of urban life has been known for a long time (Sclar & Volavka-Close, 2011, p. 561). Health (urban or rural) is the core aspect of geography of human welfare and that is why the study of health matters much at local, regional and global levels. Urban health, here, means health of the urban people. Galea and Vlahov (2005, p. 1) also stated that "urban health refers to the study of the health of urban populations". Despite several urban benefits, the continuously urbanizing world has put much pressure on urban system and has increased urban health problems.

In this context, health of the urban people stands as one of the challenging issues in a globalized regime of the twenty-first century. It is one of the central components of the Sustainable Development Goals (SDG). Millions of urban people suffer from several health problems owing to lack of proper healthcare systems in urban areas. Privatization, globalization, and commercialization of health care have turned basic health services into a market commodity and caused unequal access to people of different sections of the urban society. Trained medical personnel and medical facilities are beyond physical and financial reach of a large number of urban people. Currently, approximately 40% of the total urban population does not have proper sanitary services and sufficient potable water and suffer from infectious diseases (World Health Organization, 2021). By 2050 more than two-thirds of the total population of the world will be urban which will further increase the burden of infectious or communicable and non-communicable diseases. Therefore, from this viewpoint the global urban health continues to become a focal theme of research in the field of urban geography and in other allied subjects. This section sheds light on the historical perspective of progress in the field of urban health, changing concepts of urban health, concept and meaning of health, urban health: a resource, a problem, and a global challenge, and future perspective.

1.1.3.1 Historical Perspective of Progress in the Field of Urban Health

From antiquity urban areas all over the world have been the centres of primary focus from the point of view of development of human health. The concept of health has been changing throughout the history of the world. Historical perspective in this regard is taken into consideration as an important approach to portray the varying concepts of health that emerged through the process of advancement during different phases of time in the field of public health.

The Ancient and the Roman Period: During ancient times the concept of health was primarily related to the influence of religious faiths and the supernatural forces and hence the definition of health was not clear. People used to go to the religious institutions to appease the deities or gods with a view to receive blessings for curing diseases to achieve good health. People also believed that health problems were caused due to the action of supernatural forces and demons (Badash et al., 2017, pp. 1–2). But later with the passage of time this concept changed around the fifth century BC when Hippocrates made a significant departure from the predominantly supernatural concepts of health to a rational and scientific approach-based health and established a causal relationship between the origin of diseases and cleanliness of the environment and an individual person (Yapijakis, 2009, p. 507). Such ideas of Hippocrates came out from his beliefs that the human body consists of four fluids (humours) such as "black bile, yellow bile, phlegm and blood" and four natural or environmental elements (earth/ground, water, fire and air/wind, and their corresponding conditions or qualities such as dry, moist/humid, hot/heat and cold). In other words, health and sickness was observed as a result of the balance or imbalance condition between the said four fluids and elemental conditions. The positive state of health was found existing when these humours and elemental qualities were in balance, but imbalance or in-equilibrium among these humours and elemental qualities caused illness or negative state of health. Therefore, Hippocrates considered health as a state or condition of the body determined by the internal balance of the four bodily fluids and this balance depends on the equilibrium between external environmental forces (earth, water, fire, and air) and individual behaviours or habits relating to diet, alcohol, sex, work, leisure and exercise) (Yapijakis, 2009, p. 508;

Tountas, 2009, p. 186). The outstanding contribution of the iconic Hippocrates in the field of medical science is still very popular. After him Galen (born in 131 AD in Pergamon) became the most excellent physician during the time of the Roman Empire (beginning from first century BC to fifth century AD) and comprehensively expanded the Hippocratic concepts of health. He contended that the equilibrium among the four bodily humours play a significant role in determining the temperament and personality of an individual. The Roman period in presence of Galen became very important from the point of view of development of the concept of health. Galen (also known as Neo-platonic physician) provided a broader concept of health by incorporating mental and emotional condition. The Roman period also became important from the point of view of first attempt in organizing freshwater, sanitary, and healthcare system at societal level. In this way, the concept of maintaining health shifted from individual to population level (Badash et al., 2017, p. 3). According to Berryman (2012, pp. 210–211) Galen considered health as a bodily condition (sick, healthy, static, good to excellent condition, etc.), not merely the absence of disease and people can attain the best level of health. In fact, Galen considered health as a physiological, mental, and emotional state. Galen's concept of health was further accepted and promoted by the Christian and Muslim medical practitioners even during the medieval period.

The medieval period: This period spans between 476 AD and about 1500 AD and witnessed the constant threats of several communicable diseases (the plague, cholera, smallpox, etc.) with which the people were confronting. The occurrence of the fourteenth century (1348 AD) Bubonic plague (termed as the Black Death) was the most horrible transmissible disease that killed about twenty-five million people on the European continent (Encyclopedia.com). According to Richard (2019, p. 170) Black Death became a cause of about 20% deaths of population of the world, whereas the estimated percentage of death toll to the total population of England was in the range of 25–60%. Looking into the horrifying condition the concept of "quarantine" (the mechanism of keeping the communicable disease afflicted (fully or partially) person or persons in isolation from the contact of those not affected with the virus of infectious disease) as the new public health measure of protecting health was first introduced under practice during the fourteenth century and to be specific in 1377 to protect coastal cities people from plague epidemics (Tognotti, 2013, p. 255), though the practice of isolation was identified even during the ancient times to differentiate different diseases on the basis of nature of their ending duration. But officially the first English quarantine regulations were documented in England in 1663 as an effective public health measure (Conti, 2008, p. 456; Tognotti, 2013, p. 255). Later, it was used as the most effective public health measure tool in modern period in different countries of the world and it is proven a very effective mechanism to protect human health from Covid-19 identified first in China in 2019 and declared pandemic by WHO in 2020. Lockdown, another new public health measure concept like quarantine, was added to the history of urban health to restrict the movement of people and to keep other human activities at halt during the risk of pandemic or during certain emergency period. A formal attempt at lockdown policy was initiated by Italy in 1348 (News 18), though the known use of the term "lockdown" was made for the first time in 1973 to confine prisoners in California, USA (Lt. Hendry, 2015, p. 5) and Wikipedia. Since then it has been in practice in different countries of the world. Amid such development, the traditional and supernatural concepts of health of the Greeks were also widely practised during the medieval period and the issue of environmental purification, personal hygiene, and health education from health perspective continued as a major concern for the medieval people. They tried to reform the sanitation and health condition of the people, but could not fully succeed in achieving the goal owing to the lack of proper knowledge, information, and technology concerning public health. However, whatever the development in the public health sector took place during the middle age became the cornerstone for further development in public health in reconnaissance and modern times.

The Early Modern Period: The early modern period comprises the Renaissance period (especially sixteenth and seventeenth centuries) and post-Renaissance from 1700 to 1789 A.D. The Renaissance has been an important period for the invention and advancement of scientific knowledge in the field of medical sciences in Europe. After experiencing a huge loss of human lives from the fourteenth-century dreadful plague (termed as Black Death) the Renaissance period medical scientists made a sincere effort to renew and refine the health-related concepts and achievements of the Greeks and the Romans (Wikipedia-Medical Renaissance). Religious, supernatural, and traditional ways of treatment of a sick person continued in practice but because of important development in scientific knowledge in the field of medical sciences the influence of religious and supernatural causes of diseases began to decline during 1500–1700 AD. During this period there was scarcity of health-related national policies in the European countries, the problems of health of the urban people remained similar to that of the Middle Ages (Britannica.com). Some groundbreaking developments in the field of human anatomy by Andreas Vesalius, surgery by Ambroise Paré, circulation of blood and physiology by William Harvey are excellent contributions of this period. These are the unique examples of advancement of medical Renaissance (Toledo-Pereyra, 2015, p. 1). The search of microscopic organisms as a cause of transmissible diseases by Robert Hooke and Antoni van Leeuwenhoek between 1665 and 1683 was also an epoch-making achievement of the medical Renaissance. Such achievements became the foundation for further health-related developments beyond the seventeenth century and especially during the modern period in the history of medical and geographical thoughts. In fact, the Renaissance Period was characterized by the continuity of the ancient and Roman thoughts of health as well as by the development of new medical knowledge and scientific achievements as changes. Post-Renaissance period covers almost the entire eighteenth century except for the last eleven years. This period is noted for some important events from the point of view of urban growth and urban health. During this period European urban population speedily increased and the awareness about infant mortality and unpleasant conditions of prisons and mental hospitals also increased. As a result this period witnessed the beginning of constructing better medical institutions and enlarging health-related awareness in Europe. This period is also noted for the beginning of industrial revolution (in 1760) and for the age of enlightenment and the age of overcoming the superstitious belief concerning human health by the discovery of rational and scientific health-related ideas. This period is also noted for change from monarchy social system to the representative government by creating nation states in Europe and America. Industrialization and urbanization despite being the developmental facts became the reasons of rural–urban migration, overcrowding in poor quality housing, slum formation, filth, poor sanitation, poor living condition, air, water, and noise pollution, scarcity of safe drinking water and other urban health-related problems. These occurrences were the reasons of outbreak of epidemics like cholera, smallpox, typhus, etc. Such adverse situation deteriorated general health of the urban people and increased death rate in most of the European and American cities. Blake (1948, p. 1539) also mentioned that the eighteenth-century industrial revolution became the cause of urban health problems in England and at the same time in America. The mechanism of isolation and quarantine was in common practice to contain specific infectious diseases like plague, cholera, smallpox, and typhus. Particularly Constantinople faced epidemics many times between 1750 and 1800. Looking into the serious effect especially of the smallpox on urban health Edward Jenner, a British doctor, developed for the first time the smallpox vaccine in 1796. This was indeed an epochal contribution of his towards eradication of smallpox. Jenner's vaccine became an effective health measure against infectious smallpox disease in the modern times (nineteenth and twentieth centuries) after its inoculation and successful improvement in a thirteen year England boy in 1798 suffering from cowpox (vaccinia virus). The use of Jenner's vaccine received a big boost first in Europe, British, and Spanish colonies and later gradually in other parts of the world in the nineteenth century and towards the end of this century almost all European countries made the vaccination compulsory. By 1980 the WHO declared the world without smallpox (En.Wikipedia-smallpox) and (Biomol.com/vaccine). Leon (2008, p. 5) and Richard (2019, p. 170) in their study on urban health found that the towns or cities during the ancient, medieval, and early modern period were squalid, unsanitary, and hazardous from the point of view of outbreak and spread of dangerous communicable diseases for the health of the majority of urban population. Cities were frequently afflicted with epidemics that caused higher mortality and lower life expectancy especially in the European countries. In fact, the long-standing issue of sanitation, cleanliness of natural environment, and personal hygiene as perceived earlier remained a serious concern even during the pre modern and modern period.

The Modern Period: The long-standing issue of sanitary reform, clean physical environment, and improvement in public health system prompted the nineteenth-century health officials and organizations to take important steps with a view to protect urban health from epidemics and other diseases.

Nineteenth Century: The continuing industrialization and urbanization between 1801 and 1841 increased the urban population of London, Leeds and of other places to more than double and kept the mortality rate higher in several cities of England. The fast growth of urban population till 1850 led to severe insanitation, spread of feared contagious diseases (particularly cholera), poor health, overcrowding, shortage of housing and safe drinking water and medical facilities. Such conditions kept the general mortality higher and life expectancy near 40 or 41 in large cities of England and Wales (Leon, 2008, pp. 5–6). Owing to such conditions several European country's officials and organizations started to give more focus on the improvement in sanitation and urban public health conditions. Edwin Chadwick's work on sanitary reforms between 1832 and 1848 was one of the great strides in the field of public health. Edwin Chadwick, a noted proponent of sanitary reform, pointed out in 1838 in Britain that decay or decomposition of dead animals and human remains including organic materials, dampness, polluted water, filth in the streets and homes, compact and overcrowded housing were responsible for the outburst and spread of epidemics and higher mortality. The occurrence of smallpox in 1837, 1838, and 1840 in England is an important example of this. He stressed on sanitary reform, cleaning of streets and lanes, disposal of garbage from the streets and homes, ventilation in houses and hospitals, construction of sewers, and provision of supplying adequate potable water as the remedies of epidemics like smallpox, cholera, typhus, etc. He also pointed out that the origin and spread of infectious diseases were linked with poverty, poor standard of living, filth, and unsanitary conditions in cities of Great Britain (Chadwick, 1843, pp. 121–122). Blake (1948, pp. 1539–1543) was also of the same opinion that decay and decomposition of organic materials and filth were the major sources of contagious diseases. Chadwick's report also included financial requirement in supplying sufficient water in London and in other filthy towns for cleaning every room, house and street with a view to improve health and moral condition. Chadwick's movement of sanitation led to give England in 1848 an administration of public health (boards of health). Just before this, the discovery of chloroform by James Simpson in 1847 in Edinburgh and its use as anaesthetic substance in surgical treatment was also an important achievement during the movement of sanitary reform (Pubmed.Ncbi-chloroform). Such positive steps and the advances in European public health in the first half of the nineteenth century influenced the United States to develop effective mechanism to protect their public health from repeated epidemics like cholera, typhoid, smallpox, typhus, and the like. In 1850 Massachusetts greatly suffered from tuberculosis and very high infant mortality rate and as a result every town in Massachusetts in this year was empowered to constitute a board of health to make arrangement to clean the streets and houses, remove the filths, nuisance, garbage and control reasons of origin of diseases with a view to improve overall urban health. Recently after this, New York also suffered from smallpox and typhus because of a thick accumulation of garbage in the streets. The second half of the nineteenth century was also crucial from the point of view of several important developments in the field of urban health which took place on the continents of Europe and the United States. During this period the collective focus was given to further improvement in the city's environmental conditions, sanitary system, provision of supply of adequate and safe drinking water; research, experiment, and discovery in public health, identification and solution of urban health problems. The setting-up of the first public health department in 1866 in New York City and subsequently the board of health in California, Louisiana, Minnesota, Virginia, Columbia, Maryland, Alabama, and at many other places was an important step towards improvement in sanitary system. The discovery of antiseptics by Joseph Lister and its use in Britain in 1867 also became an effective means in surgery. In France, Louis Pasteur developed the germ theory of disease and proved the presence of bacteria in the air in

1873. In 1877 Robert Koch explained about the bacterial cause of anthrax disease and developed immunization system against this and other infectious diseases in the laboratories of the United States and Europe in 1884. With a view to develop and use scientific knowledge and skills in the direction of reforms in sanitation some governmental laboratories were established first in Massachusetts then in the city of New York, Michigan, and Providence (Rhode Island) in the last decade of the nineteenth century. During this period Sedgwick, W. T., a noted sanitation and bacteria-related scientist, detected waterborne bacteria causing typhoid and developed the technique of treatment of sewage (Institute of Medicine, 1988, pp. 58–63).

Richard (2019, p. 169–177) pointed out that poverty was the root cause of poor health and said that in the mid and second half of the nineteenth century focus was more on the specific areas of a comprehensive public health system. Health of mother and child was given priority by making provisions of health-related essentials in European countries and due to health and economy ameliorating initiatives, standard of living and health of the urban people substantially improved, and the public health was elementally changed towards the improvement in nutrition, environment, economy and in medical knowledge in the European, American and some other cities in the last quarter of the nineteenth century. In brief, he was of the opinion that measures of public health and improved living standard helped increase the expectancy of life of the urban people. Similarly, Leon (2008, pp. 5–6) was also of the similar opinion and asserted that the concept of urban health began to concentrate more towards the significance of the living standard or living condition of the urban people between 1850 and 1900. During this period urbanization and industrialization brought a significant change in social, economic, and health status of the urban dwellers in England and other parts of Europe and as a result the life expectancy at birth rose to about 46 years. All this was the reason for a decline in mortality. In fact, the late nineteenth century saw the great strides in the provision of public health and hygiene. However, there was still a lot of ill health due to the prevalence of epidemics and even by the end of the nineteenth century the life expectancy at birth remained under fifty years and the infant mortality higher, say 165 per thousand live births (Britannica.Com/public-health).

Twentieth Century: History is the witness of emergence or re-emergence of epidemics and pandemics across most of the urban tracts of the world caused by several human and nature induced factors, the mobility of people through sea, land and air routes for social, cultural, and economic reasons is attributed to contribute a lot in the spread of infectious diseases. The worldwide influenza pandemic of 1918 (known as Spanish flu) caused a great loss to the human population in the United States, European, and other countries of the world. Between 1918 and 1820 this disease infected about 500 million populations in the world (https://en.wikipedia.org/wiki/History_of_HIV/AIDS). In 1919 this influenza in India, aggravated by the 1918 famine, became the reason of about 13 million deaths of people. At the same time in 1918 the first public health was recommended in New Haven, Connecticut, USA. This happened due to poor sanitation, overcrowding, poor healthcare facilities, and human mobility from one geographical region to another. Edwin Chadwick also stressed on the link of epidemic with insanitary conditions and poor facilities of public health. Sclar and Volavka-Close (2011, pp. 561–562) stated that even at the start of the twentieth century unsatisfactory urban living, poor sanitation, and deficient healthcare facilities in the cities of industrializing countries of the west led to originate epidemics like tuberculosis, cholera, typhoid, etc. These diseases killed millions of urban people and forced the local, state, and national level governments to take necessary action in the direction of bringing improvements in the garbage disposal, sewage system, supply of water, and transportation. Such positive steps of the government were considered as a great success in the industrialized countries. Though, till 1950 the communicable diseases continued as the main reason of deaths in the urban areas of the industrialized countries. A sharp increase in the air pollution of the industrialized cities of the Western developed countries owing to continued industrialization during the middle of the twentieth century became an additional problem of urban health. This situation marked the beginning for making fresh legislative measure to control the increasing level of air pollution, though the quantity of carbon dioxide and other particulate matters in the atmosphere are still on rise in the cities of both the developed and the developing countries (McMichael, 2000, pp. 1118–1119). The development of anti-viral vaccine, the so-called polio vaccine during 1955–1960 was a significant achievement in the field of human health. In fact, the rational and scientific public health approaches adopted and progress in specific health measures made at the close of the nineteenth century played an important role in controlling the communicable diseases and in increasing the life span (longevity) of the urban people during the first half of the twentieth century. However, non-communicable diseases (the so-called primarily human lifestyle behaviour-based modern diseases) such as cardiovascular disease, diabetes, cancer, chronic respiratory, and other diseases emerged as a new challenging problem of health and became the leading reasons of deaths and disability especially in the industrialized countries. With an increase in toll, these diseases received greater attention from the health experts and government officials and the causative factors of such diseases were widely investigated in the second half of this century to study and analyse the intensity of their effects in deteriorating health of the people (Fielding, 1999, pp. xiii–xvi).

Advances in public health infrastructure, healthcare delivery system, technological and scientific approaches as well as development of effective vaccines, antibiotics, better medicines and their use at large scale during the second half of the twentieth century made a significant improvement in urban health status by controlling contagious diseases, increasing longevity and reducing mortality rate in both the developed and the developing countries of the world. HIV and AIDS evidently emerged as another urban health problem particularly during the period between 1980 and 2000, however, the cases, especially in urban areas, were suspected during the period between 1960 and 1980 also but it was not widely known. Though, its origin was first traced in Congo in 1920 as it is generally believed. The speed of mortality declined from all causes (infectious and non-infectious diseases and other reasons) and prolongation of life expectancy became much faster during the second half of the twentieth century with the introduction of proper diagnosis and proper medicines and other appropriate health measures. At the close of this century mortality from cardiovascular and other chronic diseases was reduced in most of the developed and developing nations of the world and this became possible only because of an improvement brought in the areas of diagnosis, prevention, therapy, individual lifestyle, and environmental conditions (D'Alessandro, 2020, pp. 1–8). Despite a continued sincere effort in the direction of progress of urban health, the effect of environmental hazards significantly increased the magnitude of urban health problems after 1990.

Twenty-first Century: Despite many advances, and achievements in the field of human health, urban people's health still continues to be a big global challenge for the national governments, international governmental organizations and agencies as well as for the medical experts, scientists, and researchers in the twenty-first century because of continuous urbanization and industrialization, rising social and economic inequalities, increasing urban poverty and unemployment, unsatisfactory sanitary conditions, scarcity of safe drinking water, shortage of housing, inadequate health care facilities, urban environmental pollutions, ongoing burden of infectious and non-infectious diseases. Mishra (2007, p. 509) also stressed that the first quarter of this century would be dominated by the issues of poverty, environmental pollution, child mortality, women's status, communicable and non-communicable diseases including lifestyle behaviours, etc. But these problems are also expected to continue unabated ahead in the present century. Amid all this, human health has been regarded as central to international development in the twenty-first century, for this various efforts have been made and are being made like earlier to reduce morbidity and mortality either universally or through a focus on specific population subgroups. The introduction of eight Millennium Development Goals (MDGs) in 2000 by the United Nations is considered by Ulrich and Helmut (2014, p. 1) as the first truly global health effort to improve worldwide population health. This included three health-related goals to be met globally by 2015. These three goals were related to the reduction in child (under 5 years) mortality (Goal 4); reduction in maternal mortality and access to reproductive health care (Goal 5); and reversing the spread of HIV/AIDS, tuberculosis, and malaria (Goal 6). These were instrumental in focusing on global resources for health especially in low- and middle-income countries. Though, the child and maternal mortality significantly declined all the targets of MDGs were not completely achieved by 2015 (Buse & Hawkes, 2015, p. 1).

In 2015, the MDGs were superseded by the Sustainable Development Goals (SDGs). The United Nations Sustainable Development Goals (also known as the Global Goals) relate 17 goals with 169 targets for which all the UN Member States agreed to work towards achieving them by the year 2030. They set out a vision for making the world free from poverty, hunger, and disease. SDG 3 is exclusively related to health and the only objective of this goal is to ensure good health and well-being to the people of all ages by reducing further the maternal, child, and general mortality through control of communicable and non-communicable diseases, environmental pollution, use of drug, alcohol, tobacco, and other health hazardous substances, and by providing requisite healthcare services; affordable, effective, safe, and good quality medicines and vaccines to all people across the world. All these are supported by 13 targets that cover a wide spectrum of WHO's work. The remaining other 16 goals are also indirectly related to health and their achievement in some way or other will contribute to the progress of health (https://www.undp.org/sustainable-development-goals#good-health).

The WHO, on 13 April 2021 presented the facts that each year around 71% of the total global deaths occurred only due to non-communicable diseases and continue to be a big threat to human health. The burden of non-communicable diseases is exceptionally more in less and limited resource-based low- and middle-income nations of the world. Among non-communicable diseases the relative dominance of cardiovascular diseases remains fairly high which is followed by cancer, respiratory ailments, and diabetes. The health-related sustainable development goal's objective is to reduce chronic disease-related premature mortality to one-third by 2030 by modifying or improving lifestyle behaviours regarding use of tobacco, alcohol, unhealthy diet; untimely food intake, more intake of salt and not doing physical exercise regularly, properly and adequately; by controlling metabolic related risk factors (such as high blood pressure; obesity, overweight, high level of glucose and fat in blood); improving socio-economic conditions as well as adopting required preventive measures (WHO, 2021).

Vaccine being a sort of injection or biological substance plays a great role in protecting a person from the risks of infections to be caused by viruses or bacteria of dangerous contagious diseases surviving in the atmosphere. The role of vaccines as the most effective public health measure in eradicating some of the dangerous infectious diseases (e.g. smallpox, polio) has been very commendable in the twentieth century. Vaccines are considered as the life savers because they protect the life of nearly 2–3 million people by boosting their immune system every year in the world (Delany et al., 2014, p. 708). The development of some important vaccines like Human Papillomavirus (HPV) vaccine for the protection against risks or infections of HPV, new Multicomponent Recombinant vaccine against meningococcus type B, the first therapeutic vaccine against prostate cancer, and some other vaccines (at that time already in clinical or human trials) through the application of new techniques at the start of the twenty-first century became an exemplary example of a significant contribution in the field of global public health and set the path for the discovery and development of more effective next generation vaccines against the infectious and non-infectious diseases (Siddiqui & Perry, 2006, pp. 1263–1271; Keam & Harper, 2008, p. 1; Giuliani et al., 2006, pp. 10,834–10,835; Plosker, 2011, pp. 255–256; Rappuoli et al., 2011, pp. 865–872; Delany et al., 2014, p. 717). Mascola and Fauci (2020, pp. 87–88) were of the same opinion that there is vast scope of production of most effective novel and improved vaccines in the twenty-first century to safeguard the world's people in general and the people of low-income countries in particular from the serious infectious diseases.

Truly speaking, the ongoing first quarter of the twenty-first century witnesses the worldwide prevalence of HIV and other communicable and non-communicable diseases impacting the global public health. The COVID-19 pandemic has been one of the most dreadful infectious diseases of the present century because it has dramatically led to a massive loss of human life worldwide and created an unprecedented challenge to public health, food system, employment, livelihood, working condition, land, air and water transportation system, industries, trade and commerce, governmental and non-governmental services, social connectivity, other social, cultural and economic functions and the like by keeping everything under lockdown system. The first human cases of COVID-19 were identified in Wuhan, People's Republic of China, in December 2019. The World Health Organization declared the COVID-19 outbreak a Public Health Emergency of International Concern on 30 January 2020 (WHO, 2021), and a pandemic on 11 March 2020 (WHO, 2021). Globally, as on 14 April 2022, there have been 500,186,525 confirmed cases of COVID-19, including 6,190,349 deaths, reported to WHO. Scientists around the world are working faster than ever to develop and produce vaccines that can stop the spread of COVID-19, with 21 vaccines now being rolled out in countries worldwide. There are now 137 COVID-19 vaccine candidates undergoing clinical trials and 194 candidates in pre-clinical development. According to World Health Organization as of 13 April 2022, a total of 11,294,502,059 vaccine doses have been administered globally. Even after the development of so many vaccines and other technologies we are still in the middle of the Covid-19 pandemic crisis and our health services are over burdened.

It can be summarized that during the second half of the twentieth century the general notion was widely created that communicable diseases were near an end but this was not fact (Sclar & Volavka-Close, 2011, p. 562; D'Alessandro, 2020, p. 2). Kumate (1997, pp. 155–161) had rightly pointed out that communicable diseases as an important problem of health would continue to emerge or re-emerge due to ongoing urbanization, globalization, environmental problems, use of pesticides and fertilizers, poverty and due to the effects of other health-related problems in the twenty-first century. Similarly, Bloom and Cadarette (2019, p. 1) stated that the world is expected to continue to encounter the threats of past communicable diseases to be emerged or re-emerged even despite much efforts in the development of global public health system in order to protect and boost up human health. The emergence of infectious diseases would have serious impact on socio-economic and demographic (morbidity and mortality) aspects. In fact, infectious diseases continue to exist and re-emerge in the name of new epidemic. Covid-19 is a perfect example of re-emergence of infectious disease. This was first found in Wuhan, China in December 2019, which was later recognized and declared by WHO as a pandemic in Feb. 2020. In this way, both the communicable and non-communicable diseases which were the great killers in the past are still present in different forms in the twenty-first century. Covid-19 (Corona or Coronavirus) a most infectious disease still exists around the world. The Prime Minister of India (Sunday, 24 April 2022), while addressing the nation's people through the 88th Edition of Mann Ki Baat, raised concern that Covid-19 has not finished yet rather it is still a great challenge for us. He advised people to stay cautious from the threat of the coronavirus and continue to follow Covid's appropriate behaviour. In fact, as of today (2022 AD) the global health situation in general and developing country's health situation in particular is not good and remains a serious challenge ahead in the twenty-first century.

1.1.3.2 Changing Concepts of Urban Health

In an applied sense, the concept of health (urban or rural) has undergone fundamental changes owing to certain type of rational and scientific advancement in the field of human health along with the passage of time. On the basis of practical approaches the concepts of human health as emerged through time are categorized under the following heads:

Supernatural Concept of Health: This refers to the people's beliefs in supernatural, religious, and divine forces as controlling factor of their health. In the early days, people used to believe that health of an individual was a result of the influence of supernatural forces. People used to go to the religious institutions to appease the deities or gods with a view to receive blessings for curing diseases to achieve good health. People also believed that health problems were caused due to the action of supernatural forces and demons. Though, the health related supernatural and religious faith still exists especially among those who are associated with the lower educational, socio-cultural, and economic status as several people still use to visit religious institutions with the intention of receiving blessings from God for curing diseases and maintaining good health.

Individual to Societal Concept of Health: A shift from supernatural, religious, and magical beliefs to behavioural and medicinal actions at individual level took place during the Hippocratic era of health. Around the fifth century BC Hippocrates established the fact that the origin of disease was related to the personal hygiene and environmental cleanliness and defined health of an individual as a state or condition of the body resulting from the balance between internal elemental conditions and external environmental factors including individual's behavioural habits. Such individual concept of health shifted to the societal concept of health with the scientific progress made in the field of sanitation and public health facilities during the Roman period. Therefore, sanitation and public health-related innovations during the Roman Empire lead to shift the healthcare focus from an individual level to the level of an entire society.

Quarantine/Isolation and Lockdown Concept of Health: The quarantine/isolation as a new public health measure concept was first introduced in 1377 in Dubrovnik (a Mediterranean coastal city in Croatia) to protect coastal cities people from plague epidemics which was likely to spread through mobility of persons through oceanic routes. However, the first official English quarantine regulations were documented in England in 1663 as an effective health measure to prevent infectious diseases. The root of such concept, however, is linked with the practice of isolation to differentiate different diseases on the basis of nature of their ending duration in ancient times. Lockdown, another public health measure concept like quarantine evolved in the field of urban health to restrict the movement of people and other human activities at halt during the risk of pandemic or during certain emergency period, was initiated by Italy in 1348, but it was popularly used for the first time in 1973 to confine prisoners in California. The quarantine/isolation and lockdown as the most effective mechanism of protecting health has been proved successful even during the modern times. The best example of this is ongoing Covid-19 pandemic during which different countries of the world have imposed lockdown system with a view to protect health of their people.

Rational and Scientific Concept of Health: This refers to those health-related ideas or thoughts which were rationally and scientifically revived and discovered first in the field of urban health. The period during which such progress was made is widely known for medical Renaissance as during this period the religious/superstitious beliefs (developed by the Ancients and the Romans) were revived and refined as well as some new rational and scientific knowledge were developed by Andreas Vesalius in the field of human anatomy, Ambroise Paré in surgery and by William Harvey in circulation of blood and physiology. The discovery of microscopic organisms as a cause of transmissible diseases by Robert Hooke and Antoni van Leeuwenhoek between 1665 and 1683 was also an epoch-making achievement of the medical Renaissance. The beginning for the construction of better medical institutions and an enlargement in health-related awareness in Europe together with the discovery of smallpox vaccine for the first time in 1796 by Edward Jenner (a British doctor) are the important contributions of the post-Renaissance period (popularly known for the age of industrial revolution and the age of enlightenment and the age of overcoming the superstitious beliefs of health).

Sanitary Reform Concept of Health: This refers to the condition of filth and sanitation (the result of growing industrialization and urbanization) and its connection to the infectious diseases. The movement of sanitary reform idea started between 1830 and 1850. The Chadwick's role in this regard is unforgettable. He along with others believed that stagnant piles of filth were the main reason for foul air in the atmosphere and the prolific source of infectious diseases and that is why a drive to remove filth from the cities and towns became a major focus against the contagious diseases. Filth was identified as a great enemy of human health and that is why a crusade against anti-filth was started as a result of great sanitary awakening. Ultimately in 1848, a legislative Act relating to sanitary reforms was enacted which played a considerable role in protecting human health against infectious diseases. The "filth theory of disease" originated during the mid-nineteenth century (https://www.lexico.com/definition/filth_theory) along with the sanitary movement greatly helped in improving sanitation and hygiene and in reducing infectious diseases. This theory dominated medical thoughts as it was believed that health of the people could be improved by cleanliness and sanitation at community level in the filthy urban areas (Sydenstricker, 1935, pp. 301). Popularity of the "filth theory of disease" was mainly due to the endeavour of Chadwick in England, however the work of Pettenkoffer (Germany) and Shattuck (USA) also popularized this theory (Chapin, 1902, pp. 1–2).

Biomedical Concept of Health: This refers to the concept of health which deals with the germs or bacteria responsible for causing illness or disease in humans or animals. Traditionally, good health of a person or a healthy person is perceived to be free from disease. In fact, this concept concentrates on studying the presence of germs in or on human, animal, or any other living object and its harmful or beneficial effects in causing either ill/bad or good health as some germs are beneficial to health and some others injurious (pathogenic) to health or in other words some germs help promote health and some others deteriorate health by causing infection/disease. The biomedical concept of health is based on the "germ theory of disease" developed with the epochal experimental works of Louis Pasteur and Robert Koch about the presence of germs or bacteria in the air as the cause of infectious diseases. Louis Pasteur (a French bacteriologist) proved the presence of bacteria in air in 1860 and further advanced the theory in 1873. Similarly, Robert Koch proved bacteria as a cause of anthrax disease in 1877 (Park, 2009, pp. 5 and 12). Sydenstricker (1935, p. 301) stated that the work of Pasteur and Koch passed through a golden period of bacteriology and was dominated by the idea that health of the people could be successfully attained through the action of medical measures. This theory widely dominated medical thought at the turn of the twentieth century. Germ theory of disease played a considerable role in eradicating some major infectious diseases such as plague, cholera, smallpox, etc., and saved lives of millions of people in the world. In fact, germs cause communicable diseases, not non-communicable diseases like cancer, cardiovascular disease, hypertension, diabetes, mental disorders, deficiency diseases, drug abuse, alcoholism, genetic deformities, etc. These are basically related to the lifestyle behaviours of the people. The germ theory of disease became the basis for the development, improvement, expansion, and application of modern medicines (antibiotics, sterilization kits, chemotherapy) and advanced techniques of surgery and helped in controlling and eradicating infectious diseases which are widely prevailing even today especially in the developing countries of the world (Mishra, 2007, p. 7). This theory is, however, criticized on the ground that this has reduced the role of the environmental, social, cultural, psychological determinants of health especially in urban areas. Advancements in medical, science, and social sciences indicate that biomedical concept of health was not adequate in dealing with all the causes and remedies of diseases.

Sociological, Psychological, Cultural, Economic, and Political Concepts of Health: These aspects generally seem to be neutral and not supporting health but it is not so. Different elements of these aspects directly or indirectly determine health. *Sociological concept* focuses on social structure, social cohesion, social interaction, social services, social cooperation, etc., which help promote health at the community level. Social pathology such as crime, alcoholism, drug edict, deviant behaviour, etc., evidently affects people's health to a large extent in the society. *Psychological concept* of health refers to the psychological elements which affect health of an individual. Such elements may be stress, anxiety, strain, anger, emotion, depression, hopelessness, hostility, negative feelings. These cause several physical and mental problems such as hypertension, coronary heart disease, asthma, ulcer, mental disorder, deviant behaviour, etc. Negative feelings or negative approaches forced the concerned person(s) to think that the things happening around her/him are wrong or going wrong. Such negative thoughts lead the person to perform wrong deeds/activities and even to commit suicide. *Cultural concept* of health refers to the involvement of several cultural elements such as race, religion, beliefs, spirituality, way of life (lifestyles), family and community atmosphere, working conditions, literacy, education, healthcare facilities, etc. These cultural elements are so deeply involved in the promotion, maintenance and conservation of health that health can be considered as cultural construct. In other words, health is a product of interplay of several cultural components prevailing in a culturally defined region at a given point of time. *Economic concept* of health explains how health is perceived from economic perspective. Economy and health of an area are so closely related that no one can perceive health without economy and hence health is called as a precious wealth of concerned person(s). Poverty, employment, occupation, income, standard of living, etc., shape the condition and value of health of the people of an area. *Political concept* of health is meant for those political thoughts, ideas, plans, policies, programmes, and actions involved in protecting and promoting health of both the urban and rural people. The development, production, allocation, distribution, and delivery of all public health-related resources to the people of different societies at local, regional, national, and global levels are always under the influence of governments and political power and that is why health is also considered as political phenomenon. This is also a result of the process of temporal advancement and along with that a gradual conceptual shift in the field of medical health from individual, societal/community, and medical institution to the political (governmental) levels all over the world. In fact, political concepts play a considerable role in health and well-being of the people everywhere in the world.

Geographical Concept of Health: This involves those geographical thoughts, ideas, knowledge, facts, notions, perception, information, understanding, mapping tools, and techniques which remain firmly connected with a subject under study and provide the basis to clearly understand the exact meaning, nature and contents of the concerned subject to arrive at certain convincing and informative conclusion. The geographers generally place emphasis on the application of the concept of space, time, scale, location, site, situation, proximity, connectivity, mobility, direction, interaction, and interrelationship with regard to the empirical study of a given geographic phenomenon (for instance here "health"). The role of such applied geographical concepts in finding locations of diseases, healthcare facilities, services, resources, and healthy and sick people in space at a given point of time becomes very vital from the point of view of adopting measures to prevent diffusion of diseases and to properly distribute and utilize health care services and resources for the promotion of urban and rural people's health. The geographical concepts such as the spatial concept, temporal concept, population or demographic concept, physical or natural concept, geographical information and mapping concept are applied in representing the condition of health of the people of different parts of the world.

Spatial concept-based spatial analysis helps in knowing who lives where and in what condition of health; helps in finding where and why some people are sick; helps in finding why some particular diseases are more prevalent in some portions than other portions of the world; helps in determining the position of the number of people of a particular place who were or are availing the adequate facilities of health or inversely, how many people in a particular area of a city/town or village were not or are not availing health facilities adequately. The application of spatial network analysis helps in formulating plans and policies for better delivery of healthcare facilities and services with a view to benefit the maximum people of the society at minimum cost but in case of poor people without cost. The application of cartographic and geographic information system (GIS) tools and techniques in geography for mapping diseases, heathcare system, mortality rate (infant, adult, and old people mortality rate), life expectancy at birth, food and nutritional status, malnutrition, lifestyle behaviour, health education and health awareness, poverty, unemployment, living standard, natural, human, and cultural environmental characteristics and other health indicators playing role in causing good or bad health become very vital.

The maps and other cartographic graphical representations of such variables play a great role in representing variation in the spatial pattern of the overall health of people and in turn become a source for the future spatial health planning. The geographers use the simulation model also to study the pathogenic (relating to organisms) and environmental (relating to human, cultural, and natural environmental elements) reasons of the origin and spread of diseases, especially the communicable diseases.

Temporal concept describes changes, trends, and patterns of health in specific area over a given period of time and helps in future planning for better urban and rural people's health at local, regional, national, and global levels. Demographic concept deals with variation in the pattern and trend of the intensity and spread of diseases and state of health by age, gender, social, educational, income, occupational, racial, and religious groups across different regions of the world. Physical environmental concept describes the effect of physical constituent elements such as land/relief, water, air, forest/vegetation, animals, etc., on the spatial and temporal variation of infectious and non-infectious diseases and condition of health at different spatial scales.

Ecological Concept of Health: In a wider sense, the conceptual principles of the ecologists in their "ecological concept of health" are generally the same as that of the geographers because they also consider the influence of the physical, social, cultural, economic, and political environmental conditions on disease causing pathogens and health of the people of a region over a given point of time.

Holistic Concept of Health: The above overarching conceptual principles of health including spiritual beliefs and faiths are holistic in nature and according to holistic approach health is considered as a composite result of an interplay of the variety of physical, biological, social, cultural, economic, demographic, and political elements or characteristics at individual, household, local, regional, national, and global levels.

1.1.3.3 Concept and Meaning of Health

The concept and meaning of health provide the foundational ideas to clearly understand the complex web of interconnections of multiple elements of human health. Since health (the product of genetic and environmental factors) is a very complex phenomenon, it seeks proper approach of its definition. Meaning of health here is taken into account as an important aspect of the concept of health as meaning or definition provides the basic ideas or underlying facts of health. The available literature on health provides us idea that it was Hippocrates who first considered health as a state or condition of body resulting from the balance between internal elemental conditions and external environmental factors including individual's behavioural habits. After him, Galen considered health as a physiological, mental, and emotional state of an individual's body, not merely the absence of disease. The concepts of health as developed by Hippocrates and Galen during ancient and Roman periods, respectively, were further refined and several definitions were developed by others during the transformation period between ancient and the modern. However, the definition of health as developed by the World Health Organization in 1948 became the basis of further health definitions by others. The widely accepted definition of WHO as mentioned in the preamble of its constitution is as follows: "health is a state of complete physical, mental and social well-being and not merely the absence of disease or infirmity". To be very specific, this definition was proposed by Dr. Andrija Štampar from Croatia, one of the founders of World Health Organization. Later this definition was expanded to include "socially and economically productive life" (Svalastog et al., 2017, p. 432).

After a wider recognition of health definition of WHO, several scholars during the second half of the twentieth century and at the early period of the twenty-first century made their attempt to define health in their own ways and some of them are as follows. Irwin (1964) defined health as "a state of the highest possible level of physical, mental and emotional fitness and efficiency commensurate with individual capacity". In other words, it can be stated that once a person functions at his/her highest capacity without exceeding normal physical or mental fatigue, the concerned person is considered to have optimum health. But when a person does not possess such quality of optimum health, he/she is considered either fully or partially unhealthy and hence there is always a scope for improvement in health condition. Dolfman (1974, pp. 206–209) focused on health and the environment and stated that an individual is healthy, if he/she functions adequately in a stated environment and adapts the stresses of that environment within the range of normal functioning. Conversely, an individual is not healthy if he/she does not function adequately in a stated environment and does not adapt to the stresses of that environment within the range of normal functioning. This definition revels that health is a state or condition of normal functioning of an individual in a given environment. According to Ghosh (1978, p. 266), "health suggests not only the absence of disease and weakness of the body, but the presence of the welfare of body, mind and society". Health indicates a condition of sound body and mind. This approach throws light on the disease free and perfect condition of body and mind of a person. Health is the general condition of the body with respect to the efficient or inefficient discharge of functions. It is the product of genetic and environmental factors. Ross (1982, p. 551) took into account health as a state of physical and social well-being of an individual in relation to his environment, particularly the extent of development of his abilities for survival and efficient maintenance over time. Madan (1983, p. 286) opined that health does not mean merely the absence of disease but harmonious functioning of the body and mind of an individual in relation to his physical and social environment so as to enable him to enjoy life to the fullest possible extent and to reach his maximum level of productive capacity. This definition indicates the complete adjustment of body and mind of an individual with his external environment while functioning or carrying on certain activities. Valentei et al. (1986, p. 8) stated health as an ingredient of the quality of population and held that the population health is not only the absence of mass illness and population injuries but also man's concordant physical, psychic and moral development enabling him to carry on social and labour activities freely and without restrictions. This definition is more or less similar to the above definition. Rizvi (1986, p. 236) took into consideration health as a right condition or proper functioning of the body of a person. He further mentioned that those who are regarded as healthy can take heavy food, have a good muscular body and are able to work hard. For proper health, it is necessary to be free from all kinds of mental tension too. This approach of defining health reveals the state of proper functioning capacity of the body which enables an individual to have heavy food and to work hard. Lawrence (2005, p. 481) also regarded health as a state or condition of a person or persons that results from the interrelationship of biological, chemical, social, economic, and physical environment.

After having gone through the above-mentioned concepts and meanings of health, the author is of the opinion that health is a state or condition of functioning capacity of physical and mental faculties of an individual in a given natural and human environment of a particular region and at particular point of time. Such bodily and mental faculties should also be free from disease or ailment. Health of an individual may remain constant for a certain period of time in a given environment and thereafter due to certain changes in the components of biological composition of an individual's body and also due to change in the quality of natural, human, and cultural environment and its influence on human body, health may change. In this sense, health cannot be considered as stable state of body and mind of an individual as internal structure of body changes and weakens the resistant power due to change in age and also due to change in the quality of natural, human, and cultural environment and its influence on human body. Bircher (2005, p. 336) rightly pointed out that health is a dynamic state of well-being and is interpreted to include mental, physical, and social capacity which positively support and sustain in day-to-day life activities of a person of any age and culture.

A person can remain healthy or of sound health till his or her physical and mental faculties remain free from disease or illness and also till his/her mental and physical faculties remain under the influence of congenial environment of different types. The favourable environmental condition enables a person to work adequately and efficiently without any excessive stress in the existing situations and leads him to enjoy the full length of life.

In fact, health is a state of the whole physiological and mental functioning of a person or persons or people inhabiting within a given area, environment (natural and non-natural) and time. All the physiological and mental elements of all the persons are expected to perform their functions naturally means in normal way, but this does not always happen in all the situations. Sometimes in certain situations in case of certain person's life physical and mental elements work perfectly in normal way but in another situation in case of the same person's life these do not perform their work normally because of the effects of a wide range of factors prevailing in a given space and time. Specifically, when the physiological and mental functioning of an individual is normal means positive then this state is termed as a sign of positive health and provides happiness or pleasure to him or her being unaffected by a feeling of any kind of pain or infirmity or ailment. But in reverse condition, when a person has any sort of physical and mental infirmity or illness then in this condition his or her body and mind do not function in natural or normal way and cause pain, distress, discontentment, displeasure, and unhappiness. These are the signs of negative health. Allmark (2005, pp. 1–9) rightly pointed out that where the different physiological and mental elements of the human being work normally in a positive way then they do not present any obstruction to him or her while performing daily routine activities. He was of the opinion that good life of a person is that in which he or she functions well. He further stressed that health is not an item rather it is a qualitative attribute and it is judged on the basis of health-related objective and subjective facts. Someone's good health depends on the proper functioning of different constituent parts of human being's body and mind.

Health, in a positive sense, is what we think and what we do about the happiness of a person or group of persons or people or population of a place or an area inhabiting at a given point of time and this is possible only where there is a state of perfect functioning of physiological and mental system. The state of perfect functioning of mind and body is realized in absence of feeling of any kind of pain or suffering (ranging from sweet to intense) which is caused by any type of disease or infirmity. In other words, positive health is a result of the state of normal functioning of body and mind which provides happiness to the concerned person(s) or people of an area at a certain point of time. In the same way, negative health is a result of the abnormal functioning of body and mind which provides pain to the concerned person(s) or people caused by the presence of any sort of malady. Therefore, good or positive health provides happiness and ill health unhappiness or displeasure. Similarly, happiness lowers disease risk and promotes good health and unhappiness leads to risk of diseases and impaired health.

Briefly, it can be summarized that the concepts and meaning of urban health include several dimensions that actually or potentially exist in worldwide urban areas. These are physical, mental, emotional, spiritual, social, cultural, educational, economic, demographic, political, nutritional, preventive, curative, etc. The role of these dimensions in promoting and conserving health is of paramount importance, however, their role depends to a large extent on the current state of affairs of the people of different urban regions of the world. Taking all the conceptual principles of health as mentioned above and health dimensions into account, the contemporary meaning of health can be visualized from the following perspectives. These are Biological (the state in which each and every cell and organ of a person's body function to the optimal level), psychological (the condition in which an individual feels a sense of complete well-being and completely integrated with environment), Humanistic (the state in which no one is prevented from enjoying the opportunity), Developmental (the condition in which health is perceived as an integral part of the development), social (the state in which the participation of a person in the social system remains optimum and also health is considered as a worldwide social goal), Qualitative (the state in which health is regarded as an important determinant of quality of life), and multisectoral (the condition in which health is a concern for all such as individuals, communities, medical professionals, pharmaceutical companies, governments, and agencies at state, national, and global levels).

1.1.3.4 Urban Health: A Resource, a Problem, and a Global Challenge

Health, either urban or rural, is a very vital and valuable resource, wealth, or asset of an individual, a family, community, state, or nation across the world. Lawrence (2005, p. 481) also considered health as a resource or an asset which affects every aspect of life of an individual. According to him, health is an ability of a person which he or she uses in achieving his/her potential to face challenges in everyday life. Desai (1988, p. 154) also noted health as a positive concept that emphasizes on social and personal resources as well as physical capabilities. This approach of health definition connotes health as a right condition of body and mind and not only the absence of disease. Bircher (2005, p. 335) was also of the opinion that health of an individual is a vital resource which helps satisfy the day-to-day needs of his or her life. Not only this, but also it decides his/her future life. Health in terms of physical and mental capability is required in performing any human activity at any stage of life. It increases expectancy of life, promotes labour's productivity and directly contributes to welfare of people by reducing diseases and infant mortality. Good health creates every good thing in the life of an individual and it is one of the important indicators of human development, human well-being, quality of life, prosperity, and happiness. Health like other resources has two attributes—usability and functionability in the course of producing a use value of any description to meet the needs of a person(s), a family, a community, and a society of an urban or rural area at a given point of time all over the world. Health has mainly two dimensions—physical and mental—and both are the source of physical and mental capacity which are not constant rather dynamic in nature. Health being a dynamic phenomenon has always a vast scope of improvement or promotion and from this perspective the condition of health searches opportunities for further improvement, maintenance, and conservation of health at individual, societal, institutional, and governmental levels. Health produces human resources in terms of knowledge, skills, talents, energy, stamina, and other mental and physical capability which through their efficient utilization builds a better society and contributes to better human development. Health builds a productive urban workforce, increases labour force participation rate in economic activities, fosters good livelihoods, and promotes happiness of urban people. Health is an achievable goal and each and every member of the society is expected to achieve this to the best level of health potential. From this perspective health becomes the source for creating various types of health opportunities in the towns and cities around the world to be equally availed by all individuals to achieve, maintain, or conserve their health to the optimum level. In this sense, health, like cities and its environment, is an opportunity also from the point of view of better health, well-being, and quality of life.

Health of urban people is a global problem as many people from different regions of the world suffer from several communicable and non-communicable diseases and as a result the state of health of such people is not good due to a variety of unfavourable social, cultural, economic, political, demographic, and physical environmental factors operating in towns, cities, and villages everywhere in the world. Interplay of the influence of urbanization and urban environment is responsible for ill health and for the origin and spread of infectious and non-infectious diseases, scarcity of food and nutrition, housing, water supply, sanitation, healthcare system,, etc. Urban population through its own anthropogenic activities is largely responsible for all health problems. Health Problem (as an obstacle or impediment) usually becomes an issue of debate or discussion for solution and also a global challenge concerning how to overcome the situation. Sometimes the words problem, issue, and challenge are used interchangeably.

The contemporary global urban health scenario represents intra-regional as well as inter-regional inequity and inequality in urban health across the world. Health condition of the urban people also varies at age, gender, social group, education, income, and occupational levels within all the regions of the world. Variation in health at different levels is a serious problem and becomes a major global challenge from the perspective of achieving equity and equality in urban health around the globe. Health equity and equality can be achieved by providing equal access of all types of health facilities to all the urban citizens irrespective of their social, cultural, economic, political, and spatial status as well as the health problems they face. Such health intervention would enable the urban citizens to achieve their optimal health potential.

Health of urban poor is another global challenge, as the condition of their health is worse, life expectancy of them is fairly low and they are mostly deprived of the healthcare facilities in comparison to that of the wealthier urban population. Unhealthy working and living conditions of slum people always lead them to face health hazards. One-third of the total urban population of the world live in slums and face scarcity of sufficient living space, availability of essential services, tenure security, potable water, sanitation, and healthcare facilities. Lack of these causes serious health problems in slums as well as in non-slum areas of the cities. Malnutrition among the urban children in slums is considerably higher and continues as a serious global health issue. WHO's malnutrition key facts (9 June 2021) highlighted that globally in 2020 about 45% of all deaths of children under five were attributed to under-nutrition, and the situation particularly in the developing countries is more serious as infant mortality caused by infectious diseases is exceptionally higher in poorer sections than that of richer sections of urban society (McMichael, 2000, p. 1119).

Kieny and Clos (2016, p. 8) also considered slums, the mechanism of its upgradation and capability building, environmental pollution, ageing population and migration including infectious and non-infectious diseases including HIV/AIDS, tuberculosis, as problems and challenges of the twenty-first century. Truly speaking, urban health promotion in cities depends not only on the availability of health systems but also on the conditions of the urban environment.

Among the urban poor, communicable diseases are a remaining concern. One of the greatest urban health threats is HIV/AIDS, which is associated with several social determinants. Lack of water and sanitation remains a major health threat for the urban poor. In addition, urban air pollution and traffic problems create health hazards with economic development. Accidents and road traffic injuries are also a serious concern. The poor living and working conditions create health vulnerability especially among the urban poor and vulnerable subgroups, e.g. women, infants and very young children, the elderly, and the disabled in most of the urban areas.

Occupational diseases basically caused by uncontrolled growth of small-scale and cottage industries within and in the periphery of cities are increasingly common phenomena.

The global condition of both the infectious and non-infectious diseases is alarming. Particularly, developing countries continue to experience rising rates of overweight, obesity, diabetes, high blood pressure, and dyslipidaemia among urban populations. Tobacco consumption is now universally more common among the lower socio-economic groups. Hypertension is now more common in poorer groups and diabetes is emerging as an important disease in the urban areas. Changing dietary patterns, health behaviour, and family structures in the affluent urban population have been crucial in causing heart disease, diabetes, and hypertension-related problems.

Cardiovascular diseases are major and growing contributors to urban mortality and disability across different countries of the world. Diabetes and its adverse health effects have risen more rapidly in South Asia than in any other large region of the world. India has a higher number of people with diabetes than any other country. Cancer is regarded as one of the major non-communicable diseases affecting health in South Asia. Bronchitis and asthma resulting from indoor and outdoor air pollution account for a large proportion of the burden of chronic respiratory diseasesThe prevalence of hypertension in adults remains 23% in urban areas and is found positively associated with higher socio-economic status in both urban and rural areas. Leon (2008, p. 7) also stated that the rise of non-infectious diseases especially in the middle- and low-income nations is mainly driven by enormous changes in social and economic conditions caused by continuing urbanization.

Actually, worldwide mortality (death rate) from non-communicable diseases is rising however, the rise is obviously more in the developing countries compared to that of the developed countries. According to an overview by WHO about 70% of all deaths of the world are caused by non-communicable diseases and the proportion of deaths is more in low- and middle-income nations due to poverty, more consumption of unhealthy diets, tobacco, harmful alcohol, and more sedentary lifestyle.

Worldwide rapidly increasing urban population in itself is a major challenge from the perspective of urban health for the twenty-first century as it mounts a continuous pressure on the demands for basic health infrastructure and means of improving standard of living which are generally not sufficiently available in urban areas especially in middle- and low-income countries. Rapid urbanization and fast expansion of urban areas in unplanned and haphazard way are responsible for increasing the risks of health of urban population by degrading the quality of natural and built urban environment. In fact, the overall inequities in the distribution of urban health facilities across urban space and society is a matter of grave concern as this create significant health inequalities among different sections of urban society. The poorest, migrants and other disadvantaged vulnerable groups residing in the most deprived and environmentally degraded urban areas face the hardest life for want of basic facilities of health, housing, education, employment, and other health-related requirements. Lerch (2017, pp. 1–2) also noted a swift increase in the size of urban population in general and slum population in particular especially in the cities of developing regions as a great challenge in the way of achieving the health-related SDGs by 2030. In such a fast increase in urban population as well as in the transmission of communicable diseases the role of both the national and international migration is very vital. Urban areas being socially, culturally, and economically lucrative places act as the gateways for in-migrants and immigrants.

Urban poverty, unemployment, inequity, and inequality of healthcare facilities, lack of universal urban health programmes, social vices such as crime, violence, drugs, and other deviant behaviours of health are considered as multidimensional urban health issues and challenges for the twenty-first century. All these critical issues adversely affect urban health and should be properly addressed at global level so that all urban citizens can maximize their full potential and achieve their optimal health to enjoy a happy life.

Unsustainable urban growth and poor urban governance are also the major issues for the twenty-first century because they adversely impact health of urban people. That is why, there is a need for sustainable urban growth and proper urban governance for a flourished and prosperous city with good urban health. Positively, a healthy population is essential for creating economically competitive and inclusive cities.

Briefly, the key factors affecting health of urban people belong to physical (natural), social, cultural, economic, and political environment. The apparently unstoppable urban growth puts increasing strains on food and water supplies, availability of safe housing, waste disposal, and on healthcare services. Crime and alienation, overcrowding urban and peri-urban areas, inadequate housing, solid and liquid waste management, sanitation, scarcity of electricity and water supply, air pollution from automobile and industrial sources, indoor air pollution from burning biomass and fossil fuels, surface water pollution from both municipal and industrial waste, urban coastal degradation, loss of open space and urban noise have helped increased several infectious and non-infectious diseases. Public health services in most of the urban areas particularly in the developing countries are unable to cope with overwhelming health problems. Increased frequency of HIV infection is expected to become the most serious public health problem in urban areas. Migration also contributes to certain diseases.

1.1.3.5 *Future Perspective of Urban Health*

The sheer scale and speed of the current urban expansion in the world is expected to give rise of the world's urban population to more than two-thirds by 2050. Though, the maximum rise of urban population will take place in Asia and Africa. The future urban population will mount tremendous pressure on the extra requirements of healthcare facilities, affordable housing, potable drinking water, safe and accessible transport and communication, education, employment, and sanitation-related infrastructure. So, governments of the concerned countries should adopt a multidimensional futuristic approach to deal with all these requirements in the coming future. For better health of the present and future urban population there is a need for the improvement in lifestyle behaviours concerning daily diet and dietary system, physical activities, smoking, chewing tobacco, and other health-related lifestyle behaviours with a view to control non-infectious (chronic) diseases like cardiovascular, diabetes, cancer, obesity, etc. There is also a need for developing and properly utilizing the public health-related infrastructure, medicines and vaccines, improving diagnostic and testing system, improving water supply system, sanitation, housing condition, and reducing rural–urban human migration and overcrowding to control the transmission of communicable diseases. Yoga and exercise is a powerful lifestyle-related tool for improving physical and mental health of everyone. Yoga reduces negative ideas from the mind, tiredness, and weakness and makes the person energetic by enhancing immunity power. Yoga tones up the body muscles, makes blood circulation better, and keeps a person healthier to live a good life. Hence, with a view to make sustainable urban health each country of the world must promote the practice of yoga and physical exercise. Health of slum population is another challenging urban issue of the twenty-first century and for this the local, state, and federal governments all over the world will have to pay special attention to create and provide basic housing, water, sanitation, health, and educational amenities to slum population (about one-third of the total urban population of the world) The Lancet (2015). The same requirement will be needed for both the resident and nonresident urban poor. Environmental pollution (air, water, noise, land, dust, and other pollution) stands as unquantified urban health risk and should be seriously tackled with strong urban policies now and into future at local, regional, national, and global levels to improve and maintain sound health of the present and future urban population. Capital investment in health and education is the key to open the door of peace, prosperity, well-being, and happiness for each and every country's people across the world. Looking into the future increase in urban population and need to meet the extra urban health requirements there is need for an urgency of making the worldwide cities truly smart, livable, green and sustainable. This would be a kind of great opportunity in the direction of promoting and maintaining good health, good quality of life, and well-being for the present and future urban population.

Greening cities is an innovative concept and one of the important urban planning-related issues of the twenty-first century. It enhances the environmental sustainability and liveability of cities which in turn positively impact health of urban people. It reduces the effects of urban pollutions like air, water, and noise, controls the impact of urban heat islands, diminishes the rate of soil erosion, protects biodiversity, improves the quality of air, and boosts up the psychology, comfortability, happiness, and wellness of urban inhabitants. Therefore, looking into the speed of global urbanization and urban expansion there is an urgent need of a multidimensional concrete urban development planning in order to make the cities of the world green.

Sanitation or cleanliness is of paramount importance for protecting or improving health and it should not be neglected at all. The Prime Minister of India (2022) laid emphasis on sanitation/cleanliness and said, there is direct link between health and sanitation or cleanliness. People must complete their own responsibility in keeping the cities, places, and surroundings clean or hygienic. Though, worldwide sanitation awareness has increased in the recent past, yet there is a need to increase cleanliness and environmental awareness for making the cities or other places clean for better urban health.

Improving urban health is a complicated task owing to involvement of a variety of urban environmental factors and this needs cooperation and coordination of different health-related stakeholders including residents. A sense of collective responsibility in this regard would be a better approach. Not only this, but also the overall improvement in global health system to minimize the inequity and inequality in urban health across different regions of the world must be ensured by the national governments and international agencies for good health and good life.

Despite a lot of efforts and advances in the field of public health the concept of "health for all" by 2000, originated at the assembly of the WHO in 1977 and started as a global health campaign in 1978 for securing good health, well-being, and good quality of life of the people across the world to live a life of peace, comfort, and happiness, is still at bay and remains elusive and a daunting problem (challenge) in the twenty-first century. Therefore, the need is to re-visualize the whole health-related mechanisms to produce solutions to achieve this goal in the coming future.

1.1.4 Urban Dynamics, Environment, and Health Linkages

This part includes a detailed picture of the linkages of urban dynamics, environment and health with different physical and human aspects.

Urban dynamics (urban expansion and urban growth through the process of urbanization), urban environment (the physical/natural, social, cultural, economic, and political) and urban health (a key wealth or asset and an ability or capability as a driver to thrive city) are interlinked and these three are always under the influence of geographical settings because an urban area is completely surrounded by different geographical features or attributes or elements which set the urban characteristics and urban life. In other words, geography determines the origin, growth, expansion, progress, prosperity, environment, and the complete existence of an urban area. Urban health is a product of the geographical settings and urban people's response to them. In fact, geography has a profound impact on the inter-linkages of urban dynamics, urban environment, urban health, and urban life.

The inter linkages of urban dynamics, environment, and health are multifaceted and these three vital urban components remain the major global challenges of the twenty-first century and continue to be the focal areas of research in the field of urban geography and in other allied subjects. These three themes comprise a complex and multidimensional concept and accordingly become difficult to identify, define, classify, and analyse from spatial and temporal perspective. Urban dynamics are completely linked with the environment and health aspects of urban people. The fast growing urban population and rapid expansion of the world's cities in haphazard way owing to incessant urbanization are the major causes of deterioration in the quality of environment, health, and quality of life of people. Such deterioration has created a big threat to the human society and civilization. In fact, urban areas in both the developed and developing countries have grown merely in population and environmental problems, not in prosperity. This scenario will further continue owing to continuous rise in urban population through the process of birth, death, and migration (the components of urban dynamics). Despite this the cities, on the one hand, provide the opportunity of better educational and health facilities, better human capital formation, innovation of science and technology, employment and economy for the workers, better standard of living, manufacturing and industrial growth, trade and commerce, etc., but on the other hand, unplanned and haphazard expansion of the cities and unfriendly relation between population and environment due to increasing trend of urbanization have created a big threat in the form of challenges like environmental pollution; rural–urban migration, growing slum population; loss of agricultural land, waste land, and natural vegetation; traffic congestion, diseases, and health, shortage of housing and potable water, sanitation and hygiene, headforemost use of energy, extreme economic inequalities, etc. The result of interplay of urbanization and urban environment is responsible for the ill or good health and for the origin and spread of infectious and non-infectious diseases, scarcity of food and nutrition, housing, water supply, sanitation, healthcare system, etc.

Urban dynamics, environment, and health are closely connected to each other and their interaction provides social, cultural, economic, and political opportunities for the wellness and prosperity of the urban inhabitants but side by side creates threats or risks or unfavourable circumstances to the urban people where they have to face several physical/natural, socio-cultural, economic and political consequences determining malady, hardship, unhappiness, poverty, resource crunch, etc.

Human population is central to urban dynamics, environment, and health and acts as an active factor in constructing, modifying, and changing the urban landscape. Humans through their anthropogenic activities create congenial or uncongenial environmental conditions which in turn help promote either good or poor health of the urban inhabitants.

The linkages of urban dynamics, environment, and health can be broadly explained to understand the interconnections of different dimensions of towns and cities such as physical or natural geographical features, social, cultural, economic, and political characteristics. *Linkages with the physical/natural geographical features:* Urban dynamics is closely connected with natural geographical features such as terrains, drainage, lakes/ponds/reservoirs, forests, and climate. Urban dynamics alters, transforms, or changes the geographical features into different cultural forms through the process of land use and land cover changes and gradually modify or change the natural environment of cities and in turn affect the health, well-being and quality of life of urban inhabitants. Sometimes the effects of urban dynamics on urban environment and health are positive in terms of opportunities and sometimes negative in terms of threats to urban life. *Linkages with the Social Aspects:* Unplanned urban growth increases social inequality, social exclusion, social segregation, racial, ethnic, caste, class, religious conflicts, crimes, urban terrorism, etc., and disturbs the social fabrics of urban societal environment. Though urbanization has potentiality also to develop and provide several social opportunities such as social empowerment, social amenities, social skills, social entrepreneurship, participation in the activities of social organizations which build a viable social environment and produce social values to live a life of good health, peace, comfort, and dignity in urban society. *Linkages with the Cultural Dimensions:* Cultural features (a result of the combination of natural and human resources) are the major aspects of urban life but are always under the positive and negative influences of the interaction of urban dynamics, environment, and health. In the positive sense, cities, especially the well planned ones, are the hubs of socially useful tangible and intangible cultural assets or wealth such as arts, music, education, innovative ideas, technical know-how, health, institutions, buildings, monuments, temples, mosques, churches, transports (rails, roads, airports, oceanic ports), needed services (sanitation, hygiene, water, electricity, fossil fuels, and other daily life-related materials. The role of these in thriving towns or cities is immense but poor or unplanned growth of urban centres across the world increases traffic congestion, results in creating scarcity of public services, housing, water, fuels, needed infrastructure, and other day-to-day life-related basic services. The headforemost use of nature gifted and human made energy in automobiles, industries, and in other socially, culturally, and economically productive activities in urban areas increases environmental pollution (air, water, noise, and soil pollution) which harms health, well-being, and quality of life of urban people. Excess use of technological devices and home appliances also becomes the source of environmental degradation and health problems. Similarly, other reasons of health problems, for instance, Covid-19 pandemic and other infectious and non-infectious diseases, mount pressure on cultural assets/resources like health-related infrastructure, health services, medical human resources, management and on production, demand, supply, and distribution of medicines, vaccines, etc., and often create spatial disparity which can be reduced by proper planning and management for sustainable urban growth and urban environment with a view to promote good urban health.

Linkages with the Economic Characteristics: Urban dynamics, environment, and health have a positive as well as negative relation with economic characteristics of towns or cities everywhere in the world. Changes in urban areas due to continuous rise in urban population, the result of interplay of birth, death, and migration (particularly from rural to urban), bring several opportunities of employment in the formal and non-formal economic activities belonging to the areas of industries, commerce, trade, transport, urban agriculture for the urban people to augment their income, raise standard of living and economic prosperity, but simultaneously create several economy-related problems such as urban poverty, economic inequality, unemployment, high price of land, high cost of housing, and rental spaces. Urban expansion also causes massive loss of agricultural land which is not a good sign from the point of view of future food security. If the same trend of loss of agricultural land continues, there will be a great threat of food security to human population. The study of Bren d'Amour et. al (2017, pp. 8139–8142) shows that the expansion of urban areas by 2030 will cause a loss of global agricultural land between 1.8 and 2.4% and this loss will be responsible for the worldwide loss of crop production between 3 and 4%. However, maximum loss will occur on the continents of Asia and Africa where about 80% of the total loss of global cropland will take place. Unplanned urban growth also causes loss of forests, mines, water bodies, streams, etc. the source of environmental and economic resources. Similarly, urban agriculture, forestry, industrialization, commercialization, globalization, and means of transportation become the source of air, water, noise, soil, and other environmental pollution which are ultimately responsible to increase the risks of communicable and non-communicable diseases leading to deteriorate health, quality of life and standard of living of the urban people. Ill health caused by rapid urbanization and poor quality urban environment raises poverty and economic insecurity. The relationship between migration and urban environment is very complex. Migrants bring with them human resources like skills, knowledge, talents, innovative ideas to the cities and benefit the city population as well their left-behind families by sending remittances. Migration, on the other hand, acts as the key factor for the formation of slums, for increasing overcrowding of people and for unplanned expansion of urban centres. Migration also becomes a means of transmission of several communicable diseases to harm urban health.

Linkages with the Political Activities: Urban dynamics, environment, and health aspects are closely linked with the local, regional, national, and international political activities. The expansion or growth of urban areas is always influenced by political leadership, governance and management. Good urban planning, policies, and good governance play a considerable role in making a city sustainable, smart, and livable with sustainable urban environment and better urban health. The role of governments in urban land use and in allocating lands for constructing residential, commercial, and administrative buildings, in developing transport, industries, trade, and commerce as well as in resilience building to cope with disaster risks in urban areas is immense. Rapid urbanization and fast increase in urban population due to fertility, mortality, mobility, and other urban-based pulling factors have manifold effects on urban environment and urban health. Such effects particularly in urbanizing developing countries make urban governance and urban planning a tough task. Poorly designed urban planning, poor leadership and poor urban governance lead to environment and health-related several problems. The major among them are overcrowding, scarcity of affordable housing, water, electricity, healthcare facilities. Insanitation, air, water, and noise pollution, solid and liquid waste material disposal and its management, and traffic congestion are equally challenging urban issues. Redefinition and reclassification of city boundaries by the government help urban growth and have far-reaching impact on urban environment and health. Lack of political transparency in capital investment (political economy) and clash of interest or conflicts among different political parties on urban developmental issues negatively impact urban dynamics, environment, and health. Similarly, illegal construction of buildings by builders with the collusion of government officials and in some cases the demolition or high-tech super demolition of such illegal buildings also has negative impact on urban environment and urban health. The Noida, Uttar Pradesh, India twin towers (Apex and Ceyane) high-tech supper demolition on 28 August 2022 is one of the best examples of real estate illegal business. In fact, for all these problems the local, state, and national level governments of the concerned countries are largely responsible because the solution of such problems lies in the hands of the governments.

1.1.5 The Way Forward

Looking into the detailed picture of the results of the interaction of urban dynamics, environment, and health it becomes imperative to find a way for the solution of the urban problems. In this regard, properly designed urban planning, proper policies, proper management, correct policy implementation, equitable allocation of funds and its proper utilization in all urban development-related activities with honesty and transparency would play a great positive role in controlling haphazard growth of cities, in improving the quality of urban environment and in reducing the effects of health-related risk factors to promote good health of the urban inhabitants locally, regionally, and globally. However, such approaches are not the only panacea for all the obvious reasons of urban problems.

Apart from such mechanisms, an approach to create a balanced relationship or interaction among urban dynamics, environment, and health is a prerequisite for the sustainable urban development, environment, and sustainable urban health. In this direction the vertical expansion of the world's cities (especially in the fast urbanizing developing countries) would be one of the ways of mitigating the problem of affordable housing, overpopulation, and overcrowding. The vertical expansion of towns or cities would also work in preventing the loss of precious countryside agricultural land, forest or vegetation cover, water bodies, etc. (the source of resources and ecological balance), in reducing the outdoor movement of automobiles and air pollution as well as in controlling the high price of land and buildings in urban areas. Construction and development of potentially viable underground towns or cities may also be an alternative way of protecting countryside agricultural land, forest land, waste and barren lands, and water bodies supporting urban life. Infill development in urban areas would also check to some extent the countryside expansion of cities and solve some urban problems. However, the vertical expansion of towns or cities also has some adverse impacts on urban life.

Ruralization (a novel, innovative, and counter concept of urbanization) would be another important alternative approach for making the urban areas sustainable and this is only possible when the rural areas will become self-sufficient and sustainable to support fully all the needs of the present and future rural generations. For this, there is a need for regenerating or creating the entire needed social, cultural, economic, and other opportunities in rural areas at par with urban areas. Such opportunities in rural areas will work as a centripetal or magnetic force to check the massive rural–urban migration that will lead to reducing the overpopulation and overcrowding, minimizing the mounting pressure of the demand for jobs, housing, water, sanitation, electricity, transport, fuel energy, education, health, and other quality of life-related infrastructure and services in urban areas. Better social and family cooperation, better quality of air and water, less noise and less crime, availability of pure food materials and prospect of opening of new economic opportunities in rural areas would also attract or encourage urban people to settle in rural areas through urban–rural migration. All these in turn would limit the rate of urban expansion and check the loss of countryside agricultural land, vegetation cover, waste and barren lands, water bodies, and other resources determining the environmental quality for better health and better urban life. Thus, once all the dreams of rural people are fulfilled, the urban sustainability would automatically come into existence.

A substantial reduction in air, water, land, noise, etc., pollution and hazards for better environment; control of social vices by improving law and order for better public safety; eradication of unemployment and poverty by generating employment opportunities especially for the urban poor; rise in the standard of living through the improvement in socio-economic conditions; and improvement in educational and health sectors by creating adequate education and health related infrastructure and services in urban areas globally under a well planned time bound strategy would also be the way forward for sustainable urban growth, sustainable urban environment, and sustainable urban health in the direction of social well-being and quality of life of the entire urban population.

Therefore, there is a need for worldwide concrete action plan for policy programme change in both the more developed and less developed countries towards sustainable urban development and environment for better health, prosperity, and quality of life of the present and future urban population.

1.2 Concluding Remarks

In summation, to mitigate the strains that will develop as a result of the expansion of worldwide cities in general and Asian and African cities in particular, and to maximize the potential economic opportunity that well-managed cities can offer; geographers, planners, distinguished academic experts, practitioners, and other social scientists urgently need to offer fresh, proactive, and reactive approaches in addressing the challenges of the present as well as the challenges to be appeared in future owing to ongoing urbanization, environmental degradation, and spread of a variety of diseases. Such approaches and initiatives will go a long way in spawning innovative ideas and concepts to provide the needed and required fillip to urban sustainability and sustainability of life supporting environment for better health, peace, prosperity, and human well-being.

References

Alfeld, L. E. (1995). Urban dynamics-the first fifty years. *System Dynamics Review, 11*(3). Wiley.

Allmark, P. (2005). Health, happiness and health promotion. *Journal of Applied Philosophy, 22*(1). Wiley-Blackwell Publishing

Ametepey, S. O., & Ansah, S. K. (2015). Impacts of construction activities on the environment: The case of Ghana. *Journal of Environment and Earth Science, 15*(3), 21–23. IISTE. www.iiste.org

Ashokmuthu, G. L. (2008). *Plate tectonics and climate change.* https://www.geo.umass.edu/climate/papers2

Atkinson, R., & Wulff, M. (2009). Gentrification and displacement: a review of approaches and findings in the literature. AHURI Positioning Paper No.115. Australian Housing and Urban Research Institute, Southern and Monash Research Centres.

Bacci, D. L. C., & Diniz, T. M. R. G. (2013). Mining in urban areas: methodological proposal for the identification and mediation of socio- environmental conflicts. *Rem: Revista Escola de Minas, Ouro Preto, 66*(3). Brazilian Society of Geochemistry.

Badash, I., Kleinman, N. P., Barr, S., Jang, J., Rahman, S., & Wu, B. W. (2017). Redefining health: The evolution of health ideas from antiquity to the era of value-based care. *Cureus, 9*(2). https://doi.org/10.7759/cureus.1018

Berryman, J. (2012). The art of medicine: Motion and rest: Galen on exercise and health. *The Lancet, 380.* https://doi.org/10.1016/S0140-6736(12)61205-7

Bhattacharya, B. (1991). Defilement of the Environment, Its Restoration and Management, In Jana, M.M. (1991). *Environment degradation and development strategies in India.* Ashish Publishing House.

Biomol.com/vaccine:https://www.biomol.com/resources/biomol-blog/the-first-modernvaccine

Bircher, J. (2005). *Towards a dynamic definition of health and disease. Medicine, health care and philosophy* (Vol. 8). Springer.

Blake, J. B. (1948). The origins of public health in the United States. *American Journal of Public Health, 38.*

Blanchet, D. (1998) Relationship between population growth and economic growth. In D. Paul & M. C. Geoffrey (Eds.), *The earthscan reader in population and development.* Earth Scane Publications Ltd.

Bloom, D. E., & Cadarette, D. (2019). Infectious disease threats in the twenty-first century: Strengthening the global response. *Frontiers in immunology, 10*, Article 549. https://doi.org/10.3389/fimmu.2019.00549

Bloom, D. E., Canning, D., & Fink, G. (2008). Urbanization and the wealth of nations. *Science of the American Association for the Advancement of Science, 319*(5864).

Bounoua, L., Safia, A., Masek, J., Peters-Lidard, C., Imhoff, M. L. (2009). Impact of urban growth on surface climate: A case study in Oran, Algeria. *Journal of Applied Meteorology and Climatology, 48*(2).

Bren, C. et al. (2017). Future urban land expansion and implications for global croplands. *Proceedings of the National Academy of Sciences of the United States of America (PNAS), 114*(34).

Bren d'Amour, C. et al. (2017). Future urban land expansion and implications for global croplands. In *Proceedings of the National Academy of Sciences (PNAS).* Available at: https://doi.org/10.1073/pnas.1606036114

Britannica.com: https://www.britannica.com/topic/public-health/The-Middle-Ages

Buse, K., & Hawkes, S. (2015). Health in the sustainable development goals: Ready for a paradigm shift? Globalization and Health, BioMed Central.

Cerf, V. (2012). ICTs, the internet and sustainability. In D. Souter, D. MacLean (Eds.), *Changing our understanding of sustainability: The impact of ICTs and the internet.* The International Institute for Sustainable Development (IISD). www.iisd.org

Chadwick, E. (1843). Report on the sanitary condition of the labouring population of great britain (A Supplementary Report on the Results of a Special Inquiry into the Practice of Interment in Towns). Printed by W. Clowes and Sons, Stamford Street.

Chapin, C. V. (1902). The end of the filth theory of disease. In: *Popular science monthly* (Vol. 60). McClure, Phillips and Co.

Clarke, K. C., & Gaydos, L. J. (1998). Loose-coupling a cellular automaton model and GIS: Long-term urban growth prediction for San Francisco and Washington/Baltimore. *International Journal of Geographical Information Science, 12*(7).

Clarke, K. C., Hoppen, S., & Gaydos, L. (1997). A self-modifying cellular automaton model of historical urbanization in the San Francisco Bay area. *Environment and Planning B: Planning and Design, 24*.

Conti, A. A. (2008). Quarantine through History. *International Encyclopedia of Public Health.*

Curebal, I., Efe, R., Soykan, A., & Sonmez, S. (2014). Population and settlement features of balikesir between 1940 and 2010. In: E. Recep, O. T. Tüzün, S. Igor, A. Emin, & B. Mehmet (Eds.), *Urban and urbanization.* St. Kliment Ohridski, University Press.

Daubenmire, R. F. (1974). *Plants and environment*. Wiley, Eastern Pvt. Ltd.

Davis, K. (1955). The origin and growth of urbanization in the world. *American Journal of Sociology, 60*(5), 429–437. The University of Chicago Press. https://www.jstor.org/stable/2772530

Day, L. H. (1983). *Analying population trends*. Differential Fertility in a Pluralistic Society.

De, N. K., & Bose, A. K. (1991). Sundarbans coastal region: A study in environmental development. In M. M. Jana (Ed.), *Environment degradation and development strategies in India.* Ashish Publishing House.

Delany, I., Rappuoli, R., & De Gregorio, E., (2014). Vaccines for the 21st century. *EMBO Molecular Medicine, 6*.

Demeny, P., & McNicoll, G. (1998). *The earthscan reader in population and development*. Earthscan Publication Ltd.

Demiroğlu, D., Pekin Timur, U., Ayfer, K. A., Cengiz, A. E. (2014) Ecology-based contemporary urbanism approaches. In E. Recep, O. T. Tüzün, S. Igor, A. Emin, & B. Mehmet (Eds.), *Urban and urbanization.* St. Kliment Ohridski, University Press.

Desai, V. (1988). *Rural development: issues and problems* (Vol. I). Himalayan Publishing House.

Dikshit, K. R. (1984). *A prologue to the symposium on geography and teaching of environment*. University of Pune, India.

Docampo, M. G. (2014). Theories of urban dynamics. *International Journal of Population Research*, 1–11. Hindawi Publishing Corporation.

Dolfman, M. L. (1974). Towards operational definitions of health. *Journal of School Health, 64*(4). American School Health Association. https://onlinelibrary.wiley.com/doi/https://doi.org/10.1111/j.1746-1561.1974.tb05222.x

D'Alessandro, D. (2020). Urban public health, a multidisciplinary approach. In: B. Alessandra, M. Maurizio, & I. Silvia (Eds.), *Urban health participatory action-research models contrasting socioeconomic inequalities in the urban context*. Springer Nature Switzerland. https://doi.org/10.1007/978-3-030-49446-9

Ejupi, A. (2014). Spatial and functional transformation of settlements in presevo valley. In E. Recep, O. T. Tüzün, S. Igor, A. Emin, & B. Mehmet (Eds.), *Urban and urbanization.* St. Kliment Ohridski, University Press.

En.Wikipedia-smallpox: https://en.wikipedia.org/wiki/Smallpox_vaccine

Encyclopedia.com: https://www.encyclopedia.com/science/encyclopedias-almanacs-transcripts-and-maps/public-health-middle-ages

Erdogan, N., Atak, B. K., & Nurlu, E. (2014). Modelling of land use dynamics: Case studies on urban growth in Turkey. In E. Recep, O. T. Tüzün, S. Igor, A. Emin, & B. Mehmet (Eds.), *Urban and urbanization.* St. Kliment Ohridski, University Press.

Facius, S. (2015). Ordem E progresso-precarious housing forms and the strategies of urban poor struggling against gentrification-related displacement in Sao Paulo. In H. Tulum (Ed.), *International Contemporary Urban Issues Conference Proceedings.* organized by DAKAM (Eastern Mediterranean Academic Research Center). Dakam Publishing.

Ferris, H., & Wilson, L. T. (1987). Concepts and principles of population dynamics. In J. A. Veech, & D. W. Dickson (Eds.), *Vistas on nematology: A commemoration of the twenty fifth anniversary of the society of nematologists*. Society of Nematologists.

Fielding, J.E. (1999). Public health in the twentieth century: advances and challenges. *Annual Reviews of Public Health, 20*, xiii–xvi. United States. https://www.annualreviews.org/doi/pdf/ https://doi.org/10.1146/annurev.publhealth.20.1.0

Foley, B. J. (2015). The role of plate tectonic–climate coupling and exposed land area in the development of habitable climates on rocky planets. *The Astrophysical Journal, 812*, 36. The American Astronomical Society.

Foley, J. A., DeFries, R., Asner, G. P., Barford, C., Bonan, G., & Carpenter, S. R. (2005). Global consequences of land use. *Science of the AAAS, 309*(5734).

Forrester, J. W. (1994). *Learning through system dynamics as preparation for the 21st century*. Sloan School of Management Massachusetts Institute of Technology.

Fuglestvedt, J., Berntsen, T., Myhre, G., Rypdal, K., & Skeie, R. B. (2008). *Climate Forcing From the Transport Sectors, 105*(2). PNAS, The National Academy of Sciences.

Galea, S., & Vlahov, D. (2005). *Handbook of urban health: Populations, methods, and practice*. Springer Science + Business Media, Inc.

Galster, G., Hanson, R., Ratcliffe, M. R., Wolman, H., Coleman, S., & Freihage, J. (2001). Wrestling sprawl to the ground: Defining and measuring an elusive concept. *Housing Policy Debate, 12*(4). Fannie Mae Foundation.

Ghosh, B. N. (1978). *Population theories and demographic analysis*. Meenakshi Prakashan.

Giuliani, M. M., Adu-Bobie, J., Comanducci, M., Arico, B., Savino, S., Santini, L., Brunelli, B., Bambini, S., Biolchi, A., Capecchi, B., et al. (2006). *A universal vaccine for serogroup B meningococcus* (p. 103). Proc National Academy of Sciences.

Grimm N. B. et al. (2008). Global change and the ecology of cities. *Science, 319*(5864). American Association for the Advancement of Science.

Guidebook for Evaluating Mining Project EIAs (2010). Environmental Law Alliance Worldwide (ELAW), Eugene OR 97403. https://www.elaw.org/files/mining-eia-guidebook/Chapter1.pdf

Hallett, S., Hoagland, L., & Toner, E. (2017). Urban agriculture: Environmental, economic, and social perspectives. *Horticultural Reviews, 44*. Wiley.

Hartshorne, R. (2000). *The nature of geography: A critical survey of current thought in the light of the past*. Rawat Publications.

Harvey, R. O., & Clark, W. A. V. (1965). The Nature and economics of urban sprawl. *Land Economics, a Quarterly Journal of Planning, Housing and Public Utilities, 41*(1). University of Wisconsin Press. Available at: https://www.jstor.org/stable/3144884

Heimlich, R. E., & Anderson, W. D. (2001). *Development at the urban fringe and beyond: Impacts on agriculture and rural land*. Economic Research Service, U.S. Department of Agriculture, Agricultural Economic Report.

Hui-Hui, F., Hui-Ping, L., Ying, L. (2012). Scenario prediction and analysis of urban growth uing SLEUTH model. *Pedosphere, 22*(2). Elsevier BV and Science Press.

https://en.wikipedia.org/wiki/Medical_Renaissance#:~:text=The%20Medical%20Renaissance%2C%20from%20around,many%20works%20from%20these%20societies

Institute of Medicine. (1988). *Committee for the study of the future of public health*. The Future of Public Health, National Academy Press, Washington, D.C., USA.

Ioannides, Y. M., Overman, H. G., Rossi-Hansberg, E., & Schmidheiny, K. (2008) The effect of information and communication technologies on urban structure. *Economic Policy, 54*. Available at: http://eprints.lse.ac.uk/30815

Irwin, L. W. (1964). *Health for Better Living*. Merrill Books, Columbus, Ohio, USA.

Jeanne, X. K., & Roger, E. K. (2001). *Global environmental risk*. United Nations University Press.

Jessop, R. (2012). Coinage of the term environment: A word without authority and carlyle's displacement of the mechanical metaphor. *Literature Compass, 9*(11). Blackwell Publishing Ltd. Available at: https://doi.org/10.1111/j.1741-4113.2012.00922.x

Kapluhan, E. (2014). A survey on population geography: Population mobility in Turkey and example of Uşak City. In E. Recep, O. T. Tüzün, S. Igor, A. Emin, & B. Mehmet (Eds.), *Urban and urbanization*. St. Kliment Ohridski, University Press.

Keam, S. J., Harper, D. M. (2008). Human papillomavirus types 16 and 18 vaccine (Recombinant, AS04 Adjuvanted, Adsorbed) [Cervarix]. *Drugs, 68*(3). Springer. https://doi.org/10.2165/00003495-200868030-00007

Kieny, M.-P., & Clos, J. (2016). Foreword. In *Global report on urban health: Equitable, healthier cities for sustainable development*. WHO and UN Habitat.

Kim, B., & Park, J. (2018). Effects of commercial activities by type on social bonding and place attachment in neighborhoods. *Sustainability, 10*, 1771. Multidisciplinary Digital Publishing Institute (MDPI).

Konijnendijk, C. C., & Randrup, T. B. (2004). Urban forestry. In: *Landscape and planning*. Elsevier Ltd.

Kormondy, E. J. (1991). *Concepts of ecology* (3rd ed.). Prentice Hall of India Private Limited.

Kumate J. (1997). Infectious diseases in the 21st century. *Archives of Medical Research. Summer, 28*(2). Elsevier Inc.

Lawrence, R. J. (2005). Building healthy cities: the world health organization perspective. In S. Galea, & D. Vlahov (Eds.), *Handbook of urban health: Populations, methods, and practice*. Springer Science + Business Media, Inc.

Leon, D. A. (2008). Cities, urbanization and health. *International Journal of Epidemiology, 37*, 4–8. Oxford University Press. https://doi.org/10.1093/ije/dym271

Lerch, M. (2017). International migration and city growth. UN Population Division, Technical Paper No. 10, Department of Economic and Social Affairs, United Nations.

Liddle, B. (2017). The urbanization, development, environment, and inequality nexus: stylized facts and empirical relationships. ADBI Working Paper 788. Asian Development Bank Institute. Available at: https://www.adb.org/publications/urbanization-development-environment-and-inequality-nexus

Lt. Hendry, J. A. (2015). *The origin of Lockdown*. ALICE Training Institute. Available at: https://hartlake.org/wp-content/uploads/2015/12/Origin-Of-Lockdown.pdf

Lutz, W. (2002) Population-A global challenge for the twenty-first century. In H. V. Ginkel, B. Barrett, J. Court, & J. Velasquez (Eds.), *Human development and the environment: Challenges for the united nations in the new millennium*. United Nation University Press.

Madan, G. R. (1983). *India's social problems: social disorganization and reconstruction vols*. Allied Publishers Pvt. Ltd.

Makaya, E., & Todzwo, V. (2019). Impact of urban farming on environmental sustainability: Institutional coordination dissonance. *International Journal of Environmental Science and Society, 1*.

Mascola, J. R., & Fauci, A. S. (2020). Novel vaccine technologies for the 21st century. *Nature Reviews Immunology, 20*. https://doi.org/10.1038/s41577-019-0243-3

McMichael, A. J. (2000). The urban environment and health in a world of increasing globalization: Issues for developing countries. *Bulletin of the World Health Organization, 78*(9). WHO.

Mccall, D. F. (1955). Dynamics of urbanization in Africa. *The ANNALS of the American Academy of Political and Social Science, 298*(1). https://doi.org/10.1177/000271625529800116

Michael, A. (2000). *A handbook of human resource management practice*. Kogan Press.

Mishra, R. P. (2007). *Geography of health: A treatise on geography of life and death in India* (p. 509). Concept Publishing Company.

Mitra, N. (1991). Environmental degradation on Darjeeling Himalaya In M. M. Jana (Ed.), *Environment degradation and development strategies in India* (p. 199). Ashish Publishing House.

Moore, E. G., & Mc Guinness, D. L. (1999). In K. Pandit, & S. D. Withers (Eds.), Migration and restructuring in the United States: A geographical perspective (p. 144). Rowman & Littlefield Publishers, Inc.

Nantapo, C. W. T.; Muchenje, V., Nkukwana, T. T., Hugo, A., Descalzo, A., Grigioni, G., Hoffman, L. C. (2015). Socio-economic dynamics and innovative technologies affecting health-related lipid content in diets: Implications on global food and nutrition security. In *Food research international, science direct of the AAAS* (Vol. 76, Part 4). Elsevier B.V. Available at: https://doi.org/10.1016/j.foodres.2015.05.033

Newell, C. (1988). *Methods and models in demography*. Belhaven Press.

Newmann, J. L., & Matzke, G. E. (1984). *Population: Pattern*. Prentice-Hall, Inc., Englewood Cliffs, New Jersey, U.S.A.

News 18: https://www.news18.com/news/buzz/covid-19-didnt-introduce-lockdown-to-the-world-a-look-back-at-the-history-3072605.html

Nola, I. A. (2018). Earthquakes and their environmental. *Medical and Public Health Impacts, 60*. Suplemento 1 de 2018, Salud Pública de Mexico.

Nong, Y., & Du, Q. (2011) Urban growth pattern modeling using logistic regression. *Geo-Spatial Information Science, 14*(1). Taylor and Francis, part of Informa PLC. Available at: https://doi.org/10.1007/s11806-011-0427-x

Nong, D. H., Lepczyk, C. A., Miura, T., Fox J. M. (2018) Quantifying urban growth patterns in Hanoi using landscape expansion modes and time series spatial metrics. *PLoS ONE, 13*(5).

Nwankwoala, H. N. L. (2015). Causes of climate and environmental changes: The need for environmental-friendly education policy in Nigeria. *Journal of Education and Practice, 6*(30). www.iiste.org

Open Learn Create (2017). Study Session 5 Urbanisation: Trends, Causes and Effects (pp. 9–16). The Open University, England & Wales. Available at: https://www.open.edu/openlearncreate/mod/oucontent/view.php?id=79940&printable=1

Oxford Advanced Learner's Dictionary. (2000). Oxford University Press.

Park, K. (2009). *Park's textbook of preventive and social medicine*. Banarsidas Bhanot.

Plosker, G. L. (2011). Sipuleucel-T in metastatic castration-resistant prostate cancer: profile report. *Bio Drugs, 25*(4). https://doi.org/10.2165/11207140-000000000-00000

Pubmed.Ncbi- chloroform: https://pubmed.ncbi.nlm.nih.gov/9487785/#:~:text=The%20first%20narcosis%20with%20chloroform,in%20France%20by%20Eug%C3%A8ne%20Soubeiran

Purohit, S., & Agrawal (2004). *Environmental sciences: A new approach*. Agrobios.

Rappuoli, R., Mandl, C., Black, S., & De Gregorio, E. (2011). Vaccines for the twenty-first century society. *Nature Reviews Immunology, 11*. Macmillan Publishers Limited. https://doi.org/10.1038/nri3085

Referowska-Chodak, E. (2019). Pressures and threats to nature related to human activities in European urban and suburban forests. In Forest. MDPI.

Richard, H. (1968). *Perspective on the nature of geography*. John Murray.

Richard, R. (2019). Urban public health: A historical perspective. In *Urban health*. Oxford. https://doi.org/10.1093/oso/9780190915858.003.0018

Ritchie, H., & Roser, M. (2020). *Urbanization. Our world in data*. Oxford Martin School, University of Oxford. Available at: https://ourworldindata.org/urbanization

Rizvi, S. N. H. (1986). Health practices of the Jaunsaris–A socio-cultural analysis. In B. Chaudhuri (Eds.), *Tribal health: Socio-cultural dimension*. Inter India Publicaitons.

Rodrigue, J. P., Comtois, C., & Slack, B. (2013). *The geography of transport systems*. Routledge, Taylor & Francis Group.

Ross John, A. (1982). International encyclopedia of population (Vol. Two, pp. 649, 656–663). The Free Press, a Division of Macmillian Publishing Co., Inc.

Ryabchikov, A. (1975). *The changing face of the earth*. Progress Publishers.

Sandal, E. K., Adiguzel, F. (2014). Spatial development of tarsus and the changes in land use. In E. Recep, O. T. Tüzün, S. Igor, A. Emin, & B. Mehmet (Eds.), *Urban and urbanization*. St. Kliment Ohridski, University Press.

Saxena, H. M. (1999). *Environmental geography*. Rawat Publications.

Sclar, E. D., & Volavka-Close, N. (2011). *Urban health: An overview*. Center for Sustainable Urban Development at Columbia University's Earth Institute, New York, NY, USA. Elsevier B.V.

Seto, K. C., Sanchez-Rodríguez, R., Fragkias, M. (2010). The new geography of contemporary urbanization and the environment. *The Annual Review of Environment and Resources, 35*(1). Annual Reviews.

Siddiqui, M. A. A., & Perry, C. M. (2006). Human papillomavirus quadrivalent (types 6, 11, 16, 18) Recombinant vaccine (Gardasil). *Drugs, 66*(9). https://doi.org/10.2165/00003495-200666090-00008

Silvaa, E. A., & Clarke, K. C. (2002). Calibration of the SLEUTH urban growth model for Lisbon and Porto, Portugal. *Computers, Environment and Urban Systems, 26*(6). Elsevier Science Ltd.

Singh, S. (2019). *Environmental geography*. Pravalika Publications.

Sinha, B. R. K. (2005). *Human migration: Concepts and approaches*. Foldrajzi Ertesito, Geographical Bulletin.

Sinha, B. R. K. (1987). Human Resource Utilization in Dhanarua Block, Patna Dist. (Bihar). In: Ph.D. Thesis submitted to the Department of Geography, Faculty of Science, The M.S. University of Baroda.

Souter, D. (2012). ICTs, the internet and sustainability: A discussion paper. In D. Souter & D. MacLean (Eds.), *Changing our understanding of sustainability: The impact of ICTs and the Internet*. The International Institute for Sustainable Development (IISD). www.iisd.org

Stewart, R., Korth, M., Langer, L., Rafferty, S., Silva, N. R. D., & Rooyen, C. V. (2013). What are the impacts of urban agriculture programs on food security in low and middle-income countries? *Environmental Evidence, 2*, 7.

Strahler, A. N., & Strahler, A. H. (1977). *Geography and Man's Environment* (pp. 2–3, 148, 153–157, 159). Willey.

Svalastog, A. L., Donev, D., Jahren Kristoffersen, N., & Gajović, S. (2017). Concepts and definitions of health and health-related values in the knowledge landscapes of the digital society. *Croatian medical Journal, 58*(6). https://doi.org/10.3325/cmj.2017.58.431

Swarup, R., Mishra, S. N., & Jauhari, V. P. (1992). *Encyclopaedia of ecology, environment and pollution control: An introduction to man and his environment*. Mittal Publications.

Sydenstricker, E. (1935). The changing concept of public health. *The Milbank Memorial Fund Quarterly, 13*(4). Wiley on behalf of Milbank Memorial Fund.

Tan, M., Li, X., Xie, H., & Lu, C. (2005). Urban land ex-pansion and arable land loss in China a case study of Beijing-Tianjin-Hebei region. *Land Use Policy, 22*(3).

The Compact Oxford Reference Dictionary. (2001). Oxford University Press.

The Lancet. (2015). *Editorial: Urban Health Post-2015* (Vol. 385). Available at: www.thelancet.com

The Prime Minister of India (Sunday, the 24th April 2022). The 88th Edition of Mann Ki Baat. https://newsonair.com/2022/04/27/test-track-treat-pm-modi-holds-review-meeting-amid-rising-covid-19-cases/

The Prime Minister of India. (Sunday, the 27th March 2022). The Breaking News about Sanitation at 11.10 A.M. on the News 24 an important Television Channel in India.

Tognotti, E. (2013). Lessons from the history of quarantine, from plague to influenza A. *Emerging Infectious Diseases, 19*(2), 255. https://doi.org/10.3201/eid1902.120312

Toledo-Pereyra, L. H. (2015). Medical renaissance. Journal of Investigative Surgery, 28(3). https://doi.org/10.3109/08941939.2015.1054747. PMID: 26065591

Tountas, Y. (2009). The historical origins of the basic concepts of health promotion and education: the role of ancient Greek philosophy and medicine. *Health Promotion International, 24*(2). Published by Oxford University Press.

Turk, J. (1985). *Introduction to environmental studies*. Saunders College Publishing.

Turner-Skoff, J. B., & Cavender, N. (2019). The benefits of trees for livable and sustainable communities. Plants, People, Planet 1. https://doi.org/10.1002/ppp3.39

UN-HABITAT. (2012). State of the World's Cities 2012/2013, Prosperity of Cities. United Nations Human Settlements Programme (UN-HABITAT), Nairobi.

Ulrich, L., & Helmut, B. (2014). Global health in the 21st century. *Global Health Action, 7*(1). Tayler and Francis Ltd.

United Nations. (2018). *The world's cities in 2018: data booklet* (p. 9). Economic and Social Affairs, New York.

United Nations. (2019). World Urbanization Prospects: The 2018 Revision (ST/ESA/SER.A/420). Department of Economic and Social Affairs, Population Division.

Valentei, D., Kvasha, A., & Ivanov, S. (1986). *Population and socio-economic development*. Progress Publishers.

Viana, C. M., Oliveira, S., Oliveira, S. C., Rocha, J. (2019). Land use/land cover change detection and urban sprawl analysis. In H. R. Pourghasemi & C. Gokceoglu (Eds.), *Spatial modeling in GIS and R for earth and environmental sciences*. Elsevier Inc.

Vickery, G. (2012). Smarter and Greener? Information technology and the environment: positive or negative impacts? In D. Souter & D. MacLean (Eds.), *Changing our understanding of sustainability: The impact of ICTs and the internet*. The International Institute for Sustainable Development (IISD). www.iisd.org

WHO. (2021). Noncommunicable diseases. Available at: https://www.who.int/news-room/fact-sheets/detail/noncommunicable-diseases

Webster's Encyclopedic Unabridged Dictionary of the English language. (1996). Gramercy Books.

Weeks, J. R. (1989). *Population: An introduction to concepts and issues*. Words Worth Publishing Company.

Wikipedia: https://en.wikipedia.org/wiki/Lockdown#cite_note-30

Wilson, E. H., Hurd, J. D., Civco, D. L., Prisloe, M. P., & Arnold, C. (2003) *Remote sensing of environment* (Vol. 86). Elsevier Inc.

World Health Organization. (2021). https://www.who.int/news-room/fact-sheets/detail/urban-health

Xiaoqing, Wu; Yuanman, H., Hongshi, H., Fengming, X., & Rencang, B. (2010). Study on forecast scenarios for simulation of future urban growth in Shenyang City based on SLEUTH model. *Geo-Spatial Information Science, 13*(1). Taylor and Francis, part of Informa PLC.

Yaguang, S. (2011) Development and characteristics of central business district under the philosophy of health. In *Procedia Engineering on International Conference on GreenBuildings and Sustainable Cities* (Vol. 21). Elsevier. https://www.elsevier.com/booksand-journals/procedia

Yapijakis, C. (2009). Hippocrates of Kos, the father of clinical medicine, and Asclepiades of Bithynia, the father of molecular medicine. *Vivo, 23*(4). International Institute of Anticancer Research

Zelinsky, W. (1966). *A prologue to population geography*. Prentice Hall, Inc.

Zhitin, D. V. (2011). Geographic specifics of migration flow vectors in the Russian Federation. *Russian Geographic Society Bulletin, 143*(3).

Chapter 2
Sustainable Urbanization in Southeast Asian Megacities: The Contrasting Cases of Singapore and Manila

Yves Boquet

Abstract Sustainable urbanization in recent years has become an important theme of research for the urban experts. The management of large cities to make them sustainable has become an international, well-publicized norm, but it requires efforts on several fronts such as transport and water management, which are vital for the quality of life of city dwellers. In the global context of climate change and a slow rise of sea-level waters, Asian megacities such as Singapore and Manila exhibit very different patterns of success, due to their respective levels of economic development and styles of governance. This paper first examines the dominant concepts in sustainability and urban sustainability, then proceeds to highlight some of the successes that have made Singapore an international model for urban sustainability policies, especially in regard to urban mobilities and the experiences in city greening. It then takes a critical look at the multiple failures seen in metropolitan Manila, which may be considered as almost the complete opposite of Singapore, with multiple problems such as heavy road congestion, housing woes, air and water pollution, or flooding. In Manila's case, it seems evident that sustainability must take into account the needs of the poorer segments of the population. It is because of the fact that sustainable urbanization helps provide economic and environmental opportunities for better life of such segments of urban population.

Keywords Global urbanization · Quality of life · Urban sustainability · Environmental opportunities · Peripheral areas

2.1 Introduction

Global urbanization is an irrepressible phenomenon. Urban areas now have more inhabitants than rural areas. The acceleration of urban growth, strongly correlated with globalization, the expansion of the market economy and modern technologies, is

Y. Boquet (✉)
Department of Geography, Université de Bourgogne-Franche-Comté, Dijon, France
e-mail: yves.boquet@u-bourgogne.fr

B. R. K. Sinha (ed.), *Urban Dynamics, Environment and Health*,
https://doi.org/10.1007/978-981-99-5744-6_2

particularly marked by the emergence of large metropolises and "megacities" (Kraas, 2007). The very concept of a "city" is challenged by the population explosion and urban sprawl: central cities are now less populated than their peripheries, city centres are competing for commercial activities and offices with peripheral areas, dubbed "edge cities" by Joel Garreau referring to U.S. metropolitan areas (Garreau, 1991), which can now be found in many cities around the world.

For the last thirty years, a growing proportion of Asian urbanization has occurred in very large urban areas (McGee & Wang, 1992). Many of the world's megacities are located in Asia, because of the mass of population of Asian countries (China, India, Indonesia), their density of population (Bangladesh, Philippines, Japan), a rapid economic growth symbolized by the power of their ports and airports in global networks (Singapore, Hong Kong) generating rapid urbanization (Abdullah, 2012; Dick & Rimmer, 1998; Shin & Timberlake, 2000; Tyner, 2000) fueled by industrialization (Arai, 2011; Franck, 2010).

The combination of economic growth/high rural densities/urban sprawl gave rise to large semi-rural mid-urban spaces, "large metropolitan areas" theorized as "desakota" (McGee, 1991), where "asphalt meets the rice field" (Douglass & Huang, 2007; Franck, 1993; Hara et al., 2005). Jakarta is today Jabodetabek, or even Jabodetabekpunjur, an inflation of syllables that reflects an increasing influence of the urban space on "rural" areas (Firman, 2009), a spatial growth often poorly controlled (Goldblum & Wong, 2000), and subject to heavy speculation in a context of weak urban governance and widespread corruption (Laquian, 2005a, 2005b).

While Southeast Asia as a whole shows a rate of urbanization that remains below the world average despite the presence of "megacities", situations vary greatly from one country to another (Sheng, 2010), from 77% of Brunei and 74% of Malaysia (we will leave aside the city-state of Singapore) to the low 22% of Cambodia and 33% of Laos. The strong trends observed for half a century, and those projected for the coming decades (UN data) are however clear: South East Asia is more and more urban. It is also marked by inequalities between cities, reflecting inequalities between countries. Economic and development disparities are staggering, ranging from a rich, prosperous, highly educated Singapore, to emerging countries, industrial "tigers" or "dragons" (Thailand, Malaysia), and poor, rural countries (East Timor, Laos).

The growth of cities implies an ever greater consumption of space, energy, and resources (Sheng & Thuzar, 2012). The strategic position of Asia as a major economic powerhouse accelerates the search for sustainable city models. Everyone recognizes that the contemporary Asian megacity, in developed and developing countries alike will have to be "sustainable" to be a true "global city" (Roy & Ong, 2011). These cities all exhibit similar worries about heavy flooding, as was shown in recent years in Bangkok, Jakarta, Manila, Hong Kong, and sometimes also in Singapore (Loh & Pante, 2015).

Emerging questions about sustainability, especially in the Asian context, focus on the mitigation of climate change (how to counter it with carbon sinks created through "greening" policies? (Bowler et al., 2010) how to adapt to sea-level rise in the low-lying deltaic floodplains where megacities such as Bangkok, Jakarta, Ho Chi Minh

City, Manila, Shanghai or Tokyo have been built ?), on the economic powerhouse couple (Pelling, 2003), on the "smart city" models ("knowledge-based urban development" Yigitcanlar et al., 2008, energy efficient cities, etc.…) and on the types of governance. Specificities of Asian cities (high population density, tropical monsoon climates, emerging cities in developing countries with many poor citizens) may lead to different views about what is urban sustainability. Can the Western concepts of "slow city" (Mayer & Knox, 2006; Pink, 2008), "soft mobility" (La Rocca, 2010; Le Néchet, 2012; Massot & Orfeuil, 2007) through the promotion of bicycling (Heinen et al., 2010), walking and tramways (Boquet, 2017a), pedestrian streets and walkability (Mehta, 2008; Southworth, 2005), "transit-oriented development" (Bernick & Cervero, 1996; Cervero, 1998; Cervero & Day, 2008; Ratner & Goetz, 2013) "smart growth" (Downs, 2005), "new urbanism" (Knaap & Talen, 2005) and "retracting suburbia" (Danielsen et al., 1999), "eco-neighborhood" (Bonard & Matthey, 2010; Da Cunha, 2011; Le Cadre & Mercier, 2012; Sharifi, 2016), "eco-cities" (Caprotti, 2014), find their place in a different context? Leducq et al. (2017), Ingallina (2019).

In this paper, after presenting some of the ideas about the vague notion of urban sustainability, we will contrast two cities at seemingly opposite ends of the sustainability scale, Singapore and Manila: can Singapore be a model (Chua, 2011), a blueprint, for the capital of the Philippines, which is sometimes viewed as the epitome of unsustainable urban development (Constantino-David, 2001)?

2.2 Urbanization and Sustainability

The expression "sustainable city" encompasses many dimensions of urban life and development (Theys & Emelianoff, 2001), as does "sustainable development". The standard 1987 definition (Brundtland Commission), "*Sustainable development is development that meets the needs of the present without compromising the ability of future generations to meet their own needs*", contains two key concepts: 1/the concept of "needs", in particular the essential needs of the world's poor, to which overriding priority should be given; and 2/the idea of limitations imposed by the state of technology and social organization on the environment's ability to meet present and future needs. Further developments on the idea of sustainable development have given rise to the vision of the three pillars of sustainable development, economic, environmental, and social, also informally referred to as people, planet, and profits (Purvis et al., 2018). A classic graphic representation of sustainable development shows the three pillars supporting the front pediment of a house which would collapse if one of the pillars were to fail. Another image is the intersection of 3 circles: social equity, economic viability, and environmental protection. Development is deemed viable if it is good for the economy and the environment (keeping resources such as water, land, and forests available for the future), it is bearable if the economic goals meet with the social goals, it is equitable if the environment issues and solutions coincide with social development. Where the three circles intersect, development is sustainable. The United Nations have adopted in 2015 the principle of Sustainable

Development Goals, after the 2000 Millennium Development Goals, taking into consideration the rising worries about resource depletion and climate change. Since the pioneer 1972 Meadows Report of the Club of Rome ("*The limits to growth*"), there has been an increasing recognition of the impacts of human activities on the environment on all scales, in many ways out competing natural processes leading to the idea of the Anthropocene (Barbier & Burgess, 2017; Crutzen, 2006).

The social dimension of sustainability appears the more difficult one to fully comprehend and tackle, since the logics of globalization are to make use of disparities and to increase them (Boyer et al., 2016).

Governments are facing greater challenges in providing inhabitants with a good quality of life in their cities. Urban sustainability has come forward as a rising political platform because cities are the primary sources of major environmental problems (consumption of resources and production of waste and pollutants through housing, industries, and transportation), the centres of economic and social developments, and home to more than half of the world population (Wu, 2008). The ecology of cities, often presented as the opposite of "nature", has long been neglected and understudied, but it has now taken its rightful place in the academic field, and possibly in the political arena. The carbon footprint of human activities and urban areas has now become a familiar element of the discourse of city managers and mayors. Regional leaders around the world are increasingly assuming leadership through a variety of innovative actions to fulfil the objectives of conferences such as Kyoto 1997 and Paris 2015. Some of them, as in the United States, are pushing forward despite the opposition of their national governments, and forge alliances such as the World Mayors Council on Climate Change or the Global Covenant of Mayors for Climate and Energy (Bulkeley et al., 2012; Rosenzweig et al., 2010; Schreurs, 2008). Looking at what is a sustainable city also leads us to consider what is an unsustainable city (Berger, 2014): excessive carbon footprint, dysfunctional networks (water, transportation), lack of self-sufficiency (hence the current trend to develop urban agriculture). It means examining the visions of the city of tomorrow, its place in the urban hierarchy of "global cities", detecting evolution trends, and imagining possible futures in response to the many challenges of ecological habitat, water recycling, waste recovery, non-polluting transport, preservation of biodiversity, energy efficiency and climate change, health and social justice, governance, etc. Promoting urban sustainability implies developing adequate assessment tools to measure progress (Bulkeley & Betsill, 2005). Efforts to improve the city's environment, through greening or bike paths, has the effect of increasing land values and fostering gentrification in the best areas of the city, while displacing the poorer residents unable to afford rents in "improved" neighbourhoods (Checker, 2011).

The classic indicators of economic growth, GDP per capita, Gini index, rate of poverty, used by international agencies such as the UN must be complemented with health indicators (infant mortality rate), transportation data (rate of motorization share of public transport use, availability of pedestrian paths and bicycle lanes, measurement of mobility obstacles) and environmental indicators (air quality, water quality, amount of greenery) to get a better idea of what makes a city good for all of its citizens, young and old, men, women, disabled people, local, and visitors

(Ooi, 2007). Rising incomes may be a sign of progress, but they also mean a higher consumption of energy, water, and other resources and an increasing number of motor vehicles and traffic congestion. Rapid population growth (through demography or rural-to-urban migration) also tends to aggravate urban problems such as housing. Evaluating urban sustainability is also evaluating the role of citizens in the daily life of the city: are they pawns looked upon as numbers by city executives, or are they also participants in the daily fabric of urbanization? Sustainability intersects with other social conditions, such as resilience, livability, adaptation, innovation, and reconciliation, as basic conditions of positive social change (James, 2014).

The cities must now be "smart", trying to use information technologies to better control their water, electricity, communications, garbage, and transport systems. Emergency response system and resilience are among the most crucial dimensions of smart and future cities design due to the increase in various disruptions caused by frequent man-made and natural disasters The "smart city" is hence the latest addition to the "sustainable city", "eco-city", and "green city" paradigms, as indicated by nationwide programmes like the 100 Smart Cities in Indonesia and in India (Mahesa et al., 2019; Rahmadiani, 2018), and the plans for new carbon–neutral, smart cities such as Dongtan (Shanghai, China), Masdar (Abu Dhabi in the United Arab Emirates) or New Clark City (Philippines), even if these cities, except for Songdo in South Korea (Shwayri, 2013), are still mostly in the planning stage (Evans et al., 2019; Hayat, 2016).

One of the conditions for a good functioning of cities is the existence of an efficient transport system. In a very disparate South East Asia, transport is a good indicator of development inequalities, between "informal" modes of transport (Goldblum, 2001) and efficient mobility systems (Boquet, 2009). The huge traffic jams in Manila, Jakarta, and Bangkok, which affect their productivity and image, are due to a combination of factors, mainly—urban sprawl imposing an increasing use of motorized vehicles (motorcycle or car) for the urban middle classes and feeding a rapid increase in the motorization rate (Estoque & Murayama, 2015),—the persistence of more or less informal modes of transport (Cervero, 2000; Hook & Replogle, 1996; Mateo-Babiano, 2016; Ratanawaraha & Chalermpong, 2016), that may look picturesque in the eyes of tourists (Philippine jeepneys and trisikel, Thailand tuk-tuk, cyclopousses, becak, samlo, and human-powered pedicabs). Without public transport and infrastructure worthy of the name, they are deployed at the heart of particularly dense road traffic.

It is therefore not uncommon to spend three to four hours per day in traffic. In Jakarta, apart from regional trains (crowded at peak times) and motorbikes (cheap, but not suitable for long journeys), the private car reigns supreme despite the implementation of the *Transjakarta*, a Bus Rapid Transit scheme that was inspired by experiences in Latin American cities (Curitiba, Quito, Bogota). Despite its size and the attempts to use it as a tool for Transit Oriented Development in the name of urban sustainability (Hasibuan et al., 2014), it has proven insufficient to solve the congestion problem, so there is now a new subway system being developed in the Indonesian capital. In Hanoi, Vietnam, the staggering amount of motorbike traffic is raising worries about the time when the automobile will replace the motorbike in

Vietnamese families. In Manila, the coexistence of a growing number of cars (half of the Philippine car fleet is found in the metropolitan area), an intense truck traffic linked to the presence of the commercial port in the city centre, and the uncontrolled deployment of buses, coaches from the provinces (no mainline trains in the Philippines) mingling with jeepneys and motorized tricycles contributes to inextricable congestion (Boquet, 2013a).

In stark contrast, Singapore has established itself as a model country in Southeast Asia (Henderson, 2012): safe, modern, honest, efficient, and prosperous, unlike many countries in the region plagued by endemic corruption and constant disruption. Its territory has been organized thanks to rigorous planning procedures that contrast singularly with the anarchic urban growth of other Southeast Asian cities such as Jakarta or Manila (Yuen, 1998). Can the sustainability policies of Singapore be transferred? Can other cities learn from "best practices" in urban sustainability? (Bulkeley, 2006).

2.3 Singapore: A Model City?

Singapore (720 sq.km., 5.8 million inhabitants) is the smallest state in Southeast Asia—2700 times smaller and 45 times less populous than Indonesia. In 1965, this city-state was facing the same serious problems confronting other Southeast Asian megacities today: high unemployment, poorly trained manpower, poor quality housing, crime, and filthiness. Since independence in 1965, however, Singapore has experienced a meteoric metamorphosis, following the basic orientations taken in the 60 s: authoritarian capitalism, strong state intervention, openness to the world, a search for excellence in infrastructures, technology, and education (Marginson, 2011). From a small, poor, resource-poor island, the nation-city has become one of the most stable and prosperous countries in the world. It is the richest in Southeast Asia (GDP per capita US $ 90,531 in 2017, 4th in the world after Qatar, Macau, and Luxembourg), the most developed (HDI 0.925 comparable to Germany, lowest infant mortality rate in the world 1.77 per 1000) and is regularly cited among the most economically efficient countries (ease of doing business, international competitiveness, best airline, best airport,etc.) and among the examples to follow for sustainable urban management.

The Singaporean state has tackled housing and transport issues head-on. Longtime Prime Minister Lee Kuan Yew launched an ambitious social housing programme in the early 1960s through the Housing Development Board (HDB) (Chua, 1991). At that time, the city-state was still a developing country, like its neighbours in South and Southeast Asia. Its urban part concentrated most of the population in shantytowns while the rural part was dotted with kampongs, traditional Malay villages. The goal of the HDB was twofold: to offer Singaporeans quality housing at moderate prices and to create a nation of homeowners. Today, 90% of Singaporean residents own their homes, and 80% of residents live in HDB apartments. The shantytowns of the 1960s are just a distant memory. To avoid racial tensions and communitarianism, and to foster a shared sense of national belonging, the government introduced ethnic quotas in all HDB buildings, leading to the cohabitation of various groups and the absence of ethnic ghettos (Huat, 2003; Sim et al., 2003). Today, to meet the demand of the global elite of the city, HDB also offers high-end housing in its social buildings.

To reduce the excessive density of the city centre and allow the development of a Central Business District, the Singapore territory has been divided into development zones, where residential settlements and economic activity zones are planned. A vertical urbanism in large apartment buildings in the middle of green spaces was instituted in new cities like Jurong, Tampines, and Woodlands, which welcomed people leaving the centre (Eng, 1996; Foo, 2001; Yuen et al., 2006). Everywhere in Singapore urban planning regulations specify inflexibly the land use ratios or the types of activities possible. This urban development is accompanied by a systematic effort to develop public transport, with an excellent metro network, complemented by bus lines and three local tram networks serving new cities, thus applying the principle of Transit Oriented Development (TOD) which encourages the use of public transport. Successive master plans (the most recent in 2014) prepare the evolution of the territory according to economic and demographic changes. In order to limit short-term land speculation and predict the future, the proceeds from the sale of land (or ten-year leases) add to Singapore's financial reserves, but are not included in annual budget revenues. A recent trend is the growth of tourism (Lohmann et al., 2009), with the recapture of post-port water fronts such as the banks of the Singapore River (Chang & Huang, 2005, 2011; Savage et al., 2004) and the development of recreational spaces (Sentosa Island) and casinos, for which Singapore becomes a competitor of Macao in Asia.

Singapore, true to its vision of being a high-tech vanguard country, a "Smart Island", has implemented a "biophilic" strategy of sustainable, green, ecological and intelligent cities (Calder, 2016; Gulsrud & Ooi, 2014; Newman, 2014). It aims at being a "Smart City" through the use of advanced technologies and continuous research of city management processes by the most modern means, in accordance with the technological culture that permeates Singapore, which today innovates in autonomous electric vehicles.

Urbanization has deeply affected the ecology of the island city-state, with massive coastal reclamation projects, the almost complete clearing of the original tropical lowland forests and a sharp reduction of its original biodiversity despite the early efforts of the Singapore Botanic Gardens established by colonial Britain in 1859. Since the 1980s, however, there has been an effort to include ecological aspects in the planning of Singapore, through a better understanding of its urban ecology, urban ecosystems and their role in the quality of life and sustainability of urban development (Tan & Abdul Hamid, 2014). The relatively high percentage of vegetation cover now achieved by the city due to urban greening policies shows the possibility of restoring ecological functions within a healthy urban ecosystem, even in a high-density urban environment. Singapore has proposed in 2008 an urban biodiversity index, dubbed the "Singapore index" (Rodricks, 2010), which has been adopted now by more 80 cities around the world ranging from cold Edmonton, Canada to warm Curitiba, Brazil. It is a set of indicators to measure and record how cities are growing their plant and animal species, protecting their habitats from urbanization and urban fragmentation, and include municipal bodies, local businesses, schools, and the public on its biodiversity awareness programmes.

Much of the original forest cover that has been lost due to urbanization, while the few cultivated lands have been replaced by urban green spaces, where vegetation is controlled by landscape engineers. Urban green spaces may not bring back the lost biodiversity, but they provide ecosystem services, including climate regulation through increased carbon sequestration, food, and opportunities for recreation. The quantity and structure of green space in a city may therefore have consequences for human well-being (Richards et al., 2017). In British colonial times, parks were established mostly for social and cultural aspects, defined with urban aesthetics values to improve the physical and moral condition of urban citizens (Sini, 2019). A 1958 plan reorganized green spaces and parks into a park system integrating green belts and nature reserves. However, it did not solve the issues of the fragmentation of native habitats and the rigid regimentation of watercourses and water bodies. New ideas have since emerged about considering green spaces as an ecological system rather than an eye-pleasing environment. Considerations about wildlife have resurfaced.

The country lacks resources, especially water. Although it rains heavily in Singapore, the government has implemented a multifaceted policy of water management, using the "4 taps" (Tan et al., 2009). The first, whose relative importance is to be reduced, is water imported from Malaysia via large pipes that cross the Straits of Johor. The excessive dependence on this water of the neighboring country has led Singapore to undertake storage of rainwater by reservoirs of various sizes, desalinization of seawater, an expensive technique that rich Singapore can finance, and finally a policy of maximum water recycling ("NEWater"). Today, 40% of the water used in Singapore is recycled. A major development operation, accompanying the urbanization of Marina Bay, is the Marina Reservoir, built in 2008 (Galelli et al., 2014; Sevin, 2008). With the completion of a dam close to the city centre, the reservoir, containing mainly salt water, has become a freshwater area in November 2010 after a natural desalination process (discharge of excess water to the sea after heavy rains). The reservoir provides 10% of the island's water needs. The catchment area that supplies the reservoir represents about one-sixth of Singapore's area, making it the largest of all of Singapore's reservoirs. This includes the Singapore, Kallang, Geylang, and Rochor river basins.

The policy of combating pollution and car congestion (Han, 2010) is the most energetic in the world: any car buyer must pay approximately 35,000 euros to obtain an ownership certificate valid for 10 years only, after which it must be paid anew. The number of vehicles put into circulation is severely limited (the lowest car ownership rate of all rich countries: 170 vehicles per 1000 inhabitants, against 569 in France and 811 in the United States), the driving licence is expensive, there are quotas in the number of new cars on the market, and foreigners are not allowed to drive. Taxes on imported vehicles (no car factories in Singapore) are 100%, parking spaces are scarce and expensive. Singapore is also the world pioneer in congestion charging: the Electronic Road Pricing (ERP) system, based on electronic boxes in vehicles communicating with control gates, modulates the rate of access to the city centre according to the hours of the day. The amounts collected from motorists have made it possible to finance public transport networks and to subsidize taxis, which are cheap.

Singapore's development strategies, from the beginning of Lee Kuan Yew's government, focused on greening the urban space as a key element for a quality living environment, and promoted the image of Singapore as a "Garden City" or "City in a Garden" (Goldblum, 1988; Tan et al., 2013; Velegrinis & Weller, 2012) through the implementation of "Green Plans" (Chua, 2012). It begins at Changi Airport, at terminals with numerous gardens and green spaces between boarding gates and duty free shops. While much of the city's original forest cover has been lost due to urbanization and industrialization, the Singapore State has better protected what was left of the original forests, and recreated urban green spaces through parks, public roadside afforestation and mandatory vegetation management around public and private buildings, with community gardens (Tan & Neo, 2009). Singaporean developers have developed several indicators, such as the Park Provision Ratio (PPR) which measures the number of hectares of public parks available per 1000 inhabitants (target 0.8 hectare per 1000 inhabitants), which played a key role in planning land use.

The Gardens by the Bay Park (101 hectares) (Davey et al., 2010) is located on the restored Marina Bay area and includes three gardens with large-scale green technologies and remarkable biodiversity. The "Conservatory Complex" is made up of two "Biomes", the "Flower Dome", which reproduces different Mediterranean and subtropical dry ecosystems in a dry and refreshed atmosphere, and the "Cloud Forest", which hosts in a cool and humid atmosphere Rainforest plants normally located at 2000 metres altitude. Both buildings are home to rare endangered plant species and include an indoor waterfall, a continuous flowering meadow, cascading levels of vertical planting, and raised walkways through and above the forest cover. A biomass plant uses horticultural waste to produce heat and energy, as well as fertilizer in the form of ash. The waste heat is captured to dehumidify the air before it is cooled for air conditioning of the domes. The air evacuated from conservatories is channelled to eighteen "Supertrees", futuristic tree-like structures, twenty-five to fifty metres high, which serve as breeding grounds for birds and insects. These structures are covered with a "living skin" supporting photovoltaic cells and a rainwater harvesting system. These "trees" are symbolic markers of the integration of green technologies into the urban landscape.

As the territory is limited, the preservation of "natural" spaces and the establishment of urban green spaces is not sufficient in the long term, it is also necessary to find complementary spaces in the urban landscape that can play the role of green spaces. In recent years, Singapore has focused on the concept of "vertical garden city", using already built infrastructure, such as rooftop gardens or green walls on building facades. The greenery of the terraces seems banal, but the architects in Singapore have taken it to an unprecedented level: the roof terraces of the Marina Bay Sands hotel, famous for its swimming pool resting on three buildings, or the

HDB Pinnacle Building are 50 floors high. These green roofs play a very important energetic role, lowering the temperature compared to other roofing materials, because some of the sun's energy is absorbed by plants that emit water vapour.

These energetic virtues are verified with the "vertical greenery". The "green walls", inside and outside the buildings, are composed of cassettes containing small compartments with plants. Each is connected to a watering system fed by rainwater collected, stored, and reused for irrigation of vegetation.

Singapore has therefore earned a global reputation for excellence in "Green Buildings", with a triple bottom line: a green building preserves the planet, increases the well-being of its occupants, and is more economical than an ordinary building. It must be energy efficient, use environmentally friendly materials, and take into account the health and well-being of its occupants. Singapore has developed since 2005 its own rating system called Green Mark (Han, 2019). The building candidate for certification is judged on 5 parameters: energy performance, parsimonious use of water, integration into the external environment, quality of the indoor environment, innovative character. Depending on the performance achieved, the building earns the qualification of Certified Single, Gold, Gold Plus, or Platinum. In 2012, Singapore crossed the symbolic threshold of 1,000 Green Mark certified buildings, applying its usual recipes for combining financial incentives and a binding legal framework. In addition to subsidies for environmental performance, since 2008, the law requires all new buildings of more than 2000 m^2 to be at least certified. By 2030, 80% of all buildings will need to be certified. Singapore's largest real estate developers have followed and committed to reaching legal levels 10 years in advance, with buildings "net zero", which contribute more to the environment than they consume: solar panels to generate more energy than it consumes, skillful use of natural ventilation and lighting, construction of "bioclimatic towers" (recycling of the raincoat, filtering of the sun, solar air conditioning, etc.).

2.4 How to Improve Manila?

Manila ranked poorly in the social, environmental, and economic sub-indices of the 2018 Sustainable Cities Index established by the Amsterdam-based *Arcadis* consulting firm. It was ranked 95th out of 100 large world cities, faring poorly in the three major dimensions of sustainable development: 93rd for "people", 91st for "planet", and 98th for "profit". Therefore, it "needs to reinvent itself", according to the *Arcadis* report.

The "People" pillar includes data on education, health, demographics, income inequality, affordability, work-life balance, crime, access to public transport services, transport applications and digital capabilities, cultural offerings, cost of broadband, digital public services (property tax), and wifi availability. Anybody who has been in Manila knows that "slow city" is not a positive attribute: it means heavy traffic and slow Internet.

The "Planet" pillar combines indicators about environmental exposure, green spaces, energy, air pollution, greenhouse gas emissions, waste management, drinking water and sanitation, bicycle infrastructure, electric vehicle incentives, negative emissions technologies (carbon capture and storage), and natural disaster monitoring.

The "Profit" pillar informs about transport infrastructure, economic development, ease of doing business, tourism, connectivity, employment, and university technology research.

In all cases, Singapore is ranked the best city in Southeast Asia, by far, while Manila is near the bottom, alongside Hanoi, Vietnam. The database does not include, unfortunately, the other major Vietnamese city, Ho Chi Minh City (Table 2.1).

Another set of indicators developed by Numbeo, a crowd-sourced global database of reported consumer prices, perceived crime rates, and quality of healthcare, among other statistics. Also shows major contrasts between Singapore on one side, Bangkok, Jakarta, and Hanoi on the other side. It indicates a high level of satisfaction for Singapore's residents, while Manila dwellers bemoan the daily difficulties in their city, which is the most polluted, the most exposed to pollution, has the most polluted water, is the dirtiest and noisiest of the four megacities, with mediocre green spaces and many hours wasted in traffic jams. American author Dan Brown has even described Manila as "the gates of hell" (Table 2.2).

Indicators developed by other groups or agencies confirm the dismal rankings of Manila. Air quality is catastrophic: a World Health Organization study released in May 2018 said there were about 45.3 deaths per 100,000 individuals in the Philippines due to outdoor air pollution. China ranked first at 81.5 deaths recorded while Mongolia was second at 48.8 deaths, the Philippines being the third worst country in this regard. The Philippines was in a poor 73rd spot out of 126 economies in the 2018 Global Innovation Index (GII), which ranks countries according to their capacity and success in innovation, looking into a country's institutions, infrastructure, business and market sophistication, knowledge and technology outputs, human and capital research, and creative outputs. Singapore was ranked 5th, just behind Switzerland, the Netherlands, Sweden, and the UK. The Philippines' political environment and political stability and safety ranked 93rd and 117th, respectively. The Phillipines also ranked poorly in respect to the rule of law, while Singapore was 1st, 1st and 9th best in the world in the 3 categories above mentioned. Other reports have pointed the high level of corruption (*Transparency International* places the Philippines as 99th out of 180 countries for corruption, Indonesia as 89th, Thailand 99th, Vietnam 117th, while Singapore is considered as the 3rd least corrupt in the world behind Denmark and New Zealand). In addition, the Philippines is one the most exposed and vulnerable countries of the world to natural hazards (Radtke et al., 2018): typhoons, earthquakes, and tsunami may hit most of the country, including its capital, which is also located not very far from two dangerous volcanoes, Mount Pinatubo to the north and Mt Taal to the south. Volcanoes, with their shape and height create an orographic effect and dispersive tail of rain clouds which magnify rainfall and flood hazards in Metro Manila (Lagmay et al., 2015). Urban areas such as Metro Manila have an increased vulnerability to disasters due to the number of people that are exposed to hazards.

Table 2.1 Arcadis report on the sustainability of 100 major cities in the world

People ranking			Planet ranking			Profit ranking			Overall ranking		
Rank	City	Score	Rank	City	Score	Rank	City	Score	Rank	City	Score
1	Edinburgh	78	1	Stockholm	80	**1**	**Singapore**	**86**	1	London	74
4	Taipei	73	30	Seoul	61	2	London	79	**4**	**Singapore**	**70**
7	Seoul	71	**41**	**Singapore**	**58**	3	Hong Kong	78	9	Hong Kong	67
11	Tokyo	70	50	Hong Kong	56	10	Seoul	65	13	Seoul	66
16	Macau	68	63	Tokyo	52	13	Taipei	62	24	Taipei	61
21	Hong Kong	67	64	Shenzhen	52	28	Tokyo	56	33	Tokyo	59
31	**Singapore**	**65**	71	Guangzhou	50	35	Macau	55	41	Macau	58
51	Beijing	60	72	Taipei	49	37	Kuala Lumpur	55	66	Shenzhen	51
53	Guangzhou	59	73	Macau	49	64	Bangkok	46	67	Kuala Lumpur	51
55	Shenzhen	58	76	Shanghai	46	67	Beijing	45	73	Beijing	49
59	Kuala Lumpur	56	80	Tianjin	44	70	Shenzhen	44	74	Guangzhou	47
67	Shanghai	54	83	Kuala Lumpur	43	75	Shanghai	36	76	Shanghai	45
76	Tianjin	50	84	Bengaluru	43	77	Guangzhou	33	80	Bangkok	42
82	Chengdu	48	85	Beijing	42	82	Jakarta	27	83	Tianjin	40
83	Wuhan	46	86	New Delhi	40	83	Tianjin	26	87	Wuhan	35
85	Bangkok	44	89	Mumbai	39	87	Chennai	22	88	Chennai	35
86	Chennai	43	90	Wuhan	39	89	Wuhan	21	89	New Delhi	34
87	New Delhi	43	**91**	**Manila**	**39**	90	Mumbai	20	90	Chengdu	34
91	Bengaluru	35	92	Chennai	39	91	Hanoi	19	91	Bengaluru	32
92	Hanoi	33	93	Chengdu	37	92	Bengaluru	19	93	Mumbai	30
93	**Manila**	**33**	94	Bangkok	37	94	New Delhi	18	94	Jakarta	30

(continued)

Table 2.1 (continued)

People ranking			Planet ranking			Profit ranking			Overall ranking		
Rank	City	Score	Rank	City	Score	Rank	City	Score	Rank	City	Score
94	Mumbai	32	96	Jakarta	35	96	Chengdu	16	**95**	**Manila**	**29**
96	Kolkata	31	99	Kolkata	31	**98**	**Manila**	**15**	98	Hanoi	28
97	Jakarta	28	100	Hanoi	31	100	Kolkata	8	100	Kolkata	23

Source Arcadis, Amsterdam 2018—Citizen Centric Cities The Sustainable Cities Index 2018. https://www.arcadis.com/en/global/our-perspectives/sustainable-cities-index-2018/citizen-centric-cities/

Table 2.2 Numbeo report on quality of life and pollution in 4 major Southeast Asian cities

	Bangkok	Jakarta	Manila	Singapore
Pollution index	75.41	82.85	91.37	32.36
Pollution exposition scale	133.39	147.50	165.21	55.00
Air pollution data from WHO				
PM 10	42	48	55	30
PM 2.5	24	21	17	18
PM 10 pollution level	Moderate	Moderate	High	Moderate
Air pollution	71.75 High	75.84 High	84.43 Very high	33.47 Low
Infant mortality rate p. 1000 living births	7.7 (Thailand)	18.9 (Indonesia)	19.7 (Philippines)	1.8
Water pollution	69.53 High	75.97 High	83.61 Very high	23.22 Low
Drinking water quality and accessibility	53.09 Moderate	44.76 Moderate	41.72 Moderate	87.12 Very high
Satisfaction with garbage disposal	41.46 Moderate	25.98 Low	22.00 Low	77.64 High
Dirty and untidy	63.94 High	68.21 High	80.13 Very high	18.81 Very low
Noise and light pollution	57.73 Moderate	59.38 Moderate	68.17 High	49.11 Moderate
Quality of green and parks	43.12 Moderate	31.83 Low	30.41 Low	78.91 High
Estimated annual hours spent in traffic jams (hours lost per worker per year)	208.0	124.8	233.6	89.6
Satisfaction to spend time in the city	39.37 Low	30.87 Low	22.89 Low	74.35 High

Source https://www.numbeo.com/pollution/compare_cities. Data collected on 16 July 2019

The increase of human settlements in an already vulnerable area due to rapid urbanization results in higher disaster risk rather than the increased frequency of hazards experienced (El-Masri & Tripple, 2002).

Manila's challenges are indeed many (Boquet, 2014a; Morley, 2018). It is sometimes described as a concrete jungle, with poor urban planning and very few green spaces, a city with some the highest population densities in the world, blighted by slums, traffic jams, and polluted waterways. Most of the green spaces there are located in Quezon City because of their efforts to produce lesser-density spaces and prohibitions by the government to build over public parks such as the Quezon Memorial Circle and the UP Diliman campus. There are major issues in pollution (Edelman, 2016), road traffic (Boquet, 2013a), flooding, and housing (Porio, 2011). The Pasig/

Marikina rivers' watershed is densely populated and highly polluted (Kumar et al., 2018; Qian et al., 2000), as well as Manila Bay's shores (De Castro, 2010). Even as Manila often floods, especially in poorer neighbourhoods (Zoleta-Nantes, 2000, 2002), water quality is mediocre (Hosono et al., 2010) and water availability is problematic at the end of the dry season, especially in El Niño years such as 2019, leading to rationing. The low level of water in reservoirs may limit the capacity to produce enough electricity for the city at a time there is high demand for air conditioning. Meanwhile, stagnant waters in clogged waterways are conducive to the proliferation of disease-carrying mosquitoes, leading to outbreaks of dengue and malaria. All these issues are related and systemic. Poverty and slums are part of the city's environmental problems. Encroaching shantytowns on waterways reduce the natural flow of waters. The lack of proper garbage disposal, due to insufficient investments in trash treatment facilities, is worsened by the lack of discipline of many residents (Bennagen et al., 2002; Bernardo, 2008): the esteros, small rivers in the deltaic plain where Manila has been built, are littered with debris and refuse, which is a heaven for rats, and the lack of proper restrooms forces people to defecate in the water, hence the many issues with intestinal diseases and diarrheas affecting especially children and women.

At the same time, there are well-known efforts at global-cityness in the new private developments erected by major firms such as Megaworld, Ayala, or SM in their much vaunted "townships" providing a pleasant living environment in small areas designed as "work-live-shop-and-play" enclaves (Boquet, 2016; Kleibert & Kippers, 2016), replete with manicured lawns, flower beds and fountains, designed by private developers for private profit (Mouton & Shatkin, 2019), and not as parts of a general metropolitan development plan that does not exist. Three of the ten largest malls in the world (SM Megamall, SM North Avenue, and SM Mall of Asia) are located in Metro Manila. Shopping malls play an important role in the Philippine economy: the retail industry accounts for approximately 15% of the Philippines' GDP and 33% of the entire services sector. It employs some 5.3 million people (18% of the Philippines' workforce). *SM Supermalls*, *Ayala Malls*, and *Robinsons Malls* are the major players, alongside *Walter Mart Malls*, *Gaisano Malls*, *Ever Gotesco Malls*, and *Isetann*. Most of the malls are as luxurious and comfortable as those in Singapore and Western countries. They provide an air-conditioned environment welcomed in this hot humid tropical country. However, access is controlled by guards who may reject would-be mall visitors if they do not fit the expected shopper profile. The international elites of the Philippines, if they do not choose a gated community (the first one was Forbes Park in Makati, designed in the 1950s), live close to office towers in these high-rise condominiums with many amenities: shopping, restaurants, sports facilities, etc., leading to a virtual withdrawal of the well-to-do from the poverty around them. In these condominiums, business parks, and shopping malls one does not need to bother about the noise, pollution, and poverty. Manila is clearly a dual city (Garrido, 2013) where abject poverty cohabits with luxury shopping malls (three of the ten largest malls in the world are located in Metro Manila). Authorities have built walls separating squatter areas from major thoroughfares, such as the Pinagsama area of Taguig between the airport and the BGC business centre (Michel, 2010).

One of the factors in the seemingly dysfunctional and ineffective organization of Metro Manila is the lack of strong and unified governance (Boquet, 2014b), in total contrast with the heavy and steady hand of the Singapore government. Metro Manila is made of 17 distinct municipalities with their own regulations and objectives. Because of the way the metropolis is structured, there are in fact seventeen different sets of public services, seventeen different local governments, seventeen different education systems, and basically every other public good that should be provided to the region's 12 million inhabitants replicated 17 times over. This leads to inconsistency in the delivery of public services, inefficiency and unneeded competition, corruption and political disenfranchisement, since the constituent cities of Metro Manila are effectively run as fiefdoms of the political elites (the Estrada family in San Juan, the Binay family in Makati…), who operate out of sheer self-interest without any consideration for the residents of the region as a whole. A coordinating agency, the Metropolitan Manila Development Authority (MMDA) (Manasan & Mercado, 1999) has been assigned the responsibility to assist the 17 political units in crafting and implementing an integrated development plan characterized by unity of purpose, innovation, resiliency, sustainability, adaptability and creativity in meeting the challenges of transport decongestion, climate change, waste management. However, it appears weak and unable to assert its authority over mayors jealous of their prerogatives. The directors of the MMDA are appointed by the president of the Philippines, and usually do not last long enough in their position to be able to prepare and execute long-term plans: there have been 12 directors since 1990, lasting less than 3 years on average, the longest being Bayani Fernando for 7 years, whose successor Oscar Inocentes held the position for barely 7 months. Metro Manila has grown to the point where it may need to live under one city government, not seventeen. There are examples abroad with the amalgamation of Toronto or several US cities (Indianapolis, Kansas City, Louisville), the Tokyo Metropolis government or the formation of Greater London. This issue of metropolitan governance reform is also examined in other Asian cities, such as Jakarta (Firman, 2008; Laquian, 2005a, 2005b). Unified metropolitan governance may be needed to achieve success in area-wide regional planning and to better coordinate services such as transport, water supply, sanitation, and solid waste disposal at a wider scale than the current Metro Manila area. The current Metro Manila official area covers about 600 sq. km., whereas the actual built urban area, with urban sprawl, covers about 1700 sq. km. A larger area than today could be needed to manage issues such as floods and water, transportation or housing, including Laguna de Bay, the large lake deeply impacted by the growth of Metro Manila (Guerrero, 1996). There are indeed water management problems that can only be tackled through a stronger coordination between MMDA, the LLDA (Laguna Lake Development Agency), and the PRRC (Pasig River Rehabilitation Commission).

Tools, like GIS and traffic or flooding modelling software (Badilla, 2008), and technical solutions exist to solve a number of the environmental problems of Manila. For instance, the elevation of roads or construction of well-designed drainage structures leading to the creeks would allow to make good progress on the combined problem of traffic congestion and street flooding in the many episodes of heavy rain caused by thunderstorms (Lagmay et al., 2017).

Traffic congestion is another issue that many policies have tried to tackle, without much success. There have been efforts to establish alternate days for cars with odd- and even-licence plates (Unified Vehicular Volume Reduction Program of 1995, UVVRP). This was not successful (Regidor, 2013; Salvanera et al., 2016), due to the ease of falsifying licence plate numbers, also the fact that some well-to-do people bought a second car and bribed their way to make sure the licence plate will be complementary from the first car, and also because some municipalities, especially Makati and Mandaluyong, where the largest business districts are located, refused to go along. The UVVRP has not proven an efficient tool to reduce congestion and pollution due to the fast increase in the number of vehicles on the roads, despite the improvements in per-vehicle emissions. No political leader in the Philippines has so far pushed forward any plan for a Singapore-like solution (Pay as You Go Congestion Charge), a measure that was highly successful in Singapore but goes against the automobile-bent mindset in the Philippines. In fact, the main measures to reduce traffic congestion have revolved around giving more space to automobiles with the construction of intra-urban freeways (SLEX-NLEX connector) and an effort to limit buses on main thoroughfares such as EDSA.

Buses, both metropolitan and interregional, are an important part of the Philippine transportation system (Boquet, 2012, 2013b), especially since the country has no railway system at this time, and they are for people without private cars the only way to travel from the provinces to the capital. However, the structure of the bus industry is made of a myriad of competing companies (Rimmer, 1991), where drivers are paid according to the number of passengers and not with a fixed salary (the "boundary system" in which the driver rents the vehicle from the owner and must pay back before earning for himself), leading to very aggressive driving in order to maximize ridership. The spatial effects are the proliferation of private terminals in several areas of Metro Manila, most noticeably in Pasay and Cubao. The movements of buses in and out of tiny adjacent terminals create multiple traffic jams. The current policy of MMDA is to consolidate bus terminals into 3 major facilities located in the outskirts of Manila, such as the recently opened PITX intermodal terminal in Parañaque. This has been criticized, first as a hastily planned facility, second as a measure against poorer Filipinos from the provinces, who have to transfer to city buses not equipped for people with heavy luggage. The much-touted South Korean "Gangnam model" for Seoul buses (Kim et al., 2009) or the well-conceived example of the Purabaya integrated bus terminal in Surabaya, Indonesia (Mirnasari & Suaedi, 2013; Oetomo & Ardini, 2012), have not been implemented to their full potential in Metro Manila. It has also been criticized as an anti-public transportation measure going contrary to the current trends seen in Singapore, Hong Kong, Seoul, and other cities, of favouring public transport over the private automobile. It is easy to point out that a bus, even half-full, carries more people using less space than many automobiles do.

In the same spirit, there have been increasing limits placed on the use of jeepneys, the emblematic mass transport vehicles of the Philippines since the 1940s. They have been banned from most of the EDSA semi-circular boulevard running through several municipalities of Metro Manila, and the Duterte government has pushed for their replacement with new, clean vehicles, that are too expensive to buy and maintain by the small owners controlling most of the jeepney fleet, which provides very cheap transportation to everyone. The policy aimed at reducing the high pollution caused by these ageing vehicles is deemed as anti-poor, which seems ironic when the populist president claims to be a "man of the masses" (Boquet, 2017b).

However, the Philippine administration has embraced a long overdue programme of investments in rail transportation, as well as road improvements and airport enhancement, defined by President Rodrigo Duterte as "build-build-build", with financing coming mostly from Japan and China. New railways are planned out of Manila, and several subway lines, both overhead and underground are now in construction, creating even more road traffic woes as construction is ongoing. It remains to be seen if middle-class people will embrace it and leave their cars at home in the absence of any strong Singapore-style incentive to switch modes, eschewing the famous remark by Bogota Mayor Enrique Peñalosa, "An advanced city is not a place where the poor move about in cars, rather it's where even the rich use public transportation".

The poor people in Manila are also pointed out as being part of the cause of traffic jams, with the myriad of vendors encroaching on the road pavement, hence reducing vehicular flows. It was estimated in 2005 that there were around 50,000 street vendors in Metro Manila (Bhowmik, 2005), despite MMDA and local regulations, such as in Caloocan (Illy, 1986; Recio & Gomez, 2013), prohibiting street and sidewalk vending. Regulations, as in many developing countries, are inspired by Western images of modernization largely removed from the harsh socio-economic realities of the sector, while cultural values (conflict avoidance, respect of power structures) govern the behaviour of lower-level administrative agents and street hawkers alike. As a result, regulations are mostly symbolic and ineffective, nevertheless maintaining a climate of harassment and extortion. Street vendors are part of the informal economy, making them difficult to monitor and regulate. While this allows them to dodge tax payment, this also means they lack legal protection, making them vulnerable. Street vendors are being blamed for disorder and garbage on the streets, and are even suspected of being part of crime syndicates. It may be true for some of them, but most are simply trying to make ends meet. Authorities have repeatedly launched programmes aimed at the removal of street nomads, beggars, vagrants, and homeless persons from the streets of Metro Manila. In July 2019, the newly elected Manila Mayor Francisco "Isko" Moreno Domagoso started his tenure with well-publicized street clearing operations and removal of vendors in some of the most heavily congested areas of his city, such as the Divisoria Market. His objective was to ensure an efficient flow of vehicles, reducing pollution and petty crime at the same time, but it was immediately criticized as another anti-poor measure, in the absence of an established plan to allow people continuing to earn a living through small-scale selling and street cooking (another cause of multiplied small-scale pollution due to the use of highly toxic charcoal grills). Some foreign examples may inspire Manila: India (Sinha & Roever, 2011; Schindler, 2014) has created vending zones or dedicated spaces with proper facilities for their street vendors, while Bangkok (Batréau &

Bonnet, 2016) has integrated vending into its prime tourism experiences, allowing it to continue in public spaces with strict regulation on sanitation, among other set standards.

2.5 Conclusion

It is clear that the efforts to make Manila a more manageable, more sustainable city, must also make it a more inclusive city and take into account the livelihoods of the poor. For instance, the poor should not be the victims of urban renewal (Tomeldan et al., 2014) or urban transportation policies (Loo & Chow, 2006). Despite some irreversible damage to its original natural features, Singapore was able to transform itself into a more resilient and sustainable urban area thanks to a unique combination of long-term strong political will, an eradication of corruption, a set of goals clearly defined, a law-and-order policy that attracted foreign investment fostering economic growth and allowing the country to build substantial money reserves. The emphasis on educational and technological excellence has made the use of high technology solutions to solve transportation and water supply problems. Manila is so far well behind, with heavy corruption still in place, a lack of long-term vision for the city (hence the absence of a precise blueprint for the future), and a trend towards a widening gap between rich and poor, while Singapore has been much more inclusive.

The problems have been identified for a long time (Jimenez & Velasquez, 1989), the solutions are known, the examples of better practices for a better, more sustainable, more efficient city abound in Asia, not just in Singapore, but also in Korea, Japan, Hong Kong (such as the network of pedestrian pathways above the streets of Central) or in neighboring Indonesia at an equivalent level of prosperity. Sustainable development goals should not be limited to mostly environmental issues, even if they are essential (clean air, clean water greenery, biodiversity) but also be linked with quality of life targets (Porio, 2015) for the social aspect of sustainable development and "shared growth". Education to environmental issues, especially in schools, would also provide much progress in the future (Guzman & Guzman, 2000).

The missing factor in the Philippines to implement proven solutions and adapt them to the Philippine context appears mostly political. Can the Philippines and Manila wait many more years to move towards a better, more pleasant city life for all its citizens, rich and poor, young and old?

References

Abdullah, J. (2012). City competitiveness and urban sprawl: Their implications to socio-economic and cultural life in Malaysian cities. *Procedia – Social and Behavioral Sciences, 50*, 20–29.

Arai, K. (2011). From water buffaloes to motorcycles: The development of large-scale industrial estates and their socio-spatial impact on the surrounding villages in Karawang Regency, West Java. *Southeast Asian Studies, 49*(2), 161–191.

Badilla, R. (2008). *Flood modelling in Pasig-Marikina River Basin.* International Institute for Geo-Information Science and Earth Observation, Enschede.

Barbier, E., & Burgess, J. (2017). The sustainable development goals and the systems approach to sustainability. *Economics: The Open-Access, Open-Assessment E-Journal, 11* (2017–28), 1–22. https://doi.org/10.5018/economics-ejournal.ja.2017-28.

Batréau, Q., & Bonnet, F. (2016). Managed informality: Regulating street vendors in Bangkok. *City and Community, 15*(1), 29–43.

Bennagen, M. E. G., Nepomuceno, G., & Covar, R. (2002). *Solid waste segregation and recycling in Metro Manila: Household attitudes and behavior*. Economy and Environment Program for Southeast Asia (EEPSEA) Los Baños. https://ideas.repec.org/p/eep/report/rr2002062.html

Berger, M. (2014). The unsustainable city. *Sustainability, 6*, 365–374.

Bernardo, E. (2008). Solid-waste management practices of households in Manila, Philippines. *Annals of the New York Academy of Sciences, 1140*, 420–424.

Bernick, M., & Cervero, R. (1996). *Transit villages in the 21st century*. University of California.

Bhowmik, S. K. (2005). Street vendors in Asia: A review. *Economic and Political Weekly, 40*(22/23), 2256–2264.

Bonard, Y., & Matthey, L. (2010). Les éco-quartiers: laboratoires de la ville durable. *Cybergeo: European Journal of Geography*. http://journals.openedition.org/cybergeo/23202.

Boquet, Y. (2014b). Les défis de la gouvernance urbaine à Manille. *Bulletin De l'AGF, 91*(4), 461–478.

Boquet, Y. (2014a). Metro Manila challenges : Flooding, housing and mobility. In R. B. Singh (Ed.), *Urban Development Challenges* (pp. 447–469). Springer Publishing.

Boquet, Y. (2009). Transports et mobilités dans les grandes villes d'Asie. In G. Fumey, J. Varlet & P. Zembri (Eds.), *Mobilités contemporaines. Approches géoculturelles des transports* (pp. 139–152). Ellipses.

Boquet, Y. (2012). Moving around the Philippines: Challenges and dynamics of mobility in a developing country. In *Proceedings of the 17th Hong Kong Society for Transportation Studies Conference, « Transportation and Logistics Management »* (pp. 29–36). Hong Kong University of Science and Technology.

Boquet, Y. (2013a). Bus transportation in the Philippines. In *Proceedings of the 18th Hong Kong Society for Transportation Studies conference, « Travel Behaviour and Society »* (pp. 511–518). The University of Hong Kong.

Boquet, Y. (2013b). Battling congestion in Manila: The EDSA problem, *Transport and Communications Bulletin for Asia and the Pacific*, UN-ESCAP, United Nations Economic and Social Commission for Asia and the Pacific, Bangkok, 82, 45–59, http://www.unescap.org/ttdw/Publications/TPTS_pubs/bulletin82/b82_fulltext.pdf.

Boquet, Y. (2016). MetroManila's competing business districts; Edge Cities, Philippine style? In R. Rachmawati, G. Pomeroy, & D. Mokherjee (Eds.), *Rapid urbanization and sustainable development in Asia, Proceedings of 13th International Asian Urbanization Conference* (pp. 291–302). Universitas Gadjah Mada.

Boquet, Y. (2017a). The renaissance of tramways and urban redevelopment in France. *Miscellanea Geographica–Regional studies on development, 21*(1), 5–18. https://www.degruyter.com/downloadpdf/j/mgrsd.2017a.21.issue-1/mgrsd-2017a-0005/mgrsd-2017a-0005.pdf.

Boquet, Y. (2017b). Phasing out the jeepney? A Philippine mode of transport at the crossroads of society, culture and sustainability. In A. Chen & N.N. Sze (Eds.), *Transport and society, Proceedings of the 22nd International conference of the Hong Kong Society for Transportation Studies* (pp. 471–478). Hong Kong Polytechnic University.

Bowler, D., Buyung-Ali, L., Knight, T., & Pullin, A. (2010). Urban greening to cool towns and cities: A systematic review of the empirical evidence. *Landscape and Urban Planning, 97*(3), 147–155.

Boyer, R., Peterson, N., Arora, P., & Caldwell, K. (2016). Five approaches to social sustainability and an integrated way forward. *Sustainability, 8*(9), 878. https://www.researchgate.net/publication/307894140.

Bulkeley, H. (2006). Urban sustainability: Learning from best practice? *Environment and Planning A: Economy and Space, 38*(6), 1029–1044.

Bulkeley, H., & Betsill, M. (2005). Rethinking sustainable cities: Multilevel governance and the 'urban' politics of climate change. *Environmental Politics, 14*(1), 42–63.

Bulkeley, H., Broto, V. C., & Edwards, G. (2012). Bringing climate change to the city: Towards low carbon urbanism? *Local Environment, 17*(5), 545–551.

Calder, K. (2016), *Singapore: Smart city, smart state.* Brookings Institution.

Caprotti, F. (2014). Critical research on eco-cities? A walk through the Sino-Singapore Tianjin Eco-City, China. *Cities, 36*, 10–17.

De Castro, J. A. (2010). Cleaning up Manila Bay: Mandamus as a tool for environmental protection. *Ecology Law Quarterly, 37*(2), 797–804.

Cervero, R. (1998). *The transit metropolis.* Island Press.

Cervero, R., & Day, J. (2008). Suburbanization and transit-oriented development in China. *Transport Policy, 15*(5), 315–323.

Cervero, R. (2000). *Informal transport in the developing world.* United Nations Centre for Human Settlements (Habitat).

Chang, T. C., & Huang, S. (2005). Recreating place, replacing memory: Creative destruction at the Singapore River. *Asia Pacific Viewpoint, 46*(3), 267–280.

Chang, T. C., & Huang, S. (2011). Reclaiming the city: Waterfront development in Singapore. *Urban Studies, 48*(10), 2085–2100.

Checker, M. (2011). Wiped Out by the "Greenwave": Environmental gentrification and the paradoxical politics of urban sustainability. *City and Society, 23*(2), 210–229.

Chua, B. C. (1991). Not depoliticized but ideologically successful: The Public Housing Programme in Singapore. *International Journal of Urban and Regional Research, 15*(1), 24–41.

Chua, B. H. (2011). Singapore as model: Planning innovations, knowledge experts. In A. Roy, & A. Ong (Eds.), Worlding Cities. *Asian experiments and the art of being global* (pp. 29–54). J. Wiley & Sons.

Chua, L. H. (2012). The Singapore Green Plan 2012. Ministry of the Environment. http://www.env.gov.sg/sgp2012

Constantino-David, K. (2001). Unsustainable development: The Philippine experience. *Development in Practice, 11*(2), 232–241.

Controlling nature, disciplining human nature: Floods in Singapore and Metro Manila, 1945–1980s. *Nature and Culture, 10*(1), 36–56.

Crutzen, P. (2006). The Anthropocene. In E. Ehlers, & T. Krafft (Eds.), *Earth system science in the Anthropocene* (pp. 13–18). Springer.

Da Cunha, A. (2011). Les écoquartiers, un laboratoire pour la ville durable: entre modernisations écologiques et justice urbaine. *Espaces et Sociétés*, 144–145, 193–200. https://www.cairn.info/revue-espaces-et-societes-2011-1-page-193.htm

Danielsen, K., Lang, R., & Fulton, W. (1999). Retracting suburbia: Smart growth and the future of housing. *Housing Policy Debate, 10*(3), 513–540.

Davey, M., et al. (2010). Gardens by the Bay: High performance through design optimization and integration. *Intelligent Buildings International, 2*(2), 140–157

Dick, H., & Rimmer, P. (1998). Beyond the third world City: The new urban geography of Southeast Asia. *Urban Studies, 35*(12), 2303–2321.

Douglass, M., & Huang, L. (2007). Globalizing the city in southeast Asia: Utopia on the urban edge – The case of Phu My Hung, Saigon. *International Journal of Asia Pacific Studies, 3*(2), 1–42.

Downs, A. (2005). Smart growth: Why we discuss it more than we do it. *Journal of the American Planning Association, 71*(4), 367–378.

Edelman, D. (2016). Managing the urban environment of Manila. *Advances in Applied Sociology, 6*, 101–133. https://doi.org/10.4236/aasoci.2016.63010
El-Masri, S., & Tripple, G. (2002). Natural disaster, mitigation and sustainability: The case of developing countries. *International Planning Studies, 7*(2), 157–175.
Eng, T. S. (1996). Character and identity in Singapore new towns: Planner and resident perspective. *Habitat International, 20*(2), 279–294.
Estoque, R., & Murayama, Y. (2015). Intensity and spatial pattern of urban land changes in the megacities of Southeast Asia. *Land Use Policy, 48*, 213–222.
Evans, J., et al. (2019). Smart and sustainable cities? Pipedreams, practicalities and possibilities. *Local Environment, 24*(7), 557–564.
Firman, T. (2008). In search of a governance institution model for Jakarta Metropolitan Area (JMA) under Indonesia's new decentralisation policy: Old problems, new challenges? *Public Administration and Development, 28*(4), 280–290.
Firman, T. (2009). The continuity and change in mega-urbanization in Indonesia: A survey of Jakarta-Bandung Region (JBR) development. *Habitat International, 33*, 327–339.
Foo, T. S. (2001). Planning and design of Tampines, an award-winning high-rise, high-density township in Singapore. *Cities, 18*(1), 33–42.
Franck, M. (2010). Diffusion spatiale de l'urbanisation et de l'industrialisation et formation d'une région urbaine : Le cas de Surabaya, en Indonésie. *Annales De Géographie, 671–672*, 69–92.
Franck, M. (1993). *Quand la rizière rencontre l'asphalte... Semis urbain et processus d'urbanisation à Java-Est*. EHESS.
Galelli, S., et al. (2014). Optimal real-time operation of multipurpose urban reservoirs: Case study in Singapore. *Journal of Water Resources Planning and Management, 140*(4), 511–523.
Garreau, J. (1991). *Edge city: Life on the new frontier*. Doubleday Books.
Garrido, M. (2013). The ideology of the dual city: The modernist ethic in the corporate development of Makati City, Metro Manila. *International Journal of Urban and Regional Research, 37*(1), 165–185.
Goldblum, C. (1988). Singapour (1819–1986): émergence de la ville moderne et mythe rural. *Archipel, 36*, 227–270.
Goldblum, C., & Wong, T. C. (2000). Growth, crisis and spatial change: A study of haphazard urbanisation in Jakarta. *Indonesia. Land Use Policy, 17*(1), 29–37.
Goldblum, C. (2001). Transports informels et adaptations à la métropolisation en Asie du Sud-Est. *L'Information Géographique, 65*(1), 18–32, https://www.persee.fr/doc/ingeo_0020-0093_2001_num_65_1_2731
Guerrero, R. (1996). Human impacts on Laguna de Bay, Philippines and management strategies for their mitigation. *GeoJournal, 40*(1–2), 69–72.
Gulsrud, N., & Ooi, C. S. (2014). Manufacturing green consensus: Urban Greenspace governance in Singapore. In L.A. Sandberg, A. Bardekjian, & S. Butt (Eds.), *Urban forests, trees, and greenspace. A Political ecology perspective* (pp. 77–92). Routledge.
Guzman, R. S., & Guzman, R. Z. (2000). *Environmental education for sustainable development*. Wisdom Advocates Publishing.
Han, S. S. (2010). Managing motorization in sustainable transport planning: The Singapore experience. *Journal of Transport Geography, 18*(2), 314–321.
Han, H. (2019). Governance for green urbanisation: Lessons from Singapore's green building certification scheme. *Environment and Planning C: Politics and Space, 37*(1), 137–156.
Hara, Y., Takeuchi, K., & Okubo, S. (2005). Urbanization linked with past agricultural landuse patterns in the urban fringe of a deltaic Asian mega-city: A case study in Bangkok. *Landscape and Urban Planning, 73*, 16–28.
Hasibuan, H. S., Soemardi, T., Koestoer, R., & Moersidik, S. (2014). The role of transit oriented development in constructing urban environment sustainability, the Case of Jabodetabek, Indonesia. *Procedia Environmental Sciences, 20*, 622–631.
Hayat, P. (2016). Smart cities: A global perspective. *India Quarterly, a Journal of International Affairs, 72*(2), 177–191.

Heinen, E., van Wee, B., & Maat, K. (2010). Commuting by bicycle: An overview of the literature. *Transport Reviews, 30*(1), 59–96.

Henderson, J. (2012). Planning for success: Singapore, the model city-state ? *Journal of International Affairs, 65*(2), 69–83.

Hook, W., & Replogle, M. (1996). Motorization and non-motorized transport in Asia. *Land Use Policy, 13*(1), 69–84.

Hosono, T., & al. (2010). Application of multi-isotope ratios to study the source and quality of urban groundwater in Metro Manila Philippines. *Applied Geochemistry, 25*(6), 900–909.

Huat, C. B. (2003). Multiculturalism in Singapore: An Instrument of social control. *Race & Class, 44*(3), 58–77.

Illy, H. (1986). Regulation and evasion: Street-vendors in Manila. *Policy Sciences, 19*(1), 61–81.

Ingallina, P. (2019). *Ecocity, Knowledge city, Smart city: Vers une ville écosoutenable ?* Presses du Septentrion.

James, P. (2014). *urban sustainability in theory and practice. Circles of sustainability*. Routledge.

Jimenez, R., & Velasquez, A. (1989). Metropolitan Manila : A framework for its sustained development. *Environment and Urbanization, 1*(1), 51–58.

Kim, W., Son, B., Chung, J. H., & Kim, E. (2009). Development of real-time optimal bus scheduling and headway control models. *Transportation Research Record, 2111*(1), 33–41.

Kleibert, J., & Kippers, L. (2016). Living the good life? The rise of urban mixed-use enclaves in Metro Manila. *Urban Geography, 37*(3), 373–395.

Knaap, G., & Talen, E. (2005). New urbanism and smart growth: A few words from the academy. *International Regional Science Review, 28*(2), 107–118.

Kraas, F. (2007). Megacities and global change in east, southeast and South Asia. *Asien, 103*, 9–22.

Kumar, P., et al. (2018). Evaluating future stress due to combined effect of climate change and rapid urbanization for Pasig-Marikina River, Manila. *Groundwater for Sustainable Development, 6*, 227–234.

La Rocca, F. (2010). Soft mobility and urban transformation. *TEMA. Journal of Land Use, Mobility and Environment*, *2*. http://www.classiconorroena.unina.it/index.php/tema/article/view/urn:nbn:it:unina-3558

Lagmay, A., et al. (2017). Street floods in Metro Manila and possible solutions. *Journal of Environmental Sciences*, *59*, 39–47.

Lagmay, A., et al. (2015). Volcanoes magnify Metro Manila's southwest monsoon rains and lethal floods. *Frontiers in Earth Science*, *2*, 36, https://doi.org/10.3389/feart.2014.00036/full

Laquian, A. (2005a). *Beyond Metropolis*. Johns Hopkins University Press.

Laquian, A. (2005b). Metropolitan governance reform in Asia. *Public Administration and Development, 25*(4), 307–315.

Le Néchet, F. (2012). Urban spatial structure, daily mobility and energy consumption: A study of 34 European cities. *Cybergeo: European Journal of Geography*, *580*. https://journals.openedition.org/cybergeo/24966

Le Cadre, H., & Mercier, D. (2012). Is energy storage an economic opportunity for the eco-neighborhood?. *Netnomics: Economic Research and Electronic Networking*, *13*(3), 191–216.

Leducq, D., Scarwell, H., & Ingallina, P. (2017). *Modèles de la ville durable en Asie. Utopies, circulation des pratiques, gouvernanc*. Peter Lang.

Lohmann, G., et al. (2009). From hub to tourist destination: An explorative study of Singapore and Dubai's aviation-based transformation. *Journal of Air Transport Management, 15*(5), 205–211.

Loo, B., & Chow, S. (2006). Sustainable urban transportation; concepts, policies and methodologies. *Journal of Urban Planning and Development, 132*(2), 76–79.

Mahesa, R., Yudoko, G., & Anggoro, Y. (2019). Dataset on the sustainable smart city development in Indonesia. *Data in Brief*, *25*, https://www.sciencedirect.com/science/article/pii/S2352340919304524

Manasan, R., & Mercado, R. (1999). *Governance and Urban Development: Case Study of Metro Manila*. PIDS Discussion Paper Series 1999–03. https://www.econstor.eu/handle/10419/187389

Marginson, S. (2011). Higher education in East Asia and Singapore: Rise of the confucian model. *Higher Education, 61*(5), 587–611.
Massot, M. H., & Orfeuil, J. P. (2007). La contrainte énergétique doit-elle réguler la ville ou les véhicules? Mobilités urbaines et réalisme écologique. *Les Annales de la Recherche Urbaine*, 103, 18–29. https://www.persee.fr/doc/aru_0180-930x_2007_num_103_1_2710
Mateo-Babiano, I. (2016). Indigeneity of transport in developing cities. *International Planning Studies, 21*(2), 132–147.
Mayer, H., & Knox, P. (2006). Slow Cities: Sustainable places in a fast world. *Journal of Urban Affairs, 28*(4), 321–334.
McGee, T. (1991). The emergence of 'Desakota' regions in Asia: Expanding a hypothesis. In N. Ginsberg (Ed.), *The extended metropolis: Settlement transition in Asia* (pp. 3–26). University of Hawaii Press.
McGee, T., & Wang, Y. L. (1992). La formation des mégapoles en Asie. *Mappemonde, 1992*(4), 2–3.
Mehta, V. (2008). Walkable streets: Pedestrian behavior, perceptions and attitudes. *Journal of Urbanism, 1*(3), 217–245.
Michel, B. (2010). Going global, veiling the poor global city imaginaries in Metro Manila. *Philippine Studies, 58*(3), 383–406.
Mirnasari, R., & Suaedi, F. (2013). Inovasi Pelayanan Publik UPTD Terminal Purabaya-Bungurasih. *Kebijakan Dan Manajemen Publik, 1*(1), 71–84.
Morley, I. (2018). City profile: Manila. *Cities, 72*, 17–33.
Mouton, M., & Shatkin, G. (2019). Strategizing the for-profit city: The state, developers, and urban production in Mega Manila. *Environment and Planning A: Economy and Space.* https://doi.org/10.1177/0308518X19840365
Newman, P. (2014). Biophilic urbanism: A case study on Singapore. *Australian Planner, 51*(1), 47–65.
Oetomo, H. W., & Ardini, L. (2012). SWOT analysis in strategic management: A case study at Purabaya bus station. *Journal of Economics, Business, and Accountancy Ventura, 15*(2), 171–186.
Ooi, G. L. (2007). Urbanization in Southeast Asia: Assessing policy process and progress toward sustainability. *Journal of Industrial Ecology, 11*(2), 31–42.
Pelling, M. (2003). *The vulnerability of cities. Natural disasters and social resilience.* Earthscan.
Pink, S. (2008). Sense and sustainability: The case of the Slow City movement. *Local Environment. The International Journal of Justice and Sustainability, 13*(2), 95–106.
Porio, E. (2011). Vulnerability, adaptation, and resilience to floods and climate change-related risks among marginal, riverine communities in Metro Manila. *Asian Journal of Social Science, 39*(4), 425–445.
Porio, E. (2015). Sustainable development goals and quality of life targets: Insights from Metro Manila. *Current Sociology, 63*(2), 244–260.
Purvis, B., Mao, Y., & Robinson, D. (2018). Three pillars of sustainability: In search of conceptual origins, *Sustainability Science, 14*(3), 681–695. https://link.springer.com/article/https://doi.org/10.1007/s11625-018-0627-5
Qian, X., et al. (2000). Field survey on the flow structure and water quality of Pasig river, Metro Manila. *Annual Journal of Hydraulic Engineering, JSCE, 44*, 1101–1106.
Radtke, K., et al. (2018). *World Risk Report 2018.* Ruhr University Bochum. www.WorldRiskReport.org
Rahmadiani, A. (2018). *100 Smart Cities: Between Indonesia and India*, Yogyakarta, Center for Digital Society. cfds.fisipol.ugm.ac.id/article/373/100-smart-cities-indonesia.
Ratanawaraha, A., & Chalermpong, S. (2016). How the poor commute in Bangkok, Thailand. *Transportation Research Record, 2568*(1), 83–89.
Ratner, K., & Goetz, A. (2013). The reshaping of land use and urban form in Denver through transit-oriented development. *Cities, 30*, 31–46.

Recio, R., & Gomez, J. E. (2013). Street vendors, their contested spaces, and the policy environment: A view from Caloocan. *Metro Manila. Environment and Urbanization ASIA, 4*(1), 173–190.
Regidor, J. (2013). Traffic congestion in Metro Manila: Is the UVVRP still effective ? *Philippine Engineering Journal, 34*(1), 66–75.
Richards, D., Passy, P., & Oh, R. (2017). Impacts of population density and wealth on the quantity and structure of urban green space in tropical Southeast Asia. *Landscape and Urban Planning, 157*, 553–560. https://www.sciencedirect.com/science/article/pii/S0169204616301839
Rimmer, P. (1991). A tale of four cities: Competition and bus ownership in Bangkok, Jakarta, Manila and Singapore. *Transportation Planning and Technology, 15*(2–4), 231–252.
Rodricks, S. (2010). *Singapore City biodiversity index.* http://www.teebweb.org/wp-content/uploads/CaseStudies/Singapore%20City%20Biodiversity%20Index.pdf
Rosenzweig, C., Solecki, W., Hammer, S., & Mehrotra, S. (2010). Cities lead the way in climate–change action. *Nature, 467*, 909–911.
Roy, A., & Ong, A. (2011). *Worlding Cities. Asian experiments and the art of being global.* Wiley.
Salvanera, A., Sibayan, G., Tan, M., & Rosete, M. A. (2016). The impact of the unified vehicular volume reduction program in reducing air pollution. *Review of Integrative Business and Economics Research, 5*(3), 387–408.
Savage, V., Huang, S., & Chang, T. C. (2004). The Singapore River thematic zone: Sustainable tourism in an urban context. *The Geographical Journal, 170*(3), 212–225.
Schindler, S. (2014). Producing and contesting the formal/informal divide: Regulating street hawking in Delhi. *India. Urban Studies, 51*(12), 2596–2612.
Schreurs, M. (2008). From the bottom up: Local and subnational climate change politics. *The Journal of Environment & Development, 17*(4), 343–355.
Sevin, O. (2008). Marina Bay et l'aménagement du front de mer à Singapour. *Géographie et Cultures, 62*, 79–96. https://journals.openedition.org/gc/2375
Sharifi, A. (2016). From Garden City to Eco-urbanism: The quest for sustainable neighborhood development. *Sustainable Cities and Society, 20*, 1–16.
Sheng, Y.-K. (2010). Good urban governance in Southeast Asia. *Environment and Urbanization, 1*(2), 131–147.
Sheng, Y.-K., & Thuzar, M. (2012). *Urbanization in Southeast Asia: Issues & impacts.* ISEAS Publishing.
Shin, K.-H., & Timberlake, M. (2000). World cities in Asia: Cliques, centrality and connectedness. *Urban Studies, 37*(12), 2257–2285.
Shwayri, S. (2013). A model Korean Ubiquitous Eco-City? The politics of making Songdo. *Journal of Urban Technology, 20*(1), 39–55.
Sim, L. L., Yu, S. M., & Han, S. S. (2003). Public housing and ethnic integration in Singapore. *Habitat International, 27*(2), 293–307.
Sinha, S., & Roever, S. (2011). India's national policy on urban street vendors, *Wiego Policy Brief, 2*, 12.
Sini, R. (2019). *Singapore's park system master planning A nation building tool to construct narratives in post-colonial countries.* Springer Nature.
Southworth, M. (2005). Designing the walkable city. *Journal of Urban Planning and Development, 131*(4), 246–257.
Tan, P. Y., & Abdul Hamid, A. R. B. (2014). Urban ecological research in Singapore and its relevance to the advancement of urban ecology and sustainability. *Landscape and Urban Planning, 125*, 271–289.
Tan, P. Y., Wang, J., & Sia, A. (2013). Perspectives on five decades of the urban greening of Singapore. *Cities, 32*, 24–32.
Tan, L. & Neo, H. (2009). "Community in Bloom": Local participation of community gardens in urban Singapore. *Local Environment. The International Journal of Justice and Sustainability, 14*(6), 529–539.
Tan, Y. S., Lee, T. J., & Tan, K. Y. S. (2009). *Clean, green and blue. Singapore's journey towards environmental and water sustainability.* Institute of Southeast Asian Studies.

Theys, J., & Emelianoff, C. (2001). Les contradictions de la ville durable. *Le Débat, 113*(2001/1), 122–135.
Tomeldan, M., et al. (2014). "Shared Growth" urban renewal initiatives in Makati City, Metro Manila, Philippines. *Journal of Urban Management, 3*(1–2), 45–65.
Tyner, J. (2000). Global Cities and circuits of global labor: The case of Manila. *Philippines. the Professional Geographer, 52*(1), 61–74.
Velegrinis, S., & Weller, R. (2012). The 21st-Century garden City? The metaphor of the garden in contemporary Singaporean urbanism. *Journal of Landscape Architecture, 2*(2), 30–41.
Wu, J. (2008). Making the case for landscape ecology. An effective approach to urban sustainability. *Landscape Journal, 27*(1), 41–50.
Ygitcanlar, T., Velibeyoglu, K., & Baum, S. (2008). *Knowledge-based urban development: Planning and applications in the information era*. Information Science Reference.
Yuen, B., et al. (2006). High-rise living in Singapore public housing. *Urban Studies, 43*(3), 583–600.
Yuen, B. (1998). Singapore: The planned city-state. In B. Yuen (Ed.), *Planning Singapore, from plan to implementation* (pp. 1–10). Singapore Institute of Planners.
Zoleta-Nantes, D. (2000). Flood hazards in Metro Manila: Recognizing commonalities, differences and appropriate courses of action. *Social Science Diliman, 1*(1), 60–105.
Zoleta-Nantes, D. (2002). Differential impacts of flood hazards among the street children, the urban poor and residents of wealthy neighborhoods in Metro Manila, Philippines. *Mitigation and Adaptation Strategies for Global Change, 7*, 239–266.

Chapter 3
Urbanization and Its Challenges in Nepal

Krishna Prasad Poudel and Upendra Paudel

Abstract Urbanization in Nepal is a new phenomenon. Prior to 1951, towns of Kathmandu Valley were the only confined settlements with some urban functions and infrastructure. But, out of Kathmandu Valley, few settlements were formed as hamlets or small markets comprising some commercial activities with a purpose to serve caravan traders and porters along the trade routes. In Nepali vernacular those centres named *"hatiya"*. Several such centres and also including Kathmandu Valley towns were largely practised agricultural activities during the rainy seasons and followed trade and commerce activities in off-seasons. In the long historical past, architecture of all types of market centres built as of the needs of agricultural production systems. Those towns had the similar pattern of urban morphology. The process of current urbanization in the country has been started along with the transformation of social, economic, and infrastructure development after the democratic movement of 1951. By 2020, Nepal has claimed around 60% of its total 30 million populations of the country living within the constituent urban boundaries. It is very different case of Nepal. It is also a strange to note how the urbanization process within the last 70 years transformed from almost negligible to such alarming figure. This paper aims to analyse urbanization processes, trend of urbanization, and urban facilities entertained by the people living within the constituent urban centres. It further emphasizes on associated issues and challenges of urban centres in Nepal. The method of analysis follows the review of available resources and personal observation in the context of Nepal. Based on the in-depth analysis, it is concluded that the practice of declaration of urban centres in Nepal has not been following the scientific basis and opened up a new discourse on delineating constituent urban centres.

Keywords Urbanization · Challenges · Transformation · Infrastructure · Development · Market centres · *Hatiya* · Functional agglomeration · Physiographic heterogeneity

---

K. P. Poudel (✉)
Department of Geography Education, Tribhuvan University, Kirtipur, Kathmandu, Nepal
e-mail: Kpoudel.pokhara@gmail.com

U. Paudel
Department of Geography, Prithvi Narayan Campus, Tribhuvan University, Pokhara, Nepal

B. R. K. Sinha (ed.), *Urban Dynamics, Environment and Health*,
https://doi.org/10.1007/978-981-99-5744-6_3

3.1 Introduction

Urbanization is a dynamic process of concentration of large number of people within a relatively smaller spatial extension of territory (https://www.britannica.com/topic/urbanization), and majority of people follow other than primary means of production (https://www.nationalgeographic.org/encyclopedia/urbanization/). The system of urban centre is not static. It undergoes fundamental changes with new development and changing circumstances (Shrestha, 1982: 2). In the context of Nepal, large part of it has topographically heterogeneous landscape and sparse human settlements with a national average of 179 persons km^2 (CBS, 2012). Small strip of plain area of the country is located in the southern part which comprises nearly 20% of area and around 50% of the total population (CBS, 2012). That plain landscape was under the influence of several diseases and natural calamities until the nineteenth century. In the past, large number of people of entire country was practising agrarian economy, primary production and mainly based on primitive technologies. Because of such dominant characteristics, throughout the historical past, urbanization was not a common phenomenon in the country. People used to live in scattered rural settlements and a few in compact villages even by following the agriculture activities. Prior to 1961, towns and cities were limited. There were only ten incorporated towns in the country having more than ten thousand population (CBS, 1962). Small market centres were spontaneously settled along the caravan routes. Those were developed along the main trail routes to provide services for local people as well as for the travellers and porters with main functions of small tea stalls, lodges and hotels, small groceries, clothes, and items with household needs. Many of such centres were also the rest places for porters and trekkers. Gradually, some centres also became the administrative centres after the political change of 1951. Some large market centres were accounted for zonal and district administrative headquarters. Most of those administrative centres were listed within the constituent municipalities. Because of administrative centres the flow of people increased and functions were gradually concentrated. Thus, in the initial stage urbanization in Nepal grew with administrative activities as well as increasing population with diversified needs. Because of rural and self-sufficient agrarian production systems, people hardly needed market. Salt and kerosene were the major market items. But during the farming off-season, local villagers used to fetch those items from the market centres of either from Bhote (Tibete) or from Indian border markets. They usually travelled in a folk of porters following main trail. *Hatiyas* were the location developed along the main trails with a purpose to give services to the passersby and also goods of demand of local villagers. There were also weekly market systems for the exchange of agrarian production as well as for other goods of demand of local people. Those weekly markets were movable in different locations called *Haat Bazar*. Such practice is still in some parts of the country. Gradually, *hatiyas* and *haat bazaars* used to change to permanent market centres and towns after the political change of 1951 and initiation of construction of motor roads in the country. However, the urban centres of Kathmandu Valley and its peripheries were developed in the form of *entripot* with a purpose to export and import of Tibetan goods and Indian goods (Shrestha, 1973). Some small craft industries were established in Kathmandu Valley Towns for the export purposes to both sides in Tibet and India. Similarly, the towns along the Indian border were established to facilitate transit of export and import goods. Some of those centres were later on developed for the industrial establishments. Therefore, urbanization trend is increasing rapidly after 1951 (Basyal & Khanal, 2001). In these circumstances, urbanization and its challenges have been made by reviewing publicly available documents and literatures. The analysis is mainly confined to highlight the transformation of rural-to-urban settlements, infrastructure development, urbanization processes and its challenges in Nepal.

3.2 Aims and Objectives

This paper has been analysing urbanization processes, the trend of urban development, and urban facilities and associated issues and challenges within the constituent urban centres of Nepal. More specifically, the focus has been given on tracing the urban development processes, presentation of the trend of urban development and urbanization, describing the distribution of urban facilities and services in proportion of the population as well the summarizing the issues and challenges associated with the constituent urban centres of Nepal.

3.3 Methods and Materials

Being a review paper analysis is entirely based on the published and available documents, government database, and census reports. However, the review notes are verified based on personal observation and even from the anecdotal statements.

3.4 Population Scenario and Its Concentration

Both India and China share the closest neighbour of Nepal and in terms of total population Nepal is comparatively a smaller country in the South Asia Region. For the urbanization process, both number and distribution of population play vital role. As the country first started population counting in 1911, the total figure was 5.63 million. Since then every 10 year interval counting population is operating. In 2011 it has 26.5 million total population (Fig. 3.1). At the end of 2020 it is estimated for 30 million.

Estimation of July 2020 mentioned the country has 29.1 million populations (https://www.worldometers.info/world-population/nepal-population/), and it shares 0.37% of the world total population. Crude density of population distribution in the

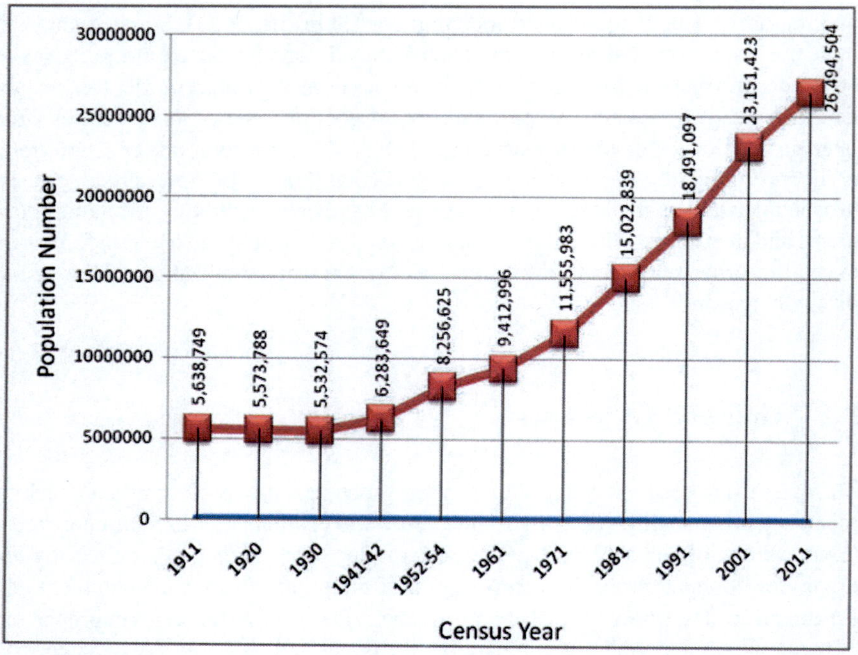

Fig. 3.1 Population of Nepal from 1911 to 2011 (CBS, 2014)

country is estimated 203 persons per square kilometre. According to the constitution of Nepal enacted in 2015, country is divided into 7 federal provinces, 77 districts, and 753 local government units or municipalities.

Nepal is a country with high physiographic heterogeneity. Therefore, the population is not evenly distributed. The 2011 census data presents Kathmandu district has the highest density (4,408 persons per km^2) and Manang district has the lowest density (3 persons per km^2). Kathmandu district has recorded the highest decadal population growth (60.93%) compared to all Nepal (14.99%) and lowest in Manang district (−31.92%). Tarai geo-ecological belt constitutes 50.15% of total population of the country, while hill belt constitutes 43.1% and mountain belt has only 6.75% (Fig. 3.2).

The distribution is highly confined in the Tarai districts and Kathmandu Valley. Because of the concentration of population it is obvious that the large numbers of urban centres are also located there.

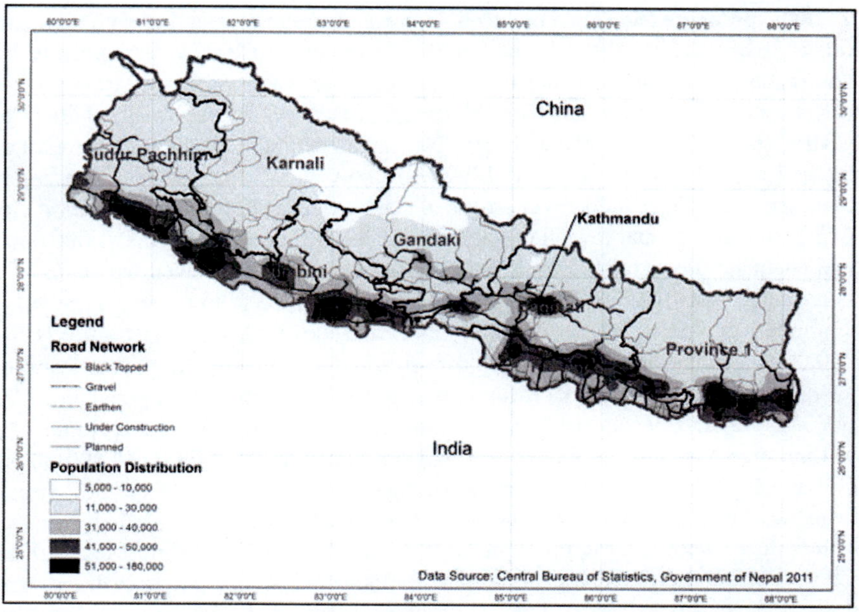

Fig. 3.2 Population distribution in Nepal (CBS, 2012). *Source* Department of Survey, Government of Nepal 2020 and population of Nepal 2011, CBS 2012

3.5 Urban Centres and Delineation Criteria

In Nepal, the first enumeration of population according to the urban centres was made in 1952/1954. In that time no any defined objective criteria was made but the government felt to designate urban centres as of the existing towns mainly with population number (Subedi, 2014: 108) with some revenue generation data from the marketing of goods. In that time only ten centres namely Kathmandu, Lalitpur, Bhaktapur, Kirtipur, Thimi, Biratnagar, Janakpur, Birgunj, Malangawa, and Nepalgunj were listed (Bastola, 1995: 244). But the first formal definition of urban settlement was started since the census of 1961. Urban centres were defined with at least 5,000 populations in a compact settlement having schools and colleges, government offices, judiciary and shops, and marketing goods. The first Municipality Act was endorsed in 1962 (Nagar Panchayat Ain 2019 B.S.) defined urban settlement (Nagar Panchayat) having 10,000 population. The fifth amendment of Nagar Panchayat Act 1962 had been made in 1976 and redefined the urban settlement. According to the amended Act 1076, the minimum requirement of population for urban centre (Nagar Panchayat) fixed for 9,000 populations. Based on those criteria the 1981 census enumerated 23 designated urban settlements in the country. After the democratic restoration in 1990, Municipality Act 1992 (Nagar Palika Ain 2048 B.S.) was endorsed by the government. According to that act some new criteria added. For the Municipality at least 20,000 population, annually around one million rupees (USD

25,000, approximately 1 USD = 40 NPR) revenue collection and urban infrastructures like health, school, road accessibility, electricity, drinking water supply, recreation centres like movie hall and theatre, etc. The act has also made three categories of urban centres as Metropolitan City (Mahanagar Palika), Sub-Metropolitan City (Upa Mahanagar Palika), and Municipality (Nagar Palika). Same categories were taken by the Local Self Governance Act 1999 (Government of Nepal 1999). According to that Act, Mahanagar Palika was an urban centre with a "minimum population size of 300,000 and annual revenue of at least Rs. 40 million rupees (USD 1,000,000), and adequate urban facilities and an area that has already received the status of a Upa-Mahanagarpalika". Similarly, Upa-Mahanagarpalika is an urban centre with a "minimum population size of 100,000, annual revenue of at least Rs. 10 million (USD 250,000), adequate urban facilities and an area that has already received the status of a nagarpalika". A nagarpalika (Municipality) requires (a) minimum population size of 20,000 in the Tarai and 10,000 in the Hill and Mountains, (b) annual revenue of 5 million rupees (USD 125,000) in the Tarai and 500,000 rupees (USD 12,500) in the Hill and Mountains and "minimum urban facilities such as electricity, motor road, drinking water, communication and other similar urban facilities". Following those criteria there were one Metropolitan City (i.e. Kathmandu), four Sub-Metropolitan Cities (i.e. Lalitpur, Pokhara, Brgunj, and Biratnagar), and 53 Municipalities as of designated urban centres in Nepal at the time of census 2011 (Fig. 3.3). Many of those located in southern plain.

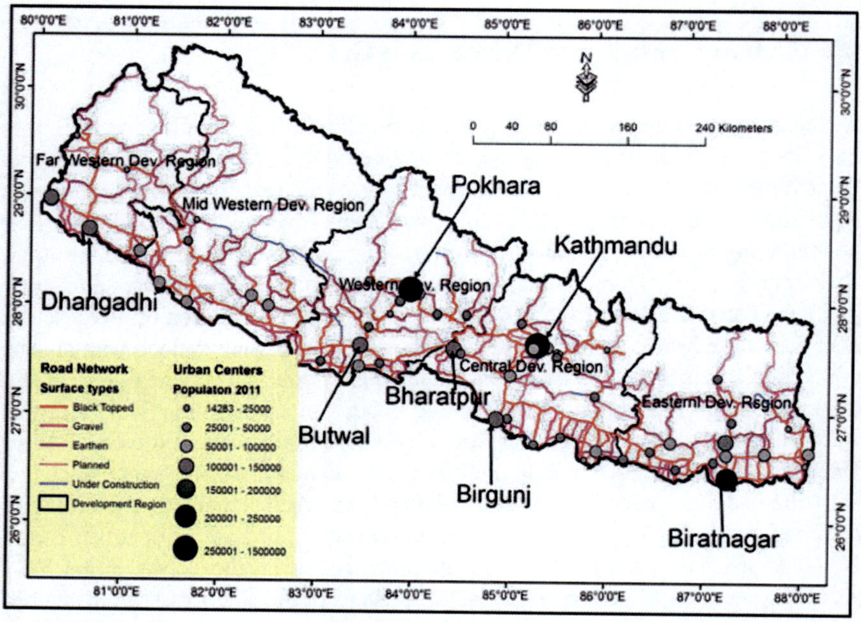

Fig. 3.3 Distribution of urban centres and population in 2011. *Source* Department of Survey, Government of Nepal 2011

After the declaration of the Federal Democratic Republic of Nepal in 2015, the Local Government Operation Act 2017 has been enacted and new criteria endorsed for the urban area delineation. The minimum population requirement for the Urban Municipalities in the Himali (mountain) districts is 10,000, annual revenue of the last five years should be 10 million (USD 100,000, approximately USD 1 = 105 in 2017), and other urban infrastructures. Minimum population requirement in the Hill districts is 40,000; for the Inner Tarai districts it is 50,000; and for the Tarai districts it is 75,000 and in the Kathmandu Valley districts it is 100,000, respectively. Similarly, revenue should be 30 million rupees (USD 300,000 around) for urban municipalities with urban infrastructures. The requirement for the Sub-Metropolitan is at least having 200,000 populations, revenue generation of 50 million rupees (USD 500,000) of the last five years average along with the physical infrastructure of college, hospital, electricity, telecommunication, and waste disposal and market services. For the Metropolitan Cities requires 500,000 populations, 100 million rupees (USD 1000,000) annual revenue, and other national and international standard urban amenities and infrastructures facilities. Based on the above criteria 293 units out of 753 local governance units are constituted within the Urban Municipalities where 6 Metropolitan Cities, 11 Sub-Metropolitan Cities, and 276 Urban Municipalities and rest 460 are within the Rural Municipalities (Fig. 3.4).

Urban centre designation procedures and criteria were quite static and similar until 1991. But after the restoration of democracy in 1990, there were abrupt changes in

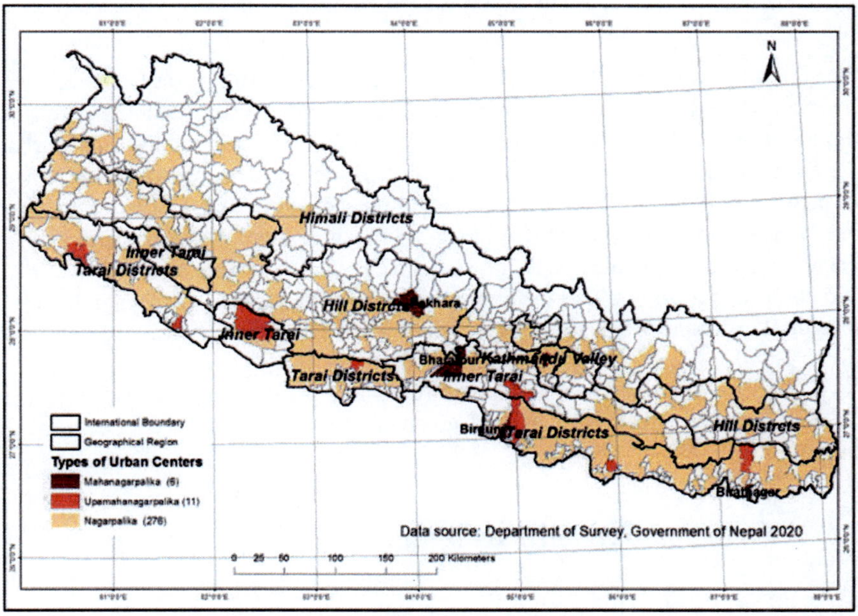

Fig. 3.4 Designated urban centres in Nepal (According to Government of Nepal, 2017). *Source* Department of Survey, Government of Nepal 2020

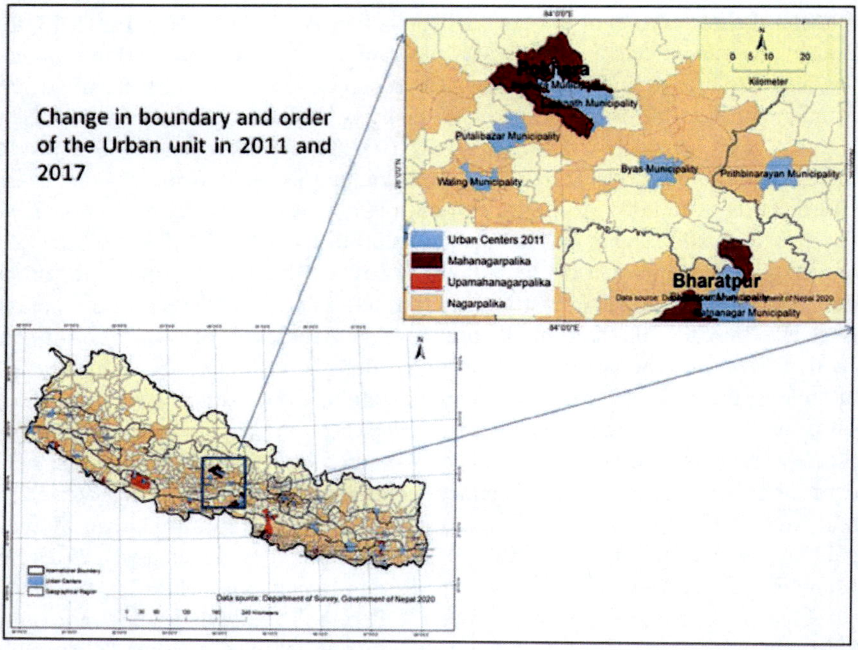

Fig. 3.5 An example of changes in urban category with expansion of boundary between 2011 and 2017. *Source* Department of Survey, Government of Nepal 2020

the systems. Because of change in criteria according to the ruling government, urban numbers as well as urban area have tremendous changes. According to 2011 census, there was only 2.2% of the total area and 17.1% of total population live in the urban area. But, massive changes had been made in 2017 in the name of structural change. Expansion of boundary even up to physically inaccessible and rural landscape had been made to meet the given criteria including both population number and revenue generation as stated in the act. By doing such changes urban area reached over 30% and comprises nearly 60% of the total population of the country. An example of the territorial expansion of previous municipalities and changes in status is presented in Fig. 3.5.

Table 3.1 shows the changes in urban units and population since 1952/1954 to 2017 as of the records of the Central Bureau of Statistics.

3.6 Urban Population Distribution

Between 2011 and 2017, 235 new municipalities have been added. The Bagmati Pradesh (Province No. 3) has the highest proportion (23.8%) of the total urban population. The province No. 1, Province No. 2, Gandaki (Province No. 4), Lumbini

Table 3.1 Trends of urban population in Nepal

Census year	Number of urban places	Total urban population	Percentage of urban population	Growth rate of urban population
1952/1954	10	2,38,275	2.9	–
1961	16	3,36,222	3.6	3.4
1971	16	4,61,938	4.0	3.1
1981	23	9,56,721	6.4	7.3
1991	33	16,95,719	9.2	5.7
2001	58	32,27,879	13.9	6.4
2011	58	45,23,820	17.1	3.8
2014 May	130	71,01,677	27.6	8
2014 Dec	191	94,03,977	35.5	–
2015	217	1,10,41,504	–	–
2017 Aug	293	16,656,057	62.9	–

Source Central Bureau of Statistics (CBS) 1995, 2003, 2011 (CBS 2001)

(Province No. 5), Karnali (Province No. 6), and Sudur-Pachhim (Province No. 7) account for 16.4, 23.2, 8.8, 13.9, 4.7, and 9.2% urban population, respectively. The urban population distribution is relatively low in Karnali and Gandaki Pradesh (Province No. 6 and 4), (Table 3.2).

Table 3.2 Distribution of urban population and density by federal sates of Nepal 2011

Province name and number	Urban area	%	Urban population	%	Density (Person/ Km^2)
1	8448.65	19.2	2,734,586	16.42	323.7
2	6753.783	15.4	3,870,159	23.24	573.0
Bagmati (3)	5488.38	12.5	3,961,963	23.79	721.9
Gandaki (4)	3967.17	9.0	1,467,240	8.81	369.8
5	6704.98	15.2	2,321,262	13.94	346.2
Karnali (6)	6014.74	13.7	774,319	4.65	128.7
Surdur Pachhim (7)	6619.03	15.0	1,526,528	9.17	230.6
Nepal Total	43,996.733	29.9	16,656,057	62.9	378.6

Source http://cbs.gov.np/image/data/2017/Population_Ward_Level_753_Local_Unit.pdf

3.7 Urbanization Processes

Sharma (2003) describes urbanization as a process of growth in the proportion of population living in urban areas. A fundamental characteristic of urbanization is the structural shifts in employment from agriculture to non-agriculture pursuits. In other words, urbanization is a territorial response to structural changes in the economy. A distinctive division of labour, technology-based production of goods, trade of a variety of goods and services, high level of spatial and economic interaction, and relatively high density and diversity of population are basic tenets associated with urbanization. In case of Nepal the process of urbanization was started lately. The definition for designated urban centres and territories is changing according to the changes of political system. Labour migration, attraction of new livelihood options, and higher rate of entering population than leaving urban areas have been contributing urbanization in Nepal (MoUD, 2015). Although, extension of municipal boundaries and declaration of new municipal areas are also main factors for urban growth and urbanization (Choe & Pradhan, 2010; Subedi, 2014: 108). Thus, urbanization in Nepal is a unique phenomenon. It is observed that urban settlements are delimiting based on political and policy level decisions rather than on functional justification, urban services, and localization of physical infrastructure. Large territory of urban units are beyond the physical suitable terrain, infrastructure access, and population concentration. Thus, the number of urban units and the share of urban population in the last decade show unnatural increment. Usually, high rate of urbanization is occurring in Kathmandu valley, Pokhara Valley, Inner Tarai, and Tarai plain at the junction of east–west highway and the north–south corridors (Muzzini & Apericio, 2013).

3.8 Transformation of Rural-To-Urban Settlements

After the restoration of multiparty democratic system in 1991, widely noted as "People's Movement I", Nepal has been experiencing several ups and downs in the various aspects of the society. The decade-long Maoist insurgency spawned a large literature, mostly of a political nature but Yadav (2016) has analysed the impacts and implications of the Maoist insurgency on contemporary socio-economic change and transformation (Sharma, 2016). Non-linear transformation of the Nepalese society had been experienced (Dahal, 2010). The discourse among the academic arena had been put forth during that period. Dahal (2010) further points out the explanation was towards the collective consciousness of Nepalese citizens, political, economic, social, and ecological trends and causes and correlation of forces provide a sound basis for the appraisal of the goal of transformation and the legitimacy of its means (Dahal, 2010). However, it is noticed that because of the so-called war against the feudalistic society along with the governance system resulted the fragmentation of civil society, polarization of corporate climate, and poor governance have eroded the political capacity of leadership to sustain non-linear transformation (Dahal, 2010). As Dahal (2010) termed the "creative destruction" of Nepal's public sector, industries, education, health, and communication has been experienced. It has been directly internalized by the market forces for the transformation of welfare state into the financial capitalism since late 1990s. That has emphasized better educational establishments, hospital, electricity, and other urban amenities at the towns and market centres with prioritizing of economic benefits. The immediate attraction of such centres spread among the dream of rural youths for better school, better health, and enjoyment of new changes. The dark, dusty, and difficult (3D) rural areas remain on backwash effect of new transformation. It is one of the major push and pull factors of rural population to the urban centres. In the mean time Nepalese citizens entered in the international labour market apart from Indian market especially in Malaysia and Gulf countries (MoL&E, 2016). The labour migration systemically broke the state's linkages with society, economy, and ecology (Dahal, 2010). Because of such changes appeared in the country the direct impacts have been experienced in the people's mobility, consumer market, and population dynamics. Unexpectedly, foreign labour migration has been swelled up in such a way that it has shifted the agricultural-based Nepalese economy towards remittance-based economy. Country receives remittances of over 25% relative to the GDP per year (Maskay et al., 2015). The government statistics shows that more than 3.8 million permits to work abroad (excluding India) were issued by the government within the period of 1993/1994 to 2014/2015 fiscal years, which represents almost 14% of the total population. According to the census data (2011), nearly 71% of the total absent population (1,921,494), or people living out of the country (including living in India) cited private and institutional jobs abroad as the reasons for leaving. Similarly, the report mentions that there has been a huge increase in the inflow of remittances, contributed 10.9 percentage share of the gross domestic product (GDP) in 2003/2004 and 27.7% in 2014/2015. The remittance flow, therefore, is a major contributor to development financing in Nepal (MoL&E, 2016). Both foreign labour migration and social transformation result the new dynamics of population concentration at the market centres, road sides, and accessible small towns. Several satellite small markets developed by the rural-to-urban migration of the people. The dynamics also enhanced by the liberal economic policy adopted by the government. Privatization on school, health, communication, transportation, and financial companies also dragged the people from the rural areas to the urban centres. That has been observed the last decade especially after "People Movement II" of 2003/2004 (movement of the people) and end of "Maoist People's (civil) War" the country is transforming from previous process to different dynamics. Several new practices have been appeared in public attitude and social behaviour. Government practices and social practices are also changed. New consumerism economy has pushed the rural-based economy, livelihood, and practices to urban-based lifestyles. The transformation or "so-called" new shift of the attitude of both public and government policy designers made happy on delimiting urban centres at adverse physical locations as well as with limited urban amenities and poor physical infrastructure. Several newly delimited urban units have even no electricity, primary healthcare services, and road access. But government is happy to report over half of the nation's population live in designated urban centres. It has raised several questions on urban geography and urbanization process. Despite the optimal expectation of the urban centres the situation of designated urban settlements in Nepal have different observed scenarios.

3.9 Infrastructure Development

Urban settlements present functional agglomeration, therefore, people living within the urban units expect different amenities and services for their happiness. Urban units are the place of hope of the people, by which people move from the rural area to the urban centres. Urban centres continuously attract the people and remain functionally active. The designated urban centres are also the major source of revenue for the government. As per the substitute of revenue collection from the urban dwellers government requires to pay back them through construction of physical infrastructures, security as well as delivery of basic services. Economic establishments generally localized at the urban centres. Similarly, other amenities and services including drinking water supply and sanitation, health services, educational institutions, road network and urban commuting services, electricity and communication are such infrastructure presents the status of urban centres. Urban safety measures for disaster, social violence, riot, and theft are also equally important. Urban land use planning, building codes, and holistic urban plan for different types of people living inside the urban territory are the required services expected by the urban dwellers as par the payment of revenue to the government. The payment of revenue by the people and construction of physical infrastructure, basic services, and urban amenities providing by the government or the city authorities are the reciprocity for the sustainable development. A healthy and smart urban centre works for the betterment of the population living within the urban territory.

3.9.1 Economic Establishment

Economic establishments within the constituent urban centres provide one of the major indicators for its standard. Economic establishments of the country were counted by the Central Bureau of Statistics, Government of Nepal in 2018. The census has provided information on number of establishments and number of persons engaged in industry wise of all the sectors (excluding unregistered agriculture, forestry and fisheries, public administration, defence, and compulsory social security (CBS, 2019). Therefore, the functional engagement of urban population besides the primary production sectors including agriculture, forestry, fisheries and other clerical, administrative, security, and service sectors were not counted.

As of the census records, the country had 936,028 establishments in total in 2018. Among those establishments 74.2% was recorded at urban municipalities and rest was in rural municipalities. Out of the total economic establishments located at the urban municipalities, 23.5% establishments were in 6 Metropolitan Cities, 10.3% establishments were in 11 sub-metropolitan cities and 66.2% of the total establishments were in the 276 urban municipalities. The economic establishments engage 10.8% of the total labour force of the country. 15.2% establishments located in the old cities of Kathmandu Valley i.e. Kathmandu, Lalitpur, and Bhaktapur, where 22.4% labour forces was engaged. The distribution of labour force engaged in the urban municipalities was 31.3% in the metropolitan cities, 10.8% in the sub-metropolitan cities, and 57.8% in the municipalities (Table 3.3).

The census data shows, economic establishments are largely concentrated in Kathmandu Valley, Pokhara Valley, Chitwan (Inner Tarai), and in the border areas with India where four out of six metropolitan cities i.e. Kathmandu, Lalitpur, Pokhara, Bharatpur Birgunj, and Biratnagar metropolitan cities are located. Economic establishments in Himali districts and other hill districts have very small figure (Fig. 3.6).

For a purpose to compute the total population and persons engaged in the economic establishments of urban municipalities had been selected all metropolitan and sub-metropolitan cities, and 25 municipalities (randomly selected from the total 276). The computed indices present quite different proportions (Table 3.4). The highest value is in Kathmandu Metropolitan City (61.79) and the lowest is in Dewahhi Gonahi Nagarpalika (3.42).

The spatial distribution of proportion of total population and person engagement in economic establishments of municipalities presents high value in metropolitan and sub-metropolitan cities whereas small urban municipalities located at remote areas have not good situation. It seems the populations in those towns still follow primary production systems (Fig. 3.7).

3.9.2 *Urban Amenities*

Besides economic establishments and the person engagements in those establishments the other urban amenities also help to observe the status of urban settlements.

3.9.2.1 Road Accessibility

Because of heterogeneous topography of the country rail network is still not in use. Owing to poor economic condition of the people air transportation is not in access to large public. Therefore, road plays a vital role on goods transport and people's movement. Road accessibility is a major criterion for the urban establishment. The distribution of road is also not properly distributed within the country. Large part of the country is still out of road access. In terms of road accessibility, Nepal has only 13,448 km road length by 2018 (Table 3.5). This shows the total road length per

Table 3.3 Economic establishments in Nepal 2018

Unit name		Total units	Total establishment	%	Total people engaged	%	Male	%	Female	%
Urban		293	694,573	74.2	2,512,204	77.6	1,579,429	78.2	932,411	76.5
	Metropolitan city	6	163,286	23.5	787,179	31.3	499,166	31.6	288,013	30.9
	Sub-metropolitan city	11	71,675	10.3	272,355	10.8	182,062	11.5	90,293	9.7
	Municipality	276	459,612	66.2	1,452,670	57.8	898,201	56.9	554,105	59.4
Kathmandu valley										
	Bhaktapur Municipality	1	6462	0.9	25,159	1.0	14,633	0.9	10,526	1.1
	Lalitpur Metropolitan city	1	18,357	2.6	97,588	3.9	60,311	3.8	37,277	4.0
	Kathmandu metropolitan city	1	80,434	11.6	440,604	17.5	278,695	17.6	161,909	17.4
	Total		105,253	15.2	563,351	22.4	353,639	22.4	209,712	22.5
Nepal		1	936,028	100	3,238,914	100	2,020,164	100	1,218,386	100

Source CBS 2018

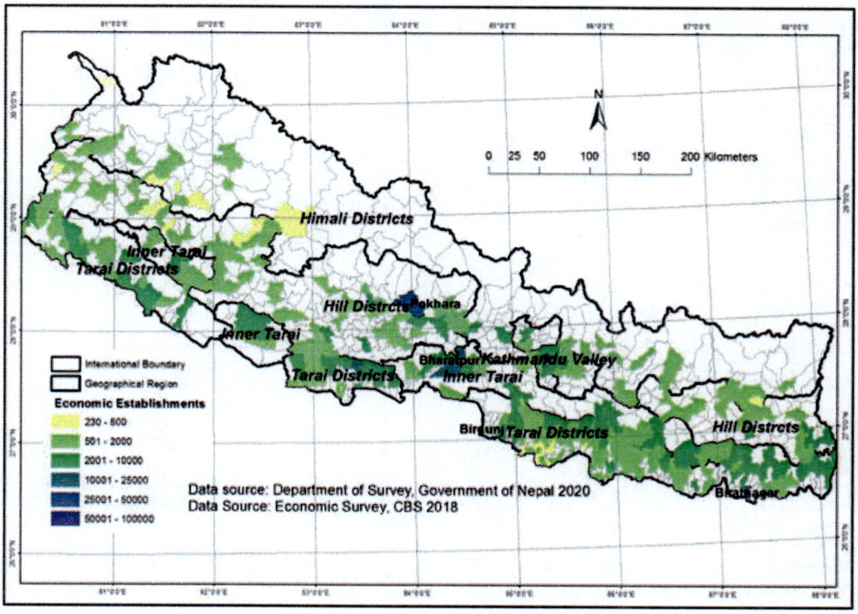

Fig. 3.6 Economic establishments in constituent urban centres in Nepal (CBS, 2019). *Source* Department of Survey, Government of Nepal 2020

100 km^2 is only 995 km. Similarly, a kilometre of road provides service for 151,655 persons. The distribution is not equal in different provinces. It shows road network connectivity is still not in better position.

The Government of Nepal has assigned urban road standards in 2019. According to the government standards, there are seven categories of urban roads like carriage way, cycle track, foot path/walk way, street light, road signals (signage), utility space, and other essential elements (including road furniture, plantation, etc.). But several cities have no cycle lane, proper footpath, road signals, and others. The country has different types of vehicles and their numbers as of the government records in 2017 (Table 3.6). But the inter-city connectivity is depending only by bus and minibus. For the goods carriage trucks are the major means.

3.9.2.2 Drinking Water Supply and Sanitation

Like other parts of the country people residing in urban areas are using drinking water from different sources. Because of the large concentration of people and low investment on drinking water supply system generally the urban areas have serious shortage of quality and adequate volume of water. Water supply problem with the urban centres of Kathmandu Valley is one of the widely noted examples for the water shortage. The Government of Nepal estimated 510 million litre water insufficiency

Table 3.4 Proportion of persons engagement in economic establishments to total population of urban municipalities

Name of municipality	Geographical region	Total population-(2017)	Total economic establishments (2018)	Total persons engaged	Proportion index*
Kathmandu Mahanagarpalika	Kathmandu_ valley	985,453	80,434	440,604	61.79
Lalitpur Mahanagarpalika	Kathmandu_ valley	284,922	18,357	97,588	51.03
Butwal Upamahanagarpalika	Tarai_ districts	138,741	10,142	38,156	43.14
Madhyapur Thimi Nagarpalika	Kathmandu_ valley	83,036	5334	22,549	42.71
Itahari Upamahanagarpalika	Tarai_ districts	140,517	8757	37,440	42.08
Pokhara Mahanagarpalika	Hill_distrcts	426,759	28,215	104,469	39.33
Dakshinkali Nagarpalika	Kathmandu_ valley	24,296	1445	5618	37.56
Bharatpur Mahanagarpalika	Inner_tarai	280,502	17,077	64,812	37.54
Nepalgunj Upamahanagarpalika	Tarai_ districts	138,951	7176	30,470	35.97
Dharan Upamahanagarpalika	Tarai_ districts	137,705	8433	30,169	35.94
Hetauda Upamahanagarpalika	Inner_tarai	152,875	9209	31,588	34.25
Mahalaxmi Nagarpalika	Kathmandu_ valley	62,172	4058	12,633	33.78
Biratnagar Mahanagarpalika	Tarai_ districts	242,548	10,392	46,796	32.35
Bidur Nagarpalika	Hill_distrcts	54,351	3010	10,480	32.33
Mahakali Nagarpalika	Himali_ districts	21,231	1169	3749	30.02
Janakpur Upamahanagarpalika	Tarai_ districts	159,468	7523	25,345	27.43
Jitpur Simara Upamahanagarpalika	Tarai_ districts	117,496	3105	18,549	27.27
Dhangadhi Upamahanagarpalika	Tarai_ districts	147,741	6089	22,909	26.85
Devchuli Nagarpalika	Inner_tarai	42,603	2054	6553	26.66
Birgunj Mahanagarpalika	Tarai_ districts	240,922	8811	32,910	24.04

(continued)

Table 3.4 (continued)

Name of municipality	Geographical region	Total population-(2017)	Total economic establishments (2018)	Total persons engaged	Proportion index*
Sanphebagar Nagarpalika	Hill_distrcts	33,788	963	4540	23.69
Mahakali Nagarpalika	Tarai_ districts	21,231	816	2468	20.83
Ghorahi Upamahanagarpalika	Inner_tarai	156,164	5219	17,846	20.51
Dharmadevi Nagarpalika	Himali_ districts	18,235	527	1972	19.52
Madhuwan Nagarpalika	Tarai_ districts	46,437	2026	4700	18.38
Buddhabhumi Nagarpalika	Tarai_ districts	64,949	1773	5992	16.89
Tulsipur Upamahanagarpalika	Inner_tarai	141,528	3688	12,157	15.82
Bhimad Nagarpalika	Hill_distrcts	31,362	1011	2638	15.52
Thaha Nagarpalika	Inner_tarai	41,623	1347	3254	14.50
Surunga Nagarpalika	Tarai_ districts	44,221	1253	3399	14.28
Ishworpur Nagarpalika	Tarai_ districts	59,986	1554	4228	13.17
Kalaiya Upamahanagarpalika	Tarai_ districts	123,659	2334	7726	11.76
Chaurjahari Nagarpalika	Hill_distrcts	27,438	682	1642	11.29
Ghodaghodi Nagarpalika	Tarai_ districts	75,586	1576	4081	10.25
Bahudaramai Nagarpalika	Tarai_ districts	39,673	686	2010	9.64
Shadananda Nagarpalika	Hill_distrcts	31,610	579	1405	8.51
Panchadewal Binayak Nagarpalika	Hill_distrcts	27,485	396	1154	8.06
Aathbiskot Nagarpalika	Hill_distrcts	33,601	604	1395	7.97
Baudhimai Nagarpalika	Tarai_ districts	36,265	586	1455	7.71
Bhangaha Nagarpalika	Tarai_ districts	46,754	782	1809	7.45

(continued)

Table 3.4 (continued)

Name of municipality	Geographical region	Total population-(2017)	Total economic establishments (2018)	Total persons engaged	Proportion index*
Mahalaxmi Nagarpalika	Hill_distrcts	62,172	748	1561	4.90
Dewahhi Gonahi Nagarpalika	Tarai_ districts	32,146	271	559	3.42

*Index is computed based on 100–{(Total Population–Total persons engagement)/(Total Population + Total persons engagement) × 100}

Source CBS (2019) (Economic Census 2018)

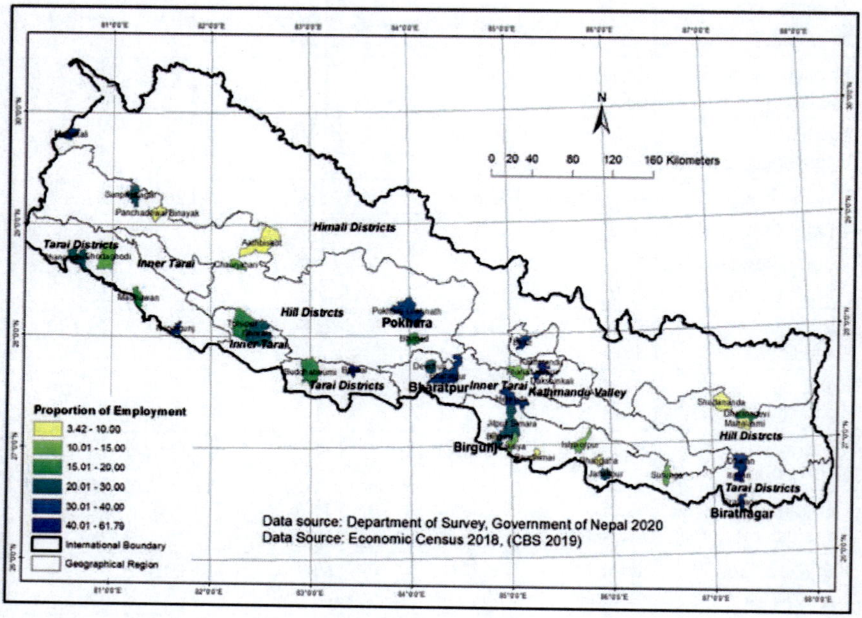

Fig. 3.7 Geographical distribution of proportion of total population and person engagement in economic establishments. *Source* Department of Survey, Government of Nepal 2020

for the people of Kathmandu Valley in early 1990s when the total population of Kathmandu Valley was estimated three million because of that the highly ambitious Melamchi Diversion Drinking Water Supply Project was designed in late 1990s. But that project is not completed till the date (NPC, 2020: 67). In reality, it seems, supply of piped water is still only periodically available in many urban areas of Nepal. Generally people living in the towns of Tarai region depend on the water of shallow tube well and ground well. The urban population facilitated by the drinking water supply types is given in Table 3.7.

Table 3.5 Nepal, length of roads by province

Name and Provinces Number	Area	Population (2011)	Total road length (km)	No of population influenced per km. road	Road density (km/100 km^2)
Province 1*	25,906	4,534,943	2,389.08	26,813	161
Province 2*	9,661	5,404,145	1,466.20	32,333	123
Bagmati	20,300	5,529,452	2,544.00	27,889	302
Gandaki	21,832	2,403,757	1,633.70	16,549	102
Lumbini	19,734	4,499,272	2,450.02	22,183	159
Karnali	30,209	1,570,418	1,343.43	10,939	70
Sudur Paschim	19,539	2,552,517	1,621.18	14,949	78
Nepal Total	147,181	26,494,504	13,448	151,655	995

*Name is not yet given
Source DOR (2018)

Table 3.6 Types of vehicles in Nepal

Types of vehicles	No. of vehicles	Percentage of vehicles
Bus	52,186	2.0
Mini bus, Mini truck	23,269	0.9
Crane, Dozer, Excavator, Truck	75,269	2.9
Car, Jeep, Van	209,199	8.0
Pickup	43,089	1.7
Micro bus	5,188	0.2
Tempo	15,348	0.6
Motor cycle	2,034,936	78.2
Tractor, Power tiller	123,797	4.8
Others	20,678	0.8
Total	2,602,959	100.0

Source Ministry of Physical Infrastructure and Transport-2017

The cities in Nepal have very measurable situation of sanitation. Even for large metropolitan cities like Kathmandu, Lalitpur, Pokhara have no properly managed sanitation. Waste disposal is major issue. In several urban centres solid wastes carried by trucks and dispose either in improperly managed landfill sites or sometimes dispose along the river and stream banks and open spaces. It can be observed that households of Kathmandu and Lalitpur Metropolitan Cities use waste pipe but the outlet goes directly to the river Bagmati and its tributaries. But in Pokhara Metropolitan City large numbers of households have septic tank on their own cost within their compound and courtyard. Survey conducted by assistance under the

Table 3.7 Water supply in urban centres of Nepal

Sources	No. of urban people	%
Tape/piped	618,664	59.2
Tube well/hand pipe	255,907	24.5
Covered well/kuwa	35,729	3.4
Uncovered well/kuwa	17,176	1.6
Spout water	30,574	2.9
River/stream	2824	0.3
Others	77,019	7.4
Not stated	7682	0.7
Total	1,045,575	100.0

Source Central Bureau of Statistics-2011

Table 3.8 Types of toilet in urban areas of Nepal

Types of toilet	No. of urban people	%
Flush toilet (Public sewerage)	316,041	30.2
Flush toilet (Septic tanks)	496,624	47.5
Ordinary toilet	129,232	12.4
Without toilet	95,030	9.1
Not stated	8148	0.8
Total	1,045,075	100.0

Source Central Bureau of Statistics-2011

ADB in three small towns-Mahendranagar, Chandrauta, and Charali indicates that the rate of access to toilet was varied as: Mahendranagar (68%), Chandrauta (75%), and Charali (88%) (ADB, 2012). Generally, in small urban areas solid waste contains 9.0%, paper, 12% plastic, and around 60–70% in organic materials (ADB, 2013). Some small towns and market centres have practice of weekly or biweekly solid waste collection by using a power tiller and hawker's cycles-van. They have no any settled landfill sites. The sanitary types for the urban population in Nepal are given in Table 3.8

3.9.2.3 Sources of Fuel and Lighting

In the urban centres basically people use electricity and LP gas. In case of Nepal the share of fire wood is still high where 25.7% urban people use fire wood and 67.7% use LP Gas. The share of electricity is very nominal (1.5%) (Table 3.9). For the light, 94.1% urban settlers use electricity and the rest 4.0% people still use kerosene (Table 3.10).

Table 3.9 Sources of fuel used by the urban population

Types of fuel	No. of urban people	%
Fire wood	268,643	25.7
Kerosene	20,990	2.0
L.P. gas	707,674	67.7
Cow dong	15,776	1.5
Bio-gas	19,121	1.8
Electricity	1255	0.1
Others	4107	0.4
Not stated	8009	0.8
Total	1,045,575	100.0

Source Central Bureau of Statistics-2011

Table 3.10 Sources of energy for lighting

Types of lighting	No. of urban people	%
Electricity	983,995	94.1
Kerosene	42,297	4.0
Bio-gas	3671	0.4
Solar	2082	0.2
Others	5438	0.5
Not stated	8092	0.8
Total	1,045,575	100.0

Source Central Bureau of Statistics-2011

3.9.2.4 Means of Communication

Means of communication is a basic requirement for the urban population. Radio, television, telephone, internet, mobile, etc., are the basic requirements. The use of those means by the urban population is not that popular. The percentage of urban population using different types of communication means is given in Table 3.11.

Based on the above described different urban facilities and services show the people living in the urban areas in Nepal are using minimal urban services. Because of that the urban centres in Nepal have not adequate level of urban infrastructure development. Even though the census data of the infrastructure distribution was collected in 2011 when there were only 58 constituent municipalities. But after 2017 restructure there are 293 urban municipalities, it is obvious that most of those newly constituted urban centres have almost none of the urban amenities. Those are mainly located in rural geographical territories. Urban centres are designated without considering the adequate urban infrastructures. Therefore, urban centres have several challenges in management and their sustainability.

Table 3.11 Mean of communication in urban population

Means of communication	No. urban population	%[a]
Radio	559,986	53.56
Television	634,332	60.67
Cable television	562,473	53.80
Computer	247,370	23.66
Internet	126,654	12.12
Telephone	236,903	22.66
Mobile	879,020	84.07
Total urban population	10,45,575	

[a]Population use multi means therefore, percentage is based on individual category
Source Central Bureau of Statistics-2011

3.10 Challenges

Adequate infrastructure and amenities are the prerequisite for urban development. In many cases infrastructures and amenities are obviously insufficient in urban centres due to burgeoning population due to several push factors from the rural peripheries either permanently or daily and short duration migrants and daily commuters to the urban centres and also due to pull factors of urban areas itself. Therefore, urban areas are usually experiencing exceptional pressure on urban infrastructure and amenities. Because of that even in the metropolitan cities infrastructural and amenities growth seems inadequate in proportion to the population.

Urbanization is an abrupt social changing system in Nepal. Declaration of urban centres as constituent urban units seems to be increased with the interest of political parties and the ruling government rather accounting to the universal urban definitions. The number of urban centres and also the share of population living in the constituent urban centres are both new phenomena in Nepal. Therefore, there are no urban centres developed by prior urban landscape planning. Unplanned development of urban centres exhibits several problems. Environmental issues and prone to different types of disasters, squatter settlements and slum inhabitants along the river banks and physically vulnerable locations, lack of urban open spaces, poor public safety and hygiene, social crimes and unemployment are some of the acute problems observed in several urban centres. Absence of good governance and proper management has been pushing to the negative impacts on all sectors, such as unscientific land use practices and haphazard layout of land, lack of sufficient water supply, and proper sanitation systems. Urban sewage and solid waste disposal is another acute problem that has been affecting public health, urban hygiene, and people's health. Encroachment of open space from illegal doers has been converting urban landscape to unplanned and unhealthy dwellings. Narrow and unplanned inner city roads seem to be major causes of traffic jam and obstruction for fire brigade and ambulances. In brief, it can be pointed out that the challenges of urbanization in the context of Nepal are as following:

- Declaration of urban settlements without considering international practices of urban definition comprises least urban functional associations and conglomerations,
- Lack of physical planning and implementation of urban standards.
- Problems of unemployment.
- Comprises poor living conditions and slums.
- Problems of environmental management.
- Social crimes.
- Improper management of urban traffic systems.
- Urban congestion and damage of urban heritage due to lack of proper planning.
- Problems of urban sprawl due to the poor and slopping topography in the large parts of the country.
- Emergence of squatter settlements in public and vulnerable spaces.
- Lack of master plan and update of urban maps.
- Political interference in planning.
- Lack of coordination and overlapping of function among different organizations.
- Weak institutional framework.
- Lack of good knowledge on urbanization.
- Poor and unscientific urban governance systems

3.11 Conclusion

The data show that in 1960s there were only ten urban centres and the number of designated towns increased by 58 with one Metropolitan City in 2011. Capital city Kathmandu was the only metropolitan city. But the figure alarmingly increased to 293 urban centres in 2017 with 6 metropolitan cities, 11 sub-metropolitan cities, and 276 urban municipalities. In terms of urban population, there was less than 20% in 2011 and it has been abruptly increased to over 60% in 2017. A country with almost 40% of its gross domestic production (GDP) used to share from the agrarian production for last several decades and at present also nearly 60% of its population directly engaged on agriculture production systems provides very little factual evidences to over 60% of its urbanized population. Simply, these figures present unscientific practice of declaration of urban centres in Nepal. It is observed that several constituent urban centres have no urban functional agglomeration, compactness of settlement, urban amenities, and basic urban infrastructures. According to the general public the declaration of urban centres was based on the interest of government to raise revenue by increasing taxation-base in people's private property and registration of functions from the interest of political leaders. The factual information presents an issue of urbanization process of the country. From the perspective of sustainable development and objective planning of urban areas it requires critical review of the functional establishments, urban amenities, physical layout of urban centres, and engagement of people living in the urban territories. The practice of the Government of Nepal on its urban centre declaration has been raising new discourse on urban planning.

References

ADB. (2012). Preparing the third small towns water supply and sanitation sector. https://www.adb.org/projects/35173-013/main. Retrieved August 24, 2020.

ADB. (2013). Solid waste management in Nepal: Current status and policy recommendations. https://www.adb.org/publications/solid-waste-management-nepal-current-status-and-policy-recommendations. Retrieved August13, 2020.

Bastola, T. S. (1995) Urbanization. In CBS (Edited), *Population monograph* (pp. 279–299). Central Bureau of Statistics, Government of Nepal.

Basyal, G. K., & Khanal, N. R. (2001). Process and characteristics of urbanization in Nepal. *Contributions to Nepalese Studies, Kathmandu, 28*(2), 187–225.

CBS. (1962). *National population census 1961*. Central Bureau of Statistics, Government of Nepal.

CBS. (2001). *National population census 2001*. Central Bureau of Statistics, Government of Nepal.

CBS. (2012). *National population census 2011*. Central Bureau of Statistics, Government of Nepal.

CBS. (2014). *Population monograph of Nepal (Vol I Population dynamics)*. Central Bureau of Statistics, Government of Nepal.

CBS. (2019). *National economic census 2018 (National Report)*. Central Bureau of Statistics, Government of Nepal.

Choe, K., & Pradhan, P. (2010). *Unleashing economic growth: Region-based urban development strategy to Nepal*. Asian Development Bank.

DOR. (2018). *Statistics of strategic road networks* (p. 2018). Department of Roads, Government of Nepal.

Dahal, D. R. (2010). Social transformation in Nepal: A personal reflection. Paper presented at a high-level dialogue on "Transformation for Peace," organized by General Federation of Nepalese Trade Unions, (GEFONT) in cooperation with Friedrich-Ebert-Stiftung (FES) on July 8–9, 2010 at Nagarkot.

Government of Nepal. (1999). *Local self governance act 1999*. Government of Nepal.

Government of Nepal. (2017). *Local government operation act 2017*. Government of Nepal.

http://cbs.gov.np/image/data/2017/Population_Ward_Level_753_Local_Unit.pdf. Retrieved August10, 2020.

https://www.nationalgeographic.org/encyclopedia/urbanization/. Retrieved August 01, 2020.

https://www.worldometers.info/world-population/nepal-population/). Retrieved August15, 2020.

Maskay, N. M., Steinkamp, S., & Westermann, F. (2015). The impact of remittances on central bank balance sheets and inflation in Nepal. *Nepal Rastra Bank Economic Review, 27*(2), 1–18.

MoL&E. (2016). *Labour migration for employment a status report for Nepal: 2014/2015*. Ministry of Labour and Employment, Government of Nepal.

MoUD. (2015). *National urban development strategy-2015*. Ministry of Urban Development, Government of Nepal.

Muzzini, E., & Aparicio, G. (2013). Urban growth and spatial transition in Nepal: An initial assessment. The World Bank, Project (TA 8346-NEP). Manila: Asian Development Bank recommendations. Asian Development Bank.

NPC. (2020). A brief introduction of national pride projects of Nepal and their progress until FY 2076/77 B.S. National Planning Commission, Government of Nepal.

Sharma, P. (2003). Urbanization and development. In CBS (Edt.) Population monograph, central bureau of statitics. *Government of Nepal, I*, 375–412.

Sharma, P. (2016). *Book Review, Resources Himalaya Foundation, Ponam Yadav (2016) Social transformation in post conflict Nepal: A gender perspective.* Routledge.
Shrestha, C. B. (1973). *Spatial integration in the Arniko Rajmarga area (Unpublished report in Nepali).* National Planning Commission.
Shrestha, C. B. (1982). Changes in the urban systems in Nepal. *The Himalayan Review, 14*, 1–11.
Subedi, B.P., (2014). Urbanization in Nepal: Spatial pattern, social demography and development. In CBS (Edts.) Population monograph (Vol. III, pp. 95–154) Central Bureau of Statistics, Government of Nepal.
Yadav, P. (2016). *Social transformation in post conflict Nepal: A gender perspective.* Routledge.

Chapter 4
Changes and Challenges in Urban Residential Space: Case Study of Bucharest

Mihaela-Daniela Preda

Abstract Bucharest's residential space has experienced transformations determined primarily by its economic characteristics, and secondly by social and administrative aspects. During the last 30 years the transformations of the Municipality' residential areas have been more significant. The aim of this study centres around understanding the changes within the dynamics of the capital city's residential space. These changes and the challenges and inherent difficulties they bring to the population are caused today by a wide range of administrative and governmental approaches in terms of urban planning in general and residential planning in particular. The characteristics of the new residential areas represent an unfortunate consequence of an unprecedented high demand for housing in the context of the rapid development of Bucharest metropolitan area, while the legislation barely managed to keep up and to correct emerging shortcomings. The methodological resources used were direct field observations, as well as the case study method, supported by the survey method, and the structured interview used as a research instrument. The objectives of the study number: highlighting the residential dynamics' typology within the sample areas, analysing the urban planning legislation, enumerating the leading dysfunctions that appeared in the new residential space, identifying the possible actors able to solve these issues. The findings observed an accelerated increase in the number and area of constructions, which expanded in the capital city's peri-urban area. Unfortunately many times, it was revealed that a large part of them do not respect current residential legislative requirements. The difficulties that the local population faces nowadays stem from trying to deal with realities of the urban infrastructure such as less than optimal housing quality, heavy traffic, a sub-dimensioned public transport infrastructure, underdeveloped technical-urban facilities, urban waste infrastructure, and an overall imbalance between expectations and the reality of a new residential area.

Keywords Residential · Urban planning · Development · Dynamic · Bucharest · Changes · Challenges · Nuclei · Urban space · Communist · Post-communist ·

M.-D. Preda (✉)
Faculty of Geography, University of Bucharest, Blv. Nicolae Bălcescu 1, 010061 Bucharest, Romania
e-mail: mihaela.preda@geo.unibuc.ro

B. R. K. Sinha (ed.), *Urban Dynamics, Environment and Health*,
https://doi.org/10.1007/978-981-99-5744-6_4

Dysfunctions · Technical-urban facilities · Residential expansion · Legislation · Living conditions · Urban development · Residential market · Policies · Infrastructure · Evolution

4.1 Introduction

Residential dynamics have become an increasingly important issue of research, mostly because of the increase in industrialization, followed by an accelerated urbanization process. In Romania, the development of cities recorded a galloping trend during the communist period. The communist authorities applied a plan for the industrialization and urbanization of the country starting with 1970. It was developed by creating industrial areas distributed homogeneously in the territory and declaring many rural localities cities, as soon as any industrial activity was located within them and they reached 5,000 inhabitants. Thus, Romania had an urban population of 23.4% in 1948, and in 1956 this percentage reached 31.3%. Between 1966 and 1977, the demographic growth was supported by governmental measures, and as such, the population living in cities increased from 38.2% to 43.6% (INSSE, 2013). Urbanization follows the same ascending trend throughout the communist era until 1990 when at the 1992 census the percentage of the urban population was of 54.3%. In the post-communist period, the urban population decreased to 52.7% in 2002. This was the result of economic restructuring marked by periods of recession or economic crises, with the smaller, mono-industrial towns being affected the most. In many cases, after the disappearance of the industrial activities, a part of these small towns asked to go back to their category of rural settlements (Mitrică et al., 2019; Preda et al., 2015). This change of status facilitated the reduction of taxes and duties for citizens, as well as the possibility to access rural development programmes financed by the European Union. The urban population percentage grew slowly so that at the 2002 census it was of 54%, and currently, it reaches the value of 56.4%. The slow urbanization of the post-communist period is the result of several economic and social factors, with cities being strongly affected by the disappearance of productive activities, and in the absence of jobs, the population migrated to Western Europe (Matei et al., 2013; Dumitrașcu et al., 2018). As a result of the reconversion capacity, large and very large cities showed sustained economic and demographic dynamics, and the capital city—Bucharest—was the one that registered positive qualitative leaps. This evolution also led to residential development, as the demand for housing was strongly dependent on periods of economic boom that alternated with economic recession ones.

Another factor that increased housing demand was related to the perception of the population vis-à-vis property, as Romanians are leading the Europe hierarchy in preference to living in self-own dwellings (96.8%), compared to the Germans where the percentage of homeowners is 51.4%, French (64.9%) or British (63.4%) (Eurostat, 2017). This is a strong trend, specific to other communist states as well because, in the past, the notion of private property did not exist during communist regimes, as the state was the unique owner. Ownership rights were reintroduced in

Romania in 1990, with the fall of the communist regime. Thus, Romanians regained the right to property, and most of them focused on buying houses with very few of them renting, and even when they rented this was usually a short-term solution adopted until they purchased their own house.

The evolution of the residential area, in Bucharest' post-communist period, is marked by a discontinuous dynamic, which saw periods of maximum development alternating with those of stagnation or regression during economic crises (Kucsicsa & Grigorescu, 2018; Mihai et al., 2015). After the 2000s, middle-class consumers especially were looking for individual homes, giving up living in blocks of apartments, typical of the communist period. Basically, the transition to a larger individual home, which did not look like the old apartments, known as "matchboxes", had to mean an improvement in the quality of housing. However, the development of residential areas inside the city or on its outskirts has not always been done in accordance to urban planning legislation, which generated the phenomenon of urban sprawl, with severe implications on the quality of life (Grigorescu et al., 2019; Nae et al., 2019; Wesołowska, 2016).

The most frequent and common dysfunctions were caused by congested traffic; old transport infrastructure—the same as the one developed during the communist years; the lack of green spaces because in many situations the residential complexes replaced areas occupied in the past by forested areas; and the lack of technical-urban facilities because local authorities did not develop or implement projects to expand the gas, water, or sewerage networks (Grigorescu & Geacu, 2017; Ianoș et al., 2010; Suditu, 2009).

Today, the city continues to grow, expanding in the peri-urban area, and blocks of flats increasingly dominate the new residential neighbourhoods. This trend is present solely as a result of the change in buyers' preference, given that the price of housing is still very high, and bank loans that cover the value of a house are more difficult to obtain. With a market that continually raised prices, Romanians who after 1990 wanted better living conditions, respectively individual villas, could afford to buy them less and less. In order to obtain financing for a home, they were made to return to the small dimensions of the "matchbox" apartments, typology specific during the communist period, because that was the only dwelling, they could afford to buy.

4.2 Data and Methodology

The study area is the country's capital, Bucharest, chosen because it represents the area with the highest socio-economic development in Romania. The capital is one of the most attractive economic poles in the country; and its development capacity produced 27.14% of the national GDP (INS, 2017). This translates into an obviously positive dynamic both from an economic but also from a social point of view, with demographic flows contributing to constant pressure on the real estate market as well as the residential space. Economic conditions lead to a high standard of living for the population, which generated an increase in consumption and change in the standard of quality of life and implicitly of housing.

Research in the studied area was performed in three stages. The first stage was a documentation one and involved, first of all, the researching of official statistical data, available on the National Institute of Statistics website. In order to establish the dynamics of the residential area, data regarding the building permits issued every month, between 2017 and 2019, exclusively for the spaces intended for housing, are centralized by the of each mayoralty of the district. As such, each of their websites was accessed online and then the selected data was processed.

The second stage was a field one and took place in September 2019 through the method of structured observation. Five categories of sample areas were established at the level of Bucharest Municipality, one for each sector, targeting the following directions: the position of the new residential nuclei in relation to the areas built during the communist period or before, houses under construction, urban infrastructure, school infrastructure, commercial facilities, service spaces, or the green spaces.

In the third stage, implemented in October–November, the survey method was used, with the interview guide as the applied/practical technique used. Five categories of sample areas were established in the study area, one for each district, targeting primarily the areas that present new housing nuclei, but also those that have appeared previously. The residents represented the target group, respectively, the people living in the selected sample areas.

A number of 100 respondents were interviewed, 20 for each district, the sampling including age categories between 21 and 49 years, as they are the ones who most often purchase homes. Most of the respondents were university graduates (53%), having incomes registered in the category 1000–2500 lei (see Table 4.1).

The field sampling targeted areas that have old blocks alternating with new ones, but also newly built areas, in order to underline the expectations and dysfunctions specific to each type of housing situation (see Fig. 4.1).

The aim of the study was to examine the opinions regarding respondents' expectations about the house they chose, but also to highlight the dysfunctions present in their living areas. Both expectations and dysfunctions were correlated with several categories of factors: economic characteristics, urban facilities, and social facilities. The respondents expressed their opinions regarding the criteria that determined

Table 4.1 Respondents profile—socio-demographic indicators (percentages)

Age		Educational status		Economic status		Marital status	
21–25 years	11.0	Secondary	3.0	<1000 lei	12.0	Married	78.0
26–30 years	18.0	High-school	44.0	1000–2500 lei	53.0	Unmarried	22.0
31–40 years	56.0	University	53.0	2500–4500 lei	18.0		
41–49 years	15.0			>4500 lei	17.0		

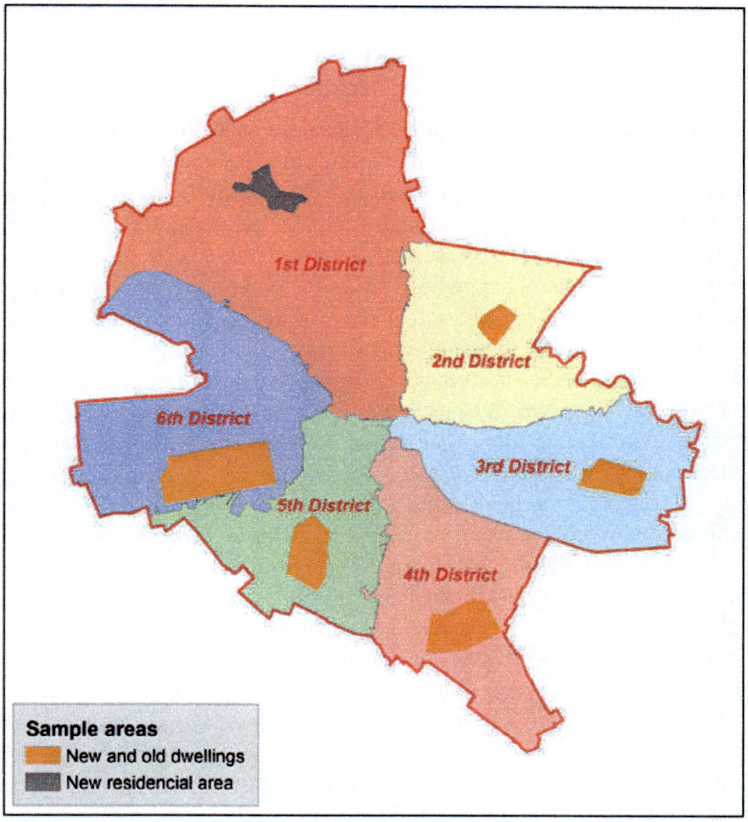

Fig. 4.1 Sample areas of the study in Bucharest

their choice of housing, from the price of housing, the quality of construction materials, traffic, transport infrastructure, education, to urban infrastructure or commercial facilities in the living area. The same variables were used to identify dysfunctions related to housing quality.

During the research, different types of analysis were used, with one of them being the method of comparing data in order to interpret the variables through an applied scale for the answers, from 1 to 5, where 1 means very low, and 5 means very high.

As per the objectives of the study, graphical methods that measure the quantitative analysis were used, the purpose being to highlight the causal typology that was the basis for the manifestation of a phenomenon. The cartographic methods aimed to spatially represent field phenomena or the results of statistical analysis, respectively, the distribution and evolution of residential space based on analyses of construction permits. ArcGIS 10.3.1 software helped map them and their spatial representation.

4.3 Characteristics of the Residential Space in the Pre- and Post-communist Period

The city of Bucharest, as the capital of Romania, manifested over time a powerful dynamic in terms of urban development, and this was accentuated more strongly and more specifically in the last two decades. The evolution of the metropolis was distinctive in two different periods: the communist period—until 1990, and the post-communist period—from 1990 until now.

The urban development happening during the communist period was a planned one, adapted to the economic and social needs of the time, with the extension of the city being made according to unambiguous criteria, and stringent urban legislation. Until 1990, we cannot talk about entrepreneurship, because the state was the principal investor, developer, and owner. Consequently, the residential space was planned in its entirety; it did not adapt to supply and demand, and was designed based on economic and functional criteria. After 1970, Romania went through a period of accelerated industrialization and urbanization, the political decision-makers encouraging rural–urban migration, in order to cover the need for labour in the newly developing cities (Sandu, 1984). The capital Bucharest was no exception to this pattern. The authorities controlled the population growth due to the migration rate, with the migration of the population to big cities being possible only in situations when the economic, social, and urban development were prepared to receive the number of inhabitants. Thus, *"The change of domicile from other localities in cities declared, according to the law, as large cities, is made under the conditions provided by the special law"* (Law no. 5/18 March 1971). It provided that, the number of those who can benefit from a change of residence is established annually, by presidential decree, and only in pre-accepted conditions such as in the interest of work or family (Decree no. 68/17 March 1976). Consequently, development targeted both the economic and the social sphere and gave birth to both iconic official administrative buildings, such as the People's House, House of the Free Press, large industrial areas, and working class residential neighbourhoods (Aldea et al., 2015). The dominant architectural style was the Soviet one, and this still broadly defines the current city.

The post-communist period is defined by a liberalization of the market, which corresponded to very high demand for housing, at the same time generating urban sprawl (Grigorescu & Dumitrescu Mitrică, 2010; Grigorescu et al., 2012). Thus, analysing the evolution of the residential space, we see two important stages: the first, between 1990 and 2000, a period of economic restructuring, and the second, from 2000 to the present. After the fall of communism, the command economy went through a period of transition towards the liberalized one, generating an economic restructuring that had dramatic consequences from a social point of view. Until the 2000s, most industrial production units had closed, unemployment was very high, and living standards depended on economic crises and hyperinflation. The already existing housing fund was the one that ensures the market needs, and the expansion of the residential area fluctuated with the successive economic crises that affected the private sector. Basically, during this period we cannot talk about an upward trend, but on the contrary, about a decrease of the number of dwellings due to a change in destination of use of specific spaces, which belonged to industrial enterprises, that previously closed or went through a reconversion process. The same situation manifested at a national level so that there were 8,006,349 dwellings in 1990, 7,682,359 dwellings in 1993, and in 1999 their number registered a slow growth (7,884,890 dwellings).

The city of Bucharest existed in the same register (see Fig. 4.2). Basically, the most substantial setback was in the early'90 s, when the workers' colony blocks, located in the immediate vicinity of industrial platforms, were abandoned because workers lost their jobs and gave up these spaces, with some buildings changing their destination and others turning into ghost buildings (see Photo 4.1).

A slight increase registered only after 1997, when Romania saw a period of economic growth, with consumption, including the one related to the real estate market, starting an ascending slope.

After the 2000s, we can talk about a stage of the continuous expansion of the residential space, as Romanians' perception of regarding living conditions changed radically. This phenomenon is present mainly in the big cities, with the capital having the highest dynamics in the country. The real estate market gained a considerable advantage, encouraged by high demand, and because the banking sector supported the purchase of housing through loans. The development of the construction sector happened in accordance with the expansion of the residential space and followed two sets of actions: by occupying the vacated land from urban or industrial space that was cleared of old buildings by demolition and by expanding outside the built-up area, using agricultural land. The upward rate was maintained until the onset of the economic crisis of 2008 when the market reacted through a stabilization/stagnation (see Fig. 4.3).

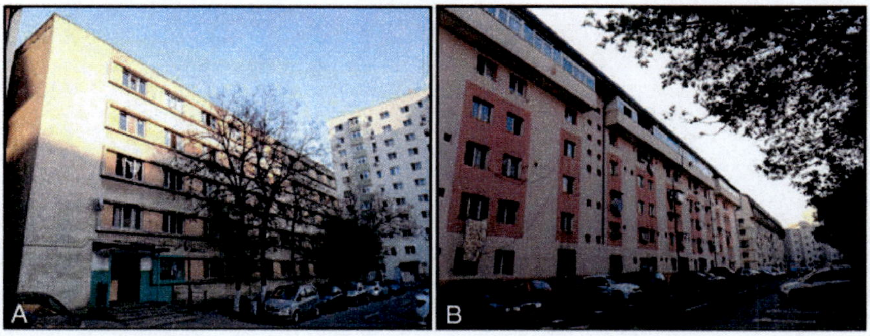

Photo 4.1 **a** Studio blocks, unrehabilitated, built during the communist era for the workers of the FAUR industrial platform, currently used for housing workers; **b** Studio blocks, built during the communist era for the workers of the FAUR industrial platform, transformed today into population housing units. *Source* Personal archive

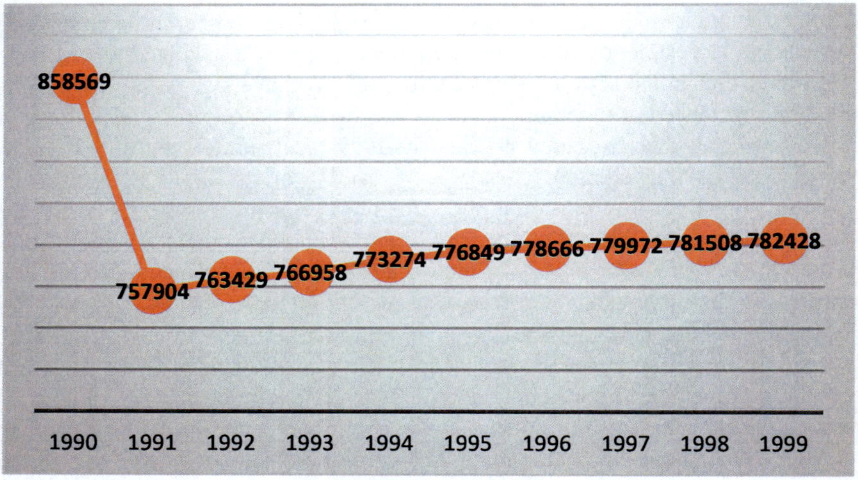

Fig. 4.2 The Evolution of the housing fund in Bucharest, between 1990 and 1999. *Source* based on INSSE statistics

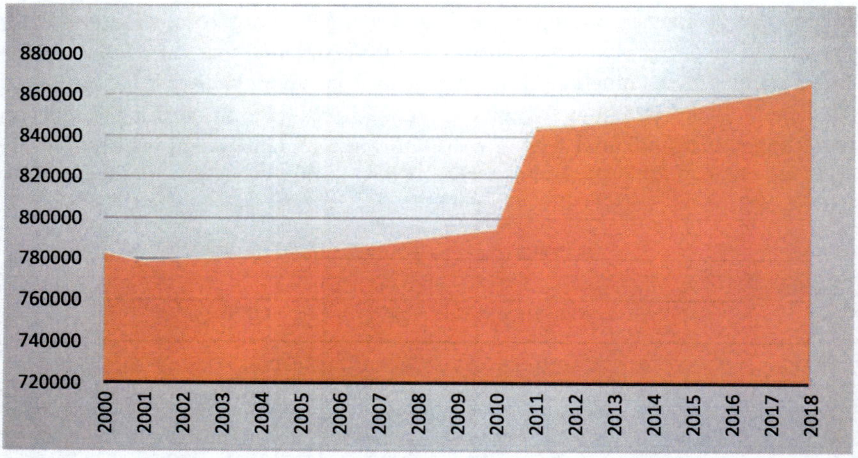

Fig. 4.3 The Evolution of the housing fund in Bucharest between 2000 and 2018. *Source* based on INSSE statistics

Between 2009 and 2011, there was a new notable increase, determined by the "First Home" governmental programme. This programme constituted the introduction on the market of state-guaranteed loans, in order to revitalize the construction sector and to offer families, especially young people, the possibility to purchase their first home.

From 2011 until now, the growth rate continued to rise, as it was dictated by supply and demand. The need for housing remained constant, being further supported by lending and also by the extension in 2020 of the "First Home" governmental programme.

4.4 How the Legislation Effects on the Current Development of the Residential Space

Bucharest' residential space has undergone eloquent changes over time, adapting to the specific socio-economic needs of each period. If at the beginning of the twentieth century, neighbourhoods made up of single-family homes predominated, in the interwar period, on the background of a sustained economic growth, areas with block-type buildings developed.

The period of a maximum explosion in terms of the evolution of residential areas was during communism, with the expansion of the city translating into an unprecedented urban restructuring, which was made according to very precise territorial planning criteria, and also involved the demolition of some historic districts. Communist authorities built blocks of flats while at the same time draw wide traffic axes, developing the urban infrastructure, the transport infrastructure, with all these being interconnected with the big new industrial platforms. The houses met the living standards of that period, though their surface was not large, but were designed precisely in order to have minimum maintenance costs and to satisfy the basic needs of the residents.

The 4 March 1977 earthquake, with a magnitude of 7.4Mw, radically changed construction safety standards. Romania registered material damages of 2 billion dollars, of which 70% in Bucharest alone; 1,578 people died, with 90% of them in the capital city. Here, 33 buildings, including high-rise blocks dating from the early twentieth century, but also three new buildings collapsed (Institutul Național de Cercetare-Dezvoltare pentru Fizica Pământului, 2017). Thus, as an immediate necessity, authorities issued on 1 July 1977, Law no. 8, which imposed standards on durability, operational safety, functionality, and quality of construction. The technical characteristics of the buildings were aimed at higher protection in relation to seismic risk established according to the zoning framework provided by law. Compliance with quality standards was required through strict control, both during the works and at their final reception (Portal legislativ, 2020a, 2020b, 2020c, 2020d). Law no. 8 imposed both administrative sanctions, such as pecuniary penalties, as well as criminal sanctions as follows: imprisonment from 5 to 15 years, for those who modify the technical specifications and as such affected the stability and strength of the building, with possible consequences of loss of life or destruction and degradation of some components of the building, imprisonment between 10 and 20 years, the prohibition of certain rights and partial confiscation of property, for those whose actions lead to the partial or total destruction of the building, affecting the health or life of several persons (Portal legislativ, 2020a, 2020b, 2020c, 2020d).

Construction quality was very easy to ensure in the conditions of a command economy, coordinated by specialized institutions, and with the state being the only investor and having the power of execution, supervision, and sanctioning. The problems appeared after the fall of the communist regime when the market was liberalized, the investors became numerous, and the legislation was no longer respected exactly. In 1995, the legislation on quality in construction was updated, with the publication of Law no. 10 of 18 January. This law will be amended and supplemented by the Government Emergency Ordinance no. 46/2015 and Law no. 163/2016.

The post-communist legislation refers to criteria related to mechanical strength, fire safety, safety, and accessibility, which are also found in Law no. 8 of 1977, but also adds current elements related to energy efficiency or sustainable use of resources. This time there was no insistence on criteria related to seismic resistance, as the article from the previous legislation was no longer present. The actors involved in quality assurance are numerous and are directly responsible for quality standards, and after construction finalizes have an obligation to offer a mandatory warranty period, between 1 and 5 years, depending on the type of building. The quality control during and at the end of the works is regulated under as an administrative circuit, and the investor has to prove it by certificates the employment awarded by accredited experts. The state started controls through a single institution, the State Inspectorate for Constructions, whose role is more one of authorizing analysis laboratories and certifying construction specialists. The legislation maintains both administrative sanctions, such as fines, and some of a criminal nature, respectively imprisonment, with sentences varying between 1 and 10 years, depending on the gravity of the facts.

Therefore, the dysfunctions related to the quality of the constructions are sanctioned most of the time with fines, as the legislation is permissive, especially in the situation in which the non-conformities appeared during the designing stage. Article 36, letter F of Law no. 163/2016 specifies that in case of elaboration of incomplete projects that lead to the non-achievement of a quality level in accordance to the legal requirements, it will proceed to "*apply the sanction and the contravention, the designer having the obligation to complete/redo the technical documentation, as the case may be, at his own expense*" (Portal legislativ, 2020a, 2020b, 2020c, 2020d). Thus, many investors decide not to comply with the legislation, using technical solutions thus affecting the criteria of quality, durability, safety, preferring to pay fines in the situation in which they received any complaints. On the other hand, consumers are not informed about the rights and obligations they have, under the same legislation, as Article 28 clarifies how non-compliances/dysfunctionalities can be notified.

The legislation on urban planning is regulated by Law no. 350 of 6 June 2001, completed by the Emergency Ordinance No. 7 of 2 February 2011. These refer to the principles of urbanism and territorial organization that must offer the population: sustainable development, the right to fair use, adequate living conditions, working conditions, service, and transport conditions that meet a diversity of needs, reduced energy consumption, protection of natural and built landscapes, conserving biodiversity and creating ecological continuity, security and public sanitation, ensuring aesthetic and architectural quality, protecting the architectural and cultural identity of urban and rural localities (Portal legislativ, 2020a, 2020b, 2020c, 2020d). This activity is managed by the Government and implemented by local authorities and other competent institutions, which have the role of elaborating urban development policies and strategies, formulating spatial planning documentation and issuing laws and regulations in this respect. Territorial planning must take into account the national, regional, or zonal land use plan. At the local level, the zonal urban plan regulates the organization of the territory in relation to the territorial, physical-geographical, economic, social, and cultural components, in order to eliminate dysfunctions. The responsibility of monitoring the principles of arrangement and development is the task of local authorities, which must ensure that the legal provisions are observed by all actors involved.

The convoluted bureaucracy, the weak collaboration between state authorities that have the role of issuing decisions or elaborating development policies, as well as the contestation of sanctions by those who do not respect the legislation, are all factors that lead to major dysfunctions, especially in metropolitan cities, as is Bucharest.

4.5 Trends and Challenges in the Dynamics of the Residential Space

The residential space has shown a very accelerated dynamic in recent years, due to the fact that the demand for housing is extremely high in the capital city. The tendency is to build a lot, the main criterion that shapes the market being the price, so houses must be cheap in order to be sold as quickly as possible. Because of this, two characteristics manifested, namely the extension to the outside of the urban area of the new neighbourhoods and the orientation of developers towards the construction of small houses or using materials of inferior quality to maintain a low price.

4.6 Highlighting the Typology of Residential Dynamics Within the Sample Areas

The evolution of the residential space has always been closely related to the economic one. Bucharest represented, until the economic crisis of 2008, the city with the highest price per built square metre, with the real estate market strongly overvalued. The conditions were favoured by loans that allowed a degree of indebtedness of 70% of the total income of the applicant, as well as by a period of economic growth correlated with an increase in incomes and implicitly in the standard of living. The real estate market has been severely affected, starting with the economic crisis of 2008 with the construction sector suffering a sharp setback. The price per built square metre sometimes decreased up to 50%, due to the decrease of living standards and purchasing power, loss of jobs, salary cuts of 25% in the public sector, diminished possibility of lending because of stricter lending rules. This strong setback was maintained until 2010, when the construction sector went through a revitalization process, determined mainly by the economic support measures implemented by the state (see Fig. 4.3). The price of housing started to rise, as the market reacting to the growing demand for housing, and newly built residential areas amplified their expansion.

The analysis of building permits issued by the urban planning office of each mayor's district office, between 2017 and 2019, showed the current trend regarding the fund of new housing. As such, in 2017, it can be seen that the largest number of building permits issued was registered in all districts in Bucharest (see Fig. 4.4). Their distribution by months shows a higher approval rate in the spring–summer-autumn months, as the schedule of the real estate developers' activity depends on the seasons with positive temperatures. In 2018 and 2019, the number of approved building permits remains constant in all five districts, except for the third district, which saw a slight decrease. We cannot speak in this case about a setback due to particular economic situations, but rather about the problems of going through specific administrative stages, which involved in 2019, the approving of a new PUZ procedure (Area Town Planning). This document had the role of setting development priorities, by extending the built perimeter in the peri-urban area, configuring new transport networks, building or arranging green spaces or extending the utility networks.

Regarding the total number of authorizations, issued during the analysed period, there is a high dynamic of constructions in the third district, mirrored by the total number of authorizations, followed by those issued in the first district (see Fig. 4.4).

The continuous expansion of the residential space in the third district was favoured by several factors. The first refers to the use of vacant lots, but also of those reclaimed through the demolition of some industrial platforms that have been going through insolvency/liquidation. This was advantageous for builders because the new homes were thusly located inside the urban area, benefiting from the infrastructure and utilities already in place since the communist period. The second factor refers to the possibility of extending the built perimeter outside the built-up area, as the agricultural areas changed their category of use. These areas, however, present multiple disadvantages related to the lack of transport or utility infrastructure, as well as the greater distance of the new neighbourhoods from the central area. The first district partially shows a similar dynamic, with a large number of building permits being determined by the availability of land used by some old communist industrial platforms, as well as the vacant lots. The peculiarity consists in the fact that within this sector an extended area represented of green space is located, respectively the Băneasa Forest as well as fruit tree plantations, whose surface diminished as a result of the appearance of residential neighbourhoods. The other districts also present an important dynamic, with the lowest change rate recorded in the second district, as this one has the smallest surface and reduced possibilities of extending towards the periphery.

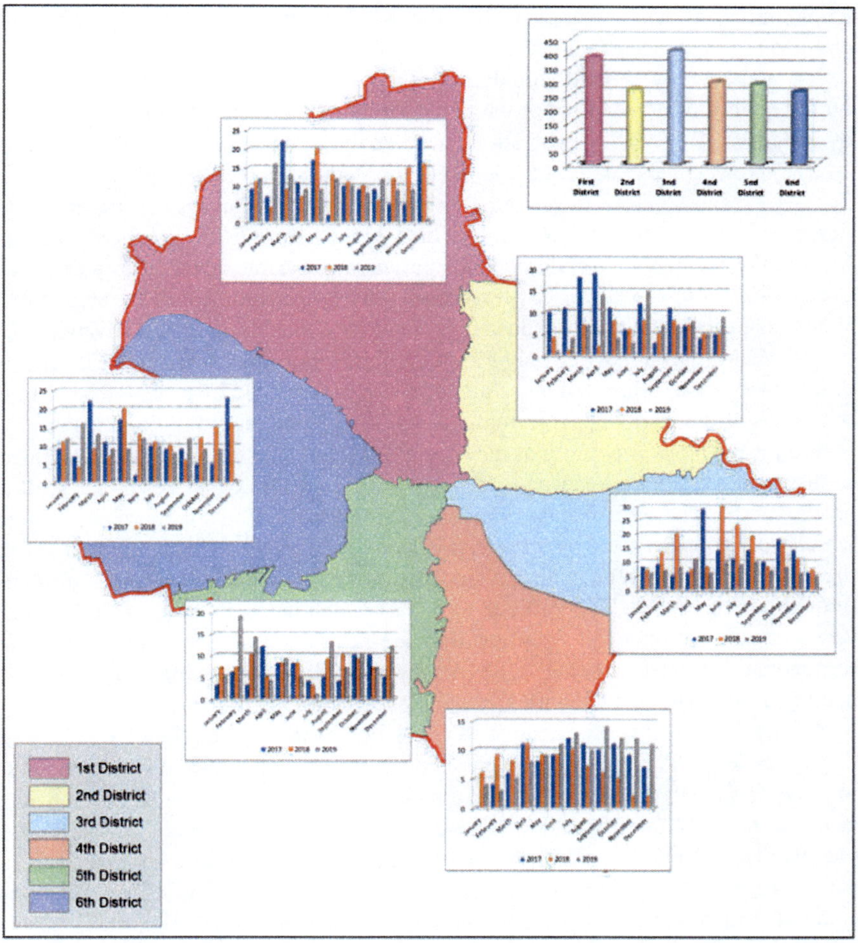

Fig. 4.4 Evolution of building permit within the districts of Bucharest between 2017 and 2019

4.6.1 Enumerating the Main Dysfunctions, and Identifying Possible Ways to Solve the Emerging Issues

The vertiginous increase in demand for housing has led, in the last two decades, to an exponential growth rate of new neighbourhoods, and also to an overestimation of selling prices. In the absence of clear regulations regarding urban planning, the consumer was forced to accept the conditions imposed by real estate developers. The pressure on the real estate market, but also imperfect information regarding a correct commercial relationship, did not allow the buyer to sanction non-conformity by, for example, a non-purchase decision. Therefore, this caused a situation in which the quality of housing in the newly built nuclei was much lower than in the communist neighbourhoods, which was clear from the field surveys, as most of the respondents claimed the existence of multiple dysfunctions.

The field survey aimed to highlight the opinion of residents in the sample areas on the quality of housing, from the perspective of two variables: expectations and dysfunctions. Residents' expectations were analysed on a scale from 1 to 5, where 1 has the lowest value and 5 the highest value. Nine variables were considered (see Table 4.2), and some of them, respectively, the price of housing, access to public transport, school infrastructure, and security being priorities for most respondents. Most of the respondents conditioned the purchase of a home on the nearby existence of schooling units (83%), security insurance (93%), or cleanliness of the area (46%). When asked about the price of housing, 64% of respondents said that an advantageous price is very important, while only 3% said that the price does not matter. The quality of the dwellings represented an acquisition criterion framed on a scale from 1 to 5, with 43% of the respondents framing it at a level 5, and only 0.50% associating it with a level 1, thus proving that they were not interested in this aspect. Technical infrastructure facilities are also prioritized by the majority, considered on the scale at levels 4 and 5. There are also two elements, namely traffic and commercial facilities, which were not prioritized by most respondents. In the case of very congested traffic, the respondents' answer can be explained by the fact that it is a widespread problem throughout the city, and it is difficult to find a residential area where traffic is fluent. Regarding the commercial facilities, they are prioritized at level 3 by 29% of the respondents, Bucharest being covered very well with commercial platforms and neighbourhood stores, which implies good accessibility at the level of all districts.

The dysfunctions were categorized in relation to the expectations expressed by the respondents. The biggest problems reported by the interviewed residents relate to heavy traffic (91%) correlated with the underdevelopment of the public transport infrastructure (77%). The residents of the newly built urban nuclei, in particular, complained about this aspect, as they have to travel to work using their personal car or have a very long waiting time when using public transport. A significant dysfunction is also the underdeveloped technical-urban facilities (63%), as most of the newly built dwellings are not connected to the public water or gas sewerage network, the streets are not asphalted, or have any public lighting or sidewalks. These are important reasons that lead to another dysfunction, namely the high price of dwelling compared to living conditions. Other elements that cause dissatisfaction are poor management of urban waste, sub-dimensioned school infrastructure, underdeveloped commercial facilities, or security-related issues (see Table 4.2).

The respondents also signalled problems that can only be solved through urban development policies implemented by the local authorities. The residential area of the Bucharest is constantly expanding. Its development should be done concurrently with the equally rapid expansion of traffic axes and implicitly the organization of an efficient public transport system that should reduce traffic, the construction of technical-urban facilities in the housing areas before the actual appearance of housing, ensuring a school infrastructure in the newly built areas, covering the needs related to correct waste management, as well as the promotion of social or commercial facilities through rigorous urban planning. According to the respondent's opinion,

Table 4.2 Residents' opinion about the expectations and dysfunctions regarding their residential area

D	EXPECTATIONS	Residents (%) 1-very low; 5-very high	DYSFUNCTIONS	Residents (%)	D
Economic	1. Dwelling prices	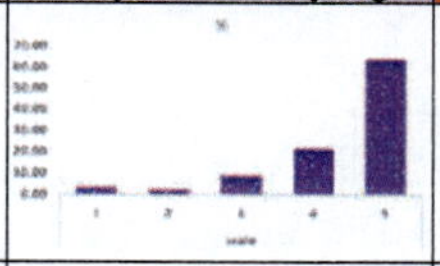	1. The high price of dwelling	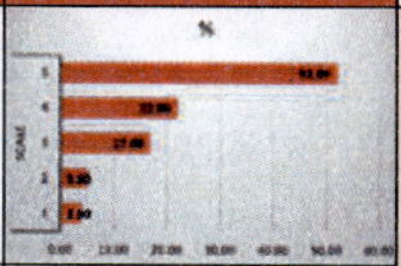	Economic
	2. The quality of the dwelling	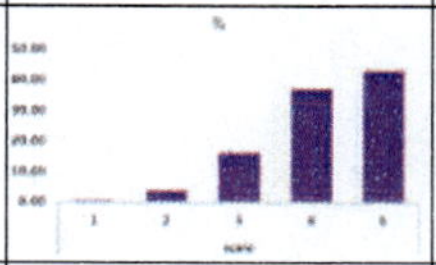	2. Poor quality of construction materials	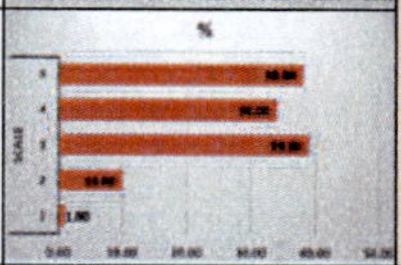	
Urban facilities	3. Public transport infrastructure	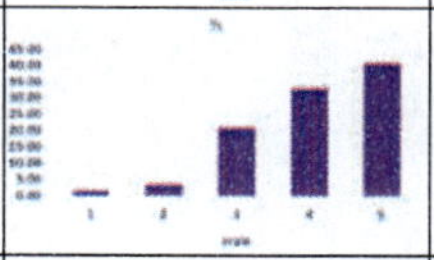	3. Sub-dimensioned public transport infrastructure	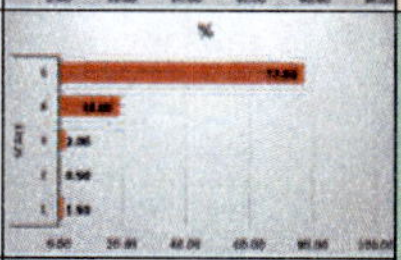	Urban facilities
	4. Traffic	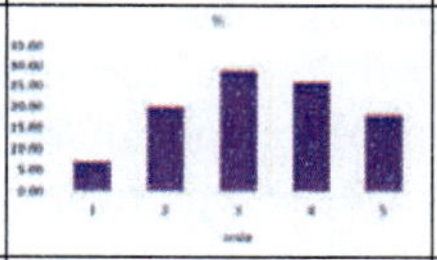	4. Crowded traffic	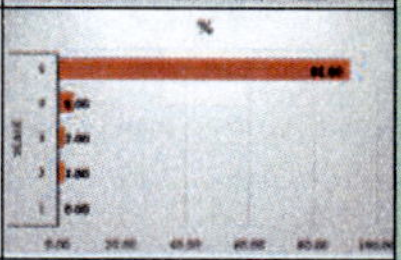	
	5. Technical-urban facilities	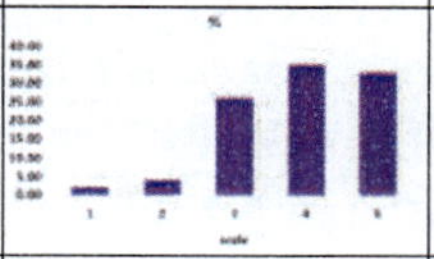	5.Underdeveloped technical-urban facilities	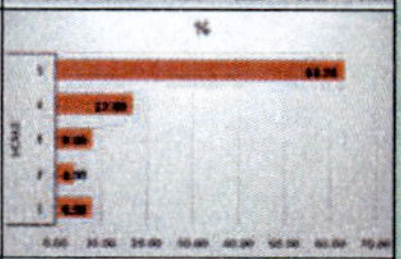	
	6. Management of urban waste	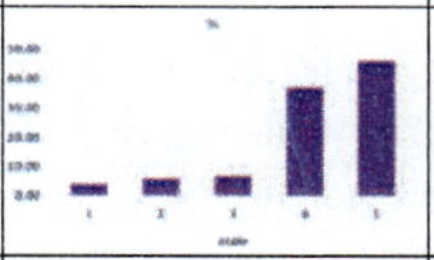	6. Poor management of urban waste	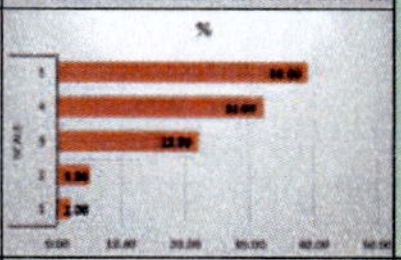	
Social facilities	7. School infrastructure	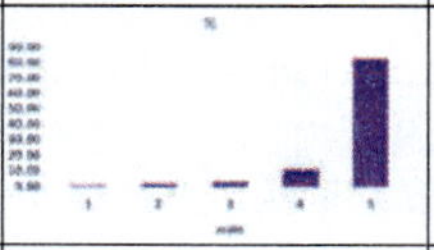	7. Sub-dimensioned school infrastructure	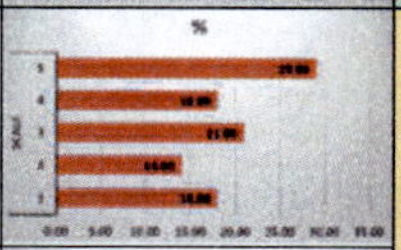	Social facilities
	8. Commercial facilities	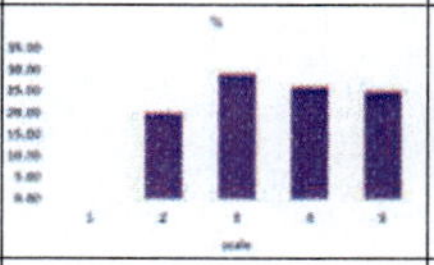	8. Underdeveloped commercial facilities	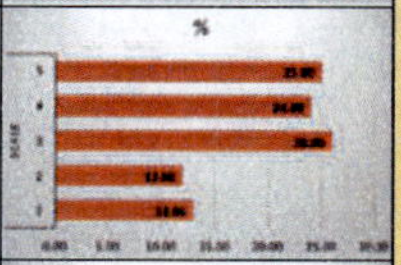	
	9. Security	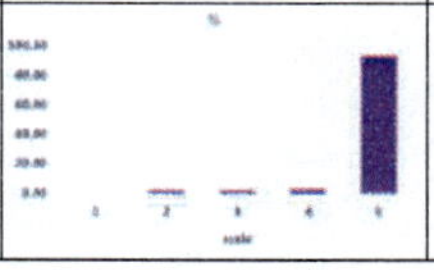	9. Security issues	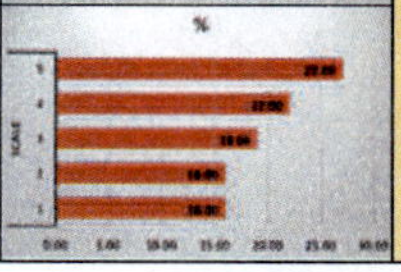	

Source Processed data survey
Note D-dimensions

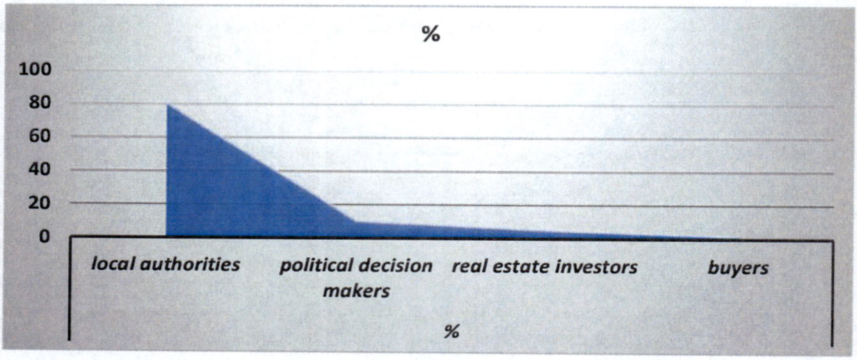

Fig. 4.5 Identifying the possible actors responsible for solving the dysfunctionalities in the opinion of the respondents. *Source* Processed data survey

the main actors responsible for solving the dysfunctionalities are local authorities (81%), which should not grant building permits until facilities such as roads, water supply, sewerage, or gas network have been developed in the area.

A second actor that has the power to solve the reported problems refers to policy makers (11%), respectively the government which, according to respondents, should issue clear legislative regulations that impose standards related to the quality of the construction materials, the minimum size of dwellings, the distance between them, the obligatory construction of parking places for each house, the obligatory existence of the green areas (see Fig. 4.5).

The third actor, mentioned by respondents, refers to real estate investors (6%) who should respect the quality standards regarding construction materials, sell the houses after their execution, and not in the project stage, practice a correct price correlated to the quality of housing, and comply with the legislation exactly.

Respondents also identified a fourth actor who could play an important role in solving problems, namely the buyer (2%), which by refusing to buy homes that do not meet the current level of housing quality standards, could require real estate investors to comply with current legislation, as well as ask local authorities for coherent spatial planning strategies.

4.7 Conclusions

The evolution of the residential space in Bucharest is marked by several stages of development, the most important being the one during the communist regime when the expansion of the city was a planned one. This involved major changes in urban morphology, when some of the historic neighbourhoods were demolished and replaced with iconic buildings that had an administrative role or with blocks of flats that served workers on the new industrial platforms. After the fall of communism, the expansion of the residential area was marked by a discontinuous evolution directly dependent on the economic evolution of the country, with periods of economic boom alternating with phases of recession.

The legislation that regulates construction quality has undergone transformations over time, with the most important being after the 1977 earthquake, when it became a necessity considering the losses registered. Thus, the communist government, the only one decision-maker and owner, imposed stringent rules regarding buildings seismic resistance norms, fire protection, but also related to their quality. The rules were followed rigorously, because the surveillance system and the coercive measures, of administrative and criminal nature, were very severe. After 1990, the legislation adapted to the installation of a competitive market, in which properties no longer belonged solely to the state, and the actors involved became private entities. The previous provisions regarding the constructions' seismic safety disappeared from the legislation, and new ones referring to energy efficiency were introduced at the same time. This meant that compliance with quality standards was not always a priority for real estate investors, and profit prevailed.

Dysfunctions regarding the quality of housing appeared in the new residential areas. They were most frequently related to traffic, correlated with the under-sizing of the public transport system, underdeveloped technical-urban facilities, underdeveloped social facilities like schooling, commercial facilities, or poor quality of construction materials all of these coming with a high price of dwellings. All this was caused by inefficient management at the local level, which hides a lack of integrated and sustainable urban planning.

This study's field survey identified local authorities as the actors that could find solutions regarding the reported problems. They could better apply the current legislation and elaborate urban development plans, and enlarge the infrastructure proportionally with the extension of the residential areas. Solutions could also come from political decision-makers who can improve the legislation in order to establish clear standards that are meant to increase the quality of housing.

In conclusion, the residential area of Bucharest still presents an expansion-oriented dynamic, as its capital status generates a rapid economic development. The new residential area is a consequence of the high demand for dwellings, but it does not fully meet the needs and quality standards of housing, requiring an immediate adjustment of urban development policies by reorienting towards functionality and sustainability.

References

Aldea, M., Petrescu, F., Parlow, E., & Iacoboaea, C. (2015). The dynamics of a city. Over 40 years of change in Bucharest and ITS detection in multitemporal satellite imagery. In *35th EARSeL symposium–european remote sensing: Progress, challenges and opportunities Stockholm*. Available at http://www.earsel.org/symposia/2015-symposium-Stockholm/pdf/proceedings/THE_DYNAMICS.pdf. Retrieved 2, February 2020.

Dumitraşcu, M., Mocanu, I., Mitrică, B., Dragotă, C. S., Grigorescu, I., & Dumitrică, C. (2018). The assessment of socio-economic vulnerability to drought in Southern Romania (Oltenia Plain). *International Journal of Disaster Risk Reduction, 27*, 142–154.

Eurostat. (2017). *Housing statistics*. Available at https://ec.europa.eu/eurostat/statistics-explained/index.php?title=Housing_statistics/ro. Retrieved 10, April 2020.

Grigorescu, I., Mitrică, B., Mocanu, I., & Ticană, N. (2012). Residential development in the metropolitan areas. *Revue Roumaine De Géographie/Romain Journal Geography, 56*(1), 43–59.

Grigorescu, I., & Geacu, S. (2017). The dynamics and conservation of forest ecosystems in Bucharest metropolitan area, *Urban Forestry and Urban Greening, 27*, 90–99. Available at http://www.sciencedirect.com/science/article/pii/S1618866716304204

Grigorescu, I., & Dumitrescu Mitrică, B. (2010). The role of metropolitan areas in the Romanian urban system. A case-study: Bucharest Metropolitan Area. In *Time-space changes of the regional and landscapes structures* (pp. 29–34) Institute of Geonics Ostrava, Brno Branch. ISBN 978-80-86407-13-5.

Grigorescu, I., Kucsicsa, G., Popovici, E. A., Mitrică, B., Mocanu, I., & Dumitraşcu, M. (2019). Modelling land use/cover change to assess future urban sprawl in Romania. *Geocarto International*, 1–19.

Ianoş, I, Cercleux, A. L., & Pintilii, R. (2010). Remarks on identity building of rural and urban communities in the bucharest metropolitan area. *Analele Universității din Oradea–Seria Geografie, Tom XX*(2), 173–183.

Institutul Național de Cercetare-Dezvoltare pentru Fizica pământului. (2017). *Cutremurele din România și efectele lor*. Available at https://www.arcgis.com/apps/Cascade/index.html?appid=b49bc333fd1b4755ae887f00db578137. Retrieved 30, March 2020.

Kucsicsa G., & Grigorescu I. (2018). Urban growth related to distance explanatory factors in Bucharest Metropolitan Area. Spatial and temporal assessment using logistic regression. *Journal of Urban Planning and Development, ASCE Library, 144*(1). Available at https://doi.org/10.1061/%28ASCE%29UP.1943-5444.0000415.

Matei, E., Braghină, C., Bogan, E., & Preda, M. (2013*). Dezvoltarea durabilă a oraşelor mici din România între potenţial şi cerinţele comunităţilor*. Editura Univerităţii din Bucureşti, Bucureşti. ISBN 978-606-16-0287-2

Mihai, B., Nistor, C., & Simion, G. (2015). Post-socialist urban growth of Bucharest, Romania–a change detection analysis on Landsat imagery (1984–2010). *Acta Geographica Slovenica, 55*(2), 223–234. https://doi.org/10.3986/AGS.709

Mitrică, B., Persu, M., Mocanu, I., Şerban, P., Grigorescu, I., & Damian, N. (2019). *Changes in the dynamics and demographic structure of the Romanian Rural Population: An overview of the post-communist period*. In J. Bański (Ed.), *Three decades of transformation in the east-central European Countryside*. Springer.

Nae, M., Dumitrache, L., Suditu, B., & Matei, E. (2019). Housing activism initiatives and land-use conflicts: pathways for participatory planning and urban sustainable development in Bucharest City, Romania. *Sustainability, 11(22)*, 6211. https://doi.org/10.3390/su11226211

Portal Legislativ. (2020a). *Legea nr. 10 din 18 ianuarie 1995 (*republicată*) privind calitatea în construcții*. Publicat în Monitorul Oficial nr. 765 din 30 septembrie 2016. Available at http://legislatie.just.ro/Public/DetaliiDocument/5729. Retrieved 30, March 2020a.

Portal Legislative. (2020b). *Legea nr. 8 din 1 iulie 1977 privind asigurarea durabilității, siguranţei în exploatare, funcționalității şi calităţii construcţiilor*. Publicat în Buletinul Oficial nr. 64 din 9 iulie 1977. Available at http://legislatie.just.ro/Public/DetaliiDocumentAfis/443. Retrieved 30, March 2020b.

Portal Legislativ. (2020c). *Legea nr. 350 din 6 iunie 2001 privind amenajarea teritoriului și urbanismul*. Publicat în Monitorul Oficial nr. 373 din 10 iunie 2001. Available at http://legislatie.just.ro/Public/DetaliiDocument/29453. Retrieved 2, April 2020c.

Portal Legislative. (2020d). *Ordonanță de urgență nr. 7 din 2 februarie 2011 pentru modificarea și completarea Legii nr. 350/2001 privind amenajarea teritoriului și urbanismul*. Publicat în Monitorul Oficial nr. 111 din 11 februarie 2011. Available at http://legislatie.just.ro/Public/DetaliiDocumentAfis/126154. Retrieved 2, April 2020d.

Preda, M., Bogan, E., & Mareci, A. (2015). Economic restructuring and unemployment dynamic and their consequences on the general population. Case study: Romania. *Analele Universității București, Seria Geografie*, 75–95. ISSN 1013-4115.

Primăria sectorului 1. (2019). *Lista certificatelor de urbanism eliberate de Sectorul 1 al Municipiului București*. Available at http://www.primariasector1.ro/info-urbanism.html. Retrieved 2, February 2020.

Primăria sectorului 2. (2018). *Liste certificate de urbanism și autorizațiilor de construire anonimizate*. Available at http://public.ps2.ro:3012/Urbanism/Liste%20CU%20si%20AC%20anonimizate/. Retrieved 2, February 2020.

Primăria sectorului 3. (2019). *Autorizații de construire*. Available at https://www.primarie3.ro/index.php/primaria/autorizatii_constructii. Retrieved 2, February 2020.

Primăria sectorului 4. (2019). *Autorizații de construire și certificate de urbanism*. Available at https://ps4.ro/urbanism/autorizatii-de-construire-si-certificate-de-urbanism/. Retrieved 2, February 2020.

Primăria sectorului 5. (2018). *Autorizații de construire. Certificate de urbanism*. Available at http://www.sector5.ro/utile/autorizatii-de-construire-certificate-de-urbanism/. Retrieved 2, February 2020.

Primăria sectorului 6. (2019). *Autorizații de construire emise*. Available at https://www.primarie6.ro/informatii-utile/autorizatii-de-construire-emise/. Retrieved 2, February 2020.

Sandu, D. (1984). *Fluxurile de migrație în România*. Editura Academiei Republicii Socialiste România.

Suditu, B. (2009). Urban sprawl and residential mobilities in the Bucharest area – reconfiguration of a new residential geography. *Human Geographies—Journal of Studies and Research in Human Geography, 3*(2), 79–93. Available at http://humangeographies.org.ro/articles/32/326_Suditu.pdf

Wesołowska, J. (2016). Urban infrastructure facilities as an essential public investment for sustainable cities—indispensable but unwelcome objects of social conflicts. Case study of Warsaw, Poland. *Transportation Research Procedia, 16*, 553–565. Available at https://pdf.sciencedirectassets.com/

Chapter 5
Rational Use of Urban Land for Sustainable Urban Growth: A Case Study of Metropolization in France

Philippe Grall, Martin Duplantier, and Camille Kamga

Abstract The worldwide sanitary crisis of early 2020 caused by the COVID-19 pandemic has greatly affected our confidence on the development of our shared urban environments. Housing and living conditions in major urban areas have been tremendously impacted by the effects of the pandemic. To protect people living in cities and reduce the spread of the transmission of the virus, government officials and public health professionals have recommended some non-pharmaceutical interventions. These measures have affected most of regular daily living habits as stay-at-home orders with restrictions on non-essential travels were enacted. Not only the majority of officials in charge of planning, regulation, supervising the development of cities, urban living conditions and environment, but also urban transportation, urban utilities and services have failed to anticipate some kind of this crisis and had tremendous difficulties in reacting to the crisis and maintain a certain level of services to inhabitants. The basic policy reaction was to forbid social life but also all kind of non-essential activity, although some adaptable activities could have been kept functioning..... Actions that have been promoted and taken revealed to be inaccurate to maintain so kind of pace of life in all major urbanized cities and metropolisis, exceptions of some few Asian countries' government which appeared to be rarely efficient. This difficult period has provided an opportunity to reassess our urban development policies and review the basic principles of smart urban growth and urbanization. A sustainable urban growth policy should rely on (1) urban land use optimized and organized to protect the quality of life and urban environment for all the inhabitants of major cities and metropolitan area, but also on (2) urban coordination and rational

P. Grall (✉)
Policy Institute for Local Officials (IFER), 112, 85 Boulevard Pasteur, 75015 Paris, France
e-mail: philippe-grall@wanadoo.fr

M. Duplantier
French Association of Architects and Contractors (AMO), 10 Rue Henri Chevreau, 75020 Paris, France

C. Kamga
Urban Transportation Research Center (UTRC) at City College of New-York (CUNY), 160 Convent Street, New Yok, NY 10031, USA
e-mail: ckamga@ccny.cuny.edu

B. R. K. Sinha (ed.), *Urban Dynamics, Environment and Health*,
https://doi.org/10.1007/978-981-99-5744-6_5

organization of urban transportation, daily commuting for workers, urban policy mix within economics, services, housing. If not successfully applied, the consequences in terms of quality of life, health, and environment are for sure bound to be extremely consequential. Using the French experience, consequences and risks of "metropolisation" are analysed, both for large cities and other urban areas in terms of quality of life, economic development, employment, land use, and infrastructure issues in urban areas. This chapter also proposes different recommendations and provides hints in terms of policy balance between sustainable development, transportation, urban development projects, and urban planning model.

Keywords Metropolization · Territorial balance · Urban congestion · Weak housing supply · Land use · Sustainable urban growth · Sustainable development

5.1 Introduction

The worldwide sanitary crisis of early 2020 caused by the COVID-19 pandemic has greatly affected our confidence on the development of our shared urban environments. Housing and living conditions in major urban areas have been tremendously impacted by the effects of the pandemic. To protect people living in cities and reduce the spread of the transmission of the virus, government officials and public health professionals have recommended some non-pharmaceutical interventions. These measures have affected most of regular daily living habits as stay-at-home orders with restrictions on non-essential travels were enacted.

The majority of officials in charge of planning, regulation, supervising the development of cities, urban living conditions and environment, but also urban transportation, urban utilities and services have failed to keep a reasonable level of urban services during the crisis. Actions that have been promoted and taken revealed to be inaccurate to help urban inhabitants undergo such an extreme situation, locked down in tiny apartments with low comfort and services around, most of the time.

This difficult period has provided an opportunity to reassess our urban development policy and review the basic principles of smart growth. A sustainable urban growth policy should rely on (1) urban land use optimized and organized to protect the quality of life and urban environment for all the inhabitants of major cities and metropolitan area, but also on (2) urban coordination and rational organization of urban transportation, daily commuting for workers, urban policy mix within economics, services, housing. If not successfully applied, the consequences in terms of quality of life, health, and environment are for sure bound to be extremely consequential. Using the French experience, consequences and risks of "metropolisation" are analysed, both for large cities and other urban areas in terms of quality of life, economic development, employment, land use, and infrastructure issues in urban areas. This chapter also discusses different recommendations and provides hints in terms of policy balance between sustainable development, transportation, urban development projects, and urban planning model.

Similarly to the rest of the world, metropolization of European cities has been accelerating in the past two decades at least. A sustainable quality of life in major cities and metropolis is clearly at risk due to increasing congestion and pollution conditions. It is urgent that urban dynamics integrate a much better land use and optimization of urban growth, in a comprehensive way of thinking about the City of tomorrow.

Urban growth has been particularly chaotic in the past three decades in France where the gap between very large cities and rural areas has been deeply widening in terms of job opportunities, housing prices, and economic development opportunities. Very large cities have been soaring to European level metropolisis, whereas some middle range cities and rural–urban have been collapsing. After having been very confident in the French urban planning system and the French way of "designing the city", researchers, architects, and policy makers are starting to wonder about the soaring demography, the growing congestion, the inefficiency of urban public transportation, the extremely concentrated of economic development and employment in limited geography areas, the lack of local housing supply, and finally the souring prices of residential real estate. To the author's personal experiences, all those issues are under scrutiny in major metropolis in Europe, Asia, and the Americas. In this chapter, the authors are sharing their thoughts about the future of cities and how to deal with it.

This chapter analyses consequences and risks of metropolization, both for large cities and other urban areas in terms of quality of life, economic development, employment, land use, and infrastructure issues in urban areas. It also demonstrates the impact and the risks for small cities and rural–urban economy, but also in terms of balance between sustainable development, transportation, urban development projects, urban demography and density, and urban economic issues. We propose to point the issues of city attractiveness and lack of urban planning using Bordeaux metropolitan area as case study. Policy recommendations in terms of foreseeing and conducting urban development in major cities and metropolis have also been discussed.

5.2 Metropolization of Cities: Attractiveness, Insufficient Housing Capacity Supply, Lack of Sustainable Urban Planning and Coordination Between Transportation, Urban growth, Economic development, and Under Pressure After the COVID-19 Crisis

For the last thirty years, French major cities have pushed to become "metropolized" in order to compete with their European counterparts. The same trend was followed by many cities in other continents. It has been a kind of frantic race for growth and attractiveness. Since 1982, in France, city mayors have discretion in decision making on all urban and construction projects within their jurisdiction. They approve construction permit together with the architectural shape, scope, and destination of real estate programmes, both commercial and residential, within national regulations. It can happen, however, that the decision-making process is concentrated at a planning level group of cities, almost like a County in some countries. The national government can enact major laws to be passed by the Parliament. But, beyond those laws, rules, and regulation are applied entirely by local elected officials with a wide range of liberty and self-administration of local bodies guaranteed by the Constitution.

Cities have been modernizing through new major real estate projects very much driven by local public governments. In the last 20 years, they committed local planning and construction agencies to create new urban areas. Major projects were putting together at the area level: sustainable development, low energy building, architectural local urbanism in the area, priority for quality of life inside the area, local community public infrastructure and services, urban transportation connection either individual or public, etc.

All of those factors led to the development of cities both very much locally oriented and promoting an attractive image that was shared with the rest of the country. In the case of the Paris capital region, this marketing campaign was sent also at an international level. As shown in the following Figs. 5.1, 5.2, and 5.3, the "urban organization" of the country has been structured around 22 cities but 12 major metropolises such as Paris capital region, Lyon, Marseille, Lille, Bordeaux, Nantes, Rennes, Toulouse, and Montpellier. Their development has been booming since the early 2010s, evidenced by the concentration and their large share of the number of essential job created within the country. The 2040 demographic forecast emphasizes the trend of urban development concentration basically in the 12 major cities in France.

Figures 5.2, 5.3 and 5.4 illustrate a major trend through a souring demographic, employment, and economic development that has no common comparison with the rest of the country. This has been a major societal break-up in the set of moral and constitutional values that keep the French nation together more than two hundred years: a united Republic. It led to a kind of social resentment from rural areas against metropolises. And metropolitan inhabitants were having in mind that they had to pay for the development of rural areas that would not really deserve the effort, since it could be considered as useless. In addition, giving the attractive credit market with the ease to obtain low interest personal loans for quite a long period, households rushed to purchase residential homes and investment while housing supply was obviously and systemically insufficient. The following map of the administrative organization of the metropolises shows a limited number of cities "chosen", thru a law, which created a metropolis status for cities in 2014. With the COVID-19 crisis, metropolitan areas have been harshly hit by the virus and more under pressure than rural areas. All of the metropolitan living conditions are questioned as not enough convenient and sufficient for inhabitants.

Current estimation of population growth has been forecasted up to 2040 and the average annual growth has been calculated. The following map is showing where there is an extremely important and regular demographic growth in the metropolises. By 2040, the Paris region will be strongly growing, in the West (Rennes, Nantes,

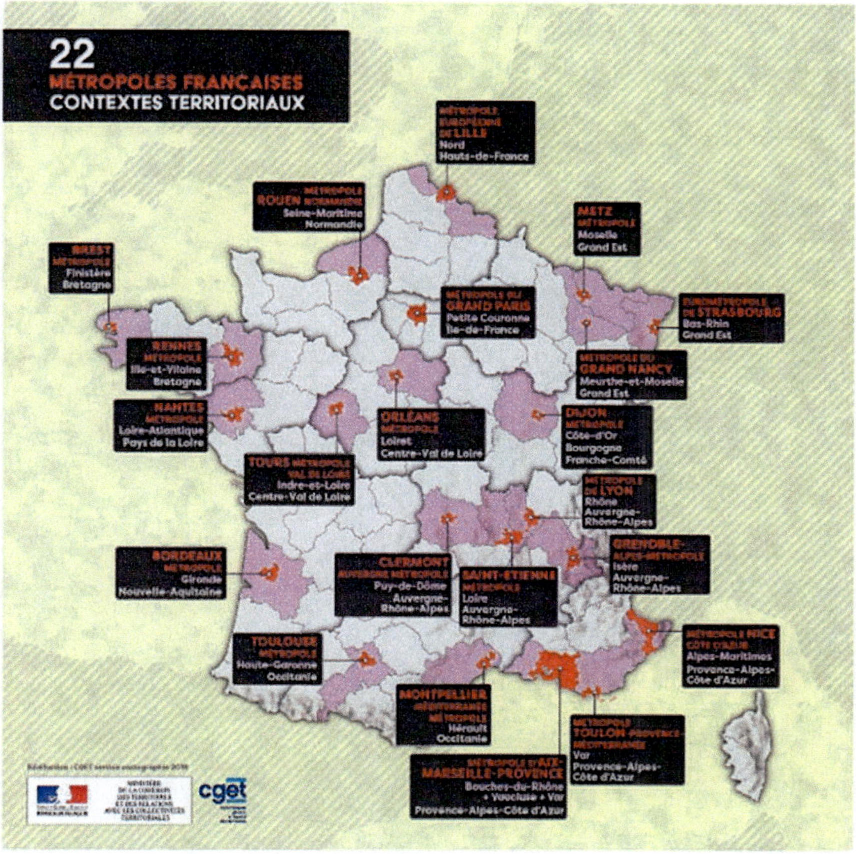

Fig. 5.1 Administrative map of French metropolises: 22 Metropolises spread throughout the country of France

and Bordeaux) and in the South (Toulouse and Lyon). The size of the limited areas is proportional to the population gained in 2040.

So far, the common approach on policy making has been to highlight the attractiveness, gain of jobs, and increased migration to the city. Almost all economists and urban planners and specialists were all advising policy makers in a unique way: France is in a world competition and its large cities could not be excluded. In order to stay globally competitive, these cities have to grow immensely to reach the size of its European and World peer cities. Economic growth is on its way, so does employment. All the public investment should go to build those metropolises since they are playing the role of "Locomotives" to drag upwards other cities in the rest of France.

As illustrated in Figs. 5.3 and 5.4, and recent figures prove it, the growth of metropolisis does not help the growth of other cities in the rest of the country. It is unfortunately not a new phenomenon. Since the financial crisis in 2008, jobs,

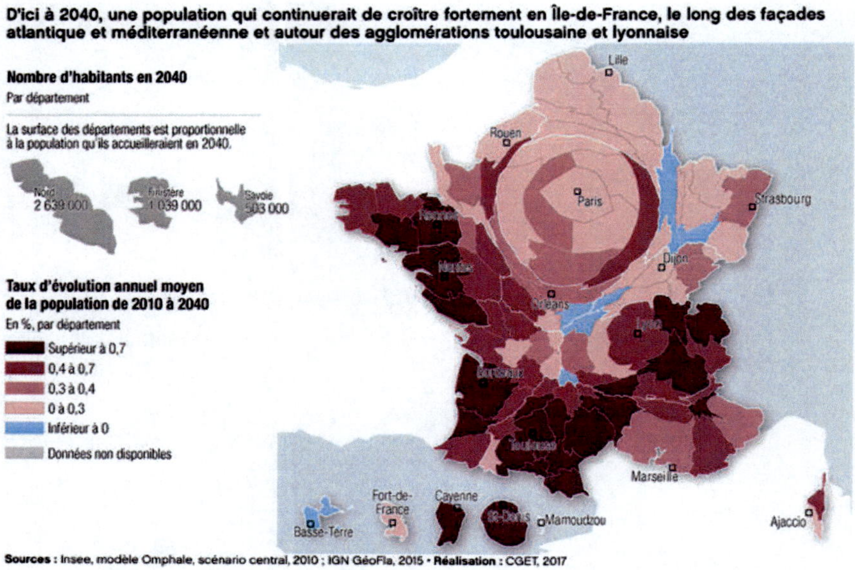

Fig. 5.2 Population annual average variation between 2010 and 2040: representing the number of inhabitants in 2040 per administrative area (Department). The surface of the latter is proportional to the population that would be in 2040. It shows obviously that the population growth will pursue its pace in Paris region, on the Atlantic and Mediterranean proximity and around Lyon and Toulouse

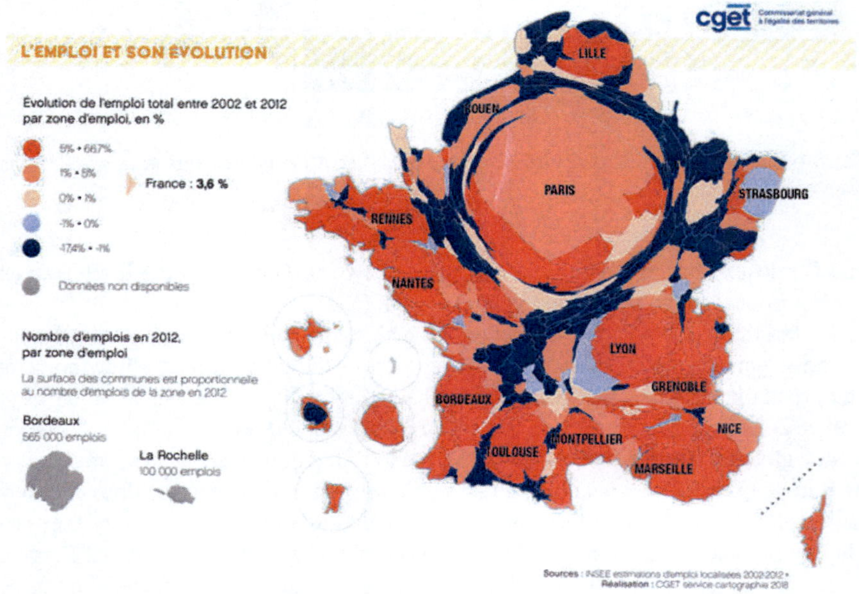

Fig. 5.3 Rate of total job growth between 2002 and 2012 per employment area. Limited areas are proportional to the number of jobs in 2012

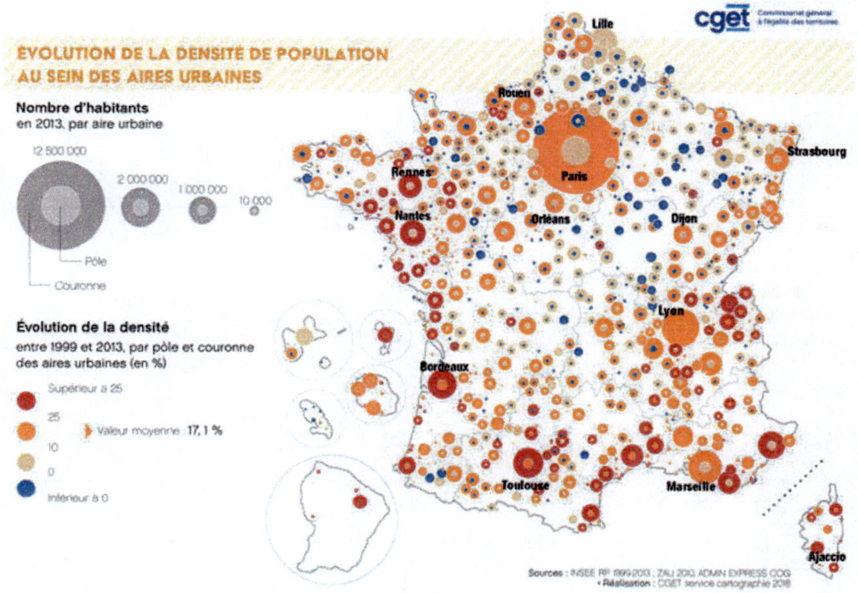

Fig. 5.4 Evolution of population in urban areas between 1999 and 2013 (size of population, rate of density). The size of the circle is the number of inhabitants per urban area, the colour is representing the evolution of the density: more than 25% between 1999 and 2013 is red, less than 0% is blue

investors, and people rushed only to the large cities, whereas, all the "medium size" or "small" cities lost population and jobs in the last 20 years. Overpopulated, those very large cities, metropolises, started around three years ago to meet serious troubles: huge congestion growing, insufficient transportation infrastructure, residential real estate prices soaring since 2012, working people moving outside metropolises since they could not afford housing anymore, while housing supply was deeply insufficient (to be detailed later).

However, small and "medium size" cities, even secondary cities around metropolises, are seriously hit by weak demography growth or loss of demography, weak job creation or even worse. "Secondary cities" gather the smaller cities that are situated around metropolises and maintain relationship with them. In the meantime, people's resentment increased in rural areas. They have the impression to be left aside.

"Housing is a basic human need, which is fundamental to our economic and social well-being. Yet housing provision is often controversial and provokes strong reactions. Seeking to promote an adequate housing supply and a more responsive provision: A weak supply of housing contributes to macroeconomic instability and hinders labour market flexibility, constraining economic growth".[1]

---

[1] Delevering Stability: Securing our Future Housing Needs, Review of Housing Supply to the Prime Minister (UK), Kate BARKER, March 2004.

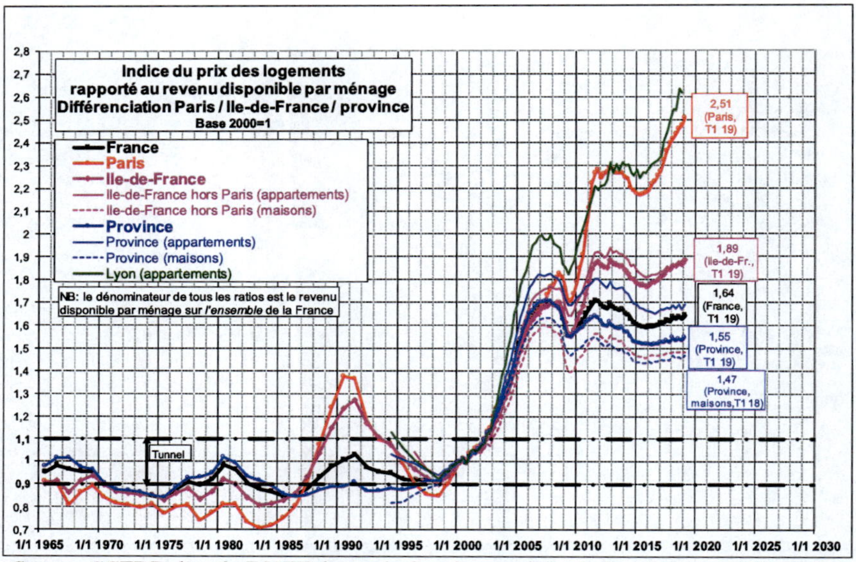

Source: CGEDD d'après INSEE, bases de donnéesnotariales, indices Notaires-INSEE désaisonnalisées.

Fig. 5.5 Housing price indexes to household income (of all French households): 1965–2018 Paris is in red, Lyon in green, Paris region in pink, France outside Paris is in blue, and black is all France. *Source* CGEDD d'après INSEE, bases de donnéesnotariales, indices Notaires-INSEE désaisonnalisées

Yet, if it is a basic need, if too many people come to the same place in the city, if prices are soaring because of credit conditions and insufficient housing supply, then people flee out of the city to find better living conditions. But, they commute much longer distance to reach their job, then congestion is soaring, and global sustainability for urban development is collapsing.

Some studies"[2] evaluated road transportation congestion to 104 h in Los Angeles a year for a road driver on peak hours, 91 in Moscow, 65 in Paris (2016 but only 45 in 2015), 29 in Bordeaux, 28 in Lyon, and 23 h in Toulouse. The average direct cost per Parisian is 2880 euros in 2013 and estimated to 4100 euros in 2030. It would cost 350 billion euros in 15 years.... Such a waste of time is not creating any penny of GDP of the country. It is a waste of time for family, loss of leisure time and working time in the city. People travel long distance to work, loose a lot of time in congestion, work in less good conditions, then the economy is suffering or just not as efficient as it should be, the global cost of saturated urban areas and metropolises is starting to explode, and the risks are high for population resentment and deep misunderstanding of what people have to bear to get a decent living.

[2] Inrix (2016).

5.3 Bordeaux Metropolitan Area: A Case Study

To illustrate the effects of metropolization thru the case of Bordeaux, it is worth gathering a few maps to get an idea of the trend. This trend has been resulting of a kind of "no" policy in terms of urban sustainable development. Urban development has been following a kind of natural evolution, since local officials just "let things going" in terms of transportation means saturation, urban construction spreading, minor prospective in terms of coordination of urban development and economic development. Although all urban regulation tools could have been used and developed in a prospective way, the structure of urban development has been following in a "laisser-faire" way. Residential real estate prices have been soaring for the last twenty years, as a consequence, low and average income workers as a result pushed out of the city and the metropolitan area.

Bordeaux is famous for its vineyards and wines. It is also a very famous and coherent City centre with its urban design and architecture dating from the eighteenth century. Metropolitan population is around 800 000 and the city of Bordeaux is only 252 000. Average annual revenue per capita is around 22 000 Euros, poverty rate is 17%, and unemployment rate is 8.2%. The city is ranked 3rd in France in terms of congestion.

Figures 5.6 and 5.7 show the dispersion of housing prices in euros. Figure 5.8 provides an idea about how resident population growth rate is spreading. For 15 years, people moved progressively to where the prices were affordable. It is not really where there are employment areas. We could also see the importance of commuting to work (Fig. 5.9). Figure 5.11 shows land consumption over 40 years in terms of urbanization.

Figures 5.9 and 5.10 show the rise of housing prices (cost of m^2 in apartment in euros) between 2004 and 2018 with a description of the few major urban and transportation development steps.

Land is a finite resource, but people have been spreading all around the metropolis, outside the employment areas, thus, causing the need of long commuting.

5.4 Policy Recommendation Hints?

After a kind of short review of the metropolises development process, let's draw a few hints for policy makers of the future of the City. Of course, every urban situation is not similar to one another, but—to our experience in Europe and in Asia—, there are some means of comparison and vigilance that we could use and learn from in order to avoid such situations that cities might get suffocated in all the sense of the word: health, economic waste, less productivity, congestion, weak quality of life, economic weak efficiency for the industry and workforce, etc.

If we refer back again to the Barker Report, although it is dating back to 2004, it remains extremely up to date for most of countries in Europe and Asia. Even if the

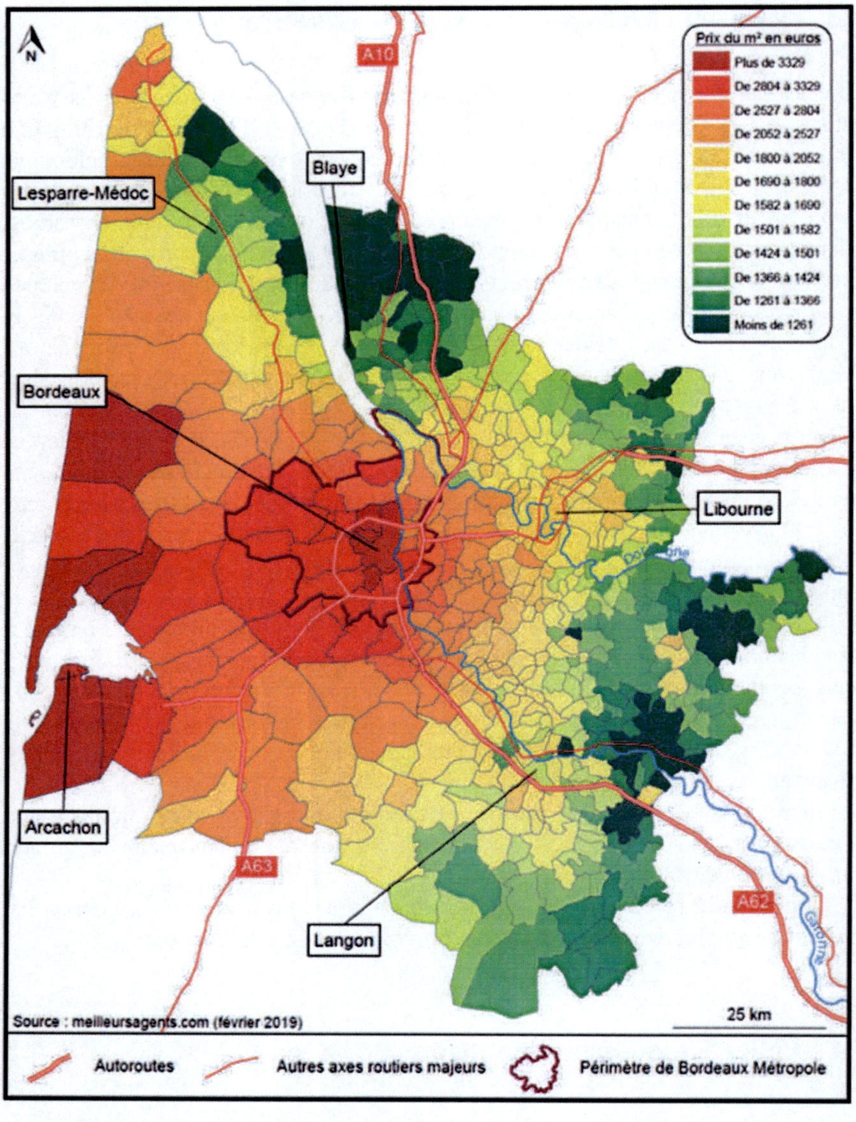

Fig. 5.6 Bordeaux residential real estate prices in euro/m^2 (Terra Nova)

recommendations are probably specific to the UK and not relevant at all for most countries, the economic and policy analyses are rather accurate. We decided to quote a brief part of the analysis of this renowned British economist, adviser to the British Government on social issues such as housing and health care. The description made is interesting, thus comparable to many other countries.

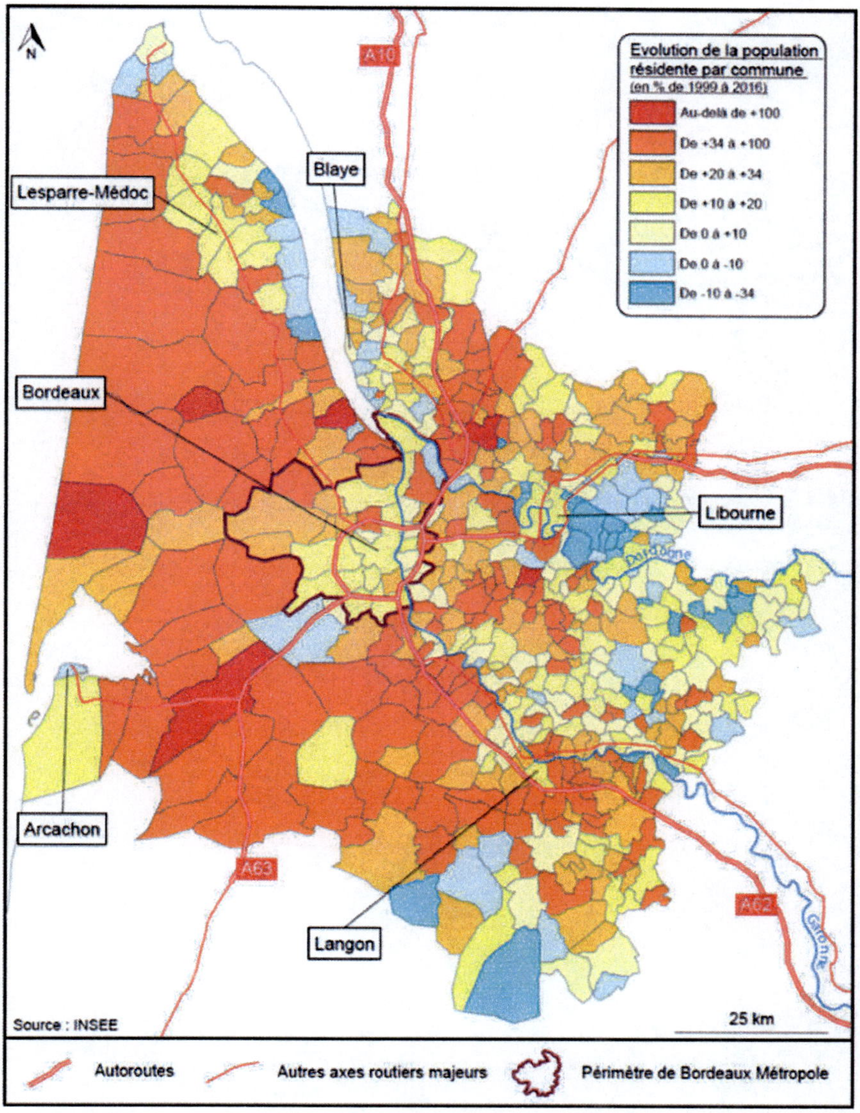

Fig. 5.7 Bordeaux rate of growth of population from 1999 to 2016 (Terra Nova)

Kate (2004), Review of Housing Supply, Delivering Stability: Securing our Future Housing Needs: "The UK should have a more flexible housing market. For many people, housing has become increasingly unaffordable over time. The aspiration for home ownership is as strong as ever, yet the reality is that for many this aspiration will remain unfulfiled unless the trend in real house prices is reduced. This brings potential for an ever widening social and economic divide between those able to access

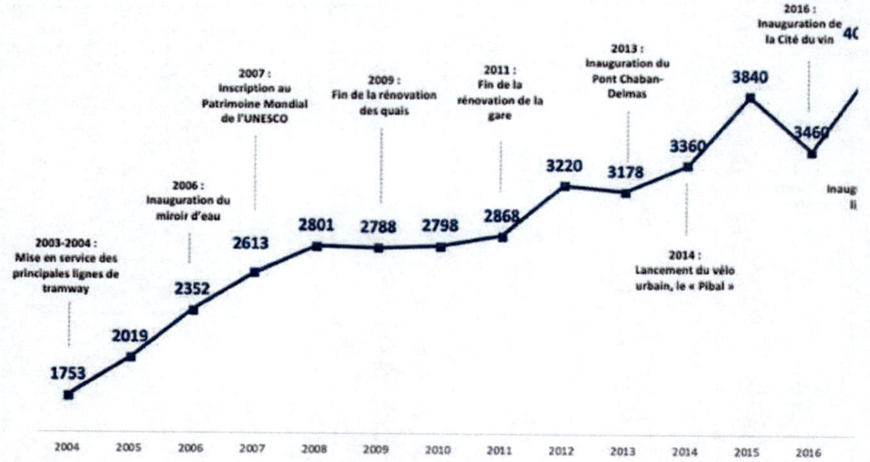

Fig. 5.8 Commuting in Bordeaux area in 2015 (DDTM Gironde): 74% of daily commute is by individual car *Source* L'Argus du logement/Challenges

Fig. 5.9 Growth of housing prices in Bordeaux (Euros) related to different steps in urban developing (Terra Nova) *Source* L'Argus du logement/Challenges

market housing and those kept out. Rising numbers in temporary accommodation is evidence of the polarization which exists today. Homes are more than shelter. They provide access to a range of services and to communities. Housing also plays a major role as an asset in household balance sheets and in household planning for their financial futures. The housing market indicates people's preferences for the types of housing, location and communities that they would like to live in, for example the

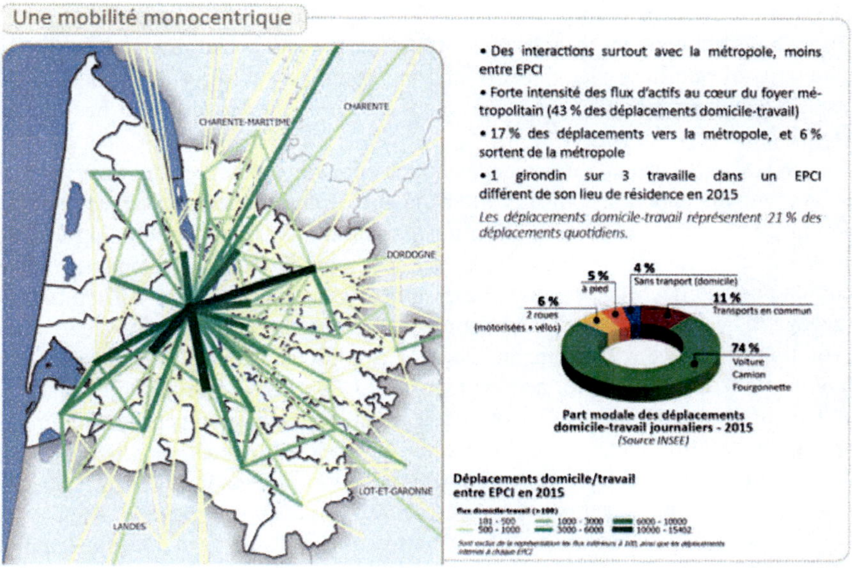

Fig. 5.10 Growth of housing prices in Euros (2004–2018) in Bordeaux, Nantes, Lille (Terra Nova)

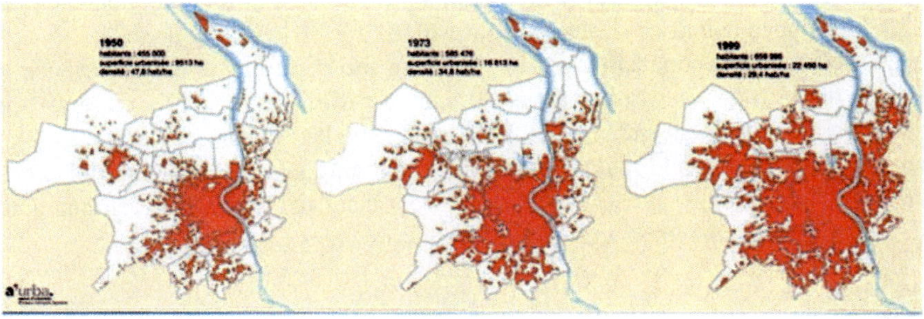

Fig. 5.11 growth of the urban area in Bordeaux Metropolis and surroundings (A'Urba, urbanism agency). Comparison of the size of the urban area in 1950, 1973, and 1999

preference for space as incomes increase. Increasing housing supply raises concerns about the environment and loss of open spaces. These considerations pose a number of dilemmas for policy makers at all levels, (national, regional and local). *There are issues around the relationship between the private sector as the main deliverer of housing and Government's objectives, which may not always accord with market pressures.* There are no easy answers and no disguising that the choices we face are difficult. Land is a finite resource, which we must make best use of. But we also need to think about other outcomes that we wish to achieve: a more flexible housing market which adapts to and reflects the needs of the economy; and a more equitable distribution of housing wealth. I do not believe that continuing at the current (low) rate of housebuilding is a realistic option, unless we are prepared to accept increasing problems of homelessness, affordability and social division, decline in standards of public service delivery and increasing the costs of doing business in the UK—hampering our economic success. Indeed, the Government has already signaled the need for action".

The UK may currently experience another economic situation, but this is not the subject of this paper, specifically after the Brexit and the sanitary crisis COVID-19. In 2004, France was not in such a situation as the UK was according to what Kate Barker is describing. Now France is undergoing the same economic and social conjuncture. Policy makers missed the shift of situation until 2017. They did not manage to avoid nor did they have serious consciousness of the general trend of the evolution of urban environment and housing conditions, which all deteriorated. It is not hopeless, if the "need for action" is seriously taken into account.

One of the major concerns in future integrated urban development is about major and structural insufficiency of affordable housing supply for inhabitants. In the insight report of June 2019 published in the context of the World Economic Forum, "Making Affordable Housing a Reality in Cities", we have chosen two quotations: "About 90% of cities around the world do not provide affordable housing or of adequate quality. The cost of housing, as well as location, prohibits people from meeting other basic living costs, threatening their employment and fundamental human rights. Based on median affordability (median price-to-income ratio), cities in less developed countries are found to be significantly less affordable (28% less affordable) than cities in more developed countries (Kallergis et al., 2018)".

As senior professionals in urban and housing and according to our experience in France but also in other European and Asian countries, we would like to draw the attention of researchers and policy makers on a few ideas to help moving to a major change in designing and making "integrated planning and development" of urban areas. We tried in this paper to demonstrate the impact and trends of development of metropolises, and the impact on other surrounding areas.

5.5 General Policy Matters

1. Integrated planning and development have to be seriously linked with the reality observed thru any statistics possible to gather, any governance and policy exchange with both the public and the private sector. In almost all countries, the statistics recorded are most of the time insufficient. But it is at least a trend and a synthetic approach that should be completed with observation of major actors.
2. Researchers and economists have to concentrate more on the urban system and on the housing and residential market, their weight, and their influence on the economy. The impact of policy measures should be also very much scrutinized. Most of the time, this sector is not sufficiently analysed and evaluated.

3. Policy makers and researchers have to investigate how the urban system (observation, planning, development, ongoing evaluation, reshaping) works and identify deadlocks, brakes to the urban market flexibility, countereffects. In the particular sense, that the primary objective of urban development has to be conducted for the flexibility of urban areas to be used by people who live, work, shop, move, and enjoy within the city boundaries.
4. Urban development and urban management should be more integrated with all what represents a large part of the GDP of a country: jobs, housing, transportation, shops location, and quality. Good leaving conditions, less transportation time wasted, working places easy reachable from homes. All this allow people to better live and better contribute to the GDP. Of course, that does not mean that housing should be on the same site of factories or office buildings!
5. Housing supply is a major contribution to help cities grow. But statistics in France and in the UK show that an increase in construction does not mean prices should diminish. Statistical elasticity is between −1 and −2 (source: Barker Report and CGEDD in France).

Governance and project management

6. Alongside with sustainable urban design, climate change mitigation, urban renewable, modern and green mobility, integrated urban development should organize "urban use" and coordinate all real estate private and public actors. Policy makers should make sure people who work don't have to cross the city to commute for example.
7. This means that policy makers should be now careful in planning the city by mixing real estate programmes: apartment, offices, shops, and even sometimes industrial sites, at least small ones.
8. Housing prices should be in some sense regulated in most of Metropolises, to avoid pure speculation. Affordable housing for the middle and lower income levels is a priority to make sure they are not forced to leave the city due to financial issues. It makes the urban area of the future, in a complete sustainable way.
9. Urban integrated development is backed also by a policy mix of macro-prudential banking measures, tax measures, and probably also land use and land management control to have the economic actors follow the government objectives. The policy model conducted in Singapore is for example worth knowing.
10. Land use and management remains essential. Land is a limited resource. Its future cannot only rely on private initiative which has usually a very short-term view. In some cases, land management should remain in the hands of public authorities. It might be more sustainable and easier to regulate construction rights allocation rather than selling urban land strategic areas to be developed.
11. Renewing obsolete or decaying area, while consolidating historic urban areas to maintain a different use of the city.
12. Urban planning and construction programme management should be optimized before developing major transportation infrastructures. Sometimes, it's too late,

and transportation should be monitored to try to ease the everyday life of commuting people. But the future of integrated urban planning is to optimize the development of the city to avoid lengthy commuting times.
13. Urban integrated planning should introduce a balanced density
14. The global real estate boom in metropolises and the soaring prices call for a transfer of the productive centres in a network of strategic urban partners. Local, regional, and national governments should organize better where activities could develop with a larger scope rather than being all at the same place.
15. The urban consequence of this should be, beyond affordable real estate (specifically residential, a higher level of life quality within cities: a higher level of green spaces, larger public spaces, available public facilities, cleaner air quality, and noise reduction.
16. A government must plan the development of satellite cities around metropolises in order to improve acceptable density and quality of the environment. Factories and businesses could be concentrated in employment areas accessible by public transport.
17. In the main city, the health system and public services for the people have to be improved. In some cases, the concentration of many super-projects simultaneously creating many high-rise buildings can disrupt the urban organization, including services to the public because of the influx of people. The COVID-19 crisis has brought up many subjects around the health policy in urban areas, but also on urban matters and urban programmes for the future. We are, yet, on the edge of a major change and a major shift in terms of definition of goals and perspectives in the future of the urban environment.
18. Indeed, urban planners and designers must carefully evaluate the risks that citizens will face such as traffic jams, lack of social services, pollution, health risks, and quality of living conditions… In every country, we should recall that the economic and financial profit for private developers should not overtake the human and community profit and interest.
19. Although the construction of public transportation is very essential, it is not sufficient itself. The problem is to logically determine the operational capacity, the intermodality between the different means (metro, bus, electric train, etc.). The stations have to be placed where people often go like the shopping centres, the university, the companies, offices areas, … Very fast transportation could be convenient if everybody could afford it and if it is relevant. It could be a solution against the bottleneck problems that the authorities are still struggling to find a way out.
20. Conventional ideas assume that proximate business is necessary to have the easiest connection to work together. But it is changing quickly, thanks to the fourth revolution. We can connect to each other thousands of kilometres away. In other words, the distance is not currently a very serious problem. In addition, the surrounding areas of metropolises have a lot of potential for economic development that we do not know as agriculture, real estate, tourism industry.

"Secondary cities"are also worth planning and developing

21. Metropolises are important. "Secondary cities" are extremely important to be enhanced and developed. They should be as much connected with metropolises, but also rely and focus on their best advantages to bring a competitive contribution to the GDP of the country. They are also a very serious link to remote areas and rural areas, very much essential to the GDP or just the current life and history of the country citizens.
22. A very much expanded idea worldwide is that large cities focus attention and public means, either political or financial. Governments should not forget to gather policy measures for the rest of the country outside metropolises. The French experience in very telling, we tried to show it in this article. Although a major very interesting programme has just been launched to try to counteract against urban devitalization, it has been after almost 40 years of no-policy for "secondary towns" and rural areas.
23. Urban policy action has to focus on urban dynamics gathering measures to boost urban quality, urban attractiveness, urban shops, urban housing, urban transportation, urban economics, and productive activities.
24. Urban integrated planning should move to a different modal to contribute more to a societal dynamics and nation spread equilibrium. The objectives are to improve urban quality of life in the whole nation and create the conditions for a better productivity of the country's economy.
25. Metropolises should contribute to develop their Hinterland, on which it could rely and reciprocally.
26. As to oppose to the "hyper-proximity syndrome" (being there and only there) a relevant answer, it looks necessary to give concrete tools for a decentralized model.
27. Spreading industrial and commercial activity not only in central metropolises, but throughout a larger regional spectrum, gives stronger sociological resilience, cheaper costs of production, and a competitive economical advantage. Secondary cities can play complementary roles to the main city. The Bordeaux study case shows that cities like Agen, Angoulême, or Libourne (around 20 to 100 kms), can activate a real partnership through back-of-house platforms, less qualified jobs but with larger real estate needs. This strategic partnership works if mobility solutions are found: fast train inter-city network, large infrastructure and digital connections.
28. Developing regional mobility system is the priority to enhance a network of cities
29. Public transport with proper network provides better regional coherence. Railway system is, in many European cities, a good testimony of a balanced regional development. The Swiss model, with its archipelago of small and medium-size cities, manages to attract and maintain economic actors, thanks to a high level of public transport and infrastructure.

5.6 Conclusion

The worldwide sanitary crisis of early 2020 caused by the COVID-19 pandemic has seriously affected the pace for urban growth and urban development as it used to be. Its economical, financial, social, and strategical consequences are both tremendous probably, but also unknown so far. The debate outraged in many major cities in Europe, in France, and also in the rest of the world, to try to gather ideas and directions to improve the daily life of inhabitants living in large cities.

Major city officials started to experiment different ideas and policy solutions, most of the time in a "rushing decision making process". This is the case of the length of bike lines that doubled in just a few weeks in Paris, to gain "a better city life". The major challenge in front of us is to contribute and manage the reduction of uncertainty and build more sustainable and integrated strategies for a more sustainable urban growth. Urban land use has to be optimized, together with planning of urban development and adequate transportation and economic growth in cities.

We strongly believe there is a serious need of rational reorientation in terms of city planning methods and urban policy towards growth. Many overaged techniques and strategies have to be questioned and reoriented to protect people and provide them with a better urban life and personal achievement.

This is the situation as of mid-September 2020, as we finalize this paper. At this point, we tried to gather experiences and ideas to contribute to a better and more sustainable urban pace of growth. We tried our best to draw a few (29) directions and recommendations to launch the debate with policy makers.

References

Commissariat Général à l'Egalité des Territoires (CGET), France: Rapport sur la Cohésion des territoires, La France après la crise: un état des lieux, (juillet 2018)

Conseil general de l'Ecologie et du Développement Durable. (2019). CGEDD, Ministère de la Cohésion des Territoires, Jacques Friggit: http://www.cgedd.developpement-durable.gouv.fr/IMG/pdf/elasticite-prix-immobilier-nombre_cle093f5d.pdf

Donzelot, J. (2011). Fondation Terra nova: Bordeaux Rive Droite: unevraieréussiteappararente, paper.

Fourquet, J. (2019). Fondation Jean Jaurès: "Pourquoi Bordeaux est-ellel'une des places fortes des "Gilets Jaunes"?, paper.

Ibid. (2019). http://www.cgedd.fr/prix-immobilier-presentation.pdf

Kate, B. (2004). *Review of housing supply, delivering stability: Securing our future housing needs.*

World Economic Forum. (2019). Making affordable housing a reality in cities, Insight report.

Chapter 6
Accessibility, Land Use, and Land Economics: A Factor Analysis Approach in Butwal Urban Area of Nepal

Chhabi Lal Chidi

Abstract Complex relationships of different urban features exist in an urban area. The objective of this study is to access the ecological pattern of urban features and their relationship. Butwal a historical town is selected as a study area which is located at the break of bulk point between Hill and Tarai region of Nepal. The factor analysis approach was applied to assess the dimension of the relationship between accessibility, urban land use, and urban land economics. Ten variables related to these features were derived from fifty randomly sampled patches. Only seven suitable variables were selected for factor analysis and two factors were derived from these variables. These two factors explain more than 65.61% of the total variance of the variables. Urban function intensity has the highest value of communality followed by land value and radial distance to the Central Business District (CBD). Factor 1 is mostly related to urban land use and radial distance to CBD. Similarly, urban economics and actual travel distance to CBD are mostly related to factor 2. However, room rent is also related to radial distance to CBD. It is concluded that urban land use pattern is mostly related to radial distance to CBD and land economics is mostly related to actual travel distance to CBD in the study area.

Keywords Accessibility · Central business district · Communality · Eigen values · Factor loading · Industrial intensity · Land economics · Marketing intensity · Radial distance · Travel distance · Urban features · Urban function

6.1 Introduction

Within the built environments of cities, human activity is primarily centred. Short-term individual decisions as well as long-term family and business decisions are significantly influenced by the location of these activity centres, particularly home, and work. (Turok & McGranahan, 2013). The patterns of land use in an urban region

C. L. Chidi (✉)
Central Department of Geography, Tribhuvan University, Kathmandu, Nepal
e-mail: chidichhabilal@gmail.com

B. R. K. Sinha (ed.), *Urban Dynamics, Environment and Health*,
https://doi.org/10.1007/978-981-99-5744-6_6

are greatly influenced by the nature of the transportation system. Land values, which are correlated with transportation costs, influence how urban land is used (Putman, 1975). The economic purpose of cities is to concentrate people and economic activity in nearby areas. The rent that is given for an urban site rises as its accessibility to popular locations does (Anas et al., 1998). Many economic sectors, including people, companies, industries, and governmental bodies, make up the urban economy. If the land market is allowed to operate as intended, the property will be occupied by the highest bidder, and the market will then decide how much land will be used for a given purpose and how intensely (McDonald, 2006). Spatial economics is not completely applied in any situation due to the variability of the spatial pattern of a physical and socio-cultural situation which can be defined as urban ecology. Urban studies are currently focused on the spatial organization and development of urban land use. For a greater understanding of the changes in urban spatial structure, complexity analysis of functional urban space evolution is highly valuable (Zhou et al., 2020). Analysis of urban land use and urban functions should be concerned with its specialized economic activities that identify suitable sites that provide them the greatest competitive advantage. This attention helps to understand the process of urban land use change and the pattern that offer the city the most competitive support (Harvey & Jowsey, 2003). Spatial variability of complex urban characteristics is commonly analysed by factorial ecology, which refers to various statistical methods using factor analysis that can synthesize several variables into a limited set of factors grounded in social theory are common (Nosoohi & Zeinal-Hamadani, 2011). The factor analysis approach is to study the differing urban ecology related to different factor combinations that can be arranged along a scale of urban development (Berry & Rees, 1969). Factor analysis depends on a large number of variables that can reduce it in an explanatory way to essential properties of a particular phenomenon of urban spaces (Gielen et al., 2017). It converts a large set of uncorrelated new variables that originally represented the majority of the information in field-derived data sets into a small set of variables that are much simpler to understand and use in subsequent analyses, making it simpler to implement effective urban land use planning (Lewis-Beck, 1994, pp. 15–20).

Urban land use planning is a rational planning process that utilizes the analysis and diagnosis of existing conditions to draw out the development needs, and land use requirements and translates them into development and land use goals and objectives (Metternicht, 2017, pp. 4–5). The Department of Urban Development and the municipal administration are in charge of planning the zoning of urban land use, however, they have very little knowledge of the land usage in many Nepalese municipalities. An unplanned urbanization process is becoming a major problem in developing countries like Nepal (Rimal, 2011, pp. 118–129). A study is on the functional ecology of municipal towns of Nepal using factor analysis (Mandal, 2005, pp. 45–57). There are very limited studies on the relationship of the various urban features under a single urban area except very few descriptive and overlay analyses. Therefore, the objective of this study is to access the ecological pattern of urban features and their relationship through the factor analysis approach of the Butwal urban area of Nepal. The output of this study will be useful information on the urban ecological pattern of various urban features of the urban area of Nepal.

6.2 Methods and Materials

This section includes the selection of the study area and its description, data source that has been collected from secondary sources and the field survey, data processing technique that was used for data management, and the use of a statistical tool like factor analysis to identify the relationship among different variables. Those have been described in the followings.

6.2.1 Selection of the Study Area

This study is limited to the urban area of Butwal Municipality of Rupandehi district of West Nepal. The geographical extension of Butwal Municipality is 83°22′52″ to 83°30′23″ east longitude and 27°40′00″ to 27°44′55″ north latitude having a total area of 69.28 km^2 (Fig. 6.1). However, this study is limited only to urbanized areas.

Butwal municipality is bordered by Palpa district to the north; Devdaha VDC to the east; Parroha VDC to the West; Salamar, Motipur, and Sankarpur VDCs to the south. This town is located at the foothill of Siwalik having slightly hot in summer and moderately cool in winter. The recorded temperature ranges from a minimum of 11.5 °C to a maximum of 35.9 °C. Butwal is crossed by East West (Mahendra) highway and Sunauli-Pokhara (Sidhartha) highway. This historical urban centre is located at the break of the bulk point of Tarai and the Hill area. So, it was developed as an urban centre in historical times.

Now link road and its location have helped for further growth of the urban area. The total population in Butwal Municipality is 118,462 with 29,662 households in 2011 (CBS, 2014) and most of them are immigrants from nearby hill districts, especially from Palpa, Arghakhanchi, Gulmi, and Syangja districts with a significant population of Tarai origin.

6.2.2 Data Source and Processing

Topographic maps of the Survey Department of the Government of Nepal (scale 1:25,000), Google Earth images, aerial photos, and city maps were used as basic information for this study. Different published and unpublished reports, data from Butwal Municipality Office, and other available sources were collected. Fifty sampled plots were selected randomly from the urban area of Butwal Municipality. The sampled

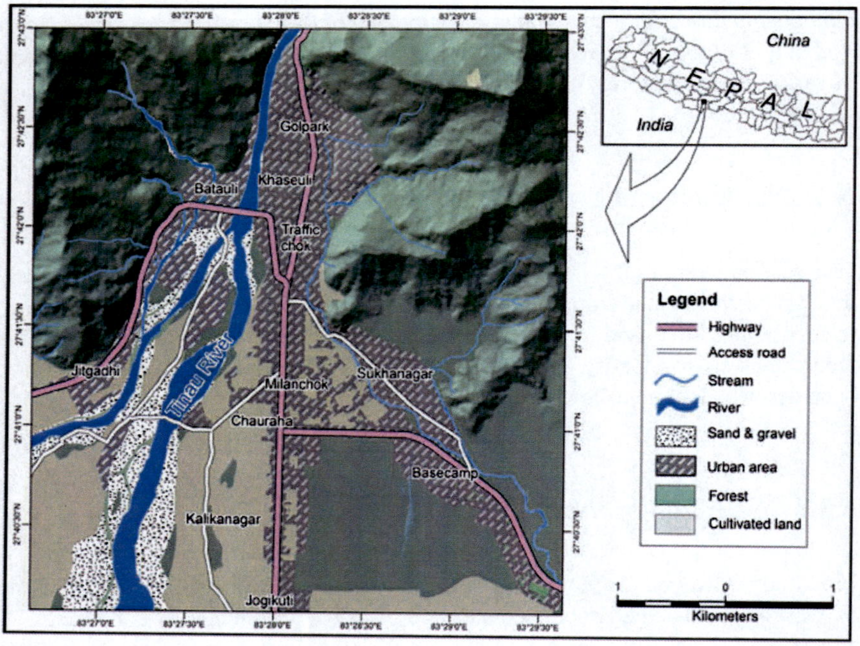

Fig. 6.1 The study area (Chidi, 2019)

area was the distinct unit area that is surrounded by streets, trails, fences, streams, etc. Plot level information such as land values land use; room rent; travel cost; and distance from road and CBD were collected from those sampled plots. Factor analysis requires a minimum of 50 observations. So, this study fixed 50 sampled plots for observation. These sample plots were selected based on snowball random sampling. The boundary of sampled plots was delineated based on unique patches separated by road, fence, wall, river, etc. So, the patch size varies with different locations. Observations were done with Google Earth images, maps, and aerial photos. Each sampled plot was visited with maps, Google Earth images, and a checklist to collect primary information from the field. Time, means, and cost of travel from CBD was collected while visiting each sampled plot. Information about the land value and room rent was collected through group discussions with local people at sampled plots. The information about differences in land value and room rent with its locational character was collected. Information on the use of rooms and floor for residential, commercial, industrial, etc., were collected through household interviews using a structured checklist from each building of the selected sampled patches. Room use of each floor and roadside/backside information were collected. Ten variables of different urban features are related to accessibility, urban function, and land economics. Required logarithmic and square root data transformation techniques were adopted for the normalization of variables.

6.2.3 Factor Analysis

Factor analysis helps to understand correlation and patterns by synthesizing data. It is used to combine a smaller number of variables (Yong & Pearce, 2013, pp. 79–94). In the mathematical model of factor analysis, p stands for the number of variables, and m for the number of underlying components. Hence, this model assumes that there are m underlying components and that each observed variable is a linear function of these factors plus a residual variate. The goal of this model is to replicate the maximum correlations.

$$X_j = a_{i1}F_1 + a_{i2}F_2 + \cdots + a_{ik}F_k + e_j$$

where, j = 1, 2, …, p.

The factor loadings are $a_{j1}, a_{j2}, \cdots + a_{jm}$ and a_{ij}, it means that a_{ij} is the factor loading of the jth variable on the first factor. The e_j stands for a particular and special component. The factor loading provides information about the contribution of the variable to the factor (Herman, 1976).

To explore the basic, the most fundamental dimension of various urban features and to examine their characteristics, factor analysis was applied by using ten different urban features of fifty patches. Derived variables of urban features are radial distance to CBD, distance to highway, travel distance CBD, land value, room rent, urban function intensity, building storey, marketing intensity, industrial intensity, and residential intensity.

Concerning the suitability of the data for factor analysis, the relatively smaller size of observations having a strong reliable correlation is also adequate (Tabachnick & Fidell, 1996). For the use of factor analysis, an empirical test shows that cases not less than 50 with correlation coefficients above 0.3 are preferable (Shoney & Pant, 1994). The correlation coefficient value of residential intensity was less than 0.3 and rejected at a 0.05 significance level. So, this variable was removed for factor analysis. Therefore, only nine variables were used for further analysis. Then, variables of distance to highway and building storey were rejected by the factor analysis process. Therefore, remaining only seven variables were used for the final analysis.

The Kaiser Meyer Olkin (KMO) measure of sample adequacy value of 0.72 demonstrates the acceptability for factor analysis, and Bartlett's test of sphericity was approved at a significance level of 0.01. Anti-image correlation matrix generated for sampling adequacy representing values more than 0.5 of diagonal and near to zero of off-diagonal elements qualifies the variables to be included in the analysis.

Principal Axis Factoring was selected to explore the basic dimensions of different urban features. Kaiser's Criterion and Scree test were used to assess the determination concerning the number of factors to be calculated and used in the model. Varimax orthogonal rotation was used to enhance the interpretability of the factor by minimizing the variables having high loading on a factor. The score of two factors for each urban feature was computed by multiplying the standard score values of variables by their corresponding regression factor score coefficients in which variance is equal to squared multiple correlations between the estimated factor score and the true factor values to analyse their spatial pattern.

6.3 Results

6.3.1 Distribution Pattern of Urban Features

In the central business district and its environs, land values are distributed more evenly. Even the highway intersection in the south seems identical. The historic Batauli region, located far east and far west of Butwal, has very poor land value. The north has land with a middling value (Khasyauli). Sukhanagar and Kalikanagar are those areas (Fig. 6.2a). Land value is significantly impacted by proximity to the CBD, highways, and access roads. Compared to the east-west highway, the north–south highway has a greater influence. The newly expanded southern portion has a higher land value than the northern portion. Moreover, Tinau river's influence has reduced land value. Road access tends to have a greater impact than the distance from the CBD.

Compared to land value, the distribution of room rent is more varied. The CBD region has the highest accommodation rent, while outside of this location, it significantly decreases. There is slightly less land value than in the CBD at the neighbouring north section and the bus park area to the south. Also, the land value along the highway heading south is higher than in other places (Fig. 6.2b). Lower room rent is available in recently expanded neighbourhoods including Kalikanagar, Sukhanagar, and Khasyauli. Rent in other locations is quite inexpensive. The distance to the CBD has the biggest influence on room rent, followed by access roads and highways. The impact of the old and new areas on land value is larger. Even close to the highway, the room rent in the Batauli area is relatively affordable. Being close to a highway, the Batauli neighbourhood boasts extremely affordable accommodation rates. Compared to the east-west route, the southwest highway has a bigger effect.

In comparison to land value and room rent, the typical building storey pattern is not consistent. The CBD area has the tallest skyscraper followed by the north of CBD, where the second highest buldings located. Throughout the place, the third building storey category has been dispersed at random. The third-category building storey place more accessible areas than other locations. Building storey height has greatly dropped outside of the city and along the road. It resembles both contemporary and historical urban regions (Fig. 6.2c). Distance from the city and road access, two important characteristics, have a bigger influence on building storey, but being close to the CBD has a greater impact than road access.

In general, but not exactly, the distribution of marketing function and building storey appears to be comparable. Outside of the city, the marketing function is more extensive. The CBD, its surroundings, and important intersections of roads are where

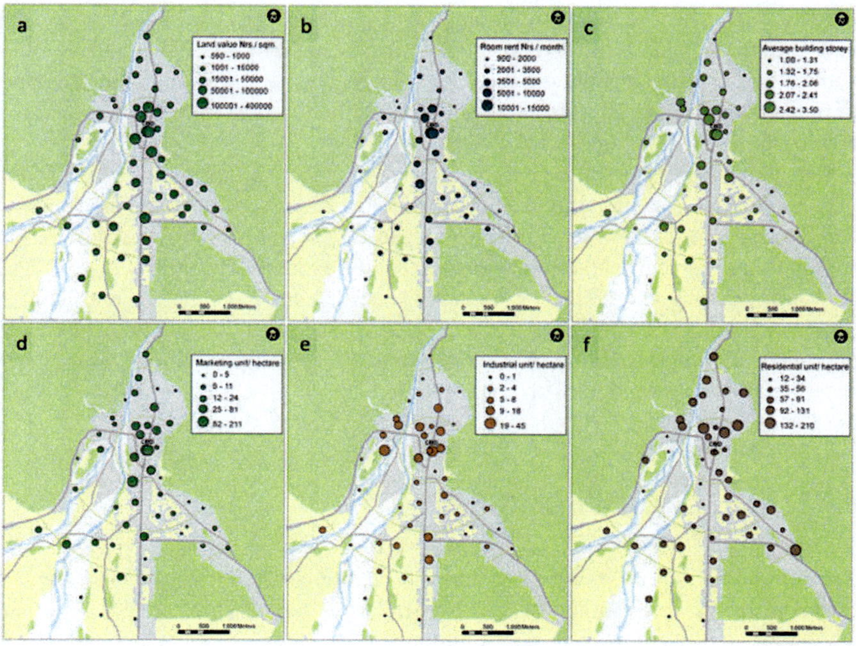

Fig. 6.2 Distribution pattern of urban features. *Source* Author's article in the Third Pole: Journal of Geography (Chidi, 2019)

the marketing activities concentrated the most. Outside of the main urban centre, there is hardly any urban activity. There are hardly any marketing operations in the recently expanded Kalikanagar and Sukhanagar regions and the old Batauli bazaar (Fig. 6.2d). The effect of distance from the CBD and road access on marketing activity is evident. Despite having a high level of accessibility, urban activity is quite low in major residential neighbourhoods. The north-south distribution of room rent shows some correlation when compared.

Around the CBD, industrial activity is highly concentrated. As the marketing function, there is much lower than in the eastern portion of the Tinau River, it has spread up to the west of the river. The southbound highway intersection is another concentration (Rajmargachauraha). Where the market function is similarly higher but fewer industrial operations are occurring along the southwest access route (Fig. 6.2e). Due to convenient access to the city and the highway junction, there are more industrial functions close by. Because more accessibility is needed than for other urban activities, there is a higher concentration at all highway intersections. The secondary road has attracted industrial activity as well, but as it gets close to the CBD, the impact of the highway junction becomes more noticeable.

Urban features are more closely clustered in a particular CBD area in terms of distribution pattern. Nonetheless, there isn't a fully monocentric area because access roads and highway connectors have attracted urban activities as well. In addition to playing a key effect in the distribution of urban features, physical barriers like hills and river basins (flood-affected areas) have also been important. There is less urban activity and more residential use outside of the larger metropolis in the little village of old Butwal. Due to physical constraints and limited access to the north, there is a dearth of interpersonal interaction.

The expansion of urban activity southward suggests that Butwal's urbanization process will move south rather than in another direction. It is due to the proximity to the Indian border and the size of the open space to the south. Along the more recent access road, more urban activities have recently expanded to the southwest. The choice of less expensive sites for residential uses is what has caused the residential area to expand outside of the city centre. The options for residential purposes are old city areas, marginal land, and recently expanded urban regions (urban fringe). Yet access to the city's centre is also crucial for it.

6.3.2 *Correlation Among Variables*

The correlation matrix shows Karl Pearson's coefficient of correlation value among seven variables (see Table 6.1). The correlation between urban function and marketing intensity has the highest and the least is between land value and industrial intensity. Travel distance to the city centre has a high relationship with many other urban features. Similarly, the urban function has also a similar relationship with others. The strength of the total relationship of urban function intensity seems higher than that of travel distance to CBD. Radial distance to CBD has weaker relation with other variables. The highest correlation between urban function and marketing intensity is because the marketing intensity itself is a major component of urban function (Ding & Zhao, 2014, pp. 83–90).

Although the urban function is the combination of marketing intensity, industrial intensity, and institutional intensity, the marketing intensity pattern has a major

Table 6.1 Matrix table of Pearson's coefficient of correlation among variables

Variables	LV	RR	UFI	MI	RDC	II	TDC
Land value (LV)	1	0.69	0.40	0.41	−0.31	0.17*	−0.56
Room rent (RR)	0.69	1	0.61	0.57	−0.34	0.43	−0.61
Urban function Intensity (UFI)	0.40	0.61	1	0.92	−0.51	0.67	−0.56
Marketing Intensity (MI)	0.41	0.57	0.92	1	−0.50	0.47	−0.47
Radial distance to CBD (RDC)	−0.31	−0.34	−0.51	−0.50	1	−0.50	0.50
Industrial intensity (II)	0.17*	0.43	0.67	0.47	−0.50	1	−0.52
Travel distance to the city centre (TDC)	−0.56	−0.61	−0.56	−0.47	0.50	−0.52	1

Note All correlation coefficient values are significant at the 0.01 level and * H_o $_{H0}$ is rejected at the 0.05 significance level. Calculated from the original field survey data dominant role in urban function intensity. Radial distance to CBD has a weaker role on the various urban feature. Instead, travel distance to CBD has a greater influence on urban land use and land economics.

6.3.3 The Variance Explained

Table 6.2 indicates the factor loading of seven urban features into two factors. These two factors can explain 65.61% of the total variance of variables. Initially, the first factor has 54.15% explanation strength but it has 38.92% after varimax rotation. Accordingly, factor 2 has a 26.69% of variance explanation capacity after rotation. The commonality value is the highest of urban function intensity followed by the land value whereas the radial distance to CBD has the least commonality value. Derived two factors can explain 97.3% of the variance of urban function intensity and the remaining unique variance 2.7% is explained by other variables. Similarly, these two factors can explain only a 36.7% variance of radial distance to CBD with a unique variance of 63.3% variance explained by others. The second most important variable is the land value which has a communality value of 0.878. It means two factors can explain 87.85% of the total variance in the unique variance is 12.2% explained by others.

Table 6.2 Factor loading (Unrotated and rotated) and estimated commonalities

Variables	Factor loading unrotated		Communalities	Factor loading rotated (varimax)	
	1	2		1	2
Urban function intensity	0.924	−0.345	0.973	0.947	0.275
Marketing intensity	0.802	−0.210	0.686	0.768	0.311
Industrial intensity	0.636	−0.289	0.488	0.683	0.149
Radial distance to CBD	−0.591	0.132	0.367	−0.553	−0.247
Land value	0.660	0.666	0.878	0.131	0.928
Room rent	0.762	0.276	0.657	0.446	0.677
Travel distance to CBD	−0.723	−0.137	0.542	−0.498	−0.542
Eigen values	3.790	0.802	4.592	2.724	1.868
Percent of the total variance	54.148	11.462	65.61	38.918	26.692

Source Calculated from the original field survey data

6.3.4 The Factor Loading

Factor loading is required to identify the strength of the relationships. Initial factor loading has indicated that factor loading of all variables is more than 0.5 in factor 1 and only land value is loaded at more than 0.35 in factor 2. Land value is highly loaded in both factors 1 and factor 2. Urban function intensity has the highest loading in factor 1 followed by marketing intensity. Since unrotated factors are unclear, factors are rotated for better understanding. The rotation aims to achieve an ideal basic structure that maximizes the number of high loadings on each variable while attempting to have each variable load on as few elements as possible (Rummel, 1970). After rotation, the factor loading of variables has changed. Variables of urban function intensity, marketing intensity, industrial intensity, and radial distance to CBD have been loaded to factor 1. These variables have very low loading in factor 2. Among them, urban function intensity has the highest loaded value. Land value is highly loaded in factor 2. Variables of room rent and travel distance to the city centre have been loaded in both factor 1 and factor 2. However, higher factors loading is in factor 2 (Table 6.3).

Here, it is clear that urban land use pattern, accessibility, and urban economics have their dimensional pattern. Urban land use is more related to radial distance to CBD rather than travel distance to CBD. However, the weak loading of radial distance to CBD indicates that it has not very strong role in determining urban land use patterns in the study area because it is also affected by topography and infrastructure (Ding & Zhao, 2014). Land economics is more related to travel distance to CBD. Land value is mostly related to travel distance to CBD rather than radial distance to CBD. Room rent is related to both radial distance to CBD and travels the distance to CBD but higher dependence on travel distance to CBD which was found in many urban studies (Jordaan et al., 2015, pp. 532–541). In addition to market influences, it is also affected by factors such as topography and infrastructure.

Table 6.3 Structure of dimension of factors: Factor 1 (Urban land use character and radial distance to CBD) and Factor 2 (Land economics and travel distance to CBD)

Variables	Factor 1		Factor 2	
	Unrotated	Rotated	Unrotated	Rotated
Urban function intensity	0.924	0.947		
Marketing intensity	0.802	0.768		
Industrial intensity	0.636	0.683		
Radial distance to CBD	−0.591	−0.553		
Land value	0.660		0.666	0.928
Room rent	0.762	0.446		0.677
Travel distance to CBD	−0.723	−0.498		−0.542

Source Calculated from the original field survey data

6.4 Discussions

The land use plan is meant to be a crucial tool for achieving the community's physical, economic, and social objectives. The plan has the potential to have a significant impact on the rate, nature, and quality of the city's physical environment expansion through its influence on public and private decisions and investments (Gallion & Eisner, 1984). The theory of urban land use became a crucial component of land economics in Butwal when the more direct statement of land use schemata was produced in the context of the land market and economics (Carter, 2010). The yield from the land for development and the costs associated with using the land will determine the rent that urban land users can afford. Hence, both improvements in yield and decreases in cost may lead to higher rents that urban land users can afford and a greater separation between their properties and the urban core. Due to the city's access road and highway connection to the southern Indian border, marketing, and industrial functions have been extended well south of the Core. The relative efficacy of different uses at various locations eventually determines how effectively land is used. Marketing and industrial activities are more concentrated close to the CBD as a result of the process of adjusting city form to the most effective land use pattern (Ritcliff, 1949). Yet, a variety of interconnections between economic, political, environmental, topographical, social, and cultural elements lead to specific patterns of urban expansion. When city populations change, these exchanges shift as well (Black & Henderson, 2003). There is a strong reaction to these mechanistic explanations because they show that abstracting a small set of economic variables and structuralizing them to act on the entire complex set affecting land use is an inadequate process. After all, most related economic variables do not fully determine how cities function and what the rent gradient looks like in the study area. The result of a very large number of distinct activity systems, each of which has unique characteristics and responds to a specific set of structures that make up a city, is the total land use area.

Even though Butwal's urban area is just a small town the urban land use evolution is similarly determined by urban infrastructures and economic activities as well. Urban economics in Butwal has been determined by the transport network and interaction of the people under the framework of overall control of landform, land cover, and disaster risk. The northern part of Butwal is the Siwalik Hill which is the barrier to the northside extension of urban function. There is the Tinau River just west side of the main urban areas and the north-south highway is on the east of the river. Thus, the urban economic activities are mostly concentrated to the east of the river. The old Butwal looks like a ghost village even though it is on the adjoining sides of the East-west highway. The situation is because of the less access to the other sides. It is now a residential place for low-class families. In general, west part of Butwal is not as urbanized as the eastern part but the extension on the eastern side is also blocked by forest areas. Thus, the highest extension of urbanization is to the south. It is not only because of the long-range connection of the transport network but also because of the physical barrier to the other side. This study revealed that urban land use, land economics, and urbanization are under the limitation of geographical condition, which develops under the local geographical features adjusting themselves to the local situation.

6.5 Conclusion

Accessibility, urban land use, and land economics pattern information derived from randomly selected patches of the Butwal urban area have revealed interrelated urban features. The factor analysis approach seems a useful tool for this analysis. The correlation matrix has given a pattern of relationship among the variables of five urban features relating to two types of accessibility i.e. radial distance to CBD and travel distance to CBD. It can be concluded that urban land use pattern such as urban functions, industrial intensity, and institutional intensity is related to both radial distance to CBD and travel distance to CBD but are more related to travel distance to CBD. The higher intensity of urban land use is nearer to CBD. In the context of land economics, land value is mostly related to travel distance to CBD rather than radial distance to CBD but some relations exist with radial distance to CBD also. Room rent is also more related to travel distance to CBD than the radial distance but some relation exists also with radial distance to CBD. Therefore, land economics is mostly related to actual accessibility and urban land use intensity is more related to radial distance to CBD. However, both types of spatial distance can not be neglected for patterning urban land use and land economics.

Some physical barriers like mountains to the north and river to the West and forests area to the east are major barriers to the regular extension of the urban area of Butwal. Thus, its urban land use and urban function are more developed towards the south along the highway. Although the west sides of the Butwal are open for urban extension the Tinauriver is a major barrier. Incomplete correlation among variables is because of the control of undefined variables in statistical analysis. Thus, it can be concluded that the relationship between the variables in such small towns is also similar to big metropolis cities although their scale is smaller. Although complex statistical analysis can address the complex relationship of different factors that determine land economics and land use, it is not a complete understanding of the complex urban environment that needs to explain undefined factors not included in statistical analysis.

References

Anas, A., Arnott, R., & Small, K. A. (1998). Urban spatial structure. *Journal of Economic Literature, 36*, 1426–1464.

Berry, B. I. L., & Bees, P. H. (1969). *Essays on commodity flow and spatial structure of Indian economy.* Department of Geography University of Chicago.

Black, D., & Henderson, V. (2003). Urban evolution in the USA. *Journal of Economic Geography, 3*(4), 343–372.
CBS. (2014). National Population and Housing Census 2011: Village Development Committee/ Municipality Rupandehi, Vol. 6. Central Bureau of Statistics (CBS), Kathmandu.
Carter, H. (2010). *The study of urban geography*. Edward Arnold.
Chidi, C. L. (2019). Bid-rent theory and urban land use of Butwal urban area, Western Nepal. *The Third Pole: Journal of Geography, 18–19*, 11–20. https://doi.org/10.3126/ttp.v18i0.27990
Ding, C., & Zhao, X. (2014). Landmarket, land development, and urban spatial structure in Beijing. In: LandUsePolicy (Vol. 40).
Gallion, A. B., & Eisner, S. (1984). *The urban pattern*. CBS Publisher and Distributors.
Gielen, E., Riutort-Mayol, G., Palencia-Jimenez, J. S., & Cantarino-Martí, I. (2017). An urban sprawl index-based multivariate and Bayesian factor analysis with the application at the municipality level in Valencia. *Environment and Planning B: Urban Analytics and City Science*, 1–27. https://doi.org/10.1177/2399808317690148
Harman, H. H. (1976). *Modern factor analysis (3rd edition revised)*. University of Chicago Press.
Harvey, J., & Jowsey, E. (2003). *Urban land economics* (6th ed.). Palgrave Macmillan.
Jordaan, A. C., Drost, B. E., & Makgata, M. A. (2015). Land value as a function of distance from the CBD: The case of the eastern suburbs of Pretoria. *South African Journal of Economic and Management Sciences (SAJEMS), 7*(3). https://doi.org/10.4102/sajems.v7i3.1363
Lewis-Beck, M. S. (1994). *Factor analysis and related techniques*. SAGE Inc.
Mandal, U. K. (2005). A functional ecological study of municipal towns of Nepal: A factor analysis approach. In M. Singh, B.K. K. C., S. Lohani, S.D. Joshi, R.D. Khatri, P. Chalise (Eds.), *Tribhuvan University Journal, XXV*(1).
McDonald, F. J. (2006). Urban transportation and land use. In: J.A. Richard & D.P. McMillen (Eds.), *A Companion to Urban Economics*. Blackwell Publishing Limited.
Metternicht, G. (2017). Land Use Planning (Global land outlook working paper). United Nations Convention to Combat Desertification.
Nosoohi, I., & Zeinal-Hamadani, A. (2011). Urban planning with the aid of factor analysis approach: The case of Isfahan Municipality. *Theoretical and Empirical Researches in Urban Management, 6*(1), 56–69.
Putman, S. H. (1975). Urban land use and transportation models: A state-of-the-art summary. *Transportation Research, 9*(2–3), 187–202. https://doi.org/10.1016/0041-1647(75)90056-8
Ratcliff, R. V. (1949). *Urban land economics*. Mcgraw-Hill.
Rimal, B. (2011). Urban growth and land use/land cover change of Pokhara Sub-Metropolitan City, Nepal. *Journal of Theoretical and Applied Information Technology, 26*(2).
Rummel, R. J. (1970). *Applied factor analysis*. Northwestern University Press.
Shoney, G. V., & Pant, M. (1994). *Statistical methods in business and social sciences*. New S. G.Wasani Macmillan India Limited, New Delhi.
Tabachnick, B. G., & Fidell, L. S. (1996). *Using multivariate statistics (3rd edition)*. Harper Collins.
Turok, I., McGranahan, G. (2013).Urbanization and economic growth: the arguments and evidence for Africa and Asia. *Environment & Urbanization, 25*(2), 465–482. International Institute for Environment and Development (IIED). https://doi.org/10.1177/0956247813490908
Yong, A. G., & Pearce, S. (2013). A beginner's guide to factor analysis: Focusing on exploratory factor analysis. *Tutorial in Quantitative Methods for Psychology, 9*(2).
Zhou, G., Li, C., Liu, Y., & Zhang, J. (2020). Complexity of functional urban spaces evolution in different aspects: Based on urban land use conversion. *Hindawi Complexity, 2020*, 9741203. https://doi.org/10.1155/2020/9741203

Chapter 7
Spatial and Temporal Changes of Land Use Pattern in Urban Fringe of Varanasi Metropolitan City, India: An Indicator of Urbanization

Chandan Kumar Kharwar and Braj Raj Kumar Sinha

Abstract Urban growth causes a significant change particularly in land use pattern of urban fringe. The use of land for different purposes such as agricultural practices, forestry, mining, industries, building construction, roads, etc. is commonly called as land use. Urban people generally intrude on the common urban fringe land and develop commercial, industrial, administrative, and residential complexes in an unplanned manner to serve their own interests. Unlike the core city, the fringe area is characterized by either poor structure or absence of the governing institutions. Therefore, haphazard development of newly built-up areas over space and time in fringe areas not only affects the land use pattern negatively by reducing the agricultural land but also creates a number of environmental problems that ultimately adversely impact health and quality of life of the inhabitants of both the fringe and City. This paper aims at dealing with the spatial and temporal changes in land use pattern of urban fringe of Varanasi city over the period of twenty years (1991–2011). For this secondary data were generated from Landsat Imageries of the respective year with the help of software like ArcGIS 10.5 and ERDAS IMAGINE 2014. Supervised classification with maximum likelihood classifier has been performed in ERDAS IMAGINE 2014 to identify the different land use categories such as built-up area, agricultural land, vegetation cover, water bodies, fallow, and waste land. The study shows that there has been a substantial decrease in the area of agricultural land, waste land, and water bodies owing to the conversion of most areas into built-up land and some proportion into fallow land and vegetation cover. All this is the result of a faster increase in the size of fringe area population owing to the dynamics of population growth and the demand of land for construction purposes to fulfil the requirements of the ongoing expansion of non-agricultural activities.

C. K. Kharwar (✉)
Department of Geography, Institute of Science, Banaras Hindu University, Varanasi, U.P. 221005, India
e-mail: chandangeobhu03@gmail.com

B. R. K. Sinha
Former Professor, Department of Geography, Institute of Science, Banaras Hindu University, Varanasi, U.P. 221005, India

B. R. K. Sinha (ed.), *Urban Dynamics, Environment and Health*,
https://doi.org/10.1007/978-981-99-5744-6_7

Keywords Land · Urbanization · Land use · Metropolitan area · Delineation · Agricultural land · Vegetation · Sustainable development

7.1 Introduction

Land is one of the most important natural resources from the perspective of overall development of a region. It covers only 29% of the total area of the earth but all is not favourable for the purpose of inhabitation of the human population. It is used to meet a variety of human needs in both urban and rural areas everywhere in the world. It is a major factor of production and plays a vital role in strengthening the socio-economic conditions of human society.

The use of land resources is very much related to the human activities which vary with the purpose it serves such as food production, provision of shelter, recreation, extraction and processing of materials and the biophysical characteristics of the land itself (Roy & Roy, 2010, p. 490). Land use is the arrangement, management, and modification of a natural environment into a cultural environment by anthropogenic activities. It is actually the product of complex interaction between the physical environment and socio-economic and cultural attributes of the society (Verma, 2017, pp. 32–33). Thus, the land use may be simply defined as the surficial use of land for the purpose of agricultural practices, forestry, mining, building houses, constructing roads, and setting up of industries. The use of land is determined by physical factors such as topography, soils, climate, minerals, availability of water, and human factors like anthropogenic activities and technology.

Today the demands of land are growing due to immense demographic pressure but the availability of land is limited. As a result of urbanization the continuous rapid growth of population has not only resulted in the change in the land use pattern of the cities but also in land use pattern of their peripheral areas or urban fringe. The land use pattern and change detection assessment have become a matter of great concern among scientists worldwide for realizing the importance of the land resource to achieve environmental security and sustainable development (Xiubin, 1996, pp. 553–558).

The urban fringe is a transitional zone between the outer boundary of the city and the countryside which is characterized by rapid land use change (Csatari et al., 2013, p. 153). According to Pryor (1968: p. 202) the term "Urban Fringe" was first used by a sociologist T. L. Smith in 1937 to signify the "built-up area just outside the corporate limits of the city" in the study of Louisiana, USA. It is characterized by regular spatial and temporal changes in terms of physical as well as cultural features depending upon the speed of urbanization beyond the main city, however, the degree of change in physical and cultural features in fringe area normally declines from the immediate city boundary to the countryside. Migration taking place from the city, as well as the countryside, plays a vital role in this direction because migration being one of the urban dynamics generally shapes the pattern of changes in the overall picture of the fringe area. However, the impact of migration determining the rate of expansion in urban characteristics also declines from the immediate city boundary to the countryside. All this happens because of the scarcity of space primarily for residential and commercial purposes within the main city area. Consequently, the area or belt developed around the city is called urban fringe and is characterized as a dynamic spatial unit.

The conversion of agricultural land, vegetation cover, wetland, barren land and water bodies of urban fringe into built-up area is a continuous phenomenon that takes place to meet the rising demand of infrastructure and industry for highly growing population in cities (Prasad & Singh, 2018, p. 191). The land lying along or adjacent to the accessible link roads beyond the main city is converted first, then the construction of new roads further helps in converting lands into built-up area in the fringe. Lal (1987: p. 98) considered accessibility, nature of the terrain, distance outside the city centre, increased transport facility, inefficient control and lack of coordination among the governing bodies as the major factors that normally accelerate the conversion of agricultural land into urban uses. Unplanned and haphazard way of conversion of land leads to imbalanced developments and creates several problems in urban fringe area and as a result there becomes a need for a sound and balanced planning for sustainable development of the fringe. From this perspective, the use of remote sensing techniques and satellite imageries can play a vital role in managing the careless conversion of fringe area land to built form. Prasad and Singh (2018, p. 191) were of the same opinion and stated that remote sensing and satellite imageries can be easily used in obtaining land use data for detecting the spatio-temporal pattern of land use changes and in getting a brief picture of their intense impacts on the fringe environment.

Like other global cities, Varanasi city has also experienced an unprecedented increase in population within the last few decades and has joined the million-plus group of cities. The million-plus cities are also called metropolitan cities.The term "metropolitan area" is applied to urban places with a population of one million and above. The census of India has defined a metropolitan as an urban agglomeration having a population of one million and above (Khullar, 1996, p. 430). The total population of Varanasi City as per the 2011 Census was 1,198,491, which is expected to reach 1,388,572 in 2022 (https://www.indiacensus.net/city/varanasi). In fact, the continuous rise of Varanasi City's population mounts pressure on its peripheral area, so-called urban fringe and becomes the reason for the conversion of fringe area productive lands into built-up land use category. Since 2011 the total population of Varanasi City already crossed a minimum threshold population of one million fixed to define a metropolitan City, Varanasi City can practically be called a Metropolitan City both for administrative as well as academic purposes.

7.1.1 *Land Use Change as an Indicator of Urbanization*

A regular transformation in the original forms of agricultural and other types of land into built-up form in the periphery of a main city for the purpose of developing the residential, industrial, commercial, transport, communication, administrative, and other secondary and tertiary activities with urban identities can be considered as an indicator of urbanization. This happens due to a gradual shift of population from the main city to its periphery caused by overcrowding, lack of open space and scarcity of social amenities within the main city. The rural–urban and urban-to-urban migration are also equally responsible for the change in the pattern of land use in the fringe area of a main city. Such process of transformation or change in land use in the periphery of a main city leads to give birth to distinct pattern of land use with urban identities. Briefly, the conversion of agricultural and other types of land of rural character into built-up form of urban nature in the periphery of a main city is an indicative of expansion of urban area. Changes in the pattern of a fringe area land use typically follow certain natural or cultural features. The conversion of rural nature of land into built-up form towards the urban character in fringe as well as in non-fringe areas initially takes place first along the already existing approaches (access roads/paths/passages) as well as along the approaches currently being developed. This process leads to creating a linear pattern of land use change and indicates a move towards linear growth of settlements with relatively dominant secondary and tertiary activities of urban identity. A similar pattern of land use change indicating urban features takes place at certain places along the rail roots, banks of a river and canal. Likewise, the fringe area also witnesses the emergence of urban characteristics around ponds, lakes, and reservoirs. In fact, land use and land cover change towards rural areas within the fringe of a city is an ongoing cultural process in the way of spatial, social, economic, political, demographic, environmental change, and urbanization. This process always involves the conversion of land from non-urban to the built up or urban category to be used for the development of new urban features and in turn changes the pattern of land use. The increasing intensity of use of land under secondary and tertiary economic functions and declining intensity of use of land in primary economic activities in the fringe area of a city regardless of its size obviously indicates the pace of urbanization. However, the intensity of land used in non-agricultural activities normally continues to decline from the adjacent outer boundary of the main city towards the countryside. Thus, the changing pattern of land use from rural or non-built nature to urban or built-up nature at the cost of agricultural and other common lands in the fringe of any city around the world indicates the process of urbanization. Therefore, the land use and land cover change in the fringe area of a city emerges as a novel subject of empirical research.

7.2 Study Area

The study area, "urban fringe of Varanasi City" as shown below (Fig. 7.1) is the same to "Rural-Urban Fringe" as employed in the Ph.D. Thesis titled "Transformation in Rural-Urban Fringe area of Varanasi City" submitted to the Banaras Hindu University, Varanasi (Kharwar, 2022). This is because of the fact that there is no difference

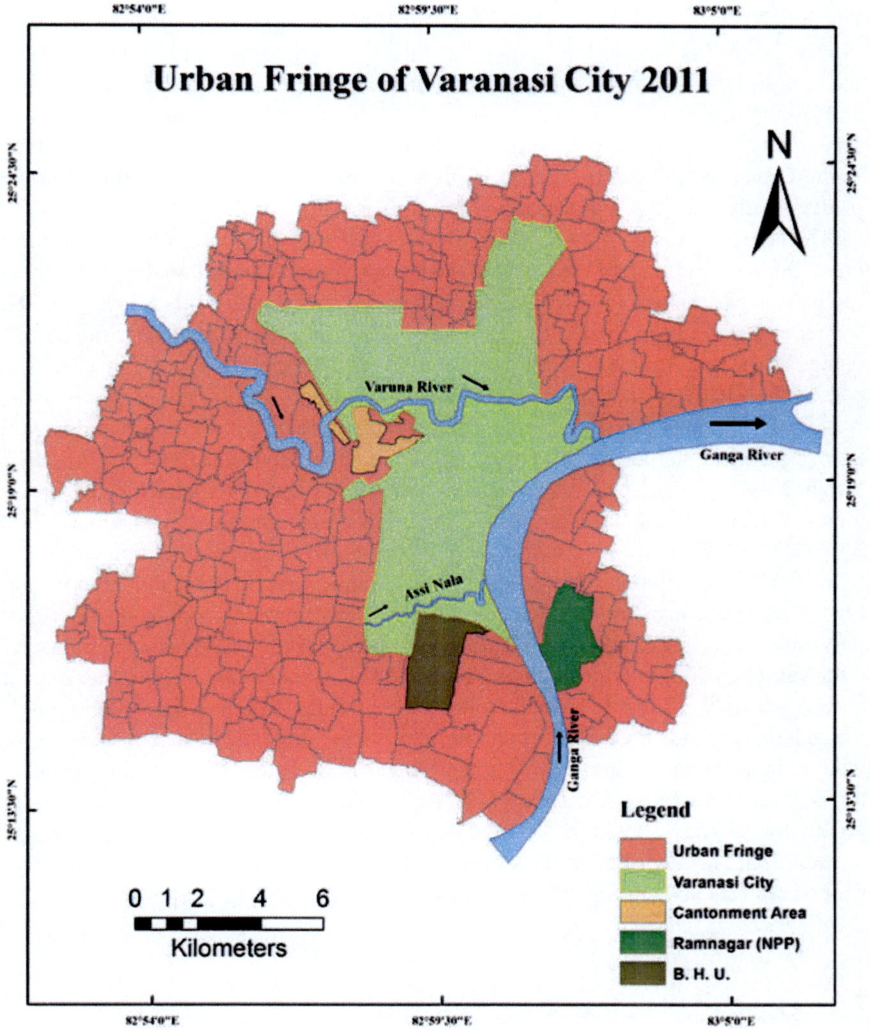

Fig. 7.1 Study area

between the area of the "Rural–Urban Fringe" and the area of "Urban Fringe". Both the terms, "Rural–Urban Fringe" and "Urban Fringe" are interchangeably used in most of the academic literature. Alpake (1942, pp. 463–478) has defined the urban fringe as "that cultural development that takes place outside the boundaries of central cities and extends to the areas of predominantly agricultural activities. Mayers and Beagle (1947) described urban fringe as "the zone between the country and the city". Firey (1946), Gist (1952), Fava (1956), Golledge (1959) and some others have used the term "fringe" instead of "rural-urban fringe" or "urban fringe". Similarly, Jutta K.

Dikshit (2011, p. 1) also considered the *"rural-urban fringe area" just equal to the "urban fringe"* or simply the *"fringe"* as this is a spatial segment that lies between the city and the vast rural countryside with their mixed characteristics of land use and population. From the above literature, it is quite clear that the terms "rural-urban fringe" or "urban fringe" are the same. Hence, keeping the above facts into mind the term "Urban fringe" has been used in place of the term "rural–urban fringe" in the present study.

The study area is an important circumferential zone of Varanasi city, the holiest city of India. This is in fact a segment of the Bhojpuri language-dominated Purvanchal region (the eastern part) of Uttar Pradesh, India. Geographically, this is located in the western margin of the middle Ganga Plain. The study area is drained by the holiest Ganga River and Varuna River. Population-wise and academically Varanasi City is one of the metropolitan cities of India as the total population of this city in 2011 was 1,198,491. It is one of the largest cities in the state of Uttar Pradesh and is well known for its numerous mystic Ghats and rich architectural heritage sites (Prasad & Singh, 2018, p. 186). The city is also called Banaras or Kashi. Urban fringe covers an area of 247.86 sq. km. and comprises 256 villages. Its latitudinal extension is from 25°11′33″ to 25°23′31″ N latitudes and its longitudinal extension is from 82°51′47″ to 83°5′52″ E longitudes. The study area falls in the region of sub-tropical monsoon type of climate consisting of three main seasons viz. summer (April to June), winter (December to March) and monsoon (July to October) (https://en.m.wikipedia.org/wiki/Varanasi). Temperature swings between 23 °C and 45 °C in the summer season and between 27 °C and 5 °C in the winter season (https://en.m.wikipedia.org/wiki/Varanasi). The study area receives an average annual rainfall of 1110 m (https://en.m.wikipedia.org/wiki/Varanasi). The urban fringe is characterized by mainly tropical mixed deciduous forest and sandy soil (DCH, Varanasi, 2011: p. 8). It has a total population of 765,480 with an average density of 3090 persons per km. sq. (Census of India, 2011). According to the Census of India, 2011, the average sex ratio in the fringe area was 904 and the literacy rate was 70.57%.

7.3 Objectives of the Study

The study aims to analyse the status of land use pattern in urban fringe of Varanasi City in 1991 and 2011 and to detect the overall changes in land use pattern between 1991 and 2011.

7.4 Methodology and Database

Research methodology is a way to systematically solve the research problem. In fact, it is a scientific approach that is used in performing research operations with certain logics, tools and techniques (Kothari & Garg, 2015, p. 6). To accomplish the above objectives the spatio-temporal data/information concerning the land use pattern of urban fringe of Varanasi city was generated from the Land Sat Images of the years 1991 and 2011 downloaded from the website (https://earthexplorer.usgs.gov). All the functions like mosaicking, stacking, clipping, and production of land use maps have been done with the help of software Erdas 10.5 and ArcGIS 10.5. Supervised classification with maximum likelihood classifier has been performed to delineate the different land use classes in the study area. The land use classes viz. agricultural land, built-up land, fallow land, waste land and water bodies have been identified in the study area. Apart from these, Microsoft Excel 2010 has also been used for different data calculations and the preparation of graphs to support the analysis.

7.5 Analysis

This section intensively deals with the spatial and temporal changes of land use pattern in urban fringe of Varanasi Metropolitan City in general and category wise such as built-up land, agricultural land, area under vegetation, fallow land, waste land, and area under water bodies in particular.

7.5.1 Changes in the Pattern of Land Use in Urban Fringe of Varanasi City (1991–2011)

The fringe area land use always remains under the impact of its main city and is visualized as a reflection of the dynamics of the concerned city, here it is Varanasi city. Generally, it is found that the greater the vicinity of a settlement with the city, the greater will be the diversification in fringe area land use. Apart from this, socio-cultural elements such as caste structure, the standard of education and the way of living jointly with physical elements like soil fertility, water supply and general character of land, play important roles in developing the types and intensity of land use (Lal, 1987: p. 110).

Continuous physical expansion of the Varanasi city outside its administrative boundary has affected the land use pattern of its fringe at a large scale as it is quite obvious from Table 7.1 and Figs. 7.2 and 7.3. This Table and Figures show a remarkable change in Land use patterns during the period of twenty years (1991–2011). Among the land use categories, the built-up area has registered the highest positive change of 28.52% followed by vegetation cover (3.41%) and fallow land (2.29%) whereas agricultural land has recorded highest negative change indicating loss of −33.78% followed by water bodies (−0.31%) and waste land (−0.13%).

The land use maps of the year 1991 and 2011 given in the Fig. 7.4 highlight the picture of spatio-temporal change in different land use categories of urban fringe of Varanasi City over the period of twenty years (1991–2011). The degree of land

Table 7.1 Changes in the pattern of land use in urban fringe of Varanasi City (1991–2011)

Land use categories	1991		2011		Change (1991–2011)		Percentage growth (1991–2011)
	Area in hectare	Percentage to total area	Area in hectare	Percentage to total area	Absolute change (in Ha.)	Percentage change	
Built-Up Land	7983.09	25.12	17,047.1	53.65	9064.01	28.52	113.5
Agricultural Land	19,335.3	60.85	8601.93	27.07	−10,733.4	−33.78	−55.51
Area under Vegetation	2575.53	8.10	3660.48	11.52	1084.95	3.41	42.12
Fallow Land	596.88	1.87	1326.87	4.17	729.99	2.29	122.30
Waste Land	483.75	1.52	440.19	1.38	−43.56	−0.13	−9.004
Area under Water Bodies	798.21	2.5	696.24	2.19	−101.97	−0.31	−12.77
Grand Total	31,772.76	100	31,772.81	100	–	–	–

Source Generated and calculated by the Authors from Land Sat Imageries of 1991 and 2011

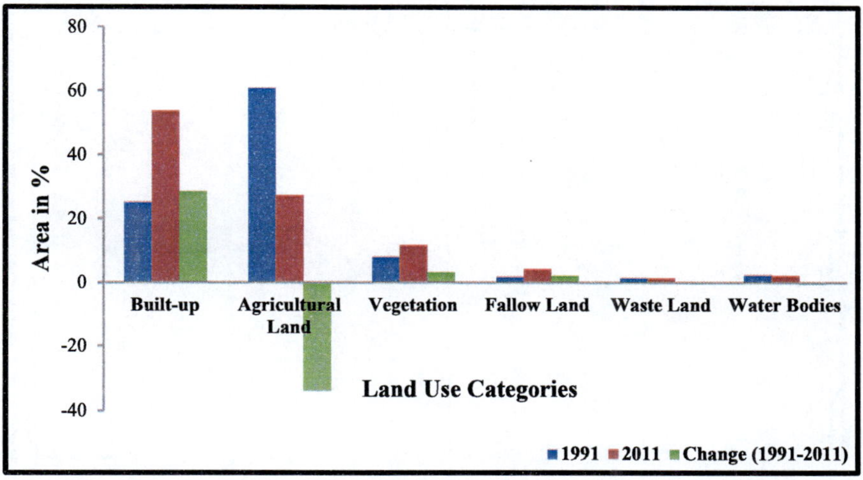

Fig. 7.2 Changes in area under different land use categories in urban fringe between 1991 and 2011(based on Table 7.1)

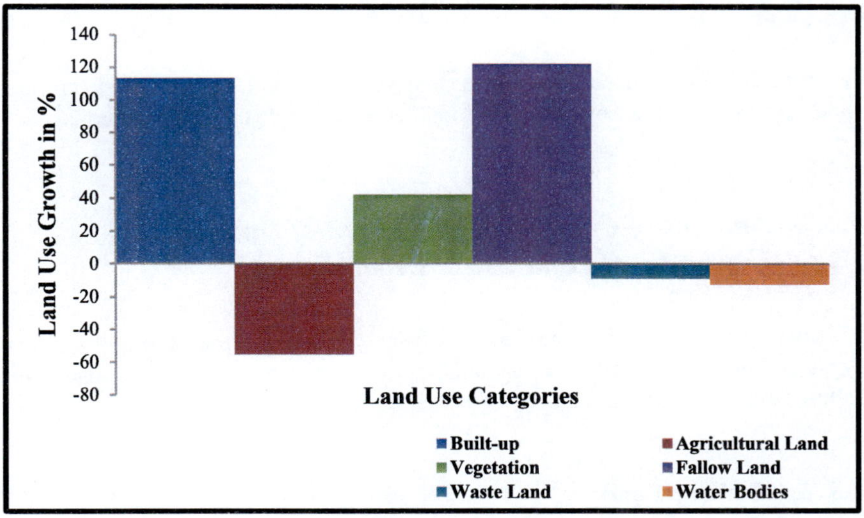

Fig. 7.3 Growth in area under different land use categories in urban fringe between 1991 and 2011(based on Table 7.1)

use changes is not the same in all parts of the fringe area rather it varies with the distance from the city. It indicates the degree of influence of the distance from the outer boundary of the City to the countryside. Due to urban expansion the fringe area closer to the main city experiences greater transformation in agricultural land into the built-up area. This is basically on the cost of precious agricultural land. The fringe

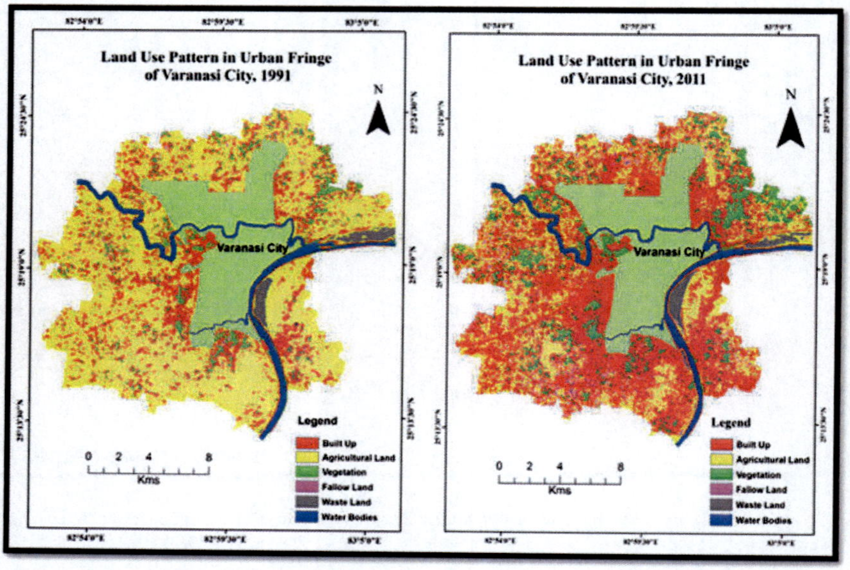

Fig. 7.4 Based on land sat imageries of 1991 and 2011

far from the city is characterized by extensive use of land because single cropped land of the past is now converted into double and multiple cropping patterns.

7.5.2 *Analysis of Changes in the Pattern of Different Categories of Land Use in Urban Fringe*

This part concentrates on the in-depth analysis of changes in the pattern of different types of land use namely built-up land, agricultural land, area under vegetation, fallow land, wasteland and area under water bodies.

7.5.2.1 Built-Up Land

The land which is covered with concrete features is called built-up land. The built-up land generally includes the land which is used for residential, industrial, commercial, transportation and other urban services (https://www.sciencedirect.com/topics/earth-and-planetary-sciences/land-use-classification). These examples of built-up land are commonly found in urban areas, that is why, the built-up land is also called as urban land. The high proportion of built-up land to total land area is generally the character of urban areas. Therefore, the maximum area of cities around the globe is found covered under the built-up land. The physical expansion of cities towards

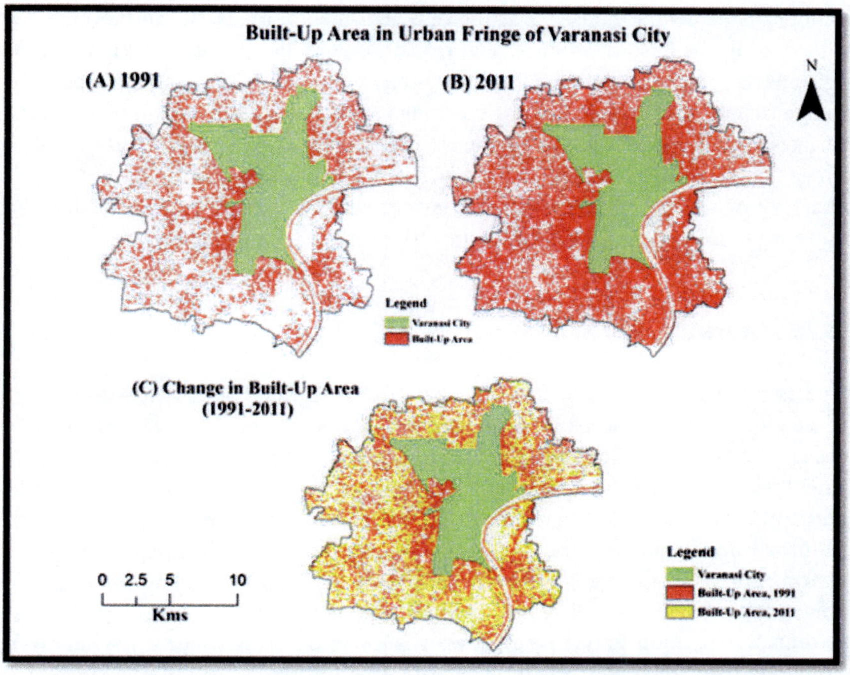

Fig. 7.5 Based on land sat imageries of 1991 and 2011

their peripheries is the main reason for this. This process leads to the conversion of non-urban areas to urban or built-up areas and makes changes in the pattern of fringe area land use. The status of built-up land in urban fringe of Varanasi city for the years 1991 and 2011 has been shown in Fig. 7.5 along with the changes occurred in area of built-up land during this period. In map (A) built-up land of urban fringe has been shown for the year 1991 while in map (B) for the year 2011. In both the maps built-up land has been illustrated in red colour. It is clear from the comparison of both maps that the area under white patches signifying for land used under other than built-up has dramatically decreased in map B only due to the conversion of chiefly agricultural land into built-up land. It shows a sharp increase in built-up land in 2011. To get a clear picture of changes in the spatial pattern of built-up land between both the years a superimposition of the maps A and B has been made to represent the map C in which built-up land of 1991 has been shown in the same red colour but the built-up land of 2011 has been displayed by yellow colour. This yellow colour representing built-up land in map C shows an increased area of built-up land between 1991 and 2011.

In absolute sense, 25.12% (7983.09 ha) to total area of urban fringe in 1991 was under the built-up category which increased to 53.65% (17,047.1 ha) in 2011 (Table 7.1). It means, over the period of twenty years there has been an increase of 28.52% (9064.01 ha) in the category of built-up land due to development of a large number of residential houses, commercial apartments, roads etc. in the periphery of the city at the cost of productive agricultural land and precious wetlands. It has registered a growth rate of 56.75% during the two decades. Thus, an unprecedented change in built-up land category of the fringe area is mainly the result of the flux of the fast-growing population of Varanasi City towards its periphery. But the rate of conversion of non-urban land into urban land is higher in the vicinity of the outer boundary of the city than that of the farther area. It is only because of the road transportation facility and smooth connectivity to the core of the city.

7.5.2.2 Agricultural Land

Land is a basic requirement for all kinds of development as it is associated with the needs of the people. Ever increasing population continues to mount immense pressure on land for different purposes and leads to diminish especially the precious agricultural land across different regions of the world. Agricultural land is the main source of production of foods to meet the increasing demand of people globally. A worldwide decline in the area of agricultural land due to urbanization particularly in the fringe areas of the cities compels the people to go for practising the double or triple cropping systems with high quantity use of chemicals and fertilizers to meets the demands of food grains of a growing population. This process has increased worldwide risks to human health by deteriorating the quality of the environment.

Here the maps (A), (B) and (C) as shown in Fig. 7.6 depict the spatial status of agricultural land in the urban fringe of Varanasi city for the year 1991, 2011 and a status of change during the period 1991–2011, respectively. A comparison between map (A) and (B) clearly shows a substantial decline in the proportion of fringe area's agricultural land in 2011, which is quite evident from map (B) where the area under yellow colour has reduced and area under white patches has increased. An increase in the patches of white colour in map B shows a loss of agricultural land in urban fringe of Varanasi City over the period of 20 years only on account of the conversion of agricultural land into built-up land.

To make the picture of above change clearer to visualize in the spatial context, the map (A) has been overlapped by map (B) to produce the picture on Map (C). In map (C) the colour green has been used to show the agricultural land of 2011 (which was earlier shown by yellow colour), while the colour yellow has been used to show the agricultural land of 1991 (which was earlier also shown by the same yellow colour). Thus, the yellow colour in map (C) represents such area which was the agricultural land in 1991, but in 2011 it was transformed into urban land for the use of different purposes. Thus, it is obvious from Fig. 7.6 that urban expansion has adversely affected agricultural land by converting this into built-up land by regular transformation in the urban fringe of Varanasi city. The picture of conversion of agricultural land into urban nature of land has not been the same throughout the urban fringe rather it has been predominantly more along the road transportation routes connecting the city.

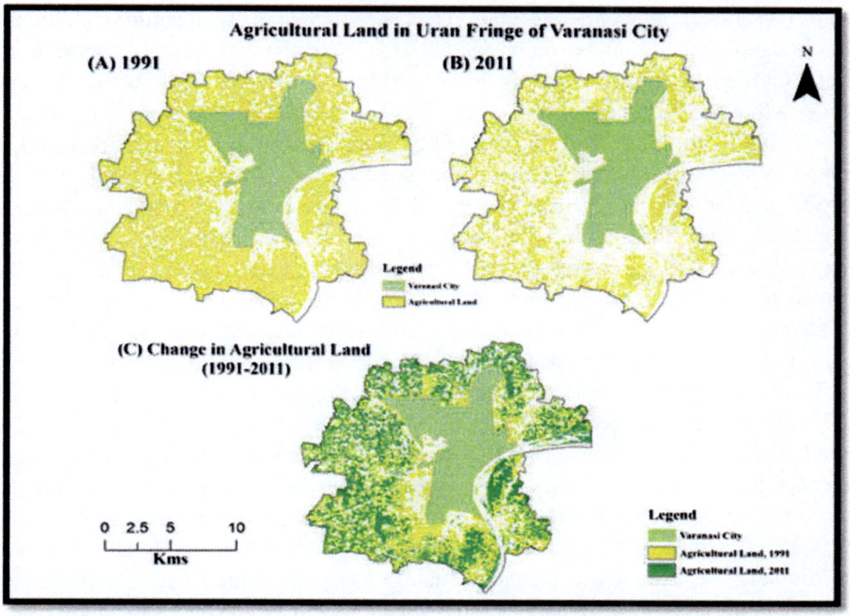

Fig. 7.6 Based on land sat imageries of 1991 and 2011

Briefly, the statistical data as shown in Table 7.1 evidently exhibits that 60.85% (19,335.3 ha) of the total area of urban fringe was agricultural in 1991 which decreased to 27.07% (8601.93 ha) in 2011. This shows a change of 33.78% (10,733.4 ha) between 1991 and 2011. But in terms of percentage growth between 1991 and 2011 the agricultural land recorded a sharp decline −55.51%.

7.5.2.3 Area Under Vegetation

Vegetation is an important natural component of the earth which makes the physical environment healthy to support human health and quality of life. But, the rapid urbanization and industrialization over space and time has influenced the vegetation negatively and has led to deforestation at a large scale. Consequent upon which, a variety of environmental problems has emerged. Globally most of the cities and their peripheries experience a decline in the vegetation cover and due to this the world cities face several problems. The scarcity of vegetation remains a challenging issue for the twenty-first century cities. However, in the case of the fringe area of Varanasi City, the picture is to some extent different which is clear from the following description.

Figure 7.7 containing map (A), map (B), and map (C) demonstrates the status of the area under vegetation in urban fringe of Varanasi City for the year 1991 and 2011 as it has been shown in map (A) and map (B) respectively. In the first two maps (A) and (B) area under vegetation has been shown in green colour while white patches indicate area under other categories of land use like built-up, agriculture, water bodies etc. After comparing these two maps it becomes obvious that the area under white patches decreased in 2011. This points out to slight increase in the area (especially in north east and east regions of urban fringe) under vegetation cover. It connotes that the area under vegetation has slightly increased in urban fringe over the period of twenty years. In order to get a clear picture of a real expansion under the vegetation cover, the map (C) has been prepared by superimposing the map (A) on map (B). In this map green colour has been used for to show the area under vegetation for the year 1991 and red for the year 2011 so that the variation in the spatial pattern of an increase in the area of vegetation may be easily detected. Thus, the red patches convey the extended area under vegetation between 1991 and 2011.

Table 7.1 depicts that area covered by vegetation (gardens and groves) was 8.10% (2575.53 ha) to total geographical area of the urban fringe in 1991 and 11.52% (3660.48 ha) in 2011. It means the area under vegetation has increased to some extent over the period of twenty years. It has registered an absolute change of 1084.95 ha and percentage change of 3.41. But in terms of percentage growth between 1991 and 2011 the area under vegetation cover recorded an increase of 42.12%. Such increase in this land use category is observed mainly along the bank of River Ganga and Varuna, in the campus of Banaras Hindu University, DLW and Railway Cantonment

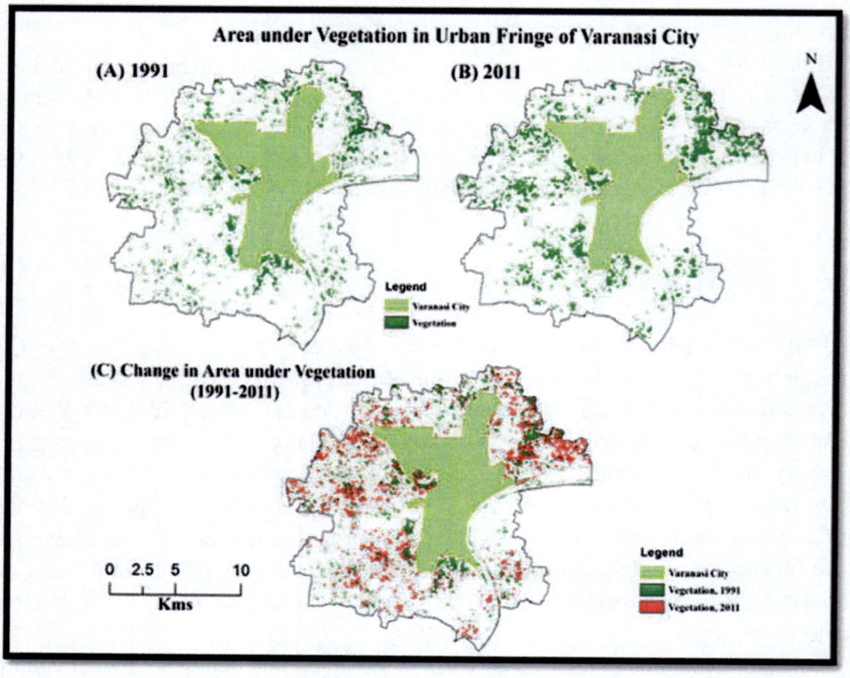

Fig. 7.7 Based on land sat imageries of 1991 and 2011

areas. Thus, it is clear from the above that in the case of urban fringe of Varanasi city the situation is different from the fringe areas of other cities because the urban fringe of Varanasi city has registered a positive change in area under vegetation. This is a healthy sign of a worldwide movement towards the concept of a green city.

7.5.2.4 Fallow Land

Fallow land is a type of arable land that is left without sowing for one or more cropping year (https://en.m.wikipedia.org/wiki/fallow) with a view to allow fallow land to recover its fertility by restoring organic matter (https://en.m.wikipedia.org/wiki/fallow). As far as fallow land in urban fringe of Varanasi city is concerned it has recorded positive change in its covered area in the last few decades. Table 7.1 reveals that in 1991, the area under fallow land to the total calculated area of the fringe was 1.87% (596.88 ha) which increased to 1326.87 ha (4.17%) in 2011 by experiencing a change of 2.29% and in absolute sense 729.99 ha over the period of two decades. This may be because of the fact that several arable plots have already been encircled by an ongoing process of development of settlements where such arable plots generally remain either water lodged or there is no facility of irrigation to grow crops.

The spatial distribution of fallow land and its changes has been demonstrated in Fig. 7.8. There are three maps which have been shown in this figure map A, map B, and map C. In maps A and B, the spatial status of fallow land in urban fringe has been displayed for the years 1991 and 2011 respectively by using the red colour, while in map C change in the area of fallow land between 1991 and 2011 has been depicted by colours red for the year 1991 and blue for 2011. The area under fallow land in urban fringe has increased from 1991 to 2011 as it is obvious from maps A and B where the area under red patches has increased from map A to map B. In other words, the area under white patches which represents the area under other land use categories has decreased from map A to map B. It means some areas of white patches have been converted into fallow land over the period of time. This change in fallow land may also be understood clearly from map C where the map (B) of 2011 fallow land has been overlapped by the 1991 fallow land map (A). Map C showing the area in red colour is the same as it was used in 2011 to represent the fallow land, but for the year 1991 the colour of fallow land has been changed from red to blue to show at a glance the spatial changes occurred in the area under fallow land between 1991 and 2011. Thus, the blue patches represent such areas that were not under fallow land in 1991 but in 2011 they have undergone fallow land due to certain reasons.

7.5.2.5 Waste Land

Waste land is generally considered as that portion of land of an area on which nothing can be grown and is no longer in use from a productive angle. But according to

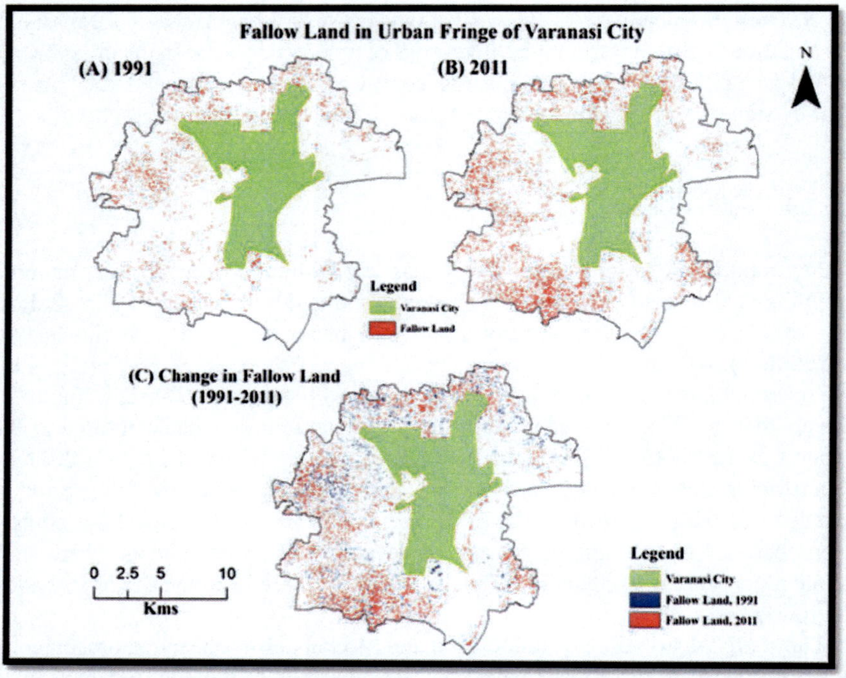

Fig. 7.8 Based on land sat imageries of 1991 and 2011

the website (https://www.yourarticlelibrary.com/wasteland/wasteland-definition-classification-and-significance-of-wasteland/29339) waste land is that land which has less than 20% of economic potential and can be made productive through anthropogenic activities. In the category of wasteland drought hit and overgrazed pasture lands, denuded forests, hilly slopes, eroded valleys, waterlogged, and marshy lands, and uncultivated land etc. are included.

Table 7.1 depicts that in the urban fringe of Varanasi City the area of waste land has decreased from 1991 to 2011. The area under waste land in 1991 was 1.52% (483.75 ha) which decreased to 1.38% (440.19 ha) in 2011 showing a negative change of 0.13% and a growth rate of −9.004%. The growing urbanization and resultant demands of new land for the purposes other than agriculture in urban fringe has been the major cause for this decrease.

Since the area under waste land in urban fringe of Varanasi City is quite less and is generally found along the River Ganga, hence no map has been prepared to show the spatial distribution of waste land. Yet, the spatial distribution of waste land in 1991 and 2011 may be seen and compared to each other in order to understand the changes occurred in the area during the period of twenty years (Fig. 7.4). It is only because the urban fringe is totally situated in the Plain of Ganga River where the soil is fertile.

7.5.2.6 Water Bodies

Water body is one of the important categories of land use and land cover and is a very significant dimension from the point of view of water resource for an urban fringe as well as for the main City. In this category, ponds, lakes, reservoirs, canals, streams, estuaries, bays, etc. are included (https://www.sciencedirect.com/topics/earth-and-planetary-sciences/land-use-classification).

For the last few decades, the water bodies in urban fringe of Varanasi City have decreased in number and size due to heavy pressure of outflow of excess population from the main city. This happens owing to rapid urbanization and commercialization taking place in the surroundings of the city and posed several environmental problems. The statistical data displayed in Table 7.1 makes it clear that the area under water bodies has declined over the period of two decades by −0.31%. The area under water bodies was 2.5% (798.21 ha) in 1991 but in 2011 it declined to 2.19%. The percentage growth (change) in the area of water bodies over the period of twenty years (1991–2011) was found −12.77%. Over this period of time this category of land use has not only decreased in terms of covered area due to the unprecedented and unplanned growth of the city but also got polluted because of the use of chemicals and pesticides in urban agriculture and horticulture and disposal of urban wastes directly into various water bodies. Since, the spatial distribution of water bodies in urban fringe has already been shown in Fig. 7.4. Hence, here a separate map in this regard has not been prepared. However, this figure shows that the water bodies in urban fringe do exist only in the form of rivers namely the Ganga and the Varuna. The comparison of both maps (Fig. 7.4) of 1991 and 2011 provides us with an idea about a mite reduction in the width of the middle portion of River Ganga indicating a small decline in the area under water bodies owing to a slight expansion of sandy waste land to the right bank. However, the width of this river has slightly increased in the northern part of the river just after meandering to the eastward. Overall reduction in the area under water bodies in the study area is not a good sign from the perspective of water resources and historical significance of the world-renowned culturally heritage City of Varanasi.

7.6 Conclusions

The above study shows a substantial change in the pattern of land use in the urban fringe of Varanasi City between 1991 and 2011. Obviously, there has been a considerable decline amounting to 33.78% in area under agricultural land. Likewise, other land use categories such as waste land and water bodies have also witnessed a decline in their areas due to a regular conversion of their land into built-up land of urban nature. Among the land use categories of positive change built-up land is the most important which has enormously increased by 28.52% for the use in different urban activities. This is followed by a slight increase in the area under vegetation and fallow land. The entire change in land use pattern of the urban fringe is attributed to the rapid population growth in Varanasi City as well as in urban fringe by human mobility. This factor has played a vital role in the expansion of built-up area at the cost of agricultural and other important common lands. A reduction in area under agriculture is a matter of great concern locally, regionally and globally as agriculture in the only source of food grains and plays a key role in the economy of the concerned area. This provides employment opportunities to a large portion of population and raw materials to the industrial sector. Regular transformation of land in urban fringe also causes environmental problems such as soil degradation, air pollution, water pollution, noise pollution and other urban problems. In such conditions, an ongoing decline in agricultural land will lead to problems of food security and employment in the long future. The conversion of agricultural land through an ongoing process of ever-increasing urban fringe across the world must be taken into account as a great threat to the human society.

The total population of the existing area of urban fringe of Varanasi city in 2011 was 765,480 which has been estimated to reach 1,706,604 by 2051 if the same trend and pattern of natural and mechanical exponential population growth rate of 0.041% between 19,991 and 2011 will remain unchanged till 2051. This is based on the personal calculation taking the 1991 Census population as the base year and the 2011 Census population as the next year. This scenario will further mount excess pressure on the earth by increasing several social, cultural, economic, demographic, political, and health cum-environmental problems to the survival of the global human society. This huge increase in future population will definitely put tremendous pressure on the remaining agricultural land of urban fringe area. It means the problems induced by decreasing agricultural land will be worse in the coming future. Thus, the estimated population is an indicative of increasing pressure upon limited resources, widening gap between the rich and the poor, poverty among the people, social crimes, communicable and non-communicable diseases, shortage of food etc.

Looking into the continuously increasing problems caused by rapid expansion of cities in their peripheral areas so-called peripheries need to be meticulously taken into consideration for concerted planning and management of land use pattern of the urban fringe. For this, the involvement of efficient scientists, planners, administrators, and academicians and their role in making the plans of sustainable development of the main City and its fringe area successful would be very important. The planning of the fringe area must be on two levels. First should be in the direction of controlling rapid growth of population in urban fringe and second should be in the direction of systematization of haphazard conversion of agricultural land to the built-up land for the development of different urban features with a hope to maintain urban environmental sustainability and urban health for the present and future urban population. As a matter of fact, planning for the fringe area is actually an urban planning of the main city. The government must strictly and honestly follow and implement all the Facts regarding uses of Urban Land (Ceiling and Regulation) as passed in the year 1976 by the government of India for the distribution of land in urban agglomerations with a view to equitable distribution of capitals in the direction of balanced development for the well-being of the present and future urban population.

References

Alpake, F. (1942). Landuse control in the urban Fringe of Portland Oregon. *Journal of Land and Public Utility Economics, 18*, 463–78.

Census of India. (2011). District Census Handbook, Series-10, Part-A, Varanasi.

Csatari, B., Farkas, J. Z., & Lennert, J. (2013). Land use changes in rural-urban fringe of Kecskemet after the economic transition. *Journal of Settlements and Spatial Planning, 4*(2), 153–159.

Dikshit, K. R. (2011). The city and the urban fringe: An introductory note. In J. K. Dikshit (Ed.), *The urban fringe of Indian cities* (p. 1). Rawat publications.

Directorate of Census Operations. Uttar Pradesh, Lucknow, pp. 1–15.

Fava, S. F. (1956). Suburbanism as a way of life. *American Sociological Review, 37*.

Firey, W. (1946). Social aspects to land use planning in the country-city fringe: The case of flint, Michigan. Agricultural Experiment Station, Michigan State College, Special Bulletin 339.

Gist, N. P. (1952). Ecological decentralization and rural-urban relationships. *Rural Sociology*, 328–335.

Golledge, R. (1959). Sydney's metropolitan fringe: A study in rural-urban relations. Australian Geographer, 243–255.

Kharwar, C. H. (2022). Transformation in rural-urban fringe area of Varanasi city. Unpublished Ph.D. Thesis, Department of Geography, Banaras Hindu University, Varanasi, 63.

Khullar, D. R. (1996). *India a comprehensive geography*. Kalyani Publications.

Kothari, C. R., & Garg, G. (2015). *Research Methodology Methods and Techniques* (3rd ed.). New Age International (P) Limited.

Lal, H. (1987). *City and urban fringe. A case study of Bareilly*. Concept Publishing Company.

Myres, R. B., & Beagle, J. A. (1947). Delimitation and analysis of the rural urban fringe. *Applied Anthropology, 6*, 14–22.

Prasad, S., & Singh, R. B. (2018). Urbanization and land use/land cover change in Varanasi city. *National Geographical Journal of India, 64*(1–2), 185–197.

Pryor, R. J. (1968). *Defining the Rural-Urban Fringe. Source, Social Forces, 47*(2), 202–215. Oxford University Press.

Roy, P. S., & Roy, A. (2010). Land use and land cover change: a remote sensing & GIS perspective. *Journal of Indian Institute of Science, 90*(4), 489–501.

Verma, N. (2017). The study of land use/land cover in ChunarTahsil, Mirjapur District using landsat data. *National Geographical Journal of India, 63*(3), 32–38.

Xiubin, L. (1996). *A review of the international researches on land use/land cover change, ActaGeographicaSinica* (pp. 553–558).

Website

https://earthexplorer.usgs.gov

https://en.m.wikipedia.org/wiki/fallow

https://en.m.wikipedia.org/wiki/urban_land_(ceiling_and_regulation)_act,_1976)

https://en.m.wikipedia.org/wiki/Varanasi

https://www.indiacensus.net/city/varanasi

https://www.sciencedirect.com/topics/earth-and-planetary-sciences/land-use-classification

https://www.yourarticlelibrary.com/wasteland/wasteland-definition-classification-and-significance-of-wasteland/29339

Chapter 8
An Evidence-Based Study on Locational Attributes, Space Use, and Operational Characteristics of Online Shopping in Dhaka City Corporation Areas

Mamata Jafrin Mouli, Ishrat Islam, and Marufa Khan

Abstract Researchers on urban dynamics are constantly working on exploring nature and trends of global phenomenon which might impact the overall spatial structure of the geographical unit. The pattern of activity centres and peoples' interaction has a significant impact on social, economic, and environmental aspects; thus, online shopping as an emerging communication technology significantly impacts shaping the city's spatial structure. To date, there is scant empirical evidence to study the prospects of online shopping from the perspective of developing countries. Hence this study has been conducted on one of the fastest-growing cities of the world, Dhaka, the capital of Bangladesh. This study aims to explore the locational characteristics and operation system of the online shops as well as their potentials and challenges in the two city corporation areas of Dhaka. The research collected information through a questionnaire survey conducted on the selected 81 online shops between 2017 and 2018. The study analyses trends of growth and nature of products sold, office and storage facilities, delivery location, and product collection and delivery system of the surveyed shops. Locational attribute findings suggest that the size of office and storage space and their locations have some differences from traditional commercial activities and few neighbourhoods have more delivery requests, where these online shops have to drop most of their deliveries. In terms of mode use for product delivery, motorcycles have the largest share while providing service within the city area. For delivering outside the study area, courier service providing agencies with large cargo or trucks are major modes. In terms of product accumulation, most of the shops procure their products from abroad except for groceries. Finally, the providers' perspective outlined the potential and challenges while running the business. In order to attain the possible positive outcomes of e-commerce induced changes in the city

M. J. Mouli (✉) · I. Islam · M. Khan
Department of Urban and Regional Planning, Bangladesh University of Engineering and Technology (BUET), Dhaka, Bangladesh
e-mail: m.jafrin.mouli@gmail.com

I. Islam
e-mail: ishratislam@urp.buet.ac.bd

B. R. K. Sinha (ed.), *Urban Dynamics, Environment and Health*,
https://doi.org/10.1007/978-981-99-5744-6_8

dynamics, it is important to analyse the forces and undertake innovative and integrated policies, rules, and national guidelines for e-commerce related business to ensure sustainable growth and efficiency of the city in the long run.

Keywords Dhaka · E-commerce · E-CAB · Online shopping · Online shop providers · Product collection

8.1 Introduction

Cities and human civilization are constantly experiencing changes. Researchers working on urban dynamics view cities as a complex system comprised of social and economic phenomena shaped by individuals and groups to attain certain goals and comforts. In recent times, communication technology has become an inseparable part of everyday life, livelihood, and lifestyle. Online shopping is one of the most visible outcomes of communication technology because of its widespread effects on not only shoppers and retailers but also on shaping the structure of a city (Cao, 2009). Online shopping is generally referred to as searching and purchasing of goods or services on the internet by consumers (Bourlakis et al., 2008). As information about products is available on the internet, and shoppers do not have to go to stores physically, accessibility is no longer impeded by time and distance of travel. As a consequence of the removal of time and space constraints, the whole process of shopping has become more malleable and diverse, thereby impacting various aspects of city planning (Golob & Amelia, 2001).

With the spreading of e-commerce based online shopping, the world's cities are experiencing a new challenge in the existence of retail stores as these online shops need less space for showcasing and direct retailing of products (Nahiduzzaman et al., 2019). One of the important planning consequences would be the decrease of retail use concentration in the Central Business District (CBD) (Nahiduzzaman & Aldosary, 2012). It can even induce decentralization; thus, people of remote and suburban areas can access their desired products (Faraget al., 2006; Visser & Lanzendorf, 2004). Online shopping also has potential impacts on urban travel as well as land use pattern (Mokhatarian, 2004). Online shopping will likely replace and reduce personal shopping trips (Cervero, 2000). Several studies claim for an increase in freight traffic as personal shopping trips are replaced by truck deliveries (Cervero, 2000; Van Loon et al., 2015; Visser & Lanzendorf, 2004; Pettersson et al., 2018). These changes are likely to bring about further changes in the land price, energy consumptions, change in amount and design of space, and a host of other factors that may alter the entire structure of a city.

The burgeoning demand for online shopping can be clear from the fact that worldwide total online sales of goods and services were 2.2 trillion USD in 2015 (E-commerce Foundation, 2016). In 2019, the sale was expected to climb to 3.6 trillion with a 60% increase just within four years. Also, within this period, 13% of total worldwide retail sales were predicted to be online sales which were 7% in 2015 (eMarketer, 2015). In the UK, the USA, and China, per capita expenditure on online shopping is 4 thousand USD, 3.4 thousand USD, and 1.8 thousand USD respectively in the year 2015 (E-commerce Foundation, 2016). In contrast, the average per capita expenditure in 2015 on e-shopping in Bangladesh was only 11–13 USD (Ministry of Foreign Affairs of Denmark, 2015). Apparently, as a developing country with limited advancement in technology, Bangladesh is still in its infancy about Business to Consumer (B2C) e-commerce and online shopping.

The E-commerce sector in Bangladesh observed slow growth in 2000–2008 as there were few e-commerce websites; however, Bangladesh had no system for online transactions, which is the prime condition for e-commerce (Islam, 2015). However, in the last ten years, with the increase of internet users being thrice, the concept of online shopping is becoming more centre-pieced and popular (BTRC, 2017). Because of being the capital and most vibrant metropolitan city in the country, online shopping is mostly seen in the core urban portions of Dhaka Metropolitan Area (DMA). The study of Ishtiaque and Sarwar (2016) shows that more than 30% of e-commerce usage is generated from here. Considering the importance in the arena of online shopping, Dhaka North City Corporation (DSCC) and Dhaka South City Corporation (DSCC), the core urban areas of DMA were selected as the study area to conduct an explorative research on online shopping. This research is aimed to examine the nature, trend, and locational characteristics and to explore the product procurement and delivery system of online shops in Dhaka City Corporations. It also focused on the prospects and problems of online shops in the concerned area.

As a very vibrant but mostly organically developed city, urban areas under DMA are changing very fast regarding land use, travel pattern, social interaction, economic activities, etc. Thus it is important to clearly understand the problems and prospects of online shopping from an urban planning perspective, but very few studies had covered this aspect in Bangladesh. Also, the lack of studies based on empirical data often brings more challenges in drawing the real scenario (Mokhatarian, 2004; Nahiduzzaman et al., 2019; Sim & Koi, 2002). The outcome of this explorative study will lead to further analysis and studies. And this may help understand how the operation of the shops and, behavioural and socio-demographic aspects of online shoppers ubiquitously affects urban city structures for a dynamic urban areas like the City Corporations of Dhaka.

8.2 Review of Literature on Online Shopping in Urban Context

The emergence of online shopping and its nexus with urban and land use planning is a new phenomenon that got attention from researchers since the twentieth century. Most of the available literature is focused on marketing and business, and spatial and urban planning and behavioural facades of online shopping have been less investigated (Farag et al., 2006). Through a review of available literature on online shopping, this section tried to understand why and how online shopping can bring changes in the course of urban planning and spatial attributes of a city.

Online shopping brings a huge change in shopping behaviour, providing an unlimited selection of products and serves convenient shopping as time and energy can be saved (Mokhatarian, 2004). Some cities experience a reduction of sales in traditional stores (Farag et al., 2006; Mokhatarian, 2004). Thus attractions of the central core of a city for retailing are deemed, and that is often replaced by residential places, which is a major shift in land use (Zhang et al., 2016). Locational preferences of online shops are different in many aspects as they mainly focus on a comfortable environment for office work, and the demand for visibility becomes a less important criterion, unlike traditional shops and stores. An important determinant for choosing the location of office space and product storage location is to minimize the transportation cost to collect the products and to deliver those to the customers (Pettersson et al., 2016; O'Healy, 2005).

Sales of online shops are found high where the shops are more concentrated in city centres in the Netherlands (Boschma & Weltrevreden, 2004). Though generally it is thought that inner cities or peripheral areas are less affected by the change, city centres face more change because of e-commerce based online shops and stores (Weltevreden & Rietbergen, 2009). However, it is also evident that online shopping also reduces the necessity of living in city centres. People living in periphery and suburban areas can order anything and can avail of them from home (Shao et al., 2016).

The number of local distribution centres or outlets may be reduced, and large or central go-down or warehouses are supposed to increase (Matthews et al., 2002). New options for delivery are identified in many cities like pick up point or parcel locker so that customers can receive the products from there, and it has also been said that those points are located in the areas which have better accessibilities (Pettersson et al., 2018). Storehouses occupied one million square feet of the new area in the last two years in Chicago. The increasing rate of online shopping is identified as the main reason behind it (Planetizen, 2016).

Studies on Bangladesh in this regard are very few. No study has been conducted yet on the spatial attributes of emerging online shops, and the city structure is never viewed as a function of the operation of these online shops. As Bangladesh is at an initial stage in the development of e-commerce, these impacts are not visible yet. The study conducted in this context is mainly for business purposes and customer-oriented.

8.3 Methodology of the Research

According to the E-commerce Association of Bangladesh (E-CAB), the official and registering organization of all the e-commerce-based shops and services of the country, 500 websites based and 2000 Facebook based companies are running e-commerce businesses in Bangladesh (e-CAB, 2018). Six criteria have been taken into considerations to select online shops for the survey, which were:

- **Registration**: Registered online shops in E-CAB were considered so that the authenticity and validity of the collected data can be ensured. About 450 online shops and service providers were enlisted on their website as the registered members of E-CAB till November 2017.
- **Modes of running the business**: There were a good number of shops that were only operated through social media (F-commerce) or mobile applications (M-commerce). Those were excluded, and online shops that have personal websites for running their business and maintain operations were considered for this study.
- **Nature of final product**: The most visible effects of online shopping are created by goods providing shops because those are needed to be transported, needs office space, and storing places that work behind reshaping a city structure. Therefore, service providing and IT-based shops were excluded from the study, and only goods providing shops were included in the survey framework.
- **Product varieties**: Some online shops sell products from different categories, whereas some online shops are focused on a single product. Both types of online shops were inclusive of the selection process.
- **Location**: As the study is aimed to explore the facades of online shopping in the North and South City Corporations of Dhaka, shops that are located and operated from the mentioned area were preferred for the study.
- **Operating policy**: Among the registered shops which maintain a Business to Consumer (B2C) operation were taken into account.

Among these registered shops, 115 goods providing shops had their websites, met the location criteria, and were B2C oriented. From the respective websites of these 115 shops, it was found that 59 of them were multi-products selling shops, and 56 were single product selling shops. Here, one thing is very important to notice that the nature of B2C (Business to Consumer) is very transforming i.e. many of the online shops stop running a business after a while or transform from B2C to B2B (Business to Business) (Suhan, 2015). So, sampling was one of the crucial parts of this study as there was a possibility that all of the selected online shops could not be available for the survey. That is why with a 90% confidence interval, and a 5% margin of error, a total of 81 shops out of 115 shops were fixed as samples considering a proportionate stratified sampling technique. Among them, 40 were single product selling shops, and 41 were multi-product selling shops.

For collecting shop owners' data, a questionnaire was prepared, and initially, two multi-products selling shops were surveyed for piloting. The final questionnaire was prepared by adjusting the drawbacks/ shortcomings identified in the questionnaire from the pilot study. Among the 81 online shops, a direct and face-to-face questionnaire survey was conducted on 61 shops. While for the rest of the 20 shops, the study team had to conduct phone interviews due to the unavailability of the respondents on time or in person. All primary data of the online shop providers were collected from October 2017 to June 2018.

8.4 Nature, Trend and Locational Characteristics of Online Shops

As online shopping is a quite new concept in Dhaka, new shops are emerging, and the growth trend is fast in recent years. The shops have some basic characteristics which are different from traditional shops and markets. Also, their locations of office and storage space often have some specialty in terms of products sold by them. These are important for viewing the prospect of online shopping.

8.4.1 Product Types and Their Service Area

Based on data collected from 115 shops, the types of products can be categorized into five groups. The first category includes clothes, ornaments, and all other fashion accessories. Automobiles and all types of electronic products are kept in another category. The third category consists of furniture, books, gifts, and necessary things for the household. Lastly, groceries have been put into two categories: perishable groceries which are fish, meat, milk, etc., and non-perishable groceries like rice, pulse, masala, pet food, etc. Among the selected 81 shops for surveying, 40 shops sell products of a single category which are addressed in the study as single products selling online shops, and 41 shops sell products of more than one category, which are named as multi-products selling shops. Figure 8.1 shows the total number of shops under these five categories and also describes how many of them sell solely one product, or they have other products.

Among the surveyed 81 shops, only 4 shops deliver their products and all of them are multi-product selling shops. 23 shops deliver products within Dhaka and its very surrounding areas. The rest 54 shops count the whole of Bangladesh as their catchment area of product delivery. In shops serving within Dhaka and within Bangladesh, both have almost proportionate numbers of single and multi-products selling shops.

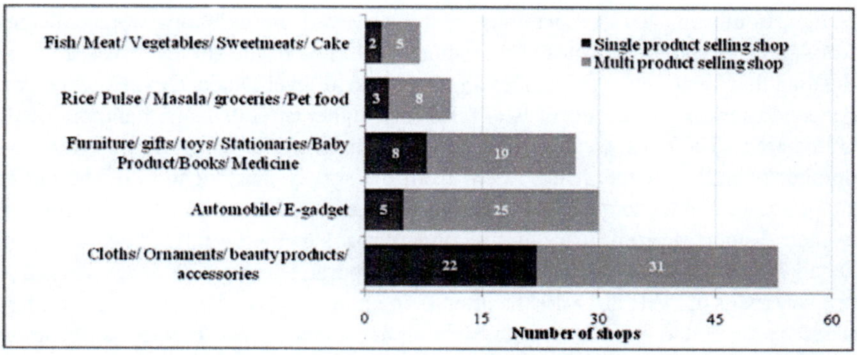

Fig. 8.1 Product types of the selected shops. *Source* Field survey (2018)

Table 8.1 Annual average increase of customers of the online shops till now from their establishment

Types of online shop	Average increase in customers (yearly) (%)	Standard deviation (%)
Single product selling	16	5
Multi-products selling	25	9

Source Field survey (2018)

8.4.2 *Growth in the Number of Shops and Customers*

The study found that only 18 shops started before 2014. In 2014, 11 shops started operating and the highest number of shops was established in 2015 which is 30% of the surveyed shops (24). Every year new shops are starting their operation in Dhaka and it especially got a boost after 2014 (e-CAB, 2018). Internet users of Bangladesh have played a dominant role in expanding online businesses as the number of internet users has become threefold in 2019 than it was ten years ago (BTRC, 2018). With the rising number of shops and internet users, online shops experience a huge increase in their customers, which is summarized in Table 8.1. the table reveals that multi-products selling shops face a higher average increase in customers than the shops that sell a single product. A higher standard deviation also demonstrates that an increase in customers is more dispersed among multi-product selling shops.

8.4.3 *Locations of Offices and Storage Facilities of Selected Online Shops*

Unlike traditional shops and markets, online shops do not need visibility and show-casing thus they do not look for frontier locations (Nahiduzzaman and Aldosary,

2012; Thompson, 2011). Locations of the surveyed online shops were collected from their respective websites and verified at the time of the survey. It is seen from the map that locations of the offices are dispersed all over Dhaka City. However, the concentration of the shop offices is relatively higher in Gulshan, Dhanmondi, and Uttara areas. The north-western and south-eastern corner of Dhaka city has the least number of online shops. It is evident from the survey that, location of the online shops shows some characteristics. The location of the shops was analysed in terms of their distance from primary and secondary roads using the road network map of Detail Area Plan (DAP), 2015 of Dhaka city corporations. Roads that are marked as primary roads in DAP, 2015 are the arterial and main roads of Dhaka city. Secondary roads are generated from these primary roads to provide access. Figure 8.2 illustrates that only 23% of multi-product selling shops (9 out of 41) and 24% single product selling shops' (9 out of 40) offices are located within the 500 m buffer along with the primary roads. It has been also seen that 88% of the shops' offices were covered within 3000 m (3 km) buffer along with primary roads. A slightly different outcome was found while working on secondary roads. The shops were found more adjacent to the secondary roads than primary roads. More than half of the shop offices are located within the buffer of 500 m (0.5 km) along the roads. It is seen that on an average 80% of shops are within 2000 (2 km) metres buffer along secondary roads.

8.4.4 *Detailing of Office Area*

To explore the locational characteristics of the office buildings, it is important to know which types of buildings are preferred for setting up offices. Among the 81 shops, 32% have their offices in buildings which have mixed land use. Almost 20% have offices in residential buildings, and the rest 48% of shops' offices are in the building which is commercial. In choosing commercial or mixed-use buildings, single products, and multi-products selling shops are almost the same in number. However, most of the single product selling shops have their offices in residential buildings, which can be further clarified from Fig. 8.2, where existing building uses in Dhaka City Corporation Areas have been illustrated.

The average office area of the online shops was 981 square feet with a standard deviation of 350 square feet. While analysing the office area, it has been found that 48% of offices have an area within the range of 501–1,000 square feet (Table 8.2). It can be generalized that very few online shops have a wide operational range and need large office spaces.

8.4.5 *Detailing of Storage Facilities*

Many cities are experiencing an increase in storage or go-down area around the world. A recent study showed that due to the increasing rate of online shopping, one

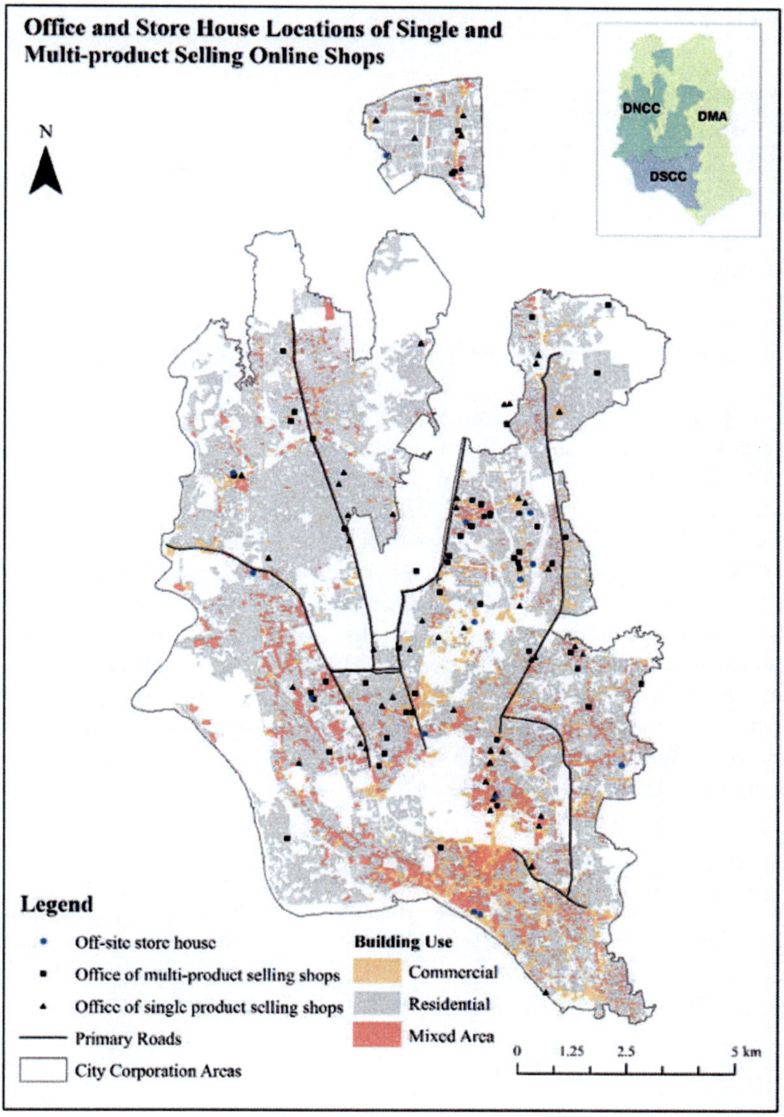

Fig. 8.2 Location of offices and off-site storehouses of the online shops with respect to building uses and primary roads in Dhaka City Corporations. *Source* Field survey (2018)

million square feet of new areas are being used as storehouses or product storage (Planetizen, 2016). It was found from the survey that the shop owners do not need to store their products, and they directly deliver the products as soon as they procure it. Grocery and perishable food item selling shops belong to this category. Because of the perishable nature of the product to the shop owners do not store those products.

Table 8.2 Office area of online shops in Dhaka

Office area	Number of shops	Percentage (%)
Less than 500 sq. feet	7	9
501–1000 sq. feet	39	48
1001–1500 sq. feet	25	31
1501–2500 sq. feet	9	11
More than 2500 sq. feet	1	1

Source Field survey (2018)

From Table 8.3 it is clear that, except for these shops, all other shops mostly store their products.

Though most of the shops need storage facilities, all of them do not have separate and designated storage space. Among these 81 shops, 59 shops have designated storage space. Others often have virtual stocks or stores unofficially in office space, even in their residences. Among the 59 shops with designated storehouses, 43 shops have designated storage space at the same locations as office sites. The remaining 16 shops have storehouses in separate locations convenient from a production, collection, and delivery point of view. Locations of storehouses that are different from office locations are pointed in Fig. 8.2. Locations of the off-site storehouses are very random. These storehouses are located in Sadarghat, Gulshan, Mirpur, Kallyanpur, Uttara, Dhanmondi, Karwan Bazar, and Motijheel areas. From the locations of storehouses, it can be said that these locations might have been chosen for convenience in collecting or delivering products as Kallyanpur and Sadarghat are two main points through which Dhaka is connected with the nearby districts.

Table 8.3 Different storing process according to product type

Type	Cloths/fashion accessories/ornaments	Automobile/e-gadgets	Books/gift/furniture/toys	Perishable grocery (rice/pulse/masala)	Non-perishable grocery (fish/meat/vegetables)
Store before delivery	29	13	11	4	14
Don't store, directly deliver	4	1	1	4	1
Both	22	20	18	8	12
Total	55	34	30	16	27

Source Field survey (2018)

8.5 Product Collection and Delivery Systems of Online Shops

Product collection and delivery systems of online shops are one of the important components of their operation. Collection and delivery procedures also vary according to product type. Other components of delivery procedure i.e. locations and logistics of delivery, can also play a role in illustrating the effects of online shopping.

8.5.1 Product-Wise Collection Procedures

According to product types, area and process of collection can be different. Surveyed online shop providers were asked to mention some major and important locations or places from where they collect their products. They could consider a place or location important for collection according to collection frequency or collection volume. More than half of the shops opined that the major share of their products is bought from abroad. The countries mainly include India, China, and the UK (Source: Questionnaire Survey, 2018). Only grocery (both perishable and non-perishable) was the category where products are collected not from abroad, rather from nearby areas. It was found that except the groceries, all categories of shops collect their major supply from abroad. So, the specific collection system of groceries was focused on. It was previously shown that 18 shops sell groceries (Fig. 8.1). Among them, 11 sell groceries that are non-perishable like rice/pulse/masala/pet food, etc. remaining seven shops sell perishable groceries: fish/meat/vegetables/milk/cake, etc.

These online grocery shops collect their products for delivery in two ways. The first one is from chain shops that have outlets at different locations in Dhaka. Online shops are often linked with chain shops. Delivery men collect what customers have ordered from the convenient outlets and then deliver. Again, online shops directly go to wholesale markets with which they maintain B2B connections. Like the previous one, deliverymen go to the selected or nearby wholesale market to collect the ordered product. Among the 18 grocery selling shops, 4 shops said they collect products from chain shops. The rest 14 shops collect products from different wholesale markets of Dhaka.

8.5.2 Procedures of Delivery

A very recent study has shown that around 80% of the total e-commerce usage of Bangladesh is from Dhaka, Chattagram, and Gazipur (Ishtiaque & Sarwar, 2016). While surveying the online shops, the percentage of delivery inside Dhaka is 77% for single product selling shops, and for multi-products selling shops, it is a bit lower

Table 8.4 Percentage of delivery inside and outside of Dhaka

Types of shop	Percentage of delivery inside Dhaka City areas (%)	Percentage of delivery outside Dhaka City areas (%)
Single product selling	77	22
Multi-products selling	68	31

Source Field survey (2018)

(around 68%). Low standard deviations of the percentage mean these statistics almost satisfy the mainly Dhaka-centric business of the online shops. Here the percentage of delivery outside Dhaka is a little bit higher for multi-products selling shops, and the possible reason behind this can be multi-products selling shops have a wider scope of spreading business (Table 8.4).

8.5.3 Locations of Delivery Within Dhaka

Providers were asked to identify the first five locations in sequential order to deliver their orders most frequently within Dhaka city. For every shop, five areas scored from 1 to 5 were named by the shop owners. The area generating the highest number of deliveries gets 5 and the area with the lowest number of delivery gets 1. 15 areas were mentioned with different scores. Then scores of these 15 areas from all the shop owners were summarized and divided by the number of shops. Thus a score was generated ranging between 1 and 5 and presents the amount of delivery that the online shops have to deliver. Malibagh, Khilgaon, Moghbazar,

And Basundhara Residential Area was not finally illustrated in the map (Fig. 8.3) as less than 10 shops mentioned them. Mirpur and Dhanmondi areas are the receivers of the highest number of deliveries from most of the shops as they have the highest scores among the twelve areas (Fig. 8.3). Gulshan and Mohammadpur areas follow them and have got close scores. Motijheel and Uttara areas are in the next place to generate almost the same amount of deliveries. Jatrabari and Old Dhaka have the lowest scores indicating that the least number of products have to be delivered among the 12 areas (Fig. 8.3).

8.5.4 Logistics for Delivery

It has been found that mainly motorcycle is used for delivering products (Fig. 8.4). Recently few product delivery service companies are operating in the city. Shops often avail this service to deliver their products to customers. 33 of 81 shops said that they regularly or frequently use these delivery companies. For delivering to the

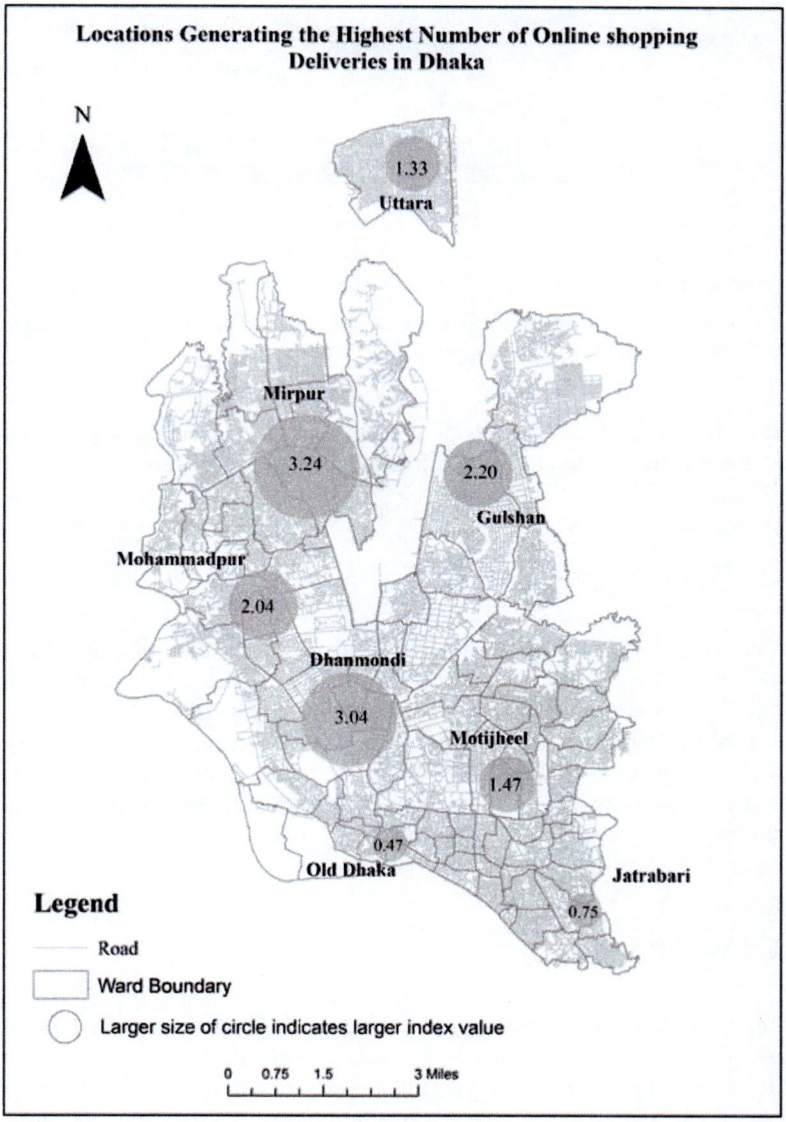

Fig. 8.3 Locations in Dhaka where maximum products of the online shops are delivered. *Source* Field survey and authors' calculation (2018)

nearest and surroundings of Dhaka City Corporation areas, like Savar, Gazipur etc., cycle and motorcycle courier is often used by the shops (Fig. 8.4).

Nevertheless, cycle or motorcycle courier cannot be used for the districts adjacent to Dhaka. For delivering products to those areas, courier service or cargo/paribahan

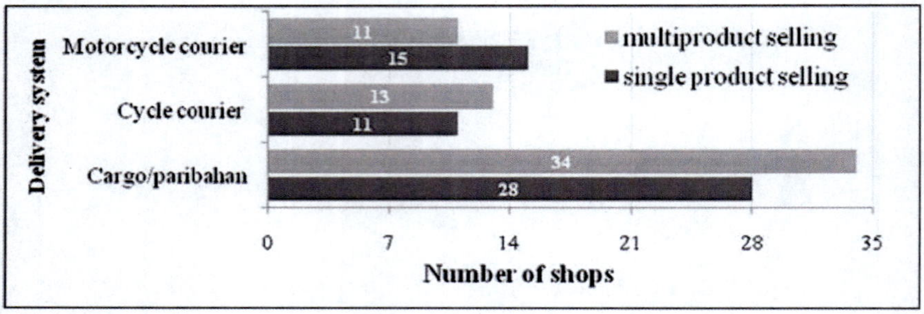

Fig. 8.4 Transportation mode or system used by the online shops to deliver products outside Dhaka. *Source* Field survey (2018) and authors' calculation (2018)

are the only way. In most cases, shops are not fixed with one single way; rather, they use all three ways according to the distance and types of products.

8.5.5 Payment System

Cash on delivery and payment before delivery both are the prevailing system of transaction. For cash on delivery, the delivery man collects the price of the products. In case of payment prior to delivery, customers have to pay the full or partial price of the products, which are mainly made through card payments or mobile banking. For delivering within Dhaka city corporations areas, customers mainly prefer cash on the delivery system since online shops still have not achieved the full trust of consumers.

8.5.6 Staff for Delivery

The number of staff for delivering products varies according to shops and sale situation. For example: at the time of sale/discount or festivals, they often hire part-time staff for delivering orders. Among the single product selling, around 32% of shops have less than 5 deliverymen, and 50% of shops have 5–10 deliverymen. The same type of data is found for multi-product selling shops. Most of the shops have 5–10 delivery men (Fig. 8.5).

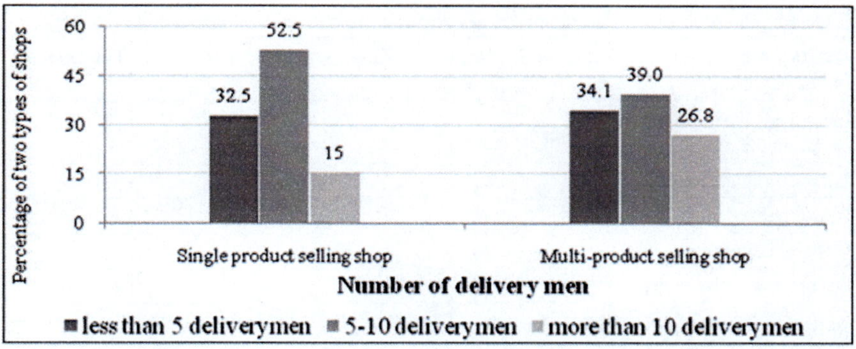

Fig. 8.5 Number of deliverymen in single and multi-product selling shops. *Source* Authors' calculation (2018)

8.6 Prospects of Online Shopping from Providers' Perspective

Providers or online shop owners are the ones who have the most comprehensive idea about the condition of online shopping in Bangladesh. About 35% of 81 shop representatives think, online shopping has a huge potential in Bangladesh, especially in Dhaka. However, 52% of them opined that it would take a long time to operate in full swing (Source: Questionnaire Survey, 2018). All 81 shop representatives were given options to choose the main reasons behind the expansion of online shopping in Bangladesh, and they had scope to choose more than one answer. 72 shop representatives agreed that increasing access to the internet and social media act as a positive agent to advertise their shop and products, which increases online shopping. The second reason appeared to be the scope of online shops to offer products from a wide range of sources than traditional shops. 37% of providers also think that online shopping will be popular day by day as it is free of time constraints and reduces the hassle of travelling and traffic jams. Other large and renowned online shopping companies abroad generally take payment before delivery. But the payment system in Bangladesh is not very updated, and people are not very familiar with it. As a result, it is thought that cash on delivery system creates huge scopes for online shopping. Day by day transportation of products and goods from abroad is getting easier and less time-consuming. Thus the product delivery time can be reduced (Table 8.5).

8.7 Challenges Faced by the Online Shops

One of the major problems of online shop owners is fraud customers. Table 8.6 summarizes the challenges faced by the online shop providers, and 61 among 81 shops said that fraud customers are a huge problem for online shopping businesses

Table 8.5 Reasons why online shopping has potential from providers' perspective

Reasons why there is a potential for online shopping	Number of shops who agree with this reason	Percentage
Increasing access to internet and social media	72	89
Larger product sources	46	57
Convenience	37	46
Cash on delivery system	26	32
Decreasing delivery time	16	20

Source Questionnaire survey (2018)

Table 8.6 Challenges faced by the online shop providers

Challenges faced by online shops	Number of shops that face these problems	Percentage
Fraud customers	61	75
Traditional thinking	55	69
Traffic jam	30	37
Lack of proper national guidelines	19	23
Unsecured payment gateway	11	14

Source Questionnaire survey (2018)

in Bangladesh. Another major problem is that the customers have traditional thinking and want to bargain about the price and the rate of returning products is also high, which is confirmed by 55 of the owners. 37% of them said that because of the traffic jam they have to incur more time both for collecting and delivering products. 19 out of 81 shop representatives were concerned about the absence of policies, as there is no such regulation or policy for e-commerce or online shopping in Bangladesh. 11 of them mentioned the unsecured payment system. A study conducted in Singapore also found that security in payment on the internet will increase the number of online shoppers by 2.7 times (Sim & Koi, 2002).

8.8 Existing E-Commerce-Related Regulations in Bangladesh

The Government of Bangladesh has enacted the "National Digital Commerce Policy, 2018" to maintain a balanced growth of e-commerce which is soon to be implemented (Moretaza, 2019). The policy set some goals and work plans which have been well accepted in the e-commerce community of Bangladesh (The Daily Star, 2020). It also puts light on almost every important for the expansion of e-commerce which protects the stake of both consumers and providers as well as focuses on strict regulations on money transfer, digital signature, copyright issues, equity of foreign ownership. It also gives importance to improving transport for delivery and expansion of e-commerce in other cities and towns (National Digital Commerce Policy, 2018, 2019). The challenges online shop providers face can possibly be wiped out through the proper implementation of this policy.

From an urban planning perspective, one of the policy's shortcomings is that there was no regulation for choosing office or storehouse locations or movement of delivery vehicles. As mentioned before, there is a practice in Bangladesh of using residential spaces for offices and storehouses, which is sometimes unsafe. Besides, the residential environment and its traffic flow can be hampered by delivery vehicles and other chaos. Several fire incidents have occurred in Dhaka in the last decade, which were initiated from storehouses and industries kept in residential buildings (Jakir, 2019). As e-commerce in Bangladesh is spreading rapidly, these issues should not be overlooked anymore and must be brought under proper regulations. Since Dhaka, as a city, is already saturated with its components, even a small anomaly or disruption in the system can affect the interaction of land use and transportation as well as the whole city structure.

8.9 Conclusion

With the spread of e-commerce based online shopping, cities worldwide are experiencing transformations in activity centres, space use, travel pattern, and the financial flow of the activities. Like developed countries, many developing countries are shifting towards e-commerce based economic activates. This research portrayed the nature of online shopping from the provider's perspective in Bangladesh and presented certain spatial aspects associated with the development of e-commerce in the capital city Dhaka. Study findings show that e-commerce is expanding at a high pace in Bangladesh. This pace is further accentuated with the increase of accessibility of the internet to a wide range of populations. Moreover, the provision of cash on delivery system helped to overcome the constraints associated with money transactions. A guaranteed and safe money transaction system will further support the spread of e-commerce. The research gap in this field will be a significant barrier to develop regulatory policies and acts to safeguard both customers and business owners. It is also important to observe the transition of traditional retail services to e-commerce from land use, transport, and revenue perspectives.

Acknowledgements The researchers like to express their gratitude to the Department of Urban and Regional Planning, Bangladesh University of Engineering and Technology (BUET), for providing all sorts of support to conduct this research work. They are also grateful to the E-Commerce Association of Bangladesh (E-CAB) for their assistance in the survey.

References

Boschma, R. A., & Weltevreden, J. (2004). B2C E-commerce adoption in inner cities: An evolutionary perspective. In M. Taylor, & S. O. Park (Eds.), *E-Commerce, E- Business and the Dynamics of Economic Development*. Ashgate.

Bourlakis, M., Papagiannidis, S., & Fox, H, (2008). E-consumer behaviour: Past, present and future trajectories of an evolving retail revolution, *International Journal of E-Business Research, 4*(3), 64–67, 69, 71–76.

BTRC. (2018). Retrieved September 26, 2018, from Bangladesh Telecommunication Regulatory Commission. http://www.btrc.gov.bd/content/internet-subscribers-bangladesh-august-2018

BTRC. (2017). Retrieved October 20, 2017, from Bangladesh Telecommunication Regulatory Commission. http://www.btrc.gov.bd/content/internet-subscribers-bangladesh-august-2014

Cao, X. (2009). E-shopping, spatial attributes, and personal travel: A review of empirical studies. *Transportation Research Board, 2135*, 160–169.

Cervero, R. B. (2000). Land use and transportation. breaking out of the box, or growing smart. In *Proceedings of the Annual Conference of the Resource Management Law Association*. Auckland.

E-commerce Foundation. (2016). *Global B2C E-commerce Report 2016: A detailed overview of the Global E-commerce Market*. Amsterdam. https://www.asendia.at/application/files/9514/8664/1553/Global_B2C_Ecommerce_Report_2016_Full_version.pdf. Retrieved 10 Oct 2018.

eMarketer. (2015). Worldwide Retail Ecommerce Sales: eMarketer's Updated Estimates and Forecast Through 2019. http://www.emarketer.com/public_media/docs/eMarketer_eTailWest2016_Worldwide_Ecommerce_Report. Retrieved 10 Oct 2019.

Farag, S., Weltevreden, J., Rietbergen, T., Dijst, M., & Oort, F. (2006). E-shopping in Netherlands: Does geography matter? *Environment and Planning B, 33*, 59–74.

Golob, T., & Amelia, A. (2001). Impact of information technology on personal travel and commercial vehicle operations: Research challenges and opportunities. *Transportation Research Part C, 9*(2), 87–121.

Islam, F. U. (2015). *E-Shopping In Bangladesh*. (Masters' Thesis). University Of Rajshahi, Bangladesh. https://www.slideshare.net/limonfnb/e-shopping-in-bangladesh. Retrieved 10 Oct 2019.

Islam, S. (2020, August 31). E-Commerce: That is why 'Cash on Delivery' is popular in online shopping in Bangladesh. *BBC News Bangla*. https://www.bbc.com/bengali/news53971502?fbclid=IwAR3p5huBqMfRC1_EVdcfOiLcEQen-V36ilwDGDWmNFgiQXKCt2ll7bw2FoY. Retrieved 1 Sept 2020.

Ishtiaque, A. S., & Sarwar, A. (2016). *E-Commerce in Bangladesh*. (Masters' Thesis). University of Liberal Arts Bangladesh, Dhaka, Bangladesh. https://ces.ulab.edu.bd/wp-content/uploads/sites/18/2016/01/CES-Final-Version-Ecommerce-Jan-28-2016.pdf. Retrieved 10 Oct 2018.

Jakir, A. (2019, February 22). Nimtoli fire replays in Chawkbazar. *The Independent*. http://www.theindependentbd.com/post/188657. Retrieved 30 July 2020.

Matthews, H. S., Hendrickson, C. T., & Lave, L. B. (2002). The economic and environmental implications of centralized stock keeping. *Journal of Industrial Ecology, 6*(2), 71–81.

Ministry of Foreign Affairs of Denmark. (2015). *Bangladesh: E-commerce Country Fact Sheet*.

Mokhatarian, P. (2004). A conceptual analysis of the transportation impacts of B2C E-Commerce. *Transportation, 31*(3), 257–284.

Moretaza, T. (2019, November 09). Revised Digital Commerce Policy yet to get Cabinet approval. *The Independent*. http://www.theindependentbd.com/post/222994. Retrieved 30 July 2020.

Nahiduzzaman, K. M., Aldosary, A.S.,& Mohammed, I. (2019). Framework Analysis of E-Commerce Induced Shift in the Spatial Structure of a City, *Journal of Urban Planning and Development. 145*.

Nahiduzzaman, K. M., & Aldosary, A. S. (2012). City structure in transition: A conceptual discourse on the impact of information and communication technology (ICT). In A. Cakir, & P. Pablos (Eds.), *Social development and high technology industries: Strategies and applications*. IGI-Global.

National Digital Commerce Policy 2018. (2019, January 30). http://e-cab.net/wp-content/uploads/2019/02/Digital-Commerce-policy-2018.pdf. Retrieved 10 Aug 2020.

O'Healy, P. (2005). *Choosing a warehouse location: Look for more than just price*. http://www.inboundlogistics.com/cms/article/look-for-more-than-just-price-when-choosing-a-warehouse-location/. Retrieved 21 Mar 2018.

Pettersson, F., Hiselius, L. W., & Koglin, T. (2018). E-commerce and urban planning—Comparing knowledge claims in research and planning practice. *Urban, Planning and Transport Research, 6*(1), 1–21.

Pettersson, F., Hiselius, L. W., & Koglin, T. (2016). Exploring the nexus between e-commerce and urban land use planning—e-commerce impacts on mobility and location strategies. *Transportation Research Procedia.*

Planetizen. (2016). Chicago's Warehouse Building Boom. https://www.planetizen.com/node/88289/chicagos-warehouse-building-boom. Retrieved 10 Oct 2017.

Shao, J., Yang, H., Xing, X., & Yang, L. (2016). E-commerce and traffic congestion: An economic and policy analysis. *Transportation Research Part b: Methodology, 83*, 91–103.

Sim, L., & Koi, S. (2002). Singapore's Internet shoppers and their impact on traditional shopping patterns. *Journal of Retailing Consumer Service, 9*(2), 115–124.

Suhan, J. (2015). Acceptance of online shopping in Bangladesh: Consumer's perspective. *IOSR Journal of Business and Management, 17*(1), 14–24.

The Daily Star. (2020, April 09). Our digital commerce policy in a nutshell. https://www.thedailystar.net/bytes/our-digital-commerce-policy-nutshell-1613056#:~:text=The%20much%20awaited%20digital%20commerce,by%20making%20safer%20digital%20tr. Retrieved 30 July 2020.

Thompson, R. (2011). E-Retailers and the Secret to A Good Location Decision. http://www.areadevelopment.com/-HighTechNanoElectronics/November2011/e-commerce-location-decision-guide-00062622.html. Retrieved 17 Oct 2017.

Van Loon, P., Deketele, L., Dewaele, J., McKinnon, A., & Rutherford, C. (2015). A comparative analysis of carbon emissions from online retailing of fast moving consumer goods. *Journal of Cleaner Production, 106*, 478–486.

Visser, E. J., & Lanzendorf, M. (2004). Mobility and accessibility effects of B2C E-commerces: A literature review. *Tijdschrift Voor Economische En Sociale Geografie, 95*(2), 189–205.

Weltevreden, J., & Rietbergen, T. (2009). The implications of E-shopping for in-store shopping at various shopping locations in the Netherlands. *Environment and Planning B: Planning and Design, 36*(2), 279–299.

Zhang, D., Zhu, P., & Ye, Y. (2016). The effects of E-commerce on the demand for commercial real estate. *Cities, 51*, 106–120.

Chapter 9
Demographic Changes and Challenges of Small Towns in Romania

Daniela Zamfir and Ilinca-Valentina Stoica

Abstract After the collapse of the communist regime (1989), Romanian small towns (under 20,000 inhabitants) followed divergent trajectories in the context of the wider political, economic, and social shift to democracy and free-market economy. These types of settlements are the basis of the urban network and are considered the most exposed to structural transformations and to the ongoing demographic changes. Among the major driving forces influencing the small town's population-related metamorphosis we can list their historical communist background, economic restructuring process, emigration flows (intensified after Romania's accession to the European Union), changes of demographic behaviour, etc. The study aims to analyse the demographic changes of small towns in Romania in the post-communism and their current characteristics and challenges through several tools: the first one being several indicators—population dynamics, vitality index, net migration rate, and old-age dependency ratio—and the last one is a composite index of demographic vulnerability (IDV), designed in an integrated perspective using five variables. The results reveal a general tendency of demographic decline and ageing, although there are also various demographic asymmetries where some small towns flourish while, at the other end of the spectrum, others experience severe population decline that is often related to a sharp deindustrialization process. Then, there are some small towns that were able to reverse their trajectories, from a downturn in the late 1990s or even later, to a prosperous perspective nowadays.

Keywords Small towns · Post-communist transformations · Emigration · Ageing · Demographic asymmetries · Demographic vulnerability

---

D. Zamfir · I.-V. Stoica (✉)
Faculty of Geography and Interdisciplinary Centre for Advanced Research on Territorial Dynamics (CICADIT), University of Bucharest, 010041 Bucharest, Romania
e-mail: valentina.stoica@unibuc.ro

D. Zamfir
e-mail: daniela.zamfir@unibuc.ro

B. R. K. Sinha (ed.), *Urban Dynamics, Environment and Health*,
https://doi.org/10.1007/978-981-99-5744-6_9

9.1 Introduction

Population decline and ageing affect many regions of Europe nowadays. Despite that, on the whole, the European Union is experiencing an increase in population, but the pace is considered too slow, and a decline is expected in the longer term (Margaras, 2019). For example, the United Nations demographic projections predict that Europe's population will decrease by 100 million by 2050 (Klingholz, 2009). This negative demographic trajectory raises other concerns such as a diminution of the share of young people, which in turn will trigger the shrinkage of the labour force, reduce the fertility rate, and lower economic growth (Akiyama, 2011; Palkina et al., 2021).

However, examination of the European Union on a macro-regional level tells a slightly different story, with major differences between regions. A study by Lutz et al. from 2019 analysed the demographic change between 1991 and 2015 in diverse parts of Europe and identified a positive trend in western and southern countries, while the eastern side was marked by a decline. One of the main reasons for this notable discrepancy was international mobility. Thus, the west and the south's developed economies, which are able to offer and sustain higher standards of living, attract immigrants from all over the world, including from Eastern Europe. Consequentially, the eastern state's significant drop in population was due to sizeable emigration correlated with very low fertility rates and negative birth rates (Batog et al., 2019; Lutz et al., 2019). This development is largely related to the fall of communism in the 1990s, and the reintegration of these countries into the global economy (Atoyan et al., 2016). Therefore, the pushing factors in Eastern European countries are connected to the systemic transformations in the post-communist period, especially the economic downturn which led to lower level of income compared to the wealthier western states. Moreover, the demographic projections anticipate this trend of population decline will persist over the coming decades (Batog et al., 2019; Lutz et al., 2017). Still, leaving aside these macro-scale typologies, urban settlements affected by the population decline in relation to different local circumstances can be found all over in the European Union.

One of the most significant outcomes of these demographic changes is population ageing. In fact, this phenomenon is common in many parts of the world, and it is estimated that by 2050, 25.1% of the global population will be 65 or older (OECD, 2015). In the European Union, the strongest impact is associated with Eastern Europe part, where the fertility rate decayed below the replacement threshold, in correlation with moderate mortality and lack of migration inflow gains (Lutz et al., 2010).

Like other eastern states, Romania registered a decrease in population from 1992 to now (−4%). But the distribution of the population losses was extremely uneven between urban and rural areas, respectively −0.11% in the former and −8.8% in the latter. This difference can be partly explained by the fact that over 60 rural units were granted urban status in this period, tipping the population scales in favour of "urban" areas. Additionally, it should be mentioned that the demographic changes were strongly influenced by the massive migration fluxes towards Western countries (Ianoș, 2016). Furthermore, the worrying demographic projections predict that Romania will be among the countries that will experience some of the largest population declines in the world (Mykhnenko & Turok, 2008). As a consequence, the share of the population over 65 years already reached 16.4% in 2018 (according to the National Institute of Statistics) and is expected to increase.

Focusing exclusively on the urban picture, several studies stress that demographic growth or decline in small and medium-sized towns are related to the context of national and regional urban systems (Servillo et al., 2014). However, in Eastern European cities, the population trajectories between 1960 and 2005 revealed that since the mid-1990s, the overwhelming majority registered a decrease in their inhabitants for the first time after the Second World War (Mykhnenko & Turok, 2008). The main drivers are post-socialist transformation, demographic changes (e.g. declining birth rates, ageing, out-migration), economic transformation (deindustrialization, downturn), suburbanization and environmental issues (e.g. natural disasters, pollution) (Haase et al., 2016; Lutz, 2010; Wiechmann & Wolff, 2013). Urban decline is often generated by the interplay of some of these processes, also depending on the local governance (Haase et al., 2016). In any case, it seems that small urban settlements registered a sharper decline than large cities (Mykhnenko & Turok, 2008), the former considered one of the most affected by the ongoing demographic changes.

Small towns represent an essential segment of the settlement network, as they serve as a liaison between urban and rural areas (Ecovast, 2013; Ianoș, 2004; Zamfir, 2007). Thus, even if small towns do not have the attraction power of metropolises, and they are not able to create agglomeration or scale economies (Noronha Vaz et al., 2013) they still fulfil a well-defined role within the urban systems (Czapiewski et al., 2016) and thus contribute to polycentric development (Servillo et al., 2014). Moreover, on the local scale, especially in regions characterized by vast rural territories unpolarized by larger urban centres, they act as development poles (Vaishar et al., 2016) and service providers for the surrounding villages (Ecovast, 2013), strengthening the regional economy (Vaishar & Zapletalová, 2009).

When looking at the features of small towns, a spatial and temporal heterogeneity across the European Union must be recognized. To that effect, several studies have described divergent trajectories of their development (Józefowicz, 2022; Servillo et al., 2014; Steinführer et al., 2016; Valtenbergs et al., 2015). Some small towns, often those located in the vicinity of a major urban centre or a metropolitan region, registered a positive trend, experiencing population growth (Wirth et al., 2016). Meanwhile, other small towns, mostly those located in remote regions, are in regression, confronted with demographic challenges like shrinking population, ageing, and out-migration of young skilled workers (Valtenbergs et al., 2015). In this regard, population decline and ageing are considered common characteristics among small towns in the European Union (Valtenbergs et al., 2015; Wirth et al., 2016).

Due to their heterogeneity and different national and regional contexts, small towns are difficult to define both from a global perspective and within the European Union (Servillo et al., 2014). In Romania, this category of settlements is characterized by a dimensional criterion in a scientific approach, designating all urban centres with less than 20,000 inhabitants as small towns. Nevertheless, these places exhibit a variety of demographic and economic circumstances, influenced by regional, historical and place-specific factors (Zamfir et al., 2009).

Against this background, the current study aims to outline the demographic characteristics of Romanian small towns in terms of population change and their current configuration and challenges. Thus, considering the demographic decline in eastern European countries and the diversity of their urban areas, the paper responds to the calls made by other researchers (Ecovast, 2013; Servillo et al., 2014; Steinführer et al., 2016) and even by the European Union guidelines (Valtenbergs et al., 2015) for more systematic research of this category of settlements. For this purpose, this study brings an overview of the Romanian situation through an in-depth analysis.

Another consideration for this study is that debates over demographic development often focus more on growing urban settlements and less on settlements facing adversity (McCann, 2017). This study contributes to filling this knowledge gap, giving insights into the entire national network. Furthermore, understanding demographic features is a prerequisite for elaborating policies and territorial development strategies, both for the sake of the small towns themselves and for the wellbeing of the surrounding rural areas.

Following this line of thought, to highlight the trajectories of small towns in Romania, the population evolution was examined under the influence of two different political regimes: communism—exhibited immediately after its fall, and the post-communism—defined by distinctive demographic and economic policies and societal context. In this regard, given that the post-socialist transformation is regarded as a cause and catalyst for urban decline (Haase et al., 2016), the analysis was focused on the period after the fall of communism (1990), investigating population decline and ageing. In order to identify the current challenges of the small towns, an aggregated index of demographic vulnerability was built, which allowed the establishment of a hierarchy, and ranked their demographic attributes.

The study is comprised of five major parts. The next section focuses on describing the relevant features of small towns and their historical background. Then, in the third part, the main methodological steps are explained. The fourth section presents the results and discussion, followed by the main conclusions.

9.2 The Contrasting Path of Romanian Small Towns

In Romania, around 2,090,000 inhabitants live in 224 small towns (with less than 20,000 persons), as revealed by the last census (National Institute of Statistics), excluding one town, which regained commune status in 2019. This tier of settlements represents 70% of the national urban network, concentrating 19% of the urban population and 10% of the total population.

Regarding the current configuration of the network of small towns, previous researches demonstrated that urban development in the post-communist period is marked by processes of path dependencies, meaning that their trajectories are influenced by their communist past (Bouzarovski et al., 2016; Nedović-Budić et al., 2006; Stanilov, 2007). This means that to understand the demographic development, an overview of previous historical periods, mainly the second half of the last century, is needed, considering that the population dynamics is strongly interrelated with political and economic legacy. In this context, 114 of the current small towns were declared urban under the communist regime, which greatly enlarged the previous network structure of 54 small towns (Stoica et al., 2020b). Thus, it can be stated that the largest part of the present ensemble of urban settlements was designated in that period. Afterwards, in the post-communist era (1990 and onwards), only 56 of the current towns gained urban status (Stoica et al., 2020a). This development attests that urbanization was considered as one of the main driving forces of the communist policy (Benedek, 2006; Ianoș, 2004). The increment of the urban population was realized in two ways, namely increasing the numbers of urban centres, and attracting new residents into the already established towns and cities. The sizeable growth was underpinned by the industrialization process, which constituted another major driving force of the communist policy (Ianoș, 2004). Newly built factories were located, predominantly, in urban nuclei, and that attracted a supplementary workforce from rural areas (Popescu, 2014). Nevertheless, for a long time, investments were directed predominantly to bigger urban centres, and as a result, the demographic increase was higher in these places, whereas in small towns it remained moderate.

In demographic terms, many of these small places acted as a relay in the rural–urban migration, gaining inhabitants from the rural areas, but at the same time losing population to larger towns and cities. In spite of this, the overall population living in small towns registered a high increase, from just a little over 440,000 inhabitants in 1948 to more than 2,020,000 in 1992 (National Institute of Statistics). This trajectory was also strongly influenced by national demographic policies regarding the proclaiming of abortions as illegal, inaccessibility to modern means of family planning, and restrictions on international migration. In fact, Romania had set one of the "toughest pronatalist policies" (Rotariu, 2019, p.16) in the communist bloc.

The radical shift from December 1989 led to a fundamental change in the regime: from the communist system characterized by central economic planning to a democracy and free-market economy. As a result, urban areas have undergone massive transformations, underpinned by economic restructuring and a deindustrialization process (Stoica et al., 2020a). In many industrial small towns, the economic downturn led to massive layoffs (especially in the manufacturing and mining industries) and unemployment. In some places, industrial activities were gradually replaced by other types of activities, like services. Other small towns maintained their previous economic functions but had to downsize them. Nevertheless, the economic decline had a major impact in the medium and long term on many of the small towns, leading to out-migration of the skilled labour force, ageing and demographic decline. Thus, a so-called "de-urbanisation" process was described, which mostly meant these towns decayed due to unemployment and out-migration (Nuțu, 2008).

Regarding the towns that gained their urban status after 1990, some of them have been known to maintain some of their rural characteristics (Stoica et al., 2020b).

Meanwhile, the complex changes (economic, political, and societal) that took place after the fall of communism have also greatly influenced the demographic behaviour of the population. The decreasing birth rate is associated with a restructuring of the fertility model in a society that is rapidly adopting the system of values and attitudes of developed countries (INS, 2012). The fertility rate in Romania has dropped significantly and is currently far below the replacement level (Ghețău, 2014). This creates a vicious circle of demographic decline, in which migrations have also played an important role. Romania's accession to the European Union (2007), which allowed intra-EU mobility, intensified the outward flows, which were mainly incentivized by the higher income that Western countries offered compared to the low wages in Romania. In fact, this problem pertains not only to Romania: it is considered that intra-EU mobility exacerbates the population decline in eastern European countries in general (Lutz et al., 2019). On the local level, the stigma of a shrinking town sometimes induces a decrease in investments and deepens the overall decline (Rybczynski & Linnemann, 1999).

Nonetheless, in the post-communist period, small towns in Romania exhibited divergent paths of transformation regarding how they adapted to market economy (Musil, 1993; Mykhnenko & Turok, 2008; Zamfir et al., 2009). Some prosperous urban settlements turned to new economic activities or maintained their traditional ones and managed to gain population. But many small towns have not fared so well, dealing with challenges like economic decline, shrinkage, population loss, ageing, etc. (Stoica et al., 2020a). Nedović-Budić et al. (2006, p. 15) label this diversity as a "mosaic of urban landscapes and experiences" that require further investigations.

9.3 Data and Methods

Statistical data used in this study were provided by the National Institute of Statistics, mainly collected from the Tempo online database, which includes information on population dynamics (population, live births, deaths, immigration and emigration) and population structure by major age groups. Based on these data, we have calculated several indicators—population dynamics, vitality index, net migration rate, old-age dependency ratio—and computed an index of demographic vulnerability, the purpose of which was to capture the post-communist transformations and the demographic perspective of Romanian small towns. Two years of reference were used, respectively 1992 as the first year with reliable statistical data after the fall of communism, expressing the effect of the former regime's policies, and 2018, reflecting the impact of implementing post-communist policies and the changes that have taken place.

The population dynamics can indicate the extent to which small towns were affected by the population decline identified on the national and urban level. With this in mind, a hierarchy allowing the classification of the population change (growth, decrease or stagnation) between 1992 and 2018 was created for each unit.

Subsequently, to identify the impact of natural increase over the population change and hints about its future trajectory, the Vitality Index was calculated as the ratio between live newborns and deaths. Following this model, values over 100% mean that the population tends to grow (if migration does not influence the process), while values below 100% indicate a sign of demographic decline in the medium and long term (Mitrică et al., 2019).

In order to investigate the impact of migration over the population dynamics, the rate of net migration was used. This indicator represents the difference between immigrants and emigrants, divided by the total population of a given small town, expressed per 1000 inhabitants. To identify the overall trend, considering inter-annual fluctuations, the average rate for the entire period of 1992–2018 was calculated.

To evaluate the population structure changes by large age groups, but also to quantify the demographic ageing, the old-age dependency ratio was calculated. This indicator reflects the level of support of elderly people (aged 65 or over) by the working-age population (aged 15–64). The value is expressed as the share of older dependents per 100 working-age population. This variable "examines the shifting population age structures for intergenerational support systems" (UNDESA, 2019, p. 1).

Then to depict the current demographic challenges of small towns in Romania in the context of post-communist changes, a composite Index of Demographic Vulnerability (IDV) was constructed based on Müller and Schiappacasse's (2009) proposal. According to the authors, the purpose of the tool was to capture the differentiation, in terms of demographic changes, between several European regions and their potential to fall in decline. Their IDV index was based on five variables: the annual growth of the total population, the annual growth of the population under 14 years of age, the annual growth of the population over 80, the life expectancy and the fertility rate. The current IDV index proposed here is adapted to the Romanian reality in terms of the components that can affect the demographic structures and lead to decline, with its imminent consequences.

The main five indicators included in the index are the following:

1. Population dynamics between 1992 and 2018.
2. Changes in the share of 0–14 years age groups—share of population aged 0–14 of the total population, calculated both for 1992 and 2018, and then the dynamics for this time.
3. Changes in the share of the population aged 65 and over of the total population, calculated both for 1992 and 2018, and then the dynamics for this time. The age structure of the population is considered an indirect driver of demographic decline (Koulov et al., 2019), the dynamics of young and elderly age groups is analysed because it reflects the shrinking generations and population ageing.
4. Vitality index dynamics (1992–2018).
5. Ageing of the working-age population—expressed as the ratio between the population aged 50–64 and the total of the working-age population (aged 15–64), for 1992 and 2018, followed by its dynamics between the two reference years. This indicator reflects the ageing degree of the workforce, with implications over productivity growth and long-term shrinking of the labour force.

Table 9.1 Description of the indicators and the division of classes

No.	Indicators name	Class 1	Class 2	Class 3	Class 4
1	Population dynamics (%)	Over 0%	0 to −10%	−10 to −15%	Below −15%
2	Changes in the share of 0–14 year's age groups (%)	0 to −5%	−5 to −10%	−10 to −15%	Below −15%
3	Changes in the share of the population aged 65 and over (%)	Below 0%	0–5%	5–10%	Over 10%
4	Vitality index dynamics (%)	Over 0%	0 to −30%	−30 to −100%	Below −100%
5	Ageing of the working-age population (%)	Below −1%	−1 to 1%	1–5%	Over 5%

Subsequently, the values of each indicator were ranked in one of four classes (score 1–4), taking significant thresholds into account (Table 9.1). For example, for population dynamics, class 1 defines an increase in the number of inhabitants between 1992 and 2018, while the split between classes 2–4 was done according to the intensity of the decline. In the next step, the individual values obtained for each indicator were cumulated (as an overall score) in the composite index (IDV). Then the ranking of small towns was done in three categories: low vulnerability (5–10), medium vulnerability (10–15), and high vulnerability (15–20).

The criteria for choosing the variables were the relevance for the demographic analysis in capturing the dynamics of the investigated processes, availability, and applicability in the Romanian context. Admittedly, there were some statistical limitations, for example: analysis of life expectancy at birth is simply not available at the local administrative level.

It is also worth mentioning that there were indicators that were calculated but proved to be irrelevant, such as demographic dependency rate. In this case, the results registered only slight differences between 1992 and 2018 because even though the elderly population has grown significantly, at the same time, there has been a substantial reduction in the young population.

9.4 Results and Discussion

9.4.1 Trends in Population Dynamics

The population dynamics of all small towns shows an overall decline of 2.9% between 1992 and 2018. This result is correlated with the downward trend registered on the national (−4%) and urban (−0.11%) levels, denoting a more significant contraction of the lower level of the urban system compared with urban areas in higher tiers. In fact, cities generally attracted more investments and population, leaving small towns far behind (Bretagnolle, 2003). Furthermore, small towns are perceived as having the lowest adaptability to the new post-communist socio-economic conditions (Benedek, 2006). The population loss, in absolute numbers, was 97,340 inhabitants, which may not seem as much, but it represents around 10% of the total decrease registered for the entire country.

Still, the category of small towns is very heterogeneous and the analysis of the population dynamics in the post-communist period revealed the following typology (Fig. 9.1):

(1) A high increase (over 15%) in the number of inhabitants was found in 8.5% of small towns. This positive development is mainly related to (a) close geographical proximity to large cities with significant population in-flows (e.g. the towns in Ilfov County, adjacent to Bucharest) (Fig. 9.2); (b) the effect of a significant stock of younger population established in some places during the communist period (e.g. Rovinari and Turceni in Gorj County); (c) traditional family views, which translate to pro-birth attitudes, among the neo-protestant religious denominations (Cace et al., 2012) like the considerable Pentecostal communities (e.g. Cajvana and Vicovu de Sus in Suceava County); (d) tourism activities (Techirghiol and Eforie in Constanța County); (e) industrial activities (Sântana

Fig. 9.1 Population dynamics (1992–2018). *Source* Author's elaboration

Fig. 9.2 Vitality index (1992). *Source* Author's elaboration

in Arad County); etc. Still, more than half of the small towns that have experienced significant population growth fall into the first group, owing their growth to their closeness to major urban centres.

(2) A moderate increase (between 1 and 15%) in the number of inhabitants was registered in 19.7% of small towns in Romania. The general explanations behind this growth are similar to the ones identified for the first category, even though, in this case, only a few of the settlements (e.g. Râşnov in Braşov County, Ungheni in Mureş County) are located near larger cities. Another group of settlements that has exhibited moderate growth is comprised of small towns registering consistent ethnic groups (like the roma population) having, traditionally, higher fertility and birth rates (e.g. Ţăndărei in Ialomiţa County, Bolintin-Vale in Giurgiu County). However, it should be noted that a conjunction of these several factors is responsible for many town's positive population trajectories, such as considerable roma communities characterized by Pentecostal religious denominations (Cace et al., 2012), or urban settlements located in the adjacent area of large cities developing different economic activities, etc.

(3) A stagnation phase manifested in 7.1% of small towns (between −1% and 1%). These can be found in every part of Romania, but there is a significant concentration of them in the south, which is where about half of all stagnant settlements can be found.

(4) Moderate decrease (between −1% and −15%) of the population is featured in half of the urban centres, over two-thirds (71%) of which have experienced population losses of over 5%.
(5) High decrease (over −15%) of the number of inhabitants was noted in 14.7% of small towns.

The causes for this behaviour of the urban centres—population decrease—are complex and varied, although we were able to identify several categories: (a) towns with a strong agricultural profile (e.g. Dăbuleni in Dolj County, Tismana in Gorj County); (b) towns affected by the reduction or disappearance of industry or related activities (e.g. Baia de Arieș in Alba County, Fieni in Dâmbovița County, Bălan in Harghita County); (c) tourist resorts (e.g. Predeal in Brașov County, Băile Herculane in Caraș-Severin County) with a high percentage of elderly population. Some of them registered early industrialization (Bănică & Istrate, 2018) from the beginning of the nineteenth century (Azuga, Sinaia and Bușteni in Prahova County) marked by a strong decline after 1990 and a reorientation to tourism; (d) some of the newly declared towns which follow the general tendency of population decline in rural areas.

The value ranges between −30.5% in the case of Bălan (Harghita County) and 206.7% in the case of Bragadiru (Ilfov County). These extreme opposites reflect two processes that took place in post-communist Romania: deindustrialization and suburbanization. Thus, for the former, the population loss is a consequence of the closure of a copper mine to which, in 1990, almost 83% of the inhabitants were related directly or indirectly (Nuțu, 2008). In the latter, the population gains are connected to the suburbanization process, as the town is located on the south-west edge of the capital city.

Several correlations were examined (with the economic profile, the dynamics of certain economic activities, religion, and ethnicity) to explain the main causes for certain dynamics between 1992 and 2018, but generally there were no well-defined patterns that were valid for the entire period. For example, even if several small towns were affected by the decline at the beginning of the period, some of them gradually changed their trajectory in a positive way. This confirms that multiple factors overlap with place-specific characteristics embedded in the local and regional context. In this regard, governance is considered one of the leading forces for urban development (Benedek, 2006). Furthermore, population changes have complex determinations, originating not only in economic, social and political aspects, but also in the decisions, attitudes, and behaviours of the individuals (Ghețău, 2014). Then, in some places, the long-term consequences of mass emigrations had a strong impact that led to population decline.

9.4.2 Vitality Index

The general tendency of the vitality index at the small town's level expresses a decreasing trajectory from an average value of 125.2% in 1992 to 86.9% in 2018. Comparing the two reference years results that in 1992, only 36% of the towns registered values lower than 100%, almost a third of which recorded values over 90%. Meanwhile, in 2018, the majority (72%) of the urban settlements presents below unity values denoting a natural population decrease. The territorial distribution reflects that immediately after the fall of communism, one of the areas that often exhibited values of over 100% was the eastern side of the country, with few exceptions (Fig. 9.2).

This area traditionally registers a higher birth rate compared to the western side, which featured lower values, even in the communist period (Cucu & Iordan, 1984). The highest values, over 300%, were recorded in four of the current small towns, namely Cajvana (Suceava County), followed by three mining towns—Motru, Rovinari (Gorj County) and Bălan (Harghita County). In fact, the vast majority of the mining towns recorded more newborns than deceased, along with towns characterized by significant industrial activities (e.g. Victoria in Brașov County, Plopeni in Prahova County). The smallest values (between 45 and 47%) were registered in the western part of the country, in Ciacova (Timiș County), Pecica (Arad County), and Geoagiu (Hunedoara County).

In 2018, the small towns with supra-unity values were spread all over the country, in several counties (Fig. 9.3). Some of them benefited from their proximity to large cities, attracting investment and young inhabitants (e.g. Bragadiru, Chitila, Otopeni in Ilfov County; Cisnădie in Sibiu County). Others carried out representative industrial activities (e.g. Rovinari in Gorj County; Sântana in Arad County) or other viable economic activities. In some places the pro-birth attitude of a part of the community led to the preservation of a high birth rate (e.g. Țăndărei in Ialomița County, Vicovu de Sus in Suceava County).

The first rank is still held by Cajvana (366.7%) which even registered a slight increase compared to 1992. The smallest values, below 30%, were found in touristic resorts (Azuga and Sinaia in Prahova County, Borsec in Harghita County, Băile Govora in Vâlcea County) and a settlement with a high share of agricultural activities (Fierbinți-Târg in Ialomița County). Additionally, the lowest value (18.2%) was registered by Sulina (Tulcea County), the only urban centre in the Danube Delta, a famously isolated area.

9.4.3 Net Migration Rate

In the post-communist period, the migration pattern registered changes that were related to the local economic turmoil, but also to the wider societal and political contexts, especially international emigration.The net migration rate depicts the overall situation for the 1992–2018 period, revealing that the attractive small towns

Fig. 9.3 Vitality index (2018). *Surce* Author's elaboration

with a positive balance (over 1.01‰ annual average value) represent just 24% of the urban settlements. The highest population inflows were recorded by small towns located around large cities, leading to the development of new residential areas as a result of the urban expansion process. The communist intra-urban pattern promoted large housing estates and as a consequence many inhabitants lived in apartment blocks. Then, mainly after 2000, a real process of suburbanization began, with people moving outside the city boundaries, in more dispersed settlements. In some parts, some economic activities relocated to the small town's territory. In any case, around 70% of the small towns which experienced inflows of inhabitants are geographically located in the surroundings of cities that are higher up in the urban hierarchy. By far, the most prosperous of these is Bragadiru, with an annual average value of 47.7‰, more than double than that of Otopeni (23.8 ‰), which occupies the second place (Fig. 9.4). Nevertheless, some of the small towns located in the adjoined area of bigger urban settlements are also developing tourism activities (e.g. Eforie and Techirghiol in Constanța County).

Other small towns that managed to attract population preserved some industrial enterprises (Rovinari and Turceni in Gorj County) or tourism services (e.g. Geoagiu in Hunedoara County) which led to an inflow of inhabitants.

The next group of towns (21%) presents a net migration rate of between −1‰ and 1‰, denoting a somewhat stationary tendency with a minimum influence of migratory flows over population change. Though affected by the decline of traditional

Fig. 9.4 Net migration rate (1992–2018). *Source* Author's elaboration

industrial activities, some of these small towns partially converted to other types of activities like tourism (e.g. Nucet in Bihor County, Baia Sprie in Maramureș County).

A moderate negative migration rate (−1.01 to 5‰) is characteristic for a large part of small towns (38%) which can be found all over the country.

The most severe outflows, with a negative average annual value of net migration under −5‰, were characteristic for 17% of the small towns. In this regard, the most repulsive areas are represented by some of the former mono-industrial towns, whose economic profile was dominated, directly and indirectly, by a single industrial branch at the end of the communist period. These places received significant inflows of employees from different parts of the country (Cucu & Iordan, 1984). Then, after the collapse of communism, the dismantling of enterprises led to retirements and layoffs in the late 90s, and many people (especially the elderly) returned to their places of origin. The highest negative values of net migration rate are registered in Bălan (Harghita County)—18.7‰, Uricani (Hunedoara County)—11.9‰, and Motru (Gorj County)—11.2‰.

Previous research concluded that mono-industrial towns "were found to be particularly vulnerable due to their narrow economic specialization" (Restrepo Cadavid et al., 2017) after the fall of communism and the subsequent dissolution of its planning policies.

The large group of towns with moderate and high negative migration rates includes different categories, like tourist resorts, urban areas affected by sharp deindustrialization, agricultural towns, etc.

9.4.4 *The Old-Age Dependency Ratio*

Despite its serious implications, demographic ageing received little attention in regional sciences and urban economics (McCann, 2017). To analyse the demographic ageing macroeconomic impact, one of the relevant indicators is the old-age dependency ratio (Aiyar et al., 2016). In the case of Romanian small towns, in 2018, the medium value of this variable was 21.9%, which is 7% higher than its value in 1992, in absolute terms. The comparative analysis over time reflects that in 2018, almost 19% of the small towns registered values of over 25%, compared to a mere 5% in 1992, while only 5% of the small towns present values below 15%, compared to 54% in 1992. These differences are partially attributed to the increase in life expectancy, which has led to a higher percentage of elderly people, but also to out-migration, which is one of the main reasons for the decrease in the working-age population.

In 2018, the highest values (over 25%) are exhibited by: (a) towns with a high share of population involved in agriculture (e.g. Dăbuleni in Dolj County, Fierbinți-Târg in Ialomița County), (b) towns that downsized their industrial activity (e.g. Zlatna in Alba County, Ocnele Mari in Vâlcea County), (c) many of the touristic resorts, etc. (e.g. Sinaia in Prahova County; Băile Govora in Vâlcea County) (Fig. 9.5).

Half of the small towns are characterized by old-age dependency ratio values of 20–25%, identified mainly as towns in the first two categories.

Fig. 9.5 Old-age dependency ratio (2018). *Source* Author's elaboration

The highest value was registered in Dăbuleni (38.3%) (Dolj County), and the lowest one was found in Rovinari (4.8%) (Gorj County). The group of towns with the lowest values (under 15%) includes some of the mining towns (Rovinari and Motru in Gorj County, Uricani in Hunedoara County), which present a large working-age population, attributed to the large inflows of young labour force in the 1980s (Ianoş, 2004) but also to the lower life expectancy and, consequently, a lower share of the older age group (Kideckel, 2001). Other small towns encountered in this category receive significant inflows of younger cohorts supporting the adult age group, correlated with a high vitality index (Țăndărei in Ialomița County, Cajvana and Vicovu de Sus in Suceava County) or favoured by the location in the surrounding area of large cities (e.g. Bragadiru in Ilfov County, Podu Ilioaiei in Iași County).

Between 1992 and 2018, the old-age dependency ratio dynamics was marked by a growth tendency in 93% of small towns. These findings correspond with the European trends, according to which a high share of the ageing population lives in relatively small towns and cities (Eurostat, 2016). The highest increases (over 10%) is specific for 27% of the urban settlements including, mainly, most of the touristic resorts and towns impacted by the dissolution of local industry (e.g. Brad in Hunedoara County, Oțelu Roșu in Caraș-Severin County) (Fig. 9.6).

The urban centres that registered an improvement of the old-age dependency rate, expressed by a decreasing value, are mainly post-communist towns (71%), half of which are located in two counties, Arad and Timiș, along the western border.

Fig. 9.6 Old-age dependency ratio dynamics (1992–2018). *Source* Author's elaboration

9.4.5 Index of Demographic Vulnerability (IDV)

Based on the post-communist dynamics of demographic processes, the IDV results are set on a hierarchic scale where the lowest value is 5 (as is the case with Pecica in Arad County and Ardud in Satu Mare County), indicating low demographic vulnerability, and the highest value is 20 (Bălan in Harghita County) denoting a high level of vulnerability.

The best-positioned category, which demonstrates low vulnerability, includes 26% of the small towns, where generally, the five indicators score in the first and the second class. In some cases, the small towns with the lowest vulnerability benefited from their proximity to large cities, attracting young adult population (e.g. the towns located near Bucharest, Tăuții-Măgherăuș near Baia Mare). Other cases rely on investments in local industry (e.g. Segarcea in Dolj County, Sântana in Arad County), tourism (e.g. Ocna Sibiului in Sibiu County), etc. (Fig. 9.7). Some of the small towns have significant ethnic or religious communities that traditionally have a pro-birth attitude (e.g. Cajvana in Suceava County, Bechet in Dolj County). However, a synergy of multiple factors can often be identified (e.g. Vicovu de Sus in Suceava County).

The medium IDV category includes the lion's share of small towns (60%). All of them present an increase in the population aged 65 and over and a predominance of class 3 and sometimes class 2 for other indicators. Although 25% of these small towns gained population after 1992, they fall in class 3 or even 4 for at least two other

Fig. 9.7 Index of demographic vulnerability—county level. *Source* Author's elaboration

indicators. This confirms previous findings, which illustrate that even if a population is not decreasing in absolute numbers, the consequences of demographic ageing are significant (McCann, 2017). Regarding this category of urban centres with medium demographic vulnerability, a particular typology associating the small towns with specific functions and urban background (for both old towns and post-communist towns) was not found.

The highest values of IDV were found in 14% of the small towns. None of them fall into the first class in any indicator, or the second class in the vitality index dynamics and ageing of the working-age population. Most of the indicators for this group of towns fall in class 3 or 4. The town with the highest index of demographic vulnerability is Bălan (Harghita County), with all five indicators in the last class, followed by two other towns (Plopeni in Prahova County and Sulina in Tulcea County) with four indicators in the fourth class.

A high IDV is characteristic for towns affected by downsizing or disappearance of industrial enterprises (e.g. Balș in Olt County, Moldova Nouă in Caraș Severin County, Victoria in Brașov County, Uricani in Hunedoara County) and longtime touristic resorts affected by a sharp decline in recent decades (e.g. Băile Herculane in Caraș Severin County, Băile Tușnad in Harghita County, Vatra Dornei in Suceava County). It is obvious that all these economic turmoils had adverse demographic repercussions. Another restrictive factor is geographical isolation (amid a shrinking economic activity) which led to outflows of population (e.g. Sulina in Tulcea County).

Correlation of the IDV results with the emergence stages of small towns reveals that the most favourable demographic profile is characteristic of post-communist towns, 70% of which present low vulnerability. Some of these urban settlements preserve certain demographic behaviours more typical of rural areas, such as a higher birth rate which in turn is reflected in other demographic indicators.

The analysis of the IDV values at the regional level outlines some differences. All four towns in the Bucharest-Ilfov Region (Bragadiru, Chitila, Măgurele and Otopeni) are in the low demographic vulnerability group, benefiting from a revival of the population, which is related to suburbanization. Next in the hierarchy are the North-East and West regions, each with 35% of the small towns in the low vulnerability category (Fig. 9.8). The causes for this are complex, like the fact that, traditionally, the North-East Region was characterized by a higher birth rate, registering late demographic transition (Trebici & Hristache, 1986). Meanwhile, the West Region is characterized by demographic rejuvenation, brought on by more significant economic development, in contrast with the early ’90s, when the region had the highest percentage of elderly people in the country (Trebici, 1996).

The region with the highest share of small towns with a high IDV is the South Region (18%), followed by the South-East Region (17%).

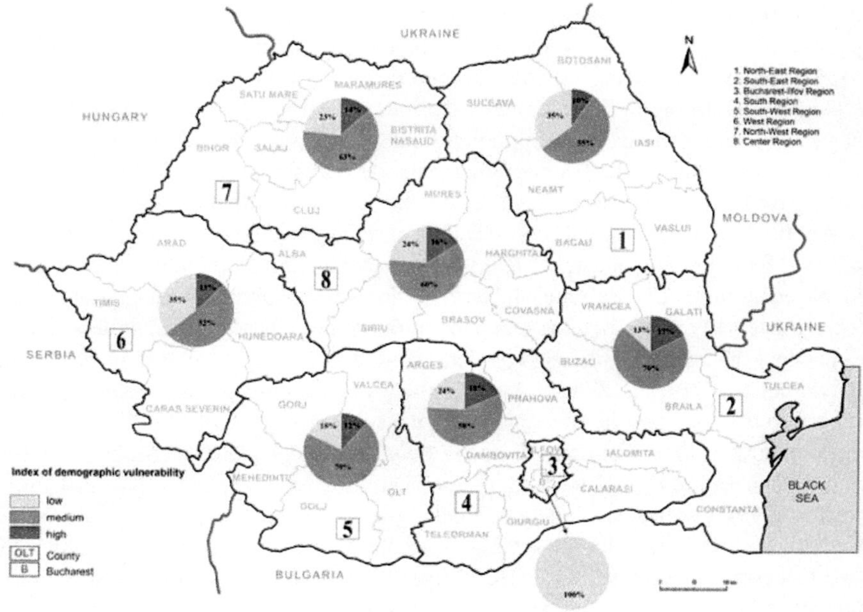

Fig. 9.8 Index of demographic vulnerability—regional level. *Source* Author's elaboration

9.5 Conclusions

Small towns are regarded as the urban category that is most vulnerable to structural transformations and ongoing demographic changes, on the background of the economic turmoil during the post-communist transition to a free-market economy. The analysis revealed major changes in the small town's demographic characteristics between 1992 and 2018, such as a general tendency of decline and ageing, given that almost 65% lost population in the time span of 28 years, while 7% registered stagnation (−1 to 1%); 80.4% presented decreases in the vitality index between 1992 and 2018, denoting a weakening of the natural demographic potential; 55% were characterized by a migratory deficit, revealed by the negative average annual value of the net migration rate (−1.1 ‰), meanwhile 21% denote a somehow stationary tendency (−1 to 1‰); 93% present a growth tendency of the old-age dependency ratio.

The composite index of demographic vulnerability pointed out that 14% of small towns register high IDV values and 60% of them register medium IDV values, but the situation can worsen if outward migration continues at the same intensity.

Setting aside this demographic overview, the urban tier of small towns is very heterogeneous, characterized by several demographic asymmetries: prosperous urban centres on one end of the spectrum, and on the other—towns gravely affected by regression. The most favourable situation is characteristic of several small towns located in the surroundings of cities higher up in the urban hierarchy, as they received significant inflows of younger cohorts, leading to a demographic revival. Others developed new economic activities (e.g. industry or tourism related) or maintained factories established in the communist period, even if it was often in a diminished capacity. These findings confirm previous studies denoting that "well-endowed and well-located cities" are far better off than others (Mykhnenko & Turok, 2008, p. 317). A positive demographic overview is also valid for towns with a pro-birth attitude of part of the local community. In contrast, demographic decay is characteristic of towns affected by a sharp deindustrialization process that lacks a sustainable conversion to viable economic activities, which leads to outflows of inhabitants, population decline, lowering fertility and birth rates, and population ageing. These types of urban centres, dependent on dwindling industries, were defined as the main "losers" of transition (Mykhnenko & Turok, 2008, p. 316). In this case, the demographic challenges are related to the uncertain future considering the subsequent reduction of the working-age population, which in turn impacts the town's economic viability (McCann, 2017) and influences the spatial reconfiguration (Dumitrache et al., 2016; Nae & Dumitrache, 2019). Other towns registering demographic depreciation are the majority of touristic resorts. The latter are defined by contradictory characteristics, falling into two categories: successful, crowded places, but with a high number of secondary residences and relatively high prices of properties; and touristic resorts defined by a downturn in touristic activities, even though they possess great unexploited potential.

In view of the magnitude of population decline and ageing in recent years, a new approach has emerged, calling to rethink growth-oriented planning and consider a downsizing perspective, which implies the acceptance of demographic changes and learning how to manage them (Hospers & Reverda, 2015; McCann, 2017; OECD, 2012).

This demographic mosaic confirms the divergent trajectories of small towns in post-communist Romania. Beyond the changes in their demographic behaviour, these trajectories were triggered by path-dependence on the communist past, the restructuring of the local economy, geographical location, opportunities in other urban areas and western European countries, and governance. This perspective emphasizes the need for designing future policies and strategies on how to deal with the challenges of demographic change.

Acknowledgements The article was partially supported by the University of Bucharest, project number UB 1322 "Integrated and sectoral analyses in trans-scalar territorial dynamics". The authors contributed equally to this article.

References

Aiyar, S., Ebeke, C., & Shao, X. (2016). The Impact of Workforce Aging on European Productivity. IMF Working Paper No. 16/238, International Monetary Fund, Washington DC. https://www.imf.org/external/pubs/ft/wp/2016/wp16238.pdf

Akiyama, R. S. (2011). Territorial future. Understanding Italian challenges and Lombardy opportunities. In L. Pedrazzini & R. S. Akiyama (Eds.), *From territorial cohesion to the new regionalized Europe*. Maggioli, Santarcangelo di Romagna.

Atoyan, R., Christiansen, L., Dizioli, A., Ebeke, C., Ilahi, N., Ilyina, A., Mehrez, G., Qu, H., Raei, F., Rhee, A., & Zakharova, D. (2016). Emigration and Its Economic Impact on Eastern Europe. IMF Staff Discussion Note 16/07. https://www.imf.org/~/media/Websites/IMF/imported-full-text-pdf/external/pubs/ft/sdn/2016/_sdn1607.ashx

Batog, C., Crivelli, E., Ilyina, A., Jakab, Z., Lee, J., Musayev, A., Petrova, I., Scott, A., Shabunina, A., Tudyka, A., Xu, X. C., & Zhang, R. (2019). Demographic headwinds in Central and Eastern Europe. International Monetary Fund, European Department (Series), Washington DC

Bănică, A., & Istrate, M. (2018). Rezilienţa oraşelor mici din Carpaţii Românesti în ultimul secol. In I. Muntele, A. Ungureanu, & C. Rusu (Eds.) *Spaţiul Geografic Românesc la 100 de ani de la Marea Unire. Edit. Universităţii*. Alexandru Ioan Cuza, Iaşi

Benedek, J. (2006). Urban policy and urbanization in the transition Romania. In *Romanian Review of Regional Studies* (Vol. 1, No. 2). Cluj-Napoca.

Bouzarovski, S., Sýkora, L., & Matoušek, R. (2016). Locked-in post-socialism: rolling path dependencies in Liberec's district heating system. In *Eurasian Geography and Economics* (Vol. 57, No. 4–5). Taylor & Francis.

Bretagnolle, A. (2003). Vitesse et processus de sélection hiérarchique dans le système des villes françaises. In D. Pumain & F. Mattéi (Eds.), *Données urbaines* (Vol. 4). Anthropos, Economica, Paris, France

Cace, S., Tomescu, C., & Cojocaru, D. (2012). The role of religion in several Roma communities. In *Revista de cercetare şi intervenţie social* (Vol. 36). Iaşi.

Cucu, V., Iordan, I., & (coord.),. (1984). *Geografia României* (Vol. II). Academiei Republicii Socialiste România, Bucharest, Romania.

Czapiewski, K., Bański, J., & Górczyńska M. (2016). The impact of location on the role of small towns in regional development: Mazovia, Poland. In *European Countryside* (Vol. 8, No. 4). Brno.

Dumitrache, L., Zamfir, D., Nae, M., Simion, G., & Stoica, I.-V. (2016). The urban nexus: Contradictions and dilemmas of (post) communist (sub) urbanization in Romania. In *Human Geographies. Romanian Journal of Studies and Research in Human Geography, 10*(1). Bucharest, Romania

Ecovast. (2013). The importance of small towns. A position paper by the European council for the village and small town. http://www.ecovast.org/papers/ECOVAST%20Importance%20of%20Small%20towns%20Position%20Paper.PDF

Eurostat,. (2016). *Urban Europe Statistics on cities, towns and suburbs* (2016th ed.). Publications office of the European Union.

Gheţău, V. (2014). Declinul populaţiei României. Două scenarii până la mijlocul secolului. In *Revista de Asistenţă Socială* (Vol. 4). Bucharest, Romania.

Haase, A., Rink, D., & Grossmann, K. (2016). Shrinking cities in post-socialist Europe: what can we learn from their analysis for theory building today? In *Geografiska Annaler: Series B, Human Geography* (Vol 98, No. 4). Uppsala, Sweden.

Hospers, G. J., & Reverda, N. (2015). *Managing population decline in Europe's urban and rural areas*. Springer.

Ianoş, I. (2004). Dinamica urbană. Aplicaţii la oraşul şi sistemul urban românesc. Edit. Tehnică, Bucharest, Romania.

Ianoş, I. (2016). Causal relationships between economic dynamics and migration. Romania as case study. In J. Dominguez-Mujica (Ed.), *Global change and human mobility, series title: Advances geographical and environmental sciences*. Springer.

Institutul Național de Statistică (INS) (National Institute of Statistics). (2012). Evoluția natalității și fertilității în România. https://insse.ro/cms/files/publicatii/Evolutia%20natalitatii%20si%20fertilitatii%20in%20Romania_n.pdf

Institutul Național de Statistică (INS) (National Institute of Statistics). (2020). Tempo online statistics. http://statistici.insse.ro/shop/?lang=en

Józefowicz, K. (2022). Small-town depopulation in the Wielkopolskie voivodeship, Poland. *Journal of Urban and Regional Analysis, 14*(1), Bucharest, Romania.

Kideckel, D. A. (2001). Labor and Society in the Jiu Valley and Făgăraș Regions of Romania, Part I: Variations in Response to the Crisis, The National Council for Eurasian and East European Research, Washington DC. https://www.ucis.pitt.edu/nceeer/2001-815-18g-Kideckel.pdf

Klingholz, R. (2009). Europe's real demographic challenge. In *Policy Review* (Vol. 157). Standford, USA.

Koulov, B., Boyadjiev, V., & Ravnachka, A. (2019) The Demographyc Draining of Bulgaria's Rural Area: A GIS-Aided Geospatial Analysis (1992–2017). In J. Bański (Ed.), *Three Decades of Transformation in the East-Central European Countryside*. Springer.

Lutz, W. (2010). Emerging population issues in Eastern Europe and Central Asia: Research gaps on demographic trends, human capital and climate change. UNFPA. https://www.unfpa.org/sites/default/files/pub-pdf/bmsablon.pdf

Lutz, W. (Ed.), Amran, G., Bélanger, A., Conte, A., Gailey, N., Ghio, D., Grapsa, E., Jensen, K., Loichinger, E., Marois, G., Muttarak, R., Potančoková, M., Sabourin, P., & Stonawski, M. (2019). Demographic Scenarios for the EU—Migration, Population and Education. Publications Office, Luxembourg.

Lutz, W., Sanderson, W. C., & Andruchowitz, S. (Eds.). (2017). *Moving frontiers in population forecasting and aging: Sergei Scherbov 65*. Annotated Reprints of Key Publications. IIASA.

Margaras, V. (2019), Demographic trends in EU regions. European Parliamentary Research Service. https://www.europarl.europa.eu/RegData/etudes/BRIE/2019/633160/EPRS_BRI(2019)633160_EN.pdf

McCann, P. (2017). Urban futures, population ageing and demographic decline. In *Cambridge Journal of Regions, Economy and Society* (Vol. 10, No. 3). Oxford, United Kingdom.

Mitrică B., Perșu M., Mocanu I., Şerban P., Grigorescu I., & Damian N. (2019). Changes in the Dynamics and Demographic Structure of the Romanian Rural Population: An Overview of the Post-communist Period. In: J. Bański (Eds.), *Three Decades of Transformation in the East-Central European Countryside*. Springer.

Müller, B., & Schiappacasse, P. (2009). The European regional demographic mosaic—A challenge for spatial development. In H. Kilper (Eds.), *German Annual of Spatial Research and Policy 2009*. Springer.

Musil, J. (1993). Changing urban systems in post-communist societies in central Europe: Analysis and prediction. In *Urban Studies* (Vol. 30, No. 6). Sage Publications.

Mykhnenko, V., & Turok, I. (2008). East European Cities—Patterns of growth and decline, 1960–2005. In: *International Planning Studies* (Vol. 13, No. 4). Routledge, United Kingdom.

Nae, M., & Dumitrache, L. (2019). Post-socialist urban changes and role strain in assessing the quality of life: The case of Bucharest, Romania. In B.R.K. Sinha (Ed.), *Multidimensional Approach to Quality of Life Issues*. Springer Nature.

Nedović-Budić, Z., Tsenkova, S., & Marcuse, P. (2006). The urban mosaic of post-socialist Europe. In S. Tsenkova & Z. Nedović-Budić (Eds.), *The Urban Mosaic of Post-Socialist Europe*. Space, Institutions and Policy. Physica-Verlag HD.

de Noronha Vaz, M. T., van Leeuwen, E. S., & Nijkamp, P. (Eds.). (2013). *Towns in a rural world*. Ashgate.

Nuțu, A. O. (2008). Life in post-socialist mono-industrial communities: the case of Bălan. http://www.aughty.org/pdf/case_of_balan.pdf

OECD. (2012). *Demographic change and local development: Shrinkage*. Organisation for Economic Cooperation and Development Publishing, Paris, France.

OECD. (2015). *Ageing in cities*. Organisation for Economic Cooperation and Development Publishing.

Palkina, M., Kislitsyna, V., & Chernyshev, K. (2021). Analysis of the relationship of investment and demographic factors in the development of depressed regions. *Journal of Urban and Regional Analysis, 13*(1), Bucharest, Romania.

Popescu, C. (2014). Deindustrialization and urban shrinkage in Romania. What lessons for the spatial policy? In *Transylvanian Review of Administrative Sciences* (Vol. 42E). Cluj-Napoca, Romania.

Restrepo Cadavid, P., Cineas, G., Quintero, L., & Zhukova, S. (2017). Cities in Eastern Europe and Central Asia: A Story of Urban Growth and Decline. World Bank, Washington DC, USA.

Rotariu, T. (2019). Notes on the demographic transformations in postcommunist Romania. In *Studia Universitatis Babeş-Bolyai Sociologia* (Vol. 64, No. 1). Cluj-Napoca, Romania.

Rybczynski, W., & Linnemann, P. (1999). How to save our shrinking cities. In *Public Interest* (Vol. 135). Washington DC, USA.

Servillo, L., Atkinson, R., Smith, I., Russo, A., Sýkora, L., & Demazière, C. (2014). TOWN, Small and Medium Sized Towns in their Functional Territorial Context (Scientific Report). Espon, Luxembourg. https://www.espon.eu/sites/default/files/attachments/TOWN_Scientific_Report_300814.pdf

Stanilov, K. (2007). Taking stock of post-socialisturbandevelopment: a recapitulation. In K. Stanilov (Ed.), *The Post-Socialist City: Urban Form and Space Transformation in Central and Eastern Europe after Socialism*. Springer.

Steinführer, A., Vaishar, A., & Zapletalová, J. (2016). The Small Town in Rural Areas as an Underresearched Type of Settlement. Editor's Introduction to the Special Issue. In *European Countryside* (Vol. 8, No. 4). Brno, Czech Republic.

Stoica I.-V., Tulla A. F., Zamfir D., & Petrişor A.-I. (2020a). Exploring the Urban Strength of Small Towns in Romania. In *Social Indicators Research* (Vol. 152, No. 3). Springer, Netherlands.

Stoica I.-V., Zamfir D., & Săftoiu L. C. (2020b). Small towns in Romania: the road from the urban dream to the rural reality. In A.H.A.-B. Steinführer, U. Grabski-Kieron & A. Reichert-Schick (Eds.), *New rural geographies in Europe: actors, processes, policies*. LIT.

Trebici, V. (1996). Demografie. Excerpta et selecta, Edit. Enciclopedică, Bucharest, Romania.

Trebici, V., & Hristache, I. (1986). *Demografia teritoriala a Romaniei*. Academiei Republicii Socialiste Romania, Bucharest, Romania.

United Nation, Department of Economic and Social Affairs (UN DESA). (2019). World Population Ageing 2019. United Nation, New York, USA.

Vaishar, A., & Zapletalová, J. (2009). Small towns as centres of rural micro-regions. In *European Countryside* (Vol. 1, No. 2). Brno, Czech Republic.

Vaishar, A., Zapletalová, J., & Nováková, E. (2016). Between urban and rural: sustainability of small towns in the Czech Republic. In *European Countryside* (Vol. 8, No. 4). Brno, Czech Republic.

Valtenbergs, V., Fermin, A., Grisel, M., Servilo, L., Vilka, I., Līviņa, I., & Bērzkalne, L. (2015). Challenges of small and medium–sized urban areas (SMUAs), their economic growth potential and impact on territorial development in the European Union and Latvia. Research report to support the Latvian EU Presidency 2015. https://www.eukn.eu/publications/challenges-of-small-and-medium-sized-urban-areas-smuas-their-economic-growth-potential-and-impact-on-territorial-development-in-the-european-union-and-latvia/

Wiechmann, T, & Wolff, M. (2013). Urban Shrinkage in a Spatial Perspective–Operationalization of Shrinking Cities in Europe 1990–2010. AESOP-ACSP Joint Congress Dublin 2013, Ireland.

Wirth, P., Elis, V., Müller, B., Yamamoto, K. (2016). Peripheralisation of small towns in Germany and Japan—Dealing with economic decline and population loss. *Journal of Rural Studies, 47*(1). Elsevier, United Kingdom.

Zamfir, D. (2007). *Geodemografia oraşelor mici din România*. Universitară, Bucharest, Romania.

Zamfir, D., Tălângă, C., & Stoica, I. V. (2009). Romanian small towns searching for their identity. *Journal of Urban and Regional Analysis, 1*(1). Bucharest, Romania.

Chapter 10
Urban "Eventscape" in Post-communist Romania: An Instrument that Shapes Tourism Destinations Reshaping Urban Identities

Daniela Dumbrăveanu

Abstract One of the widest and most recent industry, the event industry is new to Romania, with only 30 years to erase communist mentality and practice, in relation to events taking place in urban area. There is clear evidence of the development of a post-communist events industry, particularly in urban Romania. This has generated commercial event tourism, as one of the fastest growing and most dynamic forms of post-communist tourism in Romania. This paper aims first to analyse how dynamic event tourism has become, in selected urban units of the country, in relation to the events industry as a whole. Second, this study will assess the extent to which Romania, given the significant growth of its events industry, has managed to develop a particular "eventscape", and whether selected events operate as instruments to promote destination development. Third, this paper will discuss and analyse ways in which events and event tourism have contributed to shaping selected "festivalscapes" (Sighișoara and Cluj-Napoca) as post-communist tourist destinations. Further developing these themes, this paper will offer a potential explanation of the way in which festivals have contributed to defining event tourism destination urban identities in Romania, exploring the place they occupy within the Romanian post-communist "eventscape".

Keywords Destinations · Tourism · Event tourism · Landscape · "Eventscape" · "Festivalscape" · City image · Post-communism · Destination identity

The author would like to thank Dr. Anca Tudoricu, Thomas Rob Ditan and arh. Mihai Buruntia for making this study possible to come through.

D. Dumbrăveanu (✉)
University of Bucharest, Bucharest, Romania
e-mail: daniela.dumbraveanu@geo.unibuc.ro

B. R. K. Sinha (ed.), *Urban Dynamics, Environment and Health*,
https://doi.org/10.1007/978-981-99-5744-6_10

10.1 Introduction

The events taking place in Romania vary widely in content and complexity; an entire industry has grown up around them. The most significant and relevant events are the festivals and sports competitions, which have an inbuilt tourism component (Hall & Sharples, 2008; Hinch & Higham, 2011). In a globalized world characterized by fierce competitiveness and the constant need to find creative ways to resist or protect against an economic crisis, event tourism plays an essential role in local development.

In recent years, the importance of event tourism has been underlined by its ability to contribute to the development and marketing of destinations. Whether by helping to increase tourism (Smith & Jenner, 1998), extending the tourist season (Getz, 1997), or placing new destinations on the tourist map (Kotler, 2002), events have proven themselves to be an efficient tool and solution for many emerging markets.

Marked by economic changes, the Romanian tourism market has experienced growth, with a significant increase in the number, variety, and degree of sophistication of events. Although there are few official statistics to support this claim, the Romanian National Tourist Office listed 242 events in 2016, as compared with 112 in 2006 (RNTO, 2014). Recent studies (Mahika et al., 2015) have shown clear progress, with a trend for festivals far outpacing all other events in the mainstream Romanian market.

This growing post-communist event phenomenon has already had a significant impact on certain destinations, helping to define their image and offering new opportunities for developers and local authorities. To build novelty, a common goal for stakeholders in Romania, selected events have strategically incorporated aspects of local heritage into the entertainment and activities they offer.

Forced to build a new reputation starting almost from scratch, post-communist Romania had to consider many issues related to events, event tourism, and destination development. The inherited image of Romania was particularly unfortunate: "two vampires squatting on a corpse: Dracula and Ceaușescu" (Săndulescu, 2006:1). The country needed its landscape of national events, involving a variety of participants, to become a positive element in the overall process of image creation and reputation enhancement (Anholt, 2007; Getz, 2007).

10.2 Events, Event Tourism, Destination Development, and Place Promotion

Evidence suggests that event-related phenomena in Romania has developed into an industry that generates event tourism (Getz, 2007). Events are organized in accordance with precise purposes, depending on the organizer's vision and market demand. They adopt events industry mechanisms used by entrepreneurs to generate income (Araujo & Bramwell, 1999; Murphy et al., 2007). There are also events, which do not generate direct income, but are heavily involved in marketing a category of community events that raise funds, promote, or target active participation (Weber, 2004), and contribute to event tourism (Getz, 1998; Getz & Fairley, 2004).

Both providers and beneficiaries of urban tourism have advocated for events as built-in components. Travel agencies and destinations use events to stimulate economic development, create positive images, and even generate brands—while tourists see such events as attractions in their own right, or at least as extra motivators for visiting a particular place. Event marketing is essential since success may not depend on the type of event, its "star reputation", or purpose, but on how well a marketer takes advantage of collateral factors (location, weather, etc.) (Hoyle, 2002:152).

Destination development through place promotion became an important area of interest for researchers and destination marketers in the 1970s, intensifying after 1990 (Tasci et al., 2007). Destinations are intangible as such, they generate a relatively high level of perceived risk for potential customers, during the decision-making process. Since image is heavily involved in decision-making, behaviour "consists in acting in a way that is expected to realize the image of the most preferred future" (Boulding, 1961:175). The process by which one builds the image of a place involves many active factors and has been defined as a mental construct in progress, based upon "few impressions chosen from a flood of information" (Reynolds, 1965:69). Urry, states that, "Places are chosen to be gazed upon because there is an anticipation, especially through daydreaming and fantasy... or involving different senses..." (Urry, 1995:132). Whether the flood of information is obtained from primary sources or secondary sources, most commentators agree that the image is formed through a process of reasoned and emotional interpretation. Development is a complex, continuous process that follows several stages (Dann, 1996; Fishbein, 1967; Gunn, 1972; Pike & Ryan, 2004). Consequently, event tourism marketers understand that, if they are to influence affective evaluations of their destinations, both the destination's attributes and the motivations of tourists, count. (Baloglu & McCleary, 1999).

Despite having made significant progress, Romania continues to face a difficult task in selling its attractive (destination) image. The country is trapped somewhere between pre- and post-EU integration country labels. Initially, it was viewed as the country of orphans, vampires, and economic migrants; more recently has become the country of corruption and a volatile political and economic environment. Image experts (such as Simon Anholt) have suggested radical change, given that Romania has to change "one of the harshest and most unjust images a country could possibly have in the civilized world today" (Săndulescu, 2006:1). A focus on exploiting the affective component has been both suggested and considered as a possible solution. To achieve this goal, events-predominantly festivals, which Romanians tend to prefer-have been most actively exploited.

If price and climate are two common functional attributes that can be used to compare destinations, unique functional attributes can make a destination stand out; they include symbols and special events that contribute to image creation and destination identity building (Echtner & Ritchie, 1991).

Place promotion is no longer seen as an independent, self-centred phenomenon. Destinations are expected to tackle it as an integral part of their own development, in order to survive or progress within any market, let alone in an emergent tourist market, such as the post-communist Romanian context. As Ward (Ward, 1998) has described, every town, city, region, and nation, it seems, is now frenetically selling itself with assertions of its competitive place advantage due to "big changes in the social and political fabric of modern society" (Anholt, 2007:19).

The fall of communism in Eastern Europe has introduced many radical changes and demands for destinations eager to join the place—marketing game. Places have become commodities—their meanings being used to create more attractive place images to be sold and consumed. Place promotion involves the "re-evaluation and re-presentation of place to create and market a new image for localities to enhance their competitive position" in attracting or retaining resources (Short et al., 2000:318). Place promotion has been considered a way of reinventing deindustrialized or "depressed" cities in order to redefine perceptions of them, as well as an attempt to seek "new sources of wealth and new ways of stating their importance" (Ward, 1998:229).

This paper focuses its analysis on two event-based destinations selected as relevant within the Romanian context. The "Sighișoara Medieval Festival" has made an essential contribution to the city's identity as a tourist destination, as well as promoting the city to a wider target group, well beyond associating itself to the (in) famous "Dracula". Cluj-Napoca with its all year around festival agenda has succeeded in rebranding itself as the "regional cultural capital" of Transylvania.

10.2.1 Types of Events that Are Rooted in the Core of the Romanian Urban "Eventscape"

Communist Romania did not differ much from other communist countries when it came to the aims and purposes of events. Because events represent an expression of experience and exchange for participants, national leaders are able to use them as subtle expressions of authority (Picard & Robinson, 2006). During communism, this subtlety faded. Events were used, quite blatantly, to promote socialist values, ideology, the cult of personality, and festivist manifestations. They were developed into instruments designed to generate firm feelings of patriotic pride or a sense of belonging to the "right" regime. They basically fulfilled the same role at two distinctive levels: international (limited to socialist "brotherhood") and local. The number of communist events, normally attended by large numbers of people, was strictly controlled. Nevertheless, this number was drastically reduced during the last years of communism. Post-communist Romania, both as a state in transition and later, as an EU member state, took some time to reestablish the practice of holding events. Compared to other former communist countries, Romania was late to reintroduce large-scale international and national events capable of attracting a large number of recurrent "event consumers", or leading towards the development of an events industry (Tudoricu, 2008).

In recent years, the occurrence, frequency, and typology of Romanian events have progressed significantly. This pattern of development has contributed to framing a post-communist Romanian " eventscape" that is a direct consequence of the country's overall evolution, change, development, and recent history (Dumbraveanu et al., 2014; Tudoricu, 2008). Early post-communist Romania struggled to revive either pre-communist events (such as religious celebrations, traditional agrarian fairs, and rural life rituals) or communist cultural events featuring celebrities notorious for fleeing the communist regime and returning once it had collapsed. Apart from the pre-communist traditional rural fairs (fading drastically since 2014), such events had lost their significance and appeal. In most situations, they had become instruments of post-communist political manipulation during election campaigns financed by political parties or corresponding associations (Icobescu, 2014).

The major change occurred in the early 1990s, when the country opened to cultural imports, a trend that massively intensified after Romania joined the European Union in 2007. The new social, economic, and political context introduced three significant event categories into the Romanian market: traditional events with low drawing power, generally organized by public institutions; import events, designed to promote Romanian travel and tourism, and consequently generating economic benefits for their organizers; and local/regional events that emerged as a consequence of popular local activities.

Traditional events, as significant landmarks, quickly developed into hosting destinations that could attract and cater to a broad range of consumers with different motivations. Accurate figures concerning traditional events are almost nonexistent, since they are not officially recorded; there are estimated to be between 300 and 1,400 such events each year, depending on particular circumstances. They have particular themes and are specialized, mainly held in towns and cities, targeting either group of loyal local recurrent participants or non-recurrent tourists. National and international imported events (music festivals, concerts, or other cultural manifestations) make up by far the most prominent category to have developed over the past 20 years in Romania. Their significant role in generating profits has helped put the country on the international events map and has provided a frame for its own industry. The very first profit-oriented mega-event organized during the post-communist period was a Michael Jackson concert in Bucharest in 1992; the event attracted 70,000 participants. Among other achievements, this event marked the place and importance of the new political regime, creating a sense of belonging to a new society characterized by freedom and progress. This concert (equally praised and contested) had a positive impact on the image of Romania abroad for several hours, Michael Jackson turned Bucharest into the event capital of the world. As the media commented, "to Romanians, Michael Jackson meant reconnecting with the rest of the world. To the others, it was the proof Romania truly exists" (Brumă & Mincan, 2009). This concert was the starting point for a phenomenon that currently contributes to shaping the Romanian urban "eventscape".

Subsequently, several imported events arrived to launch a genuine creative industry in towns and cities. Mega-events, such as the Sibiu-European Capital of Culture (2007) or the Cluj-European Youth Capital (2015), encouraged and generated smaller events that became recurrent events, helping to develop particular destinations into "eventful cities" (Richards & Palmer, 2010), otherwise known as tourist destinations.For example, the ARTmania Sibiu Festival started in 2006 as a pre-event cultural capital initiative (Artmania festival, 2006), continues to be very popular; has added a performing arts dimension to the city identity,attracts visitors and prepares to celebrate its 15th edition in 2021.

Urban events meet the simple demand for leisure, entertainment, and similar forms of expression, ultimately providing an expression of identity for or within the local community. Events of this type developed rapidly and widely throughout post-communist Romania, particularly in the context of globalization. Within this category, the most significant and popular events are (yet again) the festivals. One good example is the ALTER-NATIVE international short film festival. This is the oldest short film festival in Romania; it has been held at Târgu Mureș ever since 1993, only one Romanian example of an event that has been held uninterruptedly year after year since it was first launched. This is a specialized and popular festival showing over 500 films from over 50 countries to loyal audiences.

10.3 The Romanian Tourist Urban "Eventscape", Destination Development, and Place Promotion

Although initially delayed and empirically developed, once in place, Romania's events industry almost immediately had an impact on place development. The Romanian post-communist events industry has made significant progress, but still has some gaps to overcome. In addition to generating profits, the events industry does attract visitors. Recently, Romania has begun to develop new and emerging forms of event tourism. Experience and tradition do not apply to Romanian event tourism since there are still no mega events or international level sports competitions directly related to Romania.

Festivals are the most familiar and popular events in Romania. Although academic theory does not link festivals with mass tourism, the Romanian experience shows that festivals can become important contemporary tourist references. The Romanian post-communist "eventscape" has rediscovered or "reinvented" traditional local events while also inventing new events, in response to social, economic, and political realities that share one main goal-to attract participating visitors. Within the category of rediscovered/reinvented events, good examples include attempts to revive communist pop music festivals, including The Golden Stag-Brasov and Mamaia Music Festival, while accommodating post-communist audience preferences; those efforts have failed (Nicolau, 2014). However, it proved easier to reinvent places as attractive destinations, using past or traditional elements. In response to a demand for tourism, different types of festivals were launched. Festivals marketed as new events have been more popular, with higher drawing power, because the local population tends to find them inspiring.

The events industry and tourism are similar in several ways. They both involve space-and time-limited activities, they are both a cause and effect of globalization, and they both have a role in reshaping social relations. They ensure that new communication networks and new public exchange processes will be implemented. In a world in which very few societies are closed to tourism, festivals (symbols of social and cultural life) promote a series of relations between tourists and stakeholders inside the tourism sector. They are seen as manifestations encompassing values that can be embraced (such as a sense of belonging or significance), contested, or renegotiated by offering a reality of their own.

In Romania, festivals and concerts are the most popular tourism-generating events (Mahika et al., 2015); they contribute to destination development and underline the importance of place promotion, provided they manage re occur regularly and long enough. This paper also focuses on two festival and place case studies, relevant in analysing how the current post-communist "eventscape" influences event tourism and destination development. Both are vivid examples of new types of festivals that have grown and developed, inventing and reinventing themselves. They have helped to turn their participants into tourists, by changing their own meaning and signification to reflect audience interpretations.

In addition to the idea of "eventscape", "festivalscape" it is equally important to be considered and argued, even if, recent and somehow not widely used as a concept. The concept of a "festivalscape" refers to the overall atmosphere generated by a unique combination of tangible-environmental and intangible-psychological elements, as experienced by participants in the festival setting. Festivals are perceived as representing both functional and affective levels (Darden & Babin, 1994). Furthermore, event tourist behaviour is influenced by inner stimuli such as taste, smell, and texture, in addition to external stimuli such as price, brand, and reputation, all of which together contribute to the tourist experience (Chang & Wildt, 1996). As a result, participant experience contributes through ambiental, spatial, and symbolic aspects towards building a "festivalscape".

Symbols are essential elements in building a destination image because they work as efficient stimuli: they generate the tourist gaze, which is "directed to features of landscape and townscape which separate them off from everyday and routine experiences" (Urry, 1995:132). Event participants, just like any other tourists, have the ability to select a variety of elements, objects, sights, or sites and to give them meaning and significance. These are the symbols that people select to keep and take away as mental constructs; for this reason, they are often associated with place promotion and increasingly used in marketing as the following case studies will argue.

10.4 The Sighişoara Medieval Festivalscape—From Emerging to Established Event Tourism Destination

Sighișoara is a town located in the historic region of Transylvania, Mureș County, well known for its medieval citadel and wide variety of authentic medieval buildings. The local community is a multi-ethnic mosaic (with Hungarians, German/Saxons, Romanians as the main groups), which can trace its complexity back to medieval times. Since the collapse of communism in 1989, Sighișoara has increasingly been associated with the medieval figure of Vlad Țepeș/Vlad The Impaler, a local leader. Despite the lack of historical facts or evidence, Vlad The Impaler (supposedly born at Sighișoara) is often associated by Western travellers in search of the "literary, historical and supernatural roots of the Dracula myth" with Bram Stoker's Count Dracula (Light, 2007:751).

During the Communist period, events and event tourism at Sighișoara were rather nonexistent. Events of the type that would be held today (and all tourism development based on anything other than socialist values), were not encouraged by the communist regime. Modest numbers of domestic and international travellers (the latter mainly from socialist countries) were encouraged but controlled, as part of a wider communist strategy during the 1970s (Răiescu, 1998). Most domestic tourism was generated through communist holiday packages, as part of the national policy to provide working-class people with holidays. The progressive decay of the social-economic and political situation from 1980 onward improved neither domestic nor foreign tourism at Sighișoara.

The collapse of communism placed Sighișoara on the modest side of tourist activity and tourism development before 2000. Today Sighișoara is one of the most popular and visited Romanian destinations, and the fourth fastest growing destination, in terms of the number of visitors per inhabitant, after Brașov, Cluj-Napoca and Sibiu (INSEE, 2019). Three main elements have contributed to its popularity and unprecedented rate of tourism development: the Sighișoara Medieval Festival ***(SMF)***, Dracula Tourism ***(DT)***, and becoming a UNESCO heritage site.

10.4.1 The Sighişoara Medieval Festival

A recent study on the behaviour of Romanian tourists, as generated by events (Mahika et al., 2015), shows that this annual festival is not only the most popular and successful festival nationwide, but also the most well-consumed event in the country. It has been held 26 times over a period of 27 years. The festival has written its own history, documenting its progress. SMF is well known as Romania's most consistent, purposeful, and long-lasting post-communist event. Despite its recurrence, importance, and significance, there have been reported controversies concerning aspects of the festival's organization,in particular, funding, fundraising and fund spending, rules and regulations, and the local authorities' approach to management. Since it was first launched, the SMF has demonstrated significant growth, development, and variety. However, it has also required organizational turn taking (the festival was initially opposed by the town hall, which then took it over, and organized and promoted it as a symbol of the city), trialled radical changes (changing its status from a performing arts event to a complex medieval festival), experienced failure (when promoted solely as a Dracula Festival in 2001), and success (when given a strong combined medieval heritage, architecture, and arts dimension in 2012). It has faced many ups and downs (in relation to visitor numbers, sponsors, and activities), faced organizational criticism (especially after deciding to charge an entrance fee), but has ultimately been widely praised by participants (for its atmosphere, quality of activities, and length). In recent years, it has become an inspiration to be copied by other Romanian destinations.

SMF grew out of a student performing arts initiative organized in 1992 in the Citadel Square, as a follow upto an earlier, spontaneous, and successful musical event, which had been highly appreciated by some enthusiastic locals. The very first festival was organized under the patronage of UNESCO Romania by Compania Studio 24 (a student association) and ODT Mureș (a youth organization) and funded both by the Romanian Ministry of Youth and Sports and by the Bucharest Youth Foundation. The event developed over time, trying several "recipes" for adding a medieval dimension to its art-related theme. Its primary goal was to raise awareness of art as a form of expression, and to address the advanced state of decay of the medieval citadel and architecture. Subsequently, the festival aimed to promote medieval culture, and to set performances in a local art-inspiring atmosphere. Starting with tens of participants in 1992, and 7,000 in 1995, the festival reached 10,000, in 1996—an unprecedented phenomenon for the Romanian events industry and tourism market, which certainly demonstrated its drawing power (Todor, 1998). This success encouraged the local authorities to consolidate the festival's double position as an organizer and beneficiary (mostly economic) of tourism. A long-running controversy has focused on their keen interest in the profits made by the festival, in comparison with their lack of interest in profits made by other local initiatives.

10.4.2 Dracula Tourism (DT)

Dracula tourism is an externally generated phenomenon in which enthusiasts of Bram Stoker's Dracula have travelled to Sighișoara in search of his roots, also attracted by the mysterious image of Transylvania promoted by international travel guides. DT, although officially labelled, "something generated by external demand that is problematic in a number of ways", (Light, 2007:748) has always been discussed in a broad context, in terms of the "dilemma of identity versus economy". In this dilemma, the past (which attracts tourists and generates revenue) collides with the post-socialist process of identity building (Light, 2000, 2007). Light has analysed the dilemma of identity versus economy and concluded that the state adopted the position of "reluctantly tolerating" Dracula tourism (Light, 2007:757) without promoting or encouraging it but managing it in its own terms. Traces of the propagandistic Dracula tourism packages launched during the communist period for the American and Western European tourist market remain. The same ambivalence is characteristic of the immediate post-socialist decade.

There is an obvious deterministic link between the Sighișoara Medieval Festival and Dracula Tourism. The increase in number of festival participants was followed by an almost identical number of tourists, their progress restrained only by the limited accommodation available. In 2001, the number of festival participants reached 10,000, while number of tourists in accommodation units was 9,393. The constant and steady growth in the number of festival participants (24,500 in 2015 and 23,000 in 2019) generated growing tourism numbers. A brief look at the statistics shows that the number of accommodation units grew from 1 hotel in 1992 to 6 units in 2001, 69 units in 2015 and 70 units in 2019. The number of places increased from 125 in 1992 to 563 in 2001, 1729 in 2015, and 2003 in 2019. The number of tourists in accommodation units has obviously followed the trend-increasing from 638 in 1992 to 9,393 in 2001, 75,008 in 2015, and 108,360 in 2019 (INSEE, 2019).

The year 2001 was a major turning point for the SMF–DT partnership. The Romanian Ministry of Tourism announced a variety of Dracula-related initiatives, including a theme park at Sighișoara, which was a novelty for the local authorities, and never to be completed. However, for Sighișoara at that time, this strategic vision was a signal to further expand and develop Dracula-related themes to incorporate into the Sighișoara Medieval Festival. This resulted in the SMF being transformed into a regular Dracula event, at the expense of the medieval festival as a cultural event (Plămădeală, 2010). The Dracula dimension steadily injected into the SMF helped to continuously increase the number of tourists registered in accommodation units until 2008 but caused them to drop in 2009. Once it became clear that a Dracula-themed festival was not financially sustainable; the first steps towards recovering the initial purpose of the festival were carried out in 2010. This took place at a time when local authorities were particularly interested in revenues and the economic crisis had led to major budget cuts. Since 2010, the SMF has shared themes, strictly related to the citadel, organized by an NGO (the Sighișoara Tourist Association). The SMF was originally financed by various donors who supported the student initiative (Art Company Studio 24) to prevent the Citadel from deteriorating (Todor, 1998). After becoming self-sufficient, sustainable, and a promising source of revenue, its funding became a priority for the local authorities, who at the moment provide one-third of the event's budget (Zaharie, 2015).

10.4.3 The Sighișoara–UNESCO World Heritage Site

Sighișoara has increased in popularity and grown into an event tourism destination with the help of UNESCO World Heritage Site status. Its Citadel was listed in 1999 (UNESCO World Heritage Centre, 2001). This has helped to attract and maintain a constant flow of international visitors, mostly Dracula enthusiasts.

After several false starts, the Sighișoara Medieval Festival managed to become the successful catalyst for a "festivalscape". Through the combination and valorization of the city's medieval attributes, both as a UNESCO heritage site and as a Dracula Tourism destination, an event tourist destination ultimately was able to emerge.

The number of accommodation units and the total number of tourists in Sighișoara were modest during the early days of the festival up until 2000. Switching the festival theme from medieval music to Dracula caused numbers to rise modestly but steadily between 2001 and 2004. The undisputed growth in numbers began in 2005, when the festival partially reverted to the theme of medievalism, while maintaining some of its previous attractive "ingredients" (music and Dracula). Heritage was reinstated as an important authentic source of supply for the medieval character of the festival. However, the Citadel itself does not attract a constant tourist flow all year long. The SMF attracts a large target group to the event itself. The questionnaire survey on the SMF shows a high recurrence rate among participants: in 2006, 47% of visitors were recurrent, as were 49.8% in 2011 and 50.1% in 2014. To be able to balance the number of tourists, the local authority strategically supports and organizes other events all year round at Sighisoara (Aldea, 2014). This further demonstrates the internally driven drawing power of the SMF and its ability to generate repeat visits. This phenomenon is caused by the SMF's capacity to generate a significant "festivalscape", which can be further developed and exploited.

Since the experience of participants is crucial in building a festivalscape, results of a series of questionnaire survey were used to put together the emerging "festivalscape" of Sighișoara during the SMF. Participant responses clearly indicated that the spatial and symbolic aspects were the most powerful elements of attraction, contributing to their experience during the event. Over 80% of interviewees chose historical items and medieval features of the destination to express their attraction. Local scenic symbols were also chosen as attraction generators. The ambiental aspects were highlighted by words related to the senses that embodied medievalism (Fig. 10.1).

The same survey suggests another function of the festival: its contribution to defining a destination image through place promotion for Sighișoara. The survey's results clearly demonstrate that *tangible symbols* related to the historical heritage, *intangible heritage* elements (such as medieval music and the performing arts), and simple, emotional elements generated by the festival and its atmosphere are all relevant image generators with drawing power.

An overall average of 20% (21% in 2006; 22% in 2009; 22% in 2011; 15% in 2014) of respondents highlighted the medieval festival setting itself as the primary element that defined the destination. More than 30% of the chosen symbols refer to elements that are not available outside the event. In effect, *the destination functions as a mirror for the image of the festival*, overtaking some of its features. Sighișoara Citadel is the mirror of the Sighișoara Medieval Festival, symbolically embodied by the Clock Tower. As Palmer (1999:316) has commented, "tourism's use of identity goes far beyond the commercial; it goes to the heart of people because it serves to define their cultural identity and to make this visible, both to themselves, and to others".

Fig. 10.1 Festivalscape symbols defining attraction to the destination: Sighișoara

The SMF Festival increased visitor numbers and the growing significance of the place by turning Sighișoara into an event tourism destination: it is therefore perceived by the local authorities and the local community as a great success. The town administration strategically added several different, smaller festivals to Sighișoara's yearly calendar in an attempt to turn it into an eventful city. In 2019, there were over 30 festivals and fairs advertised on the town hall's website and the websites of key organizations promoting the city (Shișoara Medievala, 2019).

10.5 Cluj-Napoca-"Festivalscape" Added Value to Event Tourism Destination

Cluj-Napoca is, demographically, the second largest Romanian city, located in the historic region of Transylvania, the county capital of Cluj County, well known as the "heart of Transylvania" with a citadel and wide variety of authentic medieval buildings. The local community is a multi-ethnic mosaic (Hungarians, German-Saxons, and Romanians as main groups), which can trace its complexity back to medieval times. Administratively, socially, economically, and culturally, Cluj-Napoca has been and remains the most developed, dynamic, and important city of the region, throughout history, before and after the collapse of communism in 1989.

Cluj-Napoca has been either de facto or perceived as the capital of Transylvania in all aspects, particularly cultural.

During the Communist period, events and event tourism at Cluj-Napoca were rather following the national pattern, low profiled between 1980 and 1989. Events as currently known were discouraged by the communist regime, still several strictly cultural festivals were preserved and ran with a modest number of participants. The cultural life of the pre-communist city was simply too rich and the city too important regionally to have its all cultural event agenda closed. Cluj-Napoca is particularly proud to have collaborated with "Pathe" since 1913 in the filming industry. The numbers of domestic and international travellers were encouraged and maintained around 4th highest in the country as part of a wider communist strategy during the 1970s (Răiescu, 1998). The progressive decay of the social, economic and political situation from 1980 onwards affected neither domestic nor foreign tourism at Cluj-Napoca as much as at Sighișoara. Still overall numbers particularly for foreign tourism have dropped.

The collapse of communism placed Cluj-Napoca on the rising side of tourist activity and tourism development started almost immediately after 1989. Today Cluj-Napoca is one of the most popular and visited Romanian destinations, and the second fastest growing destination, in terms of the number of visitors per inhabitant, after Brașov (INSEE, 2019).

After the collapse of communism, multiculturalism, respect for the cultural identity of "the other", freedom of expression, value sharing, and the sense of belonging to a trusting, active and diverse wide community have remained the main features and attractions of Cluj-Napoca mainly as a culture-based destination. Still it faced some changes. It was "invaded" by an increased number of tourists, among whom significantly more foreigners searching for the spirit associated with its image. People were in search of cultural diversity, creativity, freedom, and the power of expression generating waves of subtle impact, change, and development. The city has always seen itself as a "Thriving Cultural Spirit". This was redefined with a degree of cyclicality but most importantly has been reinvented. Besides its core cultural performing component, ever-present, the city steadily and continuously "capitalized" on its long-standing tradition as a cultural centre, by promoting, investing in, and gradually developing its cultural event agenda, which was then turned into an event industry of performing arts (film, theatre, music), visual arts, even poetry festivals. This was the most suitable way to adapt to change, which brings issues "assimilated as the loss and gain of…original identity" (Toma, 2005). Event tourism emerged as a reaction to destination loss of original identity, when cultural festivals ceased to be thriving by simply targeting a rather intellectual public. Promoting the "Thriving Cultural Spirit of Cluj-Napoca", among a new generation and raising awareness about preserving its past was a crucial step, culminating with the European Youth Capital 2015.

As opposed to most of the Romanian towns and cities, Cluj-Napoca has inherited, promoted, and developed a long list of festivals given its long-standing history of events, to which several have been added gradually. The city is currently hosting 32 festivals yearly. To briefly illustrate, a selection of festivals with regular, yearly occurrence since their first run ever, are considered (Fig. 10.2).

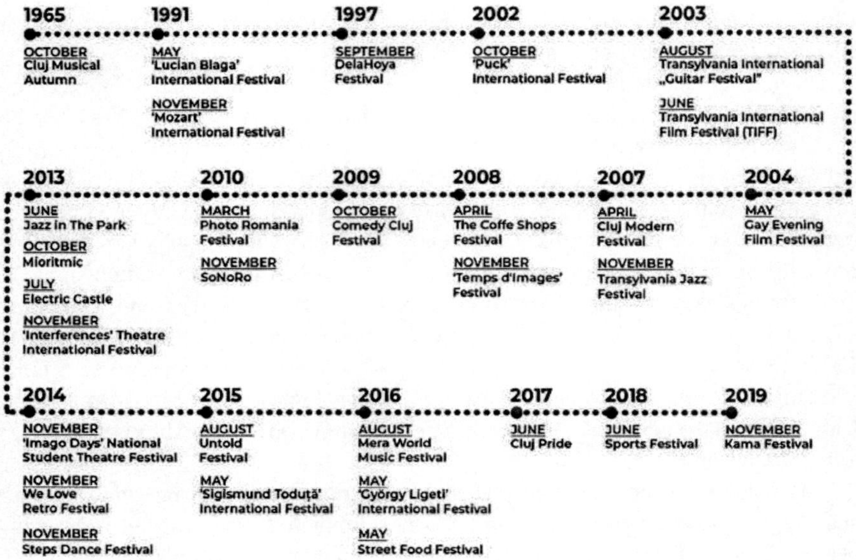

Fig. 10.2 The list of Cluj-Napoca yearly occurrence, long-standing festivals

Festivals prior to 2003 are clearly pre-communist-communist hybrid event versions of those capable of carrying on the original version of "Thriving Cultural Spirit of Cluj-Napoca". There is a clear indication that the events between 2003 and 2013 were intended as events to redefine and renew the cultural spirit of Cluj-Napoca as part of the "loss—gain" process regarding originality at time of profound social and economic post-communist change. While the events post 2013 indicate the attention paid to youth and diversity as part of reinventing the spirit of the place.

Number of participants varies widely and given the categories of festivals. Cases of huge increase in numbers are rare and as opposed to Sighișoara are irrelevant since Cluj-Napoca is aiming at a rather different strategy, more festivals with a balanced spread over the year.

Within, Cluj-Napoca "festivalscape", 4 festivals have grown to become the most important: Untold, Electric Castle, Transylvania International Film Festival (TIFF), and Jazz in the Park.

10.5.1 "Untold" Festival

It is Romania's largest modern music festival especially planned to target a large young public. The number of participants has always exceeded organizers' expectations overcoming 50,000 for each occurrence. This festival has been intended as the key summer moment for the life of the city. It had immediate success. As a result, its second occurrence has won the "Best Major European Festival" Award for the event. According to the questionnaires conducted at each occurrence, it is by far the most attractive festival in the country mainly for the youth. The participants extend their stay at Cluj-Napoca by 2, 5 days beyond the festival timetable to "visit and get to better know the city".

10.5.2 "Electric Castle Festival"

It is held nearby the city is a diverse music genre festival, aimed at attracting "all age" groups, mainly families. It includes a wide range of alternative entertainment activities with medieval themes since it takes place inside a castle. It is the second most popular festival that won "The Best Medium Sized European Festival" award in 2019. Its strategic planning feature takes advantage of the immediate vicinity of the city thus intentionally extending the popular medieval themes that spread the participants around.

10.5.3 "Transylvania International Film Festival (TIFF)"

It has become one of the most predominant European Film Event aiming to increase popularity for the industry. Local authorities are particularly interested in paying special attention to it since it is a local brand element. It is also considered as the most effective element to place Cluj-Napoca on the European map of culture. The festival is managed in close connection with the local culture developing a strong spirit of the place, enhancing the atmosphere, and extending a significant drawing power.

10.5.4 "Transylvania Jazz Festival"

It has been from its early stages foreseen as trigger image element for promoting the city. Jazz as particular music genre was aimed to attract the attention and interest of its specific public. Authorities have always aimed at involving internationally notorious musicians precisely to associate the city with something specific hence reinvent the cultural image of Cluj-Napoca. Promoting national young musicians was also part of the strategy since a side aim was to start accessing international jazz contests and festivals as part of a network.

Each of these festivals as part of a unitary strategy developed in time by the local authorities aimed to preserve on one hand the tradition of Cluj-Napoca as "bohemian medieval" place, recalling the past Austro-Hungarian Empire but on the other to add new dimensions: space for non-conformism, cultural diversity, and freedom.

Statistics clearly indicate that tourist activity increases during festival seasons, as a result of them. Prior to the collapse of communism, the city inherited a significant number of accommodation units which remained largely unchanged with a modest increase between 1991 and 2002 (20 units); it subsequently significantly increased until 2013, when it reached 91 units continuing to 116 in 2019 (INSEE, 2019). Similarly, the number of available places in the accommodation units and tourists followed the trend with one major aspect to be mentioned. Most of the new accommodation unit's location tends to correlate with the festival settings. An analysis of accommodation occupancy rates and the number of tourists in accommodation units quarterly registered shows that between 2003 and 2019: (a) during the festivals, figures increase by 160–180% compared to non-festival and unseasonal rates; (b) there is a significant difference between the number of tourists in accommodation units and festival participants. This phenomenon clearly explains the growing importance of place as an event tourism destination, contributing to building a "festivalscape".

All Cluj-Napoca festivals and particularly the 4 above-mentioned festivals have generated a continuous and unique type of "festivalscape" that clearly embeds the core shared values associated with the place. The uniqueness of Cluj-Napoca relates to the clear line between the "festivalscape" and the destination; in effect, the wholeness of the destination is the setting of the "festivalscape". These events succeed because of the existing synergy between the events and the destination. This is highlighted by the recurrence rate of participants, as shown by the questionnaire survey concerning the Untold Festival. In 2016, the Untold festival attracted 42% of recurrent participants, reaching 59% in 2017, 51% in 2018, and 53% in 2019. Most of the recurrent participants tend to have attended at least 2 other Cluj-Napoca festivals, mainly Electric Castle and TIFF.

The Untold "festivalscape" has been assessed and analysed, based on the results of the questionnaire survey. The overall response structure clearly showed that the symbolic and spatial aspects of the festival experience were its most powerfully attractive elements. 40.6% of the respondents associated the appeal of the Untold festivals with the concept of freedom expressed in a variety of ways, mostly through modern forms of music and the performing artists, to the destination as a "cool" place, "something else" than the rest of cities. Sound, variety of services- and friendship-related items were associated with attraction by 45.2% of the respondents, contributing to enhancing their experience. This reflects very well the importance of the concerts organized in the main city Square as an area for socializing. The "Thriving Spirit of Cluj-Napoca" has always been and continues to be based on socializing in a reinvented manner. 40.1% of the respondents identified ambiental aspects as factors of attraction contributing to their experience, spreading the concept over a much broader range of symbols. Ambiental aspects are expressed in terms of happiness, difference, fun, tolerance, non-conformism, crowds, and architecture. Putting together the categories of symbols that respondents use to express the drawing power of the "festivalscape", reveals a powerfully perceived symbiosis between the inner characteristics of the event and destination-related components (Fig. 10.3).

Overall, the survey showed that the Untold "festivalscape" includes an indestructible effective component having to do with the idea of freedom and tolerance that

Fig. 10.3 Festivalscape symbols that define the attraction of the destination: Cluj-Napoca

extends its drawing power in its setting, infrastructure, and entertainment opportunities; it depends on the inner characteristics of what festival is inspired by, in addition to its unique atmosphere. This argument explains how event tourists become motivated to consume the "festivalscape". It springs from the relationship between the self and the (emotionally perceived) space of Cluj-Napoca.

The interview surveys conducted on Electric Castle Festival, TIFF, and Transylvania Jazz Festival results in the particular role played by the urban heritage of the city, the recreated sense of the place, and the interest of the local authorities in events; they contributed to defining a "festivalscape" image, which in turn used them to build a "Western European" cultural image for Cluj-Napoca as destination. Tangible symbols such as old architecture, monuments, the Square, terraces, and bars, have been identified as relevant image generators widely used as cross-festival marketing symbols. In effect, Cluj-Napoca festivals function as a mirror for the destination, overtaking all of the features of the area and reflecting and asserting them. The festivals contribute to preserving and promoting an existing and inherited identity, slightly reshaping it in line with the rest of modern Central Western Europe to which it used to belong as part of the former Austro-Hungarian Empire.

10.6 Conclusions

Post-communist Romania has experienced the significant growth of its events industry (in both numbers and categories), particularly after 2000; it has become powerful, diversified, and sophisticated enough to generate an "eventscape" at the national level. After 2000, event tourism appeared and took off as the most emergent segment of the Romanian tourism market, particularly within the domestic sector. International Romanian event tourism increased modestly until becoming an EU member.

The events industry and event tourism in Romania may lack a worldwide international dimension, but not European since it has managed to create several significant national "festivalscapes". These may not massively contribute to the national economy, but they do significantly contribute to local economies. In addition, they are destination development instruments, with a massive place promotion impact. Following the successful example of Sighișoara and Cluj-Napoca, almost every Romanian tourist destination now has an event strategically planned to boost tourist numbers and revenues. This even includes places that attempted and failed to revive their communist tourist infrastructure.

Romanians are devoted consumers of events-and particularly festivals. The national "festivalscape" is abundant all year round and moderately sophisticated, in comparison with international festivals. Only a few have a developing impact on their locations. However, there are several Romanian events (mostly festivals) that have helped to develop their locations into tourist destinations of both international and national importance. They are mainly image generators that attract loyal participants, in the process of evolving into event tourist attractions. Such successes have been achieved by embedding emotions into the tourist experience and building "festivalscapes", as Sighișoara and the Cluj-Napoca Festivals have managed to do over a period of about 30 years.

Both Sighisoara and Cluj-Napoca had little to do with the events industry or event tourism before 2000.

Sighișoara, as place and destination, based its tourist attraction on its medieval citadel (which is a UNESCO heritage site), Dracula, the Transylvanian scenery, idyllic views, and a strong sense of medieval nostalgia. The Sighișoara Medieval Festival was the missing ingredient needed to turn this place into an internationally attractive destination. The SMF is Romania's oldest post-communist festival, having been held 26 times. It experienced several organizational and structural turning points before becoming capable of generating a consistent and powerful "festivalscape". The winning formula has involved a balanced combination of medievalism (celebrated in an authentic setting with elements and features of the place, such as the Citadel), historical impersonators engaged in typical festival activities, Dracula-related themes, and the performing arts (including concerts of modern and medieval music). This combination has generated one of the most powerful "festivalscapes" in post-communist Romania, attracting a large number of loyal visitors, for whom the festival is the destination. This case is unique in present day Romania, in terms of its drawing power to attract tourists. Wrapped up in an extended festival calendar, Sighișoara has turned into an "eventful city", with significant positive economic impacts. It has become an inspiration for many towns and cities (particularly in Transylvania), which have yet to prove their power.

In conclusion, by generating a powerful "festivalscape", the Sighișoara Medieval Festival manages to transfer features of the festival into the place, thus helping to develop and market it as a destination. It is a mirrored process, in which the destination mirrors the festival.

The Cluj-Napoca "festivalscape" places itself in precisely the opposite situation. Cultural events were not a novelty to the place hence the post-communism events industry first related to the pre-existing combined communist-pre-communist "cultural spirit". The place already had a strong communist-era tourist identity. It had a powerful and specific reputation, in contrast to the size of its tourist group. The events took over the main features of the place and of the destination, using them for other purposes—to preserve, promote, and develop the "Thriving Spirit of Cluj-Napoca", which originally helped to attract the events industry. This spirit was very well-defined during the communist era, when it was associated it with diversity, value sharing, intellectual resistance and militating. Since Cluj-Napoca was already a tourist destination catering for a well-defined, highly self-selected target group, event tourism offered an added dimension. The contrast between its medium tourist group and powerful destination identity disappeared, as the number of tourists increased. For places with a powerful identity such as Cluj-Napoca, festivals are instruments of expression. They immediately generated a powerful, matching "festivalscape", also "performing" as a form of emotional tourism. The Cluj-Napoca Festivals (in contrast to the Sighisoara Medieval Festival) were not in themselves primary motivators; the destination itself attracted tourists. The festivals established and developed a role as animators, embracing the "Thriving Spirit of Cluj-Napoca" and transforming it into an element of animation by taking over all of its features and characteristics to build the driving power behind the "Cluj-Napoca festivalscape".

In conclusion, the Cluj-Napoca Festivals, by generating an extremely powerful "festivalscape" in an already powerful destination, managed to further develop and promote the image of the destination by preserving, redefining, and reinventing the "Spirit of the place" as their model. This is an example of a mirrored process, in which the festival mirrors the destination.

Both Sighișoara and Cluj-Napoca are nationally relevant because their festivals have succeeded in generating "festivalscapes" capable of attracting loyal visitors and helping to develop significant event tourism. Their processes of initiating and building a destination image (Sighișoara) and developing and preserving a powerful image (Cluj-Napoca) are also relevant because of their particular details and differences. Without doubt these are powerful "festivalscapes" in the country, generating event tourism and helping to define their places as both international and national level tourist destinations. They are relevant because they have succeeded in developing cores into national "eventscapes". The Sighișoara and Cluj-Napoca "festivalscapes" are now being followed into this type of development by other "eventful cities", such as Sibiu and Alba Iulia. Although their versions of "eventfulness" are quite different, in size and typology, they have made themselves into relevant destination cases by developing powerful "festivalscapes".

Romania is currently experiencing a positive trend in the evolution of the events industry and event tourism, with clear evidence of further development. The role of these festivals as image generators and identity-builders has stimulated interest from stakeholders, who are now using them to inspire long-term tourism development strategies in a large variety of places.

References

Aldea, A. (2014) Strategia de dezvoltarea a Municipiului Sighișoara și importanța evenimentelor. In *Primăria Municipiului Sighișoara (2014) Evenimente 2014*. http://www.sighisoara.org.ro/portal/portal.nsf. Accessed 02 Feb 2019.

Anholt, S. (2007). *Competitive identity: The new brand management for nations, cities and regions*. Palgrave Macmillan.

Araujo, L. M. D., & Bramwell, B. (1999). Stakeholder assessment and collaborative tourism planning: The case of Brazil's Costa Dourada project. *Journal of Sustainable Tourism, 7*, 356–378.

Artmania Festival. (2006). ARTmania Festival 2006—Prima editie cu H.I.M. si Amorphis. http://artmaniafestival.ro/artmania-festival-2006-prima-editie-cu-h-i-m-si-amorphis. Accessed 11 Mar 2018.

Baloglu, S., & Mccleary, K. W. (1999). A model of destination image formation. *Annals of Tourism Research, 26*, 868–897.

Boulding, K. E. (1961). *The image: Knowledge in life and society*. University of Michigan Press.

Brumă, S., & Mincan, M. (2009). 1992, 1996: Concertele care ne-au pus pe hartalumii. *Adevărul, 27*(06), 2009.

Chang, T.-Z., & Wildt, A. R. (1996). Impact of product information on the use of price as a quality cue. *Psychology & Marketing, 1986–1998*(13), 55–75.

Dann, G. M. S. (1996). Tourists' images of a destination—An alternative analysis. *Journal of Travel & Tourism Marketing, 5*, 41–55.

Darden, W. R., & Babin, B. J. (1994). Exploring the concept of affective quality: Expanding the concept of retail personality. *Journal of Business Research, 29*, 101–109.

Dumbraveanu, D., Tudoricu, A., & Craciun, A. (2014). The night of museums-a boost factor for the cultural dimension of tourism in Bucharest. *Human Geographies, 8*, 55.

Echtner, C. M., & Ritchie, J. R. B. (1991). The meaning and measurement of destination image. *The Journal of Tourism Studies, 14*, 37–48.

Fishbein, M. (1967). *Readings in attitude theory and measurement*. Wiley.

Getz, D. (1997). *Event management & event tourism*. Cognizant Communication Corp.

Getz, D. (1998). Trends, strategies, and issues in sport-event tourism. *Sport Marketing Quarterly, 7*, 8–13.

Getz, D. (2007). *Event studies: Theory*. Butterworth-Heinemann.

Getz, D., & Fairley, S. (2004). Media management at sport events for destination promotion: Case studies and concepts. *Event Management, 8*, 127–139.

Gunn, C. A. (1972). *Vacation scape: Designing tourist regions*. University of Texas at Austin.

Hall, C. M., & Sharples, L. (2008). *Food and wine festivals and events around the world: Development*. Butterworth-Heinemann.

Hinch, T., Higham, J. (2011). *Sport tourism development*. Channel View Publications.

Hoyle, L. H. (2002). *Event marketing: How to successfully promote events, festivals, conventions, and expositions*. Wiley.

Icobescu, O. (2014). Alte evenimente pe bani publici politizate de PNL pentru Campania electorală. http://www.pentrutimisoara.ro/stiri_timisoara/alte-evenimente-pe-bani-publici-politizate-de-pnl-pentru-campania-electorala. Accessed 22 Dec 2019.

INSTITUTUL NAȚIONAL DE STATISTICĂ. (2019). Baza de date Tempo-Turism. http://statistici.insse.ro:8077/tempo-online/#/pages/tables/insse-table

Kotler, P. (2002). *Marketing places*. Free Press.

Light, D. (2000). Gazing on communism: Heritage tourism and post-communist identities in Germany, Hungary and Romania. *Tourism Geographies, 2*, 157–176.

Light, D. (2007). Dracula tourism in Romania Cultural identity and the state. *Annals of Tourism Research, 34*, 746–765.

Mahika, E.-C., Rădulescu, R., & Aluculesei, A.-C. (2015). The behaviour of Romanian tourists regarding the attendance at festivals. *Procedia Economics and Finance, 23*, 1239–1244.

Murphy, L., Moscardo, G., & Benckendorff, P. (2007). Using brand personality to differentiate regional tourism destinations. *Journal of Travel Research, 46*, 5–14.

Nicolau, M. (2014). Scurt Istoric al Festivalului Naţional de Muzică Uşoară de la Mamaia. http://www.topromanesc.ro/scurt-istoric-al-festivalului-national-de-muzica-usoara-de-la-mamaia. Accessed 10 Oct 2019.

Picard, D., & Robinson, M. (2006). *Festivals*. Remaking Worlds, Channel View Publications.

Pike, S., & Ryan, C. (2004). Destination positioning analysis through a comparison of cognitive, affective, and conative perceptions. *Journal of Travel Research, 42*, 333–342.

Plămădeală, M. (2010). Un festival care s-a curăţat de mici, bere şi porumb fiert. Sighişoara Medievală 2010. Observator Cultural. http://www.observatorcultural.ro/Un-festival-care-s-a-curatat-de-mici-bere-si-porumb-fiert*articleID_24054. Accessed 30 May 2019.

Răiescu, A. (1998). On Dracula tourism. In G. Erdeli, D. Dumbrăveanu (Eds.), *Romanian-British geography interferences*. Editura Corint, Bucureşti, (237 p.) ISBN: 973-9413-04-08.

Reynolds, W. H. (1965). The role of the consumer in image building. *California Management Review, 7*, 69–76.

Richards, G., Palmer, R. (2010). Eventful cities: Cultural management and urban regeneration. A Butterworth-Heinemann Title.

Romanian National Tourist Office. (2014). Romania festivals and events. http://romaniatourism.com/festivals-events-tradition-romania-2015. Accessed 20 Nov 2019.

Săndulescu, D. (2006). http://www.money.ro/simon-anholt-romania-are-imaginea-unui-cadavru. Accessed 22 June 2019.

Short, J. R., Breitbach, C., Buckman, S., & Essex, J. (2000). From world cities to gateway cities: Extending the boundaries of globalization theory. *City, 4*, 317–340.

Smith, C., & Jenner, P. (1998). The impact of festivals and special events on tourism. *Travel & Tourism Analyst, 4*, 73–91.

Tasci, A. D. A., Gartner, W. C., & Tamer Cavusgil, S. (2007). Conceptualization and operationalization of destination image. *Journal of Hospitality & Tourism Research, 31*, 194–223.

Todor, M. S. (1998). Mediovalia: Festivalul de Arta Medievala Sighisoara. http://mediovalia.blogspot.ro/p/festivalul-de-arta-medievala-sighiosra.html. Accessed 09 Sep 2019.

Toma, M. (2005). Ce-i de salvat la VamaVeche? In C. Zamfir (Ed.). Bucharest: Întrestilşi brand. Turismul alternativ la 2 Mai—Vama Veche.

Tudoricu, A. (2008). Event tourism in Romania—a tourist profile. *Human Geographies—Journal of Studies and Research in Human Geography, 2*, 95–100.

UNESCO World Heritage Centre. (2001) Historic Centre of Sighişoara—UNESCO World Heritage Centre. http://whc.unesco.org/en/list/. Accessed 12 Apr 2019.

Urry, J. (1995). *Consuming places*. Psychology Press.

Ward, S. V. (1998). Selling places: The marketing and promotion of towns and cities, 1850–2000, E & FN Spon.

Weber, S. (2004). *Reinventing a tourism destination: Facing the challenge*. Institute for Tourism Zagreb.

Zaharie, A. (2015). Sighișoara Medievală renaște pe acordul menestrelilor. http://www.zi-de-zi.ro/sighisoara-medievala-renaste-pe-acordurile-menestrelilor/. Accessed 07 July 2019.

Part II
Urban Environmental Issues

Chapter 11
Social, Cultural, and Information and Communication Technology Transformation Towards Smart Environment: Lessons Learned from the City of Magelang, Indonesia

Rini Rachmawati, Ulfah Choirunnisa, Qorih Imami, and Lely A. Nasution

Abstract Smart Environment is a terminology that appeared at the same time as the implementation of the concept of smart city. There are many elements related to environmental management. The indicators of smart environment are also varied. This paper will discuss the study of smart environment focusing on environmental management in various countries. Then also discussed in this paper will be the results of research in Indonesia, especially the research conducted in Magelang, a small city in Central Java. The data were obtained by undergoing in-depth interview with several local officials in charge of environmental management and communities. In this research the aspects of smart environment are more studied from the side of programme and activities planned and done by both the local government and communities. The result of this research shows that the implementation of smart environment in the city of Magelang has been running well, shown by the roles of both the local government and communities. There has been transformation related to social and cultural aspects, information and communication technology towards smart environment in the city of Magelang. This is shown by better service quality related to environmental management, the establishment of communities that provide support for environmental management, local people who become aware of the importance of conserving environment, and the use of information and communication technology in facilitating the activities of environmental management.

R. Rachmawati (✉)
Department of Development Geography, Faculty of Geography, Universitas Gadjah Mada, Yogyakarta, Indonesia
e-mail: rinirachma@ugm.ac.id

U. Choirunnisa
Graduate School on Geography, Universitas Gadjah Mada, Yogyakarta, Indonesia
e-mail: ulfah.c@mail.ugm.ac.id

Q. Imami · L. A. Nasution
Graduate School on Environmental Science, Universitas Gadjah Mada, Yogyakarta, Indonesia

B. R. K. Sinha (ed.), *Urban Dynamics, Environment and Health*,
https://doi.org/10.1007/978-981-99-5744-6_11

Keywords Smart environment · Social · Cultural · Information and communication technology · Transformation

11.1 Introduction

The concept of Smart City during the 1980s became a concept visualizing smart urban areas, emphasizing on the contents and digital services in urban areas to face various environmental challenges (Anthopoulos & Vakali, 2012). This concept was aimed at undergoing urban management in a modern way by applying latest technology which was environmentally friendly and met the expected results (Sikora-Fernandez & Stawasz, 2016). The ministry of Communication and Information of the Republic of Indonesia classifies the dimension of Smart City into six, namely; Smart Governance, Smart Branding, Smart Economy, Smart Living, Smart Society, and Smart Environment. The purpose of being smart in a smart city itself is the success in managing the potentials it possesses to become more efficient in facing challenges for development (Ministry of Communication & Information, 2018b).

Economic development in Asia and Pacific keeps growing. However, this growth is on the contrary with the condition of natural condition which is in danger and under pressure. Asian Development Bank (ADB) in the 2030 strategic plan focuses on the preservation of natural environment, this is by identifying efforts to tackle climatic change, creating climate, maintaining tenacity when disaster occurs, and improving environmental sustainability as the main field to obtain support. Environmental conservation becomes a prerequisite for economic growth and poverty reduction in Asia and Pacific (Asian Development Bank, n.d.). Under the control of the United Nations, many countries have so far agreed that sustainable development is the one that is intended to improve prosperity, economic opportunities, social welfare, and environmental protection (United Nation, n.d.). Environmental protection is one of the efforts to realize *Smart Environment*.

Smart environment is one of the aspects in the dimension of smart city (Ministry of Communication and Information, 2018b). The target of Smart Environment is to establish well-managed environment in a responsible and sustainable manner. Meanwhile, the programmes and indicators used are, among others; (1) developing the programme of environmental protection, (2) improving garbage and waste management, and (3) improving energy management in a responsible way (Ministry of Communication and Information, n.d.). There are three indicators used to see the condition of smart environment, namely: (1) protection for land, water, and air; (2) management of home and public garbage and waste from industries; (3) energy use emphasizing on efficiency and sustainability (Ahmadjayadi et al., 2016). Smart Environment with high value related to an underlining environmental aspect is how to manage garbage and drainage. Meanwhile, other aspects that need to be taken care of are the changing function of land for tourism interest, many people who are still found living in slum areas, no monitoring for air quality, and areas susceptible to disaster (Lanya, 2016). Local government, private sector, and local communities have an important role to play in achieving environmental management in supporting the implementation of smart cities (Rachmawati et al., 2020). One example of good environmental management is what is done by the community in Magelang City in Central Java, namely by carrying out activities such as waste processing, providing children's playgrounds, planting plants in pots using used plastic items, and creating educational gardens (Rachmawati et al., 2020).

11.2 The Overview of Smart City and Smart Environment in Different Countries

To reach a better city management, the concept of smart city can be implemented (Rachmawati, 2019). For future cities, environmental management needs to be supported with the use of ICT as an innovative strategy that is expected to be able to make it easy in the urban environmental management (Rachmawati, 2019). For example, to find out whether places for discarding garbage on the sides of the main roads have been over loaded or not, it is possible to use an ICT-based indicator. If so, a truck will come to pick up the garbage to transport it to the landfill. Another indicator can also be used to monitor air pollution whether it is dangerous for the people around the area. When somewhere on the side of the main road has been considered having been over polluted, it can be identified from the sensor installed so that it can be diverted to other road segments.

Singapore as a country that has applied the concept of smart city has always been making efforts to maintain its environmental sustainability (Ministry of the Environment and Water Resources, Ministry of National Development, 2016). In the Sustainable Singapore Blueprint, in the years between 2015 and 2030 Singapore is going to implement several programmes to support its environmental sustainability. They are, among others, (1) "Eco Smart" Endearing Towns, comprising the existence of plants, places for recreation, and liveable districts equipped with technology that is a great help for its citizens to maintain their green lifestyles and to enjoy their quality of life; (2) A "Car-Lite Singapore" is a city applying efficiency in everything and having large coverage of trains and buses, highways, and conducive lanes for pedestrians and bicycles, and smart alternatives for transportation; (3) Towards a Zero-Waste Nation, that is an effort to make Singapore free from waste by reducing the use of materials, reusing, and recycling; (4) A leading Green Economy, that is an effort to adopt greener practices in its business. That way parts of the city will become a hub for their latest sustainable businesses and that job opportunities will be created; (5) A leading Green Economy, that is an effort to maintain the spirit, commitment, and values of the citizens and that better homes can be created for good environment and better future for all; (6) Beyond Singapore, that is an effort to involve other countries to tackle environmental challenges since an environmental problem is too big to solve by one country only (Ministry of the Environment and Water Resources, Ministry of National Development, 2016).

Hong kong is also another example for sustainable environment. It can be seen from its planning document that contains efforts that will be made to realize “An Integrated Smart, Green, and Resilient Infrastructure System”, for example; Green building, energy saving, and waste-to-energy; Integrated Waste Management; Sustainable Urban Drainage and Flood Protection; Total Water Management; Infrastructure for green transport (e.g. Electric Vehicle Charging); Resilient to Landslides Flooding Typhoon, etc. (Planning Department, 2016). The ministry of Chinese Environmental Protection has also launched the programme *eco-city development.* This programme is the biggest initiative of the Chinese government to promote city development and handle a number of issues related to sustainability. Other approaches besides *eco city development* are the programme “*Low Carbon Provinces and City Program*” proposed by the Committee of Development and National reformation as well as “*Low Carbon Eco-city Program*” implemented by the Ministry of Housing and Urban-Rural Areas (Höffken & Limmer, 2019). Ecological Civilisation is one of the programmes to make China’s Eco-city come true. It is defined as human civilization that is trying to work hand in hand for all environmental problems comprehensively on industry, traffic, housing, and patterns of the society (Höffken & Limmer, 2019).

In 2015, 100 cities in India were chosen to be a “Smart City” with the help of smart technologies, that is by making use of Information and Communication Technology (ICT) (Höffken & Limmer, 2019). This is of interest for Indian people and becomes “lap-frogger” in urban development (Höffken & Limmer, 2019). Besides stressing on ICT, the aspect of being smart is also related to the connotation of environmental friendliness and it can be implemented in the form of greening (Höffken & Limmer, 2019). The same case in India, initiated by the Ministry of Communication and Information, Indonesia also launches “Movement toward 100 Smart Cities” (Ministry of Communication and Information, 2018a). The background of this movement is the complicated problem happening in urban areas, such as population growth, climate change, traffic congestion, poverty, crime, natural disaster, etc. (Ministry of Communication and Information, 2018a). The concept of Smart City has to be able to provide support of basic services for the citizens and this is adapted to the condition of local characteristics and the needs of the people in the respective region. The vision of Smart City/Regency Development is the making of unity in green city that is competitive and based on technology and that smart economy, smart people, smart government, smart mobility, and smart living can be created (Ministry of Communication and Information, 2018a).

11.3 The Study of Smart Environment in the City of Magelang, Indonesia

In relation to the movement of 100 Indonesian Smart Cities, in 2019 the city of Magelang was chosen as one of the smart cities. One of the dimensions in smart city, according to Kemenkominfo, is smart environment. The local government of the city of Magelang is considered having good management in environment. This can be shown with the achievement of that city that has achieved the award of Adipura 11 times from the State Ministry of Living Environment (Tribun Jogja, 2017). In 2014, the city of Magelang received the award Adipura Kencana since it consecutively succeeded in managing its urban environmental cleanliness sustainably. Adipura Kencana is an award given to a city that is successful in receiving the award Adipura three times in a row in the period of the last three years (Ministry of Living Environment and Forestry of the Republic of indonesia, 2018). In 2019, again the city of Magelang received the award Adipura (Tribun Jogja, 2019).

This paper will further reveal the social, cultural, and ICT transformation in supporting the achievement of smart environment. This research is aimed at (1) identifying the coverage of the implementation of smart environment; (2) undergoing the identification of the social, cultural, and ICT conditions towards smart environment, and (3) analysing the applicable strategies to achieve smart environment in the city of Magelang through social and cultural transformation and the use of ICT that is appropriate.

11.4 Methodology

This research uses descriptive-qualitative method, that is by putting forward the presentation of fact and phenomenon on field. The data collection was done by having *in-depth interview* with stake holders given authority to make policies, plans, and programmes related to smart environment in the city of Magelang, namely OPD (Organisasi Perangkat Daerah/Regional Apparatus Organization) in the city of Magelang in charge of managing environment. Besides, the data were also obtained through focus group discussion held by groups in charge of discussing the theme of smart city in the activities of technical guidance of smart city in the level of city administration held in 2019, that is the activity of making the Masterplan of Smart City as part of the Indonesian 100 Smartcity Movement organized by Ministry of Communication and Information. The OPD with which the in-depth interview was conducted comprises Unit of Nation and Politics for Community Protection, Agency for Living Environment, Agency for Communication, Information and Statistics, Agency for Agriculture and Food, Regional Development Planning Board. Meanwhile, in-depth interview was conducted with IT communities; Micro, Small, and Middle Business Community; Community of the people of kampong warna and Flower Lovers, the kampong of Tegalsari and Clean River Community as well. The technique of analysis that is used is descriptive-qualitative analysis technique.

11.5 Results

11.5.1 Environmental Condition in the City of Magelang

Environmental controlling and monitoring in the city of Magelang is implemented by the Agency for Living Environment, whose main responsibilities include monitoring, environmental control and establishment, waste, and gardening. The responsibility of controlling includes the monitoring of RTH (Ruang Terbuka Hijau/Green Open Space). 18.92% Public RTH and 19.11% Private RTH is currently found in the city. Included in the RTH are 53 city parks, one TPU (Tempat Pemakaman Umum/Public Cemetry), and one City Forest. Figures 11.1 and 11.2 show the picture of one of the city parks and city forest.

Fig. 11.1 City park

Fig. 11.2 City forest and shady trees

The Agency for Living Environment is also in charge of monitoring the quality of environment by taking river water sample at least twice a year during the dry and rainy seasons. Based on the Quality Index of Living Environment in the city of Magelang in 2018, it can be seen that the Quality Index of Water in the city of Magelang (Table 11.1) is on the index 52. This means that the condition is considered poor, meaning that it is necessary to reduce water pollution in order that the quality of water is getting better. One of the ways is by establishing Garbage Bank.

Pollution comes from the domestic waste dumped into the river by those living on both sides of the river, especially the areas in the central part of the city since it is densely populated. This is indicated by the value of fecal coli and total coliform that exceeds the quality standard. Therefore, it is necessary to make the river free from waste and garbage. In addition, it is also necessary to take a preventive action, such as giving direction and increasing the awareness of the people living around the river to keep the river clean by not dumping garbage and waste into the river (Magelang Municipality, n.d.).

In 2018 socialization to increase the quality of water environment was held by the Agency of Living Environment and Forestry of the Province of Central Java with the target of the people living around the river. Socialization was also held by the Agency for Housing and Settlement Areas with the theme Toward Magelang City Society to Stop Having Bowel Motion in the River for realizing the city of Magelang as Open Defecation Free. The city government of Magelang also runs activities of developing Local Domestic Water Waste Management System and Centralized Domestic Water Waste Management System. The Agency for Living Environment also runs the monitoring over the air quality by checking the ambient air quality in 20 sample points, that is 15 points in the location of Car Free Day/in Rindam sports field, 2 points in Kebon Bibit Senopati, 2 points in Kampong Magersari, and 1 point in Taman Atria. The parameters measured are CO, CO_2, NO, O_2, wind speed, moisture temperature, and noise. In fact, the air in the city of Magelang shows an index of 84.91 and in accordance to the category in Table 11.2 it can be considered Very Good condition (Magelang Municipality, n.d.).

Index of the Land Cover Quality in the city shows an index of 38.22 (Table 11.3), meaning that it is in the alert condition. The result of the measurement of the indexes

Table 11.1 Water quality index in the city of Magelang

Status	Total	Percentage (%)	Coefficient	Value
Appropriate	3	30	70	21
Light	5	50	50	25
Medium	2	20	30	6
Heavy	0	0	10	0
Total	10			
Value of water quality Index				52

Source Environmental quality Index of the city of Magelang (Magelang Municipality, 2019)

Table 11.2 Index of air quality in the city of Magelang

Parameter	Average	Standard quality	Parameter of pollutant index	Index of air pollution IEU	Index of air quality
		EU	IEU		
NO_2	8.44875	40.00	0.2112	0.371546875	**84.9140625**
SO_2	10.6375	20.00	0.5319		

Source Environmental quality index of the city of Magelang (Magelang Municipality, 2019)

of water, air, and land cover qualities is 56.36. This means that the condition is very lacking (Environmental Quality Index of the city of Magelang, 2018).

Regional Spatial Plan of the city of Magelang in the years 2011–2031 states that the city of Magelang has a protective zone, that is local protection zone and zone susceptible to landslide. Local protection zone includes areas on both sides of river and green open space (city forest). The city of Magelang has protective zone and city forest that is very important to meet the need for city green open space, namely conservation zone of Mt. Tidar. Education on environment is also given to those in their early age in their schools, through the implementation of the programme of living environment management in school (adiwiyata). On the field of garbage management, in fact, it has been well managed in the respective city. This can be proven with the movement of selecting garbage to discard in the trash bins provided. The selection of garbage is based on organic and unorganic one, as shown in Table 11.4.

The organic kampong programme, later called "SI IPAH", has in fact become an outstanding programme of the city of Magelang. There are 75 organic kampongs joining this programme in the city of Magelang, as shown in Table 11.5.

Garbage Bank has so far run well in the city of Magelang. There are presently 89 units of Garbage Bank (1 main Garbage Bank, 31 units in South Magelang, 28 units in Central Magelang, and 30 units in North Magelang). The distribution in details of Garbage Bank in the city of Magelang can be seen in Table 11.6.

Table 11.3 Index of the land cover in the city of Magelang

Area size (Ha)	Forest cover size (Ha)	Non-forest vegetation size (Ha)	Forest cover	Index of land cover quality
1812	110	336.384	0.172	**38.22**

Source Environmental quality Index of the city of Magelang (2018)

Table 11.4 Kinds of garbage managed in the city of Magelang

Type	No	Type	Economy			Percentage (%)	
			Low	Middle	High		
Organic	1	Leaf	13.22	9.5	9.8	9.9	47
	2	Foods	41.45	31.9	33.7	33.7	
	3	Wood/ Branches	1.78	3.9	2.8	3.4	
Inorganic	4	Paper	9.05	19.0	18.9	18.8	53
	5	Plastic	17.01	23.0	19.2	21.5	
	6	Metal	1.00	0.8	0.8	0.8	
	7	Glass/ Ceramic	2.01	3.2	4.4	3.7	
	8	Rubber/ Leather	1.58	0.7	0.4	0.5	
	9	Cloth	3.81	2.2	1.3	1.8	
	10	Others	9.09	5.9	4.7	5.9	
Total			100.00	100.00	100.00	100.00	100.00

Source Agency for Living Environment of the city Magelang (2019)

11.6 Discussion

11.6.1 The Implementation of Smart Environment in the City of Magelang

The dimension of smart environment is related to the achievement of sustainable development. The vision of smart city of the city of Magelang is "To make the city of Magelang an Economic Magnet to achieve the welfare for the people based on the typical culture" (City Government of the city of Magelang, 2019). Meanwhile, the main goal of the smart environment dimension is to make the sustainable development come true. The target is to encourage sustainable development in line with the mandate of Sustainable Development Goals. The strategies are by developing and making use of the local potencies to make sustainable development come true and environmental insight; making use of the GPS technology installed on garbage trucks so that their mobility is monitored (Magelang Municipality, 2019).

The result of in-depth interview with informants from OPDs in charge of environmental matters shows that there is difference in emphasizing to implement smart environment. Board for Political Nation Union and Society Protection aims to establish comfort, ease in services and competitiveness, public services with computerized system that is not based on cable (wifi), applicative informatics system service in each

Table 11.5 The number of Organic Kampongs in the city of Magelang based on Kelurahan *(district in a city)*

Sub-district/Village	The number of registered organic Kampongs	The number of inactive organic Kampongs
Magelang Selatan (South Magelang)	**25**	**6**
Rejowinangun Selatan (South Rejowinangun)	4	0
Magersari	3	0
Jurangombo Utara (North Jurangombo)	7	2
Jurangombo Selatan (South Jurangombo)	4	1
Tidar Utara (North Tidar)	6	3
Tidar Selatan (South Tidar)	5	
Magelang Utara (North Magelang)	**22**	**2**
Wates	4	0
Potrobangsan	5	0
Kedungsari	7	0
Kramat Utara (North Kramat)	4	2
Kramat Selatan (South Kramat)	2	0
Magelang Tengah (Central Magelang)	**28**	**9**
Keminrejo	4	2
Cacaban	6	2
Rejowinangun utara (North Rejowinangun)	2	1
Magelang	**6**	**1**
Panjang	5	1
Gelangan	5	2
Total	**75**	**17**

Source The Agency for Living Environment (2019)

unit of the Regional administration, well-managed environmental management, well-managed garbage and waste management, and responsibly managed energy management. The programmes that have been implemented related to smart environment are, in fact, not so innovative, except the one related to garbage management. Meanwhile, other applications used are, among others, Kiss Me (providing data of low-income

Table 11.6 List of Garbage banks in the city of Magelang based on Kelurahan

Sub-district/ Village	The number of registered garbage banks	Inactive garbage banks	Active customer/ Participants (Family)	Garbage volume (kg)	Omzet (Rp) turnover (Rp)
Magelang Utara (North Magelang)	**30**	**5**	**457**	**2112**	**3.270.50**
Wates	6				
Potrobangsan	6	2			
Kedungsari	4	1			
Kramat Utara (North Kramat)	6	1			
Kramat Selatan (South Kramat)	8	1			
Magelang Tengah (Central Magelang)	**14**	**6**	**209**	**729**	**1.093.700**
Kemirirejo	4	2			
Cacaban	6				
Rejowinangun Utara (North Rejowinangun)	2	4			
Magelang	**9**				
Panjang	3				
Gelangan	4				
Magelang Selatan (South Magelang)	**31**	**3**	**344**	**2294**	**2.936.363**
Rejowinangun Selatan (South Rejowinangun)	9				
Magersari	3				
Jurangombo Utara (North Jurangombo)	5	1			
Jurangombo Selatan (South Jurangombo)	4				

(continued)

Table 11.6 (continued)

Sub-district/ Village	The number of registered garbage banks	Inactive garbage banks	Active customer/ Participants (Family)	Garbage volume (kg)	Omzet (Rp) turnover (Rp)
Tidar Utara (North Tidar)	5	2			
Tidar Selatan (South Tidar)	5				
Total	**89**	**14**	**1010**	**5134.5**	**7.300.913**

Source The Agency for Living Environment (2019)

family accessible and transparent) intended to reduce the programme of reducing poverty, Si Bahenol Online (regional tax payment service-BIHTB online), demographic data, one-stop vehicle tax, e-filling, and e-budgeting (LPSE). This is proven by their easy access and of great help. However, there are other programmes that are not optimally implemented. The applications related to smart environment are Data Go, SIG (Sistem Informasi Geospasial/Geospatial Information System), Kiss Me (Kolom Informasi Kesejahteraan Sosial Masyarakat Secara Elektronik/Information Column for Social Welfare in an Electronic Way), and Si Bahenol (Sistem Informasi Biaya Perolehan Hak Atas Tanah dan Bangunan Secara Online/Online Fare Information System for the Right on Land and Building). Those applications have been running well and the local people are able to access them to obtain information and services. Yet, it is necessary to socialize them intensively to the local people.

Facilities-related smart environment initiated by The Agency of Communication, Informatics and Statistics are applications for integrated, open access, and online services. By providing virtual services through online access, it is expected to make it easy for the local people and OPDs to do coordinations so that effectiveness and efficiency can be made. The ease was realized through the launching of an application that can be downloaded by both all members of society and OPDs called "Magelang Cerdas" (Smart Magelang). This application was launched by The Agency of Communication, Informatics and Statistics in 2015 and it's a kind of online one-gate service. All kinds of activities and aspirations not only from OPDs but also from the society can be accepted by this application. Besides, any kind of transparency and information related to the City of Magelang can also be accessed freely by this application. The role of The Agency of Communication, Informatics and Statistics in implementing smart environment is more on being the facilitator for making coordinations easy among the OPDs and between OPDs and society. Besides, its other roles are providing supporting services related to the existing green open space in the City of Magelang and launching applications that are useful for establishing ease and transparency related to anything about the city of Magelang. All forms of services through an application created by Diskominsta are intended to support the implementation of smart environment. They have to be online, easy to access in an effective and efficient way, and one gate.

The Agency for Agriculture and Food plays an important role in processing compost. The organic garbage from every part of the city is collected and processed to be compost. The process itself is done by both the Agency for Agriculture and the local people working together. The compost that has been processed is collected and then distributed to groups with activities related to both agriculture and plantation. Both the individuals and groups involved in those activities get the compost freely so that it is expected that their crop is successful. Besides the Agency for Agriculture and Food that plays an important role in realizing Smart Environment, the Agency for Living Environment also plays an important role. The Agency for Living Environment of the city of Magelang provides several kinds of services, such as issuing online permit for the storage and collection of dangerous and poisonous waste, environmental setting, road sweeping, gardening, road lightening, issuing permit for cemetery, transporting garbage, absorbing latrine, tree cutting, and controlling pollution and conservation of natural resources. Implementasi Smart Environment at the Agency for Living Environment, one of the implementations related to Smart Environment that is carried out by the Agency for Living Environment is related to road lightening. In 2016–2017, there was a service provided by the Agency for Living Environment called "Siput". Siput is a kind of device that is able to give information to stakeholders when there is something wrong with light network in a place where road lightening is installed. However, road lightening is currently not running well anymore due to the factor of cost and application renewal that cannot be undergone because of the limited human resource. SI IPAH is a supreme smart environment programme that is initiated by the Agency for Living Environment in the city of Magelang. Figure 11.3 shows the concept draft of the SI IPAH programme.

SI IPAH is the result of transformation programme of Organic Kampong, that is the one that implements 3R-based garbage processing (Reuse, Reduce, Recycle). Organic Kampong has been in existence since 2013. In 2013, the city districts where kampong organic is found received some amount of fund from the mayor to run the organic kampong programme in order that it ran well. However, after 2013 the programme was not supported anymore by the local government. Therefore, in 2018 there were only 29 organic kampongs left. The percentage of the organic kampong implementation, according to The Agency for Living Environment, is only 10%. The

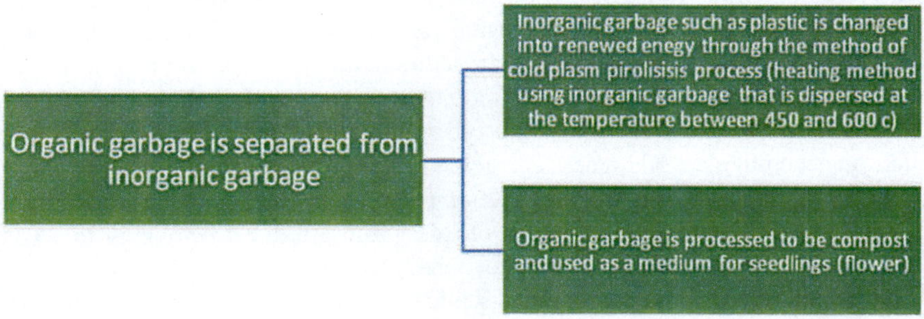

Fig. 11.3 SI IPAH scheme. *Source* Agency for Living Environment

organic kampong programme was in the stagnant condition in 2019. The thinking pattern of the people also changed. At the beginning they thought that it was meant to reduce the volume of garbage, but now they think it is only necessary to plant organic vegetables. Meanwhile, Garbage Banks found in the city of Magelang produces items such as bags made from the discarded wrapper of drink sachet, flowers made from unused plastic bags, dan some others. Unfortunately, a big problem for its sustainability is that there are no other consumers except the local government. Means of transportation for garbage is also another problem in relation to garbage management in the city of Magelang. Garbage is still mixed when transported, no separation between organic and inorganic garbage. This is because the local people do not provide organic trash bin. Besides, means of transportation for garbage provided by DLH is still limited. In relation to information, The Agency for Living Environment has owned a website that provides information on environment and it is always monitored and updated. The website is called E-Dasiku (Electronic Data and Information on Environment) with the address of http://e-dasiku.byethost9.com/profil.

The role of the Board for Regional Development Planning (BAPPEDA) in realizing smart environment is making the masterplan for Mt. Tidar. There is a plan to build a garden in Mt. Tidar and it will become a place for ecotourism and the one related to religious affairs. In the future, Mt. Tidar is not only for conservation but also a place for UKM (Usaha Kecil Menengah/Small Medium Enterprises) and tourism. BAPPEDA also plans to make the city of Magelang developed in line with the vision and misson of the mayor. Smart environment is not only related to environmental management but also related to settlement and regional facilities. Some of the local people are still in fact found to have difficulty to differentiate between grey water (waste from home) and black water (waste that needs to use septic tank). 3% of the city people in Magelang, more or less 1.165 families, are still found to have bowel motion anywhere, particularly in the river (or in Indonesian called Buang Air Besar Sembarangan/BABS). BABS itself is divided into two, namely those who have bowel motion on the side of the river and those who have already had MCK (Mandi, Cuci, Kakus/Bath, Wash, Restroom) but it is not equipped with septic tank. Instead, they flow their feces directly into the river. The Board for Regional Development Planning (BAPPEDA) is trying to solve this problem by building Local Domestic Waste Processing System, or commonly called communal septic tank. In 2018, under the coordination of the Agency for Housing and Settlement Areas of the city of Magelang, 84 communal septic tanks were built and they were funded by the Local Government through regional budget. In addition, 20 communal septic tanks were also built and they were funded by United States Agency for International Development (USAID). All those septic tanks spread in 17 city districts. One communal septic tank can accommodate 50 homes. The construction of those communal septic tanks requires the establishment of KSM (Kelompok Swadaya Masyarakat/Community Empowerment Group). The members of this group are those belonging to KPM (Kelompok Penerima Manfaat/Group of Benefit Recepient). The role of KSM is of primary importance because communal septic tank needs maintaining routinely once in 3 years. In relation to problems related to garbage, the city of Magelang is trying to reduce garbage by making use of Maggot BSF (Black Soldier Fly). This innovation is developed by the Research and Development Section of BAPPEDA. Rejowinangun Market is the one that has applied this innovation since 2018. "Gotong Royong" (Mutual Cooperation) Market will be presently applying the same innovation.

11.7 Social and Cultural Change of the Society and the Role of Information and Communication Technology

One of the challenges in development is that it is difficult to implement natural change in order that the society is able to participate in managing their living environment. Therefore, it is necessary to have the agent of change to introduce a change in a programme that is well planned and designed in order that it is able to empower society in managing living environment. The presence of communities in society becomes an important element in achieving changes related to environmental management. The communities related to smart environment are, among others, IT Community; Micro, Small, and Medium Enterprise; Community of the People of Colour Village and Flower Lovers; and Clean River Community.

IT community was established in 2009 but started its activities in 2010. This community uses ICT to create plans of innovative programmes that give benefits to the local people. They are, among others; (1) Kentongan digital (Digital Drum), an emergency system using ICT that is integrated to homes through software network. It can be accessed by using smart phone or hardware in the form of button with a certain symbol connected to people's homes. This software network is sinergized into several public services and sectors related to health, security, disaster, and it provides a warning system to the surrounding people when someone needs some help for situations, such as sickness, gas leaking, fire, theft, etc., (2) Making use of energy from garbage. This innovation can be realized by working together with garbage bank through garbage separation system. Inorganic garbage such as metal and plastic that still has economic value is collected. Meanwhile, organic garbage can be processed into *pelet*(fish food). It functions as the substitute of charcoal that is still used in *tahu* and tempe industry. ICT-based software system is very useful to help storage data and provide information on distributing garbage by garbage bank. This innovation helps reduce the use of charcoal, fossil fuel, coal, and electricity and that smart living environment can be established by managing garbage and energy. The application of the food distribution is an example of e-commerce application for food buying and selling. Moreover, Magelang is an agricultural area and becomes the one sending foods to the province of Central Java and Yogyakarta Special Region. Therefore, this application will help farmers get information in distributing their crops so that an effective, efficient, and economical way in food trade can be established. This application will make it easy for buyers to access information related to the need for food. This will also guarantee the farmers in distributing their crops so that their income will increase since the price is controllable.

The community of Micro, Small, and Medium Enterprise is the one that has received support from the government, particularly from the Agency for Investment and One-Gate Integrated Services. This agency has given trainings and guidance related to mainly living environment. They are, for example, how to process the remains of their crop produce, food materials and health, business management, and finance. An example of this enterprise related to environment is tahu-tempe(tofu-tempe) and batik industry.

The community of Colour Village and Flower Lovers is implementing the programme of Tidar Campur area as an educational place of interest by making use of the existing potentials. This flower community has done a lot to create colourful kampongs by empowering the people from the respective kampong.

The initiative of establishing colourful kampong as seen in Fig. 11.4 above is chosen as the realization of how important it is to be aware of environment. By having different colours for their houses in the respective kampong, the local people become more attentive to the cleanliness of their environment. Besides, beauty can also be created. This kampong is also regarded as an organic kampong since it is able to provide garbage bank by itself and to make use of unoccupied land for processing garbage to be compost with the fund help from the local government. This kampong also has unoccupied land as large as 1.040 m^2 for planting orchid and other grooming plants. Therefore, this community can be called to be actively participating in supporting the slogan the city of Magelang as a city with million of flowers. The Kampong of Tidar Campur also has a space for processing garbage of 3 R (Reduce, Reuse, Recycle) or popularly called TPS3R as seen in Fig. 11.5.

In relation to the environmental management by making and maintaining garden and plants on vase initiated by community, there is an area where the role of the community is high. One of the communities is found in the Kampong of Tegalsari. Previously, it was a wild piece of land where the park was developed In order to give benefits, then there was an initiative to make a park and play ground for comfort, especially to children to play around. The park and plants were managed by the local people. In this case, the awareness of the people from the Kampong of Tegalsari

Fig. 11.4 Environment established by the community of Kampong Warna (Colour Village) and flower lovers

Fig. 11.5 Garbage processing initiated by community

Fig. 11.6 The setting of environment initiated by community

can be called quite good. This can be proven with the plants found in bottles and set outside of their home walls to create beauty for their environment as seen in Fig. 11.6.

In the City of Magelang, there is also a community called Clean River Community. It focuses on cleansing the river routinely. This community has a jargon of "jeguran kali", meaning jumping into the river, which is meant to increase the environmental care from the people to the river. The target of this community is the settlement areas on both sides of the river. The community wants those living on both sides of the river to make the river as their front yard so that they are expected to keep the river clean. This community routinely runs an activity in cooperation with other groups of people and schools as a celebration especially on public holidays.

Based on the result of interview with several informants from OPDs and communities in charge of managing smart environment, there is some information that supports the environmental management implemented by the society. The result of the interview with the Board of Politic Nation Union and Society Protection reveals that social transformation that happens in society towards smart environment is related to the availability of information on social welfare of the people and the availability of integrated data so that the misuse or falsification of data can be prevented. In addition, the people who are becoming smarter and smarter and the ease to access public services also play an important role. Meanwhile, cultural transformation can take place due to the result of interaction among the society. Nevertheless, the city government of Magelang has a programme to maintain the values of local culture. The role of ICT in this case is to help provide data that can be accessed by the society, provide wifi in public areas, and provide telephone poles for the hanging cable to be installed. Meanwhile, fibre optic cable is installed inside the soil so that online system runs smoothly.

The indicator to measure the transformation that happens in society in relation to the social-cultural conditions in the city of Magelang cannot be seen directly. However, it can be felt. This can be known when the interview was conducted with OPDs related to the transformation that happens in society. This transformation can be felt in the form of the awareness of the people in keeping their environment. They really support the programme "Si Ipah" coordinated by the Agency for Living Environment. Besides, more and more garbage processing in the form of garbage bank in their own environment is evidently found. Meanwhile, based on the result of interview with the Agency for Communication, Information and Statistic, it shows that another transformation that happens in the society is the increasing number of people to access application and online services provided by the Agency for Communication, Information and Statistic. This can be seen from the statistical figures that show an increase. Such a transformation is seen to be a progress in providing services to the society. Nevertheless, transformation that is the most important is the awareness of the society in keeping their city environment clean. Such a transformation is expected to exist so that it can support the city of Magelang in making smart environment come true.

11.8 Strategies to Achieve Smart Environment

The strategies that need to be implemented to achieve smart environment are putting forward innovations to face global challenges in the future, doing synergy with stakeholders and academicians, maintaining online network system, and providing more online public services. Due to the so many challenges that appear in providing well-prepared public services electronically to society, this makes the city of Magelang, just like other cities, begin an innovation to establish a portal for *one-stop public service,* that is the improvement of an application that can be used on the grasp of hand. The innovation is called Smart Magelang, the one that is accessible by using smart phones belonging to public. This Magelang Cerdas application integrates various interoffice public services in one public portal. It is expected that the society is able to access public services in an efficient, easy, fast, and thorough way. The figures and contents shown are made as comprehensive as possible so that the size and parameter of the purpose of public services electronically are clear and detailed and that the cost to access can be made lower. Besides, it can be accessed at anytime and anywhere without any obstacles. In the future, the improvement of this Magelang Smart City application is expected to be able to integrate the management of living environment on its programme so that it can access informations related to the monitoring of living environment.

11.9 Conclusions

The implementation of smart environment in the city of Magelang has been well done by all layers of society and government institutions. The realization to make the City of Magelang able to integrate all public services as a parameter of smart city is one of the criteria for a city to be called a smart city. It has its own phase of process. This is based on the characteristics of region, society, and infrastructure available. Those three are the variables to be the basis of the consideration or concept for a city to make steps for smart environment to be true. Transformation related to the implementation of smart environment can be seen from the establishment of the existing communities and their support to manage environment. The transformation that happens in the society, awareness of the people to care for environmental issues and better services provided by the City Government of Magelang in managing environment become an important asset to make smart environment come true.

Acknowledgements We thank Graduate School on Environmental Science, University Gadjah Mada that has provided fund for this research through the 2019 Research Grant Scheme.

References

Ahmadjayadi, et al. (2016) *Melesat atau Kandas? New Indonesia: Dari Smart City Menuju Smart Nation*. H. Kusdaryanto & F. R. Kautsar (Eds.). PT Elex Media Computibdo.

Anthopoulos, L. G., & Vakali, A. (2012). Urban planning and smart cities: Interrelations and reciprocities urban planning: principles and dimensions. *Lecture Notes in Computer Science, 7281 (May 2014)*, 178–189. https://doi.org/10.1007/978-3-642-30241-1

Asian Development Bank. (n.d.). *ADB's Focus on Environment*. https://www.adb.org/themes/environment/main

Höffken, J. I., & Limmer, A. (2019). Smart and eco-cities in India and China. *Local Environment, 24*(7), 646–661. https://doi.org/10.1080/13549839.2019.1628730

Lanya, I. (2016). 'Sumber Daya Alam dan Lingkungan Cerdas (Studi Kasus Kota Denpasar)'.

Magelang Municipality. (n.d.). *E-Dasiku (Electronic Data dan Informasi Lingkungan):* Environmental Quality Index 2018. Retrieved December 15, 2019, from https://drive.google.com/file/d/1FvSHK66cDGmJzigLcROomJeEwwei-RKP/view

Magelang Municipality. (2019). Master plan smart city.

Ministry of Communication and Information. (2018a) *Gerakan Menuju 100 Smart City*. Retrieved December 14, 2019, from https://aptika.kominfo.go.id/2018/11/gerakan-menuju-100-smart-city/

Ministry of Communication and Information. (2018b) *Kemenkominfo: Penerapan Smart City Jangan Hanya Jadi Tren dan Proyek Sesaat*. https://kominfo.go.id/content/detail/13601/kemenkominfo-penerapan-smart-city-jangan-hanya-jadi-tren-dan-proyek-sesaat/0/sorotan_media.

Ministry of Communication and Information. (n.d.). *Guideplan Masterplan Smart City Gerakan Menuju 100 Smart City*. https://slideplayer.info/slide/11857474/

Ministry of Living Environment and Forestry of the Republic of Indonesia. (2018). *Peraturan Menteri Lingkungan Hidup dan Kehutanan Republik Indonesia.*

Ministry of the Environment and Water Resources, Ministry of National Development, C. for L. C. (2016). 'Sustainable Singapore blueprint message from the Prime Minister'. https://www.clc.gov.sg/research-publications/publications/urban-systems-studies/view/sustainable-singapore-blueprint.

Planning Department. (2016). Hong Kong 2030+: A smart, green and resilient city strategy (October). http://www.hk2030plus.hk/document/HongKong2030+ASGRCityStrategy_Eng.pdf.

Rachmawati, R. (2019). Toward better city management through smart city implementation. *Human Geographies-Journal of Studies and Research in Human Geography, 13.*

Rachmawati, R., Imami, Q., Nasution, L.A., Pinto, R.P.A., Pradipa, H. (2020). Urban environmental management: An effort toward Magelang smart city. *Open Access.* International Conference on Environmental Resources Management in Global Region IOP Conference Series: Earth and Environmental Science.

Sikora-Fernandez, D., & Stawasz, D. (2016). The concept of a smart city in urban management. *Business, Management and Education, 14*(1), 34–49. https://doi.org/10.3846/bme.2016.319

The Agency for Living Environment. (2019). *Informasi Layanan Publik*. Dinas Lingkungan Hidup.

Tribun Jogja. (2017). *Kota Magelang Butuh Inovasi Tambahan untuk Kembali Raih Adipura Kencana*. Retrieved January 14, 2020, from https://jogja.tribunnews.com/2017/10/13/kota-magelang-butuh-inovasi-tambahan-untuk-kembali-raih-adipura-kencana

Tribun Jogja. (2019). *Kota Magelang Kembali Raih Adipura di Tahun 2019*. Retrieved January 14, 2020, from https://jogja.tribunnews.com/2019/01/14/kota-magelang-kembali-raih-adipura-di-tahun-2019

United Nation. (n.d.). *Promoting sustainable development*. https://www.un.org/en/sections/what-we-do/promote-sustainable-development/

Chapter 12
A Bus Rapid Transit System to Help Alleviate Air Pollution in Kathmandu, Nepal

Keshav Bhattarai and Ambika P. Adhikari

Abstract In rapidly urbanizing Nepal, its bigger cities are currently experiencing transportation and environmental issues that were rarely seen before. Nepal's capital, Kathmandu Valley, is plagued by a serious issue with air pollution. This pollution is largely brought on by high traffic. Over 500,000 vehicles of various sizes and types ply each day on just a little over 2,000 kms of roads in the Valley. Because commuters spend a significant amount of time each day caught in gridlock on the highways, emissions are added each hour by high traffic and their suboptimal speeds and idling conditions. The current transportation system with low occupancy vehicles such as two- and three wheelers, personal vehicles, and taxis can move fewer people per vehicle than it would be able to if it was mostly made up of high occupancy vehicles. Our findings indicate that a considerable number of pollutants in the Kathmandu Valley can be decreased if the current traffic congestion is relieved utilizing high-capacity vehicles that can travel at ideal speed limits (above 25 km/h for metropolitan region). According to estimates, the development of a Bus Rapid Transit (BRT) system will assist cut daily automobile emissions by about 7,500 tons as BRT will move more commuters to high occupancy buses. Furthermore, the Kathmandu Valley's vehicle pollution may perhaps be decreased by more than 16,000 tons per day if these high occupancy buses were electrified. An increase in transit system efficiency will result from such a reduction in vehicle emissions, which will significantly enhance air quality. A BRT system powered by electricity would be one of the solutions to the Kathmandu transportation system that would assist in achieving the goal of this chapter in making the Kathmandu Valley a smart city.

Keywords Kathmandu valley · Transportation · Air pollution · Electric vehicles · Bus rapid transit · Smart city

K. Bhattarai (✉)
Department of Physical Sciences, University of Central Missouri, WC Morris 409 D, Warrensburg, USA
e-mail: bhattarai@ucmo.edu

A. P. Adhikari
City of Tempe, Tempe AZ 85281, USA

B. R. K. Sinha (ed.), *Urban Dynamics, Environment and Health*,
https://doi.org/10.1007/978-981-99-5744-6_12

12.1 Introduction

This paper reviews the state of vehicular transportation infrastructure and operation in the Kathmandu Valley in Nepal. It analyses some salient elements related with the existing transportation vehicles including the number of passengers served, the congestion resulting from the rapidly increasing number of vehicles on the limited road infrastructure, and the air pollution created by the vehicles. This paper analyses potential reduction in the air pollution load that can be achieved by an electric Bus Rapid Transit (BRT) system in the Kathmandu Valley. It also assesses the possible contribution of a BRT system in reducing traffic congestion, minimizing commute time, and abating the emission of pollutants and greenhouse gases. A BRT system running on electricity can be an important approach towards making the cities in Kathmandu Valley smart.

The Kathmandu Valley is experiencing major transportation problems that include the need "to catering 3,000 to 4,000 passengers per direction per hour" as per 2017 estimates (Amatya, 2017: 1). Traffic congestion is high, and the reliability of service remains poor. Because of the lack of open spaces and adequate public right-of-way in the Kathmandu Valley cities, where land is extremely expensive, it is difficult to find spaces to build an overground monorail system in the cities. A monorail transit faces other challenges too. First, it is not safe and comfortable for passengers to climb up and get down to the street level from the elevated monorail stations, especially, when the population is ageing. Second, as a monorail system can only carry limited number of passengers, it will be outdated quickly as the transit demand accelerates.

To overcome the needs to ferrying "20,000 to 50,000 passengers in each direction every hour" (Amatya, 2017) some urban planners have proposed underground metro system like in Atlanta, Georgia of the US, Asakusa and Ueno in Tokyo, Japan, Beijing, Seoul, Kolkata, Manila, and Hong Kong (Leather, 2021). They have proposed underground routes connecting east and west of the Kathmandu Valley. Such a metro-line will connect "Thankot with Sankhu and Banepa via Bhrikuti Mandap" (Adhikari & Bhattarai, 2021; Amatya, 2017). Siphoning from the main route will be another 25-km long metro-line that will connect the main metro-line with the Valley's southern city of Patan. This line will also be connected with the northern settlements of the Valley, the Budhanilkantha via Bhrikuti Mandap in the midtown Kathmandu. The old ring road will be connected by another "30-km long orbital line" (Amatya, 2017). Such a network will crisscross the Kathmandu Valley, north–south, and east–west.

The authors are in agreement with other urban planners such as Amatya (2017), Leather (2021), and Susantono et al. (2021) that the Kathmandu Valley eventually needs an underground metro system like in Bangkok and Delhi.

A metro system for Kathmandu is an attractive proposition. However, the proposed Kathmandu metro railway will face issues of cost and feasibility as it needs tunnelling work for "60-km distance including several kilometers of viaducts to smoothly operate the proposed three principal train lines" (Amatya, 2017). Further, for immediate implementation, it looks unrealistic given its high cost of construction per mile, the complex engineering required for it and the overall project management.

Another major issue is the high cost of land that would need to be acquired. Further, in the seismically vulnerable location of the Kathmandu Valley, the engineering required to design and implement a viable and safe metro system will require extensive studies, design, and increased costs. We argue that the Kathmandu Valley will need a metro-rail system from 2030 onwards as population pressure increases along with the increase in individual incomes. Changes in the economy also will help to improve various urban infrastructure. The study for a metro transit in Kathmandu should start now because the current transit system of the Kathmandu Valley cannot maintain good quality of urban life that is inefficient and has become a source of air pollution. A Bus Rapid Transit (BRT) system could be the first step towards improving Kathmandu's transportation efficiency. The next step could be a ground-level streetcar (also known as light rail) transit before the city is ready to start building an underground metro transit system.

After this brief introduction, we present materials and methods, that include the study area in the Kathmandu Valley and its existing transportation network. This is followed by conceptual framework. At the end, we present discussion and concluding remarks.

12.2 Materials and Methods

This section is divided into study area and analyses of data related to the Kathmandu Valley and its urban transit system.

12.2.1 Study Area

The bowel-shaped Kathmandu Valley surrounded by mountains and hills ranging from 2,300 to 2,800 m in elevation, extends into Kathmandu (415.5 km^2), Lalitpur (397.4 km^2), and Bhaktapur (123.6 km^2) districts among the 77 administrative districts of Nepal (Fig. 12.1). Over 85% of the total land surface of the Kathmandu district is urbanized. Despite being adjacent to the Kathmandu district, only 50% of the Lalitpur district is urbanized, while administratively, Bhaktapur is partially urbanized while many of its settlements exhibit rural characteristics (Adhikari & Bhattarai, 2021). Due to its geographic nature, very little wind flows during the winter making it highly polluted from large number of urban automobiles and other pollution sources (Bhattarai & Conway, 2021; Bhattarai et al., 2019). Contrary to the saying that the "the solution of pollution is dilution", calm wind of the valley contributes to air pollution creating several health problems for its residents.

In 2021, the population of the Kathmandu district was 2,017,532 while the population of Bhaktapur was 430,408 and Lalitpur was 548,401. The population of the urban Kathmandu Valley in 2021 was 2,033,168 as per 2021 census. The Valley being the administrative, political, educational, and health centres, its population will continue

Fig. 12.1 Map of Nepal showing the study area

to increase. A granular level data of population by location is required to plan for a more efficient transportation system. As the development in the Valley has been extremely rapid and often spontaneous, a real-time population data by administrative boundaries of Wards (neighbourhoods) will help in calculating detailed transportation demand in the Valley.

The drastic increase in the number of two-and-four-wheeler vehicles, operating old-age vehicles, existing rough roads, poor environmental conditions during construction, and ongoing factories and agricultural activities are some of the main factors turning Kathmandu into a highly polluted region. One of the most polluted cities in the world, Kathmandu has atmospheric pollutants many times greater than those considered safe by the World Health Organization (WHO) (Nepali Times, 2021) problem, preferably through the "introduction of a mass rapid transit line through an electrified metro rail" system (Amatya, 2017).

Historically, the cities and towns in the Kathmandu Valley grew organically, mostly in the form of row houses and public squares as predominantly pedestrian environments. The high density, mixed uses, and compact development patterns naturally promoted walkability. Until a few decades ago, personal vehicles in the Kathmandu Valley cities were rare, and most people were able to access urban amenities and destinations such as shopping, public spaces, temples, and even places of work on foot or bicycles within 15–20 min of travel time. Nowadays, many cities around the world such as Melbourne (Australia), Paris (France), Auckland (New Zealand), and Tempe (Arozina, USA) aspire to be 20-min cities, where most residents can reach their daily destinations and amenities to meet their essential needs within a 20-min walk, or bike, or a transit ride. The recently evolving urban pattern in Kathmandu consists of many single-family homes in the outskirts of the city. Growth of suburban type of development has accelerated, especially, after the construction of the 27-km-long Ring Road around the urban centres in the 1970s. The Ring Road provided the urban newcomers access to a large swath of open lands around the city cores for new development (Bhattarai & Conway, 2021).

Reliable and sustainable public transportation networks are only recently evolving in the newly developed urban centres of the Kathmandu Valley. In newly created urban areas, organizing a transit system is one of the challenging tasks because of the random sorting of buildings without a suitable subdivision planning system. As population grew, more motorized vehicles were added each year in the Valley to fill the gap in the demand for transportation. Transit facilities and services remained inadequate, did not cover all the urban areas, and were generally overcrowded, and sometimes even unsafe.

More than one-third of the total registered vehicles in Nepal are in the Kathmandu Valley (JICA, 2019). As the Valley has limited road network (Fig. 12.2), the existing transit system is over stretched.

In the 1990s, "vehicle fleets in Asian countries increased" at rates between 5% and 10% a year, but this annual growth rate doubled in the 2000s (Anyala & Melhuish, 2021; Leather, 2021). Many of the vehicles were not passenger cars but two-wheeled motorcycles—which are significantly riskier. For instance, in the 1990s, motorcycle traffic grew rapidly in Bangladesh, Indonesia, India, Malaysia, Pakistan, Sri Lanka, Thailand, and Viet Nam. This increase in motorcycle fleet was similar to that of the 1980s in Malaysia and Thailand. Since a motorcycle occupies small space in

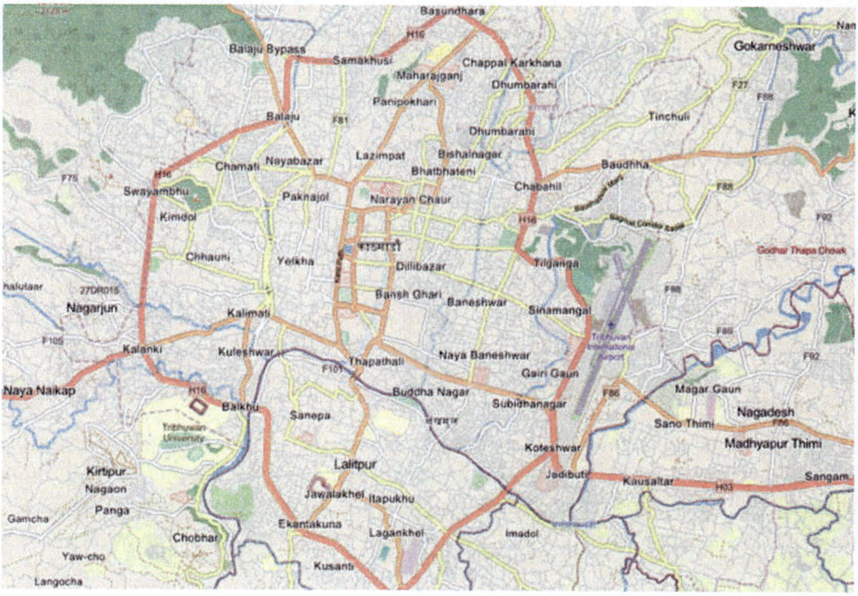

Fig. 12.2 Road network in the Kathmandu Valley (Compiled from ESRI, 2021)

Table 12.1 Average speed on various roads of the Kathmandu Valley

S. No	Name of route	Average travel speed (Km/ hour)
1	Thapathali—Kalimati–Kalanki–Thankot Road	8.78
2	Keshar–Mahal–Balaju–Modku (Trishuli Road)	10.94
3	Jamal–Lazimpat-Maharajgunj–Budhanilakantha Road	11.06
4	New Road–Paropakar–Chhauni–Sitapaila–Ramkot–Bhimdhunga Road	10.49
5	Balaju bypass–Phutung Road	10.00

Source KVDA (2017), Pokharel and Acharya (2015)

parking and requires very little spaces on the road, many Asian families are finding motorcycle as a "workhorse" (Anyala & Melhuish, 2021) and it has become a source for frequent commute, especially, for short distances.

Table 12.1 shows the average vehicular speeds on various roads in the Kathmandu Valley. Most vehicles are operating on these roads at speeds that are inefficient for controlling emissions. Overtime, the number of vehicles is increasing along with the increase in population (Table 12.2). Table 12.2 presents the increasing population densities from 1990 to 2022. The growing population density triggers the need for increased mobility and may also indicate an increase in the number of vehicles. Table 12.3 shows the increasing numbers of vehicles (by type) registered in Bagmati[1] province in different years. Their numbers are increasing in every successive census period. Table 12.4 provides the number of people travelling by various fleet modes. At present, without widening the turning points for wider buses, only buses capable of carrying 31–33 people can operate in the Kathmandu Valley. Likewise, other vehicles operating on the roads of Kathmandu Valley include minibus (ferrying 22 to 25 people), microbus (ferrying 12–20 people). Wide microbus ferries up to 20 people while a narrow microbus ferries 12 people, a Safa Tempo (electric three wheeler) ferries 7–9 people, and a taxi ferries 2–4 people. A motor bicycle ferries 1–2 people.

Due to the ease of affordability, almost every house and every job and business holders use two-wheeler vehicles, which impacts urban transit system. The lure of freedom of movement with private transport, fueled by the income from remittance and other sources, has led to the growth of cities in all directions leading to urban sprawl in the Kathmandu Valley. As a result, the Valley has been experiencing traffic congestion, environmental degradation, and road crashes, causing a decline in the quality of urban life.

The Kathmandu Valley is burdened with a high level of air pollution that is mainly caused by vehicles, factories operating under old technologies. Agricultural activities and the burgeoning construction work are adding more pollutants to the atmosphere. Many urban residents and visitors are seen using facial masks for protection against

[1] Bagmati was one of the 14 geographical zones in Nepal. The Kathmandu Valley contained most of the population of the Zone. As Nepal became a Federal State in 2015, the country was divided into seven provinces, and the administrative provision for 14 Zones were eliminated.

Table 12.2 Population density in the Kathmandu Valley (1990–2022)

Geographic area	Population density (people per hectare) in			
	1991	2001	2011	2022 (Projected)
Bhaktapur district	**14.1**	**18.2**	**25.5**	**36.4**
Bhaktapur municipality	93.2	110.6	128.0	150.1
Madhyapur Thimi municipality	28.7	43.0	75.8	141.6
Lalitpur district	**18.7**	**24.7**	**35.7**	**58.0**
Lalitpur metropolitan area	77.3	107.6	148.9	208.3
Kathmandu district	**16.2**	**25.7**	**44.1**	**78.9**
Kathmandu metropolitan area	84.8	135.9	203.6	317.6
Kirtipur municipality	21.2	27.7	44.8	76.0
Total	**16.2**	**24.1**	**37.6**	**67.2**

Source Census (2010) and JICA Study Team (This table does not include floating population)

Table 12.3 The increasing number of vehicles plying in the Kathmandu Valley for various years (Actual Data for 2009, and 2015 and Projected numbers for 2021, 2025, 2031, and 2034)

Year	Bus	Minibus	Micro bus	Car taxi	Utility	Safa tempo	Motorcycle
2009	2,314	10,738	18,153	90,868	29,638	9,714	297,747
2015	3,473	16,114	27,242	136,362	41,568	14,578	446,838
2021	5,154	23,913	40,427	193,432	58,965	20,679	565,392
2025	6,705	31,112	52,596	244,203	74,442	26,107	661,429
2031	9,950	46,169	78,052	326,799	105,598	37,034	799,027
2034	12,121	56,243	95,082	412,577	125,769	44,108	878,214

Source Department of Transport Management, Government of Nepal (2020)

Table 12.4 Daily movement of passengers by different types of vehicles (Years 2021, 2025, 2031, and 2035 are projected numbers)

Year	Bus	Minibus	Micro bus	Car taxi	Safa tempo	Motorcycle	Total
2009	76,364	264,148	223,277	198,084	70,912	387,071	1,219,856
2015	114,602	396,414	335,079	297,269	106,421	580,889	1,830,674
2021	170,067	588,270	497,250	421,682	150,960	735,010	2,563,239
2025	221,261	765,563	646,934	532,363	190,583	859,858	3,216,351
2031	328,347	1,135,767	960,036	712,422	270,346	1,038,735	4,445,652
2034	399,987	1,383,576	1,169,503	899,418	321,986	1,141,678	5,316,149

Source Department of Transportation, Government of Nepal (2020) [Does not include the floating population] the air pollution. The worsening air quality in the Kathmandu Valley has perhaps discouraged many international tourists to revisit the otherwise historic and culturally attractive old city. The tourism industry, which accounts for over 7.57% ($1,779 million in 2017) of the nation's gross domestic product (GDP), would be negatively impacted by any decline in visitor numbers (JICA, 2019).

Many premature deaths were reported in 2016 that were attributed to the environmental pollution. Of the "4.2 million" premature deaths at the global scale, many Asian countries are to blame for such incidences (WHO, 2016: 1). Looking at the case of Kathmandu, particulate matters (PM_x) along with other pollutants such as volatile organic compounds (VoCs), carbon monoxide and carbon dioxide hang in the lower tropospheric layers for a long time because the calmness in the wind-speed in the bowel-shaped Kathmandu Valley. This makes the air very unhealthy. These pollutants have contributed to several respiratory health issues. Additionally, the concentration of PM_{10} at a level higher than 579 $\mu g\ m^{-3}$ and $PM_{2.5}$ at a level higher than 160 $\mu g\ m^{-3}$ daily makes the ambient air of the Kathmandu Valley very pollutant. Particulate matters such as black carbon (BC) in the air is found at 17 per cent, sulfate at 17%, ammonium at 11%, organic carbons (OC) at 23 per cent, and nitrate at 2.5% (Shrestha et al., 2013). The Kathmandu Valley is listed among the top five most polluted cities of the world in 2015 (Thapa & Adhikari, 2016), and the third most polluted city in Asia in 2016 on the Pollution Index (Parajuly, 2016).

In Kathmandu, PM_{10} particulate count often exceeds 260 μgm^{-3} each day while taking 24 h mean average, which is much higher than the standard set by WHO which is around 50 μgm^{-3}. In 2017, the ambient air of the Kathmandu Valley had 400 μgm^{-3} PM_{10} particulate matters against the 120 μgm^{-3} set standard by the WHO (2017a, 2017b, 2018). In addition to the PM_{10} particulate matters, even smaller particulate matters like $PM_{2.5}$ are often mixed with fine dust particulates that are repeatedly crushed by vehicles plying on the road of the Kathmandu Valley. In the bowel-shaped Kathmandu Valley where the wind speed remains almost calm, these small particulates float freely in the air. These fine particles often penetrate into the blood stream during the inhaling process and contribute to the increase in carboxyhemoglobin (COHb). Once they penetrate the nervous system, they become the forerunner for the major cause for heart attacks (WHO, 2017a, 2017b) due to the blockage of arteries. Additionally, nitrogen oxide (NO_2) emitted by vehicles mix-up with other gases in the air. Inhalation of such pollutants can cause irritation in the trachea—major causes of bronchitis and broncho-pneumonia. Likewise, Sulphur dioxide (SO_2) that comes as a by-product of vehicular emission while using leaded hydrocarbon, a common practice in the Kathmandu Valley, causes eye irritation. Several conjunctivitis diseases are attributed to SO_2 gases (Li et al., 2016). These gases can become fatal as they often cause the "shortness of breath" leading to "chronic bronchitis, asthma" (Gurung et al., 2016). Further studies have shown that such respiratory illness many even lead to lung cancer (WHO, 2017a, 2017b).

Since automotive emissions are the main cause of air pollution, an improvement in urban transit will have a big impact on lowering emissions. Low-occupancy vehicles must be replaced with electrified buses for the health and safety of Kathmandu inhabitants and visitors to reduce pollution and improve public health. It is essential to reverse the trends of using single occupancy vehicles in the Valley.

In the following sections, we develop a conceptual framework to show how adoption of transit will help reduce air pollution and analyse the types and amount of emissions from various vehicles.

12.3 Conceptual Framework

In the past, leaders and professionals concerned with city planning have proposed various modes of urban transit systems to improve the quality of urban life. They have concluded that using mass transit or public transportation is the most effective way to move many of people in limited spaces such as in the Kathmandu Valley (Leather, 2021). One of the transit solutions proposed by them is a Bus Rapid Transit (BRT) system. Levinson et al. (2002), Brakewood et al. (2021), and Adhikari and Bhattarai (2021) have strongly advocated for a BRT system because of its:

a. operability at undulated topography.
b. possibility of operating BRT not only under changing socio-economic but also under dynamic demographic situations,
c. modifiable nature of BRT that can be incrementally improved as the economic conditions change, and the Kathmandu Valley becomes a perfect example for this where remittance propelled economy often influence the living conditions and commuting practices,
d. operability in different types of environments without requiring expensive new road construction and special lanes,
e. operability in in mixed traffic systems and its low cost than a railway system.

Further, as described later, a BRT can significantly reduce the air pollution levels in places like the Kathmandu Valley.

Over the century, more rail-based metro systems were introduced in densely populated Asian cities. Japan started metro-line in 1927 covering only 2.2 km as a test case. It was meant to link between Asakusa and Ueno in Tokyo. After this exemplary success in Japan, other cities also started adapting this metro system. By twentieth century, many countries have adapted a metro system, for example, Beijing in 1969, Seoul in 1974, Kolkata in 1984, and Manila in 1984. Hong Kong, China 1979 (Leather, 2021).

Efforts to replace low occupancy vehicles such as private cars, taxis, and three wheelers by high occupancy vehicles such as 35–40 passenger capacity bus can be initially challenging in Nepal because the established transportation syndicates (monopolies) and their investors often function as a cartel and have strong lobbying links with the decision-makers. Whenever the Federal and local governments try to implement strict measures to improve the Kathmandu Valley transit system, these syndicates often influence government decisions (AFP, 2017). Government decisions have been sometimes reversed under the influences of syndicates. To ensure that the future laws are favourable to them, these syndicates often provide financial support to elect their favourite politicians to the parliament. Despite these hurdles, a start of an electric operated BRT system with high occupancy vehicles (35–40 people per vehicle) at high frequency (say, every five minutes on the major routes) at optimal speed may help to reduce emission rates per person-mile travelled and help to drastically improve the environment of the Valley.

12.4 Analyses

We analysed emission data obtained from various sources for different vehicles at suboptimal (defined as less than 25 km per hour) and optimal speed (defined as over 25 km per hour for urban area) levels. Shrestha et al. (2013) published emissions status for various vehicles in the Kathmandu Valley per kilometre travelled. Their data reveals emissions from various vehicles at the engine starting phase (suboptimal or idling phase) and speed pick up (optimal or "run") phases. We have averaged the gaseous emission rates at the starting phases and pick up phases of vehicles of various categories and compared these values with the emissions values of various vehicles at suboptimal levels obtained from various literature and emission values available from different emission testing stations in the Kathmandu Valley. The cross checking was done in 2017 (summer), 2018 (summer), and 2019 (winter).

We also compared published emission values from various vehicles that operate on lead containing fuels in different geographic areas (see, Black, 2010; Chang & Morlok, 2005; Coiret et al., 2020; DART, 2018; Keuken et al., 2010; Mool et al., 2020; Tang et al., 2019; www.fueleconomy.gov, 2021) and picked up the values from those that represent the situation similar to the Kathmandu Valley. The emission values published in various publications provide information on how much quantity of emissions are exhausted from vehicles in different seasons and the time of the day under suboptimal and optimal speed levels using leaded hydrocarbons. We are using the vehicular emission values at suboptimal and optimal speed levels as published in Shrestha et al. (2013). We utilized data published in Shrestha et al. (2013), Table 12.3: Composite emission factors of different vehicle types in Kathmandu valley, 2010 (p. 586). We assumed the fuel consumption by various vehicles in run stage of the vehicle as the optimal condition and average value between start and run as suboptimal level speed. Our calculated values were then compared with the data in the published literature to compare the fuel consumption by various vehicles at optimal (>25 km/h) and suboptimal (<25 km/h). Our calculated values were within the margins of 5–10% errors with old age and new vehicles but with middle-aged vehicles, our calculated values were very close to the ground reality. Ground reality refers to a situation where drivers (three wheelers, motorcycles taxi, microbus, van, big bus) were asked in the Kathmandu Valley—how many kilometres their vehicle can go per litre of gasoline used at optimal and suboptimal levels. Their answers were considered while ascertaining the amount of gasoline consumed by various vehicles at different speed levels.

We have presented scenarios of gaseous emissions from various vehicles running at optimal and suboptimal speeds (Tables 12.5, 12.6, 12.7 and 12.8). These values are generated based on the information available at Shrestha et al. (2013). The number of vehicles given in Table 12.3 and number of passengers ferried by different transportation media such as bus, van (micro and minibus), taxi, and motorcycles as given in Table 12.4 are considered in the calculations in Tables 12.5, 12.6, 12.7 and 12.8.

Finally, we have summarized a scenario for bus, van, taxi, and motorcycle (under business-as-usual scenario) on how much of particulates and CO_2 will be emitted by the total number of vehicles per kilometre travelled to predict how chaotic the environmental conditions will be if appropriate steps are not taken soon to electrify the transit system (Table 12.9).

In these observations, we have included emission from bus (ferrying 35–40 passengers), minibus/van (22–25 passengers), microbus (ferrying 12–20 passengers), Safa-tempo (three wheeler) (ferrying 7–9 passengers), taxi (2–4 passengers), and motorcycle (1–2 passenger/s). We have calculated carbon monoxide (CO), tailpipe volatile organic carbon (VoC), oxides of sulfur (SOx), oxides of nitrogen (NO_x), particulate matter (PM_{10}), and greenhouse gas (GHG) emissions. Currently, the concentration of all particulates in the Valley is very high. If nothing is done to improve the current transit system of the Kathmandu Valley, in a few years, the Valley is likely to face more severe traffic congestions where many vehicles will operate at suboptimal speeds due to the lack of adequate road space for vehicles, and the quantity of pollutants would increase even further (Table 12.5) making the urban life miserable where the state coffer has to pour the maximum resources into health sector.

It is hoped that once the proposed outer ring road expansion (Fig. 12.3) will be completed, urban mobility level in Kathmandu may begin to improve. The Ring Road that was constructed by the help of the Government of China in the 1970s has a 2-lane carriage way. In 2011, some sections were promoted into 4-lanes (Fig. 12.3). A proposed outer ring road (Fig. 12.3) was surveyed by the Department of road (DOR) and Nepal Engineering Consultancy Services (NEPECON) in 2000. Accordingly, the ring road will be 66 km length. However, it was later corrected as 72 km long. Eight lanes are proposed, with four lanes each for the main road, four lanes for the service road on the proposed ring road and main road. The outer ring road will include the area of Harishiddhi to Lubhu area (11.3 km), Naya Thimi to Dadhikot (3.78 km), Bhatedhikuro to Thali Danchhii (10 km), Chandol to Thali Danchii (4.00 km), Satungal to Chovar (6.6 km), and Chobhar to Harishiddhi. The newly expanded length will encompass 35.68 km (Fig. 12.3). Despite the expansions of roads to the outer skirts, the pollution levels will continue to increase if the riders continue to depend on single or low occupancy vehicles for their mobility needs because the bowel-shaped geography of the Valley cannot be modified.

An efficient public transit system enables the plying vehicles to achieve higher speeds in the traffic. If a vehicle moves at optimal speed level, the engine efficiency will be better and that will help in the reduction of pollution (Tables 12.5, 12.6, 12.7 and 12.8) levels because vehicles will move at optimal speeds. In Fig. 12.4a–e the optimal (opt) level indicates improved speed (>25 kph) whereas suboptimal (subopt)

Table 12.5 Pollutants generated by the total number of buses (Table 12.3) per kilometre in kilogram while ferrying passengers (Table 12.4) at optimal and suboptimal speed limits [opt→optimal speed above 25 km/h; and subopt→ suboptimal speed below 2 km/h]

Year	CO (opt)	CO (subopt)	VoC (opt)	VoC (subopt)	NOx (opt)	NO_x (subopt)	PM (opt)	PM (subopt)	N_2O (opt)	N_2O (subopt)	CO_2 (opt)	CO_2 (subopt)
2009	17.702	39.176	4.674	9.626	80.666	89.968	23.811	23.811	0.106	0.106	6275.082	7478.177
2015	26.568	58.798	7.015	14.448	121.069	135.030	35.737	35.737	0.160	0.160	9418.047	11,223.729
2021	39.428	87.257	10.411	21.441	179.668	200.388	53.035	53.035	0.237	0.237	13,976.566	16,656.233
2025	51.293	113.516	13.544	27.893	233.736	260.690	68.994	68.994	0.308	0.308	18,182.552	21,668.616
2031	76.118	168.454	20.099	41.392	346.857	386.856	102.386	102.386	0.458	0.458	26,982.311	32,155.515
2034	92.726	205.209	24.484	50.423	422.538	471.264	124.725	124.725	0.558	0.558	32,869.607	39,171.557

Table 12.6 Pollutants generated by the total number of vans [minibuses and microbuses] (Table 12.3) per kilometre in kilogram while ferrying passengers (Table 12.4) at optimal and suboptimal speed limits

Year	CO (opt)	CO (subopt)	VoC (opt)	VoC (subopt)	Nox (opt)	Nox (subopt)	PM (opt)	PM (subopt)	CO_2 (opt)	CO_2 (subopt)	N_2O (opt)	N_2O (subopt)
2009	68.76	85.52	11.85	16.47	77.14	99.10	23.11	28.89	24,947.96	29,615.59	0.10	0.13
2015	103.19	128.33	17.78	24.71	115.76	148.71	34.68	43.36	37,438.77	44,443.37	0.15	0.19
2021	153.13	190.45	26.38	36.67	171.79	220.69	51.47	64.34	55,558.88	65,953.65	0.23	0.29
2025	199.23	247.78	34.32	47.71	223.50	287.12	66.97	83.71	72,283.53	85,807.40	0.30	0.37
2031	295.65	367.69	50.93	70.81	331.67	426.08	99.38	124.22	107,267.32	127,336.46	0.44	0.55
2034	360.15	447.92	62.04	86.26	404.04	519.04	121.06	151.33	130,672.16	155,120.23	0.54	0.67

Table 12.7 Pollutants generated by the total number of taxis (Table 12.3) per kilometre in kilogram while ferrying passengers (Table 12.4) at optimal and suboptimal speed limits

Year	CO (Opt)	CO (Subopt)	VoC (Opt)	VoC (Subopt)	Nox (Opt)	Nox (Subopt)	PM (Opt)	PM (Subopt)	CO_2 (Opt)	CO_2 (Subopt)	CH_4 (Opt)	CH_4(Subopt)
2009	3942.76	3979.11	458.88	475.24	170.83	179.92	2.70	2.75	34,767.91	35,298.58	152.66	161.75
2015	5916.75	5971.29	688.63	713.17	256.36	270.00	4.05	4.13	52,174.83	52,971.18	229.09	242.72
2021	8393.01	8470.39	976.83	1011.65	363.65	383.00	5.74	5.86	74,010.95	75,140.59	324.97	344.31
2025	10,595.97	10,693.65	1233.23	1277.18	459.10	483.52	7.25	7.40	93,436.95	94,863.10	410.26	434.68
2031	14,179.81	14,310.53	1650.33	1709.16	614.38	647.06	9.71	9.90	125,039.83	126,948.34	549.02	581.70
2034	17,901.72	18,066.75	2083.51	2157.78	775.64	816.90	12.25	12.50	157,860.21	160,269.66	693.13	734.39

Table 12.8 Pollutants generated by the total number of motorcycles (Table 12.3) per kilometre in kilogram while ferrying passengers (Table 12.4) at optimal and suboptimal speed limit

Year	CO (opt)	CO (subopt)	VoC (opt)	VoC (subopt)	NO_x (opt)	NO_x (subopt)	PM (opt)	PM (subopt)	CO_2 (opt)	CO_2 (subopt)	CH_4 (opt)	CH_4 (subopt)
2009	717.57	2009.79	267.97	768.19	47.64	125.05	41.68	77.41	19,308.89	28,936.54	157.81	259.04
2015	1076.88	3016.16	402.15	1152.84	71.49	187.67	62.56	116.18	28,977.44	43,425.95	236.82	388.75
2021	1362.59	3816.40	508.85	1458.71	90.46	237.46	79.15	147.00	36,665.67	54,947.62	299.66	491.89
2025	1594.04	4464.65	595.29	1706.49	105.83	277.80	92.60	171.97	42,893.67	64,280.98	350.56	575.44
2031	1925.66	5393.43	719.12	2061.49	127.84	335.59	111.86	207.75	51,816.90	77,653.44	423.48	695.15
2034	2116.50	5927.94	790.39	2265.79	140.51	368.85	122.95	228.34	56,952.18	85,349.23	465.45	764.05

Table 12.9 The amount of particulate matters and CO_2 projected by bus, van, taxi, and motorcycle under business as usual situation. These values are projected per kilometre. All units are in kilogram

Year	PM (opt)	PM (subopt)	CO_2 (opt)	CO_2 (subopt)	Total
Bus_2021	53.03	53.03	13,976.57	16,656.23	30,632.80
Bus_2025	68.99	68.99	18,182.55	21,668.62	39,851.17
Bus_2031	102.39	102.39	26,982.31	32,155.51	59,137.83
Bus_2034	124.73	124.73	32,869.61	39,171.56	72,041.16
MC_2021	79.15	147.00	36,665.67	54,947.62	91,613.29
MC_2025	92.60	171.97	42,893.67	64,280.98	107,174.65
MC_2031	111.86	207.75	51,816.90	77,653.44	129,470.34
MC_2034	122.95	228.34	56,952.18	85,349.23	142,301.41
Van_2021	51.47	64.34	55,558.88	65,953.65	121,512.52
Van_2025	66.97	83.71	72,283.53	85,807.40	158,090.93
Van_2031	99.38	124.22	107,267.32	127,336.46	234,603.78
Van_2034	121.06	151.33	130,672.16	155,120.23	285,792.40
Taxi_2021	5.74	5.86	74,010.95	75,140.59	149,151.55
Taxi_2025	7.25	7.40	93,436.95	94,863.10	188,300.05
Taxi_2031	9.71	9.90	125,039.83	126,948.34	251,988.17
Taxi_2034	12.25	12.50	157,860.21	160,269.66	318,129.87

Legends Bus = Bus; MC = Motorcycle; Van = microbus and minibus; Taxi = taxi and car. All calculations are based on the value published in Shrestha et al. (2013: 586)

level indicates slow speed (<25 kph). We have also examined the level of emissions from high occupancy vehicles and low occupancy vehicles. When a large section of population travels by high occupancy vehicles, per capita pollution will decrease.

If the fleet consists of high occupancy vehicles such as 35–40 seaters bus, many vehicles can operate under optimal speeds, the quantity of pollutants can be reduced by over 8,000 tons daily from the current rate of emission under the business-as-usual scenario in the Kathmandu Valley. If the number of vehicles in the fleet by high occupancy vehicle is increased by 1.5 times to replace low occupancy vehicles, it can result in a further reduction of emission by 7,518 tons daily compared to what would have been emitted under the business-as-usual scenario. Electric buses run with zero emissions (not counting the embedded carbon in the bus manufacturing process). If the urban BRT along the main road is arranged with electric buses with a higher frequency (e.g., operating in every five minutes), the air in the Valley can be made drastically cleaner.

The transit system in the Kathmandu Valley can be improved using a Transit Tracker App (TTP). A TTP can help riders to obtain real-time information about the Bus Rapid Transit (BRT) system. It helps tracking where the bus is, what time it is scheduled to arrive at a certain location, how long will it stop at certain point, and when does it arrive to the desired destination. This not only helps travellers to plan their journey, but also enhances the passenger safety (Bhattarai et al., 2019).

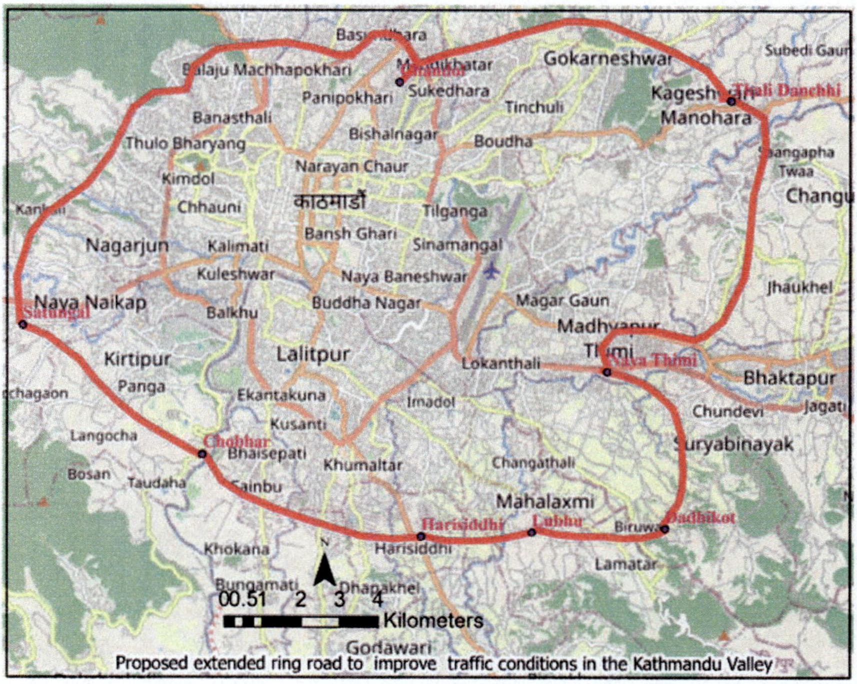

Fig. 12.3 Map of the Kathmandu Valley showing proposed and existing ring roads. *Source* Excerpted from ESRI's Street map base map 2022 and modified)

Due to changing economies of the urban dwellers especially of those families whose family members are working in foreign countries and remitting their incomes to their families, today, many of the family members have smart phones or tablets on their hands. These devices can be can handle TTA to track real-time information of the ongoing traffic and about the schedules of urban transit systems. Using the proposed TTA, riders can better plan their bus trips and bus-to-bus transfers. Once a BRT system successfully operates in the Kathmandu Valley, officials from other cities are also likely to find such a system attractive, and it might motivate the private entrepreneurs to make an investment in the urban transit system. If a BRT system is implemented in the Kathmandu Valley, in the long run, it will lower the operating cost once the system is in place and users will soon be acclimatized to the system while reducing the total amount of pollutants. However, government authorities, bus service operators including the drivers need to be educated and informed about how the system operates (Adhikari & Bhattarai, 2021; Bhattarai et al., 2019).

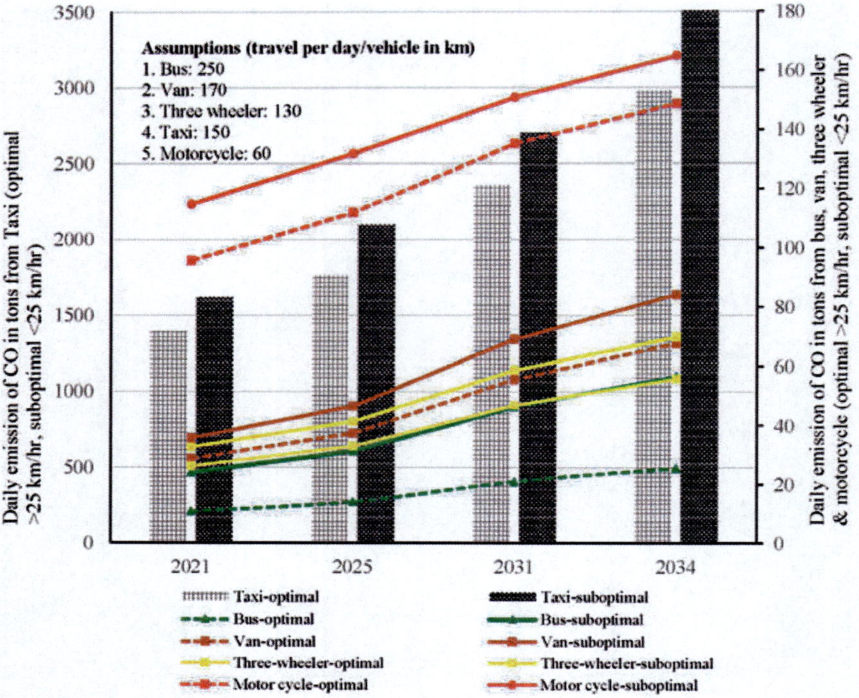

Fig. 12.4 **a** Projected Carbon monoxide emission from various vehicles operating in the Kathmandu Valley (Business as usual), **b**. Projected emission of volatile organic compounds (VoC) by bus, van, three wheeler, taxi, and motorcycle under optimal and suboptimal conditions (Business as usual scenario), **c** Projected emission of oxides of nitrogen (NO_x) by bus, van, three wheeler, taxi, and motorcycle under optimal and suboptimal conditions (Business as usual), **d** Projected emission of particulates ($PM_{10, 5, 2.5}$) by bus, van, three wheeler, taxi, and motorcycle under optimal and suboptimal conditions (Business as usual), **e** Projected emission of CO_2 by bus, van, three wheeler, taxi, and motorcycle running under optimal and suboptimal speeds (Business as usual). *Source* Graphs based on data from Shrestha et al. (2013: 586), vehicle registration (Table 12.3), speed limit, peoples' choice of mode of transit (Table 12.4)

12.5 Discussions

The increased number (projected) of buses, vans, taxis, three wheelers, and motorcycles operating in the Kathmandu Valley for years 2021, 2025, 2031, and 2034 under business-as-usual conditions is likely to make the traffic heavily congested and unsafe. In a congested environment, many vehicles would likely operate at suboptimal speeds and emit a large amount of air pollutants making Kathmandu's air quality worse. Since low occupancy vehicles carry only fewer people, obviously, the per capita consumption of hydrocarbons and per capita gaseous emission from these vehicles become higher than the high occupancy vehicles. For instance, a bus generates less emission per capita than a car and a car emits less pollutants than a

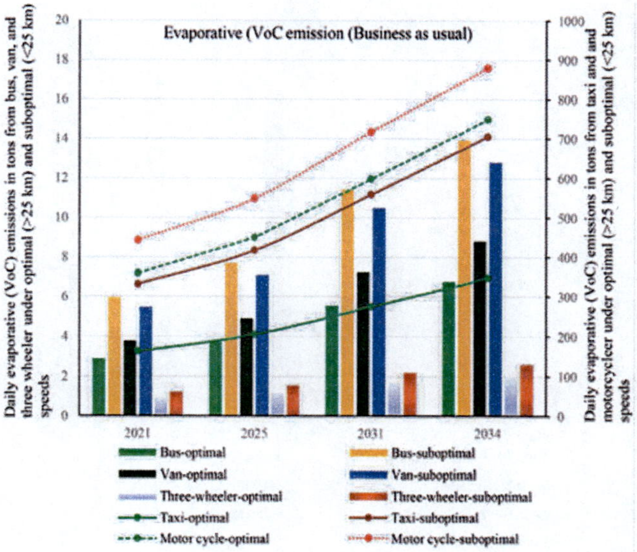

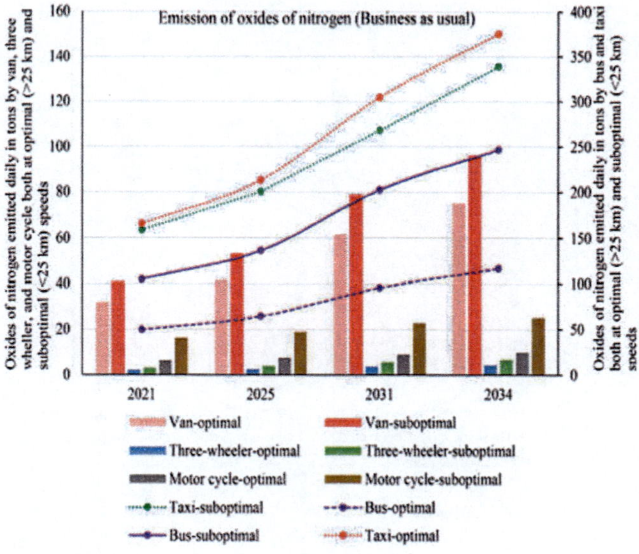

Fig. 12.4 (continued)

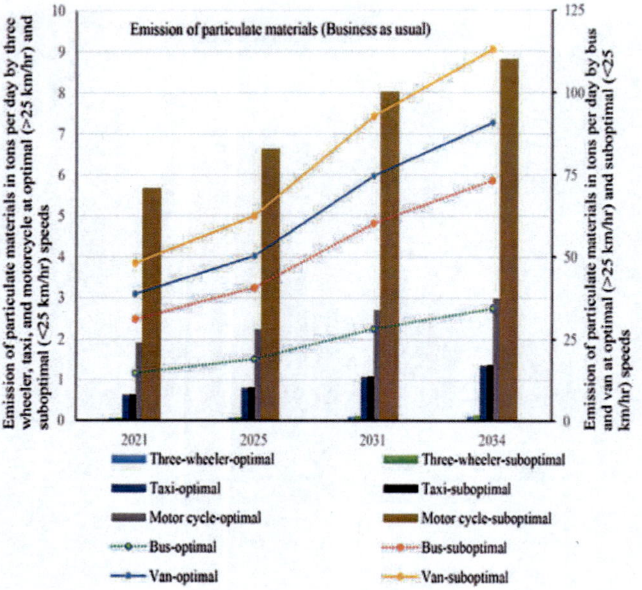

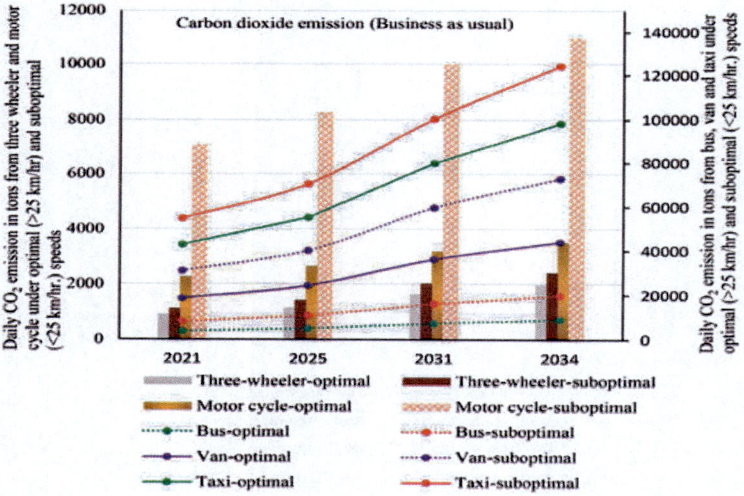

Fig. 12.4 (continued)

motorcycle. Simplifying this analogy, let us put a specific example. Only 4–6 people can travel by a car. A car with these 4–6 people emits 9.41 metric tons of CO_2 that comes close to 1.882 metric ton/person per vehicle throughout the year. This emission is lower than 6.85 metric per capita that a motorcycle contributes throughout the year (EPA, 2021). The per capita emission from a motorcycle is higher because only 1–2 person can travel by a motorcycle that generates 13.7 metric tons of emission per year. While comparing the gaseous emission by the high occupancy and low occupancy vehicles, the low occupancy vehicles contribute 3.5-fold more pollutants per rider than the low occupancy vehicle does (ACOG, 2013).

Pollutant emissions from buses and other high-capacity vehicles are lower per passenger unit. For instance, a bus can carry the same number of passengers as 17–20 motorcycles do. Use of high occupancy vehicles help smoothen the flow of traffic and reduce transportation-related accidents. After their long experiences and experimentations, Black (2010) and Sinnot have concluded that a 35–40 seaters bus will be six to seven times efficient in fuel consumption and less emission in average as compared to a single seater automobile. Since the Kathmandu Valley may not be able to have a streetcar and/or a metro train immediately due to cost, construction time, and preparation-related reasons, a BRT system seems to be the best bet for Nepal at this point.

If a proposed BRT system utilizes electric buses, emission can be reduced even more drastically (Spector, 2018). Historically, the first electric tram system in Japan started in 1895. After experiencing positive impact on the environment, by 1932, Japan had deployed electric tram systems in 65 cities (Anyala & Melhuish, 2021; Leather, 2021; Susantono et al., 2021). Because electric trams helped in improving the urban air quality, "many other Asian countries opened electric tram systems", for examples: Peking introduced electric tram system in 1899, Hong Kong in 1904, Mandalay (Myanmar) in 1904, Manila in 1905, Mumbai and Delhi in 1908. Across much of Asia, street cars become the dominant mode of mass transit during the twentieth century (Leather, 2021; Susantono et al., 2021). These systems carried more people per vehicle and emitted less pollutants.

One of the challenges of BRT system is the need to track the bus location using a Transit Tracker App (TTA) that requires wireless internet services. Likewise, an electrified BRT system needs the electric charging stations available widely. For example, California in the United States, has one public charging station for 25 to 30 electric vehicles where electric vehicle owners have relatively easy access to charging stations. In the Netherlands, the ratio is two to seven electric vehicles per public charging station (EVgo, 2021; Otonomo, 2021). However, in many Asian cities making faster charging stations ubiquitous remains a challenge because of the high cost involved for the installation of charging stations in densely populated cities. Also, faster charging during peak demand period can have a negative impact in the supply of electricity. However, using the twenty-first-century technology, this situation can be improved (Kim & Mishra, 2021). Technology advancements that produce clean electricity at a reasonable price will probably "motivate entrepreneurs" to support electric automobiles. Government tax reductions and financial incentives for the purchase and usage of entirely battery-operated or hybrid vehicles have also frequently been used to reduce emissions. In the recent decades, many countries have followed this practice. For example, China subsidizes 23% of the cost of electric vehicles, whereas Norway subsidizes 45% of the cost (World EV Outlook, 2021). Also, the price of batteries and technology are falling that has brought down the cost of electric vehicles. Together with a sharp drop in battery cost per kilowatt-hour (kWh), battery energy density and vehicle efficiency have been significantly improved. Since the operational cost per mile travelled has come down, overall transit system will be relatively cheaper with the use of twenty-first-century technology. The new batteries, which are smaller than the previous generations' batteries, can be charged at a faster rate. As a result, compared to fossil fuel vehicles, electric vehicles are becoming more energy efficient and cheaper to operate. However, one of the authors experienced that for a long distance, electric vehicle is more time-consuming; for example, it took 13 h to travel 597 miles with an electric vehicle while it was possible to cover the same distance in nine hours with a gasoline operating vehicle. Recharging takes extra time. Nonetheless, in the short distance haulage like the Kathmandu Valley, electric vehicles are good both environmentally and economically. The use of electric vehicles will become relatively cheaper with proactive checks. These vehicles use small quantity of liquids. Wear and tear of mechanical parts including brake pads replacement being relatively lower, the operation of electric vehicle will become less expensive. However, electric vehicles are heavier, and they accelerate and decelerate more quickly, their tire usage is 20–30% higher than that of the fuel operated vehicles. Since spare parts of electric vehicles are not found in all places, some spare parts may have monopolistic market making them more expensive than the spare parts of normal vehicles. Additionally, only high-skilled technicians can repair electric vehicles, maintenance employees are typically more expensive. The overall maintenance costs of electric vehicles are around 60–80% higher than conventional vehicles (Kim & Mishra, 2021). Additionally, cities using electric buses need "preferential lanes and parking access" with charging facilities (Kim & Mishra, 2021). These high expenses can be countered by initially providing additional incentive including tax reduction and banning of the uses of fossil fuel vehicles in the city centres.

Some countries have used fibre optics communication infrastructure and high occupancy vehicles to improve traffic conditions (Atha et al., 2020). To avoid traffic jams, the city of Seoul, South Korea introduced the Seoul Smart Mobility Reform in 2003 and increased the number of high occupancy vehicles. Since then, the share of bus and subway riders increased to almost 70% of the total number of vehicles on the road, and the number of cars decreased by 30%. An introduction of the improved transportation system using recent technology integrating Transit Tracker App (TTA) that captures real-time events not only decreased the traffic accidents but also helped create traffic uninterruptable (Kang, 2020). Utilizing the theory of global positioning system that uses three components—user segment, space segment, and control system, the transportation office controls the information flow at real time to monitor speed, passengers, routes, and crowd. The control segment (Seoul's Intelligent Traffic System) utilizes "Smart Transportation Card". Utilizing this concept, the Seoul transit system has greatly improved its urban transit system.

12.6 Conclusion

The Kathmandu Valley continues to experience serious transportation problems. Transport planners have faced challenges to offer efficient transportation services to commuters. Further, the use of personal automobiles for commutes, and various other factors contributing to air pollution have made the Valley one of the most polluted urban centres in the world. The public health cost due to air pollution has been very high.

Since the population of the Kathmandu Valley is rapidly increasing, an underground metro-rail with a high passenger capacity would have been a good option to cater the future traffic demands and reduce the emission of air pollution. However, a metro system does not seem feasible at present and even when it is decided upon, the planning and completion of a metro system will take more than a decade. The most problematic issues being the lack of land, dangers associated with seismic conditions, and cartels of minibus operators.

If the Kathmandu Valley introduces an electric Bus Rapid System (BRT), not only it will help improve the transit system, but also reduce the level of pollution drastically. With the introduction of BRT, a fewer number of vehicles will commute on the streets of Kathmandu Valley carrying many people as compared to low occupancy vehicles, it will help to reduce the quantity of pollutants coming from low occupancy vehicles such as motorcycles (Adhikari & Bhattarai, 2021). Since a bus can replace 16–18 motorcycles and 4–5 three wheelers, the congestions will be reduced drastically. Less congestion means more spaces are available that means vehicles will move at optimal speed limits emitting less quantity of pollutants. More importantly, switching to electric vehicles will largely reduce the overall amount of pollution. Worldwide, adoption of electric vehicles is rapidly increasing. With an expanded electric charging infrastructure, and incentives to promote the use of electric vehicles, Nepal can also join the ranks of the "Avant-Garde nations" to enter the age of electric vehicle users.

Using a BRT system with the use of a Transit Tracker App (TTP) operable on a smart phone or tablet can make the transit system user friendly. A transit tracker will help user to know bus schedule, seat availability, wait times, and the location of stops. A mobile application uses an algorithm that is capable of tracking traffic conditions in real time to provide an estimate of arrival time, speed limit, road accident, constructional activities that are happening on the road, detouring needed, and the pollution status of the ambient air (Adhikari & Bhattarai, 2021). A TTP utilizes real-time incidences. Using the real-time incidences within the overarching umbrella of a global positioning system that simultaneously interacts with the control segment, user segment, and space segment, a TTP helps in locating a vehicle. It can locate a bus and helps the users to know what time a bus leaves certain place, what are its timing to arrive and leave different stations including its parking spots. Users have been finding the TTP very useful for safe transportation planning purposes.

A transportation system gains several advantages by using electric buses for transit. The advantages include drastic reduction in air pollution and emission of greenhouse gases while enhancing the pride of the urban residents and city leadership. Although the required investments in the electrical charging station and other related infrastructure is initially high, the overall transportation system can experience significant savings in the long-run. This will also save government budget on health sector. An electricity-run BRT system for the Kathmandu Valley will support increased mobility, and transportation efficiency and help drastically reduce transportation-related air pollution in the Valley.

Authors' Contributions **Keshav Bhattarai**: Preliminary full draft including the conceptual framework, data tabulation and analysis, graph and map making.

Ambika P. Adhikari: High level concepts, analysis, and conclusion.

References

ACOG. (2013). Association of Central Oklahoma Governments (ACOG). Why transit matters: The environmental benefits of public transportation, November 14. Retrieved February 11, 2022, from https://www.acogok.org/why-transit-matters-environment/

Adhikari, A. P., & Bhattarai, K. (2021). An electric bus rapid transit system for Kathmandu valley. myRepublica. Retrieved 04 March 2022, from https://myrepublica.nagariknetwork.com/news/author/4342

AFP. (2017). Bus Mafia controlling Nepal's smog-choked capital. Retrieved January 25, 2021, from https://www.geo.tv/latest/138336-Bus-mafia-controlling-Nepals-smog-choked-capital

Amatya, B. L. (2017). Kathmandu needs metro rail. The Kathmandu Post. Retrieved April 04, 2017, from https://kathmandupost.com/opinion/2017/04/04/kathmandu-needs-metro-rail

Anyala, M., & Melhuish, C. (2021). Road safety in Asian cities. In B. Susantono & R. Guild (Eds.), *Creating livable Asian cities* (pp. 205–222). Asian Development Bank. ISBN. 978-92-9262-782-9.

Atha, K., Callahan, J., Chen, J., Drun, J., Francis, E., Green, K., Lafferty, B., McReynolds, J., Mulvenon, J., Rosen, B., & Walz, E. (2020). Chain's smart cities development: research report prepared on behalf of the U.S. -China Economic and Security Review Commission. January 2020. Retrieved February 02, 2021, from https://www.uscc.gov/sites/default/files/China_Smart_Cities_Development.pdf

Bhattarai, K., & Conway, D. (2021). *Contemporary Environmental Problems in Nepal: Geographic Perspectives*. Springer (Nature). ISBN: 3030501663. ISBN13: 9783030501662

Bhattarai, K., Yousef, M., Greife, A., & Lama, S. (2019). Decision-aiding transit-tracker methodology for bus scheduling using real time information to ameliorate traffic congestion in the Kathmandu Valley of Nepal. *Journal of Geographic Information System, 11*, 239–291. https://doi.org/10.4236/jgis.2019.112016

Black, W. R. (2010). *Sustainable transportation: Problems and solutions*. The Guilford Press.

Brakewood, C., Barbeau, S., & Watkins, K. (2021). (2014) An experiment evaluating the impacts of real-time transit information on bus riders in Tampa, Florida. *Transportation Research Part a: Policy and Practice, 69*, 409–422. https://doi.org/10.1016/j.tra.2014.09.003.AccessedJanuary10

Chang, D. J., & Morlok, E. K. (2005). Vehicle speed profiles to minimize work and fuel consumption. *Journal of Transportation Engineering, 131*(3), 173. https://doi.org/10.1061/(ASCE)0733-947X

Coiret, A., Deljanin, E., Vandajon, P.-O. (2020) Vehicle energy savings by optimizing road speed-sectioning. *European Transport Review*. Number 41, 08 June. https://etrr.springeropen.com/articles/10.1186/s12544-020-00432-8

DART. (2018). *DART first state. DART: moving forward*. Delaware Transit Corporation. Retrieved January 25, 2021, from http://www.dartfirststate.com/

Department of Transport Management. (2020). Department of Transport Management, Nepal. Retrieved July 21, 2021, from https://www.facebook.com/www.dotm.gov/

EPA. (2021). Energy and the Environment: Greenhouse Gas Equivalencies Calculator. United States Environmental Protection Agency. Retrieved November 20, 2021, from https://www.epa.gov/energy/greenhouse-gas-equivalencies-calculator

ESRI. (2021) Environmental Research Institute. Redland California. Retrieved January 10, 2021.

EVgo. (2021). Find a charger: Search for a charger near you. Retrieved January 21, 2021, from https://www.evgo.com/find-a-charger/?gclid=EAIaIQobChMIqM60zp639gIVrYFaBR3yhgEqEAAYBCAAEgLEA_D_BwE

Global Eve Outlook. (2021). More than 10 million electric cars were on the world's roads in 2020 with battery electric models driving the expansion. Trends and developments in electric vehicle markets. https://www.iea.org/reports/global-ev-outlook-2021

Gurung, A., Son, J.-Y., & Bell, M. L. (2016). Particulate matter and risk of hospital admission in the Kathmandu Valley, Nepal: A case-crossover study. *American Journal of Epidemiology, 186* (5). https://doi.org/10.1093/aje/kwx135.

JICA. (2019). *Federal democratic republic of Nepal: Data collection survey on urban transport in Kathmandu Valley: Final report*. Japan International Cooperation Agency Oriental Consultants Global Co. Ltd. Padeco Co. Ltd. Retrieved January 10, 2021, from https://openjicareport.jica.go.jp/pdf/12345484_01.pdf

Kang, M. (2020). How is Seoul, Korea transforming into a smart city? World Bank Blog, 23 January. Retrieved December 8, 2020, from https://blogs.worldbank.org/sustainablecities/how-seoul-korea-transforming-smart-city

Keuken, M. P., Jonkers, S., Wilmink, I., & Wesseling, J. (2010). Reduced NOx and PM_{10} emissions on urban motorways in The Netherlands by 80 km/h speed management. *Science of Total Environment, 408*(12), 2517–2526.

Kim, K.-J., & Mishra, R. (2021). E-mobility: Transition to sustainable transport. In B. Susantono & R. Guild (Eds.), *Creating livable Asian cities* (pp.157–164). Asian Development Bank. ISBN. 978-92-9262-782-9.

KVDA. (2017). The project on urban transportation improvement for Kathmandu Valley in Federal Demographic Republic of Nepal. Final Report, May. Kathmandu Valley Development Area.

Leather, J. (2021) Transport and urban form. In B. Susantono & R. Guild (Eds.), *Creating livable Asian cities* (pp. 135–146). Asian Development Bank. ISBN. 978-92-9262-782-9.

Levinson, H. S., Zimmerman, S., Clinger, J., Harris, D. M. J. M., & Rutherford, C. S. (2002). Bus rapid transit: An overview. *Journal of Public Transportation, 5*, 1–30. https://doi.org/10.5038/2375-0901.5.2.1

Li, Z., Bian, X., Yin, J., Zhang, X., & Mu, G. (2016). The effect of air pollution on the occurrence of nonspecific conjunctivitis. *Journal of Ophthalmology*. Article ID: 3628762. https://doi.org/10.1155/2016/3628762. Retrieved January 10, 2021.

Mool, E., Bhave, P. V., Khanal, N., Byanju, R. M., Adhikari, S., Das, B., & Puppala, S. P. (2020). Traffic Conditions and Emission Factor from Diesel Vehicles within the Kathmandu Valley: Special Session on Aerosols, Air quality and Climate Change in the Himalayan Region. Aerosol and Air Quality Research. ISSN: 1680-8584 print/2071-1409

Nepali Times. (2021). Kathmandu air pollution hits record high, January 5.

Otonomo. (2021). *EV charging networks and Mobility Intelligence drive EV Adoption.* Retrieved November 10, 2021.

Parajuly, K. (2016). Clean up the air in Kathmandu. *Nature, 533*, 321. https://doi.org/10.1038/533321e. Retrieved January 10.

Pokharel, R., & Acharya, S. R. (2015). Sustainable transport development in Nepal: Challenges, opportunities and strategies. *Journal of the Eastern Asia Society for Transportation Studies, 11*, 2015.

Shrestha, S. R., Oanh, N. T. K., Xu, Q., & Rupakheti, M. (2013). Analysis of the vehicle fleet in the Kathmandu Valley for estimation of environment and climate co-benefits of technology intrusions. *Atmospheric Environment, 81*(2013), 579–590.

Spector, J. (2018). Study: Electric buses already emit less carbon than diesel buses. In *Any state: And their climate impact will decrease as the grid gets cleaner*. Retrieved February 02, 2021, from https://www.greentechmedia.com/articles/read/study-electric-buses-already-emit-less-carbon-than-diesel-buses-in-any-Maheswaran

Susantono, B., Boarnet, M., & Guild, R. (2021). *Introduction: Opportunities for creating livable Asian cities*. Asian Development Bank. ISBN. 978-92-9262-782-9.

Tang, J., McNabola, A., Misstear, B., Pilla, F., & Alam, M. S. (2019). Assessing the impact of vehicle speed limits and fleet composition on air quality near a school. *International Journal of Environment Research Public Health, 16*(1), 149. Published online 2019 Jan 8. https://doi.org/10.3390/ijerph16010149. PMCID: PMC6339126; PMID: 30625982.

Thapa, G., & Adhikari, A. K. (2016) Kathmandu: The third most polluted city in the world. The Kathmandu Post. Retrieved January 10, 2021, from http://kathmandupost.ekantipur.com/news/2016-03-19/kathmandu-the-third-most-polluted-city-in-the-world.html

WHO. (2016). Mortality and burden of disease from ambient air pollution. World Health Organization. https://www.who.int/airpollution/ambient/en/

WHO (World Health Organization). (2017a). Nepal urban health profile. Retrieved January 05, 2020, from http://www.who.int/kobe_centre/measuring/urbanheart/nepal.pdf

WHO (World Health Organization). (2017b). An estimated 12.6 million deaths each year are attributable to unhealthy environments. Media Center. Retrieved January 05, 2020, from http://www.who.int/mediacentre/news/releases/2016/deaths-attributable-to-unhealthy-environments/en/

WHO (World Health Organization). (2018a). Global status report on road safety 2018. Geneva.

WHO (World Health Organization). (2018b) Road traffic kills 350,000 children a year. WHO Quoted in the Telegraph News. Retrieved January 10, 2021, from https://www.telegraph.co.uk/news/2018/05/25/road-traffic-kills-350000-children-year/

www.fueleconomy.gov. (2021). Driving more efficiently: Drive sensibly. Energy Efficiency & Renewable Energy. US Department of Energy. Retrieved May 1, 2021, from https://www.fueleconomy.gov/feg/driveHabits.jsp

Chapter 13
Community Participation in Mapping Poor Settlements: A Case of Chattagram City, Bangladesh

Nurul Islam Nazem and Md. Anwar Hossain

Abstract There were several attempts in identifying the urban poor communities in the cities and towns of Bangladesh in order to fix a correct target for poverty alleviation programmes. However, due to methodological problems, it was always problematic to set a correct target. It was felt by the authority and donor agencies that there was hardly any correct figure of the number of poor and realistic statistics on their characteristics. This paper highlights the methods of mapping and collection of some basic data on poor settlements. It has been possible to identify the poor easily as the poor themselves were involved in the process of identification of their own settlements. The community people, most of whom were little educated successfully conducted the task entrusted to them. This helped them understand the nature and issues of poverty which they can utilize when the actual poverty alleviation programme is launched. Although the study was conducted on 27 cities, this paper explains the mapping and poverty situation of only one city Chattagram. The exercise identified 301,527 households as poor in about 5,778 poor settlements in the Chattagram City Corporation area and estimates about 53.6% of the city's total population were poor of which 44.6% of the poor households were extremely poor.

Keywords Participatory mapping · PGIS · Poor settlements · Urban poverty · Metropolitan population · Neighbourhood groups

13.1 Introduction

In the past poverty alleviation programmes in Bangladesh were oriented mainly towards rural areas. It was perhaps due to an overwhelming majority of the people who used to live in rural areas and used to be involved in the agricultural sectors. In

N. I. Nazem (✉) · Md. A. Hossain
Department of Geography and Environment, University of Dhaka, Dhaka, Bangladesh
e-mail: nazem.info@yahoo.com

Md. A. Hossain
e-mail: anwar.geography@du.ac.bd

B. R. K. Sinha (ed.), *Urban Dynamics, Environment and Health*,
https://doi.org/10.1007/978-981-99-5744-6_13

recent years, however, various programmes in urban areas are taking root (Rahman, 2011) for several reasons. Firstly, in the last few decades, the pace of urbanization has been very rapid due to the high flow of rural migrants to the cities (Choe & Roberts, 2011; Hossain & Nazem, 2016; Jones et al., 2016).[1] Most of these migrants are poor and enter urban areas, especially in large cities to improve their livelihood conditions (Jones et al., 2016). Secondly, large cities are facing an acute problem to accommodate the growing population (Hossain & Huggins, 2021). On the other hand, incoming poor migrants cannot afford proper housing with their low income (Sultana & Nazem, 2020). A study showed that more than one-third (35%) of the metropolitan population of the country were lived in the slums and squatters in 2005 (Islam et al., 2006). Thirdly, there is also an acute shortage in the supply of basic infrastructure and environmental services (Hossain et al., 2021; Nazem & Hossain, 2019) to the poor which directly impact their social behaviour and health (Das, 2008; Diener & Suh, 1997; Eurofound, 2012; Li & Wei, 2010; Rojas, 2014). Fourthly, the importance of the poor group and their contribution to the city economy is very significant. In most of the labour-intensive industries, informal sectors are flourishing due to the availability of cheaper labourers in the city area (Bird et al., 2018; Choe & Roberts, 2011; Hossain & Huggins, 2021). Thus, it is not possible to ignore this community while we are targeting sustainable economic development where urban areas play leading roles.

Considering as one of the important parts of the urbanization process and realizing the significance of the poor and lower income community in the city's performance, poverty alleviation and upgrading the living conditions of the poor become a critical area of urban development issues in Bangladesh. However, during the operation of poverty alleviation programmes, either by the government or other agencies found several problems, such as (a) there were no usable statistics on the poor to make appropriate decisions to fix the target population; and (b) there were no comprehensive maps, to see the locations of the poor settlements (Nazem & Hossain, 2012).

It is evident that neither the characteristics of poverty are similar in all cities, nor the upgrading requirements are similar in all settlements. In such a context, a logical classification of the poor may help set the priority target groups. It was felt that a reliable baseline database for the poor of the cities and mapping their location of settlements with the participation of community people and expert groups may resolve the above-identified limitations. Under these circumstances, the present paper highlights a participatory approach of mapping poor settlements in Chattagram city as a case. The primary aim of the mapping poor settlements study was to identify the locations where the poor people live and derive their poverty status. More specifically the objectives of the study were: (a) to set a baseline to monitor future changes in the identified poor settlements in terms of their existence, (b) to explain thematic poverty indicators and aggregate poverty index at city, ward and settlement levels, and (c) to innovate a participatory mapping approach where community people and other stakeholders can use their local knowledge and perceptions and can participate in

[1] According to Population and Housing Census 2011 the growth of national population was 1.4% per year while the growth of urban population is more than 3% per year.

conducting research. All these objectives were set to improve our knowledge about the dynamics of urban poor in terms of their location pattern, size, and the magnitude of poverty and also to assess the capacity to address the problem they are facing.

13.2 Participatory Approach of Mapping: A Literature Review

Understanding both the physical and social environment is a key requirement in community-based research and planning (Chirowodza et al., 2009). In recent years, participatory research methods have become very popular and have emerged as valuable tools in social science research (Chirowodza et al., 2009; Dunn, 2007) where local community members have actively participated in formulating and conducting research (O'Fallon & Dearry, 2002). Social scientists like geographers, psychologists, or planners employ participatory tools to understand, explain, and model community environments more closely (Ungar, 2018). In this context, an alternative approach to mapping emerged in the last few years. These newer approaches are context and issue driven and seek to emphasize community participation in the production and/or use of spatial information (Dunn, 2007).

Recent mapping approaches attempt to integrate GIS and participatory techniques which encourage community participation and attachment in the production of spatial data (Casey & Pederson, 2002; Chirowodza et al., 2009; Craig & Elwood, 1998; Daniel et al., 2021). These approaches which focus on incorporating community views or understanding in GIS (Mandara, 2007) have been used in a wide range of contexts including urban planning and revitalization of neighbourhood groups (Daniel et al., 2021; Elwood, 2002); community boundary mapping (Kienberger et al., 2005); managing conflict over access to land and other natural resources (Daniel et al., 2021); land use and natural resource planning (Ventura & Niemann Jr, 2002; Walker et al., 2002); and conservation and environmental management (Mbile et al., 2003; Meredith & Yetman, 2002).

The terminologies which are often used synonymously with participatory mapping are Community mapping, Participatory GIS (PGIS), Public Participation GIS (PPGIS), and Community Integrated GIS (Daniel et al., 2021; Mandara, 2007; Weiner et al., 2002). In terms of information types, one understanding of PGIS is as a means of integrating local and indigenous knowledge with "expert" data. The inclusion of social information has helped to expose the narrowness of conventional GIS. PGIS involves local communities in the creation of information to be fed into the GIS and subsequently used in spatial decision-making which affects them. Thus, participatory mapping or PGIS offers democratic spatial decision-making (Dunn, 2007).

Participatory mapping approaches have their strengths and weaknesses. The growing movement towards integrating participatory methods and GIS highlights that neither approach alone meets society's changing information needs (Kienberger et al., 2005; Mapedza et al., 2003; Mbile et al., 2003; Weiner et al., 2002). The approach ensures that there is equal opportunity in extent and room for all relevant stakeholders to participate (Jordan, 1998) and can empower community planning and decision-making process (Mandara, 2007) by having spatial information (Alcorn, 2000). It helps to build trust and a sense of ownership as local people are the ones who provide input for information created (Poole, 1995) and help represent the local spatial knowledge (McCall, 2003). Participation should not always strive for maximum intensity, however, the intensity should be appropriate to the tasks, competencies, and specific relationships between actors in participatory spatial planning (Mandara, 2007).

13.3 Methods

13.3.1 Study Area

Chattagram is the second largest city and one of the largest commercial or business hubs of Bangladesh. Chattagram City Corporation (CCC) was selected as the study area. The city contained approximately 2.7 million populations in 2011 in an area of 168 km^2 (BBS, 2014). It is the most important city in the country after Dhaka in terms of both industrial and economic performance.

13.3.2 Methods of Mapping and Study

The study followed a participatory method to identify and assess poor settlements. The methodology involved several actors who participated in the process of activities from conceptualization to generating final maps and databases. Maps are usually prepared by specialized people but used by the authorities at the national level, city level, and also at the community level. Thus, each of the stakeholders needed to participate in making maps so that the maps could be understood easily by all of them.

The study demanded an innovative approach in the process of conducting the survey and in the preparation of maps and databases, which is a little departure from the conventional surveys. Since the whole exercise was carried out on a partnership basis, the survey team was mobilized in such a manner that it can keep the spirit of partnership. The following are the sequences of activities and the process by which poor settlements of CCC were identified (Fig. 13.1).

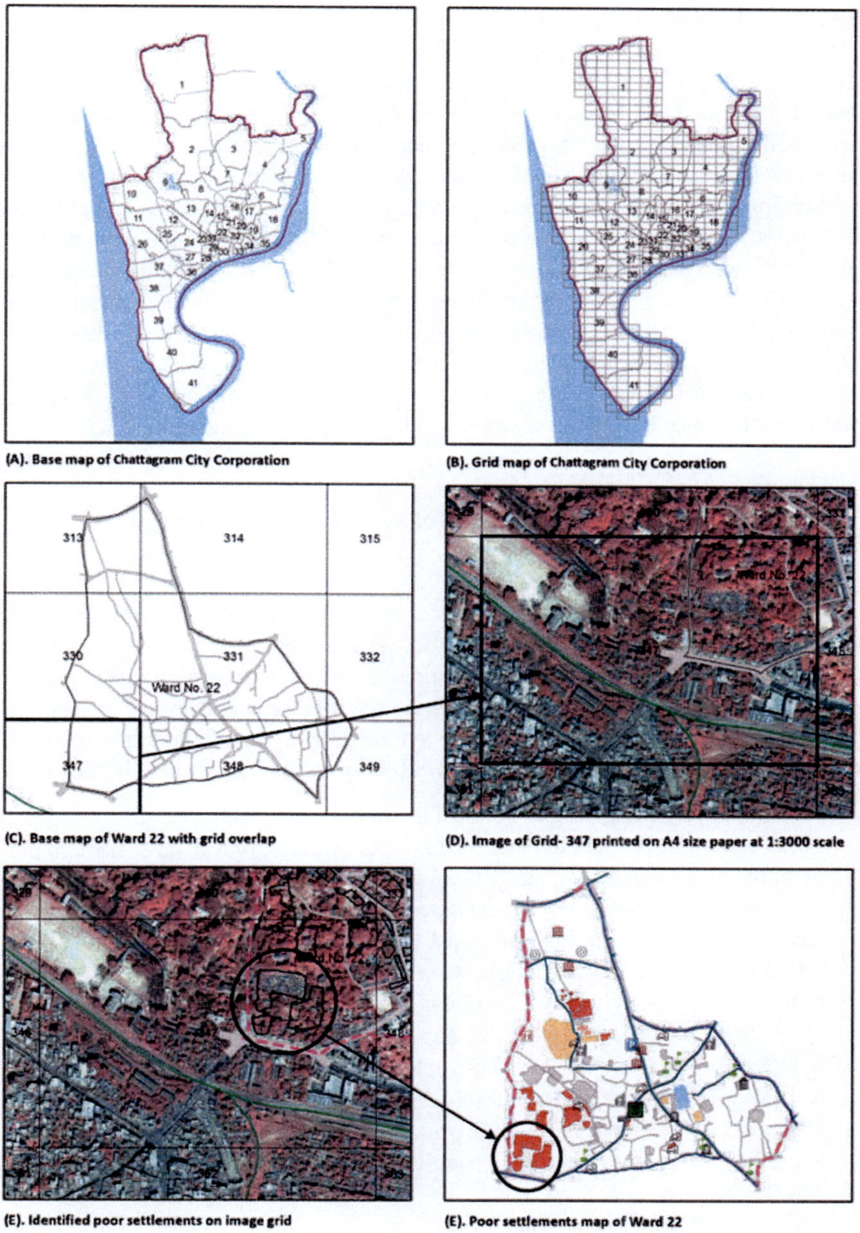

Fig. 13.1 Sequences of activities in identification of poor settlements

13.3.2.1 Base Map Preparation

The aim of preparing the base map was to produce an accurate and well-illustrated map that can be used for poor settlement mapping during the survey. Most of the cities and towns of the country lacked updated and usable maps. The city or municipal authorities have neither adequate manpower nor resources to produce and update such maps at present. Thus, it was the first step to prepare a base map before the field survey. To prepare a usable base map, first, available maps of Chattagram city were collected from the authority and were updated with the help of the most recent satellite image. Secondly, the study team consulted the city officials and surveyors regarding the level of accuracy of the map. If problems were found, these were corrected based on opinions and corrections provided by the city authority. Ground truthing was carried out during the field survey. This means that the city base map was corrected even during the survey as well as during the checking phase.

13.3.2.2 Processing and Preparation of Satellite Images for Field Survey

The purpose of obtaining the satellite images of the town was two folds: First to update the base map; and second, to identify poor settlements on the image through field survey. After collection, images were georeferenced according to Bangladesh Transverse Mercator (BTM) Coordinate System. This facilitated the proper matching of skeleton maps collected from the city authority with the obtained images of the town. Processing of images and field materials preparation include mainly four steps:

a. *Division of city into grids*: To conduct the field survey, each city had been subdivided into a number of equal-sized grids. This helped surveyors to easily comprehend the city area and can relate this with the image. The area of each grid was 100 acres. Considering the convenience, the size of the paper selected was A4 and the grid dimension represents an area of 770 m × 520 m at the ground level.
b. *Division of images into grids*: Following the division of maps into grids, the image of the same area/city was overlayed on city grids. Thus, the map grid and image were for the same area (100 acres) and the same scale. The scale of the printed image grid was 1 cm = 30 m. This helps locate settlements and delineation of its boundary on the image accurately and with proper shape.
c. *Preparation of ward level map*: Since a team of 3 persons was employed in a ward, the team was given a city map, a ward map with grids and images that correspond to grids in a ward.
d. *Printing of base maps and images*: City base map with grid overlap, ward base map and grid images were printed on A4 size paper. A folder was produced for each ward that contained these base maps and image grids.

13.3.2.3 Team Mobilization for the Survey

One Field Research Assistant (FRA)[2] and two Community Development Committee (CDC)[3] leaders formed a team to work in a ward for settlement mapping headed by a City Coordinator. A total of 40 FRAs and 84 CDCs worked in Chattagram to carry out the survey. The team collected data for each poor settlement by using a pre-designed Score Card. This three-member ward team continued the survey until they completed the work for their assigned ward.

13.3.2.4 Training and Orientation

A two-day training programme was conducted at the city level for FRAs and CDC leaders, as these two groups were responsible for the survey. The first-day training was limited to classroom lectures while on the second day, they were taken to the field to conduct a Pilot Survey to demonstrate what they had learnt on the first day. After two days of training, each (three-member) team was assigned to a particular ward to survey until the survey was completed with satisfaction.

13.3.2.5 Field Survey

Settlements were identified and drawn on the provided images. Each of the identified settlements was numbered according to ward number, block number,[4] and settlement number.[5] After identifying the settlements, the surveyors recorded the settlement characteristics in the provided scorecard consulting with the dwellers.

The study team has not compromised with the data quality. Sample sites were identified through satellite image and survey data for field checking. Approximately 15–25% of the wards were checked by the quality control team. This team updated data if required. After quality checking database was finalized for final map preparation and data tabulation.

13.3.2.6 Building Database, Processing, and Analysis

All the identified settlements were shown on the map as mentioned above and formed a database for further analysis. The study followed two approaches to analyse the

---

[2] FRA: Field Research Assistant provided by CUS.

[3] CDC: Community Development Committee. CDCs are mainly presents those City Corporation's where UPPR undertaken their projects. Where UPPR were not available during survey, two representatives provided by ward councilors worked in the places of CDC.

[4] Each ward has been subdivided into several blocks (based on community concept) and numbered from 1 for each ward.

[5] Surveyors give number to each settlement. Number started from one for each block.

data. First, a city-wise comprehensive analysis of the data; and second, ward-wise explanation to have a closer look at the scenario at ward level.

The following procedure has been used to analyse the poverty score. Each poor settlement community was asked 16 questions/issues. Each issue contained four possible answers or assessments with a value ranging from 1 to 4, respectively. Thus, the 16 questions generated a total score that varied from a minimum of 16 to a maximum of 64. This has been generated during the field survey. However, to make this score more meaningful, comparable, and analytical 16–64 values have been converted into 0–100, where 16 equals a score of 0 and 64 equals a score of 100. Poverty classes have been calculated using a quartile distribution of aggregated scores. 1st quartile represents extremely poor settlements. The second, third, and fourth quartile represents very poor, moderately poor, and marginally poor settlements, respectively (Table 13.1).

A generalized analysis has been done combining several indicators (multiple indicator analysis) to get an overall pattern in some important areas of concern such as land and housing, environmental services, infrastructure, economic and social conditions. Each of these areas accommodates several indicators. The objectives of multiple indicator analysis are to categorize settlements into good or improved and bad or worst conditions by combining several indicators. The study had done this by using the following method. In the individual indicator analysis, one can find two statuses by combining scores 1 and 2 for the worst condition and that for 3 and 4 for the improved condition. Using this generalized condition for each indicator we can find two new categories: improved and worst condition. Again, we have combined related indicators for the above-mentioned five areas of concern where each area can be explained by three interrelated indicators. Thus, each area shows 8 types of unique conditions. Derived 8 conditions by each area further combined into four generalized categories as shown in the following Table 13.2.

Then the data were processed in maps and statistical tables for the town, wards, and settlements. Each settlement has a unique identifier and has comprehensive information including a score that defines its status as poor settlements.

Table 13.1 Aggregate score and field score by poverty class (Quartile distribution) for Chattagram

Poverty status (Quartile distribution)	Assigned aggregate score: 0–100	Score assigned at field level
Extremely poor settlements (first 25%)	0.0–37.5	16–34
Very poor settlements (26–50%)	37.6–45.8	35–38
Moderately poor settlements (51–75%)	45.9–56.2	39–43
Marginally poor settlements (76–100%)	56.3–100.0	44–64

Table 13.2 Generalized categories in multiple indicator analysis

No	Indicators (e.g. Land and housing indicators)			Indicator status	Generalized condition
	Indicator 1 (e.g. land ownership)	Indicator 2 (e.g. land tenure)	Indicator 3 (e.g. housing condition)		
1	Improved	Improved	Improved	Improved condition of three indicators	Good condition
2	Improved	Improved	Insufficient/ Worst	Improved conditions of any two indicators	Poor condition
3	Improved	Insufficient/ Worst	Improved		
4	Insufficient/ Worst	Improved	Improved		
5	Improved	Insufficient/ Worst	Insufficient/ Worst	Improved conditions of only one indicator	Very poor condition
6	Insufficient/ Worst	Improved	Insufficient/ Worst		
7	Insufficient/ Worst	Insufficient/ Worst	Improved		
8	Insufficient/ Worst	Insufficient/ Worst	Insufficient/ Worst	No indicators are in improved condition	Worst condition

13.3.2.7 Limitations

The study has several limitations. Conceptually, defining poor settlements in an area and physically identifying them to record was a serious problem. Finding a poor settlement by looking at it physically or by perception was in most cases correct. However, in small towns, such settlements sometimes were not poor. Secondly, drawing settlement boundaries from images and subsequently through ground-truthing for mapping is arbitrary. It is not a legal boundary for settlement. If some other people do the same exercise may come up with slightly different results. Thirdly, obtaining information on each settlement from community people was challenging. The information is the aggregate scenario of the settlement. Especially, the use of a scorecard and the value assigned for each question is also arbitrary. Finally, the present study is unique in the sense that it involved community people for surveying along with trained expert surveyors. The community people were in most cases less educated and women. Their concentrations were in most cases poor, and conceptually weak in conducting the survey and obtaining data efficiently.

13.4 Results

The survey explores the nature of the key problems that are facing the poor at the settlement level. However, the primary focus of the survey was on the identification of the poor settlements, their location and geographic size including their basic characteristics.

13.4.1 Key Findings

The study identified 301,527 households as poor in 5,778 settlements. The study estimates that Chattagram has about 53.6% of its total population were poor. The distribution of poor by poverty class shows that 44.6% of these households are extremely poor, 20.2% are very poor, 21.2% are moderately poor, and 14% are marginally poor (Table 13.3).

Spatial distribution of poor settlements shows that in some wards like 1, 3, 7, 10, 18, 27, 38, and 39 concentration of extremely poor settlements are quite high where ward 1, 18, 27, and 38 accommodated more than a hundred extremely poor settlements each. On average, about 52.2 households were accommodated in a poor settlement, which however varies substantially among the settlements identified. Most of the poor settlements in Chattagram are new, although the average age is about 25 years. Young settlements were found very common, which means that new settlements are growing at a very rapid rate in Chattagram. The density of poor households is extremely high (about 120 households per acre of settlement area) in Chattagram. Figure 13.2 shows the distribution of poor settlements by poverty class in Chattagram City and Fig. 13.3 shows the distribution of the settlements in detail in a Ward (out of 41 wards of CCC).

Table 13.3 Settlements, households, dwelling structures by poverty class

Type of settlement	Settlements		Households		Dwelling structures		Avg size
	Number	Percentage	Number	Percentage	Number	Percentage	
Extremely poor	1574	27.24	134,595	44.64	43,592	44.00	85.51
Very poor	1360	23.54	60,896	20.20	19,692	19.88	44.78
Moderately poor	1561	27.02	63,977	21.22	20,542	20.73	40.98
Marginally poor	1283	22.20	42,059	13.95	15,252	15.39	32.78
Total	5778	100.00	301,527	100.00	99,078	100.00	52.19

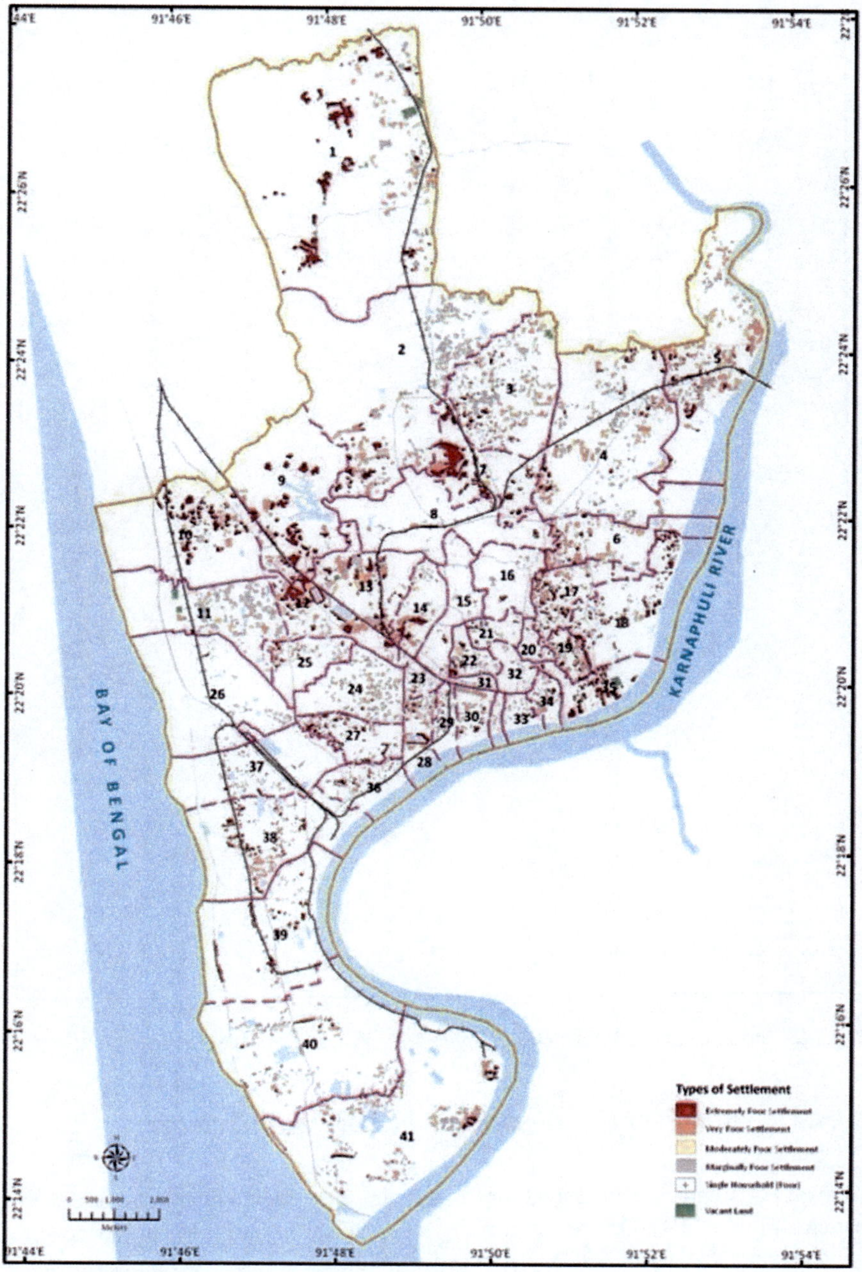

Fig. 13.2 Distribution of poor settlements in Chattagram City Corporation

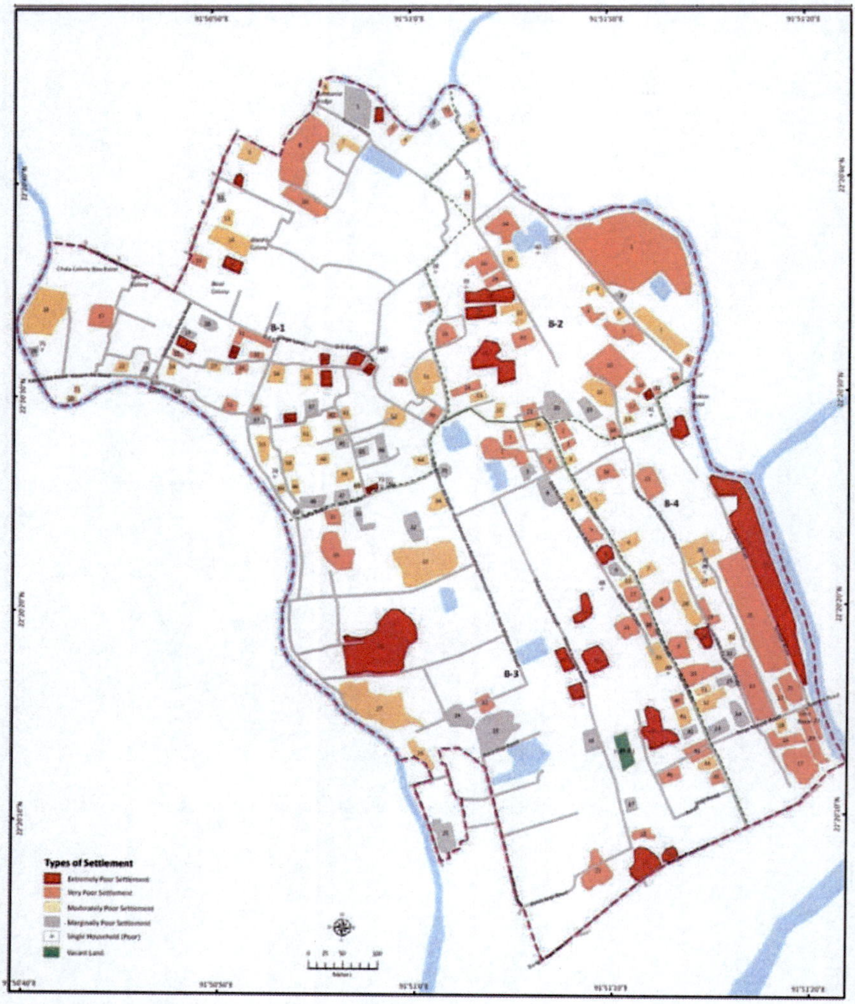

Fig. 13.3 Distribution of poor settlements in Ward 19 of Chattagram City Corporation

13.4.2 *Settlement Characteristics by Poverty Indicators*

Analyses of 16 poverty indicators for Chattagram show a vivid picture of the poverty patterns (Table 13.4). One of the striking features is the pattern of land ownership, which shows that about half of the poor were housed on their private landowner's land. Tenure status also reflects the land ownership pattern where about 20% of the poor households were living in squatters and about half of the households were tenants without any contact. Only 14.3% of households enjoy freehold rights with some variations among wards. The structural quality of housing was not satisfactory as an overwhelming majority of the settlements were composed of less than 50% permanent structures or sometimes with temporary structures. Similar conditions can be observed in environmental services such as water supply, sanitation drainage conditions, and waste management in poor settlement areas. Improved water supply, sanitation, and drainage facilities were available, respectively, in 31.6%, 34%, and 26.4% of the poor settlements. Although the electricity condition is good in Chattagram, more than one-third of the poor settlements do not have proper or improved access roads. On the other hand, solid waste management services are insufficient and dwellers of about 68.4% of the settlements (77.7% of poor households) claimed that this service is not available in their community.

The social and economic conditions in the poor settlements of Chattagram show a mixed scenario. Children's enrolment in schools is better, although one can observe variations among settlements and wards. The school enrolment of children from extremely poor settlements was really poor. Generally, income level is relatively good in all poor settlements in Chattagram as the city enjoys diverse employment opportunities. Out of about 40% of poor settlements, where the majority of the households (over 75%) earned more than Taka 4,000 per household. Savings and credit facilities in the town seem to be limited as most of the poor households do not have access to them. On the other hand, about half of the settlements, as well as households were found in a vulnerable situation both environmentally as well as economically which they cannot mitigate by themselves with their present socio-economic strengths.

13.5 Discussion

Findings of the poor settlements study of Chattagram City were discussed under five major areas of concern. The areas are (a) land and housing conditions, (b) environmental services, (c) infrastructure, (d) economic, and (f) social conditions. It has been observed that land and housing conditions in Chattagram are very poor in terms of ownership and tenure because most of the lands were privately owned by landlords and poor households living there as a tenant without any contract. The environmental services in most of the poor settlements are fragile, and so is the economic condition. Income level is relatively better than other indicators and earning opportunities are also a bit better than that of other study towns. Some physical infrastructure and social conditions were found to be improved in this port city (Fig. 13.4).

13.5.1 Land and Housing Condition

Land ownership, tenure status, and housing condition combinedly present land and housing condition of the poor settlements. Table 13.5 shows that 41.6% of the poor

Table 13.4 Poor settlements in Chattagram city by poverty indicators

Indicators	Determinants	Chattagram city				Urban area of Bangladesh (27 city average)			
		Settlements		Households		Settlements		Households	
		Number	%	Number	%	Number	%	Number	%
1. Land ownership	Private landlords	2837	49.1	144,787	48.0	14,145	32.19	370,029	32.83
	Central government	661	11.4	79,328	26.3	3940	8.97	197,629	17.53
	Local government	719	12.4	26,420	8.8	2102	4.78	66,423	5.89
	Owned by occupants	1561	27.0	50,992	16.9	23,761	54.07	493,052	43.74
2. Type of occupancy	Squatter	340	5.9	59,764	19.8	2788	6.34	153,372	13.61
	Tenant without contract	2937	50.8	145,186	48.2	16,119	36.68	421,767	37.42
	Tenant with contract	1099	19.0	53,483	17.7	3428	7.80	103,378	9.17
	Individual owner	1402	24.3	43,094	14.3	21,613	49.18	448,616	39.80
3. Housing nature	>75% Semi-permanent	506	8.8	54,415	18.0	4919	11.19	172,024	15.26
	<75% Semi-permanent	3410	59.0	180,132	59.7	23,972	54.55	635,605	56.39
	50% permanent & 50% Semi	1564	27.1	54,263	18.0	12,678	28.85	268,926	23.86
	75% permanent	298	5.2	12,717	4.2	2379	5.41	50,578	4.49
4. Water supply	No drinking water	1643	28.4	117,987	39.1	9356	21.29	308,630	27.38
	1 common tab > 15 HH	2308	39.9	129,462	42.9	15,599	35.49	474,341	42.08
	1 common tab < 15 HH	1515	26.2	46,619	15.5	15,221	34.63	275,482	24.44
	Individual pipe	312	5.4	7459	2.5	3772	8.58	68,680	6.09
5. Sanitation facilities	No toilet available	1069	18.5	95,608	31.7	10,799	24.57	328,449	29.14
	1 toilet > 15 people	2741	47.4	140,893	46.7	16,088	36.61	475,357	42.17
	1 toilet < 15 people	1482	25.6	53,157	17.6	14,419	32.81	278,343	24.69

(continued)

Table 13.4 (continued)

Indicators	Determinants	Chattagram city				Urban area of Bangladesh (27 city average)			
		Settlements		Households		Settlements		Households	
		Number	%	Number	%	Number	%	Number	%
	Individual toilet	486	8.4	11,869	3.9	2642	6.01	44,984	3.99
6. Drainage facilities	No drains, stagnant water	2372	41.1	124,735	41.4	25,258	57.47	628,568	55.77
	No drains	1879	32.5	115,783	38.4	9153	20.83	284,133	25.21
	Open drains	1201	20.8	52,250	17.3	7578	17.24	177,568	15.75
	Masonry drains	326	5.6	8759	2.9	1959	4.46	36,864	3.27
7. Access roads	No access roads	1262	21.8	72,228	24.0	13,097	29.80	345,286	30.63
	Earth or gravel roads	2323	40.2	119,277	39.6	16,225	36.92	421,487	37.39
	Paved roads, no maintenance	1271	22.0	58,400	19.4	9209	20.95	235,152	20.86
	Paved roads, maintenance	922	16.0	51,622	17.1	5417	12.33	125,208	11.11
8. Electricity supply	Not available, line is far	153	2.6	10,632	3.5	1765	4.02	41,556	3.69
	Not available, line is close	409	7.1	29,178	9.7	3683	8.38	98,205	8.71
	Available, no streetlights	3341	57.8	170,765	56.6	27,865	63.40	737,696	65.45
	Available, with streetlights	1875	32.5	90,952	30.2	10,635	24.20	249,676	22.15
9. Solid waste collection	Not available	3953	68.4	234,308	77.7	34,610	78.75	915,252	81.20

(continued)

Table 13.4 (continued)

Indicators	Determinants	Chattagram city				Urban area of Bangladesh (27 city average)			
		Settlements		Households		Settlements		Households	
		Number	%	Number	%	Number	%	Number	%
	Bins but no Reg. collection	805	13.9	35,797	11.9	3752	8.54	101,172	8.98
	Bins and Reg. collection	722	12.5	22,545	7.5	2782	6.33	60,987	5.41
	House to house collection	298	5.2	8877	2.9	2804	6.38	49,722	4.41
10. School enrolment	<25% children enrolled	604	10.5	46,435	15.4	3474	7.90	113,800	10.10
	25–50% children enrolled	1642	28.4	107,983	35.8	11,462	26.08	344,547	30.57
	50–90% children enrolled	2560	44.3	112,757	37.4	18,782	42.74	479,363	42.53
	>90% children enrolled	972	16.8	34,352	11.4	10,230	23.28	189,423	16.81
11. Employment	>50% families employed	165	2.9	16,383	5.4	2511	5.71	84,651	7.51
	25–50% families employed	1210	20.9	89,186	29.6	14,065	32.00	368,757	32.72
	>50% families self-employed	2912	50.4	136,735	45.3	20,031	45.58	523,330	46.43
	>50% families Reg. employed	1491	25.8	59,223	19.6	7341	16.70	150,395	13.34
12. Civic facilities	Not available, limited access	1199	20.8	94,188	31.2	15,443	35.14	444,612	39.45

(continued)

Table 13.4 (continued)

Indicators	Determinants	Chattagram city				Urban area of Bangladesh (27 city average)			
		Settlements		Households		Settlements		Households	
		Number	%	Number	%	Number	%	Number	%
	Not available, easy access	3597	62.3	165,174	54.8	24,589	55.95	582,209	51.65
	Available, limited access	591	10.2	27,175	9.0	2399	5.46	64,625	5.73
	Available, easy access	391	6.8	14,990	5.0	1517	3.45	35,687	3.17
13. Income	>75% HH income < TK4000	489	8.5	49,421	16.4	9294	21.15	282,843	25.09
	>50% HH income < TK4000	1244	21.5	73,022	24.2	17,642	40.14	439,514	38.99
	50–75%HH income > TK4000	1756	30.4	90,493	30.0	10,655	24.24	254,218	22.55
	>75% HH income > TK4000	2289	39.6	88,591	29.4	6357	14.46	150,558	13.36
14. Savings and credit activities	Not available	1985	34.4	62,542	20.7	9000	20.48	152,908	13.57
	<50% families participate	2476	42.9	141,195	46.8	18,849	42.89	469,762	41.68
	50–75% families participate	1052	18.2	81,296	27.0	12,209	27.78	385,396	34.19
	>75% families participate	265	4.6	16,494	5.5	3890	8.85	119,067	10.56
15. Risk and vulnerability	High risk	878	15.2	88,145	29.2	6795	15.46	264,197	23.44

(continued)

Table 13.4 (continued)

Indicators	Determinants	Chattagram city				Urban area of Bangladesh (27 city average)			
		Settlements		Households		Settlements		Households	
		Number	%	Number	%	Number	%	Number	%
	Medium risk	1597	27.6	88,080	29.2	16,338	37.18	417,562	37.05
	Low risk	2526	43.7	98,307	32.6	14,859	33.81	337,404	29.93
	No risk	777	13.4	26,995	9.0	5956	13.55	107,970	9.58
16. Social problems	>50% families face problems	720	12.5	72,619	24.1	7157	16.29	295,184	26.19
	50% families face problems	1649	28.5	112,040	37.2	13,826	31.46	388,305	34.45
	A few families face problems	2515	43.5	96,018	31.8	17,306	39.38	361,550	32.08
	Not an issue in community	894	15.5	20,850	6.9	5659	12.88	82,094	7.28
Total		5778	100.0	301,527	100.0	43,948	100.0	1,127,133	100.0

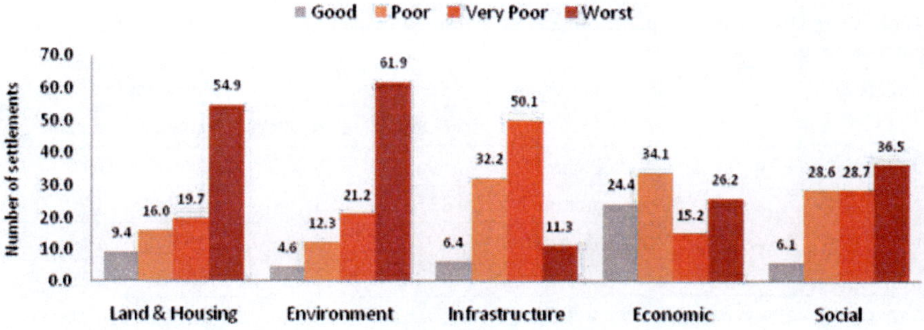

Fig. 13.4 Distribution of poor settlements of Chattagram City by major areas of analysis

settlements in Chattagram were worst in land and housing condition, where 54.9% of the households live. Only 16.8% of the settlements have good housing considering the ownership tenure and structural quality. 18.6% of settlements in Chattagram remained in very poor condition, where 19.7% of the poor households live. It is a matter of serious concern that, about 84% of the households were living in private or government land and about 68% of the poor were living in squatter settlements or without any contract with the owner, which indicates that the dwellers live in a vulnerable condition. On the other hand, more than half of the settlements' quality was semi-permanent or Katcha.

13.5.2 Environmental Services

Environmental services include water supply, sanitation, and drainage facilities. These three indicators together provide a general environmental condition of the poor settlements in Chattagram. The environmental services in the poor settlements in Chattagram show that only 8.9% are good and 19.5% are poor and 26.3% are very poor. About half of the poor settlements (45.3%) are in the worst condition where about two-thirds of the households (61.9%) live. Like other cities of Bangladesh, the environmental condition was found worst in condition. About 40% of the households do not have safe drinking water sources, 31.7% do not have sanitary toilets and about 80% do not have drainage facilities within their settlements. About half of the households use 1 tap or a toilet of more than 15 people. Due to the worst environmental conditions, poor households are highly vulnerable to health hazards.

Table 13.5 Distribution of poor settlements and households in Chattagram City Corporation by major areas of analysis

Indicator	Settlement condition	Settlement		Household	
		Number	Percentage	Number	Percentage
Land and housing	Good condition	969	16.8	28,299	9.4
	Poor condition	1332	23.1	48,283	16.0
	Very poor condition	1072	18.6	59,506	19.7
	Worst condition	2405	41.6	165,439	54.9
Environmental services	Good condition	516	8.9	13,950	4.6
	Poor condition	1127	19.5	37,225	12.3
	Very poor condition	1520	26.3	63,813	21.2
	Worst condition	2615	45.3	186,539	61.9
Infrastructure	Good condition	693	12.0	19,379	6.4
	Poor condition	1718	29.7	97,038	32.2
	Very poor condition	2914	50.4	150,948	50.1
	Worst condition	453	7.8	34,162	11.3
Economic	Good condition	1061	18.4	73,696	24.4
	Poor condition	2738	47.4	102,946	34.1
	Very poor condition	1106	19.1	45,852	15.2
	Worst condition	873	15.1	79,033	26.2
Social	Good condition	492	8.5	18,410	6.1
	Poor condition	2178	37.7	86,330	28.6
	Very poor condition	1985	34.4	86,686	28.7
	Worst condition	1123	19.4	110,101	36.5
Total		5778	100.0	301,527	100.0

13.5.3 *Infrastructure Condition*

Infrastructure includes three indicators: road, electricity, and waste management. Combinedly these three indicators explain infrastructure conditions in the poor settlements in Chattagram. Infrastructure is generally poor in all poor settlements. Only 12% of the poor settlements have good conditions having improved conditions of all three conditions. 7.8% of the settlements have no access to any infrastructure services while 50.4% of the settlements have only access to one service among the three. Ward wise infrastructure condition shows that no ward has a good infrastructural condition.

13.5.4 Economic Condition

Three indicators; employment, income, and savings were taken to explain the economic condition of the poor settlements. Table 13.5 shows that 15.1% of the settlements are in the worst condition where no one indicator is in improved status. 19.1% of the settlements have only one improved condition among the three. 18.4% of the settlements (24.4% of the households) are in good condition.

13.5.5 Social Condition

The social condition of settlements is defined by a combined status of three variables, children enrolment in school, access to civic facilities and risks. Using these indicators combined scenes improved conditions and the worst conditions have been identified. Table 13.5 shows that the overall social condition in poor settlements in Chattagram is not good. Only 8.5% of the settlements are in good condition having improved condition of three indicators. On the other hand, 19.4% of the poor settlements do not have any indicators of improved condition.

13.6 Conclusions

Most of the findings of the study go with our existing knowledge and understanding of the characteristics of the poor and their settlements. The important features of the Chattag study are that: the poor settlements are scattered all over the city with more concentration within the old, central part and industrial zones; settlements are generally smaller in size and the larger ones are worst off; and most of the poor settlements in Chattagram are comparatively new.

Poor settlements are growing at an accelerated rate in Chattagram. This indicates that the economy of the town is diverse on one hand and growing at a faster rate to attract poor migrants even from far away from the city. In fact, Chattagram though located in one corner of the country is a very dynamic city in terms of commerce, industry, education, tourism, and port facility and, therefore, the city attracts a wide range of poor and low-income people from the country's vast rural areas. This can be evident from the income pattern and economic activities of the poor.

Out of five major areas of concern (land and housing, environmental services, infrastructure, economic and social condition), housing and environmental services were found to be in critical conditions in Chattagram. Housing opportunity (land ownership and holding rights) is limited, and the management of the environment was poor.

From the above conditions of the poor settlements in Chattagram, one can identify some possible areas for interventions. The most important area is to improve housing conditions and environmental services such as water, sanitation, waste management, and drainage condition. This needs proper planning and integrated interventions. The second area of intervention is to enhance the income opportunities of the poor through careful investments in both formal and informal sectors and providing micro-credit facilities to poor households in this city.

References

Alcorn, J. B. (2000). *Borders, rules and governance: mapping to catalyse changes in policy and management* (Issue 91). International Institute for Environment and Development London.

BBS. (2014). *Population and Housing Census 2011, National Volume 3: Urban Area Report*. Bangladesh Bureau of Statistics.

Bird, J., Li, Y., Rahman, H. Z., Rama, M., & Venables, A. (2018). *Toward Great Dhaka*. The World Bank. https://econpapers.repec.org/RePEc:wbk:wbpubs:29925

Casey, L., & Pederson, T. (2002). Mapping Philadelphia's neighbourhoods. In *Community participation and geographical information systems* (pp. 65–76). CRC Press.

Chirowodza, A., Van Rooyen, H., Joseph, P., Sikoroyi, S., Richter, L., & Coates, T. (2009). Using participatory methods and geographic information systems (GIS) to prepare for an HIV community-based trial in vulindlela, South Africa (project accept-HPTN 043). *Journal of Community Psychology, 37*(1), 41–57. https://doi.org/10.1002/jcop.20294

Choe, K. A., & Roberts, B. H. (2011). *Competitive cities in the 21st century: Cluster-based local economic development*. Asian Development Bank.

Craig, W. J., & Elwood, S. A. (1998). How and why community groups use maps and geographic information. *Cartography and Geographic Information Systems, 25*(2), 95–104.

Daniel, W., M., H. T., & Craig, W. J. (2021). Community participation and geographic information systems. In *Community Participation and Geographical Information Systems* (pp. 29–42). CRC Press. https://doi.org/10.1201/9780203469484-8

Das, D. (2008). Urban quality of life: A case study of Guwahati. *Social Indicators Research, 88*(2), 297–310. https://doi.org/10.1007/s11205-007-9191-6

Diener, E. D., & Suh, E. (1997). Measuring quality of life: Economic, social, and subjective indicators. *Social Indicators Research, 40*(1), 189–216. https://doi.org/10.1023/A:1006859511756

Dunn, C. E. (2007). Participatory GIS—A people's GIS? In *Progress in human geography* (Vol. 31, Issue 5, pp. 616–637). Sage Publications. https://doi.org/10.1177/0309132507081493

Elwood, S. A. (2002). GIS use in community planning: A multidimensional analysis of empowerment. *Environment and Planning A, 34*(5), 905–922. https://doi.org/10.1068/a34117

Eurofound. (2012). *3rd European Quality of Life Survey: Quality of Life in Europe: Impacts of the Crisis*. EUR-OP.

Hossain, M. A., & Huggins, R. (2021). The environmental and social impacts of unplanned and rapid industrialization in suburban areas: The case of the Greater Dhaka region, Bangladesh. *Environment and Urbanization ASIA, 12*(1), 73–89. https://doi.org/10.1177/0975425321990319

Hossain, M. A., & Nazem, N. I. (2016). Cluster based urban economic growth in Chittagong metro region: A case of manufacturing industries. *Oriental Geographer, 57*, 1–26.

Hossain, M. A., Sultana, S., & Nazem, N. I. (2021). Spatial distribution of strategic service infrastructures that favour quality of life (QOL) in the Greater Dhaka Region (GDR), Bangladesh. *SN Social Sciences, 1*(8), 1–26. https://doi.org/10.1007/S43545-021-00219-1

Islam, N., Angeles, G., Mahbub, A., Lance, P., & Nazem, N. I. (2006). *Slums of urban Bangladesh: mapping and census 2005*.

Jones, G., Mahbub, A. Q. M., & Izazul Haq, M. (2016). *Urbanization and migration in Bangladesh*. United Nations Population Fund, Bangladesh Country Office.

Jordan, G. (1998). A public participation GIS for community forestry user groups in Nepal: Putting people before the technology. *Position Paper Presented at the Project Varenius Specialist Meeting: Empowerment, Marginalization, and Public Participation GIS, Santa Barbara, CA.* Retrieved May 03, 2006, from https://Negia.Ucsb.Edu/Varenius/Ppgis/Papers/Jordan.Pdf

Kienberger, S., Steinbruch, F., Blasckhke, T., Luis da Costa, A., & Gomes, A. (2005). The potential of community mapping and community integrated GIS: A study in the Sofala Province, Mozambique. In *10th International Conference on Information & Communication Technologies (ICT) in Urban Planning and Spatial Development and Impacts of ICT on Physical Space, Vienna University of Technology, Austria.*

Li, Y., & Wei, Y. (2010). A spatial-temporal analysis of health care and mortality inequalities in China. *Eurasian Geography and Economics, 51*(6), 767–787. https://doi.org/10.2747/1539-7216.51.6.767

Mandara, C. G. (2007). Participatory GIS in mapping local context of conflicts over pastoral resources. *A Case Study of Duru Haitemba-Babati, Tanzania*, 86.

Mapedza, E., Wright, J., & Fawcett, R. (2003). An investigation of land cover change in Mafungautsi Forest, Zimbabwe, using GIS and participatory mapping. *Applied Geography, 23*(1), 1–21.

Mbile, P., DeGrande, A., & Okon, D. (2003). Integrating participatory resource mapping and geographic information systems in forest conservation and natural resources management in Cameroon: A methodological guide. *The Electronic Journal of Information Systems in Developing Countries, 14*(1), 1–11.

McCall, M. K. (2003). Seeking good governance in participatory-GIS: A review of processes and governance dimensions in applying GIS to participatory spatial planning. *Habitat International, 27*(4), 549–573.

Meredith, T. C., & Yetman, G. G. (2002). Mexican and Canadian case studies of community-based spatial information management for biodiversity conservation. In *Community participation and geographical information systems* (pp. 205–217). CRC Press.

Nazem, N. I., & Hossain, M. A. (2012). Mapping poor settlements in 27 cities and towns of Bangladesh: UPPR experience. *CUS Bulletin on Urbanization and Development, 60–62*, 16–20.

Nazem, N. I., & Hossain, M. A. (2019). Strategic infrastructure supporting the quality of life in Dhaka. In *Multidimensional approach to quality of life issues* (pp. 209–220). Springer.

O'Fallon, L. R., & Dearry, A. (2002). Community-based participatory research as a tool to advance environmental health sciences. *Environmental Health Perspectives, 110*(SUPPL. 2), 155–159. https://doi.org/10.1289/ehp.02110s2155

Poole, P. (1995). *Indigenous peoples, mapping & biodiversity conservation: An analysis of current activities and opportunities for applying geomatics technologies* (Issue 333.95 P822). Biodiversity Support Program, Washington, DC (EUA).

Rahman, H. Z. (2011). *Urban Bangladesh: challenges of transition.* Power and Participation Research Centre.

Rojas, M. (2014). *The measurement of quality of life: Conceptualization Comes first a four-qualities-of-life conceptual framework and an illustration to Latin America.* Mexico.

Sultana, S., & Nazem, N. I. (2020). Housing affordability of ready-made Garment workers in the Dhaka Metropolitan area. *Environment and Urbanization ASIA*, 0975425320938581. https://doi.org/10.1177/0975425320938581

Ungar, S. (2018). Cognitive mapping without visual experience. In *Cognitive mapping: Past, present and future* (pp. 221–248). Taylor and Francis. https://doi.org/10.4324/9781315812281-13

Ventura, S. J., & Niemann Jr, B. J. (2002). GIS-enhanced land-use planning. In *Community participation and geographical information systems* (pp. 113–124). CRC Press.

Walker, D. H., Leitch, A. M., de Lai, R., Cottrell, A., Johnson, A. K. L., & Pullar, D. (2002). A community-based and collaborative GIS joint venture in rural Australia. In *Community Participation and Geographic Information Systems* (pp. 137–152).

Weiner, D., Harris, T. M., & Craig, W. J. (2002). *Community participation and geographic information systems.* CRC Press.

Chapter 14
Gendered Vulnerability and Role of Women in Landslide Preparedness: A Case Study of an Urban Periphery of Cox's Bazar, Bangladesh

Mumita Tanjeela

Abstract Landslides have been a recent addition to the list of natural hazards in the south-eastern part of Bangladesh particularly in some districts of the Chattagram division. Although Bangladesh is primarily characterized by flat topography, this part is hilly and mountainous, featuring numerous terraces and these hilly districts have experienced severe landslide incidents in recent years. Landslides are mostly triggered by heavy rainfall usually during the monsoon period in these areas. However, the underlying causes of landslides include deforestation, hill cutting, and unregulated and unplanned development work. Moreover, rapid urbanization has spurred hill-cutting activities, deforestation, and construction in the Chottogram Division, leading to an increasing number and impact of landslides. In particular, the influx of Rohingya refuge in Cox's Bazar is contributing to deforestation and hill cutting at an alarming rate and the risk is increasing. Poverty and landlessness also compel poor people to reside on precarious hill slopes. The nexus between climate-induced migration, poverty, urbanization, and the emergence of a new type of hazards is yet to be studied with due concern. In this context, the paper explores the landslide vulnerability and preparedness practices of the at risk community and underscores the importance of women's involvement in disaster preparedness activities. The study was conducted using a qualitative research method, adopting a case study approach of a Community-based landslide preparedness programme in Cox's Bazar district of Chottogram division, Bangladesh. Data and information were collected through focus group discussions, in-depth interviews of women living in the studied community, key informant interviews of project officials, and participant observation of their community activities. The findings clearly indicate that gender is a critical factor for shaping women's vulnerability to any hazard. The study also reveals that despite gender-differentiated vulnerabilities, women's agency and their involvement in disaster preparedness activities have a positive impact on community resilience. Thus, women's roles in emergency preparedness and responses as well as in disaster risk reduction, are equally significant as those of their male counterparts.

M. Tanjeela (✉)
Department of Sociology, East West University, Dhaka, Bangladesh
e-mail: mtanjeela@ewubd.edu; mumita_tanjeela@yahoo.com

B. R. K. Sinha (ed.), *Urban Dynamics, Environment and Health*,
https://doi.org/10.1007/978-981-99-5744-6_14

Keywords Gender vulnerability · Landslide · Women's agency · Community resiliency · Rapid urbanization · Geological hazard

14.1 Introduction

Bangladesh is ranked as the fifth vulnerable country in Asia for landslide-induced fatalities (Petley, 2012). Landslide has become a regular geological hazard in south-eastern Bangladesh, particularly in some districts of Chattagram division (Bandarban, Cox's Bazar, Khagrachhari, and Rangamati districts). This part of the country has highest average slope inclines (Ahmed, 2017). Although Bangladesh is primarily a country with a flat topography, the South-East part is hilly and mountainous, with a number of terraces covering 18% of it (Mahmood & Khan, 2008). Several of these hilly districts have experienced severe landslide incidents in recent years (Displacement Solutions, 2012). Chottogram division of Bangladesh is experiencing an increasing number and impact of landslides due to rapid urbanization, hill-cutting activities, deforestation, and construction activities. Particularly, the influx of Rohingya refuges within the camps in Cox's Bazar is contributing to deforestation and hill cutting at an alarming rate. A study by the CDMP (2012a) explains that landslides are caused by the exposure of soft sedimentary rocks due to human interventions across the slopes, combined with torrential monsoon rainfall. Increasing human activities such as hill cutting for residential development have resulted in many landslides, particularly in the CHD, putting people and properties at risk (Ahmed, 2017). A total number of 147 landslides occurred during 1995–2010 with hundreds of human fatalities in Cox's Bazar alone (CDMP, 2012b). Disaster Summary Sheet (2018) claimed that though Cox's Bazar previously had a relatively low impact of landslides, the risk is becoming increasingly higher since 2010. Additionally, rainfall pattern has undergone a significant change in recent years such as intense rainfall in a short period which is another contributing factor to prompt landslides on unstable slopes in these areas (CDMP, 2012a). Kjekstad and Highland confirm that most landslides occur due to climatic factors such as changes in rainfall patterns and human intervention and often lead to migration and unplanned new settlements (2009). The topographical map of Bangladesh is indicative of hilly slopes (Fig. 14.1).

It is evident that the hazards range from small scale to large scale depending on the flow of rain and the vulnerability of the slopes, but ultimately they all cause significant damage to lives and properties. The risk of landslides in the rapidly growing urban centres of Chattagram, particularly in Cox's Bazar requires increased attention because Cox's Bazar is the main port city of Bangladesh and the potential scale of loss and damage that can affect on the livelihood of urban poor.

Rural–urban migration is an adaptation strategy for the populations affected by environmental incidents particularly in coastal areas. Several studies show that in recent decades, rural populations have adopted migration to cities in search of new livelihood as a strategy for coping mechanism with climate-related events (Displacement Solutions, 2012; Gray & Mueller, 2012, McLeman & Smit, 2006). Roth (2009)

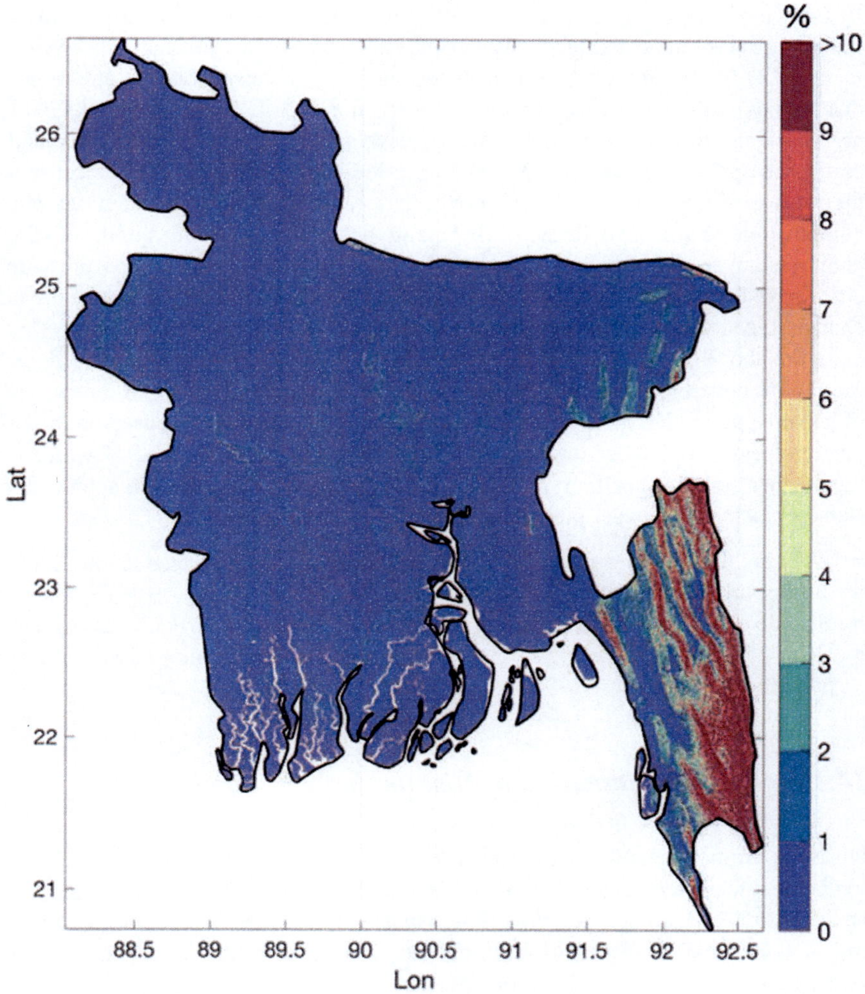

Fig. 14.1 Topographical map of Bangladesh, demonstrating slope. *Source* CYMMIT (2018). Adopted from disaster summary sheet, 2018, start fund, Bangladesh

expresses concern about the significant increasing trend of population growth concentrated in urban areas of developing countries because of the impact of environmental problems encountered by rural people. According to Penning-Rowsell et al. (2012), migration to cities is the last option for climate-displaced populations, as is the case of Bangladesh. These studies show a strong connection between unplanned urban expansion, increasing poverty, and climate-induced population displacement and migration. Others suggested that in the next few decades, it is likely that urban areas will be exposed to an increased risk of hazards which will directly impact the urban poor (Dodman & Satterthwaite, 2008; Jabeen et al., 2010).

Despite this ongoing circumstance, urban areas have yet to be considered fully at the risk of environmental impacts. The Commission on Climate Change and Development (CCD, 2009) stated, "Cities and city dwellers have received too little attention in discussions of climate change impacts and adaptation" (2009). Birkmann et al. (2010) argue that the fast growing urban areas of developing countries should be considered as hotspots of climate change risks since these cities are vulnerable because of their weak infrastructure, and their poor and unstable socio-economic demographic composition. Sharma and Tomar (2010) claim that although the cities of developing countries are becoming climate vulnerable due to increasing population pressure, cities have the potential to take mitigation measures and build resilience to climate change impacts through local governing systems.

In the above context this paper seeks to reveal the nexus among climate-induced migration, poverty, and gender vulnerabilities due to the emergence of a new type of hazards, particularly landslides in Bangladesh. The paper also focused on formal preparedness practices of the community at risk and the importance of women's involvement in this specific type of disaster preparedness activities. To achieve the stated objectives the paper poses the following research questions:

- What are the gender-differentiated vulnerabilities that women face in the context of poor urban settings?
- How do women deal with the urban environmental hazard in particular landslides?
- How youth and women can play a vital role in community-based disaster preparedness activities?

14.1.1 Gender Dimension of Disaster Vulnerability

Disasters are not a gender-neutral phenomenon. The number of female casualties in disasters is always higher than men due to their different socialization processes, social norms, and cultural practices. For instance, during the 1991 cyclone disasters in Bangladesh, 90% of the 141,000 fatalities were women (Anwarul et al., 2013). After analysing a sample of natural disasters of 141 countries between 1981 and 2002, Neumayer and Plümper (2007) claim that more women died due to their unequal social and economic position. It has been observed that apart from the higher death toll, women also bear more physical and socio-economic pressures from disasters (Cutter et al., 2003) because the issues related to their biological, social, and economic conditions position them at more risk in any natural disaster (Enarson, 2000; Enarson & Fordham, 2001; Shamim, 1995). For instance, some women remained unsafe during the Indian Ocean tsunami in 2004 as they were unable to leave their homes due to their responsibilities for the family members who were unable to move, such as the sick, the disabled, the elderly, and the children (Kottegoda, 2011). Most of the existing literature identifies the marginalized position of women, prevailing gender power relations, economic dependency, and lack of social and political power shape their disaster-related vulnerabilities (Nelson et al., 2010). A study conducted by Start Fund Bangladesh (2018) suggests that landslides heightened the risk of loss of personal goods and effects that preserve personal dignity, like clothing garments and hygiene items, also limit mobility for women and girls. Moreover, harassment and sexual assault in temporary shelters lacking adequate privacy and security particularly for young women and children, are major concern. In Bangladesh, gender power relations fix women's duties and responsibilities, and their access to, and control over, assets and property. Additionally, socio-cultural factors restrict women's mobility and potentials impacting various aspects of their lives and livelihoods. All these influencing factors contribute to women's vulnerabilities and influence their survival strategies, including how women are vital actors in country's disaster management related activities.

14.1.2 Method and Study Settings

The study was qualitative in nature and extracted from the author's Ph.D. research. The project was selected for an in-depth analysis of women's roles and contributions in formal disaster risk reduction program titled ***Community-Based Early Warning System (CBEWS) on Landslides*** from an urban periphery areas of Cox's Bazar district. Data and information collection encompass both secondary and primary sources involving separate Focus Group Discussions (FGDs) with Women Group (FGD.WG) and Men Group (FGD.MG), interviews with local level key actors (KII) and in-depth interviews of female volunteers (IDI). This community-based adaptation programme was selected for this study because it represented the first government initiative addressesing landslide hazards with the support of a development partner. Three urban communities of Jadirpahar, South Bharchara, and Mohazer Para of Cox's Bazar municipality implemented this programme were under the study.

14.1.3 Findings

14.1.4 Socio-demographical Aspects of Cox's Bazar

Cox's Bazar is a district of Chattagram division and the main sea port of Bangladesh having the world's longest (120 kms) natural sea beaches (Hassan & Shahnewaz, 2014). It is situated at the basin of the Bay of Bengal in the south and west; bordered by Chattagram district in the north, and Myanmar in the east. Cox's Bazaar district is comprised of islands, rivers, hills, and flat lands with area of 2491.86 km^2 (CDMP, 2014).

Demographic characteristics of Cox's Bazar are presented in the following Table 14.1.

Table 14.1 Demographic characteristics of Cox's Bazar district

Key indicators					
Population density	Female population	Male population	Extreme poverty ratio (%)	Average source of drinking water tube well (%)	Average source of drinking water other (%)
7579	1,120,386	1,169,604	16	85.70	10.23

Sources National statistical systems, Global Population Statistics, UNFPA Country Profiles, Population Stats, CIA World Factbook, WHO Country Statistics, World Bank Databank. Adopted from disaster summary sheet, 2018, start fund, Bangladesh

14.1.5 Development of Precarious Urban Settlements

Evidence suggests that many hilly slopes became the urban settlement for the migrated populations from the southern most islands of the country. My study also reveals that a large number of these settlers are mostly displaced from their homeland because of natural hazards like cyclones, tidal surges, sea-level rises, and river bank erosions. The residents of Jadirpahar, South Bharchara, and Mohazer Para communities of Cox's Bazar municipality describe the reason behind their settlement in such risky places:

> Our past generation had to move from our homelands due to frequent natural disasters like cyclones, tidal surges and river erosions which made their lives difficult. For homeless people, these hilly slopes were cheap lands for habitation. We have not got enough money to buy lands in plains to avoid these risky areas. Thus we struggle to survive in this unsafe place without minimal services for quality lives (FGD.WG1; FGD.MG2).

Frequent landslides are likely to result in huge population displacement from the hilly regions of Bangladesh in the near future (Displacement Solutions, 2012). A hazard like landslides is one of the outcomes of land use patterns and unplanned urban settlement from those displaced by climate change and who subsequently seek refuge in nearby urban areas (KII.2, 2014).

In most cases landslides take place suddenly without any forewarning. The landslides of these areas are mostly connected with heavy monsoon rain or approaching cyclones in the Bay of Bengal coast (ADPC, 2013). In addition, man-made activities such as illegal hill cutting, human settlement in the slopes, and continuous deforestation in Chattagram and Cox's Bazar pose a serious threat to life and the environment (IRIN, 2008). Due to a lack of awareness and understanding of the nature of pre-event symptoms, vulnerable populations cannot take precautions before the event occurs. This reduces the capacities of people who are at risk to respond in time and to make preparations to move to a safe place.

14.1.6 Landslide—A Snubbed Hazard in Bangladesh

Landslide has emerged as recent climatic hazard particularly in the Chattagram district of Bangladesh, where approximately 70,000 people at risk in 18 sub-districts of that region (IRIN, 2008). A study conducted by (CDMP, 2012a) claims that between 1999 and 2010, a total of 268 people were killed due to several landslides and in 2007 one landslide alone killed 127 people in Chattagram. Despite this alarming statistics, landslides have yet to be considered a significant hazard in Bangladesh and are excluded from the disaster risk reduction strategy of the country. The connection between climate-induced migration, urbanization and the emergence of a new type of hazards was not studied in Bangladesh until 2007. The Comprehensive Disaster Management Program (CDMP) started working on mapping urban hazards, vulnerability assessment, and loss estimation in the three cities of Bangladesh (Chattagram, Cox's Bazar, and Teknaf) in 2010 (CDMP, 2012a). This work helped to identify the vulnerability and intensity of this new type of urban hazard in the cities of the Chattagram division. Following in 2010 the Asian Disaster Preparedness Centre (ADPC) initiated a project called *Community-Based Early Warning System (CBEWS) on Landslide* focusing on disaster risk reduction in urban areas, particularly targeting landslides hazard in the two municipalities.

14.1.7 One of the first Initiatives to Address the Emerging Hazard

To address this emerging hazard, Government of Bangladesh, with the support of the Asian Disaster Preparedness Center, has initiated the community-based early warning systems in two urban municipalities, Cox's Bazar and Teknaf, as a part of landslides risk mitigation and development of community resilience. This mark one of the governments first initiatives and the project lasted for three years from March 2010 to May 2013 (ADPC, 2013, pp. 15–17).

During the study period, the respective municipalities and community volunteers are continuing the programme on a small scale as the support of the ADPC has been withdrawn. As an initial activity the project selected 60 (48 males, 12 females) volunteers from the selected communities and trained them. Three volunteers from each community (two boys and one girl) were trained to measure daily rainfall data, assess landslide probability, inform community people, maintain communication among other volunteers, and conduct landslide simulations. As part of the training the volunteers were introduced to local administration, city councillors, and relevant local government bodies to gain access to immediate support during any emergency. The project trained all councillors of the municipalities to raise awareness of the severity of the hazard and to engage local government bodies actively in risk reduction initiatives. Additionally, eight rain gauge metres were installed to monitor regular rainfall and four billboards were set up as part of behaviour change communication

(BCC) in developing community awareness. The activities were carried out in a systematically and coordination with the involvement of stakeholders. However, like many other poor urban settlements these communities have not been provided with minimum living standards facilities, so lack of proper sanitation and sources of drinking water were the main areas of distress. To address these problems (through municipality support for the communities) the project installed some tube wells and an improved sanitation system.

14.1.8 Outcomes of the Community-based Early Warning System (CBEWS)

The first activity of the project was to identify the hilly areas that were vulnerable to landslides triggered by heavy rainfall and to develop a community-based early warning system to manage the landslides. Second, the project aimed to provide a "model" landslide disaster risk reduction strategy for other hazard-prone cities of Bangladesh using its experience. Third, the project formulated a long-term landslide hazard mitigation strategy for Bangladesh by involving stakeholder institutions (CDMP, 2012c). Therefore, the programme mainly focused on identifying cost-effective methodologies and practices that can be adopted by the communities living with risks (ADP, 2010).

These efforts have helped to build community awareness and saved lives during landslides in both municipalities in June 2012, following heavy rainfall for five to seven consecutive days. 15–20 min after evacuation a major landslide occurred at Jadipahar, in the Cox's Bazar municipality. The evacuation process was a significant success for the community as a part of the project. The volunteers continue the activities even though the project has ended. During the rainy season, heavy rainfall is observed and volunteers visit the critical slopes areas several times. If volunteers observe consecutive rainfall above 150 mm they advise inhabitants to move to safer places and provide support to those who need help to evacuate (ADPC Assessment Report, 2013; project manager, personal communication, November 3, 2013). The programme was reasonably effective in involving local government bodies (as stakeholders) and received some support and services in return. Members of the community informed me that municipality councillors monitor the situation during heavy rain, give evacuation support, provide shelter, and supply food in evacuation centres (FGD.WG2). This ongoing link with the councillors has helped communities to access other services of the municipality.

14.2 Discussion

14.2.1 Urbanization, Hilly Settlement, and Vulnerable Community

The urban settlements under the study have grown over the decades due to a large number of the population migrating from their previous homelands as a way of adapting to climate-induced natural disasters. As the inhabitants of the community state:

> When the major parts of the islands Shonadiya, Kutubdiya, Sandwip, and Moheshkhali went under water due to tidal surges and river erosion, people started to move to nearby areas. Actually, this migration started a few decades back. Now the second generation of these migrant populations is living here. Over the years many more areas have gone under the sea but some new lands called 'char' have also emerged. We have heard that in the southern part of Maheshkhali a large 'char' (area of land) has emerged and can accommodate almost 20,000 to 30,000 people, whereas only 2000 to 3000 habitants are currently living there. The reason is that people do not want to go back to such places that can be inundated anytime and cause the inhabitants to lose their homes again (FGD.MG2).

However, the new settlements no longer can be considered safe. A programme coordinator from ADPC (KKI.1) has remarked that due to the increased population influx, rapid urbanization and activities such as hill cutting and settlement in hill slopes triggering landslides have become an environmental problem. Another research conducted by Ahmed (2017) found that since the government is unable to provide cheap and safer accommodation on flat lands with necessary community facilities therefore marginalized people illegally cut the hills for the development of residential houses to which are consequently making themselves vulnerable to landslides.

14.2.2 Gendered Responsibilities and Women's Vulnerabilities

There is a perceptible difference between urban and rural settlements in terms of women's roles and responsibilities. It is found that in urban settlements under the study, women are engaged in small-scale home-based informal economic sectors such as making hand made crafts, tailoring, and street vending, to support family income. Men are not always considered to be the main breadwinners of the family and female-headed households are common in this area which is utterly different from traditional rural context. As a result, traditional notions of gender division of labour are less prominent; rather women are doubly burdened with productive and reproductive roles. The following statements are from a group of women of Jadipahar Ward of Cox's Bazar municipality. When I asked about their daily activities they comment:

> All types of household duties such as cooking for family members, taking care of children and the elderly and fetching water are our responsibility. We need to walk long distances through the valleys to fetch water. Generally, in a day we need almost 30 L of water to meet our household water needs. It requires huge physical strength to collect such an amount of water every day. Our husbands cannot help us in fetching or carrying water as they work in the main town or beach side. Alongside household activities we do home-based economic activities such as sell crafts, tailoring and street vending and construction labour (FGD.WG2).

Thus urban women play vital roles in supporting the household economy. During the discussion with a women's group named "South Barchara Mohila Somitee" (FGD.WG3) in Cox's Bazar, the participants said that most of their group members are involved in home-based income-generating activities to support their families. Seashells are the main source of income since they get them from the sea without spending. They can run small craft-making businesses dependent on these natural resources without any capital. Their children collect seashells from the beach and they make ornaments, home-decorating pieces, and household materials. Several beach-side markets have flourished thanks to the resources of the Bay of Bengal where women are the main informal labour source. Local level informants reassure that women are the main labour force of informal economic sectors mostly home based such as dry fish and food preparation, cloth embroidery, and street food vending in this coastal town. According to a study by ADB (2010, p. 1) in Bangladesh women are predominant in the informal employment sector which makes up 89% of the total women's labour force. Hence, most of the women stay household chores, engage in home-based economic activities and stay inside the house putting them as a group at greater risk of sudden environmental hazards like landslide.

14.2.3 Disaster Risk Reduction: Awareness Raising and Capacity Building of Youth Volunteers

The strongest part of the CBEWS project is that trained young females volunteers have played a major role in raising awareness within the community. The findings supports Coleman and Mwingi (2013), who argue that women's visibility can bring positive impacts and generate positive attitudes and interest among women to engage in those activities that are not familiar to them. As my study found, the involvement of female volunteers can create a sense that the inclusion of women is vital to the success of any project. A female volunteer aged 20 years from the Jadipahar ward speaks about her involvement in the project:

> In 2010, along with four other boys, I was selected from our community as a volunteer for the project. I was chosen for my educational level and interest in working for the community. A three-member (one girl and two boys) volunteer team was formed in the selected wards of the municipality. We, the volunteers, received training on different climatic hazards like landslides, floods, cyclones and water logging as part of the project activities and we were taught how to deal with those. After receiving training, we started working on landslide awareness raising in our community. We trained them how to understand the first signs of landslides, when to disseminate the information to community and how to respond in a

timely way. We made them realize that we could save our lives if we moved to a safer place as early as possible (IDI.1).

Young female volunteers have expressed their interests in becoming future leaders of their community and maintaining their involvement in social activities. Sutton and Tobin (2011, pp. 899–900) found that younger people and females were more likely to engage in climate change adaptation and mitigation than the elderly and men, respectively. This project reached similar conclusions: young girls and women of the community have shown the most interest in working for a disaster resilient community.

Socio-geographic conditions also shape women's mobility and define gender roles and responsibilities. As an urban area, more women are visible in public spaces such as the market place, construction work, and informal labour sectors therefore, the overall societal attitude is not as conservative as in rural areas of Bangladesh. For this reason, female volunteers did not receive any negative response from the community when working with male volunteers. In addition, some NGO workers working in the community who are mostly women helped to shape the community's mindset about women's visibility in community activities. This finding indicates that there is a significant difference in gendered norms and values between the rural and urban societies of Bangladesh. A female volunteer from the South Baharchhara Ward narrates her experience:

We (girls and boys) usually work together during any type of disaster. We observe and recorded rainfall trends, accordingly disseminate warnings and evacuate. We have needed to rescue elderly people, children, pregnant women and people with disabilities many times. I did not get any resistance from the community as they saw we were doing work for its safety and local people know that no work can be done properly without women. That's why they support us and consider as their daughters or sisters. However, I feel that the project should train two female volunteers as male volunteers for better results (IDI.2).

14.2.4 Importance of Women's Inclusion in Community Disaster Preparedness Activities

My study findings strongly suggest that women's participation in community activities is controlled and shaped by social norms, gender division of labour, and gender power relations. This situation supports the findings of Cornwall and Gaventa (2001). Coleman and Mwingi (2013) connect women's socio-demographic status with women's engagement and involvement. Women mostly perform their community roles and duties within traditional norms and social constructs. This does not always necessarily result in a negative outcome. Ahmed (2012) argues that local practice and need-based adaptation can be more sustainable when both stakeholders (men and women) are included and can participate fully. The CBEWS project preferred to engage young girls who have fewer household responsibilities than married women or mothers. In addition, in urban areas women are more visible and it is more acceptable for them to be in public spaces as opposed to their rural counterparts. The project capitalized on this positive aspect by engaging young volunteers and the enthusiasm of youth became the driving force of the project (KII.3). Giving responsibilities to young people, particularly involving them in the development activities such as disaster risk reduction and climate change adaptation has a long-term impact for sustainable outcomes. The increased participation of women (both young and aged) will have a positive impact on the gender balance development of the future.

14.3 Conclusion

The structural discrimination rooted in women's biological, social, and economic conditions, renders them as more susceptible to environmental disasters compared to men. Empirical findings from the study areas reveal that these factors combine to shape distinct gender identities among women, which in turn influence how their roles and responsibilities are defined when it comes to responding impacts of to climate change or disasters (Tanjeela, 2016). Similarly, the different levels of engagement, involvement, and participation of women in the formal adaptation programmes hinges on their socio-economic status. Nonetheless, despite their limited participation in specific activities within the formal adaptation programmes, there has been a significant bolstering of household and community resilience. Moreover, their noteworthy contributions to community disaster preparedness and disaster management underscore the pivotal role young women play not only during disaster, but also in the preparatory phase and aftermath. Hence, it becomes imperative, to enhance the representation of women and ensure their active participation for the success of the programmes. The involvement of women in development endeavors represents a facet of their onging struggle for gender equality and signifies a gradual cultural shift from traditional patriarchal norms in society.

References

ADB (Asian Development Bank). (2010). *The informal sector and informal employment in Bangladesh: Country report–2010*. Philippines.

ADPC (Asian Disaster Preparedness Centre). (2013). *Assessment report of community based early warning system (CBEWS) on landslide: Cox's bazar and Teknaf municipality*. Dhaka, Bangladesh.

ADPC (Asian Disaster Preparedness Centre). (2010). *Concept note on community level landslide risk management in Chittagong*. Dhaka, Bangladesh.

ADPC (Asian Disaster Preparedness Centre). (n.d.). *Training manual on community based early warning system on landslide*. Dhaka, Bangladesh.

Ahmed, N. (2012). *Gender and climate change in Bangladesh: The role of institutions in reducing gender gaps in adaptation program*. Social Development Papers, a s\Summary of ESW Report no. P125705, Paper no 126, The World Bank.

Ahmed, B. (2017). *Community vulnerability to landslides in Bangladesh*. This thesis is submitted for the degree of Doctor of Philosophy (Ph.D.) at University College London (UCL).

Anwarul, A., Habiba, U., & Shaw, R. (2013). Gender and climate change: Impacts and coping mechanisms of women and special vulnerable groups. In S. Rajib, M. Fuad, & I. Aminul (Eds.), *Climate change adapatation action in Bangladesh* (pp. 165–186). Springer.
Birkmann, J., Garschagen, M., Kraas, F., & Quang, N. (2010). Adaptive urban governance: new challenges for the second generation of urban adaptation strategies to climate change. *Sustainability Science, 5*(2), 185–206.
CCD (Commission on Climate Change and Development). (2009). Closing the gaps—Disaster risk reduction and adaptation to climate change in developing countries. CCD Stockholm.
CDMP (Comprehensive Disaster Management Programme II). (2012a). Report on the landslide inventory & land use mapping, DEM preparation, precipitation threshold value and establishment of early warning devices. Ministry of Disaster Management and Relief, Government of Bangladesh (GoB).
CDMP (Comprehensive Disaster Management Programme II). (2012b). Report on the landslide hazard zonation mapping at Cox's Bazar and Teknaf in Bangladesh. Ministry of Disaster Management and Relief, Government of Bangladesh (GoB).
CDMP (Comprehensive Disaster Management Programme II). (2012c). Final report on the rainfall triggered landslide hazard zonation mapping at Cox's Bazar and Teknaf municipalities as well as introducing community-based early warning system for landslide hazard management. Ministry of Disaster Management and Relief, Government of Bangladesh (GoB).
CDMP (Comprehensive Disaster Management Programme II). (2014). Trend and impact analysis of internal displacement due to the impacts of disaster and climate change. Ministry of Disaster Management and Relief, Government of Bangladesh (GoB).
Coleman, A., & Mwangi, E. (2013). Women's participation in forest management: A cross-country analysis. *Global Environmental Change, 23*(1), 193–205
Cornwall, A., & Gaventa, J. (2001). *From users and choosers to makers and shapers: Repositioning participation in social policy.* IDS Working Paper no.127, Institute of Development Studies, Brighton, UK.
Cutter, L. S., Boruff, B. J., & Shirley, W. L. (2003). Social vulnerability to environmental hazards. *Social Science Quarterly, 84*(2), 242–261.
Dodman, D., & Satterthwaite, D. (2008). Institutional capacity, climate change adaptation and the urban poor. *IDS Bulletin, 39*(4), 67–74.
Enarson, E. (2000). Gender and natural disasters. IPCRR Working Paper no.1. International Labour Organisation. http://www.ilo.org/public/libdoc/ilo/2000/100B09_280_engl2.pdf (application/pdf)
Enarson, E., & Fordham, M. (2001). From women's needs to women's rights in disasters. *Environmental Hazards, 3*(3), 133–136.
Gray, C. L., & Mueller, V. (2012). Natural disasters and population mobility in Bangladesh. *PNAS, 109*(16), 6000–6005.
Hassan, M., & Shahnewaz, M. (2014). Measuring tourist service satisfaction at destination: A case study of Cox's Bazar sea beach. *American Journal of Tourism Management, 3*(1), 32–43.
IRIN (International Regional Information Networks). (2008). Bangladesh: 70,000 people vulnerable to landslides. News Report, July 23.
Jabeen, H., Johnson, C., & Allen, A. (2010). Built-in resilience: Learning from grassroots coping strategies for climate variability. *Environment and Urbanization, 22*(2), 415–431.
Kjekstad, O., & Highland, L. (2009). Economic and social impacts of landslides. In K. Sassa & P. Canuti (Eds.), *Landslides—Disaster risk reduction* (pp. 573–587). Springer.
Kottegoda, S. (2011). Mainstreaming gender in disaster management policy: Key issues and challenges in the Asia-Pacific region. http://www.apww-slwngof.org/index.php?option=com_content&view=article&id=88:mainstreaming-gender-in-disaster-management-policy-key-issues-and-challenges-in-the-asia-pacific-region&catid=88&Itemid=592
Mahmood, B., & Khan, H. (2008). Landslide vulnerability of Bangladesh hills and sustainable management options: A case study of 2007 landslide in Chittagong City. International Seminar on *Management and Mitigation of Water Induced Disaster*, Kathmundu, Nepal, April 28.

McLeman, R., & Smit, B. (2006). Migration as an adaptation to climate change. *Climatic Change, 76*(1–2), 31–53.

Nelson, G. C., Rosegrant, M. W., Palazzo, A., & Gray, I. (2010). *Food security, farming and climate change to 2050: Scenarios, results, policy options.* International Food Policy Research Institute.

Neumayer, E., & Plümper, T. (2007). The Gendered nature of natural disasters: The impact of catastrophic events on the gender gap in life expectancy, 1981–2002. *Annals of the Association of American Geographers, 97*(3), 551–566.

Penning-Rowsell, E., Sultana, P., & Thompson, P. (2012). The 'Last Resort'? Population movement in response to climate-related hazards in Bangladesh. *Environmental Science & Policy, 27*, 1–16.

Petley, D. (2012). Global patterns of loss of life from landslides. *Geology, 40*(10), 927–930.

Roth, M. (2009). Effects of cities on the local climate and the relationship with climate change mitigation and adaptation. *IOP Conference Series: Earth and Environmental Science, 6*(33), 332032.

Shamim, I. (1995). Women and environmental disaster: Riverine erosion and displaced women as manager. In Jahan, R., et al. (Eds.), *Environment and development: Gender perspective.* Women for Women.

Sharma, D., & Tomar, S. (2010). Mainstreaming climate change adaptation in Indian cities. *Environment & Urbanization, 22*(2), 451–465.

Solutions, D. (2012). *Climate displacement in Bangladesh: The need for urgent housing, land and property (HLP) rights solutions.* Switzerland.

Start Fund, Bangladesh. (2018). Bangladesh land slide, disaster summary sheet, June 05.

Sutton, G., & Tobin, C. (2011). Constraints on community engagement with great barrier reef climate change reduction and mitigation. *Global Environmental Change, 2193*, 894–905.

Tanjeela, M. (2016). *Untold stories: Women and climate change adaptation in Bangladesh*, Ph.D. Thesis, Centre for Public Policy, Department of Government and International Relations, Griffith University, Australia.

Chapter 15
Dhaka: A City of Development and Deprivation for Environmentally Induced Migrants

Sufia Khanom, Fiona Miller, and Sandie Suchet-Pearson

Abstract Dhaka, the capital of Bangladesh, is a city defined by inequality. It is not one city, but many. There is a city of high-rise buildings, shopping malls, and restaurants. There is also the city where people live on the streets and in informal settlements or slums (*Bastee*). This highly visible inequality, indelible in the physical and social geography, is largely the result of the dynamic push–pull factors that shape migration. This migration, together with the rapid pace of economic growth and environmental degradation, has pushed the boundaries of sustainability and liveability, creating the conditions of perceived and actual (in)security in the city. This chapter provides a context for people's everyday insecurities by discussing how the environmentally induced migrant's lives are shaped by unplanned urbanization. The objective of this chapter is to lay out the conditions of everyday and long-term (in) security experienced by these migrants in Dhaka city. This chapter analyses how Dhaka City is attracting rural environmental migrants due to the employment opportunities created by Bangladesh's role in the global economy as a source of cheap labour in the textile and construction industries. These new urban dwellers seek out shelter in abandoned places of the capital and urban planners have yet to exploit and enclose. The chapter consists of four main sections: an historical perspective; outlining the planning trajectories of Dhaka under different colonial and post-colonial periods; the construction of Dhaka as a place of hope and frustration; and critical analysis of the growth of informal settlements. The chapter concludes by explaining that the uneven geography of urban growth in Dhaka is one where development and (in) security co-evolve and indeed reflect the inequalities embedded in development.

S. Khanom (✉)
Bangladesh Institute of International and Strategic Studies (BIISS), Dhaka, Bangladesh
e-mail: sufia@biiss.org

F. Miller · S. Suchet-Pearson
Department of Geography and Planning, Macquarie University, Syndey, Australia
e-mail: fiona@mq.edu.au

S. Suchet-Pearson
e-mail: sandie.suchet@mq.edu.au

B. R. K. Sinha (ed.), *Urban Dynamics, Environment and Health*,
https://doi.org/10.1007/978-981-99-5744-6_15

Keywords Informal settlements · Environmentally induced migrants · Push–pull factors · Disaster · Historical planning · Rising cost of living

15.1 Introduction

Dhaka is a city defined by inequality. It is not one city, but many. There is a city of high-rise buildings, shopping malls, and restaurants. There is also the city where people live in skyscrapers, luxurious apartments, on streets, and in informal settlements or slums (Bastee). This highly visible inequality, indelible in the physical and social geography, is largely the result of the dynamic push–pull factors that shape migration. This migration, together with the rapid pace of economic growth; environmental degradation; and other socio-political and bureaucratic factors, have pushed the boundaries of sustainability and liveability, creating the conditions of perceived and actual (in)security in the city.

The interplay of disasters due to global environmental change and industrialization in and beyond the city acts to push and pull people towards the city, with migrants attracted because of well-paid job opportunities, health, and other daily life facilities and better educational institutions, and pushed away from their homelands due to increasing environmental changes and disasters (Islam, 1996). People have been migrating to Bangladesh for 36 years due to the climatic variabilities (Saha, 2012) and it is still one of the prime reasons for people migration towards cities. This pattern of migration, development, and urban governance has contributed to (in)security for certain groups of people who experience persistent urban poverty, pollution, traffic jams, socio-economic problems, and exposure to environmental hazards.

Nearly about eight billion people are already living in cities and towns by 2050 which will be five billion people and will share about 66% of global population (UN DESA, 2014). Global North has set up the strategies for sustainable urbanization and shrinking urban share of global population. On the other hand, Asian cities will dominate to increase the number of urban population. However, the most worrying aspect of urban growth in Global South is that the rapid growth of urbanization will increase the number of informal settlements (slums) in Asia, Africa, and to some extent in Latin America. Deprivation and poverty are very common for the poor informal settlement residents. Therefore, it will further challenge the global poverty eradication goals (Hulme, 2015). Many poverty lines indicators do not recognize the informal settlement dwellers recognize as "non-poor". Unfortunately, urban poverty is not as simple as low income but the various aspects of deprivation which are not considered by these poverty line indicators. Informal settlement dwellers can accumulate assets, increase their wealth, highly mobile but live in the precarious conditions which make them vulnerable to any shock.

In Bangladesh, the majority of the urban population of the country is concentrated in Dhaka. More than 60% of the population has migrated from the other parts of the country. It is one of the eighth fastest growing megacities in the world with a population of 16 million people. The city is projected to have more than 20 million people by 2025 and will be the third largest city in the world (World Bank, 2007). Thus, the physical characteristics of Dhaka city are changing drastically due to rapid processes of urbanization, as in other countries of the Global South.

In this chapter, we provide a context for people's everyday insecurities by discussing how their lives are shaped by unplanned urbanization, the limited capacities of urban infrastructure and service sectors, disasters, and other social vulnerabilities. The main objective of this chapter is to lay out the conditions of everyday and long-term (in)security experienced by the environmentally induced migrants in informal settlements in Dhaka. The chapter analyses how Dhaka city is attracting rural environmental migrants due to the employment opportunities created by Bangladesh's role in the global economy as a source of cheap labour in the textile and construction industries. These new urban dwellers seek out shelter in abandoned places that capital and urban planners have yet to exploit and enclose. The chapter consists of four main sections. Following the introduction, a historical perspective is provided, outlining the planning trajectories of Dhaka under different colonial and post-colonial periods. In the third part of the chapter, the construction of Dhaka as a place of hope and frustration is discussed before critically analysing the growth of informal settlements in the fourth and final section. The chapter concludes by explaining that the uneven geography of urban growth in Dhaka is one where development and (in)security co-evolve and indeed reflect the inequalities embedded in development.

15.2 Urban Planning in Dhaka City: A Historical Perspective

Although Dhaka city has a long history, it does not have a long history of urban planning. Different rulers initiated different governance, structures, and administrative steps to defend the city from natural calamities and poverty over time. Compared to the rest of the country, Dhaka was considered relatively high above sea level, and rural settlement was confined to the area between the Buriganga River and Dholai Khal (canal) until the sixteenth century. The city experienced many changes in its expansion and development under different rulers in the last 400 years. Urban planners have identified five distinct phases in Dhaka's transition from a city to megacity.[1] The following section gives a brief historical overview of significant planning initiatives that have shaped the city's form, urban character, and socio-economic conditions. Planning has been strongly shaped by the ruler's political and strategic choices rather than the common interests of the city's residents.

[1] The term "megacity" is frequently used as a synonym for words such as super-city, giant city, conurbation, and megalopolis. There has been little agreement about the size of the megacity. Megacities are defined as cities with at least ten million inhabitants by the year 2000 (World Bank, 2000).

The demand for the housing sector in Dhaka has encouraged influential people to invest in the real estate sector for huge returns. One study shows that land developers can easily make 6.7 times profit from their investments (Alam, 2014). Weak legislative authority and an absence of controlling powers by the relevant authorities are also responsible for promoting these types of investments. About 200 land developers are developing some 300 land and housing projects inside RAJUK territories (Alam & Ahmad, 2011). It has been estimated that about 700 billion BDT (approx. 11.315 billion AUD) has been invested into the housing sector, of which around one-third is foreign investment and about 20–25% came from the illegal money (Alam & Ahmad, 2011). Only the rich people can afford housing in Dhaka as the government's investment is very limited in housing projects. So, poor migrant people, the majority of the city dwellers remain excluded from their own houses.

Over the years, different authorities have shared their responsibilities for administration, planning, law, and utility services, which ultimately led to non-cooperation and undue exercise of power by certain authorities (Dewan et al., 2012). These power imbalances among different authorities at the same level create the possibility of high-level corruption in the public authorities in urban governance. A fundamental shift in the culture of the governance system is required, along with training on digitization of land records and greater investment (Ahmed & Bramley, 2015) to reduce inequalities in the distribution of resources and to reduce the associated corruption associated. The recent dissolution of Dhaka city into north and south will further complicate the situation by engaging different agencies under different ministries for planning and implementation in each jurisdiction. The following section discusses the effectiveness and functional efficiency of the key systems with special emphasis on the informal settlements.

15.3 Push–Pull Factors and Growth of Uneven Geography of Dhaka

Dhaka was a small and beautiful city of only 3.5 million people in the 1950s, with eye-soothing greenery and unpolluted air. It has now emerged as one of the megacities of the world with a population of more than 16 million (World Bank, 2007). This rapidly growing population contributes to the rapid deterioration of the city and the environmental quality of the surrounding hinterlands as well as rising competition over resources. An inability to effectively and equitably respond to this in planning and governance further affects the functionality of the urban system and undermines people's quality of life. This section focuses on the push–pull factors that have contributed to the growth of the city, social vulnerabilities, and the environmental changes associated with the wider context. The functionality of the urban system is discussed below, not from an urban planning perspective, but as it affects people's everyday lives at an individual level.

The impacts of climate change and exposure to severe "natural" disasters contribute to a huge portion of people being displaced, both temporarily and permanently. It is estimated that floods, droughts, and cyclones have displaced 47 million people and are expected to displace about 78 million by 2020 (Bhuiyan & Siddiqui, 2015, p. 15). Some 57.9 million people in the coastal areas are projected to be particularly susceptible to various "natural" disasters by 2050 (Akter, 2009). The proportion of people living under the poverty line is higher in the coastal areas than in other parts of the country (Roy et al., 2013), with environmental degradation being one of the main reasons identified for this higher incidence of poverty. A combination of a lack of capacity to prepare for disasters and failure to reduce risk, a lack of livelihood resources, and population pressures have resulted in population displacement towards the city. Figure 15.1 shows the poverty line at national, rural, and urban levels between 1991 and 2025.

Dhaka city has had one of the highest growth rates of any city in the world in the last decade, at around 1.7% per annum, resulting in a population density of 1,000 people per square kilometre (World Bank, 2007). The rapid urbanization of Dhaka city is linked with rural displacement and massive migration of the rural population towards Dhaka in search of better livelihood opportunities. The form migration takes varies along the forced-voluntary axes. A pilot study conducted in 62 villages of Bangladesh found that of total migration flows, about 75% is rural–urban, 10% rural to rural, and 24% overseas migration (Bhuiyan & Siddiqui, 2015).

In contrast to these so-called push factors, the economic development of the city is another significant factor underpinning rapid urbanization in Dhaka. The Gross Domestic Product (GDP) of Dhaka city was 11,312 million BDT in 1975 and 345,240 million BDT in recent years (Afrin et al., 2015). According to the World Bank (2007), the share of urban activity in the GDP rose from 26% in 1972–73 to 42% by 1988–89 (Mizanur, 2011). The population density in urban areas is (1800 people per square

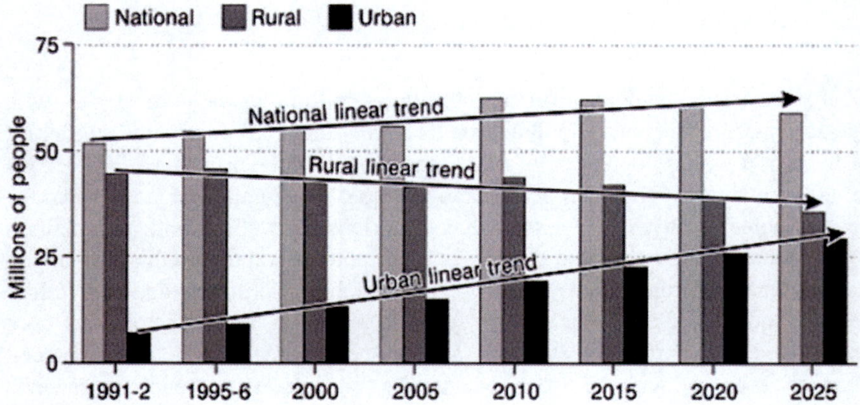

Fig. 15.1 Rural–urban population growth below the poverty line 1991–2025. *Source* Banks et al. (2011), p. 488

kilometre) in less than half in rural areas (800 people per square). However, the urban economic density (US$2.7 million per square kilometre) is eight times higher than the rural areas (US$ 32,000 per square kilometre). Therefore, the GDP per capita in urban areas (US$ 1500) is four times higher than that of rural areas (US$ 400) (World Bank, 2012, p. 164). Urbanization has a positive co-relation with the national GDP growth. City and towns play the most important role for national development through contribution to GDP. Therefore, the agricultural sector in rural areas gets prioritized over industrial and service sectors by the policy makers. Although, poverty alleviation programmes such as vulnerable group funds and others are implemented mostly in rural areas. There has been only moderate improvement in poverty reduction overall, with the number of people below the poverty line slightly decreasing in rural areas but noticeably increasing in urban areas. One can conclude that the future face of poverty is urban. Much of the wealth of the country now comes from urban rather than rural economic, activities, yet the question remains as to who is sharing in this wealth?

The majority share of GDP comes from garment industries located in and around Dhaka. About 80% of these are situated within and around the city, where more than two million people are working (Hossain, 2014). The massive development associated with the garment industries and subsidiary industries occurred within a very short time frame, attracting many people, especially women, from rural areas. Dhaka contributes about 40% of the industrial development of the country (Dewan et al., 2012), yet the benefits are not experienced evenly across society or the city.

The garment industries have spontaneously grown mainly along the major accessible routes of the city, developed through different historical stages of urban development (Hossain, 2014). During the pre-Mughal and Mughal periods, the garment (mainly muslin cloth) and other small industries developed near the Buriganga and Shitalakhya rivers due to easy transportation (Islam, 1996). Industrialization continued towards the north along the road and railway networks during the British and Pakistan periods and the core commercial areas of the city shifted, resulting in the typical concentration of this industry along "globally integrated and major thoroughfares of the city" (Hossain, 2014, p. 32).

This rapid and unplanned growth resulted in the conversion of agricultural land to industrial areas, especially along the peripheral areas. It is estimated that about 809 km^2 of agricultural land was converted annually into urban land over the last 20 years. This trend has also discouraged people from engaging in agricultural activities due to the low returns compared with income from other industries and the concomitant loss of agricultural land is increasing landlessness and the price of land. The surrounding residential areas of the urban periphery have experienced the highest growth in land prices, estimated to be about 12,000% between 1947 and 1965 (Islam, 1996). Other areas of the city have also experienced land price increases, up to 2,421% during the same period (Byomkesh et al., 2012). As a result, Dhaka city shows a mixed pattern of land use.

Apart from the garment industries, developers have converted about 10,128.33 hectares of land in the flood zone and natural water retention lakes to other uses, contrary to DMDP policy (Alam, 2014; Alam & Ahmad, 2011). This activity exacerbates urban flooding in Dhaka city and increases dependency on expensive structural measures for flood prevention such as embankments.

After the devastating flood of 1988, the Dhaka Development Authority constructed a road-cum-embankment from the middle of Dhaka towards the east and west, with assistance from the Japan International Cooperation Agency (JICA) (Bala, 2009). These development initiatives increased land encroachment inside the retention ponds, filling up the lowlands, and increasing conflicts over land. Prices for land inside the embankment have increased compared with land outside the embankment (BBS, 2010). As a consequence, the city's elites and the wealthy have gained more control over the land protected by this infrastructure. Poor migrants were further excluded from accessing urban land and were confined to exposed informal settlements. They have less freedom of choice in accessing key essential services and lack basic protection from flooding compared with richer residents, highlighting that urban flooding is an unnatural disaster. The following section discusses the key politics of growth of informal settlements as a failure of unsustainability of urban functionality in the affecting everyday lives of poor migrant people.

15.4 Growth of Informal Settlements: Understanding the Politics

In the so-called developing world, increasing inequality between cities of different sizes and economic specialization contribute to the creation of informal settlements. As Davis writes,

> If megacities are the brightest stars in the urban firmament, three-quarters of the burden of future world population growth will be borne by faintly visible second-tier cities and smaller urban areas: places where there is no or little or no planning to accommodate these people or provide them with services (Davis, 2006, p. 7).

Urban planning and growth tend to prioritize the interests of rich people and their well-being as discussed earlier in this chapter. Yet, urban growth depends on poor people's participation and contribution. An estimated two million workers are employed in the readymade garment industries in Dhaka and contribute about 85% of the country's total GDP. Most of the workers in this industry are rural migrants who live in informal settlements. Just 10% of urban housing is publicly owned and composed of low-rise buildings on expensive land (Islam, 2014) The ineffectiveness of national housing planning initiated by the government leads to the growth of informal settlements, accounting for up to 70% of the urban growth of Dhaka (World Bank, 2007, p. xv). The growth of informal settlements aligns with the increasing trend of industrialization and remains neglected by urban planners with regard to accommodating and servicing poor migrants.

Urban migrants initially concentrate on informal settlements as they are poor and lack alternatives (Hossain, 2013). Considering the high number of informal settlements, and the role poor urban planning has played in contributing to their emergence, the next section considers the politics of their growth, and who and what benefits from their presence.

The Washington Consensus emanating from the Bretton Woods Institutions (the World Bank and International Monetary Fund) led to an emphasis on neoliberal policies throughout much of the so-called developing world (Davis, 2006). Bangladesh was no exception, and successive regimes have emphasized industrialization in the cities, agricultural deregulation, and financial austerity. Davis used the term "overurbanization":

> Rather than the classical stereotype of the labour-intensive countryside and the capital-intensive industrial metropolis, the Third World now contains many examples of capital-intensive country sides and labour-intensive deindustrialised cities. "Overurbanisation," in other words, is driven by the reproduction of poverty, not by the supply of jobs (2006, p. 16).

All large South Asian cities have experienced huge growth of urban informal settlements since the 1970s. It reflects the adverse impacts of Structural Adjustment Program (SAP) on their economies. The following figure shows how fast the informal settlements of South Asian cities have grown (Fig. 15.2).

Like other developing countries, Bangladesh adopted the SAP initiated by IMF and the World Bank to keep pace with the neoliberal economic order in the mid-1980s. The country responded very positively to the call for globalization by liberalizing its trade policies and welcoming foreign investors to participate in the trillion-dollar global trade as a resource-poor nation. From 1980 to 1995, the country received a total of US$1.7 billion in financial assistance (Battacharya et al., 1999). The SAP targeted almost all sectors of Bangladesh, including agriculture, energy, manufacturing, finance, and trade, with a goal of economic growth and creating a market-oriented economy using the competitive advantages of cheap labour (Bhattacharya & Titumir, 1998). These initiatives prioritized economic growth while ignoring adverse social and environmental impacts.

The outcome of the SAP in Bangladesh has been reported as mixed (Bhattacharya & Titumir, 1998; Rahman, 1992). The measures are taken by

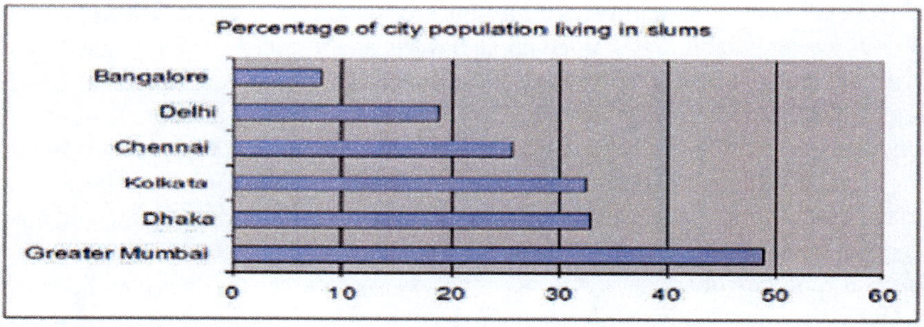

Fig. 15.2 Estimated proportion of population living in the informal settlements in some Indian cities and Dhaka city. *Source* World Bank (2007)

the SAP included price decontrol, tariff rationalization and overall trade liberalization, privatization, and de-subsidization, with the objective of reforming macroeconomic policies. These policies contributed to the exodus of migrant labour towards the cities from the rural areas and the removal of local safety nets. Rapid urban growth, currency devaluation to attract foreign investors, and state retrenchment of public service facilities aided rapid urbanization and the mass production of informal settlements (Davis, 2006). In Dhaka, Bangladesh also experienced massive growth of informal settlements, like many other developing countries in the world.

However, not only neoliberal macroeconomic policies initiated by the government and environmental factors have contributed to putting environmental migrants at the edge of urban development, but also the polarized land use pattern of the colonial and post-colonial periods. During the colonial period, the British powers dominated and dictated land use pattern according to their needs. Even after liberation from colonialism, local elites reproduced segregated zoning to defend their own class privileges and exclude others spatially (Kabir & Parolin, 2012). After the discussion on global economic policies influenced by the international organizations and colonial footprints on the urban development over the years, the following section will discuss the urban poverty situation to get an overview of the environmental migrant's existing conditions in the informal settlements of Dhaka.

15.5 Political Economy of Poverty: Urban Informal Settlement Perspectives

Poverty has a significant influence over the everyday state of insecurity which is known as "freedom from want" as human beings (Sen, 1990). There are ongoing debates on the definitions and measurements of poverty (Banks, 2015; Mitlin & Satterthwaite, 2013). In the urban context, poverty is largely conceptualized as the deprivation of resources (both monetary and non-monetary) accumulation. In doing so, the poverty assessment approaches do not capture the existing potential assets and power politics around those assets. If the power politics around the assets could be reduced their vulnerability to future poverty would be reduced (Banks, 2016). Various research findings identified urban poverty as a result of: lack of financial and physical capitals; limited skills and education; chronic illness, occupational health hazards, and malnutrition; sub-stander, poorly maintained, and insecure tenure; and unhygienic sanitation and hazardous living spaces (Baud et al., 2009; Moser, 1998). The local political-economic dimensions of household's various assets and livelihood strategies that control the access to resources are also identified as significant factors for poverty analysis (Banks, 2014). The main objective of this section is to understand political economy of access and terms of access to resources for their livelihoods.

Generally, there are two different approaches to measuring the poverty—absolute and relative poverty. According to Banks (2016), relative poverty is the best approach to understand the urban poverty whereas the absolute poverty approaches are widely used for assessing and monitoring official and formal practices. For example, poverty lines are drawn based on the price of 11 key food items (known as food bundles for minimum nutritional intake) and their consumption levels (BBS, 2004). This assessment considers the need rather than the internal power politics of accessing the resources for enhancing their capabilities. Therefore, the poverty reduction initiatives fail to resolve the problem in the long run.

Political inequalities create, maintain, and exacerbate poverty in the settlements of urban areas (Banks, 2016; Hossain & Hackenbroch, 2018; Wood, 2000). These inequalities allow a small elite group to control and access over the resources. These elite groups also set conditions or criteria for the low-income urban informal settlement dwellers to the most valuable social resources that are recognized as future social and economic pathways for them. Banks (2015) criticized the urban poverty analysis for borrowing the concepts from the rural poverty analysis such as livelihoods framework (Carney, 2001), vulnerability, assets (Moser, 1998), and capitals (Bebbington, 1999). Again, actor-oriented framework for poverty reduction approaches also overlooks the political dimension of poverty (Wood & Salway, 2000). So, the expansion of freedom of choices regarding coping strategies of informal settlement dwellers reduces the vulnerability of getting poor day by day.

High levels of casualized and informal labour; influences of intermediaries on the labour markets; limited absorptive capacities of the formal markets are the main obstacles to better employment outcomes of urban poor (Mitlin & Satterthwaite, 2013). Though, the individual and household agencies struggle to exclude and include from the labour market under the social, economic, and political structures. Therefore, the environmental migrants get limited opportunities to get rid of the adverse incorporation of patron–client relationships.

Most environmental migrants are poor as the environmental disaster hard hit upon the poor (Saha, 2012). As an urban poor, environmental migrants have rarely connected with the pro-poor urban governance system and dynamic civil society who would work for their advancement (Banks, 2015). Local people representatives (such as ward commissioner) and other government and non-government officials are almost not connected directly with the informal settlement dwellers. They maintain connections with politically affiliated local leaders to maintain the relationships. They work as a gatekeeper for both the urban migrant community and urban governance officials. These local political leaders are also connected with strong and young people (locally known as mastaans) to implement their wills. The ward commissioners consider the informal settlement dwellers as "vote bank" for elections and enforce their will by the gatekeepers by providing alternative/temporary forms of service provisions.

Urban poor are pushed back to the bottom of the urban governance system because of having no direct mandates for them. So, the limited allotted resources from the urban governance system have to pass through the several layers in the power hierarchy to reach the poor living in informal settlements. In this resource allocation processes, the interests, power, and authority of the local leaders and mastaans are protected by the government as fear the opposition party to control the urban areas which might damage their chance to re-elect. Government minimizes their risk through the chain of command among ward commissioners and local political leaders to manage and control the urban informal settlements. Environmental migrants are unable to assert their rights for formal housing and service provisions and are dependent on those gatekeepers. These "patron-client relationships are based on the economic structures of exploitations, political structures of domination, and ideological structures of consensus and control" (Lewis, 2011). The material benefit, they offer for their survival, further reinforces political and economic inequalities and strengthens the relationships among governance representatives, political leaders, and urban poor vertically (Mitlin & Satterthwaite, 2013). Migrants negotiate for their immediate security needs at the costs of long-term dependency on political leaders. The following section focuses on the location of the urban informal settlements where most of the environmental migrants in Dhaka which reveal their environmental vulnerability and abandonment by the urban governors.

15.6 Existence at the Edges: Life in Informal Settlements

Urban informal settlements are perceived as "human dumps" where urban waste and unwanted migrants end up (Davis, 2006). Similarly, the Bhola settlement is also located beside an open sewerage drain that captures all the wastes of the surrounding residential areas. Environmentally induced migrants' living conditions are overlooked by the authority and compel them to stay beside this open drain. The following figures show the location of Bhola settlement beside an open sewerage drain that should be maintained or covered up by WASA (Fig. 15.3).

Local elites, politicians, and private land developers positioned themselves as guardians of the peripheral or unused land in the urban centre for sale or rent to the poor migrants. The perceived "free land" for the poor is converted to a place for an urban latifundia[3] and crony capitalism. Therefore, poverty is not the only root cause of the growth of informal settlements, but also the uneven and unfair distribution of resources (Geeta Verma cited by Davis, 2006). Most informal settlements are located in environmentally vulnerable areas, at the precarious edges of land, between toxic factories or beside poisoned lakes and waterways. They are at the edges, in between places, in the shadows. These places are so hazardous and unattractive that they do not have the possibility of rising in value in the near future. They lack economic value in the global economy and are used by rich elites as political capital to profit from the settlement of poor migrants. Poor migrants (especially environmentally-displaced migrants) have little choice but to live with these hazards, replacing one set of hazards in the rural areas, with a new set of hazards in urban areas. The following figures show the location of the informal settlements and environmentally hazardous zones of Dhaka. They show that informal settlements are mostly located disaster-prone areas in Dhaka which make the lives of city dwellers more vulnerable.

[3] It is a Latin word. In the ancient Roman period, latfundia meant the large pieces of the agricultural estate that used many slaves or labours for production.

Fig. 15.3 Location of Bhola settlement beside WASA open drainage which represents only one of 6,400 informal settlements (or slums) in Dhaka. *Source* Ph.D. fieldwork, October 2016–April 2017

Informal settlements formed not only because of the SAP, but also because of the negligence of urban governance authorities. No provision was made for housing facilities for the urban poor. They were continuously abandoned by urban planners and started to settle in the disaster-prone areas of the city.

From an environmental restoration perspective, the city needs an alliance with the surrounding green matrix of intact ecosystems with open space and water bodies for its natural services and to recycle waste products (Geddes, cited in Davis, 2006). This waste can be a useful input for farming, energy production, and gardening for the sustainable growth of the city. Unfortunately, as has occurred in other developing countries, Dhaka is systematically polluting and destroying its crucial environmental support system in the name of economic growth. For example, Bhola has been developed by filling a water reservoir known locally as Pallabi jheel.

There is a saying that all the informal settlement dwellers are not poor, and the poor do not only live in the informal settlement. Most studies on informal settlements ignore "landlordism" or "invisible renters" (Davis, 2006) as a fundamental and divisive social relation in informal settlement life. Unlike other Asian cities, most of the settlement dwellers in Dhaka rent land from landlords or powerful elites (Ahmed & Johnson, 2014; World Bank, 2007). Tenants are usually scattered through the settlements and are often unable to exert sufficient pressure to gain access to service facilities. Because they must engage in a wide range of informal agreements with these landlords, the settlement dwellers are heavily dependent on landlords or local political elites for their existence. Therefore, they have to position themselves as political benefactors and become implicated in corrupt party machines (Piliavsky, 2014). Informal settlement development and upgrading are synchronized with the election cycle. For example, in the Bhola settlement, election years provide dwellers with leverage to develop their settlements and obtain commitments for rehabilitation (Figs. 15.4 and 15.5).

To better understand the extent and numbers of people living in informal settings, the first informal settlement survey was conducted in 1974 by the Centre for Urban Studies (CUS) and found that 275,000 people lived in informal settlements (Islam, 2005). Another survey conducted by CUS on 2,156 settlements on behalf of the International Centre of Diarrhoeal Disease and Research (ICDDRB) in 1991 found a total population of 718,143 (Islam, 2005). The Asian Development Bank (ADB)

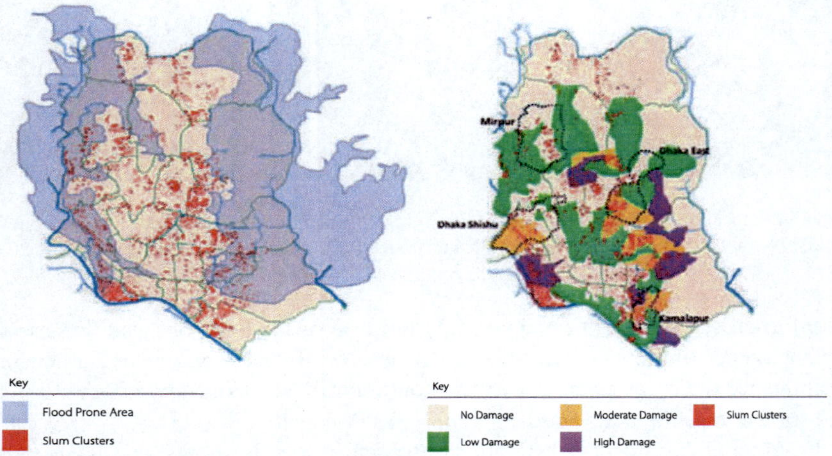

Fig. 15.4 Informal settlements and flood-prone areas. *Source* Stott and Nadiruzzaman (2014)

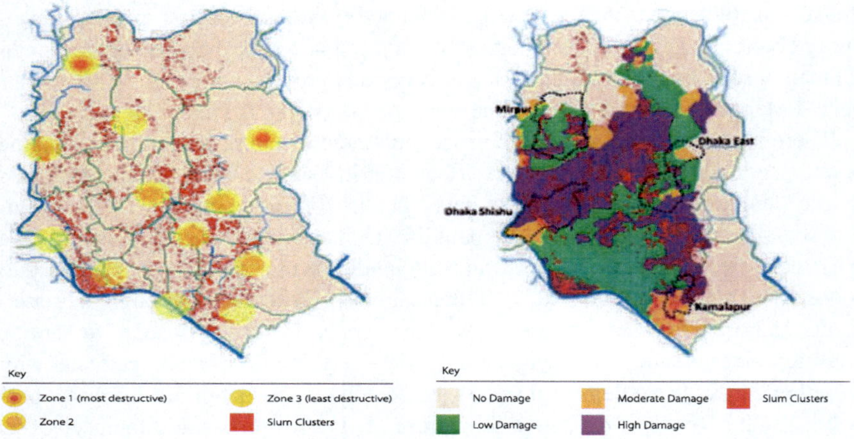

Fig. 15.5 Liquefaction zones of Dhaka. *Source* Stott and Nadiruzzaman (2014)

conducted another survey with CUS in 1996 on 3,007 informal settlements and found 1.5 million informal settlement dwellers.

The urban poor in Bangladesh remained largely invisible, statistically, until 2005. The first slum (informal settlement) census captured the size, diversity, and rapidly changing nature of the urban poor. The number of people without access to civic amenities in informal settlements has increased by about 60.43% in the last 17 years (Banks et al., 2011).

According to the Government Census, most of the settlement dwellers relocated from their place of origin after losing their land or houses to erosion or for other reasons, and because of poverty. Among them, 50.96% relocated to find jobs, 28.76 because of poverty and 7.04% after losing land due to erosion.

In regard to land ownership in urban areas, about 27.25% of settlement dwellers own their houses, while 64.87% live in rented houses and 6.99% are "floating" without any place to stay (BBS, 2014). People in informal settlements are more dependent on electricity supply facilities than other sources like kerosene and solar power. It was found that 89.65% people have access to electricity and 0.33% are dependent on solar power, while 9.7% use kerosene lamps. Availability of safe drinking water is considered one of the preconditions of health security, but about half of the population in informal settlements do not have access to safe drinking water. The report estimated that 52.48% people use tube well water and 45.21 have access to piped water supplied by the water authority (BBS, 2014). They have to use open sources of water as an alternative which causes waterborne diseases because they do not have access to piped water.

Informal settlement dwellers also lack access to hygienic toilets. About 42.19% use unhygienic toilets and 26.25% use sanitary latrines. Lack of hygienic sanitary latrines causes the spread of disease. Those who become ill become trapped in the vicious cycle of poverty as they have to take out loans for health treatment. Sickness also reduces their income as they are mostly dependent on physical labour-intensive tasks.

People in informal settlements lack skills and qualifications (Banks, 2012). Only 33.26% people are literate. In terms of occupation, 17.88% are housewives, 3.18% are students, 6.71 are recent job holders, 13.18 are garment workers, 7.58% are traders, 6.41% are domestic helper, and 6.92% are rickshaw pullers (BBS, 2014). These socio-economic statistics are similar for residents of Bhola settlement and their freedom is influenced by these socio-economic conditions.

According to their own priorities, these are the challenges faced by the people living in the informal settlements in Dhaka (see Fig. 15.6). Their priority of challenges as seen by the authorities is somewhat different. Poverty is seen by policy makers as the main challenge of the informal settlement dwellers from a materialistic perspective, and thus poverty eradication policies have received the highest attention from the government since independence. The Poverty Reduction and Strategy Paper (PRSP) has a rural bias and lacks coordination between the relevant departments and ministries. The urban poor have limited scope for raising their voice through municipal government. Local government elections in urban areas are dominated by the power, wealth, and influence of the candidates. Municipal electorates are also much

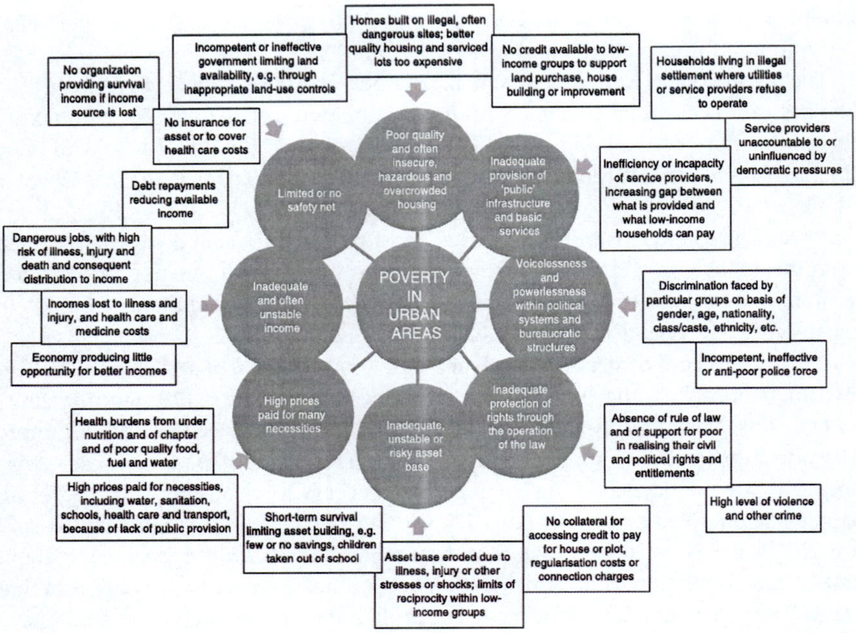

Fig. 15.6 Deprivations associated with the poor migrants living in the informal settlements. *Source* Mitlin and Satterthwaite (2013, p. 241)

bigger than that of rural electorates. The enormous size and anonymous voting blocs make leaders less accountable to the voters.

The elected political leaders are dependent on local *mastaans* (musclemen) for the distribution of goods or entitlements. The mastaans are unofficial local leaders who are affiliated with the political leaders to legitimate their power. Improved services and other benefits offered by the political leaders for the urban poor, as a vote bank, are exchanged and mobilized by the mastaans. Even though, "low-income urban residents are aware that the resources are not equally distributed and the threat of violence or retribution from hooligans is enough to prevent communities from vocalising their disapproval" (Banks et al., 2011).

It is very difficult to challenge the structure of interests necessary to deliver poverty reduction. Powerful elites distribute resources with the help of local mastaans. There is limited space for confrontation and no incentives to ensure accountability among the political leaders for delivery of services and entitlements to the urban poor. Government and NGO officials have to maintain good connections with powerful elites to run their programmes successfully in informal settlements. Therefore, poverty reduction activities are controlled by different interested groups and poor people are dependent on the elites and local mastaans. Exploitative conditions, structural inequalities, and negligence of urban planners for poor migrants leave limited scopes for challenging the relationships.

15.7 Conclusion

Urban growth in Dhaka has not brought about even economic and social development. A significant portion of city residents are still below the poverty line and this proportion is increasing compared with other parts of the country. City growth has been shaped by a massive increase in the number of informal settlements in recent decades. Historical, socio-political, structural, and environmental factors are together responsible for the emergence of the increase in the informal settlements.

Dhaka has experienced 400 years of development under various regimes. During the colonial period, the city was developed according to the needs of the rulers rather than long-term systematic planning. Partition in 1948, and its ethno-religious aftershocks, India–Pakistan war of 1964, and the 1971 secession of Bangladesh drove millions of migrants into the South Asian settlements of Bombay, Delhi, Kolkata, Karachi, Lahore, and Dhaka. After independence, the city's growth accelerated with economic development. Inadequate preparation for urban development, greedy urban elites, and improper management of planned areas has resulted in the growth of informal settlements.

Informal settlement formation in Dhaka is closely linked with rural–urban migration. Both the push and pull factors, including loss of livelihoods in rural areas, "natural" disasters, the initiation of neoliberal macroeconomic policy without proper safety nets for vulnerable groups, and job opportunities in formal or informal sectors in the city all contribute to rural–urban migration. This flow of migration hinders the functionality of urban governance with its limited resources. Urban elites are prioritized by urban governance authorities in the allocation of resources and facilities. Lack of power and influence over decision-making processes excludes the urban poor and confines them to insecure conditions.

People migrate towards Dhaka with the hope of development to find relief from the stress of their precarious rural lives due to environmental calamities. Yet, migrants find themselves living in a place which is environmentally vulnerable and invisible to urban planners and government authorities. Lack of education and formal training poses difficulties to migrants finding employment opportunities in informal, poorly paid sectors. Unhealthy and hazardous environmental conditions, the influence of landlords, mastaan, and threats of eviction further place them in an insecure state in the city.

Huge migration flows towards Dhaka and unplanned urbanization have created pressure on the environmental conditions of the city, leading to the city being ranked as the second least liveable city in the world. The stagnation of productive forces over time, the policies and measures help only the rich to get richer and result in a profound, untenable, and unjust level of inequality.

References

Afrin, S., Islam, M., & Rahman, M. (2015). Development of IDF curve for Dhaka city based on scaling theory under future precipitation variability due to climate change. *International Journal of Environmental Science and Development, 6*(5), 332.

Ahmed, S., & Bramley, G. (2015). How will Dhaka grow spatially in future? Modelling its urban growth with a near-future planning scenario perspective. *International Journal of Sustainable Built Environment, 4*(2), 359–377.

Ahmed, I., & Johnson, G. (2014). Urban safety and poverty in Dhaka, Bangladesh: Understanding the structural and institutional linkages. *Australian Planner, 51*(3), 272–280.

Ahmed, S. J., Nahiduzzaman, K. M., & Bramley, G. (2014). From a town to a megacity: 400 years of growth. In A. Dewan & R. Corner (Eds.), *Dhaka megacity: Geospatial perspectives on urbanisation, environment and health* (pp. 23–43). Springer Geography.

Akter, T. (2009). *Climate change and flow of environmental displacement in Bangladesh.* Centre for Research and Action on Development.

Alam, M. J. (2014). "The organized encroachment of land developers"—Effects on urban flood management in Greater Dhaka, Bangladesh. *Sustainable Cities and Society, 10*, 49–58.

Alam, M. J., & Ahmad, M. M. (2011). Impacts of environmental laws on land development and developers in Dhaka city, Bangladesh. *International Journal of Environment and Sustainable Development, 10*(2), 169–188.

Bala, S. K. (2009). Performance of flood control works around Dhaka city during major floods in Bangladesh. Paper presented at the 2nd international conference on water & flood management (ICWFM-2009), Dhaka, Bangladesh.

Battacharya, D., Rahman, M., & Khatun, F. A. (1999). Environmental impacts of trade liberalisation and policies for sustainable management of natural resources: A case study on bangladesh's shrimp farming industry. Geneva: United Nations Environment Programme (UNEP). Retrieved from https://www.researchgate.net/publication/237331179_FISHERIES_SUBSIDIES_IN_AN_LDC_EXPERIENCE_OF_BANGLADESH Accessed Dec 14 2018.

Bangladesh Bureau of Statistics (BBS). (2015). Population distribution and internal migration in Bangladesh. Statistics and Informatics Division (SID), Ministry of Planning, Government of the People's Republic of Bangladesh.

BBS. (2010). Population census report. Statistics and Informatics Division (SID), Ministry of Planning, Government of the People's Republic of Bangladesh.

BBS. (2014). Preliminary report on census of slum areas and floating population. Statistics and Informatics Division (SID), Ministry of Planning, Government of the People's Republic of Bangladesh.

Banks, N. (2012). Urban poverty in Bangladesh: causes, consequences and coping strategies. BWPI Working Paper 178. Brooks World Poverty Institute: University of Manchester.

Banks, N. (2014). Livelihoods limitations: The political economy of urban poverty in Bangladesh. *BWPI Working Paper, 199*, 1–27.

Banks, N. (2015). Livelihoods limitations: The political economy of urban poverty in Dhaka, Bangladesh. *Development and Change, 44*(2), 266–292.

Banks, N. (2016). Urban livelihoods in an era of climate change: Household adaptation and their limitations in Dhaka, Bangladesh. In M. Roy, S. Cawood, M. Hordijk, & D. Hulme (Eds.), *Urban poverty and climate change: Life in the slums of Asia, Africa, and Latin America* (pp. 113–129). Routledge.

Banks, N., Roy, M., & Hulme, D. (2011). Neglecting the urban poor in Bangladesh: Research, policy and action in the context of climate change. *Environment and Urbanisation, 23*(2), 487–502.

Baud, I. S. A., Pfeffer, K., Sridharan, N., & Nainan, N. (2009). Matching deprivation mapping to urban governance in three India mega-cities. *Habitat International, 33*(4), 365–377.

Bebbington, A. (1999). Capitals and capabilities: A framework for analysing peasant visibility, rural livelihoods and poverty. *World development, 27*(12), 2021–2044. UK: Pergamon

According to Kabir and Parolin (2012), Dhaka is experiencing four types of urban development. These reflect the age of the city, colonial influences, the philosophies of the planners, and the control of planning by the authorities. The city's historic core, or Old Dhaka, follows the Mughal layout or plan, with narrow lanes and compact building structures of the city's past. The planned areas, or satellite towns, have developed according to the urban plan: for example, Dhanmondi, Gulshan, Mirpur, and Baridhara were planned in 1950 and were later implemented according to the plan. RAJUK was directly involved in that development until 1980 and private housing development companies continued to adhere to the plan afterwards. The third category is a combination of the old and more recent formal patterns, occupying most parts of the city, with this development mainly influenced by private companies and individual people. In these developments, the low-lying areas in the eastern part of Dhaka were further extended and reclaimed for "services and utilities". Finally, temporary or informal settlements comprise the fourth category, where about 30% of Dhaka's city dwellers live (Banks et al., 2011). The result is a very uneven geography of urban development, resulting in multiple, divided cities.

Drainage congestion and flooding remain the major concern for both the urban planner and urban resident. This hazard is exacerbated by the actions of state elites and private development companies (Alam, 2014). Low-lying areas, wetlands, and natural drainage ditches were filled to allow the expansion of urban areas, reaching Balu and Turag rivers (Kabir & Parolin, 2012). Population growth and the advent of private housing companies led to the filling of low-lying areas to the east and west, to provide for different housing schemes for new migrants to the city. Because of land scarcity, the common pattern of development for both commercial and residential purposes is now vertical rather than horizontal (Islam, 2005). However, the informal settlements lie outside formal planning and development initiatives.

This brief review of the historical development of Dhaka city shows that urban growth has occurred with little effective planning and is on a highly unsustainable trajectory. The evidence suggests that the functionality[2] of the urban system is strongly affected as a consequence of unplanned urbanization. Planned land use changes over the years have scarcely been implemented and unplanned growth was not properly foreseen: "These deviations from the plan served the vested interests of bureaucratic or political entities, which once again highlights the severe planning and governance issues that are crippling the city's enormous potential" (Ahmed et al., 2014, p. 39).

[2] The latest Dhaka Draft Structural Plan (2015–2035) defines urban functionality in terms of making the city liveable and resilient. "Functionality" here is "the organisational or city manager's perspective that is more concerned with large systems and their functioning than an individual's experience of those systems" (RAJUK, 2015, p. 3). The essence of functionality depends on radical improvements in several key systems, such as transport, drainage, social services, utility systems, economy, energy, the natural environment and ecosystem, and urban governance. This is a formidable challenge for current and future planners, with little evidence to date that this challenge is being met or is being given the priority attention and resources it requires.

Bhattacharya, D. & Titumir, R.M. (1998). Towards demystifying a process: The structural adjustment policies in Bangladesh. Structural adjustment participatory review initiative (SAPRI). First national forum of SAPRI: Dhaka, Bangladesh.

Bhuiyan, M. R., & Siddiqui, T. (2015). Migration in the Ganga-Brahmaputara-Meghna delta: A review of the literature. DECCMA Working Paper, Deltas, Vulnerability and Climate Change: Migration and Adaptation: IDRC.

Byomkesh, T., Nakagoshi, N., & Dewan, A. M. (2012). Urbanisation and green space dynamics in Greater Dhaka, Bangladesh. *Landscape and Ecological Engineering, 8*(1), 45–58.

Carney, D. (2001). *Sustainable livelihoods approaches: Progress and possibilities for change.* Department for International Development (DFID).

Davis, M. (2006). *Planet of slums.* Verso.

Dewan, A. M., Yamaguchi, Y., & Rahman, M. Z. (2012). Dynamics of land use/cover changes and the analysis of landscape fragmentation in Dhaka Metropolitan, Bangladesh. *Geo Journal, 77*(3), 315–330.

Hossain, S. (2013). Migration, urbanisation and poverty in Dhaka, Bangladesh. *Journal of the Asiatic Society of Bangladesh (Hum.), 58*(2): 369–382.

Hossain, M. (2014). *Pro-poor urban adaptation to climate change in Bangladesh: A study of urban extreme poverty, vulnerability and asset adaption.* Heriot-Watt University.

Hossain, S., & Hackenbroch, K. (2018). Whose interest finally counts? The statutory production of urban space at the fringes of Dhaka, Bangladesh. *Planning Theory*, 1–24.

Hulme, D. (2015). *Global poverty.* 2nd ed. Routledge.

Islam, N. (1996). *Dhaka: From city to megacity: Perspectives on people, places, planning, and development issues: Urban studies programme.* University of Dhaka.

Islam, N. (2005). *Dhaka now: Contemporary urban development.* Bangladesh Geographical Society.

Islam, P. M. S. (2014). Inclusive urban development in Dhaka: Review of the national housing policy. World Town Planning Day-2014, 1–5

Kabir, A., & Parolin, B. (2012). Planning and development of Dhaka–a story of 400 years. Paper presented at the 15th international planning history society conference, cities, nations and regions in planning history, Sao Paulo.

Lewis, D. (2011). *Bangladesh: Politics, economy and civil society.* Cambridge University Press.

Mitlin, D., & Satterthwaite, D. (2013). *Urban poverty in the global South: Understanding scale and nature.* Routledge.

Mizanur, R. (2011). Rapid urbanisation and environmental degradation: the case of mega Dhaka city. Master Thesis. Centre for South and East Asian Studies, Lund University. Retrieved June 24, 2016, from http://lup.lub.lu.se/student-papers/record/2167209

Moser, C. O. N. (1998). The asset vulnerability framework: Reassessing urban poverty. *World Development, 26*(1), 1–19.

Piliavsky, A. (2014). *Patronage as politics in South Asia.* Cambridge University Press.

Rajdhani Unnayan Kottripokkho (RAJUK). (2015). *Dhaka structure plan 2016–2035* (Draft). I &II. Dhaka: RAJUK.

Rahman, S. H. (1992). Structural adjustment and macroeconomic performance in Bangladesh in the 1980s. *The Bangladesh development studies, XX* (2&3). Dhaka: BIDS.

Roy, M., Hulme, D., & Jahan, F. (2013). Contrasting adaptation responses by squatters and low-income tenants in Khulna, Bangladesh. *Environment and Urbanisation, 25*(1), 157–176.

Sen, A. (1990). *Development as capability expansion. Readings in human development.* Oxford University.

Saha, S. (2012). Security implications of climate refugees in urban slums: a case study from Dhaka, Bangladesh. In *Climate change, human security and violent conflict: Challenges for societal stability* (pp. 595–611). Springer.

Stott, C., & Nadiruzzaman, M. (2014). Disaster risk reduction in Dhaka city—From urban landscape analysis to opportunities for DRR integration. Asia Pacific HEA Community Resilience Team. Singapore: World Vision International.

United Nations Department of Economic and Social Affairs (UN DESA). (2014). *World urbanization prospects: The 2014 revision.* UN DESA.

Wood, G. (2000). Poverty, capabilities, and perverse social capital: the antidote to Sen and Putnam? In A. Khan & J. Seeley (Eds.), *Making a living: The livelihoods of of the rural poor in Bangladesh* (pp. 1–33). University Press Limited.

Wood, G., & Salway, S. (2000). Introduction: Securing livelihoods in Dhaka slums. *Journal of International Development, 12*(5), 669–688.

World Bank. (2000). *Bangladesh: Climate change and sustainable development*. Dhaka: World Bank Regional Office.

World Bank. (2007). *Dhaka: Improving living conditions for the urban poor*. Bangladesh Development Series, Paper No. 17. Dhaka: The World Bank Regional Office.

Chapter 16
Housing Quality, Affordability, and the Middle Class in Urban India

Urmi Sengupta

Abstract Since 1991 with the advent of globalization and economic liberalization, basic conceptual and discursive changes took place in the housing sector in India. The new changes suggest how housing affordability, quality, and lifestyles reality shifted for various segments of the population. The changes in lifestyles and quality have significant influence on how the concept of affordability is perceived. It is argued that it remains a dynamic concept, which is not only constantly changing but also increasingly incorporating quality and lifestyle issues. Such shift not only reflects structural patterns but also stimulates an ongoing transition process. This study highlights a twin impetus that continues to shape the ongoing transition: Expanding middle class and their wealth—a category with distinctive lifestyles, desires, and habits and the notion of nuclear family and acceptance of confined spaces and privacy of apartments. Both premises are the by-product of Westernization and their notions of privacy and ideologies. Capitalizing this, developers have consistently raised the overall quality and standards of housing (as a bundle of services) defining "lifestyle" for expanding middle class and ignoring in the process, the lifestyle preferences of the poor. This paper attempts to map the changing perception of affordability, quality, and lifestyles in Indian housing and to articulate the function of housing as a conceptualization of social reality in globalizing society in modern India. It is argued that in the context of growing agreement and inevitability of market provision of "affordable housing", the unbridled "market-defining" of housing quality and lifestyles must be checked.

Keywords Globalization and economic liberalization · Housing affordability · Distinctive lifestyles · Nuclear family · Globalizing society

U. Sengupta (✉)
School of Natural and Built Environment, Queen's University, Belfast, David Keir Building, Belfast BT9 5AG, U.K.
e-mail: u.sengupta@qub.ac.uk

B. R. K. Sinha (ed.), *Urban Dynamics, Environment and Health*,
https://doi.org/10.1007/978-981-99-5744-6_16

16.1 Introduction

Few contemporary phenomena have generated such political and commercial fervour as affordable housing in India. As the country urbanizes, the affordable housing deficit is projected to reach 38 million by 2030 from today's level of 25 million. Until 1990s, the government built large quantities of public housing for its employees and the urban poor, regarded as low-income group (LIG) and economically weaker section (EWS) in the government classification. The system however failed due to serious underinvestment and corruption in distribution. With the introduction of economic reforms, market forces and private enterprises have come to the forefront of both production and consumption of housing. This shift has benefited the affordable housing sector generally, of which two aspects stand out—availability of private capital for low-income housing through financial deregulation and engagement of government agencies in partnership with the market. There are evidence of government-developer partnership in the affordable housing sector (Sengupta, 2006, 2010) using efficient design and sustainable practices such as high-density development, mixed-income development and low-cost materials. However, these efforts are centred on "cost" and "quantity" overlooking the quality aspect of housing. The concept of affordability is viewed in isolation from household requirements and lifestyle issues.

However, India's globalizing economy is also triggering basic conceptual and discursive changes, setting the stage for fundamental change in the society. It is also a contested discourse exhibiting many variants, some of which are clearly more influential than others such as affordable housing sector. The country's rapid march to economic growth is lifting a substantial population out of desperate poverty, creating massive middle-class-centred cities. In less than a decade, between 2004 and 2015, the middle class doubled in size from 300 to 600 million. In 2015, fewer than 19 of Indians lived below the poverty line, nearly a 10% reduction from 22% in 2011. The Indian household savings rates also tripled between 2005 and 2015, with many more households having a significant disposable income according to the World Economic Forum (https://tinyurl.com/y7tysq2q). This growing middle class, as a visual urban embodiment of globalization (Farnandes, 2004) is redefining the production and consumption patterns in two complementary ways. The middle classes constitute a critical market for most goods and services, including Housing. The rising income of the expanding middle class is pushing prices and qualities of housing. Developers are also seizing this moment by focusing on building high-cost apartments, mainly catering to the middle–higher income groups. In fact, a sizable number of houses in cities remained unoccupied mostly because of the skewed supply strategies of private builders. Second, the "omnipresent" middle class, in some ways, is resulting in what Kothari (1993) terms "growing amnesia" towards the needs and preferences of the poor in liberalizing India. This dichotomy not only reflects structural patterns but also stimulates an ongoing transition process. The corporate culture is inducing Westernized concepts in residential design, enabling consumer to experience a wider suite of experiences, of lifestyles and quality altering the nature of consumption of residential space. A corollary to this view suggests design and lifestyle issues are compelling as private developers compete with each other through design and prices. In this sense, the degree to which space has been invested with the cultural semiotics of place is now shifting housing affordability, quality, and lifestyles reality for various segments of the population.

In this context, the revival of interest in affordable housing is raising fundamental questions about the changing context of social reality today. Who is buying the new homes that are mushrooming in every city? What income group does this class belong to, and how do their lifestyle preferences relate to affordability in housing? The evidence about who is consuming affordable housing in India has a distinction from such provisions elsewhere, particularly in the West. Further, the market increasingly embraces responsibility for affordable housing, and concerns for the quality have grown. This feature in contemporary Indian housing is distinctive from the previous era. Housing affordability dimension is neither absolute nor atomized, instead, defined by the performance of various components, some explicit as housing quality, availability of land and building materials, and some implicit and unexplored such as lifestyle issues. These issues are embedded in both discourse and practice to emphasize that affordability, too, needs its space and architecture, both an expression of time.

Housing in India is all about economic globalization and investigating this aphorism is where this paper is specifically situated. The paper highlights a twin impetus that continues to shape the ongoing transition: expanding middle class and their wealth—a category with distinctive lifestyles, desires, and habits and corresponding "market-defining" of affordable housing standards—to articulate function of housing as a conceptualization of social reality in modern India. The paper aims to highlight contradictions and paradoxes, and the manner in which the concept of affordability, quality, and lifestyles are embedded in both discourse and practice in India. The remainder of this paper is organized as follows. Section 16.2 outlines the macroeconomic context of housing followed by affordable housing development antecedents in Sect. 16.3. Section 16.4 explores the key effects of globalization on urban housing quality and affordability. Section 16.4 synthesizes observations on affordability, quality, and lifestyles. The final section concludes the study.

16.2 Housing and Macro-Economy: Trends and New Frontiers

This section considers macro forces responsible for housing production and consumption in India. Since independence, India has been rapidly urbanizing, creating a pattern of urbanization characterized by a continuous concentration of population in large cities and juxtaposition of poverty and affluence in equal measure. Scholars have given various interpretations of this phenomenon—"over-urbanization" (Kingsley & Golden, 1954), pseudo-urbanization (Breese, 1969), and dysfunctional urbanization (Kundu & Basu, 1998) at various stages of the process. The share of urban population has increased from 20% in the 1950s to over 30% currently in tandem with rapid economic growth during this period making "urban" India the "world's third largest country" (Bullis, 1997) by the turn of the century. Indian cities today are homes to 340 million people of diverse communities, cultures, and income bands. While unbridled urbanization, in spatial and cultural terms, has triggered India's social and economic transformation, the effects of current economic progress have manifested in two aspects of demography. First, the shape of the country's income pyramid has changed along with a dramatic rise of the middle class. In the current trend, urban households with true discretionary spending power in India are expected to increase sevenfold, from 13 million households in 2005 to 89 million households by 2025 (MGI, 2010). Two, the majority urban citizens in Metro cities continue to occupy the bottom-income segments with an earning an average of 80 (around $1.80) a day. These diverging trends are altering consumption pattern in every sector of the economy including housing, triggering a "choice-based" instead of "need-based" consumption pattern.

There are other economic factors, which are fuelling this transition. As the country's GDP grows, real estate continues to play a major role in the macroeconomic stability. In India, housing accounts for almost 80% of the real estate market in terms of volumes and has been growing at 30% annually (CRISIL, 2006). The recognition of importance of housing sector in the national economy has renewed focus on affordability and its importance as a market opportunity. A study (KPMG, 2010) puts the housing requirement for the sub-100,000 income group across seven major cities at 2.06 million units—a market size of 3,300 billion (US$ 66 billion). As a result some national developers such as Sapoorji Pallonji and Tata (of "nano car" fame) are entering this sector. The macroeconomic prudence in this focus is robust given the expanding the stock of affordable housing would bring, in addition to social benefits, substantial economic advantages through increased demand for construction, building materials, and housing finance, which in turn spur job creation and further economic growth. Further, negative ramifications of unaffordable, overcrowded, or unhealthy housing in financial, educational, and emotional well-being of individuals and families (Bratt, 2002; Dunn, 2000) are also starting to be realized. To reinforce this policy switch, some efforts to improve the quality, design, and construction standards have been observed, but Indian housing overall continues to be riddled with a complex interplay of housing appropriateness (lifestyle issues), availability, and affordability. These three issues take different forms in varied economic environments. They also interact with each other either in a trade-off, as affordability and adequacy usually do, and in concert, as availability and affordability mostly do.

16.3 Affordable Housing Development Antecedents

Housing deficit in India is estimated to be 25 million suggesting a third of all urban households unable to afford housing at market prices. Around 17 million of these households live in slums or squatter settlements with poor access to basic services of water, sewage, and sanitation and belong to the bottom two sections of the economy indicating housing affordability problem is concentrated largely with the poorest of the poor. In India's major cities, up to 54% of households are estimated to be unable to afford housing at market prices. With the demand coming from the rising wealth middle-class population eager to acquire high standard housing, the composition of homes for sale and rent on the market has been inexorably shifting towards very expensive homes.

In India, the National Housing and Urban Policy (NUHHP) 2007 identifies "affordable housing for all" as a key element to achieve sustainable urban development. The private market is considered to be the delivery vehicle for housing, although the remit of providing affordable housing remains in government domain. A number of state initiatives have taken place since pre-economic reform focused primarily on in-situ slum improvement given the size of slum population in the country made any policy consideration preclude their resettlement. It is also common knowledge that slums are politically important entities, with vested interest acting against any rehousing attempts. The Jawaharlal Nehru National Urban Renewal Mission (JNNURM) launched in 2005 is perhaps the most significant scheme with over 1.5 million low-cost homes currently under production in 65 cities for slum resettlement. Valmiki Ambedkar Awas Yojana (VAMBAY) is another centrally sponsored scheme aimed at ameliorating housing problems for the slum dwellers living below the poverty line. Since the cessation of direct construction of public housing, public–private partnership (PPP) has been a delivery vehicle for affordable housing (aimed at low-and lower mid-income group) through a system of cross-subsidy. There is also an increasing evidence of proactive development authorities such as NOIDA, Greater NOIDA, GDA, HUDA, etc., imposing a ceiling on the maximum area of the housing units alongside the requirement to provide affordable units for LIG and EWS. Despite these concerted efforts, metro cities in India continue to witness both quantity and quality problems in the provision of affordable homes. PPP for example, has been criticized for inelastic supply, causing real price appreciation of housing units and eventually pricing low-income dwellers out of the system (Sengupta, 2006). This pattern persisted in VAMBAY projects as there was simply too little affordable housing being produced relative to demand.

In hindsight, the state and market have interacted at many levels to create a structure that determines the level of quality and affordability in housing (Table 16.1) and assigns responsibility for their delivery. So far, the government-led affordable housing has primarily been low-cost single bedroom apartments in isolated suburbs, distinctly different from developer-produced higher-end homes. In the process, an informal distinction between "low cost" and "affordable" homes (Table 16.2) is being articulated, although in literature, they are often interchangeably used. Whether "low cost" or "affordable", both occupy the lower end of the market exhibiting cramped spaces and very basic services. In India as in many other countries, unit size is the principal measure of quality, although housing quality is a composite good (Fiadzo et al, 2001). Historically, the Indian government has defined basic housing units as being between 250 and 275ft^2 for a household size of 5.5 persons in EWS/LIG categories. The pre-economic reform public housing bore the dual distinction of being cramped units with very basic services, even though they boasted central locations and medium density. This historical "space squeeze", as a tool for lowering the cost, has persisted in some of the EWS homes under JNNURM. In 2008, a Task Force on affordable housing recommended raising the size threshold to 300–600 ft^2 for LIG housing, with cost of the house not more than four times annual household income and EMI not more than 30% of monthly income. However, the market forces are yet to respond to these changes. The development of *Shubh Griha*, which represents the "nano" equivalent of budget homes by Tata ranging from 283 to 465 ft^2, exemplifies developers continuing to squeeze space out of lower end homes (Fig. 16.1).

Table 16.1 Housing categories scenario in India

Housing category	Income class threshold in	Value of homes in	Average space consumption per household (in ft^2)
EWS	Under 0.2	–	Under 300
Low income	0.2–0.5	1–2	400–800
Middle income	0.5–1	2–4	800–1000
Higher middle income	1–2	3.5–8	1000–1300
High income	2–5	8–17	1250–1750
Luxury	5 plus	20 +	2500 +

Source KPMG(2010)

Table 16.2 Low cost and affordable housing in India

Parameters	Low-cost housing	Affordable housing
Amenities	Bare minimum to none	Basic
Target income class	EWS & LIG	LIG and MIG
Size	<300 ft^2	300–1200 ft^2
Location	Generally within city but can also be located on city peripheries due to high cost of land	Within city
Developer	Mostly Government agencies	Private Developers and Government
EMI to monthly income	Not exceeding 30% of gross monthly income	Not exceeding 40% of gross monthly income
Finance sources	Micro finance institutions	Traditional banking system

Source Compiled from KPMG (2010) and MGI (2010)

Fig. 16.1 F 283 ft^2 "Nano" homes by Tata currently being developed in different parts of India

When put in perspective, rising standards and quality create an upward pressure on the price affecting access to finance for the urban poor. The deregulated finance sector continues to be unabashedly neoliberal and characteristically market driven. Within the context of a supply-constrained housing market where production of affordable homes is limited, demand is very high, and regulations are stringent, there is a market bias on those with the ability to pay (Sengupta, 2014). For example, in schemes in New Town Kolkata, where up to 20% of units have been reserved for the LIG and EWS, preconditions for acquiring affordable homes are set very high. Applicants are required to deposit up to 40% of the total cost of land and pay the remaining amount in four instalments. The household income of 5000 (US$107.5) (based on official income classification for a LIG household) gives little scope for any large savings. By contrast demand for HIG plots has been very high. The demand from the upper income group coupled with pre-existing loopholes in means-testing suggests a scope for significant "down-filtering" or "negative filtering".

In a market with a constrained supply prices of housing are inevitably higher and quality is compromised. This assumption is supported by how affordability is defined in scholarly research. The term itself is a dynamic magnitude and incorporates several factors in one single bundle which represents household preferences and lifestyles. Such factors include the capital costs of housing, housing quality, household income and finance, public policies affecting housing market, and the choice that people make about how much housing to consume relative to other goods (Lee, 1990; Quigley & Raphael, 2004; Hegedus et al., 1996). This is consistent with Keare and Jimenez's (1983) definition of affordable level that includes income, housing expenditure, and propensity to consume housing. Moreover, concern for affordability is consistent with the concern for improvement in the quality of housing (Lerman & Reeder, 1987) as substandard housing can be seen as a symptom of housing unaffordability problems (Malpass, 1993).

16.4 Rising Middle Class and Its Effects on Urban Housing Consumption

This section focuses primarily on the effects of globalization and the rise of middle-class population on the housing market and the provision of affordable housing. In India, globalization refers to the increasing connectivity of our world and its cultures, driven by the exchange of people, ideas, and goods, often through key media outlets and marketing efforts (Tomlinson, 1999). This connectivity extends through economic, political, social, and cultural spheres of our lives. At a broad level globalization as an idea, aided by neoliberalism is analogous to economic growth in India. Literature is rife with the rise of the Indian middle class as a consequence of the globalization of the economy (Deshpande, 1998; Kulkarni, 1993; Lakha, 1999) engendering a new form of class division and cultural contestation. The focus is now on the lifestyles and consumption patterns of the "new middle class" as the new definer of group identity (Appadurai & Breckenridge, 1995; Scrase, 2002) articulating in the housing provision. As a new class of capitalist elites—the young urban professionals whose styles, desires, and outlook on the world are increasingly becoming hegemonic in India (Toor, 2000).

The concept of globalization has helped unlock important aspects of the changing context of housing consumption in India. In contemporary India, the consumer desire that helps precipitate the housing quality changes shows first in a frenzy of adopting a global culture on movies, music, foods, drinks, and fashion. For instance, telecommunication, unavailable in the markets for years, opened up to the mass raising connectivity. A similar revolution was witnessed in the automobile industry, especially with "nano", the cheapest car in the world. Television channels are replete with global channels, which further fuel this transition by introducing new brands of consumer goods and airing lifestyles shows. Market researchers and advertising agencies in India now commonly craft hybrid images and messages of a bi-cultural identity (one that is local and one that is global or Western) in what Mazzarella (2003) calls a "commodification" of culture. In the same vein, global media has also adapted to this change. The middle class in India is both recognizably Indian and global.

Consumption in India is nothing if not conspicuous. In Indian real estate sector today, the synonymy of progress with ever-present Westernization has whetted the appetite for higher quality houses. Developers have quickly acted to seize the opportunity to reinforce the definition of the new middle class as consumers. The cultural meanings and images intended and desired for housing have been in the heart of marketing efforts by developers attempting to sell the product with a specific and more distinctive identity. This includes an array of images targeted to stimulate the middle-class interest and meanings attributed to status symbolism. A typical housing colony is desirable only if there is a swimming pool thrown in, no matter that most of the adults in this class don't swim. Today's housing has been "new middle-class imaginaries" (Fig. 16.2) replete with fitness club, health spa, jacuzzy, steam sauna, and Turkish Bath, shopping malls to replicate Harrods ambience and products and state of the art interiors. CCTV in the lifts and lobby area re-emphasizes the safety and privacy among other features. A new city aimed at the population of over 30,000 people in Greater Noida (Delhi's outskirt) boasts of 18-hole golf course designed by Martin Hawtree, Mahesh Bhupathi Tennis Academy, a polo and equestrian club including a five star hotel, serviced apartments, a convention centre, and an exhibition hall, all packaged into one project. Such schemes are however outlandishly kitsch tribute to some of the private enclaves that have become a symbol of post-liberalization urban imagery designed to isolate super-rich from the subalterns. Aamby Valley in the outskirts of Pune, near Mumbai, is a gated and guarded township completely out of bounds to non-residents. These spatial patterns and trends can be reflective of post-colonial urban design trends—anti-urbanism and micro-urbanism (Madanipour, 1996). Their influence and inevitability as a lifestyle logic in contemporary India point to the emancipation of middle-class consumption forces fuelled by globalization. The finance sector deregulation has further fuelled this "indulgence" among the middle class. This is visible from the advertising and marketing campaigns from banks. Both governments and banks have long targeted the middle class as the socially desirable sector providing tax benefits and relaxing loan criteria. Nijman (2006, p. 773) argues that it is "consumption, not income, that has become the marker for middle-class status borne out of the explosively growing financing, marketing and advertising industries" upon finding predominantly middle-class owners in a study profiling new home buyers in 1200 newly built houses in Greater Mumbai. Ironically, up to 20% of the dwelling units in some of housing colonies are earmarked for EWS and LIG, which clearly appears incongruous with everything else.

In India, the middle class is recognized by the middle tier bulge in the population which is considered to have subgroups of range of income levels which not only categorizes an income group, but also a political and social class and a consumer market. Definitions of middle class vary but according to A McKinsey Global Institute study using National Council of Applied Economic Research (NCAER) data suggests a real annual household disposable income between 0.2 and 1 million. The importance of this segment of the population is not so much on its size but how rapidly it is growing. Within the context of globalization, the middle class is "pulling away" from the lower class (Nijman, 2006) which then makes it easier to make a distinction

Fig. 16.2 New housing colonies in Indira Puram (Delhi) and Rajarhat (Kolkata)

between two parallel lifestyle trends that exist simultaneously. The first is the "mainstream", which can, at least superficially, be argued to conceptualize lifestyle as a commodity. The second is the more humble consumer behaviour driven by the pressures of needs. This has manifested in the sort of "spatial integration". Posh areas in Mumbai such as Malad and Kandivali consisted of a large number of slum population reflecting a wider spatial disposition in the Indian cities where slums have continued their subversive coexistence alongside modern high-rise towers. However, rather than focusing on the diversity in lifestyles and spaces, the phenomenon is often articulated (by the upper class) in a simplistic binary—the slum and the multi-storied spaces in the rhetoric of built form in the city (Dwivedi & Mehoratra, 2005) and in redevelopment plans of many slum areas. Critiques of these notions find it disparaging leading to resistance to these plans. The class contestation under neoliberalism, Fernandes (2004) claims, thrives on marginalization of the poor in the remaking of the urban landscape to cater to the needs and desires of a rising, consumer-driven middle class.

16.5 Affordability, Quality, and Lifestyle: An Integrative Discourse

In India, the changes in lifestyles and quality have significant influence on how the concept of affordability is perceived. To say the least it is a dynamic concept, which is not only constantly changing but also increasingly incorporating quality and lifestyle issues. The new development in India points to three important dimensions. First, housing affordability and quality are trapped in the dilemma of perception among middle and upper-income people who as policy makers determine the "housing policy and package". For the middle class, this is associated with ever "Westernizing" package of housing and services raising housing prices to a level that is beyond the means of many households within that category. For the majority urban poor, housing provision is led by the stereotypical notion of how people live in shanties and slums—a typical 400 ft^2 home would contain a family of four, in-laws, and visiting cousins from the village who would sleep under the bed. A typical design approach for affordable housing has been to compress a home into a single room with very basic provisions and a mismatch in housing characteristics and household requirements. There are other contradictory imperatives at play, which marginalize affordable housing. Increasingly architects are exploring green building techniques to keep the cost down. Such methods range from composting toilets, rubber wood doors, mud-rammed earth, red oxide floors, solar lighting, and rainwater harvesting. The perception of what constitutes a good home for an upwardly mobile families, however poor they may be, runs counter to the broader objectives of sustainability and affordability, indicating a house means much more than a simple provision for shelter. Mud walls may be sustainable and eco-friendly, but they prefer concrete walls, proper roofs, and garish paint. This sentiment not only subscribes to what Perera (2010) calls a "packaged fantasy" of what a home should look like, influenced by globalization and Westernization, but also an inherent desire to assimilate with the mainstream. It can be argued that at conceptual level, the increasing gap between the middle class and the lower class and excessive focus on commodification of middle-class housing dream has reduced affordable houses to a marginalized sector. Such marginalization or branding of affordable housing is akin to slummification as the concept of quality and lifestyles is similarly compromised in the package. The notion of marginalization is also reinforced through the location of these colonies. High land costs in central areas drive the poor into the suburbs poorly equipped with communication and transport links. When poor households and low-quality housing are concentrated in a single area, the negative ramifications of individual housing challenges substantially increase (Galster, 2005).

Second, affordable housing components currently being provided in prestige projects are out of sync with the lifestyles of low-income population. Indian society is layered in a complex labyrinth of caste, ethnicity, and marginality contributing to divergent patterns of values, norms, and sociological practices among people. The traditional division in lifestyle preferences has manifested in spatial separation of the LIG and EWS, which has deepened with the advent of globalization. Within this context beneficiaries of these affordable homes are not able to enjoy any of the facilities provided while their lifestyle or "need-based" requirements are not addressed. For example, expensive departmental stores and the provision of golf courses mean little to low-income families engaged in a daily struggle for bread and butter. Such incongruity exists even within the provision of LIG homes indicating money is not the only consideration when building homes for the poor. In an opposite illustration, Wharton (2009) gives an example of a squatter milkman, rehoused in one of the LIG homes built by the State of Tamil Nadu went back to their coir beds under the night sky and corralled their buffaloes dung and hay intact inside their tiny one-room homes. Not only do exorbitant management fees price the poor out of the system, but the rapid property price appreciation also prompts these families to leave. The corollary to this view points to the ephemerality of the concept of "mixed communities" in today's planning triggered by "upper income invasion of affordable homes".

Third, housing today represents one of several "spaces of consumption" in which the features that draw people to the area are increasingly contextualized, and at the same time hybridized, enabling the consumer to experience them as part of a wider suite of experiences. However, the newly "reshaped" spaces in urban areas as new spaces of consumption point to increasing "standardization" of lifestyles resulting in "homogenization" of spaces through the consumption and lifestyle patterns of the middle class. A mass consumer culture can now be detected bypassing the constraints of established cultures and traditional values with a fascination for Westernized apartment blocks replete with modern facilities. The "apartment culture" entered the market as a new, stylish, and liberating concept that became increasingly synonymous with isolating, nuclear family structure at all levels in the society. People somehow liked this power of individualism, the sense of having a space, which is private. Even the lower end affordable housing has not escaped this entrapment as property developers broadly bracket it as one—to three-bedroom apartments priced below 3 million. But it is clear that such categorization does not work uniformly across different cities and income groups. Links with architecture, space, and lifestyles differ according to affordability. Affordability as a broad concept exists at every level in our society. While access to affordable housing is an acute problem households in the deprived category (annual income of less than 90,000 are unable to access basic housing across urban India. MGI (2010) estimates the affordability gap in each income segment across all tiers of cities. In Tier 1 cities such as Mumbai, the housing shortage is acute even for households with annual earnings up to 0.5 million. The shortage in the bottom two income segments is acute.

There are nevertheless some flashes of resistance and struggle evident in response to the rampant standardization of bundles of housing products and services at least in the lowest affordability category. These can be found in beneficiaries rejecting the new housing provision for being inconsistent with their lifestyle. Often, the people who live in slums or squatter settlements and those whose livelihoods are inextricably linked to the location of the place have chosen to go back to their original slum. While the external economic forces (such as price appreciation, commuting costs, etc.) influence their decision, the sense of repudiation is not only a simple case of economic prudence but gives credence to the argument that a home doesn't exist in a vacuum. It needs life, and a livelihood to sustain it. Modern slums in India have emerged from village-like habitats and have distinctive characteristics, imagery, and lifestyles that can rarely be recreated in an isolated suburb. Some resistance may be brewing in the middle-class category as well. The cheek and jowl existence of the slums and high rise in India has a legacy of socio-economic interdependence that in itself warrants a separate research. The current trend may reflect an increasing flight of the middle class to isolated suburbs of metro cities enticed by a systematic campaign in the media, how far this trend is likely to continue is questionable given it ignores their socio-economic interdependence (Hogan et al., 2010), which cannot function without the cheap labour of the service class for gardening, child care, driving, security.

16.6 Conclusion

Is it clear that quality and lifestyles do not figure as essential components to the perception of affordability and practices of affordable housing provision in contemporary India? The answer is probably yes. However, there are still traces within the scholarship that suggest new directions and possibilities for further analysis. The articulation of lifestyle studies as a post-globalization project involves the rethinking and reworking of forms of knowledge linked to affordability, quality, and consumption. The housing affordability impact of sudden swelling of the middle class (both in terms of income and aspiration) and the new forms of consumption pattern triggered by globalization and westernization of housing dream offer an area of study that sees the poor in the society being further marginalized when it comes to access to housing. Developers have consistently raised the overall quality and standards of housing (as a bundle of services) defining "lifestyle" for expanding middle class ignoring, in the process, the lifestyle preferences of the poor. The articulation of lifestyle studies as a component of housing affordability studies however requires a long view. Historically, housing affordability has been in the domain of the ultra poor in the society. But there is an acknowledgement that perhaps the concept of affordability may not be entirely exclusive and should now be understood in conjunction with quality and lifestyles at every level. The housing fantasy currently being packaged and fed through to the middle-class population has an upper middle-class bias and is set to alienate those at the lower end of the "middle-class bracket". In the context of growing agreement and the inevitability of market provision of "affordable housing", the unbridled "market-defining" of housing quality and lifestyles must be checked.

References

Appadurai, A., & Breckenridge, C. (1995). Public modernity in India. In A. Appadurai & C. Breckenridge (Eds.), *Consuming modernity: Public culture in a South Asian world* (pp. 1–20). University of Minnesota Press.

Bratt, R. (2002). Housing and family well-being. *Housing Studies, 17*, 13–26.

Breese, G. (1969). *Urbanisation in newly developing countries*. Prentice Hall.

Bullis, D. (1997). *Selling to India's consumer market*. Quorum Books.

CRISIL. (2006). Demand fundamentals support mortgage asset quality. CRISIL Insight, January 2006; ICRA. (2003). Mortgage loan market: Unmanaged growth may lead to credit concerns. ICRA Rating.

Davies, K., & Golden, H.H. (1954). Urbanisation and development in pre-industrial areas. *Economic Development and Cultural Change, 3*(1).

Deshpande, S. (1998). After culture: Renewed agendas for the political economy of India. *Cultural Dynamics, 10*(2), 147–169.

Dunn, J. R. (2000). Housing and health inequalities: Review and prospects for research. *Housing Studies, 15*, 341–366.

Dwivedi, S., & Mehrotra, R. (1995). *Mumbai—The cities within*. India Book House.

Fernandes, L. (2004). The politics of forgetting: Class politics, state power and the restructuring of urban space in India. *Urban Studies, 41*(12), 2415–2430.

Fiadzo, E. D. Houston, J. & Godwin, D. (2001). Estimating housing quality for poverty and development policy analysis: CQIQ in Ghana. *Social Indicators Research, 53*, (2), 137–162

Galster, G. (2005). Consequences from the redistribution of urban poverty during the 1990s: A cautionary tale. *Economic Development Quarterly, 19*(2).

Hegedus, J., Mayo, S. & Tosics, I. (1996). Transition of the housing sector in the East Central European countries. *Review of Urban and Regional Development Studies, 8*, 101–136.

Hogan, T., et al. (2010). Asian urbanisms and the privatization of cities.

Keare, D., & Jimenez, E. (1983). *Progressive development and affordability in the design of urban shelter projects*. The World Bank.

Kothari, R. (1993). Growing Amnesia: An essay on poverty and human consciousness. Viking.

KPMG. (2010). Affordable Housing – A key growth driver in the real estate sector? KPMG, India.

Kundu, A., & Basu, S. (1998). Informal manufacturing sector in urban areas—an analysis of recent trends. *Manpower Journal, 34*(1).

Kulkarni, V. G. (1993). The middle class bulge. *Far Eastern Economic Review, 156*(2), 44–46.

Lakha, S. (1999). The state, globalisation, and the Indian middle-class identity. In M. Pinches (Ed.), *Culture and privilege in capitalist Asia*. Routledge.

Lee, M. (1990). The affordability criteria: Inefficient, inequitable and immoral? In M. Raj & P. Nientied (Eds.), *Housing and income in third world urban development* (pp. 63–80). Aspect Publishing.

Lerman, D. L., & Reeder, W. J. (1987). The affordability of adequate housing. *AREUEA Journal, 15*(4), 389–404.

Madanipour, A (1997). Ambiguities of urban design. *The Town Planning Review, 68*(3), 363–383.

Malpass, P. (1993). Housing tenure and affordability: The British disease'. In G. Hallett (Ed.), *The new housing shortage: Housing affordability in Europe and USA*. Routledge.

Mazzarella, W. (2003). *Shoveling smoke: Advertising and globalization in contemporary India*. Duke University Press.

MGI (McKinsey Global Institute). (2010). India's urban awakening: Building inclusive cities, sustaining economic growth. McKinsey and Company.

Nijman, J. (2006). Mumbai's mysterious middle class. *International Journal of Urban and Regional Research, 30*, 4.

Perera, S. (2010). Housing complexes as packaged fantasies: A meeting of the local and the global and the standardisation of taste in Colombo. In J. S. Anjaria, & C. McFarlane (Eds) Urban navigations: Politics, space and the City in South Asia. Routledge

Quigley, J. M., & Raphael, S. (2004). Is housing unaffordable? Why isn't it more affordable? *Journal of Economic Perspectives, 18*(1), 191–214.

Scrase, T. J. (2002). Television, the middle classes and the transformation of cultural identities in West Bengal, India. *Gazette: The International Journal for Communication Studies, 64*, 323–342.

Sengupta, U. (2006). Government intervention in public private partnership in housing in Kolkata. *Habitat International, 30*(3), 448–461.

Sengupta, U. (2010). The hindered self-help: Housing policies, politics and poverty in Kolkata, India. *Habitat International, 34*(3), 323–331.

Sengupta, U. (2014). New frontiers and challenges for affordable housing provision in India. In P. G. S. M. Smets, & P. Van Der Rout (Eds.), *Affordable housing in the urban global south: Seeking sustainable solutions*. Routledge.

Toor, S. (2000). Indo-chic: The cultural politics of consumption in post-liberalization India. *SOAS Literary Review, 2*.

Tomlinson, J. (1999). *Globalization and culture*. The University of Chicago Press.

Wharton. (2009). Affordable housing: An idea whose time has come, Wharton at Work, August 27.

Chapter 17
Impact of Environmental Pollution on the Health of Workers Engaged in Informal Economy: A Case Study of Patna Metropolitan City

Anuradha Sahay and Rashmi Ranjana

Abstract The unfavourable alteration of our surroundings, wholly or largely as a by-product of man's actions, through direct or indirect effects of changes in energy patterns, radiation levels, chemical and physical constitution, and abundance of organisms is known as environmental pollution. The main reason for our concern about environmental pollution is the adverse effect it has on our health. Patna, a rapidly growing city continues to face environment and health-related problems as pollution of air, water, noise, and land is increasing. The total population of Patna in the 2011 census year was 1.698 million. A steady flight of rural and urban population to Patna in search of jobs and the failure of the formal sector to create job opportunities have raised the number of informal workers in the city. They occupy the locational spaces illegally. They are economically unsound and do not earn enough to maintain themselves or their families. This economy is growing, becoming popular and functions as the economic lifeline of the city but the progressive increase in customers has restricted the movement of vehicles creating traffic congestion, air and noise pollution. The open "Chullahs" used in tea stalls give rise to air pollution. Fast food stalls generate solid waste products and waste water creates moist and unhygienic surroundings which also adversely affects their health. They also throw their wastes on the roadside. They have to face health hazards like asthma, diarrhoea, hypertension, headache, respiratory diseases, vector borne diseases, hearing problem, etc. In this paper, an attempt has been made to highlight the impact of pollution on the health of people engaged in informal sector. Some suggestions have been given to reduce the effect of pollution on their health.

Keywords Environmental pollution · Informal economy · Traffic congestion · Health hazards · Energy pattern · Solid waste · Waste water

A. Sahay (✉)
Department of Geography, Patna University, Patna, India
e-mail: anuradha.sahay2010@gmail.com

R. Ranjana
Department of Geography, College of Commerce, Arts and Science, Patliputra University, Patna, India

B. R. K. Sinha (ed.), *Urban Dynamics, Environment and Health*,
https://doi.org/10.1007/978-981-99-5744-6_17

17.1 Introduction

The natural environment and its constituent elements like air, water, land, flora, fauna, and human beings themselves made certain places more livable and healthier than others. Any unfavourable change in the natural quality of the air, water, or soil due to anthropogenic sources leads to the environmental pollution. Pollution is a term that refers to all the ways by which people pollute their surroundings. It is the introduction of substances or energy (heat, noise, radioactivity) into the environment by humans which may result in hazards to human health, harm to living resources and ecological systems, damage to structures or amenity, or interference with the legitimate use of the environment (Unsworth, 2004–05). Environmental pollution adversely affects the physical and mental health of people and prevents them from living a healthy lifestyle. It directly kills a few per cent of people and affects many indirectly. Outdoor workers are most affected. Loss of productivity and efficiency, costs of healthcare, and decrease in working days due to innumerable reasons are other effects (Mitchell & Bates, 2011).

The increased urbanization, population growth, and urban services in Patna have led to the pollution of air, water, soil, and noise which is resulting in the occurrence of various health hazards. An increase in the rate of atmospheric pollution decreases its cleaning process which results in an increased rate of health hazards. Poor air quality or air pollutants affect our respiratory and cardiovascular system and may range from subtle biochemical and physiological changes to difficulty in breathing, coughing, and aggravation of existing respiratory and cardiac conditions. In recent years, there has been a dramatic rise in the respiratory diseases. Water pollution indirectly causes many deaths. Soil contaminants also adversely affect the health of human beings and animals. People are not much aware of the health hazards associated with various pollutants.

This study is, therefore, designed to know the impact of pollution on the health of people engaged in informal economy. "Informal economy is not taxed, monitored by government, or included in any Gross National Product (GNP)" (Becker, 2004). It has become a kind of life support system to hundreds of urban poor and slum dwellers who comprise a large part of city. Due to the massive increase in the number of street vendors and unplanned growth of Patna, excessive congestion and related problems have come up in the city. They use inputs relatively inefficiently, lack pollution control equipment, lack access to basic sanitation services such as sewers and waste disposal and have little awareness of the health and environmental impacts of pollution.

Patna has 110 slums and many squatter settlements where a large number of people live in a limited space in unhygienic conditions and pollute their surroundings which became the breeding grounds for agents that cause various diseases. They are engaged in various pollution-intensive activities such as leather tanning, bleaching, dyeing, brick and tile making, and metalworking that have significant environmental impact. The open "Chullahs" (traditional stoves) used in tea stalls, informal transportation, and dust are the main causes of air pollution. Solid waste products and waste water thrown on the roadside from fast food stalls create moist and unhygienic surroundings which also adversely affect people's health. Controlling pollution created by informal firms is complicated as they are largely unregistered, have few ties to the state, and are difficult to monitor because of their size, number, and geographical dispersion (Blackman, 2000). They have very low safety standards. Therefore, for a healthy environment and a clean city, there is a need for environmental education through demonstration, awareness, and civic sense among all sections of the society including informal workers.

17.2 Objectives

The objectives of this paper are: (i) to recognize the different types of pollution prevailing in the city; (ii) to examine the impact of environmental pollution (Air, water, soil, and noise) on the health of workers engaged in informal economy; and (iii) to list few suggestions to improve health and quality of life of the informal workers.

17.3 Database and Methodology

In this research paper, a survey of busy markets in 4 zones of the Patna Municipal Corporation area was conducted where the concentration of informal workers was found at a large scale. The source of data for the present study is both primary and secondary. The study-related data was obtained from Census Office, Bihar State Pollution Control Board (BSPCB), Patna Regional Development Authority (PRDA), and sample survey (2013). The methodology involves purposive random sampling of 100 respondents (25 from each zone) for which intensive fieldwork was conducted. Adequate analysis and interpretation of the information collected were made to fulfill the objectives. The study is supported by suitable cartographic representations like maps, bar, and pie diagrams.

17.4 Study Area

Patna, an institutional, administrative, and educational centre, is the capital city of Bihar. It extends from latitude 25°35 ′0 ″ N to 25°38 ′0 ″ N and longitude 85°5 ′0 ″ E to 85°10 ′0 ″ E. This urban centre has 72 municipal wards (Fig. 17.1) with a total area of 109 km^2 and a total population of 1.698 million persons.

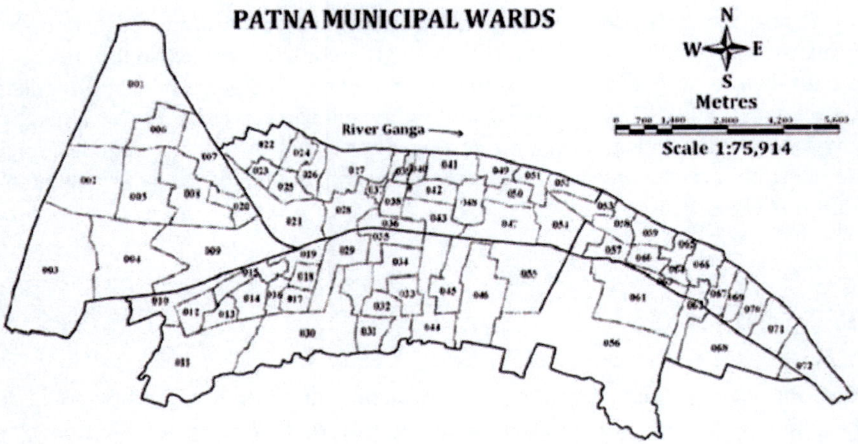

Fig. 17.1 Administrative Map of Patna City with municipal wards

17.5 Growth of Informal Economy and Informal Workers

Informal economy functions as the economic lifeline of Patna due to the lack of major industries and standard functional patterns, rural–urban migration, urban poverty, and unemployment. In the absence of major research work, Patna Regional Development Authority (PRDA) took a step to conduct a survey of informal economy in 2005 before the preparation of Master Plan, which is being implemented from 2007. On the basis of survey results, PRDA defined informal sector activities as "such kind of activities which are performed without any registration and proper approval from the authorities" (PRDA, 2006). These activities may not be illegal but they occupy the locational spaces illegally. So PRDA intends to regularize such activities by allotting suitable sites and providing proper documents for sustainable urban management.

On the basis of a personal field survey 10 markets were identified where there was a concentration of large number of informal workers (Table 17.2). A total of 8 females and 92 males were interviewed. According to the field survey, 81 respondents had an income level of less than 5000 rupees. 22 respondents were illiterate, 23 were middle school pass, 18 were matriculate, and 17 were intermediate pass while 20 respondents were graduates. Due to unemployment they were engaged in informal work. 10 respondents were engaged in vegetable/non-vegetarian items/fruits selling, 22 were in tea/fast food/betel leaf/flower/cloth selling, 15 were in selling items on Thela (push cart) or in basket on head, 8 were engaged in sweeping/rag picking, and 45 were in other various informal works (Table 17.1). Child labour was also seen in many areas. Women faced more difficulties in gaining formal employment due to illiteracy, poverty, and low wages than men. They were most vulnerable as they did not have any economic or social security and had to work in difficult working conditions.

Table 17.1 Income level, education level, and occupation of the informal workers in the study area

Income level	Less than 2000	2000–5000	5000–8000	8000–11,000	More than 11,000	Total
Total respondents	23	58	10	5	4	100
Educational level	*Illiterate*	*Middle school pass*	*Matriculate*	*Intermediate pass*	*Graduate*	*Total*
Total respondents	22	23	18	17	20	100
Occupation	Selling vegetable/ non-vegetarian foods/fruits	Selling tea/ fast food/ betel leaf/ Flower and cloth	Thela/ Feriwala*	Sweeper/Rag Picking	Others	Total
Total respondents	10	22	15	8	45	100

Source Field Survey by authors in Patna, 2013

*Refers to the person who peddles things on pushcarts or in a basket on the head for selling

Table 17.2 Locations of major informal markets, their categories, and types of principal informal activities in Patna City, 2013

Location of major informal markets				
1. Patna City	2. Ashok Rajpath	3. Sabzibagh–Khaitan Market, Birla Mandir Road	4. Bakerganj–Thakurbari Road	5. Rajendra Nagar over Bridge
6. New Market, Station Road	7. Boring Road	8. Maurya Lok, New Dak Bunglow Road	9. Raja Bazaar-Jagdeo Path	10. Lohia Nagar–Kankarbagh Auto Stand

Sl. No	*Categories*	*Types of informal activities*
1	Micro enterprise	Clay potteries, toys, confectionary, tailoring, bakery, hosiery, metal work
2	Street economy	Fast food, shoe-shining, repairing, fruits/vegetable vendors, tea stall, betal shop, milk sellers, non-vegetarian shop, magazine/book stalls, ice-cream vendors, datun sellers, flower vendors
3	Construction work	Day labourer, brick layers, electrician, carpenters, painters
4	Floating or Hawkers	Fruits/vegetable vendors, cloth/readymade garments vendors, decorative items vendors, aluminum and steel utensil items vendors

Source Field Survey by authors in Patna City, 2013

On the basis of field survey, the informal activities in various areas of Patna were identified and divided into 4 major categories and 28 types as shown in Table 17.2. The main concentrations of these informal activity-related markets were from Gandhi Maidan to Malsalami along Ashok Rajpath, from Exhibition Road to New Market

Area, and in Boring Road. Although, there were many other localities, but the rationality of geographical distribution was taken into account. During the survey, it was observed that informal workers in these areas did not have regular employment, work, and social security and were generally self-employed. They earned less, worked odd hours, and had unstable income as they did not have adequate skills and level of education required for the formal jobs. Wage employment predominated in the city. Some of them were engaged in multiple jobs to earn more. These workers were often harassed by corrupt officials, policemen, and local ruffians. For their basic necessities they were dependent on each other as the goods sold by them were cheap. Two-third of the city's total population was dependent upon these street vendors for their daily needs.

Very few numbers of market spaces have been earmarked for them by Patna Municipal Corporation. Therefore, they encroach generally on the roadside public or private areas or move from one place to another carrying their goods either on Thelas or cycles or in baskets on their heads. Their place of work is found mostly overcrowded and traffic congestion remains common. As a consequence, air and noise pollution borne diseases were found very common among them. Most of them were found residing in slums or squatter settlements where lack of sanitation and basic facilities continue to give to some serious health hazards. They are often seen as a problem in the development of city by the administration and city planners. They used to complain of being harassed by police during anti-encroachment drives. Particularly during COVID-19 pandemic they were greatly affected.

According to "Nidan", an NGO working in Patna, the total number of informal workers or vendors was approximately 28,432 in the city (Nidan, 2010). Table 17.3 indicates that 78.2% of vendors were male while only 21.8% were female. This implies that the percentage of female vendors was much less in comparison to that of male vendors. Their less number was due to the patriarchal norms, employers' preference for male labourers, and harassment from different quarters. Other than harassment, 82% of females complained about the lack of basic facilities such as toilets and crèches for their children. The percentage of static vendors and mobile vendors were 87.2% and 12.8%, respectively. Trend wise, it can be seen that out of the total 28,432 vendors, 8997 street vendors were found engaged in selling vegetables, 4414 in selling fruits, 3183 in selling non-vegetarian items, 2608 in selling readymade clothes, and 4532 in fast food vending (Fig. 17.2). Total illiterate vendors were 27%, primary educated vendors were 40%, and secondary educated vendors were 33%.

Other vendors (approximately 16.5%) were either engaged in the trade of diversified items like toys, leather goods, garlands for worship, betel, tobacco, etc., or were self-employed as cobbler, porter, barber, etc. The largest concentration of vendors was in the age group of 22–60 years. Their earnings were very low and ranged from Rs. 30–120 per day. 32% of them resided in Patna while the rest resided in the outskirts of the city or in the neighbouring districts. They used to live in out house or shades, on footpaths, religious pavements, railway platforms, etc., in very unhygienic conditions. Due to the lack of proper storage facilities they use to face the loss of their goods also. They throw solid and liquid waste in nearby areas that create a bad odour.

Table 17.3 Street Vendors in Patna, Nidan 2010

Total number of vendors		Total static vendors		Total mobile vendors	
28,432		24,805		3627	
		87.2%		12.8%	
Male	*Female*	*Male*	*Female*	*Male*	*Female*
22,246	6186	19,412	5393	2834	793
78.2%	21.8%	78.26%	21.74%	78.14%	21.86%
Trend wise total number of vendors					
Vegetable	*Fruits*	*Non-vegetarian*	*Clothes*	*Fast food*	*Others*
8997	4414	3183	2608	4532	4698
Literacy level					
Illiterate		Primary education		Secondary education	
27%		40%		33%	

Source Study on Street Vendors at Patna (Bihar) by Nidan, 2010, pp. 04
http://nasvinet.org/research-document/study%20on%20street%20vendors%20at%20patna.pdf

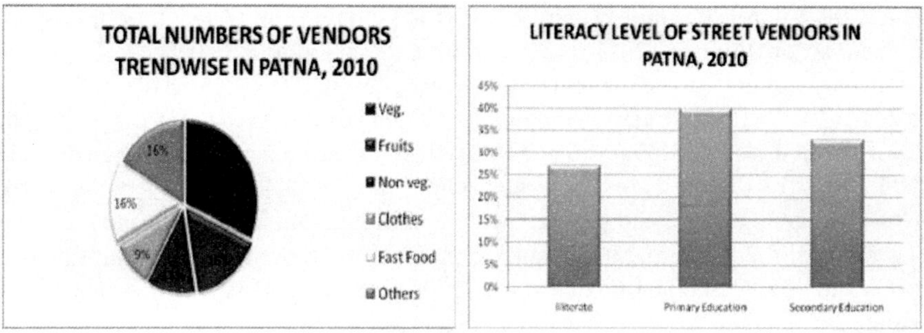

Fig. 17.2 Trend wise total numbers of vendors and literacy level of street vendors in Patna, 2010 according to Nidan (NGO)

17.6 Types of Pollution in Patna Metropolitan City and Its Impact on the Health of Informal Workers

There are four types of pollution in Patna Metropolitan City. They are: air, water, noise, and soil pollution**.** Principal pollutants in the city are gaseous and complex organic pollutants, solid waste, heavy metals, agro-chemical pollutants, inorganic chemicals and minerals, noise, heat, deposited matter, biotic components (pollen grains, spores), photo-chemical oxidants (smog, ozone), etc. All these pollutants have sometimes mild and sometimes severe effects on human health depending upon the time of exposure and immunity of the individual. It is observed that lack of specific nutrients in the diet of people harms their immunity system and makes them vulnerable to various diseases. Low purchasing power of informal workers affects their intake of quality and quantity of foodstuffs. Low immunity and long duration exposure to various pollutants easily make them susceptible to pollution borne diseases. To assess the effect of these pollutants on their health, a sample study has been done extensively.

17.6.1 Air Pollution and Its Impact on the Health of Informal Workers

Patna is now fast turning into the dustiest capital due to degraded environmental quality. The sources of air pollution in Patna are: vehicles, industrial enterprises, solid waste, fossil fuels and burning of dead bodies, solvents and paints, photo−chemical smog, water logging, construction work and generators. Bihar State Pollution Control Board (BSPCB) has installed 6 monitoring stations at Patna where pollution levels are monitored throughout the year. The main monitoring station, at Indira Gandhi Science Complex (IGSC) Planetarium, identifies the basic air pollutants PM_{10}, $PM_{2.5}$, CO, O_3, SO_2, NO_2, and benzene for Ambient Air Quality. The samples are collected as per the Central Pollution Control Board (CPCB) norms. Table 17.4 shows the ambient air quality from 2011 to 2018 at IGSC−Planetarium, Patna.

The level of particulate matter (PM_{10}) was highest in 2013 (225.78 $\mu g/m^3$), $PM_{2.5}$ in 2015 (203.83 $\mu g/m^3$), carbon monoxide (CO) in 2011 (2.35 $\mu g/m^3$), ozone (O_3) in 2011 (47.89 $\mu g/m^3$), sulphur dioxide (SO_2) in 2014 (22.65 $\mu g/m^3$), nitrogen dioxide (NO_2) in 2012 (71.69 $\mu g/m^3$), and benzene in 2011 (4.04 $\mu g/m^3$). Among them, ozone, sulphur dioxide, and benzene were within standard limits. Nitrogen dioxide was high from 2011 to 2013 and again from 2015 to 16 but within the limit for the rest of the year. Carbon monoxide was slightly high in 2011–12 but after that it was within the standard limit (Table 17.4). The most dangerous pollutants were PM_{10} and $PM_{2.5}$ in the city. They affect our lungs and respiratory system when they are inhaled. They were responsible for many airborne diseases like coughing, wheezing, asthma, bronchitis, high blood pressure, etc., among the roadside vendors and outdoor workers as they have to earn their living by working in suffocating surroundings.

Air pollution has become a major health hazard in the city. Table 17.5 and Fig. 17.3 indicate that a total of 37.75% of respondents suffered from airborne diseases in the sampled areas. Respiratory diseases were very common among children and females as they spend more time inside the house which has less or no ventilation. Maximum intensity of airborne disease was noticed in case of headache/nausea (11.45%) as these workers have to face vehicle exhaust fumes and noise daily. The airborne particles and viruses generally infect these workers and cause common cold and pneumonia (10.69%) and influenza (6.75%). The exhaust fumes contain hydrocarbons and nitrogen oxides which form ground-level ozone, the main component of smog, and are responsible for upper respiratory problems and lung diseases. It triggers asthma. Tobacco smoking among them led to bronchitis. 4.58% of workers suffered

Table 17.4 Status of ambient air quality at IGSC-Planetarium Patna (2011–2018)

Sl. No	Parameters	Annual average concentration in $\mu g/m^3$								Standards for residential, rural, and other areas
		2011	2012	2013	2014	2015	2016	2017	2018	
1	PM_{10}	186.09	194.88	225.78	211.60	199.23	–	–	–	60 $\mu g/m^3$ Annual
2	$PM_{2.5}$	–	112.86	121.90	–	203.83	122.51	116.15	118.16	40 $\mu g/m^3$ Annual
3	CO	2.35	2.02	1.98	1.87	1.67	1.55	1.49	1.43	02 $\mu g/m^3$ 8 h 04 $\mu g/m^3$ 1 h
4	O_3	47.89	37.07	27.50	34.64	21.86	34.33	24.63	35.02	100 $\mu g/m^3$ 8 h 180 $\mu g/m^3$ 1 h
5	SO_2	8.66	14.17	8.97	22.65	13.11	6.80	12.08	14.51	50 $\mu g/m^3$ Annual
6	NO_2	60.10	71.69	56.08	29.00	57.54	45.10	37.01	24.39	40 $\mu g/m^3$ Annual
7	Benzene	4.04	2.33	0.99	0.09	1.65	1.24	0.46	1.33	05 $\mu g/m^3$ Annual

Source Bihar State Pollution Control Board, Real-Time Air Quality Data, Page 1 available at http://bspcb.bihar.gov.in/AQI_data_16-17_18a.pdf

from asthma and bronchitis. 2.05% of informal workers were found suffering from heart disease, 2.07% from tuberculosis, and 0.16% from lung cancer.

High level of $PM_{2.5}$ in the city can be responsible for lung cancer but it is not medically proven. It is clearly evident that PM_{10}, $PM_{2.5}$, ozone, nitrogen dioxide, and smog were responsible for airborne diseases among these workers.

Table 17.5 Informal workers affected by Airborne diseases in Patna

Sl. No	Airborne diseases	Percentage
1	Common cold and pneumonia	10.69
2	Asthma and bronchitis	4.58
3	Tuberculosis	2.07
4	Lung cancer	0.16
5	Heart disease	2.05
6	Influenza	6.75
7	Headache/nausea	11.45
Total		N = 100

Source Field Survey by authors in Patna, 2013

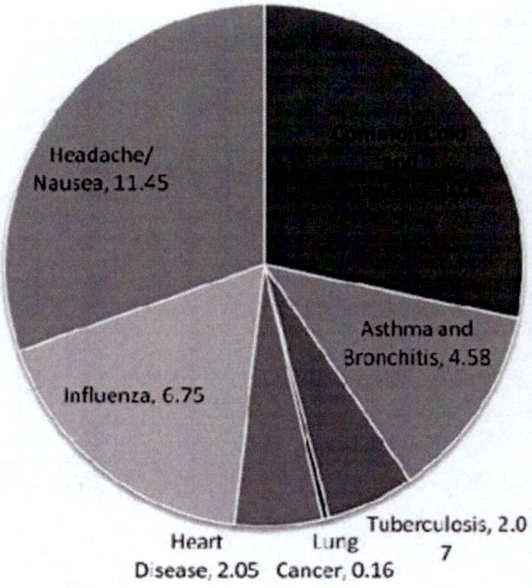

Fig. 17.3 Airborne Diseases among Informal Workers

17.6.2 Noise Pollution and Its Impact on the Health of Informal Workers

In recent years, noise pollution has become a silent killer as people are unaware of its harmful effects. Noise causes irritation which constricts arteries and increases the flow of adrenaline and forces the heart to work faster. It led to heart palpitations and high blood pressure. Noise can damage one's hearing. It also causes annoyance, aggression, high stress level, hearing loss, ischemic heart disease, vasoconstriction, and other cardiovascular problems. It is harmful to us but less harmful than air or water pollution. The main sources of noise pollution are: (i) Outdoor sources like vehicles, small−scale industries, construction work, loudspeakers, generators, etc. and (ii) indoor sources like loudly played radio, entertainment systems, electronic gadgets and appliances, lighting hum, etc.

Patna is the 8th city in India and 1st in the state where continuous noise monitoring stations were installed (BSPCB, 2013). The city has been classified into 4 different zones—silence, residential, commercial, and industrial as shown in Table 17.7. The major noise generator in the city was traffic. During the Diwali festival, maximum noise pollution was found near Patna Medical College and Hospital (PMCH) which is a silence zone at 102.1 dB, against a specified limit of 75 dB. Table 17.6 shows the location of noise pollution in Patna. Maximum noise pollution (during 2005–10) was found near Mahendru Post Office [89.2 dB (A) Leq], Boring Road Crossing, Dak Bungalow Crossing, Hanuman Mandir (Station Road), and Chirayatand overbridge.

Table 17.7 shows the ambient noise monitoring report from 2016 to 2019 at five monitoring stations. It indicates high noise pollution near Patna Medical College and Hospital (Ashok Rajpath) in recent years. Daytime noise was within the limit near Beltron Bhawan, Coca-Cola Factory, and Patna Junction Golamber (traffic circle) in 2019. Night time noise was high near Planetarium, Adalatganj (57.1 Leq dB). All

Table 17.6 Average ambient Noise level at some locations in Patna (2005–2010)

Av. ambient noise level In dB (A) Leq	Sampling location (Day time noise Between 6:00 Am. To 10:00 Pm)
60–65	IGIMS, Raja Bazaar
65–70	PMCH Campus, Ashok Rajpath
70–75	Kankarbagh Auto Stand, Dinkar Golambar (Nala Road)
75–80	Patna city railway station
80–85	Patna Market, Patna City Chowk, Kankarbagh Colony More
More Than 85	Boring Road and Dak Bungalow Crossing, Hanuman Mandir (Station Road), Mahendru Post Office **(89.2),** Raja Bazaar

Source Compiled by the author on the basis of BSPCB Reports (2005–10)

Table 17.7 Ambient noise monitoring report—NMT average annual data (5 locations) at Patna (in Leq dB) from 2016 to April 2019

Year	Area									
	Beltron Bhawan, Shastrinagar		Coca Cola, Patliputra Colony		Planetarium, Adalatganj		Patna Railway Junction Golambar		PMCH, Ashok Rajpath	
	Residential zone		Industrial zone		Commercial zone		Commercial zone		Silence zone	
	Day	Night	Day	Night	Day	Night	Day	Night	Day	Night
2016	57.0	56.0	70.0	68.9	71.0	70.5	71.3	70.9	61.0	59.9
2017	48.7	47.3	61.2	61.1	64.1	61.9	60.4	58.9	58.1	57.6
2018	47.5	43.0	58.0	57.7	63.7	56.8	59.9	55.0	51.5	48.0
2019	47.8	41.3	57.3	57.2	63.7	57.1	60.0	54.9	54.0	48.7

Source Compiled by author on the basis of BSPCB Report, 2016–19 available at http://bspcb.bihar.gov.in/AmbNoise_16_19_CORRECTED.pdf

these locations are found along major city roads where informal workers have to face noise pollution daily, however, they have become habituated to loud noise and are unaware of its ill effect.

Due to unawareness, very few respondents were able to relate noise pollution and diseases. 6.26% respondents were found suffering from hypertension (Table 17.8) but they believed that it was due to stress caused by their life situation. High sound levels caused annoyance among 5.29%, sleep disturbance in 4.64%, and hearing loss in 2.05% workers (Fig. 17.4).

Figure 17.5 shows the low, medium, and high levels of air and noise pollution zone in the study area where high pollution zones were along the major roads, road intersections near CBD, market place, and traffic pockets where informal workers were found in abundance.

Table 17.8 Informal workers affected by noise borne diseases in Patna

Sl. No	Noise borne diseases	Percentage
1	Hypertension	6.26
2	Hearing loss	2.05
3	Sleep disturbance	4.64
4	Annoyance	5.29
Total		N = 100

Source Field survey by authors in Patna (2013)

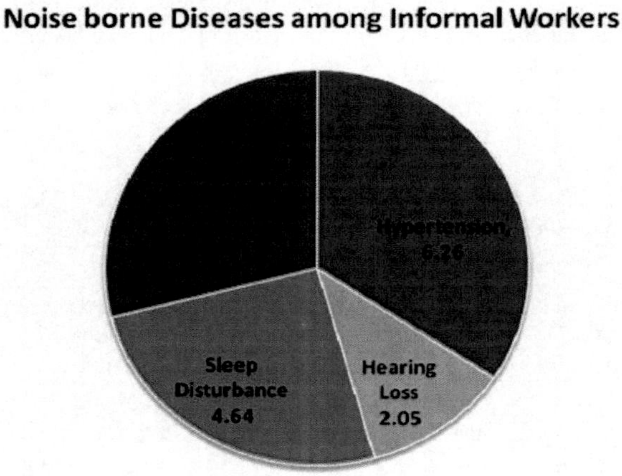

Fig. 17.4 Noise borne diseases

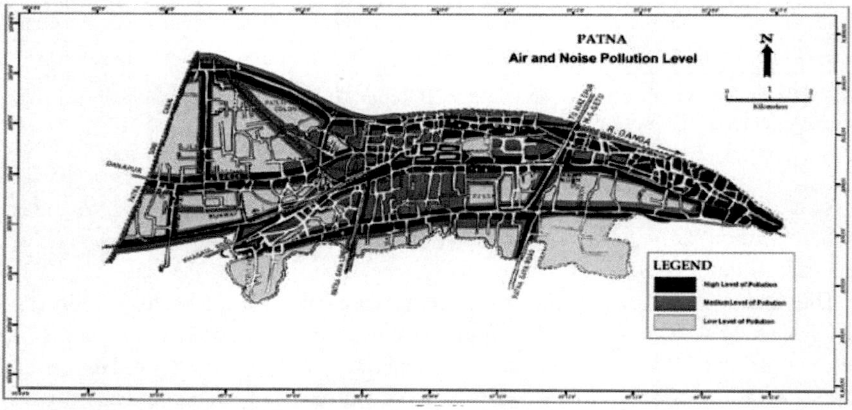

Fig. 17.5 Air and noise pollution level in Patna. *Source* By author on the basis of field survey and BSPCB Report (2011)

17.6.3 Water Pollution and Its Impact on the Health of Informal Workers

Water pollution is a major problem in Patna. Ground water is the main source of drinking water which is extracted by induction motors under the surveillance of Patna Water Board (PWB) and Public Health Engineering Department (PHED). In recent years, due to rapid urbanization and population growth the ground water is getting contaminated through seepage which contains pollutants from small–scale industries, parks, and agricultural lands (fertilizers, pesticides, and other chemicals), domestic and municipal wastes and sewage. When extracted and used by residents it causes several waterborne diseases. The construction of unplanned colonies, slums, lack of sanitation and hygiene, and improper infrastructure are further aggravating the problem of water logging which is a major source of water–related diseases.

According to Central Ground Water Board (CGWB) and Patna Regional Development Authority (PRDA), the underground water of Patna was slightly alkaline due to high level of bicarbonate ion, HCO_3 a form of carbon dioxide, which was 329 mg/l (refer to Table 17.9). pH condition was greater than 7 which also implied that ground water was slightly basic in nature. Electrical conductivity (E.C.) was 2200 μS/cm. The conductivity of the water was higher in the eastern part of Patna in comparison to the other parts of the city (1218 μS/cm). Chloride (Cl) concentration was 366 mg/l which was above its standard range of 250 mg/l but its balance was maintained by the excretion via the urine. Calcium was within range while magnesium (63 mg/l) and sodium (248 mg/l) concentration was more than the permissible limit (30 and 200 mg/l, respectively). Potassium (K) was below the permissible limit (20 mg/l). Total hardness as $CaCO_3$ was 380 mg/l (limit 200 mg/l). The ground water was generally considered good for all purposes. It was free from arsenic and fluoride as well. Arsenic is a trace element found in rocks, soils, and the water in contact with them in the aquifers up to 100 m depth in alluvial formations (CGWB, 2018). Informal workers migrated from Begusarai, Bhojpur, Buxar, Muzaffarpur, Siwan, Vaishali, etc., where arsenic contamination was found might be a sufferer of cancer and skin diseases.

Table 17.9 Analysis of water quality parameters observed in HNS wells, Patna City (in mg/l)

Location	E.C μS/cm at 25 °C	pH	CO_3	HCO_3	Cl	SO_4	NO_3	F	PO_4	Ca	Mg	TH as $CaCO_3$	Na	K
Patna City	2200	7.6	ND	329	366	–	–	–	–	48	63	380	248	12.0

Source Ground Water Information Booklet Patna District, Bihar State by Central Ground Water Board Ministry of Water Resources (Govt. of India) Mid-Eastern Region Patna, September 2013, page14

Available at http://cgwb.gov.in/District_Profile/Bihar/Patna.pdf

Table 17.10 Supply water pollution level in Patna

Quality	Zone	Affected area
I	Low contaminated water zone	Bailey Road, Jagdeo Path, Rajbanshi Nagar, Boring Road, Gardanibagh, Anisabad
II	Moderately contaminated water zone	Exhibition Road, Sri Krishna Puri, Kidwaipuri, Ashok Rajpath, Sultanganj, Baripath, Nala Road, Begumpur, Hajiganj, Khajekalan, Malsalami
III	Highly contaminated water zone	New Byepass Road, Bairiya, Patliputra Colony, Patna City, Hanuman Nagar, Kumhrar, Station Road, Mahendru, Kankarbagh, Mithapur, Jakkanpur, Rajendra Nagar, Kadamkuan, Rajiv Nagar and Mangal Talab

Source On the basis of supply water pipeline leakage, water logging, and seepage of solid and liquid waste found during field survey, 2013

The supply water also gets contaminated by overexploitation, leakage in old and unmaintained pipelines and poor maintenance by PMC. Informal workers were mostly dependent on supply water which was contaminated. Table 17.10 shows the supply water contamination zones in various parts of Patna. The highly contaminated zone comprised New Byepass Road, Bairiya, Patliputra Colony, Patna City, Hanuman Nagar, Kumhrar, Station Road, Mahendru, Kankarbagh, Mithapur, Jakkanpur, Rajendra Nagar, Kadamkuan, Rajiv Nagar, and Mangal Talab.

Due to consumption of contaminated water surrounded by unhygienic conditions, these workers were often infected with various waterborne diseases. According to Table 17.11 and Fig. 17.6, the maximum workers were found suffering from dysentery/diarrhoea (11.45%) and worm infestation (7.13%) due to their unhygienic habits. 3.67% suffered from jaundice/hepatitis A, 3.46% from typhoid, 3.19% from food

Table 17.11 Informal workers affected by waterborne diseases in Patna

Sl. No	Waterborne diseases	Percentage
1	Dysentery/diarrhoea	11.45
2	Food poisoning	3.19
3	Typhoid	3.46
4	Liver trouble	1.62
5	E. coli (UTI)	1.46
6	Jaundice/Hepatitis A	3.67
7	Malaria/Filaria	2.97
8	Worm infestation	7.13
9	Dermatitis	3.13
Total		N = 100

Source Field Survey by authors in Patna (2013)

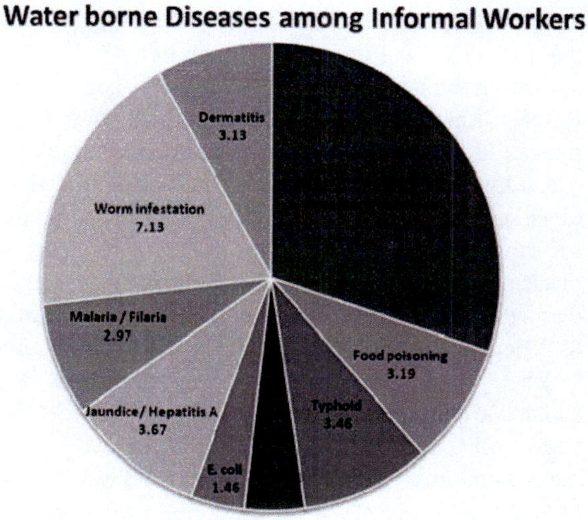

Fig. 17.6 Waterborne diseases

poisoning, 3.13% from dermatitis, 2.97% from malaria/filarial, 1.62% from liver problem, and 1.46% from E. coli. Apart from these diseases, cardiovascular diseases, water—washed diseases (skin and eye diseases), and water-insect related diseases like malaria, filarial, dengue fever, kala—azar, etc. were among the maladies transmitted by biological and microbial agents present in sewage in drinking and bathing water or by mosquitoes in water logged areas or near open Nalas.

17.6.4 Land Pollution and its Impact on the Health of Informal Workers

Most of the land area of the city shows evidence of degradation, thus affecting the productive base of the economy. Dumping of garbage in open landfills, use of pesticides, fertilizers, and biocides in agricultural land, sewage, municipal and industrial waste, etc. are major causes of land pollution in Patna. The other sources are biological agents, biomedical wastes, e—waste, loss of green cover, soil erosion, water logging, etc. They are being dumped in low—lying areas of Southern and Eastern Patna where harmful and toxic materials contaminate both the soil and ground water. Many chemicals injurious to health reach humans through the food chain. The sewerage system does not function properly, gets choked, and aggravates the problem of water logging in monsoon months. Slum and informal population generally depend on public convenience for daily personal needs. 2.5% defecate in open or along roadsides in early morning. 14.3% floating population uses public or community toilets (Urban Local Body and Line Departments Survey Formats, 2005). No secondary data is available on soil pollution as BSPCB only monitors air, water, and noise pollution in the city.

Table 17.12 shows the areas of different types of waste generation in Patna. Among them medical waste and hazardous industrial and chemical wastes are the most dangerous which have been found maximum in Patna City, Kankarbagh, Ashok Rajpath, Rajendra Nagar, and Patliputra Colony. The toxic chemicals and pathogens present in these solid and liquid wastes infiltrate the soil and pollute it along with ground water. Since soil pollution is closely linked to water pollution, many effects of soil contamination are similar to water contamination. Contact with contaminated soil can cause respiratory, liver, and skin problems, birth defects, cancer, headache, nausea, depression of the central nervous system, fatigue, eye irritation, kidney damage, and neuromuscular blockage. Children are more exposed to soil pollution as they play on the ground, inhale the dust, or sometimes accidently swallow the soil.

There is no available data on soil pollution as BSPCB only monitors air, water, and noise pollution in the city. Living and working in dirty surroundings led to many soil borne diseases among informal workers. Due to inhalation of dust and the consumption of contaminated food and water, long-term diseases like depression (2.48%), leukemia (1.78%), and kidney damage had occurred among the informal workers of the study area. Skin rash had been seen among the rag pickers (refer to Table 17.13 and Fig. 17.7).

Table 17.12 Waste generation in Patna

Type of waste	Major area
Domestic waste	All areas
Industrial waste	Patna City, Patliputra Industrial Area, Kankarbagh
Medical waste	PMCH Campus, Rajendra Nagar, Kankarbagh, Raja Bazaar, Gulzarbagh, Patliputra, Boring Road, Anisabad
Construction work waste	Kankarbagh, Byepass, Gardanibagh, Chitkohra, Anisabad, Mithapur, Yarpur, Rajendra Nagar, Bahadurpur Gumti, Gulzarbagh, all flyovers, apartments especially in Southern and Western Patna, renovation especially in Central and Western Patna, nala, and underground cabling especially in Western and Southern area, roads in almost every area
Commercial waste	Ashok Rajpath, Maroofganj, Bari Path, Station Road, Kankarbagh Main Road, Boring Road, Frazer Road, Exhibition Road, Raja Bazaar, City Chowk, Malsalami, Nala Road, Chitkohra, Mithapur

Source Field Survey by authors in Patna, 2013

Table 17.13 Informal workers affected by soil borne diseases in Patna

Sl. No	Soil borne disease	Percentage
1	Kidney Damage	0.32
2	Leukemia	1.78
3	Skin Rash	1.35
4	Depression	2.48
Total		N = 100

Source Field Survey by author in Patna, 2013

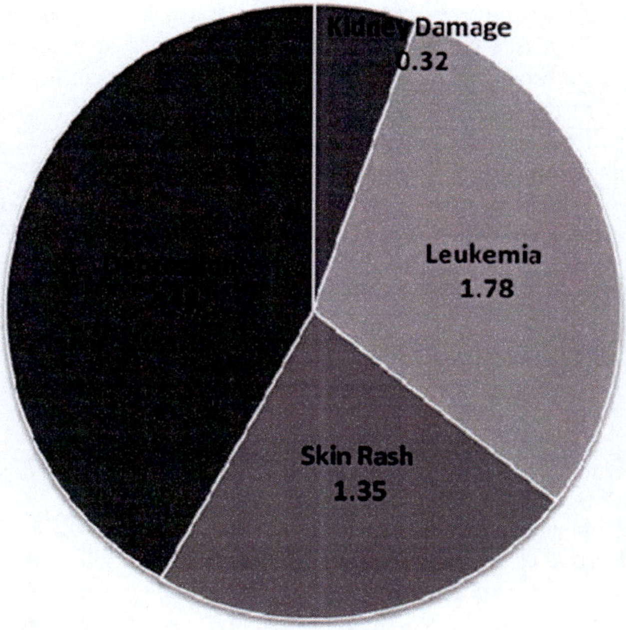

Fig. 17.7 Soil borne diseases among informal workers

Therefore, sanitation and proper sewerage system management is a major issue to be dealt with in the PMC area. Hence, it can be concluded that pollution does exist in one form or the other. The most prominent pollution existing today is air and noise pollution followed by water and land pollution.

17.7 Pollution Control Measures and Public Participation

The first ever mentioned legislative measures in India is in section 268 of the Indian Penal Code of 1860, under chapter XIV which has mentioned the offences affecting public health and defines as public nuisance all types of pollution—air, water, land, noise, etc., and section 269 of the code provides punishment for negligent act likely to spread infection or disease posing danger to life and prescribes a punishment of imprisonment which may extend to six months (Karpagam, 1991). Section 277 deals with water pollution and section 278 deals with the atmospheric pollution. The direct and exclusive treatment of the pollution problem and maintenance of ecological stability has been possible only after 1974 when the Water (Prevention and Control of Pollution) Act was passed.

In February 1972, a National Committee on Environmental Planning and Co−ordination (NCEPC) was set up in the Department of Science and Technology, which was established as the National Committee on Environmental Planning (NCEP) in April 1981. In 1981, the Department of Environment (DoE) was established for environmental conservation and sustainable national development. This later became the Ministry of Environment and Forests (MoEF) in 1985. It has also created Eco−Task Forces of ex−servicemen for eco−restoration tasks in their own areas. Reference to the environment has also been made in the Directive Principles of State Policy as well as the Fundamental Rights. Bihar State Pollution Control Board (BSPCB) was constituted in 1974 under the provisions of the Water (Prevention and Control of Pollution) Act, 1974.

The Major Acts and Rules followed by Bihar State Pollution Control Board (2021):

I. Water (Prevention and Control of Pollution) Act, 1974
II. The Water (Prevention and Control of Pollution) Cess Act 1977
III. Air (Prevention and Control of Pollution) Act 1981
IV. Environment (Protection) Act 1986
V. The Hazardous Wastes (Management and Handling) Rules 1989.
VI. The Manufacture, Storage, and Import of Hazardous Chemicals Rules, 1989.
VII. The Manufacture, Use, Import, Export, and Storage of Hazardous Micro−organism/Genetically Engineered Organisms or Cells Rules 1989.
VIII. The Chemical Accidents (Emergency Planning, Preparedness, and Response) Rule 1996.
IX. The National Environment Appellate Authority Act 1997
X. The Biomedical Waste (Management and Handling) Rules 1998.
XI. Plastic Manufacture, Sale, and Usage Rules 1999
XII. The Noise Pollution (Regulation and Control) Rules 2000
XIII. The Ozone Depleting Substances (Regulation and Control) Rules 2000
XIV. The Municipal Solid Wastes (Management and Handling) Rules 2000
XV. The Batteries (Management and Handling) Rules 2001
XVI. The Biological Diversity Act 2002.

The MoEF has also launched the National Green Corps Programme (NGC) and eco−clubs to enhance awareness among children through conferences, debates, lectures, drama, prabhat pheri, rally, etc., on public places. Tarumitra, a local organization has been networking with local groups of activists on environmental issues. World Environment Day is observed on 5th June, Bihar Paryavaran Diwas is celebrated on the occasion of the Chhath Puja, and Earth Day is observed on 22nd April every year.

17.8 Conclusion and Suggestions

These legislative measures and activities are neither perfect nor can check environmental threats alone. Many programmes and healthcare facilities are running in Patna but these programmes have failed to solve the problems of pollution and health hazard owing to the rapid growth of population, slum, and non–cooperative attitude in general. The risk of environmental pollution is increasing gradually. There is no organized health education programme and no effort for encouraging community participation. The present medical facilities are neither good nor sufficient to cater to the needs of so many people as patients come here from all over the state for better treatment. Some suggestions are put forward to improve the environmental quality and the health of informal workers:

- Public participation is must for all around development.
- Implementation of stringent and proactive population control programmes.
- Formulation of stringent pollution control legislation and effective implementation with powerful administrative machinery.
- A collaboration and commitment between agencies responsible for health, housing, and urban development is very much needed.
- Registration of authentic vendors must be done. Vending activity should be allowed within the demarcated width of footpath.
- Use of clean fuel, alternative energy, public transport, eco–friendly building materials, good ventilation in buildings, promotion of eco–labelled products, etc.
- Green technologies may also improve the quality of the work environment.
- Proper traffic management, effective mass rapid transport system, proper and regular inspection, and maintenance of vehicles.
- Sufficient width of roads should be reserved for pedestrian movement.
- Use of more efficient silencer or exhauster in building making equipment.
- Suppression of noise at the source, use of acoustic enclosures/silencers, and celebrate green festivals by avoiding firecrackers.
- Availability of water sources which is required for drinking and other vending purposes. Hands should be washed after visiting the toilet, and before preparing or eating food.
- Effective treatment of domestic sewage and modern method of sludge disposal.
- Use eco–friendly household products and non–toxic cleaning products.
- Food from street traders should be avoided unless it has been recently prepared and is served hot on clean crockery.
- Designing the water supply system to minimize stagnation and backflow and to provide temperature control for preventing bacterial growth.
- Reviewing the agricultural practices by using organic manures and reducing synthetic fertilizers, pesticides, and biocides.
- Proper collection of municipal wastes by segregation, treatment, and scientifically disposal in landfills.

- Provisions of public convenience and sanitation facilities which can be developed along the PPP model.
- Recycling and reuse of materials wherever possible.
- Proper infrastructure, better health care, adoption of healthy lifestyle, precautionary measures, nutritional and health education programmes, sanitation, etc.
- Spread of awareness among people through public platforms, newspaper, social media, NGOs, nukkad nataks, etc.
- Less cumbersome bureaucratic procedures, uniform tax schemes, and access to capital are activities which may be used to reduce the size of the informal economy.
- "Decent work" should be introduced to provide security in the workplace and to their families (International Labour Organization).
- Formalization indirectly reduces the environmental impact of the informal economy on the environment. As income and education increase, pollution intensity declines.
- Administrators and planners should prepare a short and long−term strategy in an integrated way to minimize the effect of various pollutants on the outdoor workers. This should be done on a local and regional level.

Hence, an integrated planning is the need of today. A large segment of the informal population should be motivated to improvise a healthy and clean environment for better health and quality of life. The problem of pollution should be dealt carefully with the help of NGOs, public participation, government and non−government organizations, and programmes. There should be greater emphasis on pollution and health hazard prevention rather than control.

References

Becker, K. F. (2004). The informal economy: Fact finding study. Retrieved from http://www.rrojasdatabank.info/sida.pdf

Blackman, A. (2000). Informal sector pollution control: what policy options do we have? *World Development, 28*(12), 2067–2082. Retrieved from https://www.academia.edu/14853391/Informal_Sector_Pollution_Control_What_Policy_Options_Do_We_Have

Bihar State Pollution Control Board. (2013). http://timesofindia.indiatimes.com/articleshow/18842506.cms?utm_source=contentofinterest&utm_medium=text&utm_campaign=cppst

Central Ground Water Board. (2018). Ground water quality in low aquifers in India. CGWB, ministry of water resources RD & GR, Government of India, Faridabad (pp. 18–19). http://cgwb.gov.in/WQ/Ground%20Water%20Book-F.pdf

Central Pollution Control Board. (2021). Pollution control law series, ministry of environment, forest & climate change, Government of India, Parivesh Bhawan, East Arjun Nagar, Delhi.

International Labour Organization: Advancing social justice, promoting decent work. https://www.ilo.org/global/topics/decent-work/lang--en/index.htm

Karpagam, M. (1991). *Environment economics—A text book* (pp. 245–246). Sterling Publishers Pvt. Ltd.

Mitchell, R. J., MPH, & Bates, P. MB, BS (2011). Measuring health-related productivity loss. *Population Health Management, 14*, Mary Ann Liebert, Inc. https://doi.org/10.1089/pop.2010.0014 https://www.ncbi.nlm.nih.gov/pmc/articles/PMC3128441/pdf/pop.2010.0014.pdf

PRDA, Govt. of Bihar. (2006). Final report—city development Plan for Patna under JNNURM Study on Street Vendors at Patna (Bihar) Submitted to Centre for Civil Society (CCS), New Delhi, 2010, page 04. http://nasvinet.org/research-document/Study%20on%20Street%20Vendors%20at%20Patna.pdf

Dr Rachael Unsworth. (2004/2005). GEOG3360: Urban environments: Towards sustainable cities, School of Geography, University of Leeds http://www.geog.leeds.ac.uk/courses/level3/geog3360/

Urban Local Body and Line Departments Survey Formats, Patna Nagar Parishad, 2005

Part III
Urban Health Problems

Chapter 18
The Changing Landscapes of Hospitals in the Urban Areas: Evidence from Romania

Liliana Dumitrache and Mariana Nae

Abstract Hospitals are essential entities, usually resistant to structural and societal changes, playing a vital role in shaping the healthcare landscape and designing health policies. The restructuring of the hospital sector, as an inter-related system of professions, practices, and managerial arrangements, became a challenge for governments and health authorities worldwide. In Romania, the landscape of hospitals is changing and reconfiguring in terms of hospitalization services, financing, and management performance. Despite numerous reforms, the Romanian healthcare system remains mainly hospital-based, and hospitals are important institutions that play a central role in providing medical care. This paper identifies the current issues and driving forces of the geographical distribution and reorganization of the public and private hospitals network during the past three decades. The hospital network configuration depends on the spatial arrangement, the spillover of the urban areas, the delineation of catchment areas, and the speciality care they provide. The Romanian hospital network relies on 567 hospitals, out of which 366 belong to the public sector, managed by the Ministry of Health and local authorities. The geographical distribution of hospitals in Romania is consistent with those of urban areas. Despite the large number, hospitals vary widely across the country depending on the size, specialization, and medical services they provide. Not all hospitals can provide complex care, and people address most emergency or university hospitals located in several largest cities to receive quality care. Thus, less than 50 hospital units recorded over 80% of discharges, while most municipal or town hospitals registered low occupancy rates. Compared to other European countries, Romania registered high hospitalization rates per 100 thousand inhabitants, 24.8 in 2007, decreasing slightly to 20.3 in 2016. The findings raise questions on the side effects of the healthcare structural reforms, on the challenges of the regionalization of hospital services and the necessity to reduce public

L. Dumitrache (✉) · M. Nae
Department of Human and Economic Geography, Faculty of Geography, University of Bucharest, Av. Nicolae Bălcescu 1, 010041 Bucharest, Romania
e-mail: liliana.dumitrache@unibuc.ro

M. Nae
e-mail: mariana.nae@geo.unibuc.ro

B. R. K. Sinha (ed.), *Urban Dynamics, Environment and Health*,
https://doi.org/10.1007/978-981-99-5744-6_18

hospitals discharges and spending as well as to the complementarity of a two-tier public and private healthcare system.

Keywords Public hospitals · Hospital landscape · Urban areas · Geographical distribution · Romania

18.1 Introduction

Healthcare infrastructure, relying on sophisticated organizational entities, can support and provide quality medical care to the population concerned. Hospitals are essential components of the healthcare system; to increase efficiency and reduce healthcare delivery costs, they are distributed and consolidated in multihospital networks (Dumitrache, 2014; Nolte et al., 2014).

The network of hospitals has garnered particular attention from researchers, practitioners, and policy makers during the past decade. Issues such as the ongoing shift of patient care from the inpatient to the outpatient settings, the competition between small and medium hospitals in terms of quality-of-care provision and medical staff recruitment, and the restructuring and conversion or closure of public hospitals are triggering debates and empirical research in the field. In addition, the subject of the network of public hospitals in terms of geographical distribution, infrastructure, allocation of services and quality of care has long been of concern to public authorities and health planners.

Hospitals are basic units with a local and regional geographical distribution and bureaucratic, professional structure with internal governance. Hospitals are configured depending on spatial layout, usually clustered in medium and large urban areas. In developed countries, public hospitals are considered core components of the safety net in large metropolitan areas. Indicators of population health status, healthcare use, and healthcare resources can vary by level of urbanization. The number of public hospitals is continuously declining, subject to closure or conversion, raising concerns about the increasing role of the private sector (Andrulis & Duchon, 2007).

Generally, between hospital infrastructure and services allocation, complex relationships develop. Hospitals face changing pressures in terms of the allocation of resources, disease patterns, innovation, technology, and reform policies (Jakóbczyk–Dubas et al., 2020). The allocation of hospital services worldwide also depends on patterns of use and geographical barriers to access due to changing population distribution, healthcare provision, and healthcare policies. The efficient allocation of hospital infrastructure is also a subject of debate. This allocation is often criticized, with most approaches focusing only on the spatial dimension, thus illustrating the population's accessibility to healthcare, which is regularly measured in terms of time and travel costs.

In practice, the attractiveness of a hospital infrastructure depends to a large extent on other dimensions, such as the quality or adequacy of services. Quality of hospital care is usually associated with concentration and specialization of medical services. Recent studies show that the concentration of healthcare can lead to substantial quality improvements for complex medical conditions, while the dispersion of services can increase the quality of care in chronic diseases (Porter & Ikkersheim et al., 2013; Teisberg, 2006). Therefore, the concentration of care and specialization are also considered triggering factors for increasing the efficiency of hospitals.

The size and distribution of hospitals are subjects of extensive debates. Empirical evidence highlighted a causal volume–outcome relationship; a higher case volume is usually associated with better outcomes, and larger hospitals are perceived as better quality. Hentschker and Mennicken (2018) examined for all German hospitals the causal relationship between volume and outcome for hip fracture patients; the volume had a significant positive effect on outcome (increases the number of cases treated by the hospital from 60 to 70 decreases the probability of death by 0.36 pp.).

Hospitals are subject to formal and informal internal governance arrangements. These arrangements are being reshaped and continue to be challenging, as implementing new management models interferes with hybrid professionalism (Coreia & Denis, 2016; Dent & Bode, 2014; Noordegraaf, 2007; Numerato et al., 2012). During the reform process, health professionals become involved in leadership positions. Usually, the relationship between medicine and management is perceived as contradictory or complementary. The health system is facing significant pressures both in terms of demand for services and rising healthcare costs. Hospitals are essential components of the health system, which are required to meet these challenges, thus being subject to change, reorganization, restructuring, or closure.

This paper examines the changing landscape of hospital care in Romania, emphasizing the characteristics of the hospital network from a spatial and organizational perspective.

18.2 The Space Matter and Hospital Landscape Infrastructure: The Geographical Background

Geographical space is essential for the spatial configuration of the hospital landscape infrastructure. On the one hand, space supports urban amenities and the logistics network of transport or metropolitan services. Hospitals are highly dependent on network configuration. On the other hand, hospitals are public spaces, with an increasing role in cities, as service providers for the community.

In the last decade, the design of public spaces in hospitals has become challenging. Large hospitals are an extensive, renewed, innovative part of urban integration projects. Several projects, developed in the USA, Canada, France, Italy, or Spain, include hospitals located outside cities or in low-density urban areas, designed to recreate the typical city environment, with streets, squares, green areas, and courtyards (Setola & Borgiani, 2016; Setola et al., 2013).

Moreover, public spaces in hospitals shape new social roles and symbolic environments. Current research emphasizes the therapeutic roles of the place and the need for hospital design according to various medical conditions and diseases, such as mental ones (Curtis et al., 2007; Gesler & Curtis, 2007).

In Romania, as in other Central and Eastern European countries, the transition after 1990 from the centralized model of healthcare delivery (Semasko type) to other organizational models, which proved efficient in Western European countries, required the reorganization and restructuring of the entire healthcare system. Overall, efforts focused on decentralizing the health system, strengthening the primary care services, and reforming hospital services, aiming at increasing the efficiency and quality of health services and improving the population's access to medical care.

Although primary healthcare services have significantly benefited from structural reforms, financial pressures, and constraints, on the one hand, and high hospitalization rates, on the other, have required special attention from the outset. Hospitals are considered polarizing centres of most significant importance in the organization and reconfiguration of the health system in Romania (Dumitrache et al., 2016a, 2016b; Zamfir et al., 2015).

Healthcare and hospital infrastructure support a broader organizational framework, including human resources, activities, and processes. In addition, their location in space has a legal background regulated by a central or local authority (Ministry of Health, municipalities, or health agencies).

Law regulates the location of the constructions with distinct purposes within the locality in Romania. Decision number 525/1991, republished in 1992, specifies that the university clinical hospitals can be located in the campus areas or areas adjacent to the medical faculties, while general hospitals, as well as specialized hospitals, can be located only in zones designated for health units (Government of Romania, 1991).

Also, depending on the profile of the hospitals, their location in the vicinity of pollution sources (airports, harmful industries, heavy traffic) was prohibited. However, the hospitals offering specialized care (chronic diseases, disabilities, functional recovery, psychiatric centres), must consider some of the benefits and facilities of their location outside the cities (green areas, parks, and gardens).

The hospital network configuration depends on the spatial arrangement, the spillover of the urban areas, the delineation of catchment areas, and the speciality care they provide (Fig. 18.1).

In general, the spatial distribution of hospitals in Romania follows the configuration of the urban network at the national level. The urban network is relatively balanced, with several polarized areas surrounding the largest cities.

The urban network relies on 320 cities, following a suggestive hierarchical typology, according to the demographic size: 11 large cities register over 200,000 inhabitants, 14 cities (100,000–200,000 inhabitants); 32 cities (40,000–100,000 inhabitants), 145 municipalities (10,000–40,000 inhabitants), and 118 towns (less than 10,000 inhabitants).

According to the Law 351 of 2001, urban centres in Romania are grouped into four hierarchical levels (The Parliament of Romania, 2001). The "0" rank is represented by the capital, Bucharest, with over 1.9 million inhabitants.

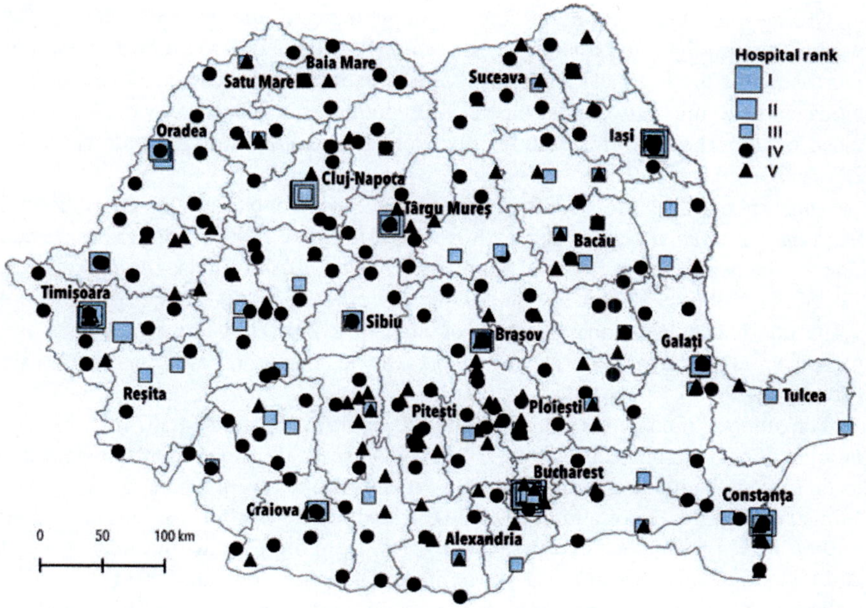

Fig. 18.1 Spatial distribution of hospitals according to the level of competency (I–V)

The first rank includes 11 cities of national importance, each with over 200,000 inhabitants. Currently, in this category, four larger cities register over 300,000 inhabitants (Iași, Timișoara, Cluj-Napoca, Constanța, Galați, and Craiova). They are considered nodal centres with direct access to major national and international transport systems. In addition, they host prestigious medical facilities and hospitals classed as category of competency II or I, along with medical faculties and research institutes.

The second-ranked cities have inter-county importance, with a balanced role in the network of localities and a population between 50,000 and 200,000 inhabitants. Among them, there is a subcategory of cities (14 in number) with a population between 100,000 and 200,000 inhabitants (Bacău, Arad, Pitești, Sibiu, Târgu-Mureș Baia Mare, Buzău, Suceava). A hospital infrastructure, with at least one county and one municipal hospital, is currently in use in each of these cities.

Another subcategory of second-rank cities consists of 20 urban centres (Târgu-Jiu, Bistrița, Focșani, Târgoviște, Tulcea, Reșița, Slatina, Călărași, Alba-Iulia, Hunedoara, Bârlad), with a population ranging from 50,000 up to 100,000, being equipped, in turn, with clinics and municipal hospitals.

Third-ranked urban centres, with a population from 5,000 to 30,000 inhabitants, generally host lower-level competency hospitals, classified as IV or V category. In contrast, in urban areas ranking 4th in the national urban hierarchy, town hospitals, and health centres, tuberculosis preventers, or palliative care centres are usually located.

During the socialist period, the state encouraged the development of the urban economy by redistributing significant investments in large urban areas, thus increasing the rural–urban dichotomy. State investments were directed to the largest cities, with administrative functions such as county capitals or the city of Bucharest, these including health infrastructures, hospitals, and health centres (Dumitrache et al., 2016a, 2016b).

This explains the current configuration of the network of public hospitals, inherited from the communist period, as new hospitals built after 1990 belong to the private sector. The healthcare network in Romania relies on 567 hospitals, 146 polyclinics, 11,252 GP offices, 10,754 specialized health offices, 14,879 dentist offices, and 9,495 pharmacies (National Institute of Statistics, 2018). In terms of ownership, most hospitals (366) are public, while most GP and dentist offices or pharmacies are private.

The number of hospitals increased after 2011, from 464 units to 567 in 2019. The number of hospital beds raises from 6.4 to 6.7 beds/per 1000 inhabitants, mainly due to the private health sector. As well, 67 public hospitals were closed in 2011. The Ministry of Health Order number 323/2011 classified Romanian hospitals according to their level of competency in five categories, from high or medium (categories I and II) to low (categories IV or V) (Monitorul Oficial al României, 2011).

The territorial distribution of high or medium-competency hospitals in Romania is consistent with those of the largest cities, and the highest categories locate in Bucharest (19), Cluj-Napoca (4), Iași (2), Timișoara (2), Târgu-Mureș (1), and Constanța (1).

While the largest cities host the highest competency hospitals providing multiple services and complex care, small towns can only benefit from the lowest competency hospitals, usually called IV or V.

Evidence shows that the coverage with medical service providers is deficient in rural areas, the urban–rural disparities deepening after 1990. For example, the degree of coverage with general practitioners is only 0.5/1000 inhabitants in rural areas, compared to 0.73/1000 inhabitants in urban areas. Other medical services suppliers are practically absent from rural areas.

This rural–urban divide is critical as the Romanian health system is hospital-based, and people address directly to hospitals even in routine investigations. As a result, Romania registered higher hospitalization rates compared to other European countries. In addition, many municipal hospitals are poorly equipped and unable to provide complex medical services. Therefore, patients are re-directed towards higher competency hospitals in large medical centres, thus increasing healthcare costs. After 1990, there were many attempts to reform hospitals; however, structural deficiencies, reduced investments, and the shortage of specialized medical personnel have significantly affected the fulfilment of the reform measures over time.

Investments in capital infrastructure and equipment in the health sector are considerably low in Romania, with 0.1% of GDP compared to European Union, 0.5% of GDP in 2014 (OECD, 2016). Also, Romania spends very little on prevention, only 18 Euros per person in 2017 or 1.7% of total health spending, compared with 3.1% in the EU (OECD, 2017). Deficient spending is associated with poor health outcomes, with Romania lagging behind all EU countries in terms of life expectancy, infant mortality, maternal mortality, cardiovascular diseases, and cancer.

The Case Mix Index (CMI) is a reliable indicator for comparing hospital performance. Usually, the index is derived from Diagnosis Related Groups (DRGs) calculated by Yale University researchers. The case mix index provides valuable information about hospital accreditation and reimbursement rates. However, it is not easy to complete international comparisons regarding hospital financing because of heterogeneous procedures in universal classification systems (France, 2003). Findings demonstrated that hospitals in the highest accreditation category have a higher Case Mix Index than those in lower categories. Marino and Quatrone (2019) highlighted geographical differences in the number of resources allocated and the funding capacity of hospitals. Findings suggested a significant disproportion between the facilities of northern Italy and central Italy, compared with those of southern Italy. Ammar et al. (2013) demonstrated that hospitals in Lebanon with more than 100 beds have a higher CMI than hospitals with less than 50 beds. The Case Mix Index and hospital accreditation remained unchanged. Mendez et al. (2014) analysed longitudinally and cross-sectionally a sizeable public database in California. They observed that CMI is associated with disease severity across hospitals with varying ownership (public versus private). CMI was lower at public hospitals than at private hospitals. The Case Mix Index (CMI) analysis, an indicator related to hospital activity, calculated as a report between the total number of weighted cases and the total number of solved cases, demonstrates differences in 2019 from 0.3 to 1.9. From 406 hospitals, 257 (63%) registered values under 1.0, meaning they could not solve complex cases, while only 18 hospitals surpassed the value of 1.5 (DRG, 2019).

Hospital discharge rates, considering the in-patients for all causes per 100,000 inhabitants, declined from 2010 to 2016, from 23,287 to 20,409 (European Commission, 2019).

Less than 50 hospitals, mainly located in the largest cities and university centres such as Bucharest, Cluj-Napoca, Iași, Timișoara, Craiova, Târgu-Mureș, Galați, Constanța, recorded over 80% of total discharges. In comparison, about 130 hospitals have a much-reduced number of cases annually (Fig. 18.2).

Moreover, given the increasing demand for healthcare services, due to demographic ageing and community-oriented healthcare policies, the distribution of RCCEs (Residential Care Centres for the Elderly) near hospitals is considered insufficient. RCCEs are built mainly on private initiative and depending on the available land reserve, and they are located on the outskirts of large cities or even in smaller urban centres. The results of previous studies on the need for care for elderly persons show that RCCEs have a diverse spatial model at the national level, with over 77% located in urban areas (Matei et al., 2018).

Hospitals, as immovable and change-resistant structures, can significantly contribute to the health of the population they serve but are, at the same time, high consumers of care budgets (McKee & Healy, 2001, 2002).

The underfunding of Romanian HCS has significantly affected the fulfilment of some reform measures over time. Total healthcare expenditure as a share of GDP varied from 3.9% in 1998 to 5.7% in 2010 and 5% in 2017, well below to EU average

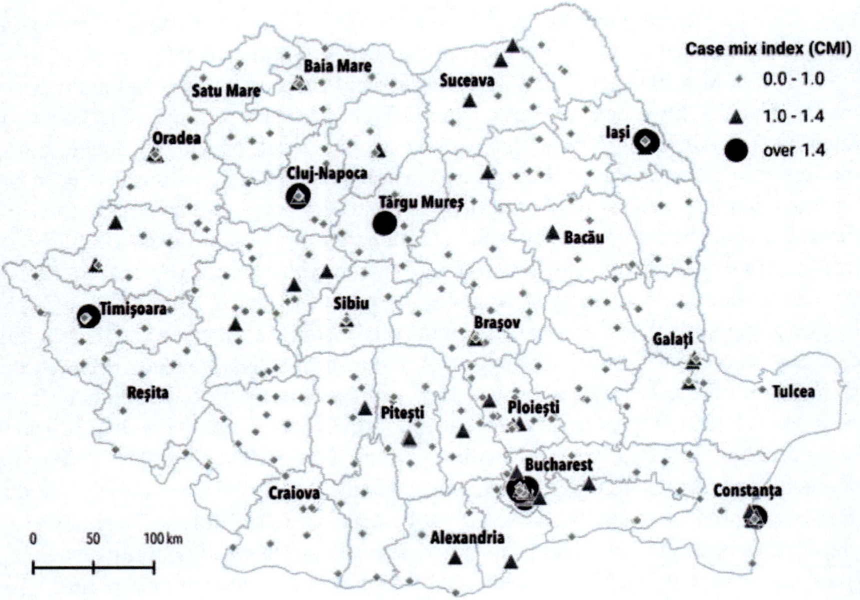

Fig. 18.2 Distribution of hospitals according to the Case Mix Index (2019)

of 9.8%, with expenditures on hospitals surpassing 42% of total health expenditure (OECD, 2017). Also, per capita health spending of 814 Euros in 2015 is the lowest in the European Union (OECD, 2019). Public sources accounted for over 79.5% of total health financing in 2017, while out-of-pocket payments cover almost 21%. Low levels of public spending on health lead to access problems and are also correlated with unmet care needs.

The hospital financing scheme is closely associated with the quality and complexity of the care they provide. Hospital care reimbursement based on the Diagnosis-Related Group systems (DRG) is used in many countries worldwide. Romania gradually introduced the DRG system starting in 1999, with 23 pilot hospitals; the process then continued with the inclusion of 195 hospitals in 2004, 276 hospitals in 2005, and 291 hospitals in 2009, presently almost all public hospitals being reimbursed under this scheme (Radu et al., 2010).

The diagnostic groups used in Romania are 499, divided into medical and surgical categories designed to cover the pathologies of acute hospital patients (INCDS, 2004, p. 1). Also, the DRG system mainly finances hospitals and divides similar patients into homogeneous diagnostic groups regarding clinical conditions and resources consumed during treatment. The hospitals are reimbursed according to a weighted case rate (TCP), which represents an average pre-calculated and weighted cost for the respective diagnostic group (Ministry of Health, Annex 1). Romania experienced the transition from the US HCFA—DRG (*Healthcare Financing Administration—Diagnosis-Related Group*) (version 18) to the Australian Refined DRG (version 5) in 2007. The changing DRG system and associated case weights have led to a slight decrease in the case mix index in hospitals (Radu et al., 2010). However, studies assessing public hospital governance and performance in Romania suggest that the current DRG cost weights do not match the reality of hospital service costs (Duran et al., 2019).

18.3 Public Hospitals and Decentralization Policy: The Normative Background

Hospitals have a legal personality and claim to have a legal personality as independent institutions classified according to the services provided. In Romania, the organization of hospitals is based on the provisions stipulated by several laws: Law 74/1995 on the Organisation of the College of Physicians; Law 145/1997 on Social Health Insurance; Law 100/1998 on Public Health; and Law 146/1999 on the Organisation of hospitals.

The *Law on the Organisation of Hospitals 146/1999* included some articles relating to hospitals classification from a double perspective: (a) depending on the medical services delivered and their specialization (general or specialized hospitals, acute or chronic hospitals, hospitals for continuing care) and (b) according to the administrative—geographical criterion, on several levels: *rural hospitals* (minimum 120 beds with internal medicine and pediatric services); *town and municipal hospitals* (at least 250 and 400 beds respectively, internal medicine, pediatrics, surgery, and gynecology-obstetrics); *county hospitals in larger cities* (ophthalmology, intensive care, orthopedics); *tertiary care in specialized units* (Institute of Oncology, Institute for Maternal and Child Care, Neurosurgery Hospitals, Institute of Pneumology).

Four healthcare units provide inpatient care: (a) long-term care hospitals for patients with chronic diseases, (b) medico-social care units managed by the local authorities, (c) sanatoriums, and (d) health centres for medical services for at least two specialties. Inpatient care is considered an unsettled issue within the care reform process.

Decentralization is a contradictory process with diverse strategies and outcomes. Nevertheless, it is generally viewed as an act of transfer of power and authority from national to regional and local levels (regions, districts, and municipalities) (Bankauskaite & Saltman, 2007).

In Romania, since 2002, decentralization of public hospitals was initiated by transferring administration of land and hospital buildings to local and regional authorities, according to Government Decision (GD) no. 866/2002, Government Decision no. 867/2002) and Emergency Ordinances (A and L 70/2002, 99/2004). Subsequently, the process continued with the Government Ordinance number 162/2008 and GD 562/2009 approving the Health System Decentralisation Strategy (Government of Romania, 2010a, 2010b). As a result, the legal framework for decentralizing 370 public hospitals out of 433 subordinated to the Ministry of Health was developed in 2010 (Government of Romania, Monitorul Oficial, 384/385/2010).

The following Law 95/2006 on Health Reform imposes different classifications of hospitals based on several criteria: regional (regional, county, local), specific to pathology (general, emergency, specialized, for chronic diseases), by the regime property (public, private, public with private sections) or from the point of view of medical education and research (clinical hospitals, institutes). The National Authority for Quality Management in Healthcare completed the first stage of accreditation on 431 hospitals from 2011 to 2016. All public and private hospitals will be gradually accredited by 2021.

Several inaccuracies in the criteria can lead to confusion (some county hospitals can be considered regional at the same time, or they are both clinical and emergency hospitals and can be public facilities with private departments). Thus, The World Bank (2018a; 2018b, p. 7) recommended the classification of hospitals into five categories according to the level of competency: tier V, with a limited level of competence, including mostly chronic diseases hospitals; tier IV, basic competence level including local hospitals with a low degree of complexity services delivered; tier III, medium competence level, including mostly county hospitals covering medical services of medium complexity; tier II, high level of competence, which includes general or specialized hospitals with a high level of equipment and specialized personnel delivering multiple and complexes medical services; and tier I, very high level of competence, accommodating hospitals able to provide medical services with a very high degree of complexity.

To maintain the national emergency system functionality, from the point of view of financing as well, the order of the Minister of Health number 1764/2006 states mainly on the classification of local, county, and regional emergency hospitals in terms of competencies, material, and human resources and their capacity to provide emergency medical care (Ministry of Health, 2014).

The issue of geographical accessibility, the allocation of health resources, and the quality of healthcare provided to the community, are key features of the changing landscape of hospitals. The equitable allocation of health resources at the national level requires decisions and political vision. According to the 2008 Presidential Commission Report, the resource allocation should be balanced within different types of health services, across different areas of the country and among different health institutions. However, territorial importance and geographical areas need to be more clearly defined. In the report mentioned above, the criteria for allocating resources were criticized as being subjective and non-transparent (especially those related to investments). Also, lacks of transparent schemes and allocation criteria and ineffective management have led to inefficient spending (Vlădescu et al., 2008).

The strategic objective of improving hospital services at the national level requires increasing coverage and equity of services and developing sustainable social policies to provide free and equal access to healthcare for the population (Ministry of Health, 2014). Thus, the regionalization process established under the Government Program 2013–2016 should have considered the necessity to reduce barriers to access to medical services of the population.

Another strategic objective resulting from the last Health Strategy (2014–2020) aims to improve health services' performance and quality. The measures focus on centralizing hospital care and creating regional reference networks, with hospitals and laboratories of different levels of competence interconnected with the specialized primary and outpatient care sector (Ministry of Health, 2014, p. 9). Seven regional hospitals are planned to be built to provide highly complex care to the population in that region.

There are also differences in the allocation of human resources, qualified medical staff, and infrastructure across the country, with many hospitals having limited capacity to provide quality, efficient, and safe medical services for the patient. Simple pathologies that currently represent "avoidable hospitalizations" could be treated in ambulatory care settings. At the same time, complex cases will be transferred to county hospitals, especially to clinical/university hospitals (Ministry of Health, 2014, p. 53).

Another measure aims to replace in-hospital care with alternative services (day-care or ambulatory care), ensuring the necessary continuity of services and strengthening long-term, cost-effective care as much as possible at the community level. Nevertheless, in 2019, the Government of Romania entrained some reforms aiming to strengthen primary healthcare access through a programme (***Results-Based Program in the Health Sector in Romania***) ratified into Law in early 2021 (Government of Romania, 2021).

Accessibility of health services, efficiency and quality of healthcare remain challenges for the national health systems. Issues related to the population's access to hospital services are still debated. The population's access to hospital services initially requires social and financial interaction between the parties involved (population, on the one hand, and health resources, on the other). The population's access to hospital services depends not only on the constraints and limits generated by travel costs but also on availability and adequacy.

Good access of the population to health resources continues to be challenging. Hospital accessibility is a contributing factor in identifying areas with a shortage of healthcare. Dumitrache et al. (2020) research findings demonstrated that the geographical distribution of the most accessible hospitals overlapped the urban spaces and highly populated territories surrounding large cities such as Bucharest, the capital of the country, Iași, Timișoara, Cluj-Napoca, and Craiova. These hospitals have an essential role at the regional level and relatively good types of equipment and access is facilitated by good road infrastructure. Conversely, areas where population has reduced accessibility to hospital care are usually located in mountainous regions or Danube Delta (Dumitrache et al., 2020).

18.4 Privatization of Romanian Hospitals: Policies and Outcomes in the European Context

In European countries, the privatization of health services was the subject of political debates, and the shift from public to private ownership is considered a way to improve the quality and efficiency of healthcare. The distinctions between public and private medical services rely on institutional and legal procedures. In many European countries, the private healthcare sector has expanded rapidly in recent years.

The privatization policies are often criticized due to their unintended effects, particularly for merging health and commercial activity (Powell, M., & Miller, 2013; 2014; Saltman, 2003, 2009; Albreht, 2009; Keaney, 2002). Depending on the socio-economic and political context of the countries concerned, there are many peculiarities in the privatization process, with multiple public and private boundaries in healthcare and shifts in public and healthcare provision (Jeurissen et al., 2016; Maarse, 2006; Waters et al., 2008). Several studies stressed the outcomes of privatization on technical efficiency, accessibility, and quality of care (Krusse et al., 2018). Findings mostly suggest that public hospitals are at least as efficient as private hospitals or an increase in efficiency after privatization (Tiemann & Schreyögg, 2012).

Privatization of health services is at the core of all stakeholders (governments, ministries, doctors, and patients), involving the formation and management of many financial and human resources. There are different ideas and views on the divergence or convergence of the two systems (public and private). On the one hand, it starts from the principle that public systems are not adapted to social requirements or the challenges of the business environment; on the other hand, the pre-eminence of the idea that the private system would solve all problems of maximum efficiency. However, opinions are divided.

In Romania, after 2010 and until now, legislative, and governmental changes have affected the functioning of the health system. Thus, depending on government policy, right-wing governments, more inclined to take measures to liberalize the medical market, have opted for more radical variants of privatization policies.

The Law on Health Reform adopted in 1996 (Law 95/2006) stipulates the continuation of the decentralization of health services, the development of the private sector, and the establishment of transparent relationships between the main actors of the health system. The 2011 Health Law draft encouraged the privatization of the health system, and private insurers prefer private hospitals and clinics. Private operators (the most efficient being hospitals and private clinics) could take over the activity of public hospitals and even participate in their privatization.

Problems related to the allocation of financial resources have worsened in the context of the economic and financial crisis of 2008. According to the National Strategy for Rationalisation of Hospitals (Ministry of Health, 2011), private hospitals could settle contracts with the National Social Insurance Fund, thus increasing competition for funds among public and private healthcare providers. There was also competition between the two sectors in terms of medical staff. Doctors practising in deficient specialities can carry out activities in public and private units that have settled contracts with the National Health Insurance House (NHIH).

The 2011 health package included measures for the liberalization of health services, the transfer of public funds to the private system, and the restructuring and closure of some medical facilities. Sixty-seven public hospitals were closed in 2011, leading to protests and social movements nationwide. Thus, in the following period, 24 out of 67 reopened as independent hospitals, 20 settings turned into nursing homes, and 11 inpatient units remained closed (Scîntee et al., 2018). Transferring public funds to private hospitals is another project that is still being debated. Often, these characteristics gave way to structural ambiguities, leading to inexplicable situations and unfair competition between the two sectors. Moreover, statistics on private healthcare providers are generally scarce.

The leading private hospitals in Bucharest are *Euroclinic* (opened in 2005 and currently owned by the operator Regina Maria), followed by *Life Memorial* (Medlife, opened in 2007). In addition, Regina Maria Maternity Hospital (opened in 2008), *West Eye Hospital* (2010), *Delta Hospital* (2011), Pediatric Hospital Medlife, *Orthopedic Hospital Medlife* (2011), and Medicover are also located in Bucharest.

New private hospitals have opened since 2004 in other cities: Polisano Hospital (Sibiu), Pelican (Oradea), ICCO (Brașov), Euromaterna, and Isis. Medical Center (Constanța), Medlife Genesis (Arad), Athena Hospital (Timișoara), and Nova Vita Hospital (Târgu-Mureş, opened in 2010). In 2011, 11 private hospitals were opened based on investments of 131 million Euros, and one year after, 22 large and medium private hospitals operated already in Romania (Domnişoru, 2012).

Currently, the medical services market is a dynamic one and is one of the most profitable sectors for private investments, registering consistent profits and a turnover estimated by specialists of over 2.2 billion Euros in 2018 (Keysfin Report, 2018). Nevertheless, the private medical services market remains concentrated around prominent players. Most private clinics and hospitals are located in Bucharest (turnover of 924.5 million Euros), and main cities as Brașov (100.2 million Euros), Timișoara (96 million Euros), Cluj-Napoca (83 million Euros), and Constanța (68 million Euros). Nevertheless, reliable data regarding private investors at the local and national levels are lacking. Therefore, a large-scale analysis of the economic performance of private healthcare providers is considered challenging and complex (Volintiru, 2018). The first three large private companies also provide hospital services and have a territorial expansion with various strategies of reinvestment (data from Keysfsin Report, 2018):

Unirea Medical Centre (CMU) is a company controlled by Royal Capital Investments and manages the private network Regina Maria, with a turnover of 68 million Euros. The CMU network consists of 3 hospitals (Bucharest) and others in the cities of Brașov, Cluj-Napoca, Timișoara, Baia Mare, and Târgu-Mureș. A network of polyclinics also operates in other cities (Bacău, Brașov Cluj-Napoca Constanța, Craiova, and Iași).

Medlife is in second place, with a turnover of over 68 million Euros. The Medlife private network consists of 12 Medlife hospitals (Bucharest, with six units), Arad, Brașov, Craiova, Iași Timișoara. The first private Medlife clinic was established in 1996, initially with activities in family planning. In 2006, *Life Memorial Hospital* was launched, the largest private hospital in Romania. Subsequently, from 2011 to 2017, the company expanded through investments in specialized hyper clinics and acquisitions of hospitals: Medlife Genesis in Arad, the Humanitas Hospital in Cluj-Napoca, and the Polisano Hospital in Sibiu.

Medlife Hyperclinics are currently available in Bucharest (8 units), Galați (2 units), Brașov (2 units), Cluj-Napoca, Constanța, Iași, Craiova, Ploiești, Arad, Râmnicu-Vâlcea, Brăila, Oradea, with one hyperclinic unit for each city. *Sanador* is a company with 100% Romanian capital, with a turnover of 52 million Euros. A Sanador hospital is currently operating in the largest city, Bucharest.

In 2020, according to data provided by the Keysfin Report (Keysfin Report, 2020), the top three largest companies by turnover remained unchanged in five recent years. *Unirea Medical Centre* remained in first place with a turnover of 149 million Euros (4.8% of the total), and Medlife ranks second, with annual revenues of 103 million Euros (3.3.% of the total). *Sanador* is third with 83 million Euros (2.7% of the total market share).

Complementarity between the two systems and achieving effective competition would be a viable solution in the medium and long term (Dumitrache, 2014; Nae, 2014). However, many issues need to be clarified regarding specific legislation, the role of private insurance, and a more precise definition of the package of covered services or access to resources (efficient development of the co-payment system).

18.5 The Clinical Activity of Hospitals and the Volume–Outcome Relationship

The clinical activity of hospitals is complex and dynamic, depending on the health workforce distribution, logistics, and various services. The evaluation of clinical activity in the DRGs system involves the development of multiple centralized indicators used to optimize the payments and for better monitoring of the quality of services provided. The hospital volume–outcome relationship has been debated for years in healthcare disciplines focusing on shortcomings in methodologies that directly affect the conclusion of a positive volume–outcome relationship (Lee et al., 2019; Morche et al., 2016). A higher volume of activities improves patient outcomes.

Usually, if the high volume is associated with better outcomes, the magnitude of the association could have significant variations (Levaillant et al., 2020). The volume–outcome relationship represents an association between the number of patients with a diagnosis treated in a hospital and patients' outcomes over time. The volume is operationalized as the number of patients assigned to a DRG.

To observe the dynamic of the clinical activity of hospitals in terms of the allocation of resources, we calculated the CAHI (Clinical Activity Hospital Index). Composite indices provide challenging methodological issues. In this case, an approach of data-driven weights techniques (Decancq & Lugo, 2013) is used. Commonly, the statistical relationships between indicators and the construction of composite indices are subject to criticism regarding the method and technique used (Greco et al., 2019; Nardo et al., 2005). The index calculation involved several stages: data selection, multivariate analysis, normalization, weighting, and aggregation (OECD, 2008; Saisana & Tarantola, 2002).

In order to determine an appropriate weighting scheme, we used the principal components analysis (PCA), which assigns factor loadings whether an indicator shares a common factor with another variable in the dataset. PCA is a multivariate technique based on linear transformations of the original data. However, we should have checked their variability and structure within the data set due to the lack of reliable data. Therefore, the robustness and sensitivity analysis for the clinical activity index was not verified.

Using the DRG database, we selected for 408 hospitals five indicators which expressed the clinical activity of the hospitals, grouped into two categories: *activity intensity indicators* (number of cases, cases at the county level, number of rural cases) and indicators *of the circumstances of hospitalization* (percentage of emergency hospitalizations, cases of family doctor referrals) Table 18.1. With a 76.5% total variance explained, four components score over 0.90 resulted (Table 18.2).

After the assigned procedure of weights scores and aggregation data, the values of the clinical activity index ranked between 445.1 and 162.9. The first ten hospitals, which recorded a powerful dynamic of clinical activity, were generally Emergency and University county hospitals located in the southern, south-eastern, eastern, and north-eastern parts of the country (Fig. 18.3).

The ten ranked hospitals according to their clinical activity are the County Emergency Clinical Hospital "St. Apostol Andrei" Galați (445.1), Ploiești Emergency Hospital (406), Craiova County Emergency Clinical Hospital (401.8), Constanța County Emergency Clinical Hospital (399.6), Râmnicu-Vâlcea County Emergency Hospital (399.1), Bacău County Emergency Hospital (397.9), Slatina County Hospital (394.5), Bucharest University Emergency Hospital (386.1), and "Sf. Ioan Cel Nou" County Emergency Hospital Suceava (376, 3).

In the northwest, west, and central parts of the country, hospitals in Cluj-Napoca, Târgu-Mureș, Oradea, Baia Mare, and Satu Mare also registered higher values of clinical activity. The hospital units, with the reduced dynamic of clinical activity, belong to healthcare centres, palliative care, mental illness, or TB (lung disease)

Table 18.1. Descriptive statistics

Indicators	N	Minimum	Maximum	Mean	Std. deviation
Nb_cases	408	43	17,737	3136,68	3474,175
Nb_ cases _county	408	17	15,637	2575,17	2945,405
Nb_cases_rural	408	10	7205	1354,11	1504,800
Nb_cases_emergency	408	0	15,067	1715,57	2345,380
Nb__cases_referral GP	408	0	11,806	685,21	1202,542
Valid N (listwise)	408				

Table 18.2 Total variance explained and the component matrix

Component	Initial eigenvalues	Extraction sums of squared loadings		
	Cumulative %	Total	% of Variance	Cumulative %
1	76.567	3,828	76.567	76.567
Extraction Method: Principal Component Analysis				
Component Matrix[a]				

	Component
	1
Z1	0.982
Z2	0.977
Z3	0.946
Z4	0.924
Z5	0.402
Extraction Method: Principal Component Analysis	
a. 1 components extracted	

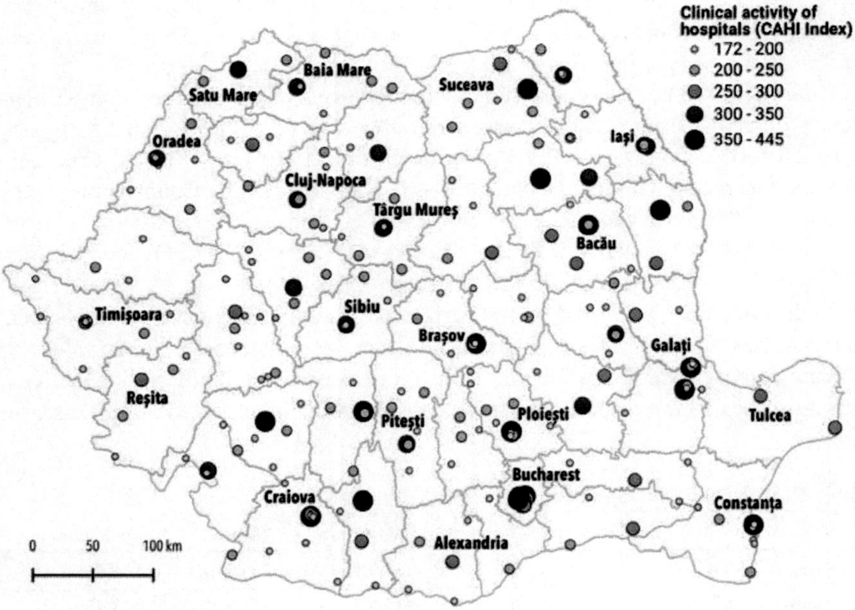

Fig. 18.3 Clinical activity of hospitals (CAHI index)

preventers. Most hospitals cannot cover complex services due to a lack of equipment or medical staff, and the patients are re-directed to higher competency hospitals.

The hospital network is considered hypertrophied, and the community-level services are deficient. Professionals claimed that hospitalizations could be avoided (on average, 7%). High rates of avoidable hospitalizations were associated with high mortality and inadequate coverage of the population with primary care (Ciutan et al., 2016; World Bank, 2011). Small-capacity hospitals, generally located in small and medium-sized cities, can deliver only essential diagnostic services, such as mild respiratory infections (Ministry of Health, 2016). In 2019, over 42% of health expenditures were directed to hospital services, compared to the EU average of 29% (OECD, 2019). Moreover, the transition to outpatient care is at an early stage.

The current strategic plans of the Ministry of Health intend to reduce the number of hospital beds for the care of patients with acute illnesses (4 beds per 1,000 inhabitants compared to 4.2 beds per 1,000 inhabitants at present) and to decrease the hospitalization rate for acute diseases, accompanied by the simultaneous increase in the number of day hospitalizations. Furthermore, regarding the total number of surgeries, 20% of them will be performed in outpatient care units.

The location of the hospitals according to the new plans acquires strategic attributes, being previously provided in the documentation (Ministry of Health, Ministerial Order number 1085/2012) and follows the configuration of the largest cities: Bucharest, Cluj-Napoca, Craiova, Iași, Timisoara, Constanța, and Târgu-Mureș.

The regional hospitals Iași, Craiova, and Cluj-Napoca, corresponding to the North-East, South-West, and North-West development regions, are currently in the design phase. For the other four hospitals (Bucharest, Timisoara, Constanța, and Târgu-Mureș), in-depth investment studies and other regional planning schemes will be provided. Another measure involves restructuring hospitals in rural areas or those with a single speciality and transforming them into other public units.

18.6 Conclusions

Hospitals play a central role in healthcare services and benefit from a substantial share of health resources. However, the determinants of their performance and quality of care require reliable assessment tools and policy interventions towards hospital services' organization. The current trends in decision-making policies aim to restructure in-hospital care and improve primary and ambulatory care.

The changes in the Romanian health system were associated, in particular, with a contradictory reform process, with significant outcomes, such as the decentralization of the system, the autonomy of the health system at a regional and local level, the improvement of financing mechanisms, and privatization of units of healthcare.

The geographical distribution of hospitals in Romania is consistent with those of urban areas. The Romanian hospital network relies on 567 hospitals, out of which

366 belong to the public sector, managed by the Ministry of Health and local authorities. However, despite the large number, hospitals vary widely across the country depending on the size, specialization, and medical services they provide. Moreover, not all hospitals can provide complex care, and people mainly address emergency or university hospitals located in several largest cities to receive quality care.

Hospitals are financed based on the diagnosis-related group system (DRGs) that provides comparable costs for patients of similar medical characteristics. Romanian hospitals received more than 45% of the total healthcare budget in 2018. However, the costs and diversity of medical services in hospitals have increased, and hospital services are generally overused.

Although the current legislation defines standards for evaluating hospitals, the planning of hospitals at the spatial level needs to be improved. There are significant differences in the self-assessment reports of hospitals in terms of the level of competence, and there is no relationship with hospitals' spatial and functional distribution.

Hospitals have benefited from several reforms in the system of organization, specialization, and management performance. In response to healthcare financing and performance management, hospitals are shifting services from inpatient to outpatient organizational settings. The reforms will continue unexpectedly from the management perspective by changing the organizational paradigm, insisting on empowerment, corporatization, and privatization according to the recommendations of the World Bank and public authorities. Assessing the reform's impact on hospital networks requires long-term analysis and reliable quantitative and qualitative data.

References

Albreht, T. (2009). Privatization processes in healthcare in Europe - a move in the right direction, a 'trendy' option, or a step back? *European Journal of Public Health, 19*(5), 448–451.

Ammar, W., Khalife, J., El-Jardali, F., Romanos, J., Harb, H., Hamadeh, G., & Dimassi, H. (2013). Hospital accreditation, reimbursement and case mix: Links and insights for contractual systems. *BMC Health Services Research, 13*(1), 1–9. https://doi.org/10.1186/1472-6963-13-505

Andrulis, D., & Duchon, L. (2007). The changing landscape of hospital capacity in large cities and suburbs: Implications for the safety net in metropolitan America. *Journal of Urban Health, 84*(3), 400–414.

Bankauskaite, V., & Saltman, R. (2007). Central Issues in the decentralization debate. In R. Saltman, V. Bankauskaite, & K. Vrangbaek (Eds.), *Descentralization in healthcare* (pp. 11–20). Open University Press.

Ciutan, M., Scîntee, S. G., Dosius, M., Mototolea, C., & Vlădescu, C. (2016). Geographical distribution of avoidable hospital conditions in Romania. *Procedia Environmental Sciences, 32*, 318–326.

Coreia, T., & Denis, J. L. (2016). Hybrid management, organizational configuration, and medical professionalism: Evidence from the establishment of a clinical directorate in Portugal. *Health Services Research, 16*(Suppl 2), 161.

Curtis, S. E., Gesler, W. M., Fabian, K., Francis, S., & Priebe, S. (2007). Therapeutic landscapes in hospital design: A qualitative assessment for staff and services users of the design of a new mental hospital inpatient unit. *Environment and Planning C, 25*, 591–610.

Decancq, K., & Lugo, M. A. (2013). Weights in multidimensional indices of wellbeing: An overview. *Econometric Reviews, 32*(1), 7–34.

Dent, M., Bode, I. (2014). Converging hybrid worlds? Medicine and hospital management in Europe. *International Journal of Public Sector Management, 27* (5).

Diagnostic Related Groups (DRG). (2019). Retrieved November 5, 2019, from http://www.drg.ro/index

Domnişoru, C. (2012). *Reforme în sistemul românesc de sănătate. Actori, mize, procese*, Friedrich –Ebert - Stiftung, România.

Dumitrache, L. (2014). Défis, incohérences et échecs de la réforme sanitaire en Roumanie. In A. Krasteva, & D. Vasilcu (Eds.), *Migration en blanc. Medecins d'est en ouest* (pp. 21–40). L'Harmattan.

Dumitrache, L., Nae, M., Dumbrăveanu, D., Simion, G., & Suditu, B. (2016a). Contrasting clustering in healthcare provision in Romania: Spatial and aspatial limits. *Procedia Environmental Sciences, ECOSMART—Environment at Crossroads: Smart Approaches for a Sustainable Development, 32*, 290–299.

Dumitrache, L., Zamfir, D, Nae, M., Simion, G., & Stoica, V.-I. (2016b). The urban nexus: Contradictions and dilemmas of (post)communist (sub)urbanization in Romania. *Human Geographies. Romanian Journal of Studies and Research in Human Geography, 10*(1), 39–58.

Dumitrache, L., Nae, M., Simion, G., & Taloș-A, M. (2020). Modelling potential geographical access of the population to public hospitals and quality healthcare in Romania. *International Journal of Environmental Research and Public Health, 17*, 8487.

Duran, A., Chanturidze, T., Gheorghe A., & Moreno, A. (2019). Assessment of public hospitals governance in Romania: Lessons from 10 case studies. *International Journal of Health Policy and Management, 8(4),* 199–210.

European Commission, European Core Health Indicators. Retrieved December 3, 2019, from https://ec.europa.eu/health/indicators

France, F. H. R. (2003). Case mix use in 25 countries: A migration success but international comparisons failure. *International Journal of Medical Informatics, 70*(2–3), 215–219. https://doi.org/10.1016/S1386-5056(03)00044-3

Gesler, W., & Curtis, S. (2007). Application of concept of therapeutic landscapes as to the design of hospital in the UK: The example of a mental health facility in London. In: Williams. A. (Ed). *Therapeutic landscapes* (pp.149–164). Ashgate Publishing.

Government of Romania. (1991). Hotărâre nr.525 din 27 iunie 1996 pentru aprobarea Regulamentului general de urbanism. Anexa 1.7. Constructii pentru sănătate, republicat în Monitorul Oficial nr. 856/27/2002. Retrieved May 18, 2020 from legislatie.just.ro/Public/DetaliiDocumentAfis/9499

Government of Romania.(2010a). Hotărârea Guvernului nr. 529/2010a pentru aprobarea Listei spitalelor pentru care se menţine sau se transferă managementul asistenţei medicale la autorităţile administraţiei publice locale şi la Primăria Municipiului Bucureşti, *Monitorul Oficial al României*, Partea I, nr. 385/10.06.2010a. *Monitorul Oficial* Publishing House, Romania.

Government of Romania.(2010b). Ordonanţa de Urgenţă a Guvernului nr. 48/2010b pentru modificarea şi completarea unor acte normative din domeniul sănătăţii în vederea descentralizării, *Monitorul Oficial al României,* Partea I, nr. 384/10.06.2010b. *Monitorul Oficial* Publishing House, Romania.

Government of Romania.(2021). Law no 1/2021 for the ratification of the loan agreement regarding the results based program in the health sector in Romania, between Romania and IBRD, signed on September 23rd, 2019, and updated on July 31st, 2020, and August 26th 2020; 2021. Retrieved August 20, 2021, from http://legislatie.just.ro/Public/DetaliiDocument/235938

Greco, S., Ishizaka, A., Tasiou, M., & Torrisi, G. (2019). On the methodological framework of composite indices: A review of the issues of weighting, aggregation, and robustness. *Social Indicators Research, 141*, 61–94.

Hentschker, C., & Mennicken, R. (2018). The volume-outcome relationship revisited: Practice indeed makes perfect. *Health Services Research, 53*(1), 15–34.

Ikkersheim, D., Tanke, M., van Schooten, G., Bresser, N., & Flueren, H. (2013). Modelling hospital infrastructure by optimizing quality, accessibility, and efficiency via a mixed integer-programming model. *Health Services Research, 13*, 220.

INCDS (Institutul Național de Cercetare și Dezvoltare în Sănătate). (2004). *Sistemul de finanţare bazată pe caz rezolvat a spitalelor* (sistemul grupelor de diagnostic—DRG), întâlnire de lucru INCDS cu mass-media, 16 ianuarie 2004.

Jakóbczyk, D. K., Albreht, T., Behmane, D., Bryndova, L., Dimova, A., Dzakula, A., Habicht, T., Murauskiene, L., Scîntee, S. G., Smatana, M., Velkey, Z., & Quentin, W. (2020). Hospital reforms in 11 central and Eastern European countries between 2008 and 2019: A comparative analysis. *Health Policy, 124*(4), 368–379.

Jeurissen, P., Duran, A., & Saltman, R. (2016). Uncomfortable realities: The challenge of creating real change in Europe's consolidating hospital sector. *Health Services Research, 16*(2), 168.

Keaney, M. (2002). Unhealthy accumulation: The globalization of healthcare privatization. *Review of Social Economy, 60*(3), 331–357.

Keysfin Report. (2018). Romanian seasonal illnesses and the public healthcare crisis lead the private medical services to a new record in 2018. Retrieved June 20 from https://www.keysfin.com/EN#!/Pages/News/NewsDetails&title=bolile-romanilor-si-criza-din-spitalele-publice-duc-piata-serviciilor-medicale-private-la-un-nivel-record

Krusse, F., Stadhouders, N., Adang, E., Groenewoud, S., & Jeurissen, P. (2018). Do private hospitals outperform public hospitals regarding efficiency, accessibility, and quality of care in the European Union? A literature review. *The International Journal of Health Planning and Management, 33*, e434–e453.

Lee, K. C. L., Sethuranam, L., & Yong, J. (2019). On the hospital volume and outcome relationship does specialization matter more than volume? *Health Services Research, 50*, 36.

Levaillant, M., Marcilly, R., Levaillant, L., Vallet, B., & Larmer, A. (2020). Assessing the hospital volume-outcome relationship in surgery: A scoping review protocol. *British Medical Journal Open, 10*, e038201.

Maarse, H. (2006). The privatization of healthcare in Europe: An eight—country analysis. *Journal of Health Politics, Policy and Law, 31*(5), 981–1014.

Marino, D., & Quattrone, G. (2019). A proposal for a new index to evaluate hospital resource allocation: The case of Italian NHS rationalisation. *European Research on Management and Business Economics, 25*(1), 23–29. https://doi.org/10.1016/j.iedeen.2018.10.001

Matei, E., Saghin, D., Dumitrache, L., & Nae, M. (2018). Multidimensional approach on sustainability of ageing in Romanian residential care centres for elders. *Sustainability, 10*, 3336.

McKee, M., & Healy, J. (2001). Hospitals in transition in central and eastern Europe and central Asia. *European Observatory of Healthcare Systems, 7*(3), 1–90.

McKee, M., & Healy, J. (2002). The significance of hospitals. An introduction. In M. McKee, J. Healy (Eds.), *Hospitals in a changing Europe,* (pp. 3–13). Open University Press.

Mendez, C. M., Harrington, D. W., Christenson, P., & Spellberg, B. (2014). Impact of hospital variables on case mix index as a marker of disease severity. *Population Health Management,* 17(1), 28–34. https://doi.org/10.1089/pop.2013.0002

Morche, J., Mathes, T., & Pieper, D. (2016). Relationship between surgeon volume and outcomes: A systematic review of systematic reviews. *Systematic Reviews, 5*, 204.

Ministry of Health. (2014). Strategia Națională de Sănătate 2014–2020, București. Retrieved July 28, 2019, from Strategia Nationala de Sanatate 2014–2020 – Ministerul Sănătății (ms.ro)

Ministry of Health. (2016). Regional health service plans. *Monitorul Oficial* No. 988 bis/8 December 2016.

Ministry of Health. (2011). Strategia Națională de Raționalizare a Spitalelor (varianta actualizată). Retrieved September 22, 2019, from https://www.srcv.ro/pgs/strategia%20nationala%20de%20rationalizare%20a%20spitalelor.pdf

Monitorul Oficial al României no.274/2011. 2011. Ministry of Health Order no.323/2011, Monitorul Oficial Publishing House 2011, Bucharest, Romania.

Nae, M. (2014). Le système de santé et sa réforme: représentations dans le cyberespace. In A. Krasteva, & D. Vasilcu (Eds.), *Migration en blanc. Medecins d'est en ouest* (pp. 45–65), L'Harmattan, coll. Global & Local.

Nardo, M., Saisana, M., Saltelli, A., Tarantola, S., Hoffman, A., & Giovannini, E. (2005). *Handbook on constructing composite indicators*. OECD Publishing.

National Institute of Statistics. (2018). Tempo Online Statistics. Retrieved April 13, 2020, from http://statistici.insse.ro:8077/tempo-online/#/pages/tables/insse-table

Nolte, E., Pitchforth, E., Miani, C., & McHugh, S. (2014). *The changing hospital landscape: An exploration of international experiences*. Santa Monica, California. Rand Europe, Cambridge, UK. Retrieved May 15, 2020, from https://www.rand.org/content/dam/rand/pubs/research_reports/RR700/RR728/RAND_RR728.pdf

Noordegraaf, M. (2007). From 'Pure' to 'Hybrid' professionalism. Present-day professionalism in ambiguous public domains. *Administration and Society, 39*(6), 761–785.

Numerato, D., Salvatore, D., & Fattore, G. (2012). The impact of management on medical professionalism. A review. *Sociology of Health and Illness, 34(4),* 626–644.

OECD. (2008). *Handbook on constructing composite indicators: Methodology and user guide*. OECD Publishing.

OECD/EU. (2016). *Health at a Glance: Europe 2016—State of health in the EU cycle*. OECD Publishing. Retrieved January 19, 2020, from https://doi.org/10.1787/9789264265592-en

OECD, European Observatory on Health Systems and Policies. Romania: Country Health Profile, 2017 OECD/European Observatory on Health Systems and Policies. Romania: Country Health Profile 2017, State of Health in the EU, OECD Publishing, Paris/European Observatory on Health Systems and Policies, Brussels. Retrieved January 29, 2020, from https://doi.org/10.1787/9789264283534-en

OECD, European Observatory on Health Systems and Policies. Romania: Country Health Profile, 2019 OECD/European Observatory on Health Systems and Policies. Romania: Country Health Profile. (2019). State of Health in the EU, OECD Publishing, Paris/European Observatory on Health Systems and Policies, Brussels. Retrieved January 29, 2020, from https://doi.org/10.1787/f345b1db-en

Parliament of Romania. (2001). Law No.350/2001 on territorial planning and urbanism; modified and completed by OG 27/2008; Monitorul Oficial 373/2001; Monitorul Oficial Publishing House. Retrieved June 18, 2020, from http://legislatie.just.ro/Public/DetaliiDocument/29453

Porter, M. E., & Teisberg, E. O. (2006). *Redefining healthcare*. Boston Harvard Business School Press.

Powell, M., & Miller, R. (2013). Privatizing the English national health service: An irregular verb? *Journal of Health Policy, Politics and Law, 38*(5), 1053–1061.

Powell, M., & Miller, R. (2014). Framing privatisation in the English national health service. *Journal of Social Policy, 43*(3), 575–594.

Radu, C. P, Chiriac, N. D., & Vlădescu, C. (2010). Changing patient classification system for hospital reimbursement in Romania. *Croatian Medical Journal, 5*(1), 250–258.

Radu, C. P, Chiriac, N. D., Vlădescu, C., & Shah, J. (2010). România în contextul internațional al sistemelor de clasificare a pacienților. *Management în sănătate, XIV/1/2010*, 15–17.

Saisana, M., & Tarantola, S. (2002). State-of-the-art report on current methodologies and practices for composite indicator development. European Commission, Joint Research Centre, Institute for the Protection and the Security of the Citizen, Technological and Economic Risk Management Unit, Ispra, Italy.

Saltman, R. (2003). Melting public—private boundaries in European health systems. *European Journal of Public Health, 13*, 24–29.

Saltman, R. (2009). The rise of pragmatism in state/market debate. *Health Economics, Policy and Law, 4*, 509–511.

Scîntee, S. G., Vlădescu, C., & Sagan, A. (2018). The unexpected outcomes of the closure of 67 inpatient care facilities in 2011 in Romania. *Health Policy, 122*, 1161–1164.

Setola, N., & Borgiani, S. (2016). Interface space between city and hospitals. In S. Nicoleta & B. Sabrina (Eds.), *Designing public places in hospitals* (pp. 10–15). Routledge.

Setola, N, Borgiani, S., Martinez, M., & Tobari, E. (2013). The role of spatial layout of hospital public spaces in informal patient - medical staff interface. In by Y.O. Jim, H.T. Park, & K.W. Seo (Eds), *Proceedings of the Ninth International Space Syntax Symposium.* Sejong University, Seoul.

Tiemann, O., & Schreyögg, J. (2012). Changes in hospital efficiency after privatization. *Health Care Management Science, 15*, 310–326.

Vlădescu, C., Pascu, O., Astărăstoae, V., Verboncu, I., Anghel, R., Stănescu, A., Molnar, G., Oslavszky, V., & Irimia, C. (2008). *Un sistem sanitar centrat pe nevoile cetăţeanului.* Comisia Prezidenţială pentru Analiza şi Elaborarea Politicilor din Domeniul Sănătăţii Publice din România, București.

Volintiru, M. (2018). The internalization context of private health services providers in Europe: Romanian market case study. In *Proceedings of the 12th International Conference on Business Excellence* (pp. 1038–1047). De Gruyter Open.

Waters, H. R., Hobart, J., Forrest, C., Siemens, K. K., Pittman, P., Murthy, A., Vanderver, B. G., Anderson, G., & Morlock, L. (2008). Trends: health insurance coverage in central and Eastern Europe: Trends and challenges. *Health Affairs, 27*(2), 478–486.

World Bank. (2011). România—Analiză funcțională: Sectorul sanitar. Retrieved March 22, 2019, from http://sgg.gov.ro/docs/File/UPP/doc/rapoarte-finale-bm/etapa-II/MS-RO-FR-Health-Sector-ROM

World Bank. (2018a). Romania public finance review: Enhancing the efficiency of public spending in the health sector. Washington, DC: World Bank. Retrieved July 28, 2019, from http://sgg.gov.ro/docs/File/UPP/doc/rapoarte-finale-bm/etapa-I/06_RO%20FR_Public%20Finance_%20Final%20Report.pdf

World Bank. (2018b). Romania regional hospital analysis study. Regional Referral Networks in Romania. June 28, 2018b. Retrieved July 20, 2019, from https://documents.worldbank.org/en/publication/documents-reports/documentdetail/798931530245678590/romania-regional-hospital-analysis-study-regional-referral-networks

Zamfir, D., Dumitrache, L., Stoica, I.-V., & Vîrdol, D. (2015). Spatial inequalities in healthcare provision in Romania: Milestones for territorial sustainable development. *Carpathian Journal of Earth and Environmental Sciences, 10*(3), 177–188.

Chapter 19
Influence of Social Environment and Unconventional Substance Consumption on Young People's Sexual Health Behaviour in Lagos Metropolis, Nigeria

Onipede Wusu and R. A. Soniyi

Abstract This study examines the effect of social environment and unconventional substance consumption on sexual health behaviour among young people in Lagos metropolis. The cross-sectional survey design was used to elicit data among 1020 young people (10–24 years old). Statistical analysis involved univariate, bivariate, and multivariate tools. Chi-square analysis suggested school and ethnic environment were associated with modern contraception to prevent sexually transmitted infections (STIs) ($P < 0.05$) and pregnancy ($P < 0.01$) among females. Similarly, family environment was significantly associated with modern contraception for STIs prevention among sexually active males ($P < 0.01$). Also, ever use of any unconventional substance and contraception for STIs prevention were significantly associated in the total sample of males ($P < 0.001$) and females ($P < 0.001$). Adjusted logistic regression models indicated that out-of-school environment significantly predicted contraception for STIs prevention (OR = 6.2, $P < 0.01$) and unwanted pregnancy (OR = 5.7, $P < 0.001$) compared to those in schools among females and sexually active males in monogamous family (OR = 0.9, $P < 0.05$) relative to single parenthood family environment among sexually active males. Also, never use of unconventional substance significantly predicted contraception to prevent STIs among males (OR = 0.2, $P < 0.001$) and unwanted pregnancy among their sexual partners (OR = 0.2, $P < 0.001$). Thus, modern contraceptive use to prevent STIs and unwanted or mistimed pregnancy is likely common among young males but less likely among females in school environment. Sexually active males in monogamous family environment are less likely to report modern contraceptive use for STIs prevention. Surprisingly, those who ever consumed unconventional substance were more likely to use modern contraceptives to prevent STIs and unwanted pregnancy compared to those who never used. Therefore, policies that could awake the sexual subjectivity of young people in school and family environments, and neutralise patriarchal stereotypes shaping

O. Wusu (✉) · R. A. Soniyi
Department of Sociology, Lagos State University, Lagos, Nigeria
e-mail: onipede.wusu@lasu.edu.ng

B. R. K. Sinha (ed.), *Urban Dynamics, Environment and Health*,
https://doi.org/10.1007/978-981-99-5744-6_19

sexual subjectivity in the environments are imperatives. Also, everything necessary must be done to discourage consumption of unconventional substance use among young people in Nigeria.

Keywords Social environment · Unconventional substance use · Unwanted pregnancy · Sexually transmitted infections · Contraceptive use

19.1 Introduction

Lagos metropolis is a result of a continuous process of urbanisation traceable to the combined effects of high fertility and rapid rate of migration. As a result, the city is characterised by an interesting interconnection of more urban and a network of suburban settlements. Wealth of knowledge on those explicit characteristics of the metropolis exists but little is known about the influence of the interaction of existing social environment (SE) and the culture of unconventional substance consumption on the sexual health of young people in the metropolis. Incidentally, young people constitute about 53% of the entire population resident in the metropolis. Sexual health challenge remains a critical public health concern facing young people in sub-Saharan Africa (Wusu, 2013). Nigeria with a youthful population has multiple sexual health problems confronting the country at local, state, and federal government levels because of poor contraception behaviour. One of such sexual health problems is low contraceptive use, which increases young people's risk of contacting STIs and having unwanted or mistimed pregnancy (Wusu, 2016).

The pathetic sexual health situation among young people in Nigeria (National Population Commission [Nigeria] and ICF Macro International, 2019) calls for further research towards building better empirical understanding of the determinants of poor use of contraception. In light of this need, it is imperative for rigorous consideration of some of the factors that have not been well explored is a significant research goal. This paper considered two of such factors, namely social environment (SE) and a growing component of this environment—unconventional substance use—that have become popular in the society (Jubril, 2019; Wusu et al., 2019). Social determinants of health (SDH) that have been reported as a significant predictor of sexual health among young people (Maness & Buhi, 2016) and the association between unconventional substance use and sexual health behaviour among them deserves more attention in Nigeria. Also, sex as means of survival is a common practice among young people in sub-Saharan Africa (Hampanda, 2013; Svinurai & Makhubele, 2018; Wusu, 2010), and substance use and sex for monetary benefit have been reported to be significantly related (Svinurai & Makhubele, 2018). Therefore, it is important to examine the role of substance use (especially unconventional substance) in sexual health.

Previous studies have shown that components of SDH (SE is used in the context of this study as synonym of SDH) such as perception of peers and peer networks have been reported to exert stronger influence over contraceptive behaviour among young people than their individuals' attitude (Jones et al., 2017; Ochako et al., 2015;

Viner et al., 2012). The studies suggest that the individual attitude towards use of contraceptives for prevention of sexual health challenges such as unwanted pregnancy and STIs are likely subject to that of the peers or the dominant contraceptive attitude in their peer network (Ochako et al., 2015). Research has also demonstrated that supportive families as well as safe and supportive schools (a SE indicator) with effective sexual health education programmes are significantly related to improved sexual health among young people (Chernick et al., 2019; Maness & Buhi, 2016; Srijaiwong et al., 2017; Viner et al., 2012). Family structure and living arrangement of young people will either enhance or endanger their sexual health. It is expected that a homogenous or nuclear family living arrangement is likely to be more supportive and proper monitoring of young people is likely to be a norm, and young people's basic needs are better provided for. Consequently, young people's sexual health is likely positioned to improve in such settings. In contrast, young people who live in polygynous or single parenthood family settings are most likely not properly monitored and their needs may not be well taken care of, and consequently such young persons' sexual health may end poor. Also, supportive schools that are safe and socially cohesive, and which expose their students to sexual health tips from time to time are SE where young people can experience sexual subjectivity (connotes consciousness about personal and others' sexual right) that predisposes them to improved sexual health (Lys et al., 2019; Maness & Buhi, 2016). The reverse will be the case among those in unsafe and unsupportive schools and neighbourhoods (Decker et al., 2018).

Moreover, substance use has become an integral component of the global human society and studies have shown that substance consumption is capable of promoting poor sexual health among young people (Adekeye et al., 2015; Elkington et al., 2010; Lys et al., 2019; Svinurai & Makhubele, 2018). Some of the consequences of substance use for sexual health identified include increased sexual risk behaviour owing to poor sexual subjectivity, drunkenness or intoxication, weakening of sense of judgement and suppressing the feeling of fear and reducing the perception of risk. However, recently a new set of substance use culture has emerged among young people in Nigeria, which is coined here as unconventional substance use (Wusu et al., 2019). Some of such substances include animal dung, hydrogen sulphide gas (sewer gas), premium motor spirit, rubber solution, nail polish, pawpaw leave, pawpaw seed, moringa leave, gun powder, and gutter from toilet (Danjuma et al., 2015; Wusu et al., 2019).

In the light of all the studies highlighted above, the role of social environment in the young people sexual health remains largely obscure in Nigeria. Also, it is apparent that knowledge on the influence of the interaction between social environment and unconventional substance use on sexual health of young people in the global literature is either non-existent or shallow. These are the considerations that underscore the significance of this research. This study employed the positivist methodology to test the hypothesis that the interaction between SE (indicated as school environment, religious environment, family environment, and ethnic environment) and unconventional substance use (defined here as unusual substances used as drugs) is likely to influence the sexual health indicator (use of modern contraceptives to prevent STIs and mistimed or unwanted pregnancy) among young people in Lagos metropolis.

19.2 Objectives of the Study

This study aims to: examine the influence of SE indicators and unconventional substance consumption on the indicators of sexual health among young people in Lagos metropolis, Nigeria; find the association between SE indicators (school, family, and religious environments) and sexual health indicators (modern contraception to prevent STIs and unwanted or mistimed pregnancy) among young people and those sexually active; determine the key influencer(s) of sexual health indicators among the SE indicators among young people and those sexually active; and identify the association between ever use of unconventional substance and sexual health indicators among young people and those sexually active.

19.3 Data and Methods

This study used the data collected through a survey conducted among young people (10–24 years of age) in Lagos metropolis in 2019. Lagos State has a population of over 20 million and it is the commercial hub of Nigeria. The location of Lagos State on the Nigerian landscape is represented in Fig. 19.1. Lagos State has 18 out of its 20 Local Government Areas (LGAs) interconnected into a mega city. The state houses a megacity that is situated between Badagry and Epe, which are the two remote largely rural LGAs at the two horizontal ends of the state. The metropolis emerged from a continuous process of urbanization traceable to the combined effects of high fertility and rapid rate of migration. As a result, the city is characterized by an interesting interconnection of more urban and a network of suburban settlements.

The survey employed a stratified and a modified cluster sampling strategy in selecting the respondents. The research team stratified the study population into in-school and out-of-school. On the one hand, the in-school survey was conducted in two purposively selected Junior and Senior Secondary Schools in Alimosho and Ikorodu LGAs, and two tertiary institutions located in Lagos Mainland LGA. The data collection instrument was administered to all Junior and Senior Secondary School Students of classes selected with the assistance of their class teachers, after due approval of the Head of Administration. The respondents sampled in the two tertiary institutions were randomly selected and the data collection instrument was administered to them.

On the other hand, the out-of-school sample was selected through a modified cluster sampling procedure because the substance users were considered as special population that may be difficult to reach. First, the team carried out acquittance visit through the metropolis to identify the areas that are notorious for substance use. Through this visit, the research team listed various locations and assigned them

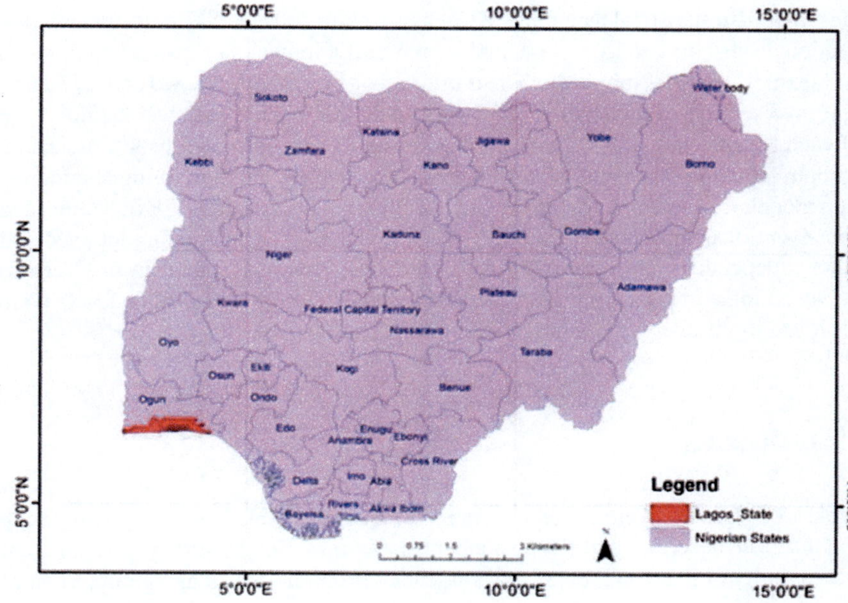

Fig. 19.1 Map of Nigeria highlighting Lagos State. *Source* Onwuadiochi et al. (2021)

numbers. Random selection process was adopted to select the locations where the data collection instrument (questionnaire) was administered on the survey respondents in the areas where young people who were unconventional substance consumption gather in the evenings to relax and for social interactions. The research assistants negotiated access through the leaders of the groups after explaining the purpose of the study to them.

For ethical reasons the trained field assistants who administered the questionnaire made it clear to every potential respondent that their consent to participate in the study was needed, that it was a voluntary exercise. They could decide not to participate and they could withdraw their consent at any stage of the interview. Interview was only conducted after a verbal consent to participate in the study was expressed by the respondents.

Data analysis employed the Statistical Package for Social Sciences (SPSS) version 25.0 in the conduct of three levels of data analysis. The three levels of analysis focused on all respondents sampled and those who were sexually active within the three months before the survey. Nevertheless, the bivariate and logistic regression analyses results presented exclude sexually active females and modern contraceptive use for pregnancy prevention because there was no significant association between the independent and dependent variables, and owing to the inability to construct a statistically significant model for females and on contraception for pregnancy prevention. First, univariate analysis employed descriptive statistical tools to describe the background characteristics of all respondents. Second, bivariate analysis entails deployment of the chi-square statistical technique to test association between SE indicators, unconventional substance consumption, and sexual health indicators separately among all and sexually active respondents. The third level of analysis was in two parts. The first part involved use of the logistic regression statistical technique to test the influence of each SE indicator and unconventional substance use on sexual health indicators without taking into account other independent variables (unadjusted logistic regression models). The second part examined the influence of each of the SE indicators and unconventional substance use on sexual health indicators while taking into account other independent and modifying characteristics (age was crucial in this regard) (adjusted logistic regression models). All possible tests of significance levels were deployed in the analysis, including 0.001, 0.01, and 0.05.

19.4 Results

Table 19.1 shows the univariate description of all the variables involved in this study and the basic description of the respondents interviewed in the survey. It is apparent that majority of male and female respondents were in the 15–19 age group. Most of the respondents were adolescents. And a vast majority of them were schooling. Also, majority of the adolescents were in Christian religious environment and majority of them indicated being more religious. Monogamous family environment was very popular among the respondents and a vast majority were of the Yoruba extraction in Nigeria. With respect to ever use of unconventional substance, although majority of the respondents claimed they never consumed any, a higher proportion of females (81%) than males (56%) reported they never used any unconventional substance. Obviously, most of both male and female respondents reported never used modern contraceptives to prevent STIs and pregnancy, but a higher proportion of males indicated ever use to prevent STIs and pregnancy (among their sex partners). In a similar vein, only a small proportion reported ever engaging in sexual acts (males = 49.1% and females = 16.1%). The proportion that reported engaging in sex three months before the survey was higher among males (35%) than that of females (11.5%).

Table 19.2 presents the description of the background characteristics of sexually active respondents in the study survey. One hundred and sixty-six (16.3% of the total sample) males and 61 females (6% of the total sample) reported they were sexually active in the three months before the survey. Obviously, majority of both sexes were in adolescent age group and, just like in the total sample, a vast majority indicated they were schooling. In terms of religious environment, majority of both males and females indicated Christianity and while males reported being more religious, majority of females reported they were less religious. Also, monogamous family environment predominated among both male and female respondents. With respect to ever use of any unconventional substance, majority of males indicated ever use of the substance but majority of their female counterparts reported never use of any unconventional substance. In addition, majority of males reported use of modern

Table 19.1 Percentage distribution of respondents by selected socio-demographic characteristics by sex

Characteristics	Male (n = 478)	Female (n = 541)
	%	%
Age		
< = 14	8.8	14.2
15–19	69.5	74.9
20–24	21.8	10.9
School environment		
No	7.6	4.3
Yes	92.4	95.7
Religious environment		
Christianity	58.8	57.4
Islam	38.7	40.9
Others	2.5	1.7
Religiosity		
Less religious	31.2	27.1
More religious	68.8	72.9
Family environment		
Monogamous	58.1	60.2
Polygynous	23.1	19.3
Single Parenthood	18.8	20.5
Ethnic environment		
Yoruba	74.6	78.4
Others	25.4	21.6
Ever use of unconventional substance use		
No	55.9	80.7
Yes	44.1	19.3
Modern contraceptive use for STIs prevention		
No	66.1	92.6
Yes	33.9	7.4
Modern contraceptive use for pregnancy prevention		
No	64.2	88.7
Yes	35.8	11.3
Ever engaged in a sexual act		
No	50.9	83,9
Yes	49.1	16.1

Source Survey in Lagos Metropolis in 2019

contraception for STIs prevention but majority of their female counterparts reported otherwise. However, in the case of modern contraception for pregnancy prevention, majority of both males and females reported use.

In Table 19.3 the bivariate association between the independent and dependent variables (showing only those who reported modern contraceptive use to achieve

Table 19.2 Percentage distribution of sexually active respondents by selected socio-demographic characteristics and sex

Characteristics	Male (n = 166)	Female (n = 61)
	%	%
Age		
< = 14	4.2	4.9
15–19	62.0	52.5
20–24	33.7	42.6
School environment		
No	12.0	18.0
Yes	88.0	82.0
Religious environment		
Christianity	65.2	73.8
Islam	32.3	24.6
Others	2.4	1.6
Religiosity		
Less religious	44.0	59.0
More religious	56.0	41.0
Family environment		
Monogamous	59.0	67.2
Polygynous	23.0	11.5
Single Parenthood	18.0	21.3
Ethnic environment		
Yoruba	71.5	55.7
Others	28.5	44.3
Ever use of unconventional substance use		
No	30.3	55.0
Yes	69.7	45.0
Modern contraceptive use for STIs prevention		
No	31.3	60.7
Yes	68.7	39.3
Modern contraceptive use for pregnancy prevention		
No	27.7	34.4
Yes	72.3	65.6

Source Survey in Lagos Metropolis in 2019

Table 19.3 Percentage distribution of young people by selected indicators of social environment, ever use of unconventional substance and sexual health indicators and by sex in Lagos Metropolis

Characteristics	Use of modern contraceptives to prevent STIs		Use of modern contraceptives to prevent pregnancy	
	Male (Yes, n = 162)	Female (Yes, n = 40)	Male (Yes, n = 171)	Female (Yes, n = 61)
Age				
< = 14	6.8(11)	5.0(2)	6.4(11)	3.3(2)
15–19	64.2(104)	52.5(21)	64.9(111)	55.7(34)
20–24	29.0(47)	42.5(17)	28.7(49)	41.0(25)
Chi-square	08.0*	45.0***	8.3*	66.2***
School environment				
No	6.2(10)	22.5(9)	7.2(12	18.0(11)
Yes	93.8(152)	77.5(31)	93.0(159)	82.0(50)
Chi-square	0.68	35.1***	0.11	31.8***
Religious environment				
Christianity	59.6(96)	75.0(30)	60.0(102)	75.4(46)
Islam	37.3(60)	25.0(10)	37.1(63)	23.0(14)
Others	3.1(5)	–	2.9(5)	1.6(1)
Chi-square	0.5	5.7*	0.4	9.3*
Religiosity				
Less religious	37.9(61)	50(20)	38.2(65)	49.2(30)
More religious	62.1(100)	50(20)	61.8(105)	50.8(31)
Chi-square	5.1*	11.5**	6.1*	17.0***
Family environment				
Monogamous	51.9(81)	61.5(24)	52.7(87)	50.0(36)
Polygynous	26.9(42)	17.9(7)	27.3(45)	16.7(10)
Single Parenthood	21.2(33)	20.5(8)	27.3(33)	23.3(14)
Chi-square	3.7	0.1	3.4	0.5
Ethnic environment				
Yoruba	75.2(121	65.0(26)	74.1(126)	63.9(39)
Others	24.8(40)	35.0(14)	25.9(44)	36.1(22)
Chi-square	0.0	4.6*	0.1	8.5**
Ever use of unconventional substance use				
No	32.1(51)	57.9(22)	31.0(52)	59.3(35)
Yes	67.9(108	42.1(16)	69.0(116)	40.7(24)
Chi-square	55.4***	13.6***	66.0***	19.4;***

Note * significant at $P < 0.05$; ** significant at $P < 0.01$; *** significant at $P < 0.001$; a few unequal summations of cases were observed due to non-response to some questions
Source Survey in Lagos Metropolis in 2019

clarity and neatness) is presented. Among the variables of interest to the study, school environment, religious environment, and ethnic environment as well as every use of unconventional substance were significantly associated with contraceptive use for prevention of STIs and unwanted or mistimed pregnancy. Respondent's age was also significantly associated with the indicators of the dependent variable. First, school environment was only significantly associated with contraception for STIs and pregnancy prevention among females ($P < 0.001$). A vast majority of females in schools reported contraception to prevent STIs (77.5%) and mistimed or unwanted pregnancy (82.0%) while only a negligible proportion of their counterparts not in any school environment reported modern contraceptive use for STIs (22.5%) and pregnancy (18.0%) prevention. Second, religious environment was only significantly associated with use of modern contraceptives to prevent mistimed or unwanted pregnancy among females ($P < 0.05$). While a vast majority of Christians reported modern contraceptives use to prevent pregnancy, minority of Muslims and followers of other religions reported same. Moreover, a general pattern among both males and females was that a higher proportion of those who indicated being more religious reported use of modern contraception to prevent STIs and pregnancy than their less religious counterparts.

Third, ethnic environment was also only significantly associated with modern contraception for STIs ($P < 0.05$) and pregnancy ($P < 0.01$) prevention among female respondents. At least 63% of female respondents of the Yoruba extraction reported use of modern contraceptives to prevent STIs or pregnancy. Ever use of unconventional substance was significantly associated with modern contraception for prevention of STIs and pregnancy ($P < 0.001$). The analysis revealed that a vast majority of males and at least 41% of females who indicated ever use of unconventional substance reported modern contraception to prevent STIs and pregnancy.

In Table 19.4, results of bivariate association between each of SE indicators, unconventional consumption, and modern contraception for STIs prevention among sexually active male respondents are presented. Again, there was no significant association observed in the sexually active female data and modern contraception for pregnancy prevention among both sexually active males and females. Hence, Table 19.4 shows only bivariate analysis for males and modern contraception for STIs prevention. The analysis revealed that school, religiosity, and family environment were significantly associated with modern contraception for STIs prevention. Nearly all sexually active males in school environment indicated use of modern contraception to prevent STIs ($P < 0.01$). Also, majority of sexually active more religious males indicated contraceptive use for STIs prevention ($P < 0.05$). Similarly, the highest proportion of male respondents who indicated contraceptive use (50.5%) were in monogamous family environment ($P < 0.01$).

Table 19.5 presents the unadjusted logistic regression odds ratios of the association between social environment indicators along with ever use of unconventional substance use and contraception for prevention of STIs and unwanted/mistimed pregnancy. It is apparent that school environment and ever use of unconventional substance use were significantly associated with the two indicators of dependent variable. Male respondents who were not in any school environment were less likely

Table 19.4 Percentage distribution of young people by selected indicators of social environment, ever use of unconventional substance and sexual health indicators by sex among sexually active young males in Lagos Metropolis

Characteristics	Use of modern contraceptives to prevent STIs	
	(No, n = 52)	(Yes, n = 114)
Age		
< = 14	1.9(1)	5.3(6)
15–19	59.6(31)	63.2(72)
20–24	38.5(20)	31.6(36)
Chi-square	1.5	
School environment		
No	23.1(12)	7.0(8)
Yes	76.9(40)	93.0(106)
Chi-square	8.7**	
Religious environment		
Christianity	72.5(37)	61.9(70)
Islam	25.5(13)	35.4(40)
Others	2.0(1)	2.7(3)
Chi-square	1.7	
Religiosity		
Less religious	55.8(29)	38.6(44)
More religious	44.2(23)	61.4(70)
Chi-square	4.3*	
Family environment		
Monogamous	76.9(40)	50.5(55)
Polygynous	13.5(7)	27.5(30)
Single Parenthood	9.6(5)	22.0(24)
Chi-square	10.2**	
Ethnic environment		
Yoruba	65.4(34)	74.3(84)
Others	34.6(18)	25.7(29)
Chi-square	1.4	
Ever use of unconventional substance use		
No	26.9(14)	31.9(36)
Yes	73.1(38)	68.1(77)
Chi-square	0.4	

Note * significant at $P < 0.05$; ** significant at $P < 0.01$; *** significant at $P < 0.001$; a few unequal summations of cases were observed due to non-response to some questions
Source Survey in Lagos Metropolis in 2019

Table 19.5 Unadjusted Odds ratios of Logistic Regression (with 95% CI) of the influence of social environment and ever use of unconventional substance on indicators of sexual health by sex among young people in Lagos Metropolis

Characteristics	Use of modern contraceptives to prevent STIs		Use of modern contraceptives to prevent pregnancy	
	Male	Female	Male	Female
School environment				
No	0.5(0.2–0.3)***	0.1(0–0.3)	0.6(0.3–1.4)	7.7(2.9–20.2)***
Yes	rc	Rc	Rc	rc
Religious environment				
Christianity	0.4(0.1–1.6	a	0.5(0.1–1.8)	1.4(0.1–13.4)
Islam	0.4(0.1–1.6)	a	0.5(0.1–1.9)	0.5(0.1–5.4)
Others	rc	a	Rc	rc
Family environment				
Monogamous	0.8(0.4–1.3)	0.9(0.4–2.4)	0.9(0.5–1.5)	0.9(0.4–1.9)
Polygynous	0.9(0.5–1.8)	1.0(0.4–2.4)	1.1(0.6–2.0)	0.8(0.3–2.1)
Single Parenthood	rc	Rc	Rc	rc
Ethnic environment				
Yoruba	1.0(0.6–1.7)	1.5(0.7–3.2)	0.9(0.6–1.6)	0.8(0.4–1.5)
Others	rc	Rc	Rc	rc
Ever use of unconventional substance use				
No	0.2(0.1–0.1)***	3.2(1.5–6.6)**	0.2(0.1–0.3)***	0.3(0.2–0.5)
Yes	rc	Rc	rc	rc
Model Chi-square P value	59.3***	27.8***	68.6***	43.9***

Note * significant at $P < 0.05$; ** significant at $P < 0.01$; *** significant at $P < 0.001$; a: variable not included in model; rc: reference category
Source Survey in Lagos Metropolis in 2019

to report contraceptive use for STIs prevention ($P < 0.001$) relative to those who were in school environment. Conversely, females who were not in a school environment were nearly 8 times more likely to report contraceptive use to prevent unwanted/ mistimed pregnancy compared to their counterparts in school environment ($P < 0.001$). Also, males who never used any unconventional substance were less likely to report contraception to prevent STIs and unwanted/mistimed pregnancy (for sex partners) compared to those who ever used unconventional substance ($P < 0.001$). Their female counterparts who indicated never used any unconventional substance were over 3 times more likely to report use of modern contraceptives to prevent STIs relative to those who ever used any unconventional substance ($P < 0.01$).

Moreover, Table 19.6 shows the adjusted logistic regression odds ratios on the influence of SE indicators and ever use of unconventional substance on contraceptive use for STIs and unwanted/mistimed pregnancy prevention. When age and religiosity

Table 19.6 Adjusted Odds ratios of Logistic Regression (with 95% CI) of the influence of social environment and ever use of unconventional substance on indicators of sexual health behaviour by sex among young people in Lagos Metropolis

Characteristics	Use of modern contraceptives to prevent STIs		Use of modern contraceptives to prevent pregnancy	
	Male	Female	Male	Female
School environment				
No	0.5(0.2–1.1)	6.2(2.0–18.8)**	0.5(0.2–1.2)	5.7(1.8–17.3)**
Yes	rc	rc	rc	rc
Religious environment				
Christianity	0.5(0.1–2.1)	a	0.6(0.1–2.3)	0.8(0.1–9.2)
Islam	0.5(0.1–2.2)	a	0.6(0.1–2.5)	0.4(0.0–4.4)
Others	rc	a	rc	rc
Family environment				
Monogamous	0.8(0.4–1.3)	1.0(0.4–2.7)	0.9(0.5–1.5)	0.9(0.4–1.9)
Polygynous	1.0(0.5–1.8)	0.9(0.3–3.0)	1.1(0.6–2.1)	0.8(0.3–2.2)
Single Parenthood	rc	rc	rc	rc
Ethnic environment				
Yoruba	1.0(0.6–1.7)	1.2(0.5–2.8)	0.9(0.6–2.1)	1.2(0.6–2.6)
Others	rc	rc	rc	rc
Ever use of unconventional substance				
No	0.2(0.1–0.4)***	0.5(0.2–1.7)	0.2(0.1–0.3)***	0.5(0.3–1.1)
Yes	rc	rc	rc	rc
Age	1.1(1.0–1.2)	1.5(1.2–1.7)*		1.5(1.3–1.7)***
Religiosity				
Less religious	0.9(0.6–1.5)	2.5(1.2–5.4)	1.1(1.0–1.2)*	2.6(1.3–4.9)**
More religious	rc	rc	rc	rc
Model Chi-square P value	62.3***	59.9***	72.2***	100.9***

Note * significant at $P < 0.05$; ** significant at $P < 0.01$; *** significant at $P < 0.001$; a: variable not included in model; rc: reference category
Source Survey in Lagos Metropolis in 2019

were adjusted for school environment still significantly predicted contraception for prevention of pregnancy among female respondents ($P < 0.01$). Among males, ever use of unconventional substance use significantly predicted contraception for STIs prevention ($P < 0.001$). Female respondents who were not in any school environment were over 6 times and about 6 times more likely to report contraception for the prevention of STIs and unwanted/mistimed pregnancy respectively compared to those who were in schools. Among their male counterparts, those who never used any unconventional substance were less likely to report contraception for STIs prevention relative to those who ever used.

Table 19.7 Unadjusted and adjusted Odds ratios of Logistic Regression (with 95% CI) of the influence of social environment and ever use of unconventional substance use on indicators of sexual health indicators among sexually active male respondents in Lagos Metropolis

Characteristics	Use of modern contraception to prevent STIs	
	Unadjusted OR	Adjusted OR
School environment		
No	0.3(0.1–0.7) *	0.3(0.1–0.9)
Yes	rc	Rc
Religious environment		
Christianity	0.6(0.1–6.5)	0.5(0.1–5.4)
Islam	0.8(0.1–8.8)	0.6(0.1–7.0)
Others	rc	rc
Family environment		
Monogamous	0.3(0.1–0.9) *	0.9(0.1–0.9) *
Polygynous	0.8(0.3–3.1)	0.9(0.2–3.4)
Single Parenthood	rc	rc
Ever use of unconventional substance use		
No	1.3(0.6–2.8)	1.1(0.5–2.4)
Yes	rc	rc
Age	–	0.9(0.8–1.1)
Religiosity		
Less religious	–	0.6(0.3–1.4)
More religious	–	rc
Model Chi-square P value	17.6**	21.2**

Note * significant at $P < 0.05$; ** significant at $P < 0.01$; *** significant at $P < 0.001$; a: variable not included in model; rc: reference category
Source Survey in Lagos Metropolis in 2019

Table 19.7 shows the odds ratios of unadjusted and adjusted logistic regression models on the influence of indicators of SE, unconventional substance use on modern contraception practice for STIs prevention among sexually active male respondents. The unadjusted model revealed that school environment and family environment significantly predicted contraceptive use for STIs prevention ($P < 0.05$). Sexually active males who were not in any school environment were less likely to report modern contraceptive use compared to those in a school environment. Similarly, sexually active male respondents who were in a monogamous family environment were less likely to report modern contraceptive use compared to their counterparts in single parenthood family environment. However, after adjusting for age and religiosity in the adjusted model, the SE indicator that significantly predicted modern contraceptive use for STIs prevention was family environment ($P < 0.05$). Similar to the behaviour of the indicator in the unadjusted model, sexually active male respondents in monogamous family environment were less likely to report modern contraceptive use for STIs prevention relative to those who indicated single parenthood family environment.

19.5 Discussion

This study has examined the influence of SE indicators and ever use of unconventional substance consumption on modern contraceptive use for the prevention of STIs and unwanted or mistimed pregnancy among young people (10–24) using a cross-sectional survey data collected in Lagos metropolis in 2019. SE indicators were significantly associated with sexual health indicators and the association could be described in two major ways. First, influence of individual SE indicators and unconventional substance use on sexual health indicators in the entire sample and among those who were sexually active three months before the survey was statistically significant. Second, influence of SE indicators and unconventional substance consumption taking into account other social, economic, and demographic variables was likely to impact on sexual health. In the first case, SE indicators significantly associated with sexual health indicators separately include school, religious, and ethnic environments before taking into account another SE indicator. Conversely, when the analysis was limited to sexually active respondents, only school and family environments, and religiosity were the SE indicators that were significantly associated with modern contraceptive use for STIs prevention only among males. Thus, at the bivariate level of analysis, this study demonstrated that it is consistent with previous studies that indicated that school environment, family environment, and religion are the SE indicators that are statistically significant factors influencing sexual health among young people (Jones et al., 2017; Maness & Buhi, 2016; Srijaiwong et al., 2017; Viner et al., 2012).

However, in the unadjusted and adjusted logistic regression models, it was apparent that school environment was the only SE indicator that sustained its significance as a predictor of the sexual health indicators in the total sample when other SE indicators and ever use of unconventional substance were taken into account. The association was positive among males in the unadjusted model but negative among females. Thus, on the one hand, male respondents in schools were more likely to report modern contraceptive use relative to those outside school. On the other hand, females outside the school walls were more likely to report modern contraception use compared to those in schools. Moreover, when the analysis was limited to sexually active respondents, school and family environments significantly predicted contraception for STIs prevention in unadjusted model only among males. But in the adjusted model, family environment became the only SE indicator that significantly predicted sexual health indicators only among males. There was no SE indicator that predicted sexual health indicators among sexually active females. Nevertheless, the findings that school and family environments significantly predicted the sexual health indicators support previous studies that have supported that school and family environments are important influencers of sexual health among young people (Chernick et al., 2019; Srijaiwong et al., 2017). However, in the population studied, the role of school and family environments in the sexual health of sexually active young females is unclear. Further studies are required to explore this area.

Furthermore, it is important to observe the twist to the influence of school environment on sexual health indicators in this sample. At the bivariate level of analysis, higher proportion of females in school environment compared to those out-of-school reported use of modern contraceptives for STIs and mistimed or unwanted pregnancy among both total sample and among the sexually active male respondents. Conversely, in the unadjusted and adjusted logistic models among females, those out of school environment were more likely to report modern contraceptive use compared to those in school in the total sample. Among their male counterparts, out-of-school environment were less likely to report modern contraceptive use for STIs and pregnancy prevention (females and sex partners of males) in total study sample among those sexually active, those out-of-school were also less likely to report modern contraception for STIs prevention but was not a significant predictor in the adjusted model. It thus suggests that the school environments were more supportive to young males to use modern contraceptives to prevent STIs and unwanted or mistimed pregnancy but not very supportive to females.

The contrast in the role of school environment in young people's sexual health between males and females may be related to the fact that in a patriarchal society such as Nigeria where premarital sexual engagement among young females is frowned at less societal disapproval is given their male counterparts may be an active influencer in schools (Ntoimo & Odusina, 2019; Wusu & Adedokun, 2018). Also, females' sexual subjectivity in patriarchal societies may be weak and make it difficult for them to take responsive sexual health decisions while in school (Lys et al., 2019; Wusu & Isiugo-Abanihe, 2008). As such, schools may be more supportive to males but appear less supportive to females in the adoption of contraceptives for the prevention of STIs and mistimed or unwanted pregnancy. In addition, the result suggests that females in the total sample outside the school environment have greater freedom to adopt modern contraception to ensure their sexual health.

In the light of the objective of this study, which set out to determine SE indicators that are significant predictors of sexual health among young people, it is important to emphasize the role of family environment among sexually active males as revealed in this analysis. Family environment appears the most important predictor of sexual health among sexually active male respondents. Thus, this finding supports earlier study that reported family structure as the most significant predictor of sexual health among young people (Jones et al., 2017; Srijaiwong et al., 2017). Those from monogamous family environment were less likely to report modern contraceptive use for STIs prevention. A plausible explanation is that young people from monogamous family may suffer from the general challenge of poor sexual health education aimed at exposing them to available modern contraceptive methods. Their ignorance owing to the popularity of considering sex-related discussion with children as "no go area" by parents in most part of Nigeria. Lack of proper knowledge may be a barrier to adoption of contraception to prevent STIs. Monitoring of young people by parents may be very strong in monogamous families to the extent that such children's involvement in sexual activity may be just casual and unprepared and because they may be mostly sexual novice. In such cases of sexual contact, modern contraceptive use negotiation with sex partner is likely a difficult task, which makes them vulnerable to STIs. Conversely, young persons in single parenthood are less likely to operate in monitored social environment, they may be more sexually informed with higher level of sexual subjectivity that could facilitate contraceptive use negotiation with sex partners.

The second independent variable in the analysis of significant predictors of modern contraceptive use was unconventional substance consumption. Ever use of unconventional substance use significantly predicted modern contraceptive use among both male and female respondents at bivariate and multivariate levels. Males who ever used unconventional substance were more likely to report modern contraceptive use compared to those who never used. Conversely, their female counterparts who ever used unconventional substance were less likely to report adoption of positive sexual health. These results are consistent with the outcome of an earlier analysis that reported that young males who ever used the unconventional substances were more likely to report positive sexual health behaviour (Wusu et al., 2019). It is surprising to note that males who used unconventional substance reported preventive behaviour but not so among their female counterparts. This result could be explained in terms of young males' involvement in unconventional substance use gained consciousness of the risks of STIs and unwanted pregnancy among their sex partners whereas their female counterparts tend to disregard such consciousness. Conversely, unconventional substance consumption failed to predict modern contraception in the models constructed for the sexually active sample. Thus, the outcome suggests that unconventional substance consumption may not be a key predictor of modern contraception for STIs and unwanted or mistimed pregnancy prevention in the sexually active sample. However, a previous study observed significant association between unconventional substance use and sexual risk behaviour (Srijaiwong et al., 2017). It is likely that the contrast between this study and the 2017 study is predicated on the fact that the previous study did not take into account the interactive effect of SE indicator.

19.6 Conclusion

Overall, the key SE indicators that predicted sexual health among young people sampled in this study were school and family environments. Religious and social environment were not a significant predictor of sexual health behaviour in this sample. Moreover, modern contraceptive use to prevent STIs and unwanted or mistimed pregnancy was likely common among young males but less likely among females in school environment. Also, sexually active males in monogamous family environment were less likely to report modern contraceptive use for STIs prevention. Also, surprisingly, those who ever consumed unconventional substance were more likely to use modern contraceptives to prevent STIs and unwanted or mistimed pregnancy compared to those who never used. Therefore, policies that could address the sexual subjectivity of young people in school and family environments and neutralise patriarchal stereotypes shaping sexual subjectivity in the social environments are imperatives. Also, everything necessary must be done to discourage consumption of unconventional substance use among young people in Nigeria.

Acknowledgements The authors are indebted to the following final-year students of the Department of Sociology, Lagos State University, Lagos, Nigeria, who were actively involved in the collection of the data analysed here. The affected students include Z. A. Kazeem, R. Y. Abogunloko, A. R. Animashaun, O. M. Balogun, B. E. Ajibade, J. O. Otike, and O. M. Adeoye.

References

Adekeye, O. A., Adeusi, S., Chenube, O. F., & Sholarin, M. A. (2015). Assessment of Alcohol and substance use among undergraduates in selected private universities in Southwest Nigeria. *IOSR Journal Of Humanities And Social Science, 20*(3), 01–07. http://www.iosrjournals.org

Chernick, L. S., Siden, J. Y., Bell, D. L., & Dayan, P. S. (2019). A qualitative assessment to understand the barriers and enablers affecting contraceptive use among adolescent male emergency department patients. *American Journal of Men's Health, 13*(1), 1557988319825919.

Danjuma, A., Taiwo, A. I., Omoniyi, S. O., Balarebe, S. A., Kolo, S., Solomon, L. S., & Yashi, G. N. (2015). Nonconventional use of substance among youths in Nigeria: Viewpoints of students in Nigerian tertiary institution. *Journal of Nursing Care, 4*(3). https://doi.org/10.4172/2167-1168.1000311

Decker, M. J., Isquick, S., Tilley, L., Zhi, Q., Gutman, A., Luong, W., & Brindis, C. D. (2018). Neighborhoods matter. A systematic review of neighborhood characteristics and adolescent reproductive health outcomes. *Health & Place, 54*, 178–190.

Elkington, K. S., Bauermeister, J. A., & Zimmerman, M. A. (2010). Psychological distress, substance use and HVI/STI risk behaviour among youths. *Journal Youth Adolescence, 39*, 514–527.

Hampanda, M. K. (2013). The social dynamics of selling sex in Mombasa, Kenya: A qualitative study contextualizing high risk sexual behaviour. *African Journal of Reproductive Health, 17*(2), 141–149.

Jones, J., Salazar, L. F., & Crosby, R. (2017). Contextual factors and sexual risk behaviors among young, Black men. *American Journal of Men's Health, 11*(3), 508–517.

Jubril, U. Y. (2019). The effect of drug abuse among youths in Nasarawa Local Government, Kano State, Nigeria (2014–2019). *Journal of International Politics, 01*(3), 13–25.

Lys, C., Gesink, D., Strike, C., & Larkin, J. (2019). Social ecological factors of sexual subjectivity and contraceptive use and access among young women in the Northwest Territories, Canada. *The Journal of Sex Research, 56*(8), 999–1008.

Maness, S. B., & Buhi, E. R. (2016). Association between social determinants of health and pregnancy among young people: A systematic review of research published during the past 25 years. *Public Health Reports, 131*(86–99).

National Population Commission [Nigeria], & International, I. M. (2019). *Nigeria demographic and health survey 2018*. National Population Commission [Nigeria], ICF Macro International.

Ntoimo, L., & Odusina, E. K. (2019). Family structure, women autonomy and utilisation of maternal health care services in Nigeria. In E. Nwokocha, & O. Fayehun (Eds.), *Concise demography of Nigeria*. BumbleBee Publishing.

Ochako, R., Mbondo, M., Aloo, S., Kaimenyi, S., Thompson, R., Temmerman, M., & Kays, M. (2015). Barriers to modern contraceptive methods uptake among young women in Kenya: A qualitative study. *BMC Public Health, 15*, 118. https://doi.org/10.1186/s12889-015-1483-1

Onwuadiochi, C. I., Ijioma, M. A., & Ozoemene, M. L. (2021). Comparative analysis of the measured and extrapolated wind shear distribution at Muritala Mohammed and Port Harcourt International Airports, Nigeria. *Journal of Environment and Earth Science, 11*(5), 63–78. https://doi.org/10.7176/JEES/11-5-06

Srijaiwong, S., Sindhu, S., Ratinthorn, A., & Viwatwongkasem, C. (2017). Factors influencing sexual behaviors among Thai adolescents. *Journal of Population and Social Studies [JPSS], 25*(3), 171–193.

Svinurai, A., & Makhubele, J. C. (2018). The nexus between substance abuse and commercial sex work in Musina border town, Limpopo Province, South Africa. *Gender & Behaviour, 16*(3), 12152–12166.

Viner, R. M., Ozer, E. M., Denny, S., Marmot, M., Resnick, M., Fatusi, A., & Currie, C. (2012). Adolescence and the social determinants of health. *The Lancet, 379*(9826), 1641–1652.

Wusu, O. (2010). Reproductive health issues among undergraduates. In L. Olurode (Ed.), *Reciprocal impact: A study of LASU and its neighbours*. Department of Sociology, Lagos State University.

Wusu, O. (2013). Exposure to media content and sexual health behaviour among adolescents in Lagos metropolis, Nigeria. *African Journal of Reproductive Health, 17*(2), 157–168. http://www.ncbi.nlm.nih.gov/pubmed/24069761

Wusu, O. (2016). A re-assessment of drivers of contraceptive acceptance, use and discontinuation among young females in Nigeria. In J. Oucho, G. Rutaremwa, & J. B. Nyakaana (Eds.), *The demography of Uganda and selected African Countries* (pp. 88–105). Fountain Publishers.

Wusu, O., & Adedokun, O. A. (2018). Young women emerging preferences: Exploring what it holds for family reconstruction in Southwestern Nigeria. *Gender and Behaviour, 16*(3), 1596–9231.

Wusu, O., & Isiugo-Abanihe, C. U. (2008). Understanding sexual negotiation between marital partners: A study of Ogu families in Southwestern Nigeria. *African Population Studies, 23*(2), 155–171.

Wusu, O., Soniyi, R. A., Kazeem, Z. A., Abogunloko, R. Y., Animashaun, A. R., Balogun, O. M., Ajibade, B. E., Otike, J. O., & Adeoye, O. M. (2019). Unconventional substance use among young people in urban Lagos and the implications for sexual and reproductive health. *The Nigerian Journal of Sociology and Anthropology, 17*(2), 78–98.

Chapter 20
Dealing with Mental Health Issues in Urban Romania—A Study on Youth's Perception of Depression

Alina Mareci and Anca Tudoricu

Abstract Multiple studies prove that poor mental health is usually correlated to other health and behavioural shortcomings. Mental health issues are more and more common, especially among young people. One of the main problems in investigating mental health in general, apart from a limited number of mental health professionals and their sometimes-limited qualifications, is the considerable patient stigma that comes with this subject. This study aims to breach the issue of depression and mental health issues among teenagers and young adults (15–35 years old) in urban Romania. The objectives of the study included obtaining an image of what depression means for this target group and how the target group usually acted when faced with one symptom that is mostly associated with depression: sadness. The authors also measured how the target group would deal with their own or someone else's depression. The methodology had several steps: constructing a radiography of the Romanian health system and emphasizing its shortcomings, organizing focus groups with mental health professionals as well as with members of the target group, creating and disseminating a questionnaire to more than 1.000 respondents. The results show that the target group's members have a solid understanding of the importance of the issue, many times due to their own experiences or because this phenomenon affected someone from their family or circle of friends. However, the study shows that the target group mostly lacks the mechanisms to address depression, be it in themselves or others.

Keywords Depression · Health system · Mental health · Youth · Perception · Urban · Romania

A. Mareci (✉) · A. Tudoricu
Faculty of Geography, University of Bucharest, N. Bălcescu, 1, 010041 Bucharest, Romania
e-mail: alina.mareci@unibuc.ro

A. Tudoricu
e-mail: anca.tudoricu@g.unibuc.ro

B. R. K. Sinha (ed.), *Urban Dynamics, Environment and Health*,
https://doi.org/10.1007/978-981-99-5744-6_20

20.1 Literature Review

Depression is one of the most significant public health concerns worldwide due to its incidence and its devastating impacts on individuals, families, and society. Although the international literature covers many issues related to depression, few studies are analysing the perception of young people of this condition, and even fewer that analyse it relating to the Romanian youth.

According to the American Psychiatric Association (Parekh, 2017), depression is a common and serious illness that negatively affects how you feel and act, and the way you think. The American National Institute of Mental Health also remarks its commonality and emphasizes that living with this illness hinders the person's ability to perform daily routine activities (National Institute of Mental Health, 2018), while the World Health Organization also names it "a major contributor to the overall global burden of disease" (World Health Organisation, 2020). Its presence dates back to antiquity when the concept of *melancholia* was used to cover feelings of sadness, dejection, and despondency, and often fear, anger, delusions and obsessions. For example, Hippocrates, the initiator of the humoral theory, associated melancholia with autumn, coldness, and dryness, considering that it was caused by an excess of black bile (Tacchi & Scott, 2017). Aristotle was the first one to associate this condition with the intellect and creativity itself (Akiskal & Akiskal, 2007), while the physician Soranus of Ephesus connected it with the loss of reason, suggesting that mental health could be improved through physical activity and psychological interventions (Toohey, 1990). In the Middle Ages, mental disorders were perceived as gifts or punishments from god(s); therefore, melancholia was either an experience to be admired or to be cured. The twentieth century brought significant contributions in the field of psychology; Sigmund Freud stated the difference between sadness and depression, being the one who opened the path towards modern psychiatry. The importance of the condition is also marked by the fact that the very first Diagnostic and Statistical Manual of Mental Disorders (DSM-I), the book that changed the world of practitioners, contained the concept of "depressive reaction" (American Psychiatric Association, 1952).

Worldwide, about 1 billion people were diagnosed with a mental disorder, out of which more than 264 million people of all ages suffered from depression, making it the most common mental disorder (James et al., 2018). In 2014, 7.1% of the EU-28 population (2.7% males and 4.1% females) reported having chronic depression (Ritchie & Roser, 2018). Romania was one of the countries with the smallest percentage (1.5%), but in 2017 it reached 5%, showing alarming rates among the elderly (2%) and women (2.7%) (Ministry of Health, 2019).

Even if sometimes depression can be mistaken with sadness, both its symptoms and its effects have a different intensity. As Truschel (2019) mentioned, the difference does not necessarily lie in the gravity with which a person feels down, but rather in a unique blend of factors which determine the duration of these negative feelings, other psychophysiological symptoms, and the effect upon the individual's ability to function in daily. Therefore, in order to standardise the diagnosis of depression, the

1. Depressed mood most of the day, nearly every day.
2. Markedly diminished interest or pleasure in all, or almost all, activities most of the day, nearly every day.
3. Significant weight loss when not dieting or weight gainor decrease or increase in appetite nearly every day.
4. A slowing down of thought and a reduction of physical movement (observable by others, not merely subjective feelings of restlessness or being slowed down).
5. Fatigue or loss of energy nearly every day.
6. Feelings of worthlessness or excessive or inappropriate guilt nearly every day.
7. Diminished ability to think or concentrate, or indecisiveness, nearly every day.
8. Recurrent thoughts of death, recurrent suicidal ideation without a specific plan, or a suicide attempt or a specific plan for committing suicide.

Fig. 20.1 Symptoms specific to depressive disorders. (American Psychiatric Association, 2013)

DSM-5 (2013) outlines that over a 2-week period the individual must be experiencing five or more symptoms out of which at least one is either depressed mood or loss of interest or pleasure (items 1 and 2 from Fig. 20.1).

20.1.1 The Healthcare and Political Entanglements in the Study Area

Before 1990 the Romanian Healthcare System was organized according to the Semasko model: totally centralised, state-owned, financed by the state budget, and coordinated by the Ministry of Health. The reform of the Healthcare System started in 1990 and hoped to decentralize it; restructure primary healthcare and separate it from secondary care and tertiary care; change the doctor-patient relationship by introducing the concept of family doctor/general practitioner (GP); improve the quality and efficiency of sanitary services. As presented in Fig. 20.2 the main actor of the healthcare system is still (per the Governmental Decision 244/1997), the Ministry of Health (The Romanian Government, 1997). In this hierarchy, directly under the ministry function several lower-level units with the most influent being the county-level Departments of Public Health which hold the authority at the local level. The public and private units that provide healthcare services are placed on the last level (Amnesty International, 2004).

The Ministry of Health ensures the general management and the regulatory oversight at a national level while the public health local authorities are responsible for delivering the actual services. Similarly, the National Health Insurance House administers and regulates the health and social insurance system through its territorial branches which actually contract healthcare services. Officially, all medical services are free for the entire population, but in reality, some costs are directly or indirectly transferred and are met by patients. These include the informal payments of medical personnel (Carol Davila University of Medicine and Pharmacy, 2018). In

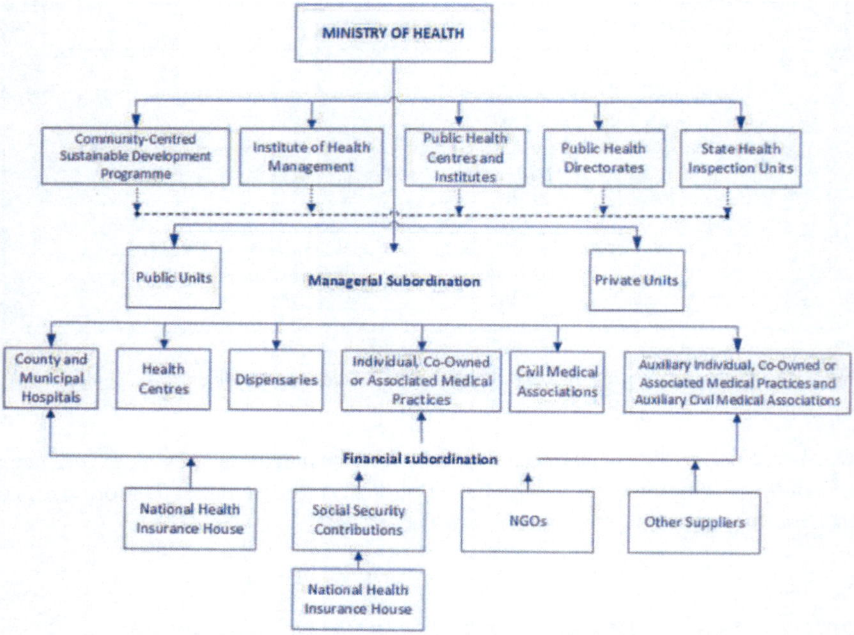

Fig. 20.2 General organizational chart of the Romanian healthcare system—according to the Governmental Decision no. 244/1997 and Law no. 100/1998 after http://www.ms.ro/wp-content/uploads/2019/09/Organigrama.pdf

Romania, out-of-pocket payments represent 21.3% of the total health spending of the population, and are presently increasing from a level of 18.5% measured in 2005. Most of them are targeted towards pharmaceutical products (70.8% compared to an EU average of 44.2%). The out-of-pocket payments also include direct payments for services that are not covered by the social health insurance system, but also unofficial payments made directly to the medical personnel. The amplitude of out-of-pocket payments is difficult to evaluate accurately but was expected to decrease starting with 2014 when additional sanctions were introduced for those who accepted under the table payments (Secretariatul General al Guvernului, 2011).

Between 1990 and 2020 there have been 30 changes at the leadership level of the Romanian Ministry of Health. Mathematically this would allow for each minister (interim or not) one year to manage a system struggling both financially and structurally. Law no. 95 from 2006 on Health Reform was modified 67 times between its promulgation and 2020 (Guvernul României, 2015). Many of the changes were done through emergency ordinances. This type of governmental act is meant to be used in extraordinary situations because the nature of the decision they include cannot be postponed. They represent a conduit that gives the government the power to legislate while bypassing the parliament. As such, emergency ordinances are ad-hoc

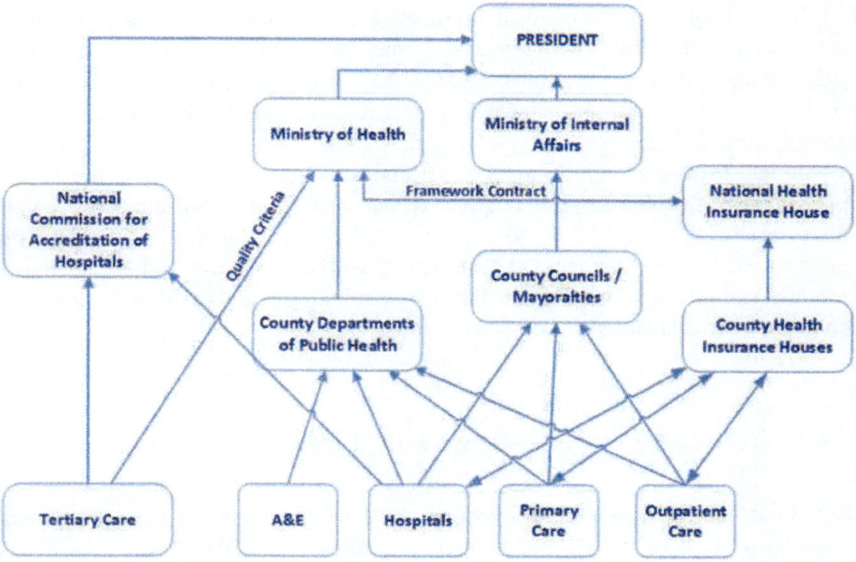

Fig. 20.3 Responsibility relationships framework of the Romanian healthcare system—after Banca Mondială Regiunea Europa și Asia Centrală (The World Bank) (2011), Analiza Functională a Sectorului Sănătate în România, Secretariatul General al Guvernului, https://sgg.gov.ro/docs/File/UPP/doc/rapoarte-finale-bm/etapa-II/MS-RO-FR-Health-Sector-ROM.pdf

changes, done without preliminary discussions and analysis. In the case of healthcare, legislating without public forums and feasibility analysis only increased the chaos of the system, both from an organisational and financial point of view, as well as in terms of managing accountability.

A 2012 report performed by the World Bank, which included discussions with personnel from national and local health care units as well as hospital staff, concluded that the system is way too complex. One of the main problems emphasized was the difficulty with which all the involved actors were able to identify the managing or leading entity. The different and sometimes contradictory answers reflect the lack of information and knowledge even at the level of the more important decisional forums. The uncertainty about who is responsible for precisely what is conducive to chaos and in turn, ineffective management and low productivity (Secretariatul General al Guvernului, 2011) (Fig. 20.3).

In 2019, Romania ranked last in the European Union in terms of the gross domestic product spent on health care. While the average for other European countries was of 9.6%, Romania only allocated healthcare half of that—4.84%. The healthcare minister at the time presented it as a historical increase (Neagu, 2019). Still, this figure lags behind other European, but not EU member countries, such as Albania (6.8%), North Macedonia (6.1%), or Montenegro (5.9%) (Organizația pentru Cooperare și Dezvoltare Economică—OECD, 2017).

One of the problems more difficult to solve and/or overcome is that of the social insurance system. The system remained centralized despite recent efforts of decentralizing its regulatory functions. Factors that hinder access to healthcare in this country are numerous and subject to numerous studies (Dumitrache et al., 2016; Zamfir et al., 2015).

Similar to most countries in the UE and also around the world, the highest percentage of people who declared themselves unsatisfied with healthcare are more impoverished people (11.1%) who live in rural areas. As such, it is expected that in urban areas, the coverage and accessibility of the population to both healthcare services and information is better. This is one important reason the authors focused on presenting the results of their study on urban areas.

20.2 Managing and Treating Mental Health in Romania

Romanians suffered severe deprivations in all fields during the time the country functioned as a communist dictatorship (between 1965 and 1989). This also affected the mental health branch of the country's healthcare system. The country's 1,000 psychiatrists at that time, and many of their patients, were victimized by the political system. Many became willing or unwilling participants in the abuse of their profession. During this period, psychiatric hospitals were used for the repression of political dissidents and others who were deemed too problematic to simply be jailed. Other abuses included: biased interpretation of the laws in detaining persons not suffering from mental illness; false diagnoses and treatment; detentions in secret facilities. In addition, the field of clinical psychology—an essential complement to psychiatry in the treatment of mental illness—was marginalized to the point where mental health care was covered entirely following a biological model of care (Adler et al., 1993). Unfortunately, this legacy of marginalization will prove difficult to shake, and it is still moulding opinions and behaviour of the population.

Since 1989, Romania's psychiatric system functioned under the same framework of economic stagnation and institutional challenges that have plagued most of the country's state-provided services. The people who are involved in the treatment of patients with mental illness face severe obstacles to the efficient treatment of their patients. Another significant challenge for the Romanian mental health system is that, on the background of alternating economic problems facing the country and consequently its financing, psychiatric hospitals recorded increasing numbers of patients—particularly, with major depressive disorder and various anxiety disorders. Specialists associate this increase in patients precisely with the economic hardships felt after 1990, when economic reforms began to take their greatest human toll, specifically with increasing unemployment and stress levels (Friedman, 2006).

The mental health care system receives a tiny share of the healthcare budget, just 3%, seeing that mental health problems (excluding dementia) account for 12% of the overall burden of disease (World Health Organization, 2001). The sums received for mental health treatment and care account for 35–45% of the sums budgeted for most other medical specialities, strengthening the message that mental illness not only carries a social stigma for patients but is also stigmatized in the broader healthcare system. Apart from this, it has been especially difficult for the medical community to receive greater support from the state due to the lack of quality epidemiological data for mental health illnesses' identification and tabulation.

According to a 2005 study, individuals looking for mental health treatment in Bucharest are seen in the emergency and outpatient wards of the largest general psychiatric hospital in Romania (Gater et al., 2005). In 2017 there were 11.44 psychiatrists per 100,000 inhabitants, and while this value has seen a continuous growth it is only comparable with very few surrounding countries: Bulgaria reached the value of 7.77 psychiatrists per 100,000 inhabitants in the same year, and Poland 9.23. A vast majority of the EU states were better ranked in 2017 (Eurostat, 2020). People living with mental illness are stigmatized almost everywhere, even in the twenty-first century. In Romania especially, it may be more difficult compared, for example with countries from the western part of the continent. From a cultural perspective, any mental health inadvertences are filed under "madness". This perspective sometimes manifests as a combination of evil (moral and/or religious), genetic weakness, and/or stupidity on behalf of the patient (a conflation of mental illness with learning disabilities). There are reported cases of priests advising families and patients that their mental illness is some form of devil workings. Even within the professional field of social work, young social worker trainees tend to avoid choosing mental health as a sub-specialization because they can experience working with people with mental illness as somehow "dangerous" and "tainting".

Many countries from the eastern part of the continent are still trying to shed their past more conservative views and increase their development ranking and standard of living. This goal sometimes proves to take more time and effort than expected, primarily due to the increased bureaucracy and centralized organization of all their institutions. So, in many situations, there is a tendency to present, at least statistically, a more positive picture of reality. Mental healthcare falls into a category that still needs much work and, as such, many studies sustain that existing data should be considered with much reluctance. For example, more than 70% of the institutions suggest that group and individual psychotherapy is a common practice, that psychological evaluation is offered as one of the regular services and that occupational therapy is delivered in more than 60% of the facilities. However, few of the professionals working in hospital wards are trained psychotherapists (official data from 2006 account for around 500 psychotherapists working in public units). Since there is no formal training for occupational therapy, such reports are optimistic at best (Junjan et al., 2009a, 2009b).

Nowadays, many of those diagnosed with mental health illnesses are placed and treated in social assistance institutions of certain types, or psychiatric hospitals (under the management of the Ministry of Health). People with less severe mental health issues are treated in psychiatric wards of general or university hospitals while those with more serious and/or severe conditions that require long-term treatment are hospitalized in units designated for chronic psychiatric illnesses. According to a 2014 Eurostat survey, the number of hospital beds in psychiatric hospitals reached 16,503 and was decreasing slightly starting in 2010. Most of these beds are found in psychiatric hospitals rather than psychiatric wards of general hospitals, with hospitals' capacity varying from 50 to 1,250 beds, and half of them are located in the biggest hospitals of the country, thus still perpetuating the centralized framework of the entire system (Mental Health Europe, 2018; World Health Organization, 2011). The psychiatric wards of general hospitals, which could handle emergencies or less critical cases, are considerably smaller (with 21 to 300 hospital beds). Out of the total number of hospital beds specifically designed for patients with mental health issues, two-thirds are occupied by people with acute issues and the rest with people who need long-term care (Junjan et al., 2009a, 2009b). A 2014 report from the Institute for Public Policy Bucharest regarding the implementation of the UN Convention on the Rights of Persons with Disabilities showed that the main reason for people getting out of the social assistance system is death (76%) and this percentage increases every year as the population ages. Only 16% of patients left these facilities in order to (re)integrate into a family-type setting (Mental Health Europe, 2018). Data available at the end of 2016 shows that most of the social assistance institutions host a heterogeneous mix of vulnerable people from elderly to adults, from people with intellectual/learning and mental disabilities to people with other types of disabilities. This makes monitoring and treating specific mental health issues virtually impossible. Moreover, it could also lead to the alienation of people suffering from mild mental health problems. This, in turn, translates into them spending their entire lives in specialized institutions. In 2018, 80% of those suffering from intellectual or mental disabilities were hospitalized (Ministry of Health, 2019).

Romania's mental healthcare system is still based on patients' institutionalization even though there are numerous propositions, both from NGOs and governmental bodies, to adopt an approach based on deinstitutionalized practices. An issue such as depression does not need institutionalization; as such people who deal with it benefit the most from out of hospital treatment. Unfortunately, these exact practices are still not common place in this country. Feasibility studies were done in order to bring forward such proposals, and they pinpointed the many vital issues that Romania needs to deal with to achieve this goal: a lack of qualified personnel such as psychiatrists, auxiliary medical staff, psychologists, social workers; a weak and insufficiently endowed infrastructure; lack of training for those dealing with people with mental health issues; the stigma towards people dealing with mental health issues displayed even by specialists and medical personnel; human rights abuses; discrimination against people with mental health issues; inadequate or even lack of implementation of the existing legislation regarding mental health (Mental Health Europe, 2018).

20.3 Methodology

The aim of this study is to obtain an image of depression as seen from the perspective of the younger aged cohorts of the population in urban Romania. The authors also aimed at contextualizing it in the broader framework of the Romanian healthcare and mental health care specifically. The research hypothesis that the authors started with was that urban dwellers and especially young urban dwellers would have an overall perspective on this issue that mirrors to a high degree the specialized description of the illness, issues, and obstacles that people living with it would face. We based this hypothesis on several lines of thought applicable to the urban younger generation: they would be less burdened by the historical baggage the illness comes with; they would be better informed because of the easier available sources on this topic; they would have an open-minded attitude on issues of mental health.

The main objectives of the study included obtaining an image of what depression means for this target group. For another objective, the authors measured how the target group usually acted when faced with one symptom that is mostly associated with depression: sadness. The authors also measured how the target group would deal with their own or someone else's depression and tried to identify the possible obstacles this illness poses to the population.

The research itself employed both quantitative and qualitative methods. This was necessary because we dealt with both a demographic and social topic which is difficult to target with a single type of methodology. The main instrument used for this endeavour was a questionnaire with both open and closed-ended questions. In constructing the questionnaire, there were organized two focus groups with people working in the field, NGO representatives, and people from the target group's age bracket. The discussions were free but were focused on the idea of how young people saw, dealt with, and reacted to mental illness and depression specifically. The results of these discussions were compiled in a questionnaire, which for two weeks was tested on a group of 35 people within the same age bracket. This phase aimed to measure the questionnaire's readability, addressability, accessibility and solve issues regarding the clarity of certain questions. The final product was disseminated online with a total of 1,212 people filling it in, out of which 1,156 were declared valid. The rest were eliminated for any of the following reasons: they were ineligible due to their age; they did not respond to all questions, thus making comparisons ineffective; or their answer was purposely malicious. The authors considered this important point to mention as the last category because it is statistically insignificant and shows that the researched topic still remains controversial within it. This study only presents the analysis results of the answers coming from urban areas. The demographic characteristics of the respondents are presented in Table 20.1. Each of the age brackets represents different educational levels of the Romanian youth: high school, university, and young adulthood that could signify post-university studies or insertion on the labour market. The high prevalence of female respondents is supported by previous research that shows that women use verbal expressions of emotions more often than

Table 20.1 Age and gender of the respondents

	Age brackets			Total
	15–17 years old	18–23 years old	24–35 years old	
Females	301	301	83	685
Males	25	51	24	100
Total	326	352	107	785

men (Goldshmidt & Weller, 2000), and they are more inclined to fill in online questionnaires, regardless of topic (Preda et al., 2020). This research was conducted in accordance with the Declaration of Helsinki, and the existing protocols of the Ethics Committee of the Faculty of Geography, the University of Bucharest.

The final data was compiled using Dimensions v7, which enabled a comprehensive representation of the data as well as allowing us to have a quantitative measurement of each response and to check all possible correlations between responses and/or correlate responses with the socio-demographic construct of the respondents.

When the respondents were asked to define the concept of depression and explain their opinion and perspective on how it would manifest itself, what obstacles people living with it might face, or how they themselves would deal with it, they used images, (pre)conceptions, or shortcuts that they have created from own experience or that they bowered or inherited from the society they live in and were moulded by. As such the authors considered that the symbolic interaction theory (Aksan et al., 2009) was a good way to analyse and interpret the responses to these open-ended questions. This methodology was extensively used when discussing depression and especially in studies targeting the same population group: studying the low esteem aspect it causes in adolescents (Plunkett et al., 2007); its connection with abuse from living with it Karp (1994); but also, in studying depression in general, from living with it Solomon (2014) to understanding it Lewis (1995). The principles of the symbolic interaction theory were used in all open-ended questions sometimes to create specific categories for a question in order to allow comparison or just to simplify and clarify specific ideas, for example in constructing the word cloud designating what depression means.

20.4 An Attempt to Define Depression

In order to construct a baseline for our results, the questionnaire's first request was to provide an open definition or description of what depression was according to the respondents' opinion. Figure 20.4 exemplifies a word cloud of the most frequently used words in this endeavour. As expected, two of the words that were extensively used were *sad(ness)* and *mental*, even if non-dysphoric depression is considered to be just as important as major depression (Gallo et al., 1997). As mentioned before, even if depression and sadness have been used interchangeably in several studies of

Fig. 20.4 What depression is in the respondents' opinion?

the twentieth century, it is commonly agreed that sadness is a regular emotion, while depression involves a state of pathology (Stiver & Miller, 1988).

Sadness leads most of the time to self-isolation, an attitude associated by 26.5% of our respondents to people suffering from depression. Previous research shows that social isolation has potential implications for well-being and health in general (Caspi et al., 2006), with teenagers and young adults being a risk group in this respect (Qualter et al., 2015). Even if isolation and loneliness do not necessarily co-occur, due to the high value attached to close friendship and romantic relationships during adolescence, they have a robust association with depressive symptoms. The perception of being excluded from social groups makes teenagers feel rejected and unwanted, triggering a range of responses (behavioural, cognitive, and physiological) geared towards self-protection (Matthews et al., 2016). This is also highlighted by the opinion of the target group (14.3%) who considered lack of support from friends and family one of the main problems that people with depression deal with, multiplying its harmful effects. Words such as *social, struggle, wall, misunderstood, ignored, helpless,* and *indifferent* present in the cloud above underline this aspect.

Other physical difficulties met while dealing with this condition, as identified by the target group, are related to the inability to work and interact normally (21.7%), inability to perform everyday tasks (9.3%), and permanent physical and mental fatigue (21.9%). These consequences of depression have been extensively analysed (Fuhrer & Wessely, 1995), being present in case of all ages, genders, social statuses, or work profiles (Puffer & McShane, 1992). Words such as *exhaustion, fatigue, insomnia, drained,* or *unable* highlight their incidence among young people and their acquaintances.

Of course, even if the issues presented above have been presented as having physical implications, they all have psychological roots. Depression is, by far, associated with loss. It can be triggered by physical, tangible loss (a member of the family, a partner, a property) and it generates both tangible and intangible loss (loss of appetite, of various abilities, of self-esteem). Solomon (2014:21) opens his book with a brilliant affirmation: "Depression is the flaw in love. To be creatures who love, we must be creatures who can despair at what we lose, and depression is the mechanism of that despair". In this respect, the respondents chose many words related to despair *suffering, grief, distrust, disappointment, restlessness, terror, unrest, and persecution.*

An overwhelming number of words chosen to define depression have negative connotations because not only the victims of this disease are strongly affected, but also the ones close to them. According to Beck (1967, 1979), people suffering from depression perceive the world through a depressogenic schema; therefore they tend to think negatively about themselves and others, which has a severe impact on cognitions about help-seeking. Whether they do not believe in the positive outcomes of the specialized help or if they do not have the will to look for social support, the negative cognitive bias increases as the condition worsens (Haaga et al., 1991; Lienemann et al., 2013; Siegel et al., 2014). Negative thinking sometimes leads to suicidal thoughts and even actions, which makes depression a silent killer. Suicide is the second leading cause of death in 15–29-year-olds (World Health Organisation, 2020); therefore, it is essential to generate awareness regarding the impact of this condition on health and well-being, especially among young people.

In order to determine the target groups' perception regarding the gravity of this disease, the authors categorized the results into two major categories: those that defined depression as a clear-cut illness and those that categorized it as a mood or emotional state (as something transitory, or a phase). A pattern became visible, a pattern that repeated itself when analysing the target group from all its socio-demographic perspectives. The latter group (those that saw depression as a mood or emotional state) being almost every time slightly more numerous (56.43% compared to 41.40%) and 2.17% of those questioned in the urban areas admitted that they could not describe depression as they do not have enough information or knowledge about it. There was no significant difference in terms of males or females in categorizing depression, as most of the respondents in both groups saw it as an emotional state rather than an illness (56.64% in the case of the females and 55% of the males). Nevertheless, there was an important disparity when analysing the target groups' interpretation of what depression is, based on their age. The authors think that the first important point worth emphasizing is that the percentage of those that answered they do not have enough information to have an opinion decreased while the age of the respondents increased (2.76% for those aged 15 to 17, 1.99% for those aged 18–23, and 0.93% for those aged 24–35). One finding could be that no matter the quality of the information, the target group felt that they received enough of it to form an opinion. While not necessarily definitive, there is a noticeable distinction in

Table 20.2 Depression and its effects—a respondents' opinion on categories

	Physical (%)	Psychological (%)	Both (%)	DK/NA (%)	Total (%)
Female	2.19	75.47	18.69	3.65	100
Male	0	84	11	5	100

defining depression based on the age of the respondents. The youngest age bracket sees depression overwhelmingly as an emotional state (60.34%), while only half of the other two age groups of the respondents categorize it as such. The same pattern was observed when splitting the group based on their marital status. However, a deviance occurred when analysing the occupation of our respondents: over 60% of the unemployed categorized depression as being an illness, thus strengthening the relationship between economic strife and depression.

The authors also constructed a multidimensional correlation between this question and the rest of the questionnaire in order to find out if, apart from the socio-demographic distribution, other patterns or abnormalities in the answers of the target group could be pointed out. As such, those that described depression as an illness also supported specialized treatment when dealing with it (62.77%) compared with only 53.95% of those that see it as an emotional state. They were also more numerous in identifying the issues that someone suffering from depression faces daily as being both psychological and physical (Table 20.2) and having both an inner and outer effect (affecting both themselves and the surrounding people be they close friends and family or work colleagues) as opposed to favouring one over the other.

20.5 Studying Healthcare Behaviour When Dealing with Depression

Regardless of whether they see depression as an illness or emotional state, a majority would seek specialized treatment (described as individual or group therapy or medication) if they suffered from it, with 56.94% supporting this course of action. There is an unequivocal acceptance of specialized treatment for the older cohort of the target group and consequently by those with a higher level of education. Only a small size of the target group rejected specialized help, and a third of the interviewees said they did not know whether they would use it or not. Other studies show that less than half of the people with major depressive disorder receive adequate treatment (Keeler & Siegel, 2016; Kessler et al., 2003) so, it is essential to motivate youngsters to seek help when needed. This insecurity regarding approaching specialized help when dealing with depression is directly correlated with the powerful stigma depression symbolizes for the Romanian population. Even though the research hypothesis was that urban dwellers would be well informed about depression and open about approaching both the subject and the causing factors and treatment opportunities, it seems that this particular issue still needs time and may be better information sources and methods. This study brings additional evidence that the negative perception that depression brings with it, continues to weigh on the population.

When asked to enumerate possible reasons that would prevent the interviewees from seeking out specialized treatment or help, the most commonly used answer pointed to stigma as the most important barrier. As this question was once again an open-ended one, the coding of opinions created six major categories: stigma, financial reasons, lack of trust in the medical system/specialized personnel, the belief that they would cure themselves, lack of time, or refusal to deal with the issue altogether. A substantial number of people (12.87%) declared that they saw no barriers in seeking specialized help, while almost 6% did not answer. The most often cited reason for avoiding specialized treatment in treating depression was, as mentioned before, the stigma that comes with acknowledging it. Almost half of this group was made out of the younger cohort of interviewees (15–17 years old) with the rest of the socio-demographic features corroborating their age. There is no distinction at the level of those naming stigma as a very important obstacle when analysing their defining of depression with an almost equal three-way split between it being an illness, an emotional state, or the DK/NA group. Social exclusion from the peer group (lack of attention, being ignored, exclusion from activities) has the worst impact on teenagers suffering from depression (Dolphin & Hennessy, 2014), leading to isolation in particular and destructive behaviour in general. However, according to the attribution theory (Weiner et al., 1988), people stigmatize when they perceive the behaviour as controllable. Therefore, people tend to judge depending on the level of responsibility attributed to a behaviour; the easier they identify an external cause for depression, the better their attitude is towards the people dealing with the illness.

The next reason identified by the weight of responses relates to financial issues. In Romania, a psychotherapy session costs approximately 6.28% of the average wage,[1] it means that the monthly amount to be spent on mental health would be around 25%. When dissecting this group of respondents, the following features stand out: they are predominantly part of the older cohorts of our target group, specifically those aged 24–35; all of them offered concrete opinions when asked to describe depression (although they are split between categorizing it an illness and an emotional state) and three-quarters of them declared that they would be willing to seek specialized help if needed. This is an important finding that even in the case of overcoming the inherent stigma of this illness; people who would be willing to seek specialized help are unable to receive it due to financial constraints. When analyzing the age structure of the responding group, most of those naming lack of financial resources as an obstacle are those aged between 18 and 23. This group is positioned somewhere between still relying (if only partially) financially on their family and starting to earn their income and become financially independent. One important economic reality this age group faces presently in Romania is the fact that in urban areas living costs are high and increasing yearly, and most of the entry-level jobs come with an income that would cover only the basic necessities. As such, young people, even if employed,

[1] In 2020 the average net salary in Romania was 3176 RON and a 50 min psychotherapy session around 200 RON.

still have to seek financial help from other sources. It is still very difficult to receive continuous treatment, even if employed, given the high costs of seeking specialized help in Romania. This is supported by the fact that 25% of those naming financial issues as an obstacle in treating depression are in fact, employed.

The last most often quoted obstacle in seeking specialized treatment was the lack of trust in the medical system/specialized personal. This could stem from the very complicated relation of responsibility and subordination of the Romanian Healthcare System that creates problems for patients and doctors alike, from the convoluted and heavily bureaucratized system that is difficult to navigate even for the simpler medical needs, from the very nuanced nature of mental health issues, and also from the feeling of the stigma that often seeps into the medical practice of this nature. There is a correlation between not trusting specialized personnel and defining depression, as 60% of this group do not see depression as an illness but instead as an emotional state.

20.6 Identifying the Incipient Markers of Depression

Due to the fact that depression is very often correlated with a state of intense sadness, which the respondents also pointed out in Fig. 20.4, they were also asked about the usual actions they performed when faced with such a situation. A state of sadness could be the first sign of a more serious condition or just a momentary setback. Measuring the response to such a situation could be an indicator of their predisposition to acting the same in a more severe position. Sometimes multiple possible actions were identified by the respondents, given that this was also an open-ended question. The answers were codified into main categories; some very clear including a more varied range of behaviours. As such the respondents declared that when faced with an episode of sadness they would: try to socialize; engulf themselves into different hobbies (with reading being named most often but also writing, painting, cooking, and others); go for long walks or other sporting activities; seek professional advice; meditate; completely ignore it and continue with their routine; eat; isolate themselves; continuously think about the reasons they are in that state and analyse their situation; listen to music; watch movies (series, or other shorter forms of entertaining either online or from their own collections); cry or sleep for long periods. One person usually named more than one type of possible behaviour, but the weight of each action is presented in Fig. 20.5. The most often cited courses of action are neither constructive nor destructive necessarily but could be correlated to an attempt to isolate the person: listening to music, watching things on their laptops or TV, crying, and/or sleeping. A total of 48% of the times, the answers describe one or more of these courses of action. This range of behaviours was cited by both the males and females participating in the study, with females citing them slightly more often (half of the females that filled in the questionnaires named neutral actions in their answers while males only cited neutral actions in 38% of their answers). In terms of more constructive actions (trying to socialize, performing different hobbies,

Fig. 20.5 Activities performed while being sad

sports, meditating, and/or trying to speak to someone trained in this field), found that these were considered more often by males than females, as these actions were found in 35% of the males' answers and only named in 29% of cases by females.

20.7 Conclusions

There is a vested interest of the public for the issues discussed in this paper, given a large number of responses to the online questionnaire. This study showed that women are more willing to talk about depression than men, even if the results proved that it affects all categories, no matter the age, gender, or social status.

The detailed and sometimes euphemistic descriptions of how the target group sees depression to support the hypothesis that young urban dwellers are at least willing to share their experience or to talk about the issue. This study proves that the incoming generation desires to have a dialogue on the topic of mental health and confirms the importance of seeking specialized help in dealing with depression. At the same time, young people are inclined to distrust the current healthcare system, as it continues to be plagued with problems like underfunding, increased bureaucracy; an overbearing network of responsibility and dependency relations among its institutions; reduced and under-prepared medical personnel. It is encouraging that in describing depression the urban youth identified elements used by the professionals to diagnose patients, which proved their solid knowledge and interest in the topic. However, the majority considered it an emotional state, not an illness and associated it with a temporary physio-psychological state.

Moreover, the target group highlighted most of the difficulties faced by people suffering from depression, both from an inner and an outer perspective. Negative thinking and suicidal thoughts, loss of self-esteem and self-isolation, permanent fatigue, inability to communicate efficiently and perform everyday tasks, financial issues, etc. are just some of the categories that were taken into account in analysing the impact of the disease on its subjects.

The most important conclusion of this study is that young urban dwellers prove to have an increased awareness of the existence of depression and its negative effects on those suffering from it; however, they still use a lot of stereotypes and shortcuts when asked to discuss this issue more deeply. This could perpetuate and even strengthen the stigma of this illness and also cause the alienation of those suffering from it.

Further research should thrive to analyse the reasons behind the incongruence between knowledge and behaviour and propose concrete actions towards improving the accessibility to both information about depression and means to specialized care.

Acknowledgements The chapter is the result of the research conducted by the authors for Universitur Association throughout the project *Mind the Gap* coordinated by Philmys Cultural Association, co-financed by The Administration of the National Cultural Fund (AFCN). The authors would also like to thank Prof. BRK Sinha for his continuous support during this process.

References

Adler, N., Mueller, G. O. W., & Ayat, M. (1993). Psychiatry under tyranny: A report on the political abuse of Romanian psychiatry during the Ceausescu years. *Current Psychology, 12*(1), 3–17. https://doi.org/10.1007/BF02737088

Akiskal, H. S., & Akiskal, K. K. (2007). In search of Aristotle: Temperament, human nature, melancholia, creativity and eminence. *Journal of Affective Disorders, 100*(1), 1–6. https://doi.org/10.1016/j.jad.2007.04.013

Aksan, N., Kısac, B., Aydın, M., & Demirbuken, S. (2009). Symbolic interaction theory. *Procedia—Social and Behavioral Sciences, 1*(1), 902–904. https://doi.org/10.1016/J.SBSPRO.2009.01.160

American Psychiatric Association. (1952). *Diagnostic and statistical manual of mental disorders I* (1st ed.).

American Psychiatric Association. (2013). Depressive disorders. In *DSM library. Diagnostic and statistical manual of mental disorders*. xxx10.1176/appi.books.9780890425596.dsm04

Amnesty International. (2004). *Memorandum to the Romanian government concerning inpatient psychiatric treatment*. Brussels.

Beck, A. T. (1967). *Depression: Clinical, experimental, and theoretical aspects*. Hoeber Medical Division, Harper & Row.

Beck, A. T. (1979). *Cognitive therapy of depression*. Guilford Press.

Carol Davila University of Medicine and Pharmacy. (2018). *Course for 5th-year students, Management and public health*. Retrieved March 20, 2020, from https://umfcd.ro/wp-content/uploads/C-10-Sisteme-de-sanatate.pdf

Caspi, A., Harrington, H., Moffitt, T. E., Milne, B. J., & Poulton, R. (2006). Socially isolated children 20 years later: Risk of cardiovascular disease. *Archives of Pediatrics & Adolescent Medicine, 160*(8), 805–811. https://doi.org/10.1001/archpedi.160.8.805

Dolphin, L., & Hennessy, E. (2014). Adolescents' perceptions of peers with depression: An attributional analysis. *Psychiatry Research, 218*(3), 295–302. https://doi.org/10.1016/j.psychres.2014.04.051

Dumitrache, L., Nae, M., Dumbrăveanu, D., Simion, G., & Suditu, B. (2016). Contrasting clustering in health care provision in Romania: Spatial and aspatial Limitations. *Procedia Environmental Sciences, 32*, 290–299. https://doi.org/10.1016/j.proenv.2016.03.034

Eurostat. (2020). Physicians by medical speciality. Retrieved March 23, 2020, from https://appsso.eurostat.ec.europa.eu/nui/show.do?dataset=hlth_rs_spec&lang=en

Friedman, J. R. (2006). The challenges facing mental health reform in Romania. *EUROHEALTH-LONDON-, 12*(3), 36.

Fuhrer, R., & Wessely, S. (1995). The epidemiology of fatigue and depression: A French primary-care study. *Psychological Medicine, 25*(5), 895–905. https://doi.org/10.1017/S0033291700037387

Gallo, J. J., Rabins, P. V., Lyketsos, C. G., Tien, A. Y., & Anthony, J. C. (1997). Depression without sadness: Functional outcomes of nondysphoric depression in later life. *Journal of the American Geriatrics Society, 45*(5), 570–578. https://doi.org/10.1111/j.1532-5415.1997.tb03089.x

Gater, R., Jordanova, V., Maric, N., Alikaj, V., Bajs, M., Cavic, T., Dimitrov, H., Iosub, D., Mihai, A., Szalontay, A. S., Helmechen, H., & Sartorius, N. (2005). Pathways to psychiatric care in Eastern Europe. *British Journal of Psychiatry, 186*(6), 529–535. https://doi.org/10.1192/bjp.186.6.529

Goldshmidt, O. T., & Weller, L. (2000) "Talking emotions": Gender differences in a variety of conversational contexts. *Symbolic Interaction*. John Wiley & Sons, Ltd., *23*(2), 117–134. https://doi.org/10.1525/si.2000.23.2.117.

Haaga, D. A., Dyck, M. J., & Ernst, D. (1991). Empirical status of cognitive theory of depression. *Psychological Bulletin, 110*(2), 215.

James, S. L., Abate, D., Abate, K. H., Abay, S. M., Abbafati, C., Abbasi, N., & Murray, C. J. L. (2018). Global, regional, and national incidence, prevalence, and years lived with disability for 354 diseases and injuries for 195 countries and territories, 1990–2017: A systematic analysis for the Global Burden of Disease Study 2017. *The Lancet, 392*(10159), 1789–1858. https://doi.org/10.1016/S0140-6736(18)32279-7

Junjan, V., Micluţia, I., Popescu, C., Ciumăgeanu, M., Sfetcu, R., & Ghenea, D. (2009a). Overview of the Romanian inpatient psychiatric services. *Romanian Journal of Psychopharmacology, 3*, 133–138.

Junjan, V., Micluția, I., Popescu, C., Ciumăgeanu, M., Sfetcu, R., & Ghenea, D. (2009b). The provision of psychiatric care in Romania–need for change or change of needs? *Transylvanian Review of Administrative Sciences, 5*(26), 81–93.

Karp, D. A. (1994). Living with depression: Illness and identity turning points. *Qualitative Health Research, 4*(1), 6–30. https://doi.org/10.1177/104973239400400102

Keeler, A. R., & Siegel, J. T. (2016). Depression, help-seeking perceptions, and perceived family functioning among Spanish-Dominant Hispanics and Non-Hispanic Whites. *Journal of Affective Disorders, 202*, 236–246. https://doi.org/10.1016/j.jad.2016.05.017PMID-27280309

Kessler, R. C., Berglund, P., Demler, O., Jin, R., Koretz, D., Merikangas, K. R., Rush, A. J., Walters, E., & Wang, P. S. (2003). The epidemiology of major depressive disorder results from the National Comorbidity Survey Replication (NCS-R). *JAMA, 289*(23), 3095–3105. https://doi.org/10.1001/jama.289.23.3095

Lewis, S. E. (1995). A search for meaning: Making sense of depression. *Journal of Mental Health, 4*(4), 369–382. https://doi.org/10.1080/09638239550037424

Lienemann, B. A., Siegel, J. T., & Crano, W. D. (2013). Persuading people with depression to seek help: Respect the boomerang. *Health Communication, 28*(7), 718–728.

Matthews, T., Danese, A., Wertz, J., Odgers, C. L., Ambler, A., Moffitt, T. E., & Arseneault, L. (2016). Social isolation, loneliness and depression in young adulthood: a behavioural genetic analysis. *Social Psychiatry and Psychiatric Epidemiology, 51*(3), 339–348. Retrieved from https://doi.org/10.1007/s00127-016-1178-7 PMID - 26843197

Mental Health Europe. (2018). *Mapping and understanding exclusion in Europe.* Retrieved March 18, 2020, from https://mhe-sme.org/wp-content/uploads/2018/01/Mapping-and-Understanding-Exclusion-in-Europe.pdf

Ministry of Health. (2019). *Ziua Europeană Anti-Depresie 1 octombrie 2019 Analiză de situație.*Retrieved February 5, 2020, from https://insp.gov.ro/sites/cnepss/wp-content/uploads/2017/04/Analiza-situatie-ZEAD-2019-.pdf.

National Institute of Mental Health. (2018). Depression. Retrieved from https://www.nimh.nih.gov/health/topics/depression/index.shtml

Neagu, A. (2019). The Economist: Sistemul de sănătate din România, cel mai slab din Uniunea Europeană. Retrieved March 18, 2020, Retrieved 5 February, 2020, from https://www.hotnews.ro/stiri-sanatate-23509000-the-economist-sistemul-sanatate-din-romania-cel-mai-slab-din-uniunea-europeana.htm

Organizația pentru Cooperare și Dezvoltare Economică—OECD. (2017). *Romania: Profilul Sănătății în 2017.* Retrieved from https://books.google.ro/books?id=ahpADwAAQBAJ

Parekh, R. (2017). *What is depression.* Retrieved 15 April, 2020, from https://www.psychiatry.org/patients-families/depression/what-is-depression

Parlamentul României. (2015). LEGE nr. 95 din 14 aprilie 2006 (republicată) privind reforma în domeniul sănătății. Retrieved February 23, 2020, from http://legislatie.just.ro/Public/DetaliiDocument/71139

Plunkett, S. W., Henry, C. S., Robinson, L. C., Behnke, A., & Falcon, P. C. (2007). Adolescent perceptions of parental behaviors, adolescent self-esteem, and adolescent depressed mood. *Journal of Child and Family Studies, 16*(6), 760–772. https://doi.org/10.1007/s10826-006-9123-0

Preda, M., Mareci, A., Tudoricu, A., Taloș, A. M., Bogan, E., Lequeux-Dincă, A. I., & Vijulie, I. (2020). Defining the concept of family through the lens of fertile-aged women in Bucharest, Romania—between traditionalism and inclusion. *Sustainability, 12*(7), 2691.

Puffer, J. C., & McShane, J. M. (1992). Depression and chronic fatigue in athletes. *Clinics in Sports Medicine, 11*(2), 327–338.

Qualter, P., Vanhalst, J., Harris, R., Van Roekel, E., Lodder, G., Bangee, M., Maes, M., & Verhagen, M. (2015). Loneliness across the life span. *Perspectives on Psychological Science, 10*(2), 250–264. https://doi.org/10.1177/1745691615568999

Ritchie, H., & Roser, M. (2018). *Mental health.* Retrieved from https://ourworldindata.org/mental-health

Secretariatul General al Guvernului. (2011). *Analiza Funcţională a Sectorului Sănătate în România.* Retrieved February 5, 2020, from http://sgg.gov.ro/docs/File/UPP/doc/rapoarte-finale-bm/etapa-II/MS-RO-FR-Health-Sector-ROM

Siegel, J. T., Lienemann, B. A., & Tan, C. N. (2014). Influencing help-seeking among people with elevated depressive symptomatology: Mistargeting as a persuasive technique. *Clinical Psychological Science, 3*(2), 242–255. https://doi.org/10.1177/2167702614542846

Solomon, A. (2014). *The noonday demon: An atlas of depression.* Simon and Schuster.

Stiver, I. P., & Miller, J. B. (1988). *From depression to sadness in women's psychotherapy.* Retrived April 10, 2020, from https://www.wcwonline.org/pdf/previews/preview_36sc.pdf

Tacchi, M. J., & Scott, J. (2017). Depression: A very short introduction. In *Very short introductions.* Retrieved April 10, 2020, from https://books.google.ro/books?id=sGfODQAAQBAJ

The Romanian Government. (1997, June 3). Decision no. 244/ 30.05.1997 regarding the organisation and functioning of the Ministry of Health. *Monitorul Oficial.*

Toohey, P. (1990). Some ancient histories of literary melancholia. *Illinois Classical Studies, 15*(1), 143–161. Retrieved April 12, 2020, from http://www.jstor.org/stable/23064398

Truschel, J. (2019). Depression definition and DSM-5 diagnostic criteria. Retrieved April 12, 2020, from https://www.psycom.net/depression-definition-dsm-5-diagnostic-criteria/

Weiner, B., Perry, R. P., & Magnusson, J. (1988). An attributional analysis of reactions to stigmas. *Journal of Personality and Social Psychology, 55*, 738–748. https://doi.org/10.1037/0022-3514.55.5.738

World Health Organisation. (2020). Depression. Retrieved February 15, 2020, from https://www.who.int/news-room/fact-sheets/detail/depression

World Health Organization. (2001). The world health report 2001: Mental health : New understanding, new hope. In *Public health series.* Retrieved April 12, 2020, from https://books.google.ro/books?id=GQEdA-VFSIgC

World Health Organization. (2011). *Mental Health Atlas 2011*. Retrieved from https://doi.org/10.1093/bja/aes067

Zamfir, D., Dumitrache, L., Stoica, I.-V., & Vîrdol, D. (2015). Spatial inequalities in health care provision in Romania: Milestones for territorial sustainable development. *Carpathian Journal of Earth and Environmental Sciences, 10*(3), 177–188.

Chapter 21
Active Ageing and Health Status of the Aged Population of Bucharest City, Romania

Ana-Maria Talos

Abstract This paper focuses on the specific problem of the elderly population in the context of the demographic ageing process that Romania, as well as other European countries, is facing. Experts working in the area of ageing population consider the concept of active ageing very important as Europe is getting older. In light of this, the present study aims at evaluating the impact of lifestyle on general health status of the elderly population of Bucharest. The paper also intends to evaluate the degree of satisfaction of the elderly with their personal health status; highlight the needs of elderly population in the perspective of active ageing and assess the role of lifestyle in meeting these needs. The analysis and interpretation of the study is based on the elderly population-related statistical data gathered from National Institute, as well as on the information collected from 400 respondents over 65 years old through a personal health survey conducted in Bucharest during the period of June–September 2018. Results of the study show that lifestyle, being an important driving factor of active ageing, determines the elderly population's health status. The elderly persons who follow a healthy and active lifestyle present better health and hence lifestyle becomes a reliable indicator of health. Income and education levels are also found correlated with the differences in health and lifestyle of the elderly.

Keywords Elderly population · Lifestyle · Health status · Active ageing · Romania

21.1 Introduction

Because of demographic changes, the world faces major socio-economic transformation. In European countries, people are already relatively old compared to the rest of the world and this trend is likely to continue. The people of 60 and above years

A.-M. Talos (✉)
Department of Human and Economic Geography, Faculty of Geography, University of Bucharest, Boulevard Nicolae Balcescu No.1, Bucharest, Romania
e-mail: ana.talos@geo.unibuc.ro

B. R. K. Sinha (ed.), *Urban Dynamics, Environment and Health*,
https://doi.org/10.1007/978-981-99-5744-6_21

are growing faster than that of any other age group and the proportion of them is estimated to rise from 14% to nearly 25% of the total population by 2050. In other words, by the year 2050 there will be 2 billion people aged 60 and above and 80% of them are expected to live in the developing countries. All these will place huge pressure on labour markets, social protection systems and the environment (WHO, 2002).

Ageing is a continuous and unavoidable process, but it is very important for older persons to remain socially integrated and active in the social and family life because healthy and active ageing encourages a dynamic balance between body, cognition, and affection. They need special care concerning diet, physical activity, healthy habits, and correct medical behaviour. A successful ageing is considered to be related with the 4Ds: dependency, disease, disability, and depression. In this sense, it becomes important to combine longevity, happiness, and health for a successful ageing (Dias & Couceiro, 2017).

The elderly population is often lumped together to include everyone over the age of 65, but according to specific age the characteristics of the elderly people clearly differ. So, it is very useful to divide the elders into groups to understand them better. According to Mulan, 2000; Bartram & Roe, 2005; Werner, 2011, the older population can be divided into three life-stage subgroups: the young-old (from around 65 to 74 years of age), the middle-aged (75–84 years of age), and the old-old (over 85 years of age).

It is expected that in Romania the proportion of people aged 60 or above will increase from 24.4% in 2015 to 36.4% by 2050, as the life expectancy is projected to increase in 2050 to more than 80 years, on average (UN, 2015).

The term "active ageing" was adopted by the World Health Organization (WHO) in the late 1990s to define the process of "optimizing opportunities for health, participation and security, in order to enhance quality of life as people grow old". Active ageing aims to extend healthy life expectancy and quality of life for all people as they get older. The concept is related to a longer and healthier life, a high rate of employment for the elderly population, an active social and economic life, low rate of dependency, and better healthcare services (WHO, 2013).

There are several researches which conclude that individuals with better health tend to retire at an old age compared to that of less healthier persons as better health has several positive effects on delaying the timing of retirement. For example, Bloom et al. (2007) found that individuals of all ages that are healthier and live longer lives determine an increase in the optimal retirement age. In Austria, Staubli and Zweimüller (2013) found that an early retirement was more popular among less healthy and low-wage workers looking for disability or unemployment benefits. Moreover, Coile and Levine (2007) discovered that older workers with health problems have a greater probability of becoming unemployed or wishing to retire as soon as possible, being constrained by poor labour conditions.

An adequate physical activity is very important for a healthy ageing. There are several theories that talk about a balance between genetics and environment. With age, elderly population becomes wiser, since wisdom can only be acquired with experience. There are also several myths and stereotypes against elder people like senility, lack of productivity, deterioration of intelligence, isolation, and lack of serenity (Serra e Silva et al., 2015). In general, certain diseases are more frequent in the elderly population. Here, there seems to be a socio-pathological complex created largely by stress, lifestyle, diet and lack of activity. That is why prevention is better than treatment in the event of an active ageing (Minett, 2012).

When one talks about the association between health and retirement, it becomes necessary to take into account the preventive health programmes. Those individuals who have the highest levels of productivity seem to have a greater interest in extending their working life and also in investing in preventive health, compared to the rest of population (Aisa et al., 2015). In Europe, Carrieri and Wuebker (2018) showed that mammography and blood tests are popular among high-wage individuals, despite their substantially lower diagnostic needs. Also, in the USA, Ross et al. (2007) discovered that the adults over 50 are not making use of preventive care if they are lower income workers. Courtemanche and Zapata (2014) demonstrated that the 2006 Massachusetts Healthcare reform, oriented to guarantee universal access to healthcare, has led to positive effects on overall health, especially among those with low incomes.

Several studies examined a number of indicators covering health, economic, and social sectors and the results reflected the dynamics of gender, family structure, health, economics, and social conditions. It is important to mention that not all of these factors have the same effect across age groups, meaning that the same group may be more likely to involve in one set of activities, but not others. For example, a more educated older person is likely to participate in social gatherings and activities, compared to someone with no formal education (Dommaraju, 2016).

Active Ageing Index (AAI) is a composite index used in the European Union since 2012 to compare the extent and progress of active ageing in EU countries. It is considered that AAI has a major role in monitoring and implementing the guiding principles of the ageing policies at the European and national levels (European Commission & United Nations Economic Commission, 2015).

In Romania, as in other European countries, the epidemiological profile is characterized by the prevalence of circulatory system diseases, cancer, and accidents. More exactly, 86% of the total mortality cases in Romania is represented by cancer, heart diseases, respiratory diseases, and diabetes. The number of obese people and disabilities associated with diabetes is increasing, also the problems caused by the lifestyle's characteristics (smoking, alcohol consumption, medical behaviour, etc.) are more widespread. All these health problems are affecting the ageing society, having long-term consequences on the healthcare system. The frequency of type 2 diabetes, hypertension, and obesity is higher among elderly population, as well as the risk of comorbidity, while most preventive measures are needed during childhood and adulthood (National Institute of Statistics, 2021).

21.2 Background Information—Elderly Population in Europe, European Union, and Romania

In Europe, the statistics indicate two categories of aged population: 65–79 years old and 80 and above. In 2020, there were 14.6% people between 65 and 79 years old and 5.9% people 80 and above in Europe. These average values are higher in some countries (Bulgaria, Italy, Finland, Croatia) indicating an accentuated process of population ageing (Fig. 21.1).

Healthy Life Years (HLY) at birth measures the number of years that a person at birth is still expected to live in a healthy condition. HLY is a health expectancy indicator which combines information on mortality and morbidity. The data required are the age-specific prevalence (proportions) of the population in healthy and unhealthy conditions and age-specific mortality information. A healthy condition is defined by the absence of limitations in functioning/disability. This indicator is calculated separately for males and females and it is also called disability-free life expectancy (DFLE) (Eurostat, 2021b).

According to Fig. 21.2, the average age for healthy life is 64.6 in the European Union, with variation between the European countries: from 73.3 in Sweden to 53.1 in Latvia, due to lifestyles (nutrition, physical activity, smoking), environmental conditions, working and living conditions, poverty, genetic predisposition and differences in healthcare systems. Moreover, some national data suggest that the expectancy of healthy life years can vary across socio-economic groups measured using education, working category, or income.

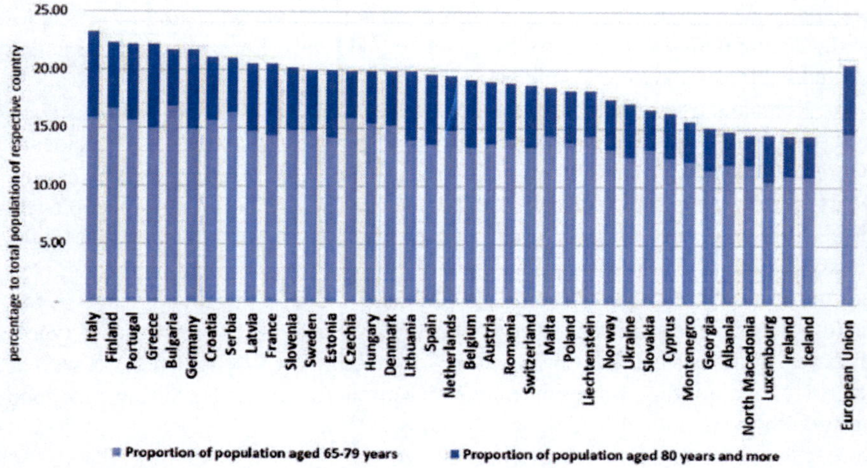

Fig. 21.1 Proportion of elderly population in the European countries and European Union in 2020. (*Source* Eurostat, 2021a)

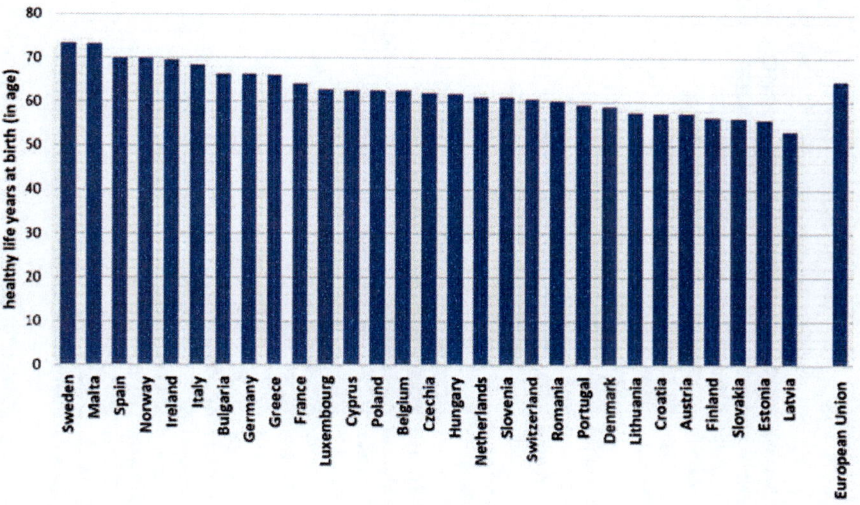

Fig. 21.2 Healthy life years at birth in 2019 in European countries and European Union. (*Source* Eurostat, 2021b)

Healthy Life Years (HLY) at age 65 measures the number of years that a person at age 65 is still expected to live in a healthy condition. HLY at age 65 is like the one presented before, just referring to population aged 65 years (Eurostat, 2021c).

According to Fig. 21.3, the average age for healthy life at 65 is 10.3 in the European Union, with variation between the European countries: from 16.2 in Sweden to 4.7 in Latvia and Slovakia. These discrepancies are due to differences in the perception of these problems and in health expectations, but also to differences in exposure to health problems and in their management (for example, differences in mortality and morbidity).

One important factor that can explain these inequalities is income, and an indicator was calculated to evidence the risk of poverty among aged population (Fig. 21.4). This indicator is the share of persons with an equivalized disposable income below the risk-of-poverty threshold, which is set at 60% of the national average equivalized disposable income (after social transfers) (Fig. 21.5).

In 2020, Romania registered one of the highest rates of poverty among people over the age of 65 compared to other countries of the EU, reaching 24.5%. Similar values were recorded by Malta (26.3%), Croatia (31%), Lithuania (36%), Bulgaria (38.3%), Latvia (40.9%), and Estonia (41.4%), Fig. 21.4.

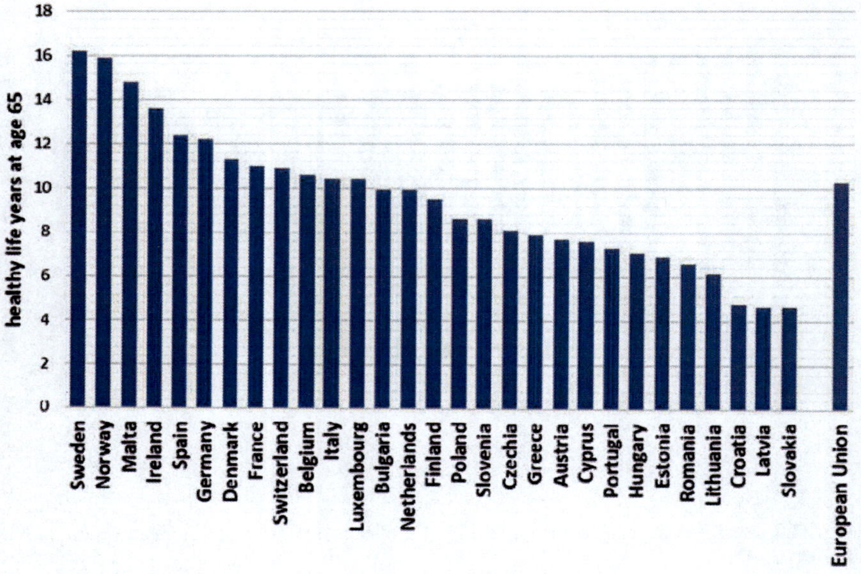

Fig. 21.3 Healthy life years at age 65 in 2019 in European countries and EU. (*Source* Eurostat, 2021c)

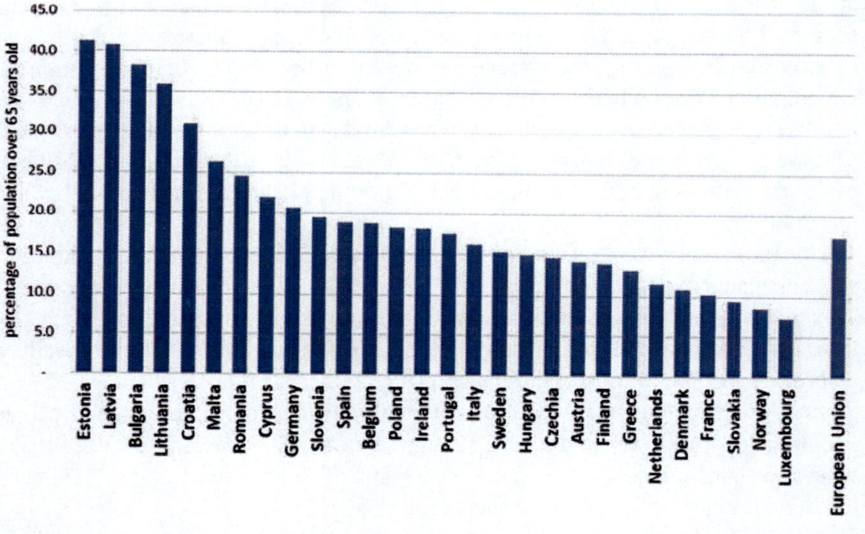

Fig. 21.4 At-risk-of-poverty rate of people aged over 65–EU-SILC survey 2020. (*Source* Eurostat, 2021d)

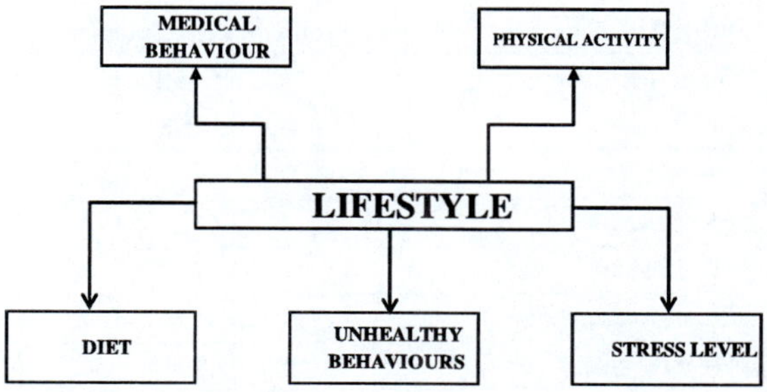

Fig. 21.5 Lifestyle components considered by the author

21.3 Data and Methodology

While analysing the concept of active ageing, it becomes important to consider some other significant elements such as population health status, dependency rate, and lifestyle. It is important to analyse the elderly population's health status from the perspective of active ageing as health is an important element in ensuring an active ageing. A healthier, more productive, participatory, and independent life is essential for both the elderly and young people. Ageing is a premise for a long life and it is important to pay attention to preventive measures, early detection of disease, especially chronic disease.

Bucharest (the capital of Romania) is one of the important cities in the East European region and is facing the problem of the ageing population, so it was chosen as the study area to fulfil the objectives of this research. More exactly, in 2014 the aged population (over 65) represented 14.8% of the total population (National Institute of Statistics). Many researchers have already carried out their research works in Bucharest City but most of them have focused on the economic impact (effects on labour market) or simply on demographic characteristics. Only a few of them have tried to assess the problem of active ageing in Bucharest. So, this study exclusively concentrates on examining the problem of ageing population in Bucharest.

The study is based on the statistical information collected from 400 respondents over 65 years old living in Bucharest City by conducting personal field survey between June and September 2018. The sampling was done following the aim of the research, the respondents were chosen from a varied spectrum in terms of age, economic, and educational background, from all of Bucharest's six sectors (Table 21.1).

The descriptive statistics of the independent variables in our analysis are presented in Table 21.1. The share of aged population between 65 and 70 years is the highest (49.5%) in which there is preponderance of females. In the selected sample, the number of females is more in all age groups except in the age group of 70–75 years. Regarding the educational background, high school studies are the most common among the respondents, and by gender, the educational status varies. More than 50% of the respondents live with less than 1000 RON per month (<200 €) and men tend to earn more money than women.

Table 21.1 Socio-demographic profile of the respondents

Socio-demographic indicators	Gender		Total
	Male	Female	
Age			
65–70 years	43.2	55.3	49.5
70–75 years	57.7	41.3	25.9
75–80 years	45.6	54.4	14.2
Over 80 years	48.4	51.6	7.7
Educational status			
Secondary	41.9	55.8	10.7
Highschool	51.6	46.6	40.1
Professional school	45.7	52.2	11.5
University	45.7	54.3	26.2
Economic status			
Under 1000 RON	47.1	51	51.7
1000–2000 RON	59.7	40.3	19.2
Over 2000 RON	59.3	40.7	6.7

Source Author's field research

In the present study, the cartographic graphical method and survey have been used to achieve the objective as stated above. The values of share of elderly population, healthy life years at birth, healthy life years over 65, and risk of poverty as illustrated in respective Figs. 21.1, 21.2, 21.3, and 21.4 for the EU Member States became useful in shaping the background of ageing phenomenon in Europe. Each indicator extracted from the Eurostat website explains the existing differences in terms of healthy ageing and poverty among the EU states.

The statistical data concerning total population, share of elderly population were gathered from National Institute of Statistics. Other required data were collected through questionnaire (containing 15 questions, both open and closed) by conducting personal field survey. The open questions included the information about the personal beliefs and opinions, whereas the closed questions were related to quantify the respondents age, gender, education level, monthly income, satisfaction degree regarding personal health and living conditions, frequent use of medical services, family help in need, etc. The data collected through questionnaire were processed by using specific software (SPSS 20) in order to obtain some territorial perspectives. Collected data were analysed through descriptive analysis, crosstab, Pearson correlation, and cluster analysis.

SPSS is important and very useful software which is widely used by the social and behavioural scientists in order to process the data for preparing the statistical tables to analyse the given facts. The descriptive statistical analysis is used for presenting data which becomes very useful in recognizing patterns. On the other side, Pearson correlation coefficient is useful for measuring the degree of association between two variables, showing the power and direction of that relation. Moreover, regressions offer the possibility of evaluating or predicting the weight when there is a relation between two variables, and crosstabs analysis combines two variables and emphasizes the distribution of data for each of the questions (Howitt & Cramer, 2010).

The lifestyle profile of the older population was obtained by applying cluster analysis, under non-hierarchical procedure based on the K-means algorithm. Thereafter, three clusters were defined: the first cluster was used to define "healthy lifestyle" that includes satisfaction regarding health, activity in the family, independence in travel aid, and technological skill. The second cluster corresponds to "moderate healthy lifestyle" defined by medium scores which were the averages of each of the chosen variables. Finally, the cluster of poor scores defined the "unhealthy lifestyle" of the aged population.

21.4 Analysis

21.4.1 Health Satisfaction Among Elderly Population

The approach to health in geographical studies has as a starting point in the recognition that health is the result of multiple factors, including social factors (Dumitrache, 2004). Health is a resource for everyday life and constitutes a positive concept that emphasizes social and personal resources as well as physical capabilities (Walker, 2010).

Healthcare reform in the majority of countries is driven by the desire to control costs, which have been constantly increasing above the rate of inflation due to a number of factors, including advances in medical science (drugs, technology, and procedures), population demands, and demographic change (Santerre & Neun, 2010).

The Healthy People 2020 programme includes health promotion for older people in Taiwan with a particular focus on the prevention and management of chronic conditions including diabetes, hypertension, and asthma (Ti, 2010).

The bivariate distribution of the independent variables (educational level and economic status) with the level of satisfaction regarding personal health is presented in Table 21.2. The highly educated older persons and the individuals with higher monthly income are happier with their personal health status (48.6 and 55.6%). This could be explained by the fact that those respondents who have higher odds in maintaining more social ties are more likely to participate in social activities compared to the poorest and less educated from the community. We also noticed that those with a lower level of satisfaction regarding personal health are more likely to maintain social ties as some of them are dependent on others in terms of their daily activities. In other words, such persons need someone to help them with the transport, with the shopping, with their periodical examination, etc.).

Table 21.2 Crosstab correlation between socio-demographic indicators and satisfaction level of health

Socio-demographic indicators	Level satisfaction				
	Very happy	Happy	Moderately happy	Un Happy	Very unhappy
Educational status					
Secondary	9.3	37.2	27.9	20.9	4.7
Highschool	6.2	37.9	32.9	21.1	1.9
Professional school	4.3	30.4	41.3	19.6	4.3
University	6.7	48.6	25.7	14.3	2.9
Economic status					
Under 1000 RON	4.8	41.3	35.6	15.9	1.9
1000–2000 RON	10.4	42.9	27.3	14.3	3.9
Over 2000 RON	18.5	55.6	7.4	18.5	0

Source Author's field research

In terms of medical services, 52.9% older persons preferred to use public health care services, 6% older persons relied on using only private services, and 41.4% older persons chose to use both types of medical services in case of prevention or treatment. There is a small difference between those respondents whose monthly income is not enough (46.3%) and those satisfied with their revenue (52.5%). It is important to mention that older persons with lower income (less than 1000 RON) use only public healthcare services when needed but in case of persons whose income is higher, the preferences are changing, and they choose private services or both of them combined. Older persons earning more money are least likely to be unhappy with their personal health status whereas those with lower income are the opposite. Given that older persons' satisfaction with health status is strongly influenced by their economic status, it is of interest to see whether these correlations are seen to be statistically significant in the Pearson analysis that follows (Table 21.3).

Table 21.3 Correlation between healthcare preferences and socio-economic variables

		Health care services	Satisfaction with health status
Economic status index	Pearson correlation	0.134	0.282*
	Sig. (2-tailed)	0.007	0.000
	N	400	400
Education level index	Pearson correlation	0.348*	–0.072**
	Sig. (2-tailed)	0.000	0.152
	N	400	400

*. Correlation is significant at the 0.05 level (2-tailed)
**. Correlation is significant at the 0.01 level (2-tailed)

Most employed seniors have a low-income level. More than 50% of seniors declared that they are currently working but earning less than 1,000 lei/month (approx. 205 Euros). This percentage decreases as the income level increases. As expected, the most active category in terms of employment is the youngest segment of the sample as two-thirds of respondents who declared to be still employed belong to the age group of 65–70 years.

Seniors need to work because they do not have the freedom to exit the labour market because of their low pension income. Taking this into account, seniors realize that their revenues are limited. A large percentage of respondents (46.3%) declared that their income is limited and does not cover the cost of their current medication. This explains that a large number of seniors are employed, even though they could have officially been retired. In fact, 59% of the respondents declaring their monthly income level too low to cover the costs of their medication are still employed. This reflects the high poverty risk rate for a significant percentage of the senior population in Romania. This is also clear from Fig. 21.4 where the percentage of population over 65 in Romania is slightly less than 25%.

21.4.2 Aged Population Needs in the Context of Active Ageing

The aged population needs are generally related to the medical assistance, transport aid for personal needs (shopping, medical checkup), social and community integration, household services (cleaning, food), technology support, etc.

This study reveals that respondents are generally quite independent, but in case of transport aid 34.1% respondents were found dependent, 82.3% were active in the family activities, 29.9% were less active in the community and 7.7% were active in politics. 21.9% of the population needs health care services at home, 11.2% of population needs cleaning services, and 6% needs various services at home. 31.2% of the population uses a laptop or computer and 67.1% uses a mobile phone. The lack of solitude positively influences a person's ability to be active, as the family and/or partner satisfy their emotional needs and represent the base of a supportive and mutual helping relationship. Most of the seniors (62.4%) do not call for help outside their immediate family or partners when they have to go somewhere, and 34.1% rely on their close circle of friends or neighbours.

Normally, the seniors prefer not to externalize their problems and choose, firstly, to ask for support to their family members. A smaller percentage declared that they go for help to their neighbours or friends. Seniors ask for mobility support for several reasons: health (transport to a clinic or doctor; or to buy medicines), groceries, and walks in the park or other recreational activities.

In terms of using specialized services for medical transport to the hospital, GP, or other adjacent services, 77.9% of seniors declared they do not use them. Only 22.9% of seniors requested a personalized health-paid service if the state services, free of charge, are not available or cannot solve a situation. Seniors rely on state-provided free-of-charge medical services, not necessarily because they are satisfied by their quality, but because they cannot afford private healthcare services. In fact, about 20% of seniors, regardless of their income level, appeal to paid private medical services proving the insufficiency of the public sector.

Table 21.4 shows the bivariate distribution of the needs of the elderly population with each of the explanatory variables used in the analysis to examine local problems. It is important to highlight some of the main aspects that emerge from the table. First, gender-based differences in the needs of the elderly are not very significant, as the resulting values are very similar. However, men are less dependent on transport aid compared to women; they are less active in the family actions, more active in the community, less proficient in the use of technology (laptops and mobile phones), and call for medical services at home more frequently. In terms of age differences, it is observed that as age increases, the dependency on travel is higher (29.6% in case of elders in the age group 65–70 and 54.8% in 80 + age group), they call for cleaning services at home more often (16.1% for those aged over 80) and they are using less technology (laptops and mobile phones).

In this bivariate distribution, it can be noted that education has a strong influence on older persons' technology skills, social engagement, and specialized home services. 88.9% older persons with higher income and 74.3% with higher education use technology more frequently, 40.7% older people with higher income and 19% older people with higher education ask for cleaning services at home more frequently, and 51.9% older persons with higher income and 35.2% older persons with higher education are more active in the community actions. Older persons with lower educational background and lower income are less likely to participate in community meetings, but are very active in the family works.

21.4.3 Role of Lifestyle in Meeting the Needs of the Elderly

Lifestyle includes several components like diet, physical activity, stress level, unhealthy behaviours, and medical behaviours. Lifestyle has a profound impact on population health status and that is why health status of the elderly remains precarious even though they adopt a healthy lifestyle (Taloș, 2017).

Older people may be particularly prone to the negative effects of a sedentary lifestyle; it is well established that sedentary lifestyles contribute to chronic illnesses (Byrne et al., 2018) including obesity, diabetes mellitus, non-alcoholic fatty liver disease, and cardiovascular disease (Al-Dayyat et al., 2018).

Social interaction, especially in later life, is important for brain health during ageing. An active and socially integrated lifestyle may protect against dementia (Fratiglioni et al., 2004).

Moreover, there is a strong link between early-life education and the risk of dementia, suggesting that people with fewer years of educational schooling may be at a higher risk of dementia (Sharp & Gatz, 2011).

Table 21.4 Bivariate distribution between elderly needs and socio-economic factors

	Transport aid	Weekly active in the family	Weekly active in the community	Medical services at home	Cleaning services at home	Technology support
Male	29.9	75.8	35.6	25.4	11.4	64.8
Female	38.1	89.1	24.8	18.3	11.4	69.3
Age						
65–70 years	29.6	84.4	30.7	17.7	9.1	79.3
70–75 years	33.7	77.9	32.7	27.9	13.5	61.5
75–80 years	40.4	86	26.3	22.8	14	47.4
Over 80 years	54.8	71	29	19.4	16.1	48.4
Living alone	34.9	68.8	34.9	23.9	15.6	54.1
Educational status						
Secondary	69.8	74.4	30.2	18.6	4.7	48.8
High school	57.8	82	30.4	21.7	8.7	65.8
Professional school	69.6	91.3	23.9	15.2	6.5	71.7
University	68.6	81	35.2	25.7	19	74.3
Economic status						
Under 1000 RON	33.2	81.7	28.8	22.7	9.7	64.7
1000–2000 RON	28.6	79.2	42.9	19.5	15.6	80.5
Over 2000 RON	29.6	88.9	51.9	25.9	40.7	88.9

Source Author's field research

In the present study, the first group consists of older individuals with healthy lifestyle, which is essential for an active ageing. It is found that the persons who are satisfied with their health status are active in their family tasks, but are not seeking help in terms of travel assistance and frequently use technology (laptops and mobile phones) for various reasons. This group/cluster includes the largest group of respondents (57%) which reveals the positive aspects that elderly people need to adopt a healthy lifestyle in order to have a healthy ageing. The second cluster is the smallest (20% of the group), constituted by people who are moderately satisfied with their health, who mostly perform passive activities, need travel assistance, and use mobile phone less than others. The third group (21%) is identified as having an "unhealthy lifestyle", represented by people who are below average in the activities in question.

21.5 Conclusion and Suggestions

After having gone through the above analysis of active ageing and health of the aged population of Bucharest city it has been found that the proportion of the city's elderly population above the age of 65 is gradually increasing and their percentage has reached about 24–25% in 2020. Age-wise variation shows that the percentage of aged population was highest from the age group of 65–70 years in which the share of females was more. The number of females has been found more in all age groups except the age group of 70–75 years. Levels of education vary between men and women, although there is a predominance of respondents with high school studies. Similarly, more than 50% of the respondents were found living with less than 200 Euros per month. Statistical evidence shows that most elderly people with higher levels of education and higher monthly income were happier with their personal health status.

In terms of assistance or services 52.9% older persons preferred to use public health care services, 6% private health services and 41.4% older persons chose to use both types of medical services in case of prevention or treatment. Similarly, older persons with low income use only public health care services, but persons with higher income prefer to use private services or they use both public and private. Older persons with higher income were happier than those whose income was low. More than 50% elderly stated that they are currently working but their earning is less than 1,000 lei/month.

In terms of medical assistance, transport aid for personal needs, social and community integration, and household services the responses of the elderly are not the same rather vary from one type of assistance or service.

Education and income have been found influencing the older persons' technological skills, social engagement, and specialized home services. 74.3% older population with higher education and 88.9% older persons with higher income use technology more frequently.

Lifestyle being an important factor has a strong positive as well as negative influence on the health status of the city's elderly population. In fact, the effect of lifestyle varies from one group of people to another in terms of their level of satisfaction with their health. The elderly who follow a healthy and active lifestyle have better health status. Therefore, the elderly are suggested to follow a healthy lifestyle from the perspective of improving and maintaining their health.

An assessment of the degree of satisfaction of the elderly with their personal health status shows that the majority of the elderly have a moderate level of satisfaction with their health status.

Health is an important component of active ageing and the Romanian retired people are interested in health, social happiness (family life), and social involvement (active in their family, less than other communities). In this context, public policies and programmes are needed for the training and integration of the elderly population into social life in order to prevent harm to the health of the elderly and to educate their priorities.

Besides gender, socio-economic variables such as education and income have an influence on lifestyle and elderly needs. Health conditions are in strong connection with family activity, dependency rate, social life, and daily routine.

To achieve long-term healthy and productive life, there is need for investment in health and education-related sectors for early prevention and detection of diseases and to increase the lifestyle-related education and awareness of older persons. There is also need to revise its pharmaceutical policies regarding the provision of medical care to the elderly. Promoting a healthy lifestyle among the elderly population needs the provision of making the preventive measures and services available at the primary health centres in the community. This will generate a positive result towards a more active, healthy, and socially integrated life of the elderly population in Bucharest.

References

Aisa, R., Larramona, G. & Pueyo, F. 2015. Active aging, preventive health and dependency: Heterogeneous workers, differential behavior. *Journal of Economic Behavior & Organization, 117*, 1-9.

Al-Dayyatal., H. (2018). Non-alcoholic fatty liver disease and associated dietary and lifestyle risk factors. *Diabetology and Metabolic Syndrome, 12*, 569–575.

Bartram, L., & Roe, B. (2005). Dependency ratios: useful policy-making tools? *Geriatrics & Gerontology International, 5*, 224–228.

Bloom, D., Canning, D., Mansfield, R., & Moore, M. (2007). Demographic change, social security systems and savings. *Journal of Monetary Economy, 54*, 92–114.

Byrne, H. et al. (2018). Self-directed exercise programmes in sedentary middle-aged individuals in good overall health; a systematic review. *Preventive Medicine Journal, 114*, 156–163.

Carrieri, V., & Wuebker, A. 2013. Assessing inequalities in preventive care use in Europe. *Health Policy, 113*, 247-257.

Coile, C., & Levine, P. (2007). Labor market shocks and retirement: Do government programs matter? *Journal of Public Economy, 91*, 1902–1919.

Courtemanche, C., & Zapata, D. (2014). Does universal coverage improve health? The Massachusetts experience. *Journal of Policy Analysis and Management, 33*, 36–69.

Dias, G. & Couceiro, M. 2017. Active ageing and physical activity- guidelines, functional exercises and recommendations. In *Springer briefs in well-being and quality of life research, Switzerland.* s.l.:s.n.

Dommaraju, P. (2016). Perspectives on old age in India. *Contemporary Demographic Transformations in China* (pp. 293–308). Springer.

Dumitrache, L. 2004. *Starea de sănătate a populației României. O abordare geografică.* Univers Enciclopedic.

European Commission & United Nations Economic Commission. (2015). *Report of the fifth meeting of the Expert Group on the Active Ageing Index,* s.l.: s.n.

Eurostat. (2021a). *Population by age.* http://appsso.eurostat.ec.europa.eu/nui/show.do?dataset=tps00010&lang=en.

Eurostat. (2021b). *Healthy life years.* https://ec.europa.eu/eurostat/statistics-explained/index.php?title=Healthy_life_years_statistics.

Eurostat. (2021c). *Healthy life years at age 65.* https://ec.europa.eu/eurostat/statisticsexplained/index.php?title=Healthy_life_years_statistics#Healthy_life_years_at_age_65.

Eurostat. (2021d). *At-risk-of poverty rate by detailed age group—EU-SILC survey.* http://appsso.eurostat.ec.europa.eu/nui/show.do?dataset=tessi120&lang=en.

Fratiglionial., L. (2004). An active and socially integrated lifestyle in late life might protect against dementia. *The Lancet Neurology, 3*, 343–353.
Howitt, D., & Cramer, D. (2010). *Introduction in SPSS for psychology.* 2 ed. s.l.:Polirom.
Minett, C. (2012). Prevention better than cure for optimized active ageing. *Quality in Ageing and Older Adults, 4*(2), 291–300.
Mullan, P. (2000). *The imaginary time bomb. Why an ageing population is not a social problem.* IB Tauris.
National Institute of Statistics. (2021). *Data base.* http://statistici.insse.ro:8077/tempo-online/.
Ross, J., Bernheim, S., Bradley, E., & Teng, H. (2007). Use of preventive care by the working poor in the United States. *Preventive Medicine Journal, 44*, 254–259.
Serra e Silva, P., Dias, G., Branquinho, M., & Mendes, R. (2015). Envelhecimento activo. In: *Envelhecimento activo e actividade fisica* (pp. 1–21). s.n, Coimbra.
Sharp, E., & Gatz, M. (2011). The relationship between education and dementia an updated systematic review. *Alzheimer Disease and Associated Disorders Journal, 25*, 289–304.
Staubli, S., & Zweimüller, J. (2013). Does raising the early retirement age increase employment of older workers? *Journal of Public Economy, 108*, 17–32.
Shu, T. C. (2010). *Health promotion in the elderly—an Asian perspective.* Gastein, Paralel forum on healthy ageing at European Health Forum.
Talos, A. (2017). *Stilul de viață și impactul acestuia asupra stării de sănătate a populației. Studiu de caz: județul Ialomița.* University of Bucharest.
UN. (2015). *United Nations. World population prospects: The 2015 revision population database.* s.l.: s.n.
Walker, A. (2010). The emergence and application of active ageing in Europe. *Soziale Lebenslauf-politik* (pp. 585–601). VS Verlag für Sozialwissenschaften. Wiesbaden.
Werner, C. (2011). *The older population: 2010.* U.S. Census Bureau. http://www.census.gov/prod/cen2010/briefs/c2010br-09.pdf.
WHO. (2002). *World Health Organisation. Active ageing: a policy framework.* s.l.: s.n.
WHO. (2013). *Health 2020:A european policy framework and strategy for the 21st century.* s.l.: s.n.

Chapter 22
Evaluating Displaced People's Life Satisfaction in Urban Resettlement Community in Nanjing City, China

Zhu Qian

Abstract China's unprecedented urbanization has led to prevalent rural collective-owned land conversion to state-owned land to meet the demands in urban expansion and economic growth. One of the critical socio-economic consequences is the resettlement and life transformation of affected rural villagers after land conversion. The overall well-being of resettled people has societal and economic implications in Chinese cities. This paper assesses displaced people's life satisfaction in government-designated urban resettlement communities after land acquisition to advance our understanding of the levels, characteristics, and determinants of landless people's well-being after their resettlement due to land acquisition. From the theoretical perspective of subjective well-being, the study explores the relations between the overall life satisfaction of displaced people and their subjective satisfaction with three dimensions in their post-resettlement life—(1) material living conditions, (2) social security and employment support, and (3) attitudinal perceptions, social relations, and social participation. Survey questionnaire and statistical analysis were employed in this research and the valid survey results of 98 resettled households from two urbanized towns of Nanjing, Jiangsu Province were collected for descriptive statistical and logistic regression analyses. The study shows that while compensation and resettlement policy reforms have improved resettled people's material living conditions, they still struggle with urban life transformation and adaptation from the social and attitudinal perspectives. An integrated resettlement approach is proposed to facilitate better accessibility to social security programmes and non-agricultural employment opportunities, and to address issues in identity adaptation, lifestyle transformation, and social activity participation.

Keywords Life satisfaction · Resettlement · Land expropriation · Displaced people · Living conditions

---

Z. Qian (✉)
School of Planning, University of Waterloo, Waterloo, ON N2L 3G1, Canada
e-mail: z3qian@uwaterloo.ca

B. R. K. Sinha (ed.), *Urban Dynamics, Environment and Health*,
https://doi.org/10.1007/978-981-99-5744-6_22

22.1 Introduction

Rapid industrialization and massive urbanization in China's economic reform era have created enormous demand for urban construction land. The pace of urbanization has accelerated since the 1990s, leading to rural population decrease and loss of agricultural land to urban development in rural land expropriation. An increasingly large number of rural villagers has been resettled to urbanized communities and lost all or part of their rural land during the process. Resettled people then have to face numerous challenges in their new urban life and often find themselves in a disadvantaged position in urban society. As passive participants in the process, resettled people typically have no alternative but to obey decisions from local authorities. Consequently, some resettled people may develop a resistant attitude towards resettlement policies and processes (Tang et al., 2016). Meanwhile, urban residents' perception of resettled people and the sense among villagers of being inferior to urban citizens formed by the long existing urban-rural dualism might reinforce their limitations in competing in the labour market and discourage their active participation in urban social activities. Once deprived of their rural property rights and land collective ownership, resettled people encounter financial challenges in their daily life as well as face marginalization in the labour market and urban social security programmes such as pension plans, healthcare, low-income subsidies, and unemployment insurance (Liang & Zhu, 2015; Ma & Hu, 2014). All these influence resettled people's attitudes towards compensation and resettlement policies and their overall life satisfaction with urban life. Over the past decade, multiple compensation and resettlement approaches have been devised to address these issues but with mixed consequences.

Existing empirical studies on resettled people's well-being have not sufficiently addressed whether resettled people consider policy reforms effective and how they evaluate their life quality and well-being after land expropriation and resettlement. Research on resettled people's life satisfaction in urban society from their own opinions will contribute to local government's understanding and planning of resettlement programmes in their policy revisions. It will also provide an opinion platform to resettled people and facilitate a bottom-up mechanism in assessing resettlement and compensation systems to give more considerations to the rights and interests that resettled people care about. It is imperative to investigate how resettled people's life has been impacted by resettlement and compensation programmes and how (un)satisfied these former villagers feel about the resettlement and compensation institution in bettering their life experience in urban setting.

China's compensation principles are focused mainly on direct economic loss, leaving insufficient considerations to indirect losses associated with rural land conversion to urban uses such as agricultural and environmental amenity changes, new urban life assistance, tradition and culture transformation, and lifestyle changes (Ma & Hu, 2014). Yet, resettled people's life quality after land expropriation can experience tremendous changes resulting from material living conditions, social security coverage, re-employment, social network reconstruction, and urban lifestyle adaptation. Studies have examined various aspects of resettled people's satisfaction with their life quality (Chen et al., 2013; Duan & Chen, 2009; Hu et al., 2014; Luo & Fan, 2008), but few have employed the perspective of subjective well-being and incorporated multiple dimensions to assess resettled people's satisfaction level with new urban life from their own opinions. The subjective well-being discourse emphasizes on individual's own evaluation that can be cognitive, experiential, and attitudinal (Diener, 2000). This research examines the practicability and sustainability of resettlement and compensation programmes by investigating resettled people's self-assessment of life quality and well-being in their resettlement communities designated by local government. Life quality satisfaction assessment serves to provide more empirical evidence of how resettled people evaluate their life quality, and clarify what life aspects they are satisfied or unsatisfied with, and understand what they really expect and highly value in current resettlement and compensation institutions.

22.2 Conceptual Discourse on Subjective Well-Being and Resettlement

Subjective well-being is an approach to defining how good or happy people report their quality of life, and is often seen as "the science of happiness" (Diener, 2000). Hu et al. (2014) point out that subjective well-being can serve as an empirically adequate and valid evaluation for individually experienced welfare and happiness. The theory of subjective well-being closely relates to and emphasizes people's own evaluation that is cognitive, experiential, emotional, and attitudinal (Diener, 2000). Inglehart (1990) believes that people's subjective well-being would become increasingly valuable after their basic material needs are fulfilled.

Measurement of subjective well-being typically contains multiple components, and overall life satisfaction and satisfaction with important aspects of life are two major indicators (Diener, 1994, 2000; Diener et al., 2003). Self-reported satisfaction is a popular tool of analysis, and with its strong psychometric properties, it has adequate reliability, factor invariance, and sensitivity to subjective well-being (Diener, 1994; Pavot & Diener, 1993). Economic, social, and cultural factors can influence subjective well-being in many ways. For instance, cultural elements affect people's perceptions of well-being from the psychologistic perspective (Diener, 2000). Economic situations can determine long-term satisfaction levels, while social and cultural factors explain floating variances in subjective well-being (Diener et al., 2003).

Arguments against the inclusion of subjective indicators into policy decision-making systems have existed for long. Their reasons are that the subjective assessment is irrational and only specific to individual judgements, and that the assessment criteria are not transparent or precise (Böhnke, 2005). Diener (2000) notes that the measurement of subjective well-being can possibly be undermined by prejudices, biases, temporary moods, and other situational factors. Veenhoven (2002) points out that policy instruments should not take account of situation-based preferential opinions, which simply reflect media-driving values, current moods, or personal sensitivities. Nevertheless, Böhnke (2005) advocates the informative and analytical value of the subjective well-being theory, which brings about indicators of life quality open to individuals' genuine satisfaction level about their living conditions, preferences, anticipations, and adaptations. For policy makers, an overview of what is likely to improve life quality and what determines the self-reported satisfaction level of a certain group of population can be provided based on the assessment of subjective well-being. From the perspective of social stability and cohesion, opinions and perceptions about policy implementation and the subjective well-being of vulnerable groups help identify conflicts and prevent potential harms stemming from such conflicts. Life satisfaction measurement can provide policy-implicated evidence such as how people evaluate their life quality, what life aspects they are satisfied and dissatisfied with, what they really expect from policy mechanisms, and what they highly value for their life well-being (Böhnke, 2005).

Subjective well-being is influenced by economic elements such as employment rate and social capital accumulation (Böhnke, 2005). Subjective well-being contains interdependent preferences of individuals which depend on their relative situations in the society. In income distribution, one's subjective well-being considers income as a comparable indicator of one's income relative to others', median income, and income gaps in the society (Bellani & D'Ambrosio, 2011). Hence, economic conditions impose a great impact on people's satisfaction with life quality (Li & Raine, 2014). Many approaches to enhancing life satisfaction level are far beyond any individual capability, and largely dependent on policy interventions, such as minimum income standards and social security programmes.

Subjective well-being theory is central to this life quality research. Life quality refers to the overall well-being of individuals and is typically defined by multiple value-based indicators such as equal distribution of resources, living standards, access to employment opportunities, and social security systems (Bellani & D'Ambrosio, 2011; Böhnke, 2005). It entails material conditions, social relationships and participation, subjective satisfaction perception, and psychological experience (Böhnke, 2005; Diener, 2000). Allardt (1993) summarizes life quality in a three-pillar concept of "having, loving and being" in which "having" refers to material living conditions, "loving" refers to social relationship networks, and "being" refers to a sense of belonging and recognition as well as integration into the new society. Besides material conditions, subjective well-being also considers social interactions and psychological factors such as attitudes and perceptions.

Subjective well-being theory concerns about individual's self-reported satisfaction with multiple dimensions of life quality (Böhnke, 2005; Diener et al., 2003). From the subjective well-being perspective, life satisfaction can be assessed in three dimensions: (1) material living conditions, (2) social security and employment support, and (3) attitudinal perceptions, social relationships, and social participation. Each of the three dimensions may exert influence on other dimension(s).

Fig. 22.1 Three dimensions of life satisfaction and influencing factors of subjective well-being

Factors in material living conditions include living environment, income, expenditure, consumption capability, and disposal of monetary compensation from land expropriation; factors in social security and employment support include social security types, coverages, professional skill training, employment information availability, and accessibility; factors in social relationships, participation, and attitudinal perceptions include participation in community activities, knowledge of compensation policies obtained by social interactions, attitudes towards urban life, and use of social networks (Fig. 22.1).

Subjective well-being has been strongly influenced by material living conditions (Böhnke, 2005). Material living conditions relate to financial burden, consumption capability, and living environment (Bellani & D'Ambrosio, 2011). Substantial relationships exist between a variety of influencing factors of material living conditions and people's self-reported satisfaction (Bayram et al., 2012; Böhnke, 2005; Liu et al., 2008; Shui et al., 2014). Typically, those with limited access to the labour market is more vulnerable to unemployment risks and insufficient social security system (Silver, 1994). Long-term unemployment means not only depletion of income and material sources, but also loss of social status, self-identity, and social network (Böhnke, 2005). For social relationships and attitudinal perceptions, research has found that active participation in social interactions, either formal or informal, is positively related to life well-being and happiness (Helliwell, 2001; Putnam, 2001). Displaced villagers usually are faced with challenges in self-identity recognition and adaptation to urban life (Zhang & Tong, 2006), because of factors such as institutional inequality and attitudinal perception of rural being inferior than urban (Wu & Qin, 2008). Some displaced people still consider themselves as traditional rural villagers in psychological, cultural, and social aspects, despite of their urban household registration status (Liang & Zhu, 2015).

22.3 Research Methodology

Since the 2000s, Nanjing has witnessed two major resettlement and compensation policy reforms, aiming to improve displaced villagers' life quality and experience in urban society. The 2004 policy reform tremendously increased compensation rates and diversified approaches, in order to provide resettled people with better living standards (Zhou et al., 2014). For instance, monthly monetary compensation replaced the lump-sum compensation payment. About 70% of the compensation and resettlement fund went directly to displaced people's personal accounts instead of their rural collective's. However, the 2004 policy reform did not grant resettled villagers the same social security package that was enjoyed by urban residents. The 2011 policy reform introduced a social security fund and a wide range of social insurance for resettled people's long-term social welfare. It also ushered in a comprehensive compensation mechanism that is based on a zoning compensation system, an advanced social security arrangement (such as endowment insurance, healthcare, and a unemployment insurance), and a resettlement fund. The compensation standard takes location and differential land rent into consideration, in which rural land is classified into three acquisition zones with differentiated acquisition and compensation prices. Compensation rates were also raised from the 2004 standards.

The case study was conducted in two urban fringe sites in Nanjing—Tiexinqiao and Longtan, which are located respectively in south and north to the city core (Fig. 22.2). Both sites accommodate resettled people in government-designated urban resettlement communities. Tiexinqiao was built in the early 2000s and Longtan in the mid to late 2000s. Tiexinqiao is closer to the city center than Longtan. Structured survey was conducted during March and May 2017. Among 98 valid questionnaire returns, 55 were from Tiexinqiao (Chunjiang New Community) and 43 were from Longtan (Longtan Community). Sampling participants were randomly selected and survey unit was resettled household. The survey questionnaire includes five categories: economic and living environment indicators representing the dimension of material living conditions; social security coverage and employment situation indicators representing the dimension of social security and employment support; and indicators representing the dimension of social relationships, participation, and attitudinal perceptions. A Likert Scale was used ranging from 1 for very unsatisfied, 2 for partly unsatisfied, 3 for neutral, 4 for partly satisfied, to 5 for very satisfied; or 1 for strongly agree, 2 for partly agree, 3 for neutral, 4 for partly disagree, 5 for strongly disagree. This study used an ordinal logistic model to assess life satisfaction, expressed in two equations:

$$Y = \log P/(1 - P) = \alpha + \beta_1 X_1 + \beta_2 X_2 + \ldots + \beta_n X_n \qquad (22.1)$$

$$f(Y_1 = 1) = \log(\theta_1) = \alpha_1 - \beta_1 X_1,\ where\,\theta_1 = prob(score \leq 1)/(1 - prob(score \leq 1)) \qquad (22.2)$$

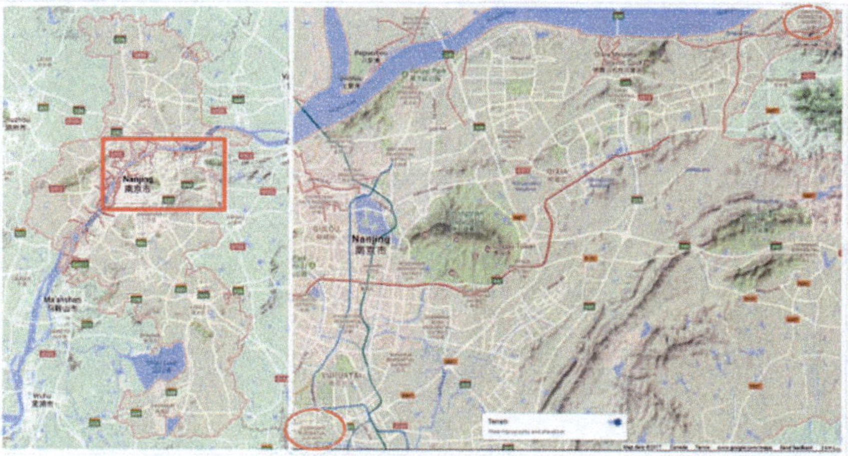

Fig. 22.2 Nanjing and the two urban fringe sites—Tiexinqiao (in southwest) and Longtan (in northeast)

P represents the probability of resettled people's life satisfaction scale, qualitative variable, or a variable with a binary feature after linearization. Resettled household's life satisfaction is classified into five levels from very unsatisfied, partly unsatisfied, neutral, partly satisfied to very satisfied. Given P as the probability that a resettled household is satisfied and the probability of unsatisfied is "1-P", $0 < P < 1$. Let α be the constant value, χ_n are independent variables that denote various factors influencing resettled people's satisfaction level, and β_n is the partial regression coefficient in the logistic regression (Table 22.1).

22.4 Research Findings

22.4.1 Descriptive Analysis

Among the 98 resettled residents, 60 were male and 38 were female, with a mean age of 51 years old. 84.5% of these resettled residents already converted their rural household registration status to urban, sponsored by the local government. The status made them eligible for urban social security welfare. More than half of these surveyed residents were resettled before 2004, and only about 10.7% were relocated after the 2011 policy reform, leaving the rest displaced in the years between 2004 and 2011. Resettled residents with elementary school education or illiterate had the largest share (46.6%), and those with college/university education or above only had a mere share of 4.8%. Before land expropriation, 73.8% of resettled residents worked in agricultural sector, and 9.7% were in local private industries, 6.8% seasonal migrant workers,

Table 22.1 Define latent and explanatory variables with 24 variables

Variable	Define variable	Assigned value of variable	Mean
Y1	Satisfaction level with overall life quality	1 = very unsatisfied, 2 = partly unsatisfied, 3 = neutral, 4 = partly satisfied, 5 = very satisfied	2.52
Y2	Satisfaction level with compensation standard and approach	1 = very unsatisfied, 2 = partly unsatisfied, 3 = neutral, 4 = partly satisfied, 5 = very satisfied	2.38
Y3	Satisfaction level with material living conditions	1 = very unsatisfied, 2 = partly unsatisfied, 3 = neutral, 4 = partly satisfied, 5 = very satisfied	2.33
Y4	Satisfaction level with social security	1 = very unsatisfied, 2 = partly unsatisfied, 3 = neutral, 4 = partly satisfied, 5 = very satisfied	2.43
Y5	Satisfaction level with employment support	1 = very unsatisfied, 2 = partly unsatisfied, 3 = neutral, 4 = partly satisfied, 5 = very satisfied	2.00
X1	Age	1 = 18–30, 2 = 31–45, 3 = 46–60, 4 = 61–70, 5 = > 70	3.33
X2	Education level	1 = elementary school and below, 2 = middle school, 3 = equal to high school, 4 = college and above	1.81
X3	Work status before and after acquisition	1 = agricultural sector, 2 = private industry, 3 = migrant worker, 4 = self-employed, 5 = government/government-owned enterprise, 6 = student, 7 = unemployed, 8 = retirement	before: 1.64 after: 5.60
X4	Length (year) of resettlement	before 2004 = 1, 2004–2011 = 2, after 2011 = 3	1.67
X5	Change in monthly income	1 = significantly decrease, 2 = decrease, 3 = similar, 4 = increase, 5 = significantly increase	3.63
X6	Change in monthly expense	1 = significantly decrease, 2 = decrease, 3 = similar, 4 = increase, 5 = significantly increase	4.41
X7a	Overall living environment	1 = worse than before, 2 = generally similar, 3 = better than before, 0 = not sure	1.99
X7b	Housing area	1 = worse than before, 2 = generally similar, 3 = better than before, 0 = not sure	1.52
X7c	Housing quality	1 = worse than before, 2 = generally similar, 3 = better than before, 0 = not sure	1.86
X7d	Utility facilities	1 = worse than before, 2 = generally similar, 3 = better than before, 0 = not sure	1.99
X7e	Transportation services	1 = worse than before, 2 = generally similar, 3 = better than before, 0 = not sure	2.48
X7f	Community safeguard	1 = worse than before, 2 = generally similar, 3 = better than before, 0 = not sure	2.18
X7g	Telecommunication services	1 = worse than before, 2 = generally similar, 3 = better than before, 0 = not sure	2.33

(continued)

Table 22.1 (continued)

Variable	Define variable	Assigned value of variable	Mean
X7h	Convenience and richness of daily activities	1 = worse than before, 2 = generally similar, 3 = better than before, 0 = not sure	2.47
X8	Satisfaction level with endowment social insurance	1 = very unsatisfied, 2 = partly unsatisfied, 3 = neutral, 4 = partly satisfied, 5 = very satisfied	2.24
X9	Satisfaction level with urban healthcare insurance	1 = very unsatisfied, 2 = partly unsatisfied, 3 = neutral, 4 = partly satisfied, 5 = very satisfied	2.26
X10	Satisfaction level with new rural cooperative healthcare insurance	1 = very unsatisfied, 2 = partly unsatisfied, 3 = neutral, 4 = partly satisfied, 5 = very satisfied	2.97
X11	Satisfaction level with unemployment insurance	1 = very unsatisfied, 2 = partly unsatisfied, 3 = neutral, 4 = partly satisfied, 5 = very satisfied	1.00
X12	Whether vocational training was offered or funded by town government or rural collective	1 = Yes, 0 = No	0.15
X13	Whether it was helpful or necessary	1 = Yes, 0 = No	0.19
X14	Do you have organizational access to job market information	1 = Yes, 0 = No	0.18
X15	Do you receive government assistance for employment	1 = Yes, 0 = No	0.42
X16	Feel hard to adapt to urban lifestyle and environment	1 = strongly agree, 2 = partly agree, 3 = neutral, 4 = partly disagree, 5 = strongly disagree	2.38
X17	Prefer rural housing and living environment	1 = strongly agree, 2 = partly agree, 3 = neutral, 4 = partly disagree, 5 = strongly disagree	2.02
X18	Prefer farming lifestyle and feel psychologically close to land	1 = strongly agree, 2 = partly agree, 3 = neutral, 4 = partly disagree, 5 = strongly disagree	2.35
X19	Expect better living quality and security with rural land	1 = strongly agree, 2 = partly agree, 3 = neutral, 4 = partly disagree, 5 = strongly disagree	1.54
X20	Prefer to keep rural registration status and land rather than receive any compensation	1 = strongly agree, 2 = partly agree, 3 = neutral, 4 = partly disagree, 5 = strongly disagree	2.48

(continued)

Table 22.1 (continued)

Variable	Define variable	Assigned value of variable	Mean
X21	Reduce frequency to contact with old friends and relatives	1 = strongly agree, 2 = partly agree, 3 = neutral, 4 = partly disagree, 5 = strongly disagree	2.38
X22	Add frequency to meet with new friends and neighbours	1 = strongly disagree, 2 = partly disagree, 3 = neutral, 4 = partly agree, 5 = strongly agree	3.17
X23	Participation in community activities and development	1 = strongly disagree, 2 = partly disagree, 3 = neutral, 4 = partly agree, 5 = strongly agree	2.58
X24	Awareness of public notice about relevant compensation policies	1 = barely concern, 2 = know a little bit, 3 = somewhat familiar, 4 = fully knowledgeable, 0 = no policy notice	1.68

7.8% self-employed, and only 1.9% in government or town and village enterprises (TVEs). After resettlement, 43.6% of resettled residents chose to retire (which was well reflected by the median age of 51 in the survey), 22.8% were in local private industries, and 10.9% were small business owners. Resettlement changed villagers' employment and many were pushed to work in non-agricultural sectors or become unemployed. The employment changes were closely related to educational attainment—none of those with post-secondary or higher education attainment, typically young- and middle-aged, engaged in agricultural work prior to land expropriation.

After land expropriation, more than 60.0% of resettled residents were unsatisfied with their material living conditions, and more than 70.0% of resettled residents expressed strong dissatisfaction with their current employment situations, while only 11.4% held a positive attitude towards life quality. On one hand, about 90.4% of resettled residents who felt dissatisfied with their overall life quality also expressed dissatisfaction about their material living conditions. About 88.5% of resettled residents who felt dissatisfied with their overall life quality also expressed dissatisfaction about social security programmes, and 73.1% of resettled residents who felt dissatisfied with their overall life quality also expressed dissatisfaction about employment assistance from the local government. On the other hand, among those who were satisfied with their overall life quality, 63.9% showed a positive attitude towards material living conditions, and 66.7% were satisfied with social security programmes, but only 11.1% appeared to feel satisfied with employment assistance from the local government.

Among various compensation approaches, monetary compensation was the dominant one in all resettlement cases. About 41.6% of resettled residents also received in-kind apartment unit compensation, and 31.7% of resettled residents received social insurance such as endowment fund, healthcare, unemployment insurance, and/or low-income subsidies. Only 2.0% of resettled residents agreed that they received employment arrangement or vocational training as part of their compensation package.

Resettled villagers shared common choices in spending their monetary compensation fund—45.5% of them used most of the compensation fund to purchase discounted apartment units exclusively available for them, and 30.7% of them used part of the compensation fund for furniture and interior decoration. About 24.8% of resettled residents chose to use part of the compensation fund to purchase social security package. Only 11.9% of resettled residents were able to reserve some monetary compensation for financial investment. While about 55.6% of resettled residents reported that their monthly income increased after land expropriation, all surveyed residents complained that their monthly expenditure grew significantly after land expropriation.

About 40.8% of resettled residents indicated that they were satisfied with living environment, most of whom lived in Longtan, while about 37.8% of resettled residents expressed very low satisfaction with living environment, most of whom lived in Tiexinqiao. In-kind apartment unit was part of the compensation package, the average area of in-kind apartment unit floor area was 103 m^2 before 2004, 114 m^2 between 2004 and 2011, and 246 m^2 after 2011. About 74.0% of resettled residents were allocated only one apartment unit, almost all of them received the compensation before 2011. Those households that received more than three apartment units were typically resettled after 2011. A few villagers who were resettled after 2011 in Longtan admitted that they received several apartment units as part of their compensation packages. While the total floor areas of these multiple units were large, each unit's floor area was insufficient to accommodate a traditionally large household of five to six family members. Therefore, many resettled households preferred to have large-sized apartment units.

Resettled villagers demonstrated improved satisfaction with social insurance policy. About 70.0% of the residents who were resettled before 2004 expressed strong dissatisfaction with their social security arrangement. About 66.7% of the residents who were relocated after 2004 felt partly satisfied or very satisfied with their social security arrangement. Among those who were displaced after 2011, almost 80.0% of them were at least partly satisfied with their social security package. About 80.6% of resettled residents reported that they were covered by endowment social insurance programmes, and 87.8% of resettled residents claimed that they were covered by urban healthcare. However, only 23.5% of these surveyed were covered by unemployment insurance. The survey demonstrates the lack of employment assistance programmes in both communities. Only 14.6% of resettled residents received vocational training after land expropriation, and 11.7% of resettled residents agreed that the vocational training was supported by local township administration. Many resettled residents opted out training workshops for various reasons such as age, time and financial restrictions, course contents, and lack of practical objectives of training programmes. A few residents found the vocational training that they received from the local government was not helpful in finding a position in the job market. Instead, those who chose to attend technical skill-related training programmes by themselves found them helpful. Social capital and social media played important roles in resettled villagers' job seeking. About 30.6% of surveyed residents admitted that they relied on social networks of relatives and friends in finding jobs and social media was also widely used in finding jobs (39.8%). The conventional way of resorting to employment intermediary agencies became less popular (16.3%). Close to 60.0% of resettled residents groused about receiving little assistance from government in their search for employment.

Social participation was reflected by former villagers' intention to involve in community activities and their engagement in community development. Resettled residents' awareness of compensation and resettlement policies was also an indication of their social participation. Up to 55.4% of resettled residents showed little interest in community affairs and activities and 55.1% of residents had very limited knowledge of compensation- and resettlement-related policies. When asked about their interactions with others, 60.0% of resettled residents admitted that they reduced contacts with rural relatives, neighbours, and friends after resettlement. Meanwhile, 52.0% of resettled residents became acquainted with new neighbours and friends after resettlement. A large share of resettled villagers showed a strong inclination to rural living environment and lifestyle—about 60.0% of residents claimed that they would rather give up their new urban household registration status and compensation package in exchange for their (lost) rural land.

22.4.2 *Ordinal Regression Analysis*

Cronbach's Alpha reliability analysis was performed to test independent variables in order to examine whether the multiple questions measure a same dependent variable. The Cronbach's Alpha of each dimension is summarized in Table 22.2. The negative Alpha value for personal socio-economic attributes was calculated from a negative average covariance among the variables, which was against the assumption of the reliability model. This was because of a non-consistent scale of measurement among the variables (X1–X4). It indicates that these variables did not have good fitness when being incorporated into a regression model. However, it does not mean this dimension had little importance to life satisfaction assessment. All values in Table 22.4 indicate a high level of internal consistency. But the factors of personal socio-economic attributes and employment support appeared to be not correlated or consistent. Therefore, these variables may not have good fitness in measuring resettled people's satisfaction in a continuous regression model.

Table 22.3 shows that the overall life quality was positively correlated with the material living condition and social security dimensions, with a significance value lower than 0.01. The overall life quality seems less relevant to the employment dimension and the significance value did not support a convincible statistical relationship between the two. The dimension of material living conditions was closely correlated to the social security dimension and the employment dimension. In addition, there

Table 22.2 Alpha reliability analysis of influencing factor clusters

Influencing factors	Cronbach's Alpha	N of variables
Personal socio-economic attributes	Negative	X1–X4
Material living conditions	0.681	X5–X7
Living environment	0.747	X7a–X7h
Social security	0.739	X8–X11
Employment support	0.392	X12–X15
Perceptive and social dimension	0.828	X16–X24

was significant relationship between the social security and the employment dimensions. All discussed positive relationships were accepted with a high significance level (p-value < 0.05).

Then, the overall life satisfaction (Y1) and its sub-dimensions (Y3–Y5) was correlated against measures of satisfaction with assorted compensation policies (Y2a–Y2c). The positive, relatively large parameter and high significance level in Table 22.4 demonstrate that displaced people's satisfaction level with the specific compensation policies was dominantly related to the level of their overall life satisfaction. Moreover, it is observed that the compensation package received by resettled residents had a significant influence on their satisfaction level with material

Table 22.3 Correlations between life quality satisfaction and levels of satisfaction in sub-dimensions

Correlations

		Y1: Life quality	Y3: Material conditions	Y4: Social security	Y5: Employment
Y1: Life quality	Pearson correlation	1	0.720**	0.663**	0.193
	Sig. (2-tailed)		0.000	0.000	0.058
Y3: Material conditions	Pearson correlation		1	0.684**	0.319**
	Sig. (2-tailed)			0.000	0.001
Y4: Social security	Pearson correlation			1	0.347**
	Sig. (2-tailed)				0.000
Y5: Employment	Pearson correlation	0.193	0.319**	0.347**	1
	Sig. (2-tailed)	0.058	0.001	0.000	

**Correlation is significant at the 0.01 level (2-tailed)

Table 22.4 Correlations between life satisfaction dimensions and compensation policies

Correlations

		Y2a: Resettlement	Y2b: Compensation fees	Y2c: Insurance
Y1: Life quality	Pearson correlation	0.775**	0.747**	0.701**
	Sig. (2-tailed)	0.000	0.000	0.000
Y3: Material conditions	Pearson correlation	0.638**	0.857**	0.625**
	Sig. (2-tailed)	0.000	0.000	0.000
Y4: Social security	Pearson correlation	0.601**	0.632**	0.478**
	Sig. (2-tailed)	0.000	0.000	0.000
Y5: Employment	Pearson correlation	0.072	0.244*	0.098
	Sig. (2-tailed)	0.481	0.017	0.353

**Correlation is significant at the 0.01 level (2-tailed)
*Correlation is significant at the 0.05 level (2-tailed)

living conditions, especially in relation with the standards of compensation fees. Additionally, the aspect of social security was positively correlated with a series of compensation policies, influenced by the social insurance policy and relevant to the monetary and resettlement compensation. However, the aspect of employment was not necessarily taken into consideration of the current compensation policies, despite a significant yet relatively weak statistical relationship with the standard of compensation fees.

The analysis using the ordinal logit model for the overall life quality shows that the satisfaction with material living conditions and social security had a positive influence on the overall life satisfaction. The dimension of material living conditions was a stronger contribution to former villagers' life quality satisfaction than the social security dimension. Ordinal logit model regression can also be conducted for material living conditions, social security, and employment support, as well as social relationships, participation, and attitudinal perceptions to reveal more detailed relationships. For instance, resettled villagers' frequency in interacting with their rural relatives, friends, and neighbours, and their preference for rural lifestyle and traditional attachment to rural land indicated that the more they clung to their rural lifestyle, the lower their overall life quality satisfaction was.

22.5 Conclusions

This study reveals that despite several national and regional reforms of the compensation system, the current directives of compensation standards and approaches are still restricted to the losses of economic well-being associated with land productive functions, while non-economic well-being associated with land non-productive functions fails to be reflected in the compensation mechanism. In fact, even the economic well-being is insufficiently quantified, only via policy prices instead of true market value.

Among the variables with constant significant relationships, the positive influencing factors on displaced people's life satisfaction include: year of compensation and resettlement (i.e. policy reform periods), change of income, overall living environment, housing floor area, satisfaction level with urban healthcare insurance, adaptation capability, expectation of life security with land, frequency of contacts with rural networks, and cognition level of compensation policies. The negative factors influencing their life satisfaction include: satisfaction level with pension insurance, preference to rural lifestyle and land, and attitude towards compensation. Overall, resettled people are likely to face potential marginalization and exclusion in welfare programmes and employment status in the urban society. In Nanjing, resettled people after the 2011 reform are more satisfied with their urban life, especially in terms of monetary compensation standards and social security policies.

The study shows that resettled people's higher adaptation capability in living in urban environment and less preference for rural lifestyle and rural amenity can facilitate their satisfaction with various aspects in urban living. Several policy implications can be proposed. First, reformed compensation and resettlement should give consideration to both economic well-being and non-economic aspects to reflect the multi-function of rural land that is lost to urbanization. Second, compensation and resettlement approaches can include more alternatives for employment support, extended endowment security insurance, and healthcare insurance. Unemployment insurance coverage needs to be expanded so that more resettled villagers can benefit from the programmes. Third, local government, in collaboration with non-government interests, needs to pay more attention to resettlement community management to address concerns in housing conditions, public facilities, community safety, and community activities to create a better living experience for resettled people.

Acknowledgements The early version of this article was published as a research paper titled "Evaluating Villagers' Life Satisfaction in Resettlement Community: A Pilot Study of Suburban Nanjing, China", pp. 1-8, vol. 29, 2020 in *Journal of Research in Architecture and Planning*. I appreciate the journal publisher's permission to include the materials. Revisions have been made and new contents added since then.

References

Allardt, E. (1993). Having, loving, being: An alternative to the Swedish model of welfare research. *The Quality of Life, 8*, 88–95.

Bayram, N., Aytac, S., Aytac, M., Sam, N., & Bilgel, N. (2012). Poverty, social exclusion, and life satisfaction: A study from Turkey. *Journal of Poverty, 16*(4), 375–391.

Bellani, L., & D'Ambrosio, C. (2011). Deprivation, social exclusion and subjective well-being. *Social Indicators Research, 104*(1), 67–86.

Böhnke, P. (2005). *First European quality of life survey: life satisfaction, happiness and sense of belonging*. Office for Official Publications in the European Communities.

Chen, D., Tang, K. K., Zhao, L., & Zhang, Y. (2013). Will China's cooperative medical system fail again? Insight from farmer satisfaction survey. *Health Promotion International, 30*(2), 251–261.

Diener, E. (1994). Assessing subjective well-being: Progress and opportunities. *Social Indicators Research, 31*(2), 103–157.

Diener, E. (2000). Subjective well-being: The science of happiness and a proposal for a national index. *American Psychologist, 55*(1), 34.

Diener, E., Oishi, S., & Lucas, R. E. (2003). Personality, culture, and subjective well-being: Emotional and cognitive evaluations of life. *Annual Review of Psychology, 54*(1), 403–425.

Duan, J. H., & Chen, J. B. (2009). Statistical investigation and analysis on satisfaction degree of Chinese urban and rural people. *Statistics & Information Forum, 24*(4), 79–84. (in Chinese).

Helliwell, J. (2001). Social capital, the economy and well-being. The review of economic performance and social progress. In A. Sharpe & K. Banting, Di (Ed.), *The review of economic performance and social progress 2001: The longest decade: canada in the 1990s*, Vol. 1, Centre for the Study of Living Standards. The Institute for Research on Public Policy.

Hu, W., Zhang, X., Song, Y., Shen, L., Liu, J., & Zhang, A. (2014). Life satisfaction approach to farmers' compensation for land acquisition: Empirical study from the suburbs of Wuhan City. *Chinese Journal of Population Resources and Environment, 12*(4), 316–323.

Inglehart, R. (1990). *Culture shift in advanced industrial society*. Princeton University Press.

Li, J., & Raine, J. W. (2014). The time trend of life satisfaction in China. *Social Indicators Research, 116*(2), 409–427.

Liang, Y., & Zhu, D. (2015). Subjective well-being of Chinese landless peasants in relatively developed regions: Measurement using PANAS and SWLS. *Social Indicators Research, 123*(3), 817–835.

Liu, Y., He, S., & Wu, F. (2008). Urban pauperization under China's social exclusion: A case study of Nanjing. *Journal of Urban Affairs, 30*(1), 21–36.

Luo, Y. M., & Fan, L. M. (2008). The economic effects of rural infrastructure and farmers' satisfaction research based on 44 administrative villages in Shandong Province. *Inquiry on Economic Issue, 12*, 67–73. (in Chinese).

Ma, K., & Hu, L. (2014). The research of landless peasants social risks management which is based on the perspective of social security. *International Integration for Regional Public Management (ICPM)*, 126–130. Atlantis Press.

Pavot, W., & Diener, E. (1993). Review of the satisfaction with life scale. *Psychological Assessment, 5*(2), 164.

Putnam, R. (2001). Social capital: Measurement and consequences. *Canadian Journal of Policy Research, 2*(1), 41–51.

Shui, W., Bai, J., Zhang, S., & Chen, Y. (2014). Analysis of the influencing factors on resettled farmer's satisfaction under the policy of the balance between urban construction land increasing and rural construction land decreasing: a case study of China's Xinjin county in Chengdu City. *Sustainability, 6*(12), 8522–8535.

Silver, H. (1994). Social exclusion and social solidarity: Three paradigms. *International Labour Review, 133*, 531–578.

Tang, S., Hao, P., & Huang, X. (2016). Land conversion and urban settlement intentions of the rural population in China: A case study of suburban Nanjing. *Habitat International, 51*, 149–215.

Veenhoven, R. (2002). Why social policy needs subjective indicators. In M. Hagerty, J. Vogel, & V. Møller (Eds.), *Assessing quality of life and living conditions to guide national policy* (pp. 33–45). *Social Indicators Research Series, 11*. Springer.

Wu, S., & Qin, Q. (2008). Research progress of the identity recognition of land-lost farmers during the process of urbanization. *Journal of Anhui Agricultural Sciences, 36*(23), 193–196. (in Chinese).

Zhang, H., & Tong, X. (2006). Self-identity of the passive-urbanized group in the process of obtaining urban adaptability and modernity: An empirical study on 561 land-displaced peasants in Nanjing. *Sociological Studies, 2*, 84–106.

Zhou, J., Yao, P., Xu, L., & Chen, Z. (2014). Land requisition system reform on the background of urban-rural integration: reform practice and performance of Nanjing. *Modern City Studies, 08*, 25–30. (in Chinese).

Chapter 23
Socio-economic Status of Mothers and Stunting Among Their Children of Under-Five Years Age in Urban Areas of DKI Jakarta

Titik Harsanti and Azka Ubaidillah

Abstract Currently, one-third of developing countries in the world face with stunting phenomenon. Various policies focus on poverty and food shortages in rural areas. So, the solution is provision of food and facilities. Even though Special Capital Region of Jakarta (Daerah Khusus Ibukota or DKI Jakarta) is an urban city with easy access, facilities, and a low percentage of poor people, the cases of stunting children are still found. So, the provision of food and facilities is not the answer. This study analyses the maternal factors that influence under-five years stunting children in the urban area of DKI Jakarta. The major factors here are mother's education, working status of the mother, share of family income, and purchase decision of durable goods. Using 136 mothers of non-poor families, Bayesian binary logistic regression processing was applied and regression results show that all of the maternal factors are associated with stunting children. More educated mothers, working mothers, and mothers with the largest share in family income decrease the probability of stunting among the children. These variables show that the mother's care through socio-economic role in the family brings better health and nutrition for their children. However, the mother's authority to purchase durable goods in the family increased the probability of stunting children. Durable goods are generally expensive, and their purchase may reduce spending on nutrition for toddlers due to budget constraints on the mother.

Keywords Stunting children · Socio-economic status · Bayesian binary logistic regression

---

T. Harsanti (✉) · A. Ubaidillah
Politeknik Statistika STIS, Jakarta, Indonesia
e-mail: titik@stis.ac.id

A. Ubaidillah
e-mail: azka@stis.ac.id

B. R. K. Sinha (ed.), *Urban Dynamics, Environment and Health*,
https://doi.org/10.1007/978-981-99-5744-6_23

23.1 Introduction

Stunting is a chronic condition reflecting poor nutrition and health status, more common in developing countries. The proportion of children under five who are stunted mostly occurs in developing countries as a form of acute nutrition problem which is currently a major concern, because it has long-term effects. It is estimated that 195 million children have stunted growth due to stunting, and the number of children under five who experience nutritional problems in the form of stunting reaches one-third of the total number of children under five worldwide (Rah et al., 2010). Childhood stunting will become a complex problem because it risks child morbidity and death (Black et al., 2008). Based on data from a 1990 born South Africa urban cohort study combined with an analysis of longitudinal data from the Philippines, Jamaica, Peru, Indonesia, and Brazil, revealed that stunting between 12 and 36 months of age predicted poor cognitive performance and/or lower school grades attained in middle childhood (Victora et al., 2008). In the long run, stunting children under five years of age decreased school performance, decreased productivity, had low wages, and increased chances of becoming poor in adulthood.

Indonesia is one of the countries in ASIA facing stunting and malnutrition problems. Despite numerous government efforts to tackle child under-nutrition problems, the levels of children under five years of age remain high, with stunting conditions estimated at 37.2% in the year 2013 and 30.8 per cent in the year 2018 (Indonesia Basic Health Research data). Although it continues to decline, this figure is still higher than the world average of 22% in 2018. Indonesian Health Ministry's target to reduce the cases and stunting could become the next issue of the SDGs targets in 2025.

Most studies indicate that stunting issues focus on people living in rural areas, low socio-economic status, poor household conditions, limited access to food, and harmful environmental conditions (Saniarto & Panunggal, 2004; Fernald & Neufeld, 2006; Taguri, 2008). Based on National Social & Economic Survey data (BPS, 2014), as a capital city of Indonesia, Jakarta has good access to health facilities, a small population living in poor condition, and most of the population has good access to drinking water. However, 26% of under-five-year-old children are stunting and not only occur in low-income families. Why stunting under five Years of age Children still a phenomenon in DKI Jakarta?

This study aims to investigate a mother's role in determining under-five children's nutritional status in the urban area of the Special Capital Region of Jakarta (Daerah Khusus Ibukota (DKI) Jakarta. Geographical conditions, population, and children under five in DKI Jakarta are described in advance in order to obtain an overview of the situation of the research location. Furthermore, the study will discuss stunting and its causes, nutritional and health status for children under five years of age, and socio-economic status of mother in DKI Jakarta.

23.2 Overview of the Province of DKI Jakarta

Based on law number 29 of 2007 DKI Jakarta Province as the state capital has a special status and is given special autonomy so that all policies regarding government and budget are determined at the provincial level. Jakarta is the capital city of the Republic of Indonesia, located in Java island, consists of 1 district, namely Kepulauan Seribu, and 5 cities: North Jakarta, Central Jakarta, East Jakarta, West Jakarta, and South Jakarta. DKI Jakarta Province has a land area of 661.52 km^2 and oceans covering an area of 6.977,5 km^2, recorded $\pm$ 110 islands scattered in the Kepulauan Seribu (Fig. 23.1).

Jakarta Province is bordered by Banten Province in the west and the eastern, southern borders with West Java Province, and to the north is bordered by the Java Sea. Some parts of the area are below sea level. In general, the climate is tropical with an average temperature between 28 and 310 C.

Based on population census data in 2000, 2010 and National Economic and Social Survey 2018, DKI Jakarta is the province with the highest population density in Indonesia, which is 12,592; 14,518, and 15,804 people per km^2 (BPS-Statistics

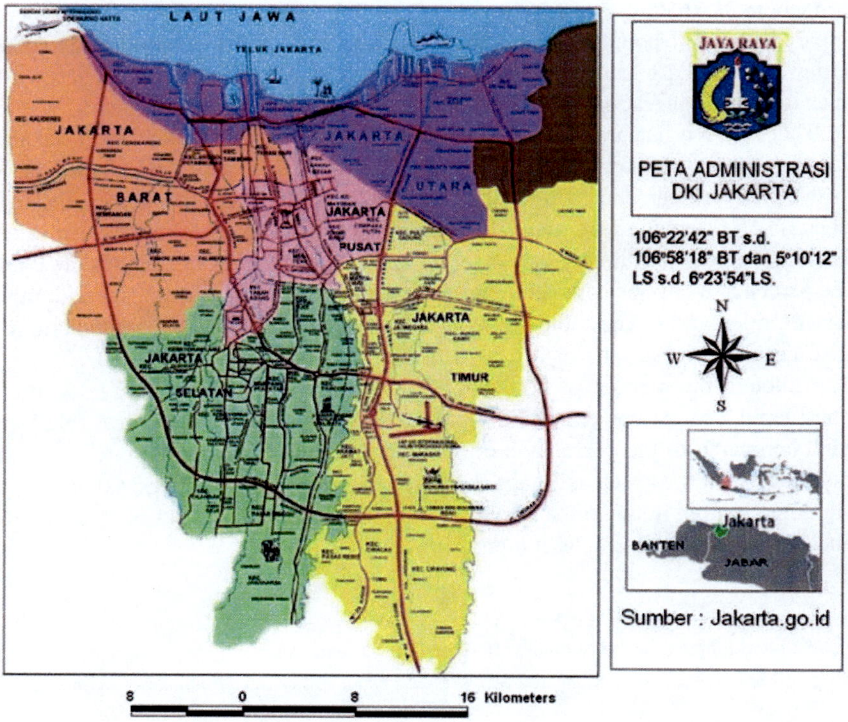

Fig. 23.1 Map area of DKI Jakarta. *Source* Jakarta.go.id

Table 23.1 Percentage of poor population based on Districts/Cities in DKI Jakarta, 2015–2017

District/cities	Percentage of poor population		
	2015	2016	2017
Kepulauan Seribu	11,4	12,58	12,98
South Jakarta	3,41	3,27	3,38
East Jakarta	3,24	3,19	3,31
Central Jakarta	4,16	3,91	3,78
West Jakarta	3,64	3,38	3,25
North Jakarta	5,91	3,57	5,59
DKI Jakarta	3,93	3,75	3,77

Source From the 2015–2017 National Socio-Economic Survey, March

Indonesia). Compared with other cities in the South East Asia, Manila is the world's most densely populated city with 42,857 people per km^2. This density is much higher than that of New Delhi city 29,259.12 people per square mile, Mumbai (23,000 per km^2), Paris (20,150 per km^2), and Tokyo (10,100 km^2).[1] Besides, other cities in the South East Asia are Bangkok 6,584 per km^2 (Year 2010–2011),[2] Woodland East, Singapore 38,763 km^2 (year 2020),[3] Ho Chi Minh City 4,292 and Hanoi 2,398 km^2 (year 2009).[4] The population density of DKI Jakarta is quite high compared to other cities in South East Asia and the world. This causes various problems such as waste management, congestion, pollution due to motorized vehicles, floods, etc.

There are 5 urban areas in DKI Jakarta including North Jakarta, Central Jakarta, East Jakarta, West Jakarta, and South Jakarta. Kepulauan Seribu district as the region with the highest percentage of stunted children under five is excluded from the study because of different environmental problems such as the availability of clean water, health services, and limited access as the major problems of the islands area, and the percentage of poor population in this district relatively high compare with other urbans/cities areas. The table below shows the percentage of poor population in districts and cities in Jakarta, 2015–2017 (Table 23.1).

Although the number of poor population in the urban area of DKI Jakarta is considered low, the percentage of stunting children in the year 2016 was relatively high compared to the percentage of stunting children in the capital city of neighbourhood countries such as Bangkok, Thailand, and Kuala Lumpur, Malaysia. The stunting children in the cities were 6,5% in 2016 in Bangkok (Okubo et al., 2020) and 10.5% in 2018 in Kuala Lumpur (Tay, 2019).

[1] https://worldpopulationreview.com/world-cities/manila-population.

[2] https://shared.ontariotechu.ca/shared/faculty-sites/sustainability-today/publications/compendium_entries/bangkok.pdf .

[3] Woodlands East (Subzone, Singapore) - Population Statistics, Charts, Map and Location (citypopulation.de).

[4] Ho Chi Minh City Population in 2021 (Demographics, stat) (rentapartment.vn).

23.3 Stunting and Its Causes

Stunting is one of the major health problems facing developing countries. A stunted child is often assessed through taking anthropometric measurements. For identification of stunting children, anthropometric measurements use height for age. Using these measurements, data on children's age, sex, and length/height is required. To determine if a child can be categorized as having stunting problems, the height for age is compared to the height for the age of the reference population using the z score. If the height is below –2 standard deviations then the child can be categorized as stunted (WHO, 1992).

This research uses height or body length with the consideration that height or body length can be used as a measure of bone mass growth that occurs as a result of nutritional intake. Height can be used as an anthropometric parameter to describe linear growth. The increase in height or length occurs over a long period of time, therefore problems related to height or body length are an indicator that indicates a chronic nutritional problem. This can occur due to a history of disease in the past, lack of appetite, food intake that is not appropriate for their age and inadequate care for the needs of children's growth and development which results in low nutritional intake (Chikungu, 2013).

To determine the variables that affect the survival of infants, children, and toddlers, especially in developing countries, theories that are widely used are developed from analytical framework Mosley and Chen (1984). The framework describes the relationship between socio-economic variables operating through proximate determinants to exert an impact on mortality. Proximate determinants are variables that directly affect a child's health and mortality status, consist of maternal, environmental, nutritional adequacy, accidents, and individual disease control factors. Advanced research on social policy and medical intervention used the framework to improve child survival. Nutrition research focuses on breast feeding, dietary practices, and food availability, related to nutritional status. This study is grouped to see how socio-economic factors impact on the nutritional status of children under five in the form of stunting based on several previous studies. Stunting can be directly caused by the proximate determinants and indirect causes, namely socio-economic determinants. Furthermore, based on previous research, the direct causes divide into causes that come from within and from outside.

Direct causes that come from within are caused by the vulnerability of individual toddlers who can be at risk of stunting, namely low birth weight, male sex (Taguri, 2008), the age of the child (Teshome et al., 2009), malnutrition during pregnancy, stunting mother (Lindtjørn and Alemu, 2002). Besides, the external causes were a history of diarrhea (Teshome et al., 2009; Taguri et al., 2008), dietary behaviour (Saniarto & Panunggal, 2004; Teshome et al., 2009), access to food, cold weather, hygiene-related problems with water supply, parasitic infection, history of diarrhea, colostrum deprivation, duration of breastfeeding, pre-lacteal feeding, type of food, age at first introducing supplementary feeding and feeding method (Teshome et al., 2009), lack of micronutrients, inadequate protein intake, insufficient food consumption, illness, sanitation and hygiene conditions (Assis, et al., 2003).

Meanwhile, the indirect causes are socio-economic variables. Based on previous research, socio-economic variables include the father's low level of education (Taguri et al., 2008; Saniarto & Panunggal, 2004); low psychosocial stimulation (Taguri et al., 2008), poverty (Taguri et al., 2008; Saniarto & Panunggal, 2004), many occupants (Lindtjørn and Alemu, 2002); maternal age, education, intelligence, and low maternal height (Fernald, & Neufeld, 2006) low wages, insufficient food consumption, illness, inadequate sanitation and hygiene conditions associated with health and nutritional status (Assis et al., 2003).

This study sees nutritional problems as one of the proximate determinants that need attention because they have an impact on the health of toddlers, and in the long term can cause growth faltering and death. Various health research shows that by improving the nutritional status of children under five, the health status of children under five will be better and death can be prevented through the work of socio-economic factors. Socio-economic determinants must operate through more basic proximate determinants that in turn influence the risk of disease and the outcomes of the disease processes. Child survival might be clarified factors that are involved in the family's production of healthy children to provide a foundation for formulating health policies and strategies.

Nowadays, many developing countries are still facing under nutrition and health problems. Therefore, studies that discuss socio-economic variables are increasingly developed. For example, study of Smith et al. (2003) and Bold et al. (2013) discuss about how women influence their children's nutritional status through their effect on childcare practices.

UNICEF's (1990) conceptual framework showed that child survival, growth, and development are influenced indirectly by food security, healthcare and a healthy environment, and care. Focus on nutrition and health, to perform care practices, the caregiver needs sufficient education, knowledge, and beliefs. Education has an important role in improving the health status of toddlers, because mothers who have a good education also have good knowledge in choosing and utilizing health facilities, maintaining a clean environment and are able to seek the latest knowledge and technology for the health of their children. Mothers who have a better level of education are also more adaptable to science, so they have the ability to explore their potential in maintaining and creating a cleaner and healthier home environment as a form of commitment to improving children's health status.

23.4 Nutritional Foods, Nutritional Status and Stunting

Based on Mosley and Chen's analytical framework, Chikungu (2013) develops the socio-economic factors consisting of family, household, community, and government or state variables which influence basic proximate determinants. The family variables consist of mother or father variables such as skills, education, health, and time, household level variables such as income/wealth, water, preventive care and community variables such as ecological setting and health system. Using Chikungu conceptual framework, this study proposes to analyse socio-economic status of mothers and stunting among their children of under-five years age in urban areas of Jakarta.

Chikungu (2013) divides stunting problems based on the national level in the form of programmes and policies, which at the community level depend on credit, social safety nets, education and information, access to health services, and others. Urban areas are different from rural areas because of the different characteristics and the facilities available in urban areas characterize the differences between rural and urban areas. This study chose DKI Jakarta as a city area as well as the capital city of Indonesia which already has good facilities but has a fairly high percentage of stunting, especially compared to neighbouring countries such as Thailand and Malaysia.

Differences in characteristics at the regional level have an impact on the status of toddlers at the family level, where individuals who play an important role in the toddlers' care are mostly the responsibility of the mother. As explained in the framework of Chikungu (2013), background characteristics of mothers such as education, etc. will affect access to consumption, adequate food, child care, etc., which has an impact on the child's nutritional status in the form of stunting. In general, in Indonesian society, mothers still play an important role in caring for children under five, even in big cities. So even though mothers work to help meet the family's economic needs, the task of caring for toddlers remains the mother's responsibility. When mothers live in the DKI Jakarta city area, seeing the socio-economic characteristics of mothers associated with the nutritional status of children under five becomes important because the burden that must be borne by mothers becomes greater. In addition to the educational status and occupation of the mother, this study also looks at the Durable Goods Purchase Decision and Share of Family Income of the mother in the family and their impact on the child nutritional status, which have not been widely studied. Therefore, the framework of this research as mentioned below in Fig. 23.2 clearly explains the facts of socio-economic mother and child nutritional status.

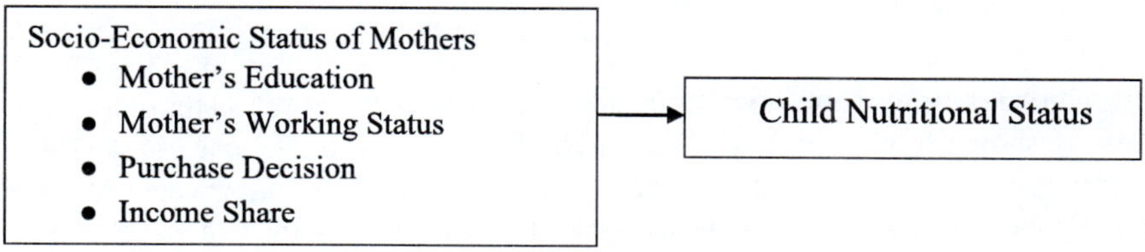

Fig. 23.2 Framework for socio-economic mother and child nutritional status

23.5 Socio-economic Status of Mothers

Socio-economic status shows certain inequalities, where members of society have different level of income, and several individuals have greater access towards jobs of higher status, different level of education, greater access towards better education, different on their economic resources, and different level of power to affect society (Santrock, 2007). Socio-economic status describes a person's condition or a society in terms of social and economics such as level of education, income, working status, and so on.

The fundamental problems of children's nutritional status depend on their family, which functions to ensure its members' health and well-being. In the family, the mother's role is defined by the social norms or expectations of nurturance and protection of her young. The more precise definition of any person occupying biological motherhood is that the person performs the role in a socially acceptable way (White & Klein, 2002). Many studies show that the mother plays a vital role in their children in the family. Infants treated by the non-biological mother significantly increase the chances of becoming stunted (Bloss et al., 2004).

Gender inequality, poor empowerment of women, and poor decision-making powers adversely influence socio-economic status and purchasing power which in turn impact on women's status with implications on rate of child stunting. Women influence their children's nutritional status through their effect on childcare practices (Bold et al., 2013). This ability is determined by the human resources possessed by the mother in the form of nutritional status and physical ability of the mother as well as factors that influence the empowerment of the mother in the family such as education, knowledge, and ability to make decisions to control, obtain, and empower existing resources. (UNICEF 1990; Engle et al., 1997).

Based on theory and previous studies, this research analyses the mother's care and mother's socio-economic variables that affect the nutritional status of children under five years of age such as mother's education, working status of the mother, share of family income, and purchase decision of durable goods.

23.5.1 Mother's Education

Education is an important factor to consider in relation to the health status of toddlers. This is shown by empirical data in research conducted by Liaqat et. al (2007) in Pakistan. The results of this research show that almost all uneducated mothers experience stunting problems. Meanwhile, mothers with primary education have a greater tendency to be stunted compared to mothers with secondary education. Education is related to many things that have an impact on children's health status, for example education is a good tool to increase mother's knowledge about nutrition and nutrition, increase understanding of the importance of utilizing modern medicine, increase capacity to access knowledge, and increase ability to make decisions and opportunities in utilizing professionals in child care.

23.5.2 Working Status of the Mother

Nowadays, the economic needs of contemporary families are not entirely the responsibility of the husband. In Indonesia, the WorkForce Participation Rate (WFPR/TPAK) of women, which has increased from year to year, shows mothers' role in the family economy increases, especially in urban areas. The amount of a mother's income has a positive relationship with childcare. According to Raley et al., (2012), the greater and the continuous participation of the female workforce, the more work experience she accumulates, and the more her income. As the income increases, the possibility of an economic contribution to the family increases as the economic costs available to care for children increase. Mothers with high earning potential have a better ability to share the responsibility for caring for their children with others, especially with their spouses. When women earn more, they become better able to organize a more equitable division of domestic labour. In addition, the higher their income compared to their spouse, the better their ability to negotiate and decide what they want to meet the needs of their children's care.

23.5.3 Durable Goods Purchase Decision (Women Autonomy)

In addition to education and employment, decision-making power by women in the family is important in empowering women, but has not been widely discussed in policy making or research that discusses maternal and child health care. In several literature studies, decision-making power is mentioned as a form of women's autonomy. For example, Woldemicael (2007) mentions that decision-making power in making large household and daily purchases is a form of women's autonomy besides freedom of movement to visit families. Therefore, researchers use this term as the freedom of women in the family to make important decisions when determining the purchase of durable goods. So when women have the ability to determine the purchase of durable goods in the family, then women will have a bargaining position in determining other important decisions in the family that can improve the welfare of their children, especially related to the family economy.

Bloom et al. (2001) proved in their study that women's autonomy has an important role in addition to educational and economic variables in determining the choice to use healthcare facilities in Uttar Pradesh, North India. The link between women's decision-making autonomy and maternal and child healthcare was also investigated by Woldemicael (2007). He uses different dimensions to investigate women's decision-making autonomy and child healthcare based on Demographic and Health Surveys of Ethiopia and Eritrea data. The research showed that women's decisions in the family were influenced by socio-economic factors that had an impact on child healthcare utilization.

Although women's autonomy is widely referred to in many studies, the concept of women's autonomy is still unclear and its relation to demographic problems has not been studied much (Presser & Sen 2000). For example, women's autonomy is concluded as the power to manipulate their environment to control the resources and information in solving their own or close family members' problems (Basu 1992; Dyson & Moore 1983). In this case, women's autonomy can also be interpreted as the ability and freedom to be able to carry out activities according to personal wishes independently without intervention from other people. For example, a mother does not need to ask permission from her husband or other family members to take her sick child to the hospital. So, women's autonomy can be interpreted as the ability to make important decisions in the family even though the husband or other family members are against her wishes (Mason, 1995).

A research using data set from demographic and health surveys of 36 countries in the three developing regions from South Asia, sub-Saharan Africa and Latin America and the Caribbean concluded that women's decision-making power plays an important role in improving the nutrition status of their children. The importance of equal status between women and men in making family decisions is shown by empirical data to reduce malnutrition status in children by 13 per cent or roughly 13.4 million children (Smith et al., 2003). Meanwhile, research by Shroff et al. (2009) concluded that in South Asia, childcare practices such as breastfeeding initiation, complementary feeding practices, actions taken when children are sick, immunizations, etc. are highly dependent on how child healthcare practices are carried out. Based on a study conducted in India, women who have better autonomy, for example, have higher access to money or have the freedom to choose to go to market tend to have healthier children and not experience with stunting (Shroff et al., 2009).

The importance of empowering women to food adequacy, calories, and good feeding processes for infants is shown in studies conducted in Pakistan, Bangladesh, and India. It is shown that in line with the younger age of marriage for women, also with lower levels of education and lower levels of income and purchasing power, the potential for their children to become stunted in South Asia increases (Bold et al., 2013). Furthermore, Smith & Haddad (2015) concluded that gender issues are things that can be considered in reducing stunting rates in South Asia.

In Demographic and Health Survey (DHS), questions about specific decision making in the household are meant to estimate the degree of decision-making autonomy by men/women in their household. Information collected about this includes several types of decisions, including: household purchases for daily needs, health check/care, and household purchases for the need for durable goods. This research proposes women purchasing durables goods to investigate affect of mother's autonomy through nutritional status of the child in the family. For the same reason, this research proposes mother's share of family income as mother's autonomy.

23.5.4 Share of Family Income

In Indonesian culture, women have a dual role because apart from having the nature to give birth, women also have domestic tasks in the family. This is in contrast to men, behind the nature they carry, women still cannot leave their domestic role. Since ancient times, the majority of Indonesian people have believed for generations that domestic work is a woman's inherent obligation as a natural and determined by God (Harun, 2015). So, socially, wives are second class, inferior citizens under men's domination and naturally become economically dependent on men (husbands).

Economic functions in contemporary families are essential because decision making in families is often based on economic power. The problem of power relations between husbands and wives in most families in Indonesia which is studied based on economic power shows that when a woman owns at least 25% of household assets, the impact is that the woman or wife has power or influence in decision-making by the wife or women in the family (Beegle et al., 2001). When mothers have empowerment in the economy, mothers are freer in providing good food and facilities, and the impact is to solve the problem of the child nutritional status.

Currently, women work in order to increase the resources they need to meet the economic needs of their families and communities, closely related to how they manage time and money. With the ability to manage time and money well, they can increase the human resources they have and their children. This can affect their level of welfare (Folbre, 2006).

23.6 Material and Methods

This research was conducted in 5 urban area regions of DKI Jakarta, consisting of Jakarta Utara, Jakarta Pusat, Jakarta Timur, Jakarta Barat, and Jakarta Selatan. The sampling unit of the study is chosen using a 2-stage sampling. The first step chooses Posyandu were selected by means of Simple Random Sampling without Replacement (SRS-WOR). The next step is a family with children under five years of age who are underweight. Posyandu is a form of Community Based Health Efforts (UKBM) which are managed and organized from, by, for, and with the community in the implementation of health development in order to empower the community and provide facilities for the community in obtaining basic social health services to accelerate the reduction of the Maternal Mortality Rate and Infant Mortality Rate. One hundred thirty-six respondents, mothers who are not poor, and have a child under five years of age, are interviewed.

23.6.1 Variables in the Study

The dependent variable used is the nutritional status of children under five years old. It is coded as "1" for those children who are stunting while "0" for those children who are not stunting. The independent variables are maternal factors, such as mother's education, working status of the mother, family share income, and durable goods purchase decision. All the variables are divided into two categories. Mother education is divided into two categories that is high categories if at least graduated high school and low categories if maximum graduated junior school. The working status of the mother consists of work and not work. Family share income is a family member who provides the largest financial share to meet the daily needs of the family, with open answers from father or mother. The last is durable goods purchase decision variables with open answers, father or mother, family member who makes the main decisions in purchasing durable goods.

23.6.2 Method of Data Analysis

This study uses Bayesian Binary Logistic Regression Model methods. Binary Logistic Regression Model is used to explain the probability of a binary response variable as a function of some covariates. The dependent variable in this study (Yi) is considered as dichotomous. Code 1 is for those who are stunting, and 0 is for those who are not stunting.

A binary logistic regression model with a Bayesian approach is applicable to examine the effect of the predictors on the response variable, the nutritional status of children. For binary response variable Y, a binary logistic regression model with logit function:

$$Y = log\left(\frac{P_i}{1 - P_i}\right) = \beta_0 + \beta_1 X_{i1} + \beta_2 X_{i2} + \cdots + \beta_p X_{ip}$$

where, Yi = the observed status of nutrition for the ith children, Pi = the probability that the ith children to be stunting, $X_1, X_2, \ldots X_p$ are the set of independent variables and the $\beta_0, \beta_1, \beta_2, \ldots \beta_p$ are unknown binary logistic regression parameters. In the classical approach, the analysis of the logistic regression model is based on estimating parameters of the model through Maximum Likelihood Estimation (MLE). MLE in small samples encounters severe inferential problems. However, in the Bayesian method, inference about the model parameters is conducted based on their posterior distribution, which is the combination of the likelihood function of observed data and information from previous studies or personal experiences known as prior distribution (Howson & Urbach, 1993). This approach is applied to a binary logistic regression model taking account of the uncertainty in the parameters to identify the determinants of children's nutritional status. The relationship between the posterior distribution and the prior and likelihood distributions can be written as follows,

$$\text{Posterior distribution} \propto \text{likelihood} \times \text{prior distribution}$$

On the Bayes theorem, if there is a parameter θ given by the observation data y, then the probability distribution for the posterior θ on the y data will be proportional to the multiplication between the prior distribution θ and the likelihood function θ given by the y data. It can be written mathematically as follows:

$$f(\theta|y) = \frac{f(y|\theta)f(\theta)}{f(y)}$$

$$f(\theta|y) \propto f(y|\theta)f(\theta)$$

where $f\ (\theta|y)$ is the posterior distribution which is proportional to the multiplication of the likelihood function $f\ (y|\theta)$ and the prior distribution $f\ (\theta)$. This posterior distribution is a complex function of a parameter, and a numerical method is needed to obtain the marginal posterior distribution for each model parameter. The most popular method is the simulation of the posterior distribution using the Markov Chain Monte Carlo (MCMC) simulation method.

The basic idea of MCMC is to generate sample data from the posterior distribution according to the Markov Chain process by using an iterative Monte Carlo simulation in order to obtain convergent conditions to the posterior (Ntzoufras, 2009).

These conditions must fulfil the characteristics of a strongly ergodic Markov Chain (Ubaidillah et al., 2017).

Irreducible, meaning that the sample parameters generated through the Markov Chain Monte Carlo (MCMC) process are random. This is shown from the Autocorrelation Function (ACF) plot, which only cuts off at the 0th lag. It means that the MCMC iteration process fulfils the random nature.

Aperiodic, meaning that the generated parameter sample does not have a periodic pattern in a particular value domain. It is also shown by the ACF plot, which shows that the sample results from the MCMC process do not have a periodic pattern.

Recurrent, meaning that the sample parameter changes occur stably in a particular value domain. It is shown from the plot history and plot trace where the pattern of sample changes occurs stably in certain value domains. It also shows that specific sample values can be recovered after a particular iteration period.

23.7 Results and Discussion

Based on primary data processing, the sample in this study shows that 47.06 per cent of children under five years old are not stunting, and the rest, 52.94%, are stunting. Mother characteristics and nutritional status of the children can be seen in Table 23.2. The table shows that percentage of stunting children is low if the mother at least graduated from senior high school education.

Variable of the mother's role in the economic field in the family is work status, share income in the family, and decisions on purchasing durable goods. The percentage of stunting children is lower if the mother's status is working compared to mother's status is not working. Another variable that describes the economic role of the mother is the share income in the family. The percentage of stunted children is lower if the mother shares the most considerable income in the family than if the person who shares the largest income in the family is the father. The last variable that describes a mother's role in the family economy is purchasing durable goods. The result shows that empirical data differ with the statement before about empowerment in the decision to purchase durable goods and the association with the nutritional status of children. The percentage of stunting children is higher if the mother decides to purchase durable goods in the family and the percentage of stunting children is lower if the father is the person who decides to purchase durable goods in the family.

This study uses Bayesian binary logistic regression analysis with the R open-source software programme to test the variables that significantly influence the nutritional status of children under five. Figure 23.3 shows density plots of significant

Table 23.2 Socio-economic status of mothers and nutritional status of children under-five years old of age

Mother's status	Nutritional status of children	Total		
	% Stunting	% Not Stunting	n	%
High	41.3	58.7	63	47.7
Low	65.2	34.8	69	52.3
Mother working status				
Work	35.7	64.3	14	10.6
Not work	55.9	44.1	118	89.4
Income Share				
Mother	25.0	75.0	4	3.0
Father	54.7	45.3	128	97.0
Purchase decision				
Mother	65.0	35.0	20	15.2
Father	51.3	48.7	112	84.8

Source Own compilation

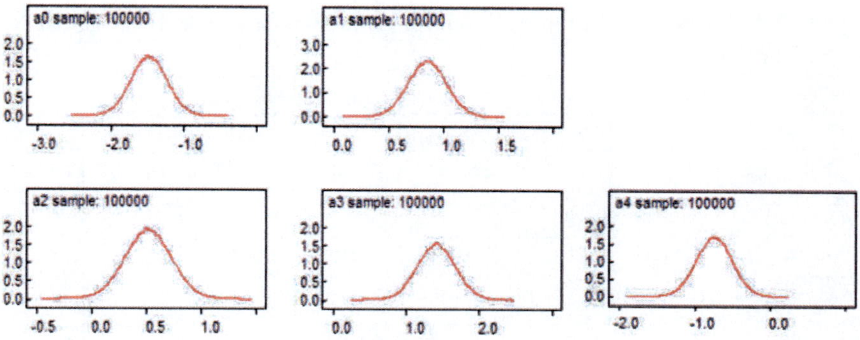

Fig. 23.3 Density plots for only statistically significant regression coefficients in the model

regression coefficients in the model and simulated samples from posterior distribution for each regression coefficient are smooth, uni-modal shape of posterior marginal distribution indicating that simulated parameter value indicates convergence to the target distribution.

Figure 23.4 shows the historical plots of the regression coefficients in the model and this option produces iteration number on x-axis and parameter value on y-axis. The plots look like a horizontal band, with no long upward or downward trends indicating that the chains have converged.

Table 23.3 shows that all value of Mean lies between credible intervals of 2.5% and 97.5%. It can be concluded that all coefficients of model estimation are significant. Besides, the estimation obtained has already met the convergent requirements and the characteristics of the Markov Chain, which are strongly ergodic.

Equation Model of Bayesian Binary Logistic Regression,

$$Y = -1.480 + 0.851X_1 + 0.5118X_2 + 1.41X_3 + -0.7455X_4$$

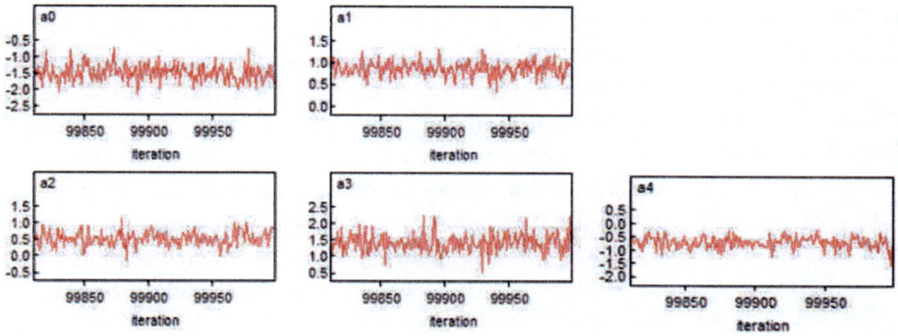

Fig. 23.4 Time series (history) plots for only statistically significant regression coefficients in the model

Table 23.3 Results of bayesian binary logistic regression-based on variables in the equation

Variables in the equation

Node	Mean	SD	MC error	2.50%	Median	97.50%	$\widehat{\beta}$
Constant	−1.480	0.245	7.87E-04	−1.960	−1.481	−0.9971	0,2276
Mother education (X1)	0.851	0.1734	5.31E-04	0.512	0.8513	1.1920	2,3420
Mother working Status (X2)	0.5118	0.211	6.27E-04	0.0989	0.5120	0.9236	1,6683
Income Share (X3)	1.41	0.2625	8.12E-04	0.8968	1.4100	1.9280	4,0960
Purchase Decision (X4)	−0.7455	0.2365	6.77E-04	−1.210	−0.7444	−0.2818	0,4745

Source *Own compilation*

β is the constant value, which is equal to the Exponential value of Mean of the second column at Table 23.3. The most significant tendency for children under-five malnutrition children is income share variable. Mothers who provide the most significant income share in the Family income tend to have children with good nutritional status, 4.0960 times compared with income share given by husband. This study shows the importance of the economic empowerment of mothers associated with children's nutritional needs. The economy is an important part of the lives of young couples who have toddlers in big cities. The role of mothers in helping to increase family income has become commonplace, in order to fulfill the need for quality food consumption, which is generally quite expensive.

The important variable in this study is maternal education. The result of this study shows that the probability of less educated mothers is 2.3420 times to have stunting children compared with more educated mothers. The results of this study are consistent with Strauss (1990) and Thomas et al. (1991) which show that maternal education significantly contributes to child health. More educated mothers tend to have a better ability to find a better job, have a better salary, negotiate with husbands, and so on, than less educated mothers. It is related to mother role and empowerment in the paternal culture, which can reduce the probability of stunting children.

Another variable that reflects the economic role of the mother is work status. The working mother is 1,6683 times to have children with better nutritional status compared with non-working mother. This result shows that the mother economic role in the family is an important variable associated with stunting children. This study noted that such opportunities and the associated increase in women's incomes are, in principle, beneficial for nutritional outcomes for under-five-year-old children. A study using household-level data in the Philippines in 1985 and 2003–2004, which compared between time and income, concluded that instead, income effects predominate than time to care. Although time dedicated to care may decline in some cases, children's welfare is generally increased when mothers augment their income with employment outside the home (Salazar & Quisumbing, 2009; ADB, 2013: p. 69).

The purchase of durable goods is a significant variable, but the coefficient is negative. If the mother decides to purchase durable goods in the family, the probability of children having good nutritional status is 0,4745 compared with if the father is deciding to purchase durable goods in the family. This result contradicts the previous hypothesis that empowering mothers by deciding on purchasing durable goods will improve the nutritional status of children under five years old. But this finding is in line with and consistent with Mahastanti and Wiharjo (2012). In economics, individuals consume goods or services not only based on their needs, but also influenced by the community environment in which the individual lives. Besides, gender and income have a relation with doing shopping. Her study shows that in doing shopping more women have a higher impulsive shopping nature than men. This happens because when doing shopping women prioritize feeling and pleasure than logic. In addition, individual consumption behaviour is influenced by income earned, how much income is consumed, and how much is saved. Family consumption decisions affect overall behaviour, both short and long term to meet the needs and desires for consumption, individuals make decisions to finance their consumption by saving or making loans depending on the income they receive (Mahastanti & Wihardo, 2012). Budget constraints owned by mothers and the desire to shop for durable goods will reduce spending on nutritious food needs and the health of the child.

23.8 Conclusion

This study shows the importance of the role of socio-economic status of mothers in determining the nutritional status and in checking stunting of under-five year of age children, especially in the economic aspect of the family. Indonesia is a country with a patriarchal culture, that husband has the duties and responsibilities of earning a living. However, the economic needs of contemporary families are not entirely the responsibility of the husband, especially in urban areas. Economic functions in contemporary families are essential because decision making in families is often based on economic power. In this study, mother's role in the economy is the mother who has a large share income in the family and work. These two variables indicate that mothers who do not have a role in economic capital tend to have stunting children. It is a fundamental nutritional problem for under-five children, even though the family is not poor.

Another variable in this study is mother education. Education can improve mother knowledge and change culture. Education is an important variable that can decrease the probability of stunting children. This study suggests empowering mothers through the economic role of mothers, such as mother education, mother-share income in the family, and working mothers. Government's regulation is needed to support working mothers while they care for the children, for example, provide a lactation room at work.

The last significant variable affecting stunting children in this study is the mother who decided to purchase durable goods in the family. If the mother is the one who decides on the durable goods purchasing, the probability of under-five children to become stunting tends to increase. This may be because mothers are faced with many consumptive goods in the big city such as Jakarta. Budget constraints owned by mothers and the desire to shop for durable goods could reduce spending on nutritious food needs and the health of the child. Researchers suggest a more in-depth study on the consumption behaviour of mothers in big city life associated with their socio-economic status which can have an impact on the nutritional status of children under five.

Acknowledgements The authors thank to family, friends, members of the Politeknik Statistika STIS and Universitas Indonesia for supporting the research work. Thanks to all mothers who become respondents of this research.

References

ADB. (2013). Gender *equality and food security: women's empowerment as a tool against hunger*. Mandaluyong City, Philippines: Asian Development Bank.

Anjullo, B. B., & Haile, T. T. (2018). *Bayesian binary logistic regression approach in identifying factors associated with exclusive breastfeeding practices at Arba Minch Town.* Department of Statistics, Arba Minch University, Ethiopia.

Assis, A. M. O., Prado, M. S., Barreto, M. L., Reis, M. G., Pinheiro, S. M., Conceição, P., & I. M. and Blanton, R. E. (2003). Childhood stunting in Northeast Brazil: The role of Schistosoma mansoni infection and inadequate dietary intake. *European Journal of Clinical Nutrition, 2004*(58), 1022–1029. https://doi.org/10.1038/sj.ejcn.1601926.

Basu, A. M. (1992) Culture, the status of women, and demographic behaviour: illustrated with the case of India. Clarendon Press, Oxford. ISBN: 9780198283607.

Beegle, K., Frankenberg, E., & Thomas, D. (2001). Bargaining pwer within couples and use of prenatal and delivery care in Indonesia. On-line working paper series, University of California. https://escholarship.org/uc/item/180736vt.

Black, R. L., Allen, Z., Bhutta, L., Caulfield, M., de Onis, M., & Ezzati, C. (2008). *Maternal and Child Undernutrition: Global and Regional Exposures and Health consequences.* Lancet

Bloss, E., Wainaina, F., & Bailey, R. C. (2004). Prevalence and predictors of underweight, stunting, and wasting among children aged 5 and under in Western Kenya. *Journal of Tropical Pediatrics, 50*(5). Oxford University Press.

Bold, M. V., Quisumbing, A. R. & Gillespie, S. (2013). *Women's empowerment and nutrition: an evidence review.* IFPRI. Discussion Paper 01294, Transform Nutrition, International Food Policy Research Institute.

Bloom, S. S., Wypij, D., & Gupta, M. D. (2001). Dimensions of women's autonomy and the influence on maternal health care utilization in a north indian city. *Demography, 38,* 67–78. Springer.

Bolstat, W. M. (2010). *Understanding computational bayesian statistics.* John Wiley & Sons, Inc.

BPS–Statistics Indonesia. (2014). *Expenditure for Consumption of Indonesia by Province.* Based on Susenas, March 2014. Book 3.

BPS—Statistics Indonesia. (2022). *Sirusa.* https://sirusa.bps.go.id/sirusa/index.php/indikator/index. (Acces 16 Mar 2022).

Chikungu, L. C. (2013). *Modelling under-nutrition in under-five children in Malawi.E-Prints Soton.* University of Southampton Research Repository.

Dyson, T., & Moore, T. (1983). On kinship structure, female autonomy, and demographic behaviour in India. *Population and Development Review, 9*(1), 35–54.

Engle, P. L., Menon, P., & Haddad, L. (1997). *Care and nutrition: Concepts and measurement*. California Polytecnic State University.

Fernald, L. C., & Neufeld, L. M. (2006).Overweight with concurrent stunting in very young children from rural Mexico: prevalence and associated factors. *European Journal of Clinical Nutrition.*

Folbre, 2006. Measuring care: gender, empowerment, and the care economy. *Journal of Human Development 7*(2). University of Massachusetts Amherst, USA.

Fujiati, D. (2014). Relasi Gender Dalam Institusi Keluarga dalam Pandangan Teori Sosial dan Feminis. *Muwazah Jurnal, 6*(1). IAIN Pekalongan, Indonesia.

Harun, M. Q. (2015). Rethinking peran perempuan dalam keluarga. Karsa, Pamekasan, Indonesia.

Hendrawan, D. (2005). Kualitas Air Sungai dan Situ di DKI Jakarta. *Makara, Teknologi, 9*(1), 13–19.

Howson, C., & Urbach, P. (1993). *Scientific reasoning: The Bayesian approach* (2nd ed.). Open Court.

Lindtjørn, B., & Alemu, T. (2002). Year-to-year and seasonal variations in stunting among preschool children in Ethiopia. *Journal of Health, Population and Nutrition - JSTOR.* Rerearchgate.net.

Mahastanti, L. A., & Wiharjo, K. K. (2012). *Mental accounting dan variebl demografi: sebuah fenomena pada penggunaan kartu kredit*. National Seminar Proceedings Menilai Kinerja Bisnis & Ekonomi Indonesia: Problematika, Perspektif, & Prospek. Universitas Atmajaya, Yogyakarta, Indonesia.

Manary, M. J., & Solomons, N. W. (2009). Gizi kesehatan masyarakat, gizi dan perkembangan anak. penerbit buku kedokteran EGC. In Gibney M. J., Margetts, B. M., Kearney, J. M., & Arab, L. B (Eds.), Translate by Public Health Nutrition. Publishing Ltd, Oxford.

Mosley, W. H. & Chen, L. C. (1984). An analitical framework for the study of child suvival in developing countries. *Population and Development Review 10* (Suppl), 25–45.

Mason, E. S. (1995). Gender differences in job satisfaction. *The Journal of Social Psychology, 135*(2).

Ntzoufras, I., Fouskakis, D., & Draper, D. (2009). *Bayesian decision theory in biostatistics: The utility of utility*. Department of Applied Mathematics and Statistics, University of California, Santa Cruz, USA.

Okubo, T., Janmohamed, A., Topothai, C., & Blankenship, J. L. (2020). Risk factors modifying the double burden of malnutrition of young children in Thailand. *Maternal & Child Nutrition. 16*(S2). Willeyonlinelibrary.com/journal/mcn. Bangkok, Thailand.

Presser, H., & Sen, G. (2000). Women's empowerment and demographic processes: Moving beyond cairo. Oxford University Press. UK.

Rah, J. H, Akhter, N, Semba, R. D, Pee, de S, Bloem, M. W, Campbell, A. A, Pfanner, M. R, Sun, K, Badham, J., & Kraemer, K. (2010). Low dietary diversity is a predictor of child Stunting in rural Bangladesh. *European Journal of Clinical Nutrition,* 1393–1398. http://www.sightandlife.org/fileadmin/data/Publications/EJCN.64.2010.Low_dietarydiversity_is_a_predictor_of_child_Stunting_in_rural_Bangladesh.pdf [diakses 23 Januari 2013].

Raley, S., Bianchi, S. M., & Wang, W. (2012). When do fathers care? mothers' economic contribution and fathers' involvement in child care. *American Journal of Sociology, 117*, 1422–1459.

Salazar, L., & Quisumbing, A. (2009). *Assessing the Impact of Gendered Labor Markets in the Rural Philippines*. Paper for the FAO, IFAD, and ILO Workshop on Gaps, Trends and Current Research in Gender Dimensions of Agricultural and Rural Employment: Differentiated Pathways out of Poverty. Rome. 31 March–2 April.

Saniarto, F., & Panunggal, B. (2004). *Pola makan, status Sosial Ekonomi Keluarga dan Prestasi Belajar pada Anak Stunting Usia 9–12 Tahun di Kemijen*. Semarang Timur. Universitas Diponegoro.

Shroff, S., Griffiths, P., Adair, L., Suchindran, C., & Bentley, M. (2009). Maternal autonomy is inversely related to child stunting in Andhra Pradesh, India. *Maternal and Child Nutrition, 5*, 64–74. Blackwell Publishing Ltd. https://doi.org/10.IIII/j.1740-8709.2008.00161x.

Schiffman, L. G., & Kanuk L. L. (2008). Consumer behaviour (7th edn). Prentice-Hall, Inc. Upper Saddle River, New Jersey 07457. Translate by PT Indeks, Indonesia.

Smith L. C., Ramakrishnan U., Ndiaye A., Haddad L., & Martorell, R. (2003) The importance of women's status for child nutrition in developing countries. International Food Policy Research Institute (IFPRI) Research Project 131.

Smith, L. C., & Haddad, L. (2015). Reducing Child Undernutrition: Past Drivers and Priorities for the Post-MDG Era. *World Development, 68*, 180–204. Elsevier. https://doi.org/10.1016/j.worlddev.2014.11.014.

Strauss, J. (1990). Households, communities, and preschool children's nutrition outcomes: evidence of rural cote d'Ivoire. *Economic Development and Cultural Change, 38*, 23161.

Sunaryanto, H. (2015). Dampak fertilitas terhadap kebutuhan dasar keluarga (studi kasus pada keluarga dengan jumlah anak lebih dari dua di desa Pondok Kelapa, kabupaten Bengkulu Tengah). *Jurnal Sosiologi Nusantara, 1*(1). Universitas Bengkulu, Indonesia.

Taguri, A. E. et al. (2008). Risk factors for stunting among under-fives in Libya. *Public Health Nutrition 12*(8), 1141–1149. US National Library of Medicine National Institutes of Health.

Tailor, H. M., & Carlin, S. (1998). *An introduction to stochastic modelling* (3rd ed.). Academic Press.

Tay, C. (2019). *Malaysia's Childhood stunting shock: Rates higher than Palestine and several African countries* – new report. https://www.nutraingredients-asia.com/Article/2019/06/12/Malaysia-s-childhood-stunting-shock-Rates-higher-than-Palestine-and-several-African-countries-new-report. (AccesSed 10 Mar 2021).

Teshome, B., Kogi-Makau, W., Getahun, Z., & Taye, G. (2009). Magnitude and determinants of stunting in children underfive years of age in food surplus region of Ethiopia: The case of West Gojam Zone. *Ethiopian Journal of Health Development, 23*(2). eISSN:1021–6790. https://doi.org/10.4314/ejhd.v23i2.53223.

Thomas, D., Strauss, J., & Hendriques, M. H. (1991). How does mother's education affect child height? *The Journal of Human Resources, 26*(2), 183–211. University of Wisconsin Press. https://doi.org/10.2307/145920, https://www.jstor.org/stable/145920.

Ubaidillah, A., Kurnia, A., & dan Sadik, K. (2017). Generalized multilevel linear model dengan pendekatan bayesian untuk pemodelan data pengeluaran perkapita rumahtangga. *Jurnal Aplikasi Statistika & Komputasi Statistik 9*(1). Politeknik Statistika STIS, Jakarta.

UNICEF (United Nations Children's Fund). (1990). *Strategy for improve nutrition of children and women in developing countries.* UNICEF.

Uyun, Q. (2002). *Peran Gender dalamBudayaJawa.* Psikologika No. 13 Tahun VII, 2002. Universitas Islam Indonesia, Yogyakarta.

Victora, C. G., Adair, L., Fall, C., Hallal, P. C., Martorell, R., Richter, L., & Sachdev, H. S. (2008). Maternal and child undernutrition: Consequences for adult health and humancapital. *The Lancet, 371*, 340–357.

White, J. M., & Klein, D.M. (2002). *Family Theories.* (pp. 68). SAGE Publication, Inc. Thousand Oaks, California.

Woldemicael, G. (2007). *Do women with higher autonomy seek more maternal and child healthcare? Evidence from Ethiopia and Eritrea* (p. 28). Digitala Vetenskapliga Arkivet (DiVA). Sociologiska institutionen, Stockholm University.

Chapter 24
Role of Educational Attainment of Mothers in Child Delivery, Post Natal Check-Up, and Child Immunization in Urban India: A Study in the Context of Child Health

Prabhakar Nishad and Braj Raj Kumar Sinha

Abstract Child birth, postnatal check-up, and immunization are essential components of child health. In this context, the role of educational attainment of mothers becomes very important, especially in the developing countries like India. A considerable percentage of mothers are either illiterate or less educated even in urban areas, despite the availability of suitable educational and health facilities. In fact, there is well known relationship between mother's education and maternal-cum-child health. The mother is considered as the most important household-level health worker. The performance of a mother as a household-level health worker depends mainly on her educational attainment and health-related awareness. It is believed that educated mothers are generally more aware in taking care of their health, which is prerequisite for delivering a healthy child. Place of delivery (institutional and home) is linked with the health of the mother as well as the child. Home birth increases the risk of maternal and neonatal mortality because the delivery is assisted by less skilled attendants. Still, in urban India, a substantial percentage of deliveries take place at home. Immunization of children against vaccine-preventable diseases can reduce child mortality to a great extent. Educational attainment of mothers plays an important role in proper compliance of immunization schedules. The first 2 days after delivery are crucial for the mothers and new-born babies. Most of the neonatal deaths occur during this period. In light of the above, this paper aims at: establishing relationship between educational attainment of mothers and the place of delivery of child in urban India; assessing the impact of educational attainment of mothers on immunization of children aged 12–23 months; and investigating the effect of mother's educational

P. Nishad (✉)
Department of Geography, D. A. V. Post Graduate College,
Siwan, J. P. University, Chapra, Bihar, India
e-mail: prabhakaragna@gmail.com

B. R. K. Sinha
Former Professor, Department of Geography, Institute of Science, Banaras Hindu University,
Varanasi, U.P, India

B. R. K. Sinha (ed.), *Urban Dynamics, Environment and Health*,
https://doi.org/10.1007/978-981-99-5744-6_24

attainment on the postnatal check-up of mother and new-born baby. This paper is based on the secondary data obtained from the National Family Health Survey 4 (NFHS-4), 2015–16. The findings show that the educational attainment of mothers is highly correlated with the place of delivery. The institutional delivery is preferred by educated mothers. Furthermore, improving maternal education is important for increasing child vaccination uptake and coverage. The role of educational attainment of mothers in postnatal check-up is also evident.

Keywords Education · Immunization · Child delivery · Postnatal check-up · Urban health · Child health

24.1 Introduction

Approximately 34 children per 1000 live births do not survive till the age of five years in India (World Bank, 2019) due to various factors such as neonatal infections, low birth weight, pneumonia, and diarrhoea, among other reasons. Cross-country comparisons using large data sets, such as the World Fertility Survey and the Demographic and Health Surveys, have shown that education in general and female education in particular exerts a very strong influence in reducing child morbidity and mortality and improving the health of child. In a study done by Racine and Joyce (2007, p.1770), it was found that 15% of the poorest mothers had problem with basic literacy and it was estimated that each additional year of schooling would increase a women's income by 10–20%, which in turn has a direct impact on the child's health. Maternal education was found to play an important role in improving the child survival and health in developing countries (Caldwell, 1979, p. 410). This can be attributed to the fact that educated mothers have a better knowledge about health care, nutrition and may provide healthy sanitary habits and a safer environment for their children (Strauss & Thomas, 1998, p. 812). The National Research Council and Institute of Medicine (2004) has defined child health as the extent to which children are able or enabled to develop and realize their potential, satisfy their needs, and develop the capacities that allow them to interact successfully with the biological, physical and social environment. In a study done by Barrera (1990, p. 82) it has been predicted that maternal education has the greatest impact on child health especially among the children of age group 0–2 years. Improving female education is important for reducing child mortality rate. This is an important reason for governments to emphasize upon the education of girl child. Studies have shown that the use of basic health facilities that relate to childhood survival has a direct correlation with mother's education. Hence the aim of the study was to analyze the impact of maternal education on nutritional status, morbidity factors—diarrhoea, respiratory infections, and personal hygiene.

One of the major public health problems in India is neonatal mortality. According to National Health Profile Report 2010, it has been estimated that close to one million babies die every year before completing their first month of life. The neonatal mortality rate in India was 32 per 1000 live births in the year 2010, a high rate that has not declined much in the last decade (RGI, 2009). The neonatal mortality rate is significantly higher in rural India (38/1000 live births) compared to that of urban areas (22/1000 live births) as per the National Family Health Survey-4 (IIPS, 2017). The place of delivery of the child is an important factor responsible for neonatal mortality. Delivery in a health facility with a skilled doctors/nurses reduces early neonatal mortality and infant mortality. Skilled birth attendance and an institutional environment capable of providing effective obstetric and neonatal care are needed to significantly reduce maternal deaths, still births, neonatal, and infant deaths (Gouder et al., 2015, p. 13).

Immunization is a proven tool for controlling and eradicating infectious diseases. According to World Health Organization (WHO), immunization prevents 2–3 million infant deaths globally every year. Though child immunization has been very successful, vaccine-preventable diseases remain a major health problem among children in low and middle-income countries. WHO has estimated that vaccine-preventable diseases are still responsible for 1.5 million child deaths each year due to inadequate vaccination coverage. It includes those children who were partially vaccinated, completely unvaccinated, or received vaccine later than scheduled (WHO, 2005). It has been hypothesized that children of mothers who are educated are more likely to complete their vaccination because maternal education helps a mother to follow health-seeking behaviours that enhance child vaccination (Balagon et al. 2017 p. 68). To assess the determinants of child immunization in general and the association of maternal education and immunization coverage in particular, studies have been carried out in different settings and different populations. A survey conducted in Eritrea in 2017 among children aged 24–35 months to assess the national Expanded Programme on Immunization coverage found that 92% of children of mothers with middle or higher education were fully immunized compared to only 78% of children of mothers with no education (MoH, 2017). Other studies conducted in Nigeria and Ethiopia found that children of educated mothers were more likely to be fully immunized than children of uneducated mothers (Fatiregun & Okoro, 2012, p. 736). Conversely, a study conducted in Uganda in 2010 found that infants, whose mothers had a secondary education, were at least 50% less likely to miss scheduled vaccinations compared to those whose mothers had only primary education (Animaw et al. 2014, pp. 36–38). Maternal education may influence a child's health in different ways. The information transfer hypothesis (Baker, 2010) suggests that women enrich their health knowledge base through direct exposure to school curricula covering disease classification, germ theory, health and hygiene, and biology, among other subjects.

24.2 Objectives

The objectives of this paper are to:

i. find out the relationship between educational attainment of mothers and place of child delivery in urban India;
ii. assess the impact of educational attainment of mothers on immunization of children aged 12–23 months; and
iii. investigate the effect of mother's educational attainment on the postnatal check-up of the mother and new-born baby.

24.3 Data Source and Methodology

The study is based on secondary data collected from the most recent round of the National Family Health Survey (NFHS-4) 2015–2016, which is a nationally representative, multi-round, population-based survey conducted in all the states and union territories of India under the direction of the Ministry of Health and Family Welfare, Government of India. Descriptive approach of analysis was adopted to determine the role of educational attainment of mothers in child birth, postnatal check-up, and child immunization in urban India. The term "educational attainment" as used in this paper connotes the same meaning of the term "years of schooling" of the person concerned.

24.4 Analysis

24.4.1 Role of Educational Attainment of Mothers in the Place of Child Birth

Monitoring deliveries in health facilities is essential to ensure that women receive quality care and deliver babies in a set up equipped to meet emergency situations. Giving birth to child in a health facility can increase survival prospects for mother and baby through access to appropriate equipment and supplies that are available on site or through immediate referral to a higher-level facility. However, it also remains essential to ensure that the delivery is carried out by skilled health personnel who are capable of anticipating or detecting signs and symptoms of complications. In addition, it is critical that water, sanitation and hygiene services are available in health care facilities. This is the key for infection prevention and control, patient safety, and child and maternal health, particularly during the first few days of child birth.

Table 24.1 Percent distribution of live births to women in the age group of 15–49 in the five years preceding the survey by place of delivery according to year of schooling in Urban India, 2015–16

Years of schooling	Health facility			Home
	Public hospital	NGO/trust hospital	Private hospital	
No Schooling	49.5	0.3	11.8	38
< 5 years complete	54.4	0.4	15.2	29.7
5–7 years complete	58.4	0.5	21.2	19.5
8–9 years complete	60.9	0.7	23.7	14.4
10–11 years complete	53.8	0.7	37	8.4
12 or more years complete	41.2	0.7	52.8	5.1

Source National Family Health Survey (NFHS-4) 2015–16

According to UNICEF (2022) a large proportion of babies are delivered in health facilities in many countries, particularly middle-income and high-income countries. Over 9 out of 10 births occurred in health facilities in Eastern Europe and Central Asia (98%), East Asia and the Pacific (91%), and Latin America and the Caribbean (94%). In contrast, 59 and 63% of births were delivered in health facilities in West and Central Africa and Eastern and Southern Africa respectively, where the burden of maternal and new-born deaths is highest. In sub-Saharan Africa, there was an urban–rural gap of over 25% points (78 vs. 52%) and a staggering 48% point gap between women in the richest (88%) wealth quintiles and those in the poorest (40%) wealth quintiles.

Increasing institutional deliveries is an important factor in reducing maternal and neonatal mortality. Seventy-nine percent of live births in the five years before the survey were delivered in a health facility (Table 24.1). According to the NFHS4 data, the most common reason for not going to a health facility for delivery was that the woman did not consider it necessary (40%). Eighteen percent of women said that the institutional health facility was too far or there was no transportation, another eighteen percent said that the husband or family did not allow them to have the delivery in a health facility, and 16% said that the cost of institutional delivery was too high. The mother's educational status is highly correlated with the place of delivery. Ninety-five percent of births to mothers with 12 or more years of schooling were delivered in a health facility, compared with 62% of births to mothers with no schooling.

24.4.2 Skilled Assistance During Delivery

It includes births delivered with the assistance of doctors, midwives, nurses, Dai, and ASHA workers. Assistance during childbirth can influence the birth outcome and the health of the mother and the newborn. A skilled attendant can manage complications of pregnancy and delivery or refer the mother and/or the baby to the next level of care.

Table 24.2 shows the percent distribution of live births to women in the age group of 15–49 in the five years preceding the survey by person providing assistance during delivery, percentage delivered by caesarean section (C-section), according to mother's year of schooling in Urban India, 2015–16. The majority of births were attended by doctors (56%), followed by ANMs, nurses, midwives, LHVs (25%), and Dais (TBAs) (11%). The mother's educational status is highly correlated with skilled delivery. Ninety-five percent of births to mothers with 12 or more years of schooling were delivered by a skilled attendant, compared with 66% of births to mothers with no schooling.

A caesarean section, or C-section, can be a lifesaving intervention and is an essential part of comprehensive emergency obstetric care, preventing maternal and neonatal mortality and morbidity when medically justified. According to UNICEF data, approximately 18% of newborns were delivered by C-section in 2021 on a global level. In Latin America and the Caribbean, C-sections accounted for 44% of all deliveries, almost 10 times higher than in sub-Saharan Africa (5%). High percentage of C-sections in Latin America and the Caribbean may suggest over-medicalization of childbirth; conversely, the low percentage of C-sections in sub-Saharan Africa (5%) is alarming, suggesting a dire lack of access to this potentially lifesaving intervention.

Access to caesarean section (C-section) can reduce maternal and neonatal mortality and complications such as obstetric fistula. However, a caesarean section without a medical need can put a woman at risk of short and long-term health problems. WHO advises that caesarean sections should only be done when medically necessary, and does not recommend a target rate for countries to achieve at the

Table 24.2 Percent distribution of live births to women in the age group of 15–49 in the five years preceding the survey by person providing assistance during delivery, percentage delivered by caesarean section (C-section), according to mother's year of schooling in Urban India, 2015–16

Year of schooling	Person providing assistance during delivery						Percentage delivered by caesarean section
	Doctor	ANM/ nurse	Other health personnel	Dai	Friends/ relatives	Others	
No schooling	35.9	28.9	1.1	20.4	11.1	2.5	6
< 5 years complete	48.8	24.3	1	15.6	8.8	1.5	10.6
5–7 years complete	56.1	26	0.6	10.3	5.9	0.9	14.7
8–9 years complete	59.7	27	0.6	7.5	4.5	0.6	17.4
10–11 years complete	72.2	19.7	0.4	3.8	3.2	0.6	25.7
12 or more years complete	76.6	18	0.3	2.4	2.3	0.4	33.6

Source National Family Health Survey (NFHS-4) 2015–16

population level. More educated women are much more likely to undergo caesarean deliveries. The caesarean rate for deliveries for women with 12 or more years of schooling is 34%, compared with 26% for women with 10–11 years of schooling, 15% for women with 5–7 years of schooling, and 6% for women with no schooling.

24.4.3 Postnatal Health Check for Mothers

Postnatal care is a specialized job for the mother that begins immediately after delivery up to a period of roughly eight weeks. The care includes prevention, elimination, early detection and cure of health complications, counselling for breastfeeding, immunization, an interactive session on the importance of birth spacing and maternal nutrition, and so on. After delivery, mothers go through several emotional and physical changes, as they learn to take care of their newborn. Therefore, postnatal care involves proper rest, nutrition, and vaginal care for the mother, facilitating a quicker and healthier recovery.

The postnatal period is critical for the health of the mother and the child. According to WHO, most maternal and infant deaths occur in the first month after birth: almost half of postnatal maternal deaths occur within the first 24 h and 66% post-delivery maternal deaths occur during the first week. It is estimated that 2.8 million new-born babies died within one month of their birth in 2013 and close to one-third of death happened on the first day itself. The World Health Organization (WHO) recently updated global guidelines on postnatal care for mothers and newborns through a technical consultation process. These guidelines address the timing and content of postnatal care for mothers and newborns with a "special focus on resource-limited settings in low and middle-income countries". These guidelines are in addition to the existing guidelines and are aimed at reducing complications emerging immediately after child birth.

24.4.4 Timing of First Postnatal Check-Up for the Mother

As mentioned above, large proportion of maternal and neonatal deaths occur during the first 24 h after delivery. For both the mother and the infant, prompt postnatal care is important to treat complications that arise from delivery and to provide the mother with important information on caring for herself and her baby. The Ministry of Health and Family Welfare (MoHFW) recommends that all women who deliver in a health facility must receive a postnatal health check-up within the first 24 h after delivery and that women giving birth outside of a health facility should be referred to a health facility for a postnatal check-up within 12 h after giving birth. The 2015–16

NFHS found among women aged 15–49 giving birth in the five years before the survey, 65% of mothers had a postnatal check-up during the first two days after birth. Thirty percent of mothers did not receive any postnatal check-ups.

Table 24.3 shows the percentage distribution of women in the age group of 15–49 giving birth in the five years preceding the survey on the basis of mothers receiving a postnatal check-up after their most recent live birth and the timing of the first postnatal check-up in the first two days after birth by year of schooling in urban India. From Table 24.3 it is clear that only 18% of women with 12 or more years of schooling have no postnatal check-up while this percentage is far greater (44%) among the women who have no schooling. From the above table we can also conclude that timing of first postnatal check-up among the women having 12 or more years of schooling is highest while it is lowest among the women having no year of schooling. About 67.4% of women having 12 or more years of schooling had their first postnatal check within 4 h of delivery while this percentage is only 43.6 among the women having no year of schooling. So we can say that there is a positive relationship between the level of education of mother and postnatal check-up among the women in the 15–49 year age group.

Table 24.3 Percentage distribution of women aged 15–49 giving birth in the five year preceding the survey by timing of postnatal check-up for the mothers by their year of schooling in Urban India, 2015–16

Years of schooling	Timing for first postnatal check-up for the mother					No postnatal check-up
	Less than 4 h	4–23 h	1–2 days	3–41 days	Don't know	
No schooling	43.6*	2.8	4.6	3.8	0.8	44.5
< 5 years complete	49.6	3.9	5.9	5.2	0.9	34.5
5–7 years complete	55.6	3.7	5.8	4.3	0.7	29.9
8–9 years complete	58.4	4.3	5.6	3.9	0.9	26.9
10–11 years complete	63.3	4.1	6.1	4.0	0.7	21.7
12 or more years complete	67.4	4.4	5.8	3.5	0.8	18.1

Source National Family Health Survey (NFHS-4) 2015–16

* Decimal figures are in percentage to total women aged 15–49 giving birth in the five year preceding the survey

24.4.5 Type of Health Provider of First Postnatal Check-Up for the Mother

Ensuring that all births are assisted by skilled health personnel is critical for reducing maternal and new-born deaths. Skilled health personnel, including medical doctors, nurses, and midwives, are trained to handle normal deliveries safely. They are also able to recognize warning signs for complications and refer mothers to emergency care in case of complications like haemorrhage and sepsis, which are the leading causes of death among mothers during and after childbirth. It is important to see how education and knowledge are important for reducing maternal morbidity and mortality.

Skilled attendants (people with midwifery skills, such as midwives), doctors, and nurses, who have been trained to manage normal (uncomplicated) pregnancies, childbirth and the immediate postnatal period, and who are able to identify, manage, or refer complications in the woman and newborn, are best placed to ensure the survival and safety of pregnant women and their infants. Whatever their professional title, health professionals functioning as skilled attendants should be able to identify early signs of complications and offer first line emergency obstetric care (including emergency new-born care) when needed.

Table 24.4 shows the percentage distribution of women in the age group of 15–49 giving birth in the five years preceding the survey by types of provider of the mother's postnatal check-up for the last live birth, according to mother's years of schooling in urban India, 2015–16. Among the various types of postnatal health providers, doctors are considered as most skilled while Dai and ASHA workers are comparatively less skilled in handling various postnatal check-ups. From the table it is clear that around 61.7% of women having 12 or more years of schooling are attended by doctors and this percentage decreases as the years of schooling decreases. The less skilled healthcare providers like Dai and ASHA workers were highest among the women having no year of schooling while this percentage becomes less than 1% among the women having 12 or more years of schooling. So we can say that highly educated women are attended by high skilled postnatal healthcare providers and less educated women are more attended by less skilled postnatal healthcare providers.

24.4.6 Postnatal Health Check-Ups for Newborns

The first 48 h of life is a critical phase in the lives of new-born babies and a period in which many neonatal deaths occur. Lack of postnatal health check-ups during this period can delay the identification of complications in the new born and it may delay the initiation of appropriate care and treatment. Identifying conditions requiring additional care, and giving proper counselling to parents on the requirement of a new-born baby to be taken to a health facility is critical. Families should be counselled on identification of danger signs, understanding the care that both the woman and newborn

Table 24.4 Percentage distribution of women aged 15–49 giving birth in the five years preceding the survey by type of provider of the mother's postnatal check-up for the last live birth, according to mother's years of schooling in Urban India, 2015–16

Years of schooling	Types of health provider of new-born's first postnatal check-up					
	Doctor	ANM/nurse	Other health personnel	Dai	ASHA	Others
No schooling	23.8	24.3	0.4	2.9	3.1	0.3
< 5 years complete	35.1	22.9	0.4	2.3	3.6	0.3
5–7 years complete	41.8	23.4	0.3	1.7	2.2	0.1
8–9 years complete	44.7	24.4	0.2	1.1	1.8	0.1
10–11 years complete	56.9	18.8	0.2	0.7	0.9	0.1
12 or more years complete	61.7	18.1	0.1	0.4	0.7	0.0

Source National Family Health Survey (NFHS-4) 2015–16
* Decimal figures are in percentage to total women aged 15–49 having postnatal checks for the last live birth

need, and where to reach services when needed. Basic care for all newborns should include promoting and supporting early and exclusive breastfeeding, if possible, keeping the baby warm, increasing hand washing, and providing hygienic umbilical cord and skin care.

In India, according to NFHS4, only 27% of newborns had a first postnatal check-up within the first 2 days after birth. Sixty-four percent of newborns did not receive any postnatal health check-up. There is a positive relationship between the mother's level of education and a postnatal check-up in the first two days after birth. Thirty one percent of babies born to mothers with 12 or more years of schooling received a postnatal check-up within two days, compared with 22% of babies born to mothers with no schooling.

Table 24.5 shows the timing of postnatal check-ups of the newborn by their mother's year of schooling in urban areas. Around 25.7% of new-born babies had postnatal check-ups within 4 h among the women having 12 or more years of complete schooling. This share declines with a decline in mother's year of schooling. Only 16.9% of new-born babies get their first postnatal check-up within 4 h of birth among the women having no schooling. Overall, almost 30% new-born babies had their first postnatal check-up within 2 days of birth among women having 12 or more years of complete schooling. This share declines to 20% among the women having no year of schooling. This indicates that education of mother plays an important role in post-natal check-up of the newborn. From the Table 24.5 we can say that percentage of postnatal check-up of new-born within 2 days increases considerably with increase in the years of schooling of mother (Table 24.6).

Table 24.5 Percentage distribution of new-born by the timing of first postnatal check-up according to mother's year of schooling in Urban India 2015–16

Years of schooling	Timing for first postnatal check for the new born					No postnatal check
	Less than 4 h	4–23 h	1–2 days	3–41 days	Don't know	
No schooling	16.9	1.1	3.9	7.5	0.4	70.2
< 5 years complete	20.9	1.5	4.6	10.1	0.4	62.5
5–7 years complete	21.4	1.3	4.6	9.6	0.3	62.8
8–9 years complete	23.2	1.2	4.4	9.9	0.3	61
10–11 years complete	24.4	1.4	4.3	9.2	0.4	60.4
12 or more years complete	25.7	1.2	3.7	8.3	0.3	60.9

Source National Family Health Survey (NFHS-4) 2015–16

Table 24.6 Percent distribution of last births in the five years preceding the survey by type of provider for the newborn's first postnatal check during the two months after the birth, according to background characteristics, India, 2015–16

Years of schooling	Types of health provider of new-born baby's first postnatal check						No postnatal check
	Doctor	ANM/nurse	Other health personnel	Dai	ASHA	Others	
No schooling	9.7	11.9	0.3	2.2	5.1	0.6	70.2
< 5 years complete	15.5	13.2	0.3	1.7	6.3	0.6	62.5
5–7 years complete	17.2	12.7	0.2	1.3	5.3	0.4	62.8
8–9 years complete	19.4	12.8	0.2	0.9	5.3	0.4	61
10–11 years complete	23.9	10.9	0.1	0.6	3.6	0.5	60.4
12 or more years complete	26	9.3	0.1	0.5	3	0.3	60.9

Source National Family Health Survey (NFHS-4) 2015–16

The type of health provider for postnatal check is an important indicator in the context of the child's health. Doctors are considered as most skilled health providers. The percentage of postnatal check of new-born babies by doctors is highest among the women having 12 or more years of schooling. About one fourth postnatal check-ups of new-born babies were taken by doctors in the category of 12 or more years of schooling of mother. This share is only about 10% in the case of no schooling of mother. The percentage share of Dai for postnatal checks of newborn was highest among the women having no year of schooling while it was lowest among the women having 12 or more years of schooling.

24.4.7 *Role of Educational Attainment of Mothers in Child Immunization*

Immunizing children against vaccine-preventable diseases can greatly reduce childhood morbidity and mortality. Despite sincere efforts of the government to immunize children, there are many sections of the society where child immunization is far from satisfactory. The role of education of mothers is critical in the immunization of the babies. It has been widely regarded that the education of women must be considered as an important factor behind immunization of children. Deasai and Alva (1998, p. 79) have critically analyzed the causal relationship between child health and female education in a study encompassing twenty two developing countries. The dependent variables for the study are child mortality, immunization status, and height for age whereas the independent variables are mother's education, age of children, education of father, area of residence (rural/ urban), marital status, and some environmental factors. The primary education of mother is closely linked with data of infant mortality in 11 countries while the secondary education of mother is correlated with infant mortality in 15 countries. The pioneer work of Cleland and van Ginnekan (1988) made the researchers aware of the strong link between the health of the child and the education of mothers.

Children aged 12–23 months who received specific vaccines at any time before the survey (according to vaccination card or the mother's report). To have received all basic vaccinations, a child must receive at least: one dose of BCG vaccine, which protects against tuberculosis, three doses of DPT vaccine, which protects against diphtheria, pertussis (whooping cough), and tetanus, three doses of polio vaccine and one dose of measles vaccine.

Table 24.7 Percentage of children aged 12–23 months who received specific vaccines at any time before the survey by mother's level of educational attainment in Urban India, 2015–16

Years of schooling	Types of vaccination					All basic vaccinations
	BCG	Hepatitis B1	DPT	Polio	Measles	
No schooling	86.4	53.5	68.3	85.7	72	51.5
< 5 years complete	92	64.8	78.8	91.6	80.5	63.2
5–7 years complete	92.5	62.1	78.4	91.5	81.4	61.2
8–9 years complete	93.8	65.3	82	92.7	84.8	65.9
10–11 years complete	94.5	68.4	82.8	92.8	85.6	67.6
12 or more years complete	95.5	69.5	86.1	94.1	88.1	69.7

Source National Family Health Survey (NFHS-4) 2015–16

Table 24.7 shows the coverage for each of the basic vaccinations among children aged 12–23 months. Coverage was highest for the BCG vaccine and lowest for the third dose of polio vaccine. Vaccination coverage increases with increasing level of mother's schooling. 70% of children aged 12–23 months whose mothers have 12 or more years of schooling have received all basic vaccinations, compared with 52% of children whose mothers have no schooling. It is apparent that the educated mothers are more concerned with the timely vaccination of their babies.

24.4.8 *Spatial Distribution of Immunization Among Children Aged 12–23 Months in Urban India*

India is a land of diversity with thousands of dialects, languages, cultural practices, and religious beliefs. One can identify geographic clustering of certain groups in certain regions. Therefore, regional variation is well marked in terms of child vaccination due to the differential implementation of health policies in different states/regions. The religious and caste status of individuals also play an important role in these diversities.

Figure 24.1 shows the state-wise variation in the pattern of vaccination of children in the age group of 12–23 months in 2015–16. It is obvious from the figure that the southern, eastern, and northern regions of the country have better immunization coverage than the western, central, and north-eastern regions of India. The socio-cultural and economic diversity is the chief factor behind this regional variation.

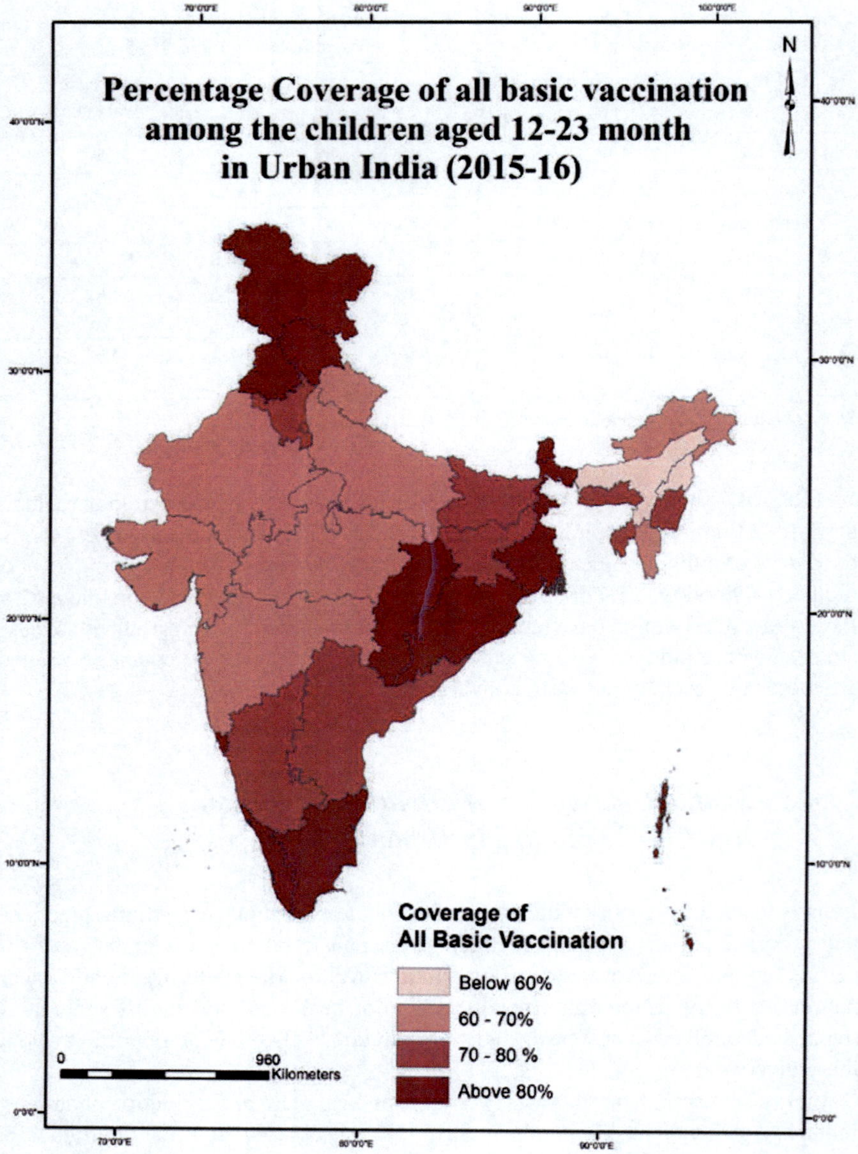

Fig. 24.1 Coverage of all basic vaccination among the children aged 12–23 months in urban India (2015–16)

24.5 Conclusion

The mother's educational attainment is highly correlated with the place of delivery. Ninety-five percent of births to mothers with 12 or more years of schooling were delivered in a health facility, compared with 62% of births to mothers with no schooling. More educated women are much more likely to undergo caesarean deliveries. The caesarean rate for deliveries for women with 12 or more years of schooling is 34%, compared with 26% for women with 10–11 years of schooling, 15% for women with 5–7 years of schooling, and 6% for women with no schooling.

There is a positive relationship between the mother's level of education and a postnatal check in the first two days after birth. Thirty-one percent of babies born to mothers with 12 or more years of schooling received a postnatal check-up within two days, compared with 22% of babies born to mothers with no schooling. The percentage of postnatal check of newborn by doctors is highest among the women having 12 or more years of schooling/education. This share is only about 10% in the case of no schooling of mother. The percentage share of Dai for postnatal checks of newborn was highest among the women having no year of schooling while it was lowest among the women having 12 or more years of schooling.

Vaccination coverage increases with increasing level of mother's schooling. 70% of children aged 12–23 months whose mothers have 12 or more years of schooling/education have received all basic vaccinations, compared with 52% of children whose mothers have no schooling. The analysis of data proves that the effect of maternal education on child vaccination is significantly positive.

One may draw substantive conclusion from the above that education is an empowering tool for women. An educated mother is more aware of her own well-being as well as the health of her child. The education of women is of utmost importance in developing and under-developed countries where the level of female education remains alarmingly low despite global consensus to work proactively in this field.

References

Animaw, W., Taye W., Merdekios, B., Tilahun, M., Ayele, G. (2014). Expanded program of immunization coverage and associated factors among children aged 12–23 months in Arba Minch town and Zuria District, southern Ethiopia, 2013. *BMC Public Health*, 14–464.

Baker, D. P. (2010). *Facts, attitudes, and health reasoning about HIV and AIDS: explaining the education effect on condom use among adults in sub-Saharan Africa AIDS Behav.*

Balogun, S. A, Yusuff, H. A., Yusuf, K. Q., Al-Shenqiti, A. M., Balogun, M. T., & Tettey, P. (2017). Maternal education and child immunization: the mediating roles of maternal literacy and socioeconomic status. *The Pan African medical Journou, 26* (2017)..

Barrera, A. (1990). The role of maternal schooling and its interaction with public health programs in child health production. *Journal of Development Economics, 32*, 69–91.

Caldwell, J. (1979). Education as a factor in mortality decline: An examination of Nigerian data. *Population Studies., 33*, 395–413.

Cleland, J. G., & van Ginneken, J. K. (1988). Maternal Education and Child Survival in Developing Countries: The Search for Pathways of Influence. *Social Science and Medicine., 27*(12), 1357–1368.

Desai, S., & Alva, S. (1998). Women education and child health: Is there a strong casual relationship? *Demography, 35*(1), 71–81.

Fatiregun, A. A., & Okoro, A. O. (2012). Maternal determinants of complete child immunization among children aged 12–23 months in a southern district of Nigeria. *Vaccine, 30*(4), 730–736.

Goudar, S. S., Goco, N., Somannavar, M. S., Vernekar, S. S., Mallapur, A. A., Moore, J. L., Wallace, D. D., Sloan, N. L., Patel, A., & Hibberd, P. L. (2015). Goldenberg Institutional deliveries and perinatal and neonatal mortality in Southern and Central India. *Reproductive Health, 12*, S13.

International Institute for Population Sciences (IIPS). (2017). *ICF. National Family Health Survey (NFHS-4), 2015–2016*. India IIPS. http://rchiips.org/NFHS/NFHS4Reports/India.pdf

Ministy of Health (2017). *Eritrea National EPI coverage survey*.

Racine, A., & Joyce, T. (2007). Maternal education, child immunisations and public policy: Evidence from the US national immunisation survey. *Social Science and Medicine, 65*, 1765–1772.

Registrar General of India. (2009). *Compendium of the Vital Statistics, in India (1970–2007: Sample Registration System)*, pp. 1–212. RGI, Ministry of Home Affairs: New Delhi, India

Strauss, J., & Thomas, D. (1998). Health, nutrition and economic development. *Journal of Economic Literature, 36*(2), 766–817.

UNICEF/WHO. (2022). *Database of skilled health personnel, based on population based national household survey data and routine health systems.*

World Bank (2019) Level and trends in child mortality. *World Bank*, UN Inter-agency Group

WHO. (2005). *Global immunization vision and strategy 2006–2015* [Internet]. WHO. http://www.who.int/immunization/givs/en/

Chapter 25
A Study of Lifestyle Behaviours of In-Migrants in Asansol City, West Bengal, India: An Aspect of Urban Health

Priyanka Thakur and Braj Raj Kumar Sinha

Abstract Lifestyle generally signifies for a particular habit, attitude, behaviour, practice, and a way of life adopted by a particular person or persons in day-to-day life. Rapid urbanization and modernization have a great impact on the way of life of people in both the developed and developing regions. In-migrants in the process of adjustment of their life to the milieu of recipient or destination place often change their food, drinking, smoking and chewing habits, and the habits or practices of wearing cloths, using leisure time, using cosmetics, mode of entertainment, involvement in physical activities, using types of vehicle, etc. All these lifestyle habits play both the positive and negative role in health and well-being of the concerned person(s). Therefore, lifestyle of in-migrants, now-a-days, emerges as an important subject of research in the field of geographic discipline.This study aims to examine the pattern of lifestyle of In-migrants across different age groups, social groups, educational attainment categories, occupational and income categories by gender in the context of spatial location and time. The study is mainly based on the lifestyle-related primary data collected through questionnaire from purposely selected 300 respondents during a field survey conducted in 2019–2020. Appropriate statistical methods and relevant cartographic techniques have also been used for analyzing the pattern of lifestyle of the in-migrants. Expected findings show that particular lifestyle varies through demographic, social, cultural, and economic conditions of the surveyed in-migrants. This study is expected to set a new direction in the field of further research on lifestyle, health, well-being, and quality of life of the people in both the developed and developing regions.

Keywords Lifestyle habits · Rapid urbanization · Unhealthy lifestyle · Chronic diseases · De-addiction centres

P. Thakur (✉)
Department of Geography, Banaras Hindu University, Varanasi 221005, India
e-mail: priyanka.thakurbhu@gmail.com

B. R. K. Sinha
Former Professor, Department of Geography, Banaras Hindu University, Varanasi 221005, India

B. R. K. Sinha (ed.), *Urban Dynamics, Environment and Health*,
https://doi.org/10.1007/978-981-99-5744-6_25

25.1 Introduction

In recent decades, lifestyle has become one of the focal themes of research worldwide in the area of health studies as most of the chronic diseases, which are on the rise, are caused by the lifestyle behaviours of the people. Lifestyle simply means the "living style of an individual or society" which changes from place to place in a given environment over time. Lifestyle of an individual is a matter of expressing choice, habit, belief, leisure activity, opinion, and interest (Ginsberg, 1975, p. 668; Jensen, 2007, p. 68; Jagadeeswari, 2005, p. 21; Utama, 2021, p. 19). Yadav (2015. p. 11) opined lifestyle as an attitude, behaviour, practice, and a way of life that a person follows in their daily life. The words Habit and Lifestyle are used interchangeably. When a person used to do or practice life-related activity again and again then it becomes a regular habit which leads to a certain situation in which the concerned person leads his life. Jensen (2007, pp. 65–68) considered habit as a repeated practice of any activity or a way of habitual life that become a style or style of living. Katz-Gerro and Shavit (1998, p. 370) after studying the works of Rojek (1985) and Featherstone (1991) suggested lifestyle as an expression of the pattern of one's food consumption, use of apparels (western or traditional), travelling, spending leisure time, way of performing certain works, games and sports, chewing, smoking, drinking, and the like. Unhealthy lifestyle leads to several lifestyle-borne diseases. Lifestyle is also related with the type of physical activities (Yoga, exercise, dancing, running, walking, jogging etc.) performed or adopted by a particular person or persons in daily life. In other words, a pattern of lifestyle is derived from the existing and potential resources and opportunities of social, cultural, economic, political, and environmental types that encourage or discourage the individual's motives, action, attitude, desire, and needs to migrate from one place to another to enjoy life. Lassetter and Callister (2008, p.100) rightly pointed out that an opportunity of employment and promotion in it generally leads the person(s) to migrate which in turn enhances the socio-economic status and lifestyles of the migrants. Hence lifestyle and migration are intertwined. Rapid expansion of urbanization or westernization and growth of automobile industries to meet the economic demand of cities modify lifestyles of the migrants as well as increase the emission of carbon dioxide to pollute the air to lead to outbreaks of several diseases. All this results in an increased risk of mortality from chronic lifestyle-related diseases (Allender et al., 2008, p. 939; Mishra & Ganda, 2007, p. 697). Thus, migration turns to shift in access to better lifestyle, including diet, physical activity, and ultimately health of in-migrants (Wandel et al., 2008, p. 376).

In the opinion of Mead (1962) eating habits are generally considered to be habitual decision or a response to social and cultural pressures to choose, consume and use the food available to an individual or group of people, in Singh (2013, p. 22). Physical activity is an important factor that helps in reducing the risks of cardiorespiratory disease and promotes muscle strength, flexibility, motor skills and agility (Nowicka and Flodmark, 2007, p. 39). Caspersen et al. (1985, p. 126) also stressed on physical activity such as yoga and exercise as beneficial for health.

The habits of chewing betel leaf with or without tobacco and smokeless tobacco locally known as Khaini, Gutkha (generally made up of betel nut, slaked lime, paraffin wax, catechu with sweet or savoury flour) etc. are very common in Asansol city. In this study, a detailed analysis of lifestyle habits such as chewing, smoking, drinking, and physical activity has also been done. Smoking hemp and Beedi (a thin cigarette filled with tobacco flakes and wrapped in a Tendu leaf with a string or adhesive at the end) affects the health of consumers in terms of heart, lung, and kidney failure. The habit of consuming alcoholic beverage prepared by the fermentation of grains, fruits, or other sources of sugar is also prevalent in the city. It contains ethanol and is very harmful to health of the consumers.

25.2 Concept and Meaning of Lifestyle

Sinha (2005, p. 403) opined that concept provides us ideas or facts about the given subject and sets a new direction for further study. Katz-Gerro and Shavit (1998, p. 370) used the concept of lifestyle to denote a unique or distinctive style of life of a specific group of people. Pulkkinen and Kokko (2000, p. 450) are of the opinion that the concept of lifestyle describes about how to live a life or how someone wants to live his/her life or the way of life. The word lifestyle or way of living is derived from the Latin expression *modus Vivendi* the meaning of which is a way or the manner of living. It should be taken in terms of value and attitude (Jensen 2007, p. 63). Further in 1924, Alfred Adler coined the term "Style of life" and further elaborated that when people set a goal for himself/herself and follow certain methods to achieve that goal, it is called lifestyle (Jensen, 2007, p. 67). In 1978 Max Weber emphasized that there was a link between people's social status, consumption patterns, needs, and desires and according to him lifestyle is the way people live their lives, usually based on membership of a status group (in Veal, 2001, p. 360). It is Bourdieu who combined cultural aspects of lifestyle and in 1984 made an attempt to link it with geography (in Chowdhury (2017, p. 4). In short, lifestyle is a unique pattern of living of people which is determined by social, cultural, economic, political, environmental, and psychological factors. According to Dima-Cozma et al. (2014, p. 111) lifestyle is a complex behavioural strategy or routine, attitude, norm, and value of life that an individual or group of persons use to adopt in the society.

Jensen (2007, p. 67) mentioned the definition of lifestyle of Giddens as propounded by him in 1991. Giddens suggested lifestyles as routine practices included in way of dressing, eating, and acting in a given environment conducive to encounter others; but the routines followed remain clearly open to change in light of the dynamic nature of individual.

World Health Organization (1998) defined lifestyle as a way of living based on recognizable patterns of behaviour in which personal characteristics of an individual, social interactions, socio-economic and environmental living conditions play a significant role. Yadav (2015, p. 11) adopted the definition of lifestyle as given by the National Family Health Survey which includes working or living conditions of people, their food habits, smoking, using alcohol, chewing tobacco, spending leisure time in video parlours or travelling. In general, a person's lifestyle is a unique way of living in terms of opinions or choices, attitudes, behaviours, physical activities, food habits, dressing, chewing, smoking, drinking, and spending time in leisure activities. Most of such unhealthy lifestyle behaviours harm the health of the urban people.

25.3 Determinants of Lifestyle

The determinants of lifestyle behaviour are another important aspect of this study. Determinants of lifestyle are those factors which normally influence the nature or behaviour of a person in his/her regular life activities. Lifestyle factors are generally related to the social, cultural, economic, demographic, political, and environmental conditions in which a person lives. Such conditions, not necessarily in combination, but also individually play a significant role in determining the lifestyle of a person or persons or people of a community in a given space and time. In broader sense, ***social aspect*** includes social interaction, social events, social groups, social status, marital status (unmarried, married, widow/widower, divorced/separated), and types of society such as primitive and modern society. Fan et al. (2015, p. 5) stated that living alone or remaining unmarried, and spending widowhood also affect subsequently cognitive deterioration and lifestyle. Such social components determine the nature, attitude, and behaviour of people in general and in-migrants in particular. ***Cultural attributes*** such as literacy and education, language, means of Information and Communication Technologies (ICTs) such as television, radio, computer, internet, mobile, religion and religious tradition, custom, belief, taboo, medical facilities, etc. significantly affect and modify the lifestyle of the people. For instance, Thorpe et al. (2019, p. 2) found that a person having a higher level of education is more aware and likely to be accustomed of improved dietary pattern. Higher education is directly related to the wages of the individual and better income and in turn related to better living style. After studying the same situation in China, Yang et al. (2015, pp. 10–11) also supported the above statement by saying that better education leads to higher income, entertainment, and other home services and lifestyle changes. In general, people with poor education and low economic status were viewed as smoking, chewing tobacco, and drinking alcohol (Palosuo, 2000, p. 1329), low consumption of fruits and vegetables, and almost no interest in physical activity (Palosuo, 2000, p. 1330) compared to highly qualified and wealthy individuals (Kim et al. 2004, p. 190). Employment status is linked to a healthier lifestyle for people. Daily wage workers have the highest frequency of alcohol consumption compared to white-collar workers (Yang et al., 2015, p. 11; Palosuo, 2000, p. 1329). Similarly, a particular religion normally has a deep spiritual feeling and faith in his religion-based customs which help in promoting a healthy lifestyle. ***Economic aspects*** such as occupation, income, resources, unemployment, poverty etc. also contribute a lot to day-to-day lifestyle of a person. In other words, economic status is the most important factor that determines lifestyle of a person in the society.

Frankish et al. (1998, p. 293) rightly noted that a better job and higher income also change the lifestyle of a person. Apart from these, political aspects such as plans, policies, programs of the governments also modify and promote people's lifestyle. In the same way ***political stability*** provides a safe and peaceful place to live during the wars or rights, which ultimately improves lifestyle of the concerned persons. Floods, droughts, earthquakes, tsunamis, landslides etc. adversely affect the lifestyle of the people. Similarly, ***demographic aspects*** like age and gender among other demographic components play a very significant role in promoting certain lifestyle. Lifestyle depends to a great extent on the age brackets and gender. For instance, sometimes persons (especially children below fifteen years) learn and opt for bad habits or lifestyle after coming in association with wrong companions in the society. The youth in the age group of 15-30 years are more eagerly changing their food habits from traditional to unhealthy food like cold drinks, snacks, pizza, pasta, and noodles as compared to the older age group (Cockerham, 2000, p. 1315). Women are generally more aware than their counterparts about healthy eating habits such as fresh and seasonal fruits, green leafy vegetables, getting physical exercise, and less consumption of any form of tobacco including alcohol (Vari et al., (2016, pp. 160–161). Generally consumption of unhealthy diet, excess physical activity and use of excess alcohol, drugs and narcotics, stress-anxiety-depression, obesity, wrong association in the society, poverty, etc. also influence the lifestyle of the people as well as migrants.

25.3.1 Migration–Concept and Meaning

Sinha (2005, p. 403) was of the view that concept provides us the underlying ideas or facts about a given subject and sets a new direction for further study. The term "migration" is derived from the Latin word "migrare", which means to change in one's place of residence. The UN (1984) expressed the concept of migration and stated that when people permanently move from their place of birth or from the place of their last residence to another geographically defined place, it is called migration. A place or space can be administratively defined in terms of geographical or political/ administrative line or boundary, for instance, a village, a town, or a city within the country of residence or outside the country and movement in it for more than a year is considered migration (Khan, 1981, p. 2). Other eminent scholars from different regions have tried to define migration from their point of view. Some of them are given below. Similarly, Ross, (1982, p. 448) was of the same opinion about the concept of migration. He explained migration as an act of geographic mobility of human beings and the change of their residential place between defined political areas of different types at a given period of time. A well-known scholar called Weeks (1989, p. 186) had the similar opinion about the definition of migration. He discussed that migration is the process of the permanent transfer of an individual from a geographically defined dwelling place to another geographically defined area and in this process a person due to social or economic motives crosses one or more than one political boundary of village or state or nations. Zelinsky (1971, p. 223) considered migration, not just an unequivocal biological event that occurs from one geographic space to another but here he also tried to make bondage among other socio-cultural influencing factors that determine the volume and pace of migration. Clarke (1965, p. 123) contended that there is no unanimity over the meaning of migration, though many consider it as a movement involving a change for residence for substantial duration.

25.4 Urban Lifestyle, Migrants and Health: An Interrelationship

Urban lifestyle, migrants, and their health are the three major dimensions in the field of migration studies and these three are completely intertwined with each other as shown in Fig. 25.1. The people who enter into a town or city from its surrounding as well as from other distant areas with the purpose of permanent settlement are called In-migrants and those who go outside from the concerned town or city with the same declared choice of permanent settlement are called out-migrants. Urban Lifestyle or lifestyle in general is now-a-days a burning global issue of the twenty-first century as most of the non-communicable diseases worldwide are found as the result of unhealthy lifestyle behaviours of the people in general and in-migrants in particular. The major components of lifestyle are food habits, drinking, sleep, smoking, chewing tobacco, physical activity, etc. These greatly affect an individual's health in which in-migrants are also included. In-migrants or out-migrants sometimes work as important carriers of infectious viruses and help spread the communicable diseases. The worldwide spread of COVID-19 is an exemplary example of this. Leaving aside the case of communicable diseases through migration, one can say that health is positively/negatively related to the migrant's lifestyle behaviour. In fact, migration and lifestyle are very much related to each other and to understand the role of migration in the context of lifestyle is important because migration plays a great role in determining the way of life of the migrants moving from their original place of abode to the new destination of their declared choice (Benson & O'Reilly 2015, pp. 12–13). Both migrants and their lifestyles are not constant rather they are dynamic in nature. Ottesen and Wandel (2012, p. 1) after studying changes in the dietary habits of immigrants and its consequences on health, they found that the immigrants after coming to the place of their destination normally change their earlier food habits and other lifestyles in course of time by adopting the culture of the majority of the host people through the ongoing mutual exchange of ideas and habits with them. This in case of some immigrants becomes the reason of obesity harming health. Casali et al. (2015, pp. 4–5) were of the similar opinion in case of particularly women immigrants and changes in their food habits and other lifestyles after reaching to the place of their new destination for permanent living. The study of Yang (2015, p. 10) clearly exhibits that there is a positive relation between migration and lifestyle as migration augments the income of the migrants which helps in promoting either unhealthy

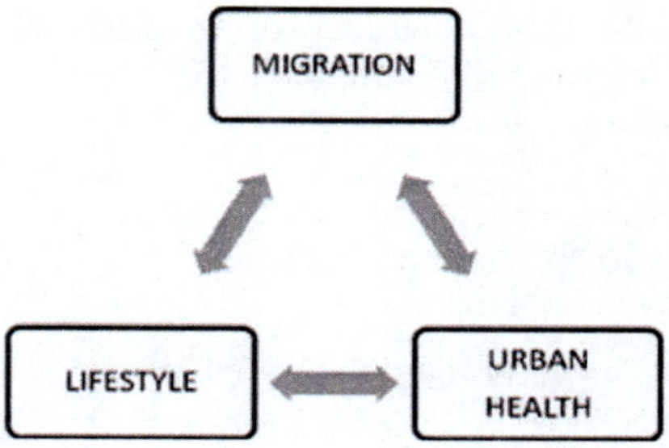

Fig. 25.1 Synergetic relationship among migration, lifestyle and urban health

or healthy lifestyle. In the same way, Wandel et al. (2008, p. 376) also mentioned about the relationship between in-migrants and changes in their food habits. The in-migrants often adopt the food habits and other lifestyles in the line of the food habits of the local people of the place of their destination after coming in contact with them.

In support of the above statement Mishra and Ganda (2007, p. 697) after studying several developed and developing countries, came to a conclusion that migration leads to sedentary lifestyle, which increases the risk of chronic diseases, metabolic disturbances, and lifestyle-related diseases. Ottesen and Wandel (2012, p. 1) studied the works of Abate et al. (2001) and Jenum et al. (2005) and found that some ethnic groups were suffered from obesity, Type 2 diabetes and cardiovascular diseases after migration. It means, healthy or unhealthy habits adopted by the people in their daily life or living become the reasons of several chronic diseases.

Both lifestyle and health are closely related to each other. The status of health of a person is always determined by his/her healthy/unhealthy lifestyle. The World Health Organization also stated that health-related quality of life highly depends on one's lifestyle behaviour (Ziglio et al., 2004, pp. 204–206). For instance, practising regular physical activities is likely to control several chronic diseases and metabolic disturbances. Cancer, cardiovascular diseases, diabetes, stress and anxiety, obesity, hypertension etc. can be checked to a large extent by avoiding chewing tobacco and smoking, alcoholic consumption, unhealthy eating habits, irregular sleep, and by avoiding other bad habits. Kruger and Chang (2008, p. 892) found the same situation when studying the health behaviours and lifestyle of the US population, that high levels of physical inactivity, higher consumption of cigarettes and alcohol increase the effects of stress and anxiety and several chronic diseases. Taking comparable example from the Moscow and Helsinki, Palosuo (2000, p. 1329) found that lifestyle-behaviours relating to smoking, drinking, diet, physical activity, leisure time, sleep and the like seriously affect the health of the urban people.

In fine, the status of health is directly or indirectly influenced or maintained by the good or bad lifestyle of an individual. In this study, we exclusively concentrate on dealing with the lifestyle of in-migrants and its impact on their health in an urban setting which could be partially different from the lifestyle and health of the rural people globally and in context of the developing countries like India in particular as the lifestyle of the urban people is much unhealthier than that of the rural people. In this context the study of the relationship among urban lifestyle, migrants, and health can obviously considered as a perfect example of an aspect of urban health.

25.5 Study Area

The study area "Asansol City" is a part of Burdwan District of West Bengal in Eastern India. It is located between 23° 24′ and 23° 53′ North latitude and 86° 48′ to 87° 32′ East longitude. The total population of this City as per 2011 Census was 563917. The name Asansol is a combination of "Asan", one of the species of tree that typically grows to about thirty metres tall, found on the bank of the Damodar River; and "sol" refers to land. It is a segment of Rarh region of West Bengal. It is about 210 km. North West from proper Kolkata City of India. Asansol is quite economically developed as it is surrounded by mines of several minerals mainly coal, manganese, iron-ore, lime-stone etc. The lifeline of communication and transportation is NH 19 which connects Asansol with Kolkata and New Delhi. Asansol is mainly dependent on its steel and coal industries, railways, trade, and commerce.

25.6 Objectives

The objectives of this study are to examine the pattern of lifestyle of in-migrants: by age group and sex; social group and sex; level of educational attainment and sex; income group and sex; across economic activities and sex and types of lifestyle by disease and sex respectively.

25.7 Methodology

Methodology is an important part of any research as it helps in collecting information, classifying the data; interpreting and analyzing the facts and logically and scientifically arriving at certain conclusions. It involves field survey, sources of data, method of survey and sampling design, use of tools and techniques in achieving the defined objectives. In this study, the sample units for the study were families who had migrated to Asansol City during the last 50 years. However, purposive random sampling technique was adopted in selecting the respondents. Therefore, the study is mainly based on the primary data collected through questionnaire in 2019–2020 at the household level. Asansol City consists of 50 wards but to accomplish the objectives all 50 wards were divided into five zones- north, south, east, west, and centre. From each zone two wards were randomly chosen and while doing so a total of ten wards from all five zones were selected. After this, 30 respondents (one respondent from one household) by covering 06 respondents from each social group namely General (higher), Other Backward castes (OBC), Scheduled Caste (SC), Scheduled Tribe (ST), and Muslim were randomly selected. While selecting respondents only those persons were taken into consideration whose age was 15 years or more. Thus, a total of 300 in-migrant respondents (141 males and 159 females) from 300 households were covered who presented information about their five types of lifestyle such as food habits in terms of single response; chewing tobacco, smoking, drinking alcohol, and regular physical activity in terms of single as well as more than one response. The study based on the information of 300 respondents will be the representative of the whole study area. The data were analyzed using the SPSS package version 22. The entire study is based on the percentage statistical technique.

25.8 Analysis

This section deals with the pattern of lifestyle behaviours of in-migrants in Asansol City at age group, gender, social group, and educational level as well as at income and occupational level.

25.8.1 Pattern of In-Migrant Respondents by Age and Sex

Table 25.1 evidently shows the pattern of percentage distribution of in-migrant respondents at >15, 15–34, 35–59, and 60 and above year age groups for the year 2019–2020. Out of 300 total in-migrant respondents 47% were males and 53% females. Slightly higher percentage of females might be due to marriage migration. A high percentage of both males (53.2) and females (53.5) to total male or female respondents in the age group of 35–59 years in comparison to that of other age groups is observed as a result of greater mobility of persons to the City in the earlier decades when the opportunity of employment in different sectors of the City was more in comparison to the recent decades. This is on the basis of respondent's response. This may also be due to greater length of time in this age group compared to that of other age groups. It has also been observed that the rate of in-migration in the recent years has declined due to overall reduction in the production of coal in the collieries and the closure of the coal plants. No new investors find better opportunities of opening new firms, industries, or business establishment that can attract the migrants to the City. The second higher percentages of male and female immigrants are in the age bracket of 15–34 years. The picture of the pattern of such percentage in this age group in-migrants may be due to some certain reasons such as many people from surrounding rural areas send their children with the purpose of achieving better education in the City, family moved from rural to urban or urban to urban areas, as well as more female migration through marriage at this age group. 60 and above year age group people had in-migrated before sixty years either alone or with their family in search of jobs or starting commercial activities.

Table 25.1 Pattern of Age-sex composition of in-migrant respondents in asansol City

Age group	Sex of In-migrants		Total
	Male	Female	
15–34	49	66	115
	34.8*	41.5	(38.3)
35–59	75	85	160
	53.2	53.5	(53.3)
60 and Above	17	8	25
	12.0	5.0	(8.3)
Grand total	141	159	300
	(47.0)	(53.0)	(100.0)

Source Authors Field Survey 2019–2020
Note * Unbracketed lower figures are the percentages to grand total males (141) or females (159)
+ Bracketed lower figures are the percentages to Grand Total of males and females

25.8.2 Pattern of Lifestyle-Related Responses of In-Migrant Respondents by Age Group and Sex

This section deals with the pattern of lifestyle-related responses furnished by the in-migrant respondents during the field survey at age group and sex level. The lifestyle-related responses have been categorized into two major groups, namely food-related habits (persons having breakfast+ 2 meals/day and persons having only 2 meals / day) as well as other habits/lifestyles comprising chewing tobacco, smoking, drinking alcohol, and practising regular yoga and exercise as mentioned in Table 25.2. The former group includes single response of each respondent regarding food habits, while the later group includes one or more than one response given by the concerned respondent regarding other than food habits. After looking into the detailed picture of the pattern of lifestyle-related responses it is found that 76.6% males and 86.8% females to total male or female respondents used to have breakfast + 2 meals/day, whereas 23.4% males and 13.2% females were found to eat only 2 meals /day. This difference may be partly explained by the fact that most of the females generally remain in the house as house wife and take more care of their health as well as the health of their family members and consume breakfast + 2 meals/day in comparison to their male counter parts who are generally reluctant in having timely food rather they are more worried to complete their different economic activity related day-to-day works within or outside their houses. The age group and sex wise picture in the categories of respondents eating breakfast + 2 meals/day and only 2 meals/day is different. The percentage of respondents who used to take breakfast+ 2 meals/day are normally more from 35–59 and 15–34 year age groups and less in the 60 and above year age bracket. However, the percentage of females is highest in 15–34 year age group and the percentage of males is highest in 35–59 year age group. Similarly, the percentage of respondents/persons whose habit is to take only 2 meals per day is highest among those who belonged to 60 and above year age and less in the 15–34 and 35–59 year age groups. The lesser percentage of old persons/respondents in the category of breakfast+ 2 meals/day and higher percentage in the category of only 2 meals/day is observed due to poor appetite and digestive system caused by their ageing factors and certain type of negligence at the part of the old people's family.

After looking into the pattern of percentage of responses of the respondents in the category of lifestyle other than food habits it is found that the percentage of males having habits of chewing tobacco is generally more in 15–34 year age group and less among the males of older and old age groups, whereas the percentage of 35–59 year age group females in chewing tobacco is more but it is less among the older and old women respondents of the study area. The higher percentage of responses of female respondents from 35-59 year age group in chewing tobacco may be due to greater involvement in their day-to-day hard and low-paid economic activities especially in non-formal sectors where women normally adopt the habits of chewing tobacco from

Table 25.2 Pattern of lifestyle-related responses of in-migrant respondents by age group and sex

Types of lifestyle		Age group and sex						Total responses	
		15–34		35–59		60 and Above			
		M	F	M	F	M	F	M	F
Food habits	Breakfast + 2 meals/day	39	63	63	71	6	4	108	138
	Only 2 meals /day	79.6*	95.5	84.0	83.5	35.3	50.0	(76.6)+	(86.8)
		10	3	12	14	11	4	33	21
		20.4	4.5	16.0	16.5	64.7	50.0	(23.4)	(13.2)
Grand total respondents		49	66	75	85	17	8	141	159
		(34.8)	(41.5)	(53.2)	(53.5)	(12.0)	(5.0)	(100.0)	(100.0)
Other habits	Chewing tobacco	23	5	56	24	11	1	90	30
		44.2	20.8	34.6	43.6	35.5	20.0	(36.73)	(35.71)
	Smoking	15	2	58	6	12	–	85	8
		28.8	8.3	35.8	10.9	38.7	–	(34.70)	(9.52)
	Drinking Alcohol	7	1	34	1	–	–	41	2
		13.5	4.2	21.0	1.9	–	–	(16.73)	(2.38)
	Regular yoga and exercise	7	16	14	24	8	4	29	44
		13.5	66.7	8.6	43.6	25.8	80.0	(11.83)	(52.38)
Grand total responses		52	24	162	55	31	5	245	84
		(21.2)	(28.5)	(66.1)	(65.4)	(12.6)	(5.95)	(100.0)	(100.0)

Source Authors Field Survey 2019–2020

Note * Unbracketed lower figures are the percentages to total responses of males or females of respective age group calculated separately in food habits and other habits

+Bracketed lower figures are the percentages to grand total male or female respondents in food habits, and to grand total responses of male or female in-migrant respondents in other habits other women in the working society. Such women are generally socio-economically poor and are illiterate or have less education. The same reasons are applied in case of the higher percentage of 15–34 year age group males having habits of chewing tobacco. In case of smoking the percentage of responses of both male and female respondents normally decreases from the old (60 and above years) to the older (35–59 years) and the younger age group (15–34 years). In case of drinking alcohol the percentage of responses of the male respondents in the younger and the older age groups is relatively more than that of their female counterparts. In case of practising regular yoga and exercise the percentages of female respondents are higher than that of the male respondents in all the age groups, however the higher percentages of both male and female respondents in the old age group are due to ageing factor as most of the aged are advised to continue regular yoga and exercises to maintain their health. But higher percentages of only female respondents in the younger and older age groups is due to the facts that such women are more conscious in practising regular yoga and exercises just to keep their body fit and healthy.

In respect of lifestyle Cockerham (2000, pp. 1318–1319) believed that there is a significant impact of age on lifestyle and stated that younger individuals are more likely to use or chewing tobacco and alcohol, eat junk foods, and practice exercises in their spare time and such practices are associated with older age too. However, the pattern of such lifestyles in this study area is to some extent different due to some certain reasons. He further asserted that the frequency of chewing tobacco, smoking, and drinking alcohol generally decreases after the age of 50 and beyond due to health issues as most of the old people suffer from several chronic diseases and practice yoga and exercises.

In brief, the total number of actual male and female respondents in the study area was 141 and 159 respectively, but in the lifestyle category of other than food habits, "means in other habits", the grand total responses of 141 male respondents became 245, while the grand total responses in case of 159 female respondents remained only 84. This indicates that most of the male respondents get themselves involved in more than one category of other habits or lifestyles but the picture in case of actual female respondents is quite different as only 84 (52.8%) female respondents were found involved in some categories of lifestyle other than food habits. This is a kind of positive sign in case of better health of the females in comparison to that of the males. Also, out of the grand total responses of male or female respondents the majority of responses were from the older working age group (34–59 years) which is followed by the younger working age group of 15–34 years and by the old people of 60 and above years.

25.8.3 Pattern of Lifestyle-Related Responses of In-Migrant Respondents by Social Group and Sex

Pattern of lifestyle-related responses of in-migrant respondents also varies across social group and sex as it is evident from Table 25.3. The social group wise detailed picture of variation in the pattern of single response with respect to food habits submitted by the in-migrant males or females exhibit that the percentage of responses of in-migrant males or females generally decreases from the General to the OBC, SC, ST, and to the Muslim social groups, whereas the percentage of responses in case of the category of only 2 meals/day generally increases from the General to the OBC, SC, ST and to the Muslim social groups. This is a reflection of the socio-cultural and economic status of the respondents of different social groups. It is believed that the General and OBC people are socio-economically comparatively better off than the SC, ST, and the Muslims and normally prefer to take Breakfast + 2 meals/day while the SC, ST, and the Muslims owing to their poor socio-economic conditions and their compulsion to complete their assigned works mostly as informal workers normally skip their breakfast in the morning and have only 2 meals/day. The highest percentage of skipping breakfast in the ST females and Muslim males is observed due to their weaker socio-economic status as well as the pressure of completion of their assigned jobs mostly in case of the informal sectors. This may also be because of the facts that most of the ST females were found engaged in daily wage labour and had to leave for the workplace early in the morning. Most of the Muslim male in-migrants survive on cottage industries such as preparing paper bags, embroidering, tailoring, and other arts and crafts activities. That is why Muslim male in-migrants have very little time to have breakfast and two meals daily. In case of other social groups skimping of breakfast may be due to negligence and overburden of their works in different economic activities.

After having a close view on the social group and sex wise pattern of percentage of responses of the respondents in the category of lifestyle other than food habits (where the concerned respondent submitted one or more than one or no response in different lifestyle categories) it is found that the percentage of responses of both male and female respondents in the category of chewing tobacco (Paan (betel and betel nut), Gutkha and Khaini) generally increases in case of both males and females from the General to the OBC, SC, ST and to the Muslim social groups except few cases in exception. This pattern also indicates that chewing tobacco is more prevalence among the respondents of comparatively poor socio-economic conditions. The smoking of Cigarette, Ganja, and Beedi is prevalent in today's culture and is considered as a part of leisure activities. Smoking percentages were found higher among Muslim males (38.6), OBC males (37.9), General males (34.3), SC males (29.5), and ST males (29.5). The percentage of smoking was found higher among the women of ST (11.8) and Muslim (22.7). They generally smoke Beedi (a kind of mini cigarette locally prepared by using flake of tobacco and rolled in a leaf of east Indian ebony tree) instead of cigarettes as cigarettes are costly.This is also an indicative of poverty. In the category of drinking alcohol the percentage of in-migrant male respondents

Table 25.3 Pattern of lifestyle-related responses of in-migrant respondents by social group and sex

Types of lifestyle		Social group and sex										Total responses	
		General		OBC		SC		ST		Muslim			
		M	F	M	F	M	F	M	F	M	F	M	F
Food habits	Breakfast +2 meals/day	25	30	30	24	20	27	15	31	18	26	108	138
		83.3	100.0	85.7	92.3	74.1	84.3	71.4	79.5	64.3	81.2	(76.6)+	(86.8)
	Only 2 meals/day	5	–	5	2	7	5	6	8	10	6	33	21
		16.7	–	14.3	7.7	25.9	15.7	28.6	20.5	35.7	18.8	(23.4)	(13.2)
Grand total respondents		30	30	35	26	27	32	21	39	28	32	141	159
		(21.3)	(18.9)	(24.8)	(16.3)	(19.1)	(20.1)	(14.9)	(24.5)	(19.9)	(20.1)	(100.0)	(100.0)
Other habits	Chewing tobacco	12	6	19	1	17	6	18	6	24	11	90	30
		34.3	27.3	32.8	10.0	33.3	46.2	40.9	35.2	42.1	50.0	(36.73)	(35.71)
	Smoking	12	1	22	–	16	–	13	2	22	5	85	8
		34.3	4.5	37.9	–	31.4	–	29.5	11.8	38.6	22.7	(34.70)	(9.52)
	Drinking alcohol	3	–	10	–	12	–	9	2	7	–	41	2
		8.6	–	17.2	–	23.5	–	20.5	11.8	12.3	–	(16.73)	(2.38)
	Regular yoga and exercise	8	15	7	9	6	7	4	7	4	6	29	44
		22.8	68.2	12.1	90.0	11.8	53.8	9.1	41.2	1.0	27.3	(11.83)	(52.38)
Grand total responses		35	22	58	10	51	13	44	17	57	22	245	84
		(14.3)	(26.2)	(23.7)	(11.9)	(20.8)	(15.5)	(18.0)	(20.2)	(23.3)	(26.2)	(100.0)	(100.0)

Source Authors Field Survey 2019–2020
Note * Unbracketed lower figures are the percentages to total responses of males or females of respective social group calculated separately in food habits and other habits
+ Bracketed lower figures are the percentages to grand total male or female respondents in food habits, and to grand total responses of male or female in-migrant respondents in other habits

was higher in scheduled caste (23.5) and scheduled tribe (20.5) social groups in comparison to that of the other backward cast and general social groups. In drinking alcohol only scheduled tribe females (11.8) reported drinking alcohol or palm wine (an alcoholic beverage produced locally from the sap of various trees such as palm trees, coconuts, and date palms). These are also cheaper than that of the more refined company-based or foreign alcohols. This also shows that poor people drink more such alcoholic material, however, all these are harmful to health. The percentage of responses of male in-migrants in the category of practising regular yoga and exercise declines from the socio-economically better groups to the socio-economically poor groups. Comparatively higher percentages among the respondents of general castes and OBC are noted just because of their less physical activities in their daily life. But the picture among the women in-migrants is quite different than that of their male counterparts as the percentage of responses of female in-migrants is found fairly higher in all social groups which indicate greater awareness among the urban females towards maintaining their health. This is a positive sign from the perspective of keeping their mind and body fit.

The percentage of overall responses to total male or female in-migrants of respective social group and sex was found higher in the OBC males (23.7), Muslim males (23.3) than that of the SC males (20.8), ST males (18.0), General (14.3) whereas in case of female in-migrants the percentage of responses was higher among general casts and Muslims than that of the females of the rest social groups.

25.8.4 Pattern of Lifestyle-Related Responses of In-Migrant Respondents by Level of Education and Sex

Level of education is also an important factor that determines lifestyle behaviours of an individual and that is why the day-to-day lifestyle of persons varies from one level of education to other levels of education. It is quite evident from Table 25.4 that 100 per cent females having under graduate and post graduate and above education were found having their breakfast and two meals/day, while 94.9% in-migrant graduate males and 85.7 % post graduate males were found having breakfast and two meals per day. Such percentage of responses is followed by the respondents of higher secondary, secondary, and primary level of education. 60% illiterate females were also found taking breakfast and two meals per day. In case of only two meals/day it was found that the proportion of illiterate (40.0 %) and primary (24.2%) passed women was higher as compared to that of the women of secondary (10.0%) and higher secondary (2.9%) education who used to eat only two meals per day while the percentage of male respondents in case of only two meals per day was fairly higher towards lower level of education and lower towards the lower level of education. Briefly, the educational level wise detailed picture of variation in the pattern of single response with respect to food habits given by the in-migrant males or females clearly shows that the percentage of responses of in-migrant males or females generally increases from lower levels of education to upper or higher levels of education. Similarly, the percentage of responses of in-migrant males or females in the category of only 2 meals/day generally decreases from the lower level of education to upper and higher levels of education. Wandel et al. (2008, p. 382) stated that skipping breakfast can be a result of the daily work schedule of oneself or other family members.

Chewing tobacco, betel and betel nut, khaini, gutkha and smoking all are harmful to human health and leads to several chronic diseases. After having gone through the detailed picture across the levels of education and sex with respect to other day-to-day habits/lifestyles behaviours it is found that the percentage of responses of the illiterate respondents or the respondents with lower levels of education in the lifestyle category of chewing tobacco and smoking is fairly higher in both sexes, however, the percentage of responses of both male or female in-migrant respondents having middle or upper levels of education is generally lower leaving aside few cases in exception. This indicates the impact of levels of educational achievements of the concerned person's on the habits of chewing tobacco and smoking. Palosuo (2000, p. 1329) was also in favour of the similar opinion that illiterate and less educated women consume more tobacco whereas, Palosuo (2000, p. 1335); Cockerham (2000, p. 1319); and Nierkens et al. (2015, p. 387) opined that less educated people are more prone to smoke and use more smoking. But there is a different pattern in case of drinking alcohol as the percentage of responses of in-migrant male respondents having middle or upper level of education is generally higher which slightly decreases towards lower levels of education but again the percentage is exceptionally high among the illiterate respondents. It has been observed that illiterate and less educated persons drink especially locally produced alcohol which is quite cheaper, but the persons of upper level of education having better source of income drink especially branded alcohol which is quality wise better and costlier. With respect to practising regular yoga and exercises the percentage of responses of female in-migrant respondents generally increases from lower levels of education to upper and higher levels of education by indicating a positive relationship between the regular yoga-exercises and levels of education. More or less similar trend is noticed in case of male in-migrants too, however, the exact pattern is to some extent different. But in general, it can be hypothesized that lower is the level of education lower will be the percentage of responses and higher is the level of education higher will be the percentage of responses of both in-migrant males and females in case of practising regular yoga and exercises. Palosuo (2000, p. 1330) also found a positive association between education and exercise during his studies in Moscow and Helsinki.

25.8.5 Pattern of Lifestyle-Related Responses of In-Migrant Respondents by Per Capita Monthly Income and Sex

This section does not deal with the pattern of lifestyle-related responses of all in-migrant respondents because all of them were not economically active, means all

Table 25.4 Pattern of lifestyle-related responses of in-migrant respondents by level of education and sex

Types of lifestyle		Level of education and sex												Total responses	
		Illiterate		Primary		Secondary		Higher secondary		Under graduate		Post graduate and above			
		M	F	M	F	M	F	M	F	M	F	M	F	M	F
Food habits	Breakfast + 2 meals/day	–	12	10	25	29	36	26	33	37	28	6	4	108	138
		–	60.0	55.6	75.8	64.4	90.0	83.9	97.1	94.9	100.0	85.7	100.0	(76.6)+	(86.8)
	Only 2 meals/day	1	8	8	8	16	4	5	1	2	–	1	–	33	21
		100.0*	40.0	44.4	24.2	35.6	10.0	16.1	2.9	5.1		14.3	–	(23.4)	(13.2)
Grand total respondents		1	20	18	33	45	40	31	34	39	28	7	4	141	159
		(0.7)	(12.6)	(12.8)	(20.7)	(31.9)	(25.1)	(22.0)	(21.4)	(27.6)	(17.6)	(5.0)	(2.5)	(100.0)	(100.0)
Other habits	Chewing tobacco	1	10	14	7	38	5	16	5	18	2	3	1	90	30
		33.3	55.6	42.4	50.0	43.2	31.2	30.2	35.7	32.7	11.1	23.1	25.0	(36.73)	(35.71)
	Smoking	1	3	10	2	32	2	19	–	19	1	4	–	85	8
		33.3	16.6	30.3	14.3	36.4	12.5	35.8	–	34.5	5.5	30.7	–	(34.70)	(9.52)
	Drinking alcohol	1	–	5	1	14	–	9	–	9	1	3	–	41	2
		33.3	–	15.2	7.1	15.9	–	17.0	–	16.4	5.5	23.1	–	(16.73)	(2.38)
	Regular yoga and exercise	–	5	4	4	4	9	9	9	9	14	3	3	29	44
		–	27.8	12.1	28.6	4.5	56.3	17.0	64.3	16.4	77.8	23.1	75.0	(11.83)	(52.38)
Grand total responses		3	18	33	14	88	16	53	14	55	18	13	4	245	84
		(1.2)	(21.4)	(13.5)	(16.7)	(35.9)	(19.1)	(21.6)	(16.7)	(22.5)	(21.4)	(5.3)	(4.8)	(100.0)	(100.0)

Source Authors Field Survey 2019–2020

Note * Unbracketed lower figures are the percentages to total responses of males or females of respective educational group calculated separately in food habits and other habits + Bracketed lower figures are the percentages to grand total male or female respondents in food habits, and to grand total responses of male or female in-migrant respondents in other habits were not engaged in producing economic goods or services. This part of the study analyzes the pattern of lifestyle-related responses of only those in-migrant male or female respondents who were engaged in any type of economic function and were earning money. There were only 116 males and 30 females among the in-migrant respondents. Table 25.5 evidently shows variation in the pattern of percentage of responses provided by the concerned in-migrants male and female respondents during the field survey in the study area. After having a close look at the pattern of percentage of responses of the concerned in-migrant respondents by their monthly income regarding breakfast and two meals per day it was found that the percentages of both male and female in-migrant respondents clearly increases from the lower categories of monthly income to the categories of higher levels of income or vice versa. But opposite is the pattern with regard to only two meals per day as the percentage of responses of the concerned in-migrant male respondents obviously increases from the categories of higher levels of monthly income to the lower levels of monthly income, whereas all the six female in-migrants in this category of food habits reported their monthly up to rupees 10000. Such variation in the pattern of responses of both males and females in the categories of breakfast + two meals as well as only two meals per day clearly indicates that the types of food habits and its relative dominance in terms of percentages are largely determined by the monthly income of the respondents/persons or in other words, income can be considered as one of the important factors in determining the types of food habits and its relative dominance in the city. The prospects of earning better income in the city encourage the worldwide rural folk to migrate to the cities. Kruger and Chang (2008, p. 890), and Zhang and Xiang (2019, p. 6) also mentioned that income is positively related to the standard of living and lifestyle resources that migrants can access in times of their need. Per capita monthly income category-wise overall picture in the category of food habits shows that out of the grand total respondents the majority of male respondents belonged to the income category of above rupees 15000, whereas the majority of female respondents to grand total female respondents belonged to the income category of below rupees 15000. Such per capita monthly income variation also indicates the positive impact on food habits of the respondents.

In the category of other habits comprising chewing tobacco, smoking, drinking alcohol, and regular yoga and exercise the percentage of responses of males to the grand total responses of males in each category of per capita monthly income was exceptionally higher in the habits of chewing tobacco and smoking. The next higher percentage of responses of males to the grand total responses of them in almost each category of per capita monthly income was in the habit of drinking alcohol. In comparison to the responses of males the percentage of responses of females to the grand total responses of females in chewing tobacco, smoking, and drinking alcohol was less and they belonged to only three categories of per capita monthly income as shown in Table 25.5. In respect of regular yoga and exercise the percentage of responses of females to the grand total responses of females in respective category of per capita monthly income was exceptionally higher in comparison to that of the males. However, the majority of males who practice yoga and exercise were from among higher income categories that is from more than 15000 rupees while

Table 25.5 Pattern of Lifestyle-related responses of in-migrant respondents by per capita monthly income and sex

Types of lifestyle		Categories of Per capita monthly income												Total responses	
		<5000		5000–10000		10000–15000		15000–20000		20000–25000		>25000			
		M	F	M	F	M	F	M	F	M	F	M	F	M	F
Food habits	Breakfast + 2 meals/day	–	2	12	11	9	3	14	2	13	–	39	6	87	24
		–	66.7	48.0	68.7	64.3	100.0	77.8	100.0	81.2	–	92.9	100.0	(75.0)+	(80.0)
	Only 2 meals/ day	1	1	13	5	5	–	4	–	3	–	3	–	29	6
		100.0*	33.3	52.0	31.3	35.7	–	22.2	–	18.8	–	7.1	–	(25.0)	(20.0)
Grand total respondents		1	3	25	16	14	3	18	2	16	–	42	6	116	30
		(0.9)	(10.0)	(21.6)	(53.3)	(12.1)	(10.0)	(15.5)	(6.7)	(13.8)	–	(36.2)	(20.0)	(100.0)	(100.0)
Other habits	Chewing tobacco	1	–	23	4	13	–	14	1	10	–	26	1	87	6
		50.0	–	46.0	44.4	37.1	–	56.0	50.0	38.5	–	31.7	16.7	(37.8)	(33.3)
	Smoking	1	–	17	–	13	–	13	–	9	–	28	1	81	1
		50.0	–	34.0	–	37.1	–	52.0	–	34.6	–	34.1	16.7	(35.2)	(5.6)
	Drinking Alcohol	–	–	8	–	7	–	5	–	3	–	18	1	41	1
		–	–	16.0	–	20.0	–	20.0	–	11.5		22.0	16.7	(17.8)	(5.6)
	Regular yoga and exercise	–	1	2	5	2	–	3	1	4	–	10	3	21	10
		–	100.0	4.0	55.6	5.7	–	12.0	50.0	15.4	–	12.2	50.0	(9.1)	(55.5)
Grand total responses		2	1	50	9	35	–	25	2	26	–	82	6	230	18
		(0.9)	(5.6)	(21.8)	(50.0)	(15.2)	–	(10.9)	(11.1)	(11.3)	–	(35.7)	(33.3)	(100.0)	(100.0)

Source Authors Field Survey 2019–2020

Note * Unbracketed lower figures are the percentages to total responses of males or females of respective monthly income category calculated separately in food habits and other habits + Bracketed lower figures are the percentages to grand total male or female respondents in food habits, and to grand total responses of male of female in-migrant respondents in other habits the majority of females who practice yoga and exercise were from among lower income categories. Here, the statement of Palosuo (2000, p. 1336) that the high-income migrants practice more yoga and exercise than low-income groups, but in case of this study his statement is more applicable with regard to males rather than females. The grand total responses show that the total response in male category was 230 and in female category it was only 18. The comparative picture in this respect exhibits that the majority of responses in male category to the grand total responses of males were towards more than 15000 rupees per capita monthly income, whereas the majority of responses in case of females were towards less than 15000 rupees per capita monthly income. This also shows that the majority of females possess less income in comparison to the income of males.

25.8.6 *Pattern of Lifestyle-Related Responses of In-Migrant Respondents by Economic Activity and Sex*

Occupation or economic activity is also one of the factors that determine a particular habit or lifestyle behaviour of in-migrants. Table 25.6 exhibits the pattern of lifestyle of in-migrants by their economic activities and sex. Frankish et al. (1998, p. 293) rightly considered economic activity as one of the factors of lifestyle. After having a close look on the pattern of the percentage of responses of male or female in-migrants to total male or female in-migrants in the respective type of economic activity of in-migrants was in almost all the economic activities except livestock and fishing whereas, the majority of responses of female in-migrants in case of having breakfast and 2 meals/day were not in all economic activities. The majority of them were in teaching activities, professional and technical works, household industrial activities, others, and commercial activities in descending order. In case of only 2 meals/day male respondents were only in livestock and fishing, household industrial activities, medical professions, others, transport and communication activities, real estate activities and commercial activities whereas, female respondents were only in a few economic activities such as commercial activities, others and household industrial activities in descending order. Overall picture shows that 75.0% males and 80% females to total male (116) or female (30) in-migrant respondents were in the category of breakfast and 2 meals/day while, 25.0% males and 20% females to total male or female in-migrant respondents were in the category of only 2 meals/day. It means the majority of male and female respondents used to have breakfast and 2 meals/day.

The pattern of responses of males and females to grand total responses of males or females shows that the percentage responses of males to the grand total responses of males in the category of chewing tobacco, smoking, drinking alcohol, and in regular yoga and exercise was fairly higher than that of the responses of females in most of the economic activities of the study area. The responses of female in-migrants in the category of chewing tobacco, smoking, drinking alcohol, and in regular yoga

Table 25.6 Pattern of lifestyle-related responses of in-migrant respondents by economic activities and sex

Types of economic activity	Types of lifestyle														Total responses	
	Food habits				Total respondents		Other habits									
	Breakfast+ 2 meals/day		Only 2 meals/ day				Chewing tobacco		Smoking		Drinking alcohol		Regular yoga and exercise			
	M	F	M	F	M	F	M	F	M	F	M	F	M	F	M	F
Livestock and fshing	–	–	1	–	1	–	1	–	1	–	–	–	–	–	2	–
	–	–	100.0	–	(0.9)	–	50.0	–	50.0	–	–	–	–	–	(0.9)+	–
Household industrial activities	2	4	2	1	4	5	3	–	2	–	1	–	–	2	6	2
	50.0*	80.0	50.0	20.0	(3.4)	(16.7)	50.0	–	33.3	–	16.7	–	–	100.0	(2.6)	(11.1)
Construction works	4	–	–	–	4	–	2	–	4	–	3	–	1	–	10	–
	100.0	–	–	–	(3.4)	–	20.0	–	40.0	–	30.0	–	10.0	–	(4.3)	–
Commercial activities	30	5	15	4	45	9	37	3	25	1	10	1	6	1	78	6
	66.7	55.6	33.3	44,4	(38.5)	(30.0)	47.4	50.0	32.1	16.7	12.8	16.7	7.7	16.7	(33.9)	(33.3)
Transport and communication activities	9	–	5	–	14	–	13	–	13	–	7	–	–	–	33	–
	64.3	–	35.7	–	(12.0)	–	39.4	–	39.4	–	21.2	–	–	–	(14.3)	–
Teaching activities	8	8	–	–	8	8	7	1	5	–	4	–	3	4	19	5
	100.0	100.0	–	–	(6.8)	(26.7)	36.8	20.0	26.3	–	21.1	–	15.8	80.0	(8.3)	(27.8)
Professional and technical works	19	4	–	–	19	4	11	1	13	–	12	–	6	2	42	3
	100.0	100.0	–	–	(16.2)	(13.3)	26.2	33.3	31.0	–	28.6	–	14.3	66.7	(18.3)	(16.7)

(continued)

Table 25.6 (continued)

Types of economic activity	Types of lifestyle															
	Food habits				Total respondents		Other habits								Total responses	
	Breakfast+ 2 meals/day		Only 2 meals/ day				Chewing tobacco		Smoking		Drinking alcohol		Regular yoga and exercise			
	M	F	M	F	M	F	M	F	M	F	M	F	M	F	M	F
Administrative services	5	–		–	5	–	3	–	6	–	1	–	1	–	11	–
	100.0	–		–	(4.3)	–	27.3	–	54.5	–	9.1	–	9.1	–	(4.8)	–
Medical professionals	1	–	1	–	2	–		–	2	–		–	–	–	2	–
	50.0	–	50.0	–	(1.7)	–		–	2.5	–		–	–	–	(0.9)	–
Real estate activities	5	–	2	–	7	–	5	–	5	–	1	–	3	–	14	–
	71.4	–	28.6	–	(6.0)	–	35.7	–	35.7	–	7.1	–	21.4	–	(6.1)	–
Others	4	3	3	1	7	4	5	1	5	–	2	–	1	1	13	2
	57.1	75.0	42.9	25.0	(6.0)	(13.3)	38.7	50.0	38.7	–	15.4	–	7.7	50.0	(5.6)	(11.1)
Grand total responses	87	24	29	6	116	30	87	6	81	1	41	1	21	10	230	18
	(75.0)	(80.0)	(25.0)	(20.0)	(100.0)	(100.0)	(37.8)	(33.3)	(35.2)	(5.6)	(17.8)	(5.6)	(9.1)	(55.6)	(100.0)	(100.0)

Source Authors Field Survey 2019–2020

Note * Unbracketed lower figures are the percentages to response of males or females of respective economic activity calculated separately in food habits and other habits

+ Bracketed lower figures are the percentages to grand total male or female respondents in food habits, and to grand total responses of male or female in-migrant respondents in other habits and exercise were found only in some certain economic activities such as household industrial activities, commercial activities, teaching activities, professional and technical works and in other types of economic activities, however, their number and percentages were more in the category of chewing tobacco and regular yoga and exercise and obviously less in the category of smoking and drinking alcohol. The overall picture exhibits that the percentage of responses of males to grand total responses of males in descending order was in the categories of chewing tobacco, smoking, drinking alcohol, and in regular yoga and exercise whereas, the percentage of responses of females to grand total responses of females in descending order was in the category of regular yoga and exercise, chewing tobacco, and one (5.6%) each in smoking and drinking alcohol. Hence, the majority of females used to practice regular yoga and exercise and some had habits of chewing tobacco. In the words of Frankish et al. (1998, p. 293) both male and female in-migrant respondents from teaching activities, professionals and technical works, household industrial activities, commercial activities, real estate activities, and some other economic activities used to practice more regular yoga and exercise by sparing some times daily in their homes. This can be considered as a healthy sign for good health. The above description explains that certain occupation/economic activities have varying impacts on the lifestyles of the both the male and female in-migrants in the study area.

25.8.7 Pattern of Lifestyle-Related Responses of In-Migrant Respondents Concerning the Types of Disease by Sex

The types of lifestyle and its relation with chronic diseases of migrants (in-or-out-migrants) are now-a-days the matter of great interest for the researchers across the world because most of the chronic diseases are the results of unhealthy lifestyle. WHO (2011) rightly pointed out that non-communicable diseases such as diabetes, cancer, cardiovascular disease are caused by irregular food habits, eating unhealthy food, use of substances such as chewing tobacco (khaini, betel leaf, and betel nut with and without tobacco), smoking of cigarette, consumption of branded or non-branded alcohol and lack of physical activities. Such unhealthy lifestyles can lead a person to gastric, obesity, cancer, heart disease, diabetes, asthma, joint-pain, blood pressure (hypertension), etc. Reversely, a healthy lifestyle or healthy and timely diet are generally associated with fitness of the body and lighter body weight and control the increased prevalence of non-communicable diseases (Palosuo, 2000, p. 1331; WHO, 2008). Importantly, psychological stress and low-self-esteem in-migrants are also responsible for the increased frequency of smoking, alcohol consumption, unhealthy dietary practice, and physical inactivity that can inadvertently exacerbate several chronic diseases (Parrott, 1999, p. 1155; Salmon 2001, p. 33; Dima-Cozma et al., 2014, p. 116).

In this context, Table 25.7 presents the detailed picture of the pattern of responses of the surveyed in-migrants regarding their types of lifestyle (healthy or unhealthy) and its relation with non-communicable diseases. From this table it is found that the majority of percentage of responses of male or female respondents to their grand total responses in the lifestyle categories of breakfast+ 2 meals/day, only 2 meals/day, chewing tobacco, smoking, drinking alcohol, and regular yoga and exercise in descending order is in the gastric, blood pressure, arthritis, obesity and diabetes except in male column of regular yoga and exercise where the percentage of responses of male in-migrants was found highest in blood pressure and it was followed by gastric, obesity, arthritis and diabetes. In other words, it can be said that the majority of both male and female respondents in descending order suffer from gastric, blood pressure, arthritis, obesity, and diabetes except few cases in exception. However, some respondents suffer from other diseases such as heart disease, cancer, and asthma. Among the migrant respondents, there was only one male and one female who reported suffering from asthma.

The work of Van Dam et al., (2013, p. 1156) on food habits and its association with a number of health problems as mentioned in the below table affirmatively supports the relation of food habits and types of disease. Similarly, the statement of Vincet et al. (2002, p. 431) and Yadav (2015, p. 183) that most of the chronic diseases result from the interplay of genetic susceptibility and environmental factors, including diet, chewing tobacco, smoking, drinking alcohol and physical inactivity also supports the above facts as mentioned in the above analysis and Table 25.7. Chewing tobacco, khaini, and betel leaf and betel nuts lead to many oral diseases like cancer and other chronic diseases (Johnson, 2001, p. 330; Kuper et al., 2002, p. 455). The studies of Brandhagen et al. (2012, p. 311); Kuntsche et al. (2009, p. 1267) and Casali et al., (2015, p. 8) also supports the facts as mentioned in the above analysis and Table 25.7 because they also found that consuming beedi, ganja, and cigarette in an empty stomach causes gastric problems. Obesity, blood pressure, arthritis, etc. are all due to less physical activities. More fatty food, excess intake of sugar and salt etc. also cause obesity, diabetes, and blood pressure. It can be seen from the above analysis that chewers, smokers, and drinkers were suffering from gastric diseases, blood pressure, and obesity diseases. Practice of regular yoga and exercise always helps in improving and maintaining health.

The overall picture of the types of lifestyle and types of disease of male and female in-migrants in Asansol City shows that the percentage of responses of males to grand total responses of male in-migrants was found highest in the lifestyle category of chewing tobacco which was followed by the lifestyle category of smoking, breakfast + 2 meals/day, regular yoga and exercise, only 2 meals/day and in the category of drinking alcohol. But the pattern in case of responses of the female in-migrants was different as the percentage of responses of females to grand total responses of female in-migrants was found highest in the lifestyle category of breakfast + 2 meals/day, which was followed by regular yoga and exercise, chewing tobacco, only 2 meals/day, smoking and drinking alcohol. This indicates that the majority of male respondents in descending order were in the lifestyle category of chewing tobacco, smoking, and breakfast + 2 meals/day, while in case of female in-migrants the majority were in the lifestyle category of breakfast + 2 meals/day, regular yoga and exercise and chewing tobacco.

Table 25.7 Pattern of the types of lifestyle and diseases of in-migrants by sex

Types of disease	Types of lifestyle												Total responses	
	Food habit				Other habits									
	Breakfast+ 2 meals/day		Only 2 meals/ day		Chewing tobacco		Smoking		Drinking alcohol		Regular yoga and exercise			
	M	F	M	F	M	F	M	F	M	F	M	F	M	F
Diabetes	12	14	6	4	13	6	13	2	3		12	12	59	38
	9.8*	9.1	10.0	11.8	9.7	8.4	9.8	10.0	6.2		15.2	13.5	(10.2)+	(10.3)
Cancer	1	1			1				1		1	1	4	2
	0.8	0.6			0.7				2.9		1.3	1.1	(0.7)	(0.5)
Asthma	1	1										1	1	2
	0.8	0.6										1.1	(0.2)	(0.5)
Arthritis	19	27	12	8	23	13	23	4	7	1	13	17	97	70
	15.5	17.5	20.0	23.5	17.2	18.3	17.3	20.0	14.6	50.0	16.5	19.1	(16.8)	(18.9)
Obesity	18	27	9	4	18	12	19	4	7		14	15	85	62
	14.6	17.5	15.0	11.8	13.4	16.9	14.3	20.0	14.6		17.7	16.9	(14.7)	(16.7)
Heart Disease	7	4	3		7	3	8	1	1		8	3	34	11
	5.7	2.6	5.0		5.2	4.2	6.0	5.0	2.1		10.1	3.4	(5.9)	(3.0)
Blood Pressure	25	35	12	6	27	18	27	4	11		16	19	118	82
	20.3	22.7	20.0	17.6	20.2	25.3	20.3	20.0	22.9		20.2	21.3	(20.4)	(22.2)
Gastric	40	45	18	12	45	19	43	5	18	1	15	21	179	103
	32.5	29.2	30.0	35.3	33.6	26.8	32.3	25.0	37.5	50.0	19.0	23.6	(10.2)	(27.8)

(continued)

Table 25.7 (continued)

Types of disease	Types of lifestyle													
	Food habit				Other habits								Total responses	
	Breakfast+ 2 meals/day		Only 2 meals/ day		Chewing tobacco		Smoking		Drinking alcohol		Regular yoga and exercise			
	M	F	M	F	M	F	M	F	M	F	M	F	M	F
Grand total responses	123	154	60	34	134	71	133	20	48	2	79	89	577	370
	(21.3)	(41.6)	(10.4)	(9.2)	(23.2)	(19.2)	(23.0)	(5.4)	(8.3)	(0.5)	(13.7)	(24.1)	(100.0)	(100.0)

Source Authors Field Survey 2019–2020
* Unbracketed lower figures are the percentages to total males or females of respective lifestyle category
+Bracketed lower figures are the percentages to grand total responses of in-migrant males or females

Similarly, with respect to diseases the majority of male in-migrants were found suffering from blood pressure, arthritis, diabetes, and gastric, whereas the majority of female in-migrants were found suffering from gastric, blood pressure, arthritis, and obesity.

25.8.8 Conclusion

The overall picture that emerged from the above analysis of lifestyle behaviour-related data of 300 in-migrants shows that lifestyle behaviours of in-migrants are more or less positively related to their age, sex, social group, educational attainment, income, and economic activities. The summarized results of the study in this respect are as follows.

The main aim of this study was to find out the possible relationship between lifestyle behaviour, health-status, and socio-economic condition of the in-migrant population in the study area. This study involved data on selected lifestyle such as food habits, chewing, smoking, drinking yoga and exercise of in-migrants across age, sex, social group, education, income, economic activity, and diseases. Out of the 300 respondents, 159 (53%) were female and 151 (47%) were male in-migrants. The slightly higher percentage of female in-migrants than that of male in-migrants might be due to the preponderance of female in-migration through marriage.

Majority of both male and female respondents were from the age group of 35–59 years, which was observed as a result of greater mobility of people to the City in the earlier days when the job opportunities were more in various economic activities of the City compared to that of the recent times. Lifestyle category-wise responses show that in the food habits category of lifestyle the total responses of males and females remained just equal to the number of respective respondents as food comes under the essential item of everyday requirements for all. But in other categories of lifestyle, the grand total responses of 141 male respondents became 245, while in case of 159 female respondents it remained only 84. This indicates that larger number of males usually get engaged themselves in more than one category of lifestyles regardless of its risk factors, but a smaller number of females get engaged themselves in lifestyles other than food habits as they are more health conscious. This is a kind of positive sign from the perspective of better health of females in comparison to that of males.

The percentage of responses in only 2 meals/day category increases from the General to the OBC, SC, ST, and to the Muslim social groups. This is observed due to variation in the socio-cultural and economic status of the respondents of different social groups. Skipping breakfast may be due to negligence and overburden of their works in different economic activities. The percentage of responses of both male and female respondents in chewing tobacco, Betel leaf or betel nut), Gutkha, and Khaini generally increases from the General to the OBC, SC, ST, and to the Muslim social groups in both male and female categories except few cases in exception. Percentage of smoking was higher among the Muslim males and it was followed by males of OBC, General caste, SC, and ST. The percentage of smoking was found higher among the women of ST and Muslim social groups. They generally smoke Beedi instead of cigarettes as Beedi is quite cheaper than cigarette. In the drinking alcohol category of lifestyle the percentage of in-migrant male respondents was found higher in scheduled caste and scheduled tribe social groups in comparison to that of other backward cast and general social groups. The percentage of responses of male in-migrants in practising regular yoga and exercise declines from the socio-economically better to poor groups.

Educational category-wise pattern of breakfast cum two meals per day exhibits that the percentage of responses of in-migrant males or females increases from lower levels of education to higher levels of education. Likewise, the percentage of responses of in-migrant males or females in only 2 meals/day category generally reduces from the lower level of education to higher levels of education. But in other habits such as chewing tobacco and smoking the percentage of responses of the illiterate respondents or the respondents with lower levels of education was fairly higher in both sexes. This is an indicative of the impact of educational achievements of the concerned person's on the habits of chewing tobacco and smoking. Persons with illiteracy and less education drink more the locally produced cheaper alcohol, while the persons with upper level of education drink more expensive branded alcohol. The percentage of responses of female in-migrant respondents in regular yoga and exercise increases from lower levels of education to higher levels of education. This reveals a positive relationship between the regular yoga-exercises and levels of education.

In case of monthly income and breakfast cum two meals per day category of lifestyle it was found that the percentages of both male and female in-migrant respondents clearly increase from the lower categories of monthly income to the categories of higher levels of income or vice versa. In other habits such as chewing tobacco, smoking, drinking alcohol, and regular yoga and exercise the percentage of responses of males to their grand total responses in each category of per capita monthly income was exceptionally higher in the habits of chewing tobacco and smoking. The next higher percentage of responses of males to their grand total responses in almost each category of per capita monthly income was in the habit of drinking alcohol. In respect of regular yoga and exercise the percentage of responses of females to their grand total responses in respective category of per capita monthly income was exceptionally higher in comparison to that of the males.

Economic activity wise pattern of lifestyle manifests that the number males in all the economic activities is larger than that of females. Not only this, but also there is predominance of male in-migrants in almost all categories of lifestyle and economic activity, whereas female in-migrants are fewer and they were engaged in certain economic activities such as commercial, teaching, household, and technical/professional activities and lifestyle categories and economic activities. Percentage wise variation shows that the majority of both male and female in-migrants were in food habits. The pattern in other than food habits, the number and percentage of female in-migrants to their grand total responses were more in the regular yoga and exercise as well as in chewing tobacco. The pattern of responses of male and female in-migrants clearly show that the percentage of responses of males to their grand total responses in the chewing tobacco, smoking, and in drinking alcohol was fairly higher than that of the responses of females in most of the economic activities of the study area, but the responses of female in-migrants were more in the regular yoga and exercise and most of them were engaged in teaching, technical/professional, commercial and others.

The pattern of the types of lifestyle and types of disease shows that the majority of male respondents in descending order were in chewing tobacco, smoking, and breakfast + 2 meals/day categories of lifestyle, while in case of female in-migrants the majority were in breakfast +2 meals/day, regular yoga and exercise and chewing tobacco categories of lifestyle. Likewise, with regard to diseases the majority of male in-migrants were found suffering from blood pressure, arthritis, diabetes, and gastric, whereas the majority of female in-migrants were found suffering from gastric, blood pressure, arthritis, and obesity.

Suggestions: From the above it is quite clear that all the in-migrant respondents do not have proper food habits in eating breakfast and two meals per day as well as several persons among them were addicted to chewing tobacco, smoking, and drinking alcohol and as a result of this the concerned in-migrants suffer from some of the lifestyle-related chronic diseases.

Those including the respondents who don't take breakfast plus two meals per day timely and regularly should be motivated and educated about proper food habits with a view to maintain good health.

Similarly, those including the respondents who don't practice yoga and exercise regularly should be encouraged through social media to practice yoga and exercise to improve and maintain their health. This can be more effective when the governments and non-governmental agencies establish yoga centres at suitable places in the city. Frequent organization of camps for yoga and exercise would also be a step forward in awakening awareness about the health benefits of the physical activity. In the same way, for those including the respondents who chew tobacco, smoke, and drink alcohol there should be the establishment of a required number of de-addiction centres in the study area by the government and non-governmental agencies to control the addictive habits. They should be provided some kind of incentives too so that the addicted persons can leave their addiction. Such harmful habits to health can also be checked through online social media programs and support groups. The contents concerning lifestyle-related bad habits and its negative impacts on health should be included in the curriculum activities of the formal educational system. This would also be a positive move towards controlling addictive habit or behaviour. Therefore, a more rigorous research involving larger sample sizes and extended variables related to lifestyle is also needed to achieve a better scenario of health-related problems at local, regional, national, and global levels.

References

Abate et al. (2001). Ethnicity and T2D: focus on Asian Indians, *Journal of Diabetes and its Complications, 15*(6), 320–327.

Allender, S., et al. (2008). Quantification of urbanization in relation to chronic diseases in developing countries: A systematic review. *Journal of Urban Health: Bulletin of the New York Academy of Medicne, 85*(6), 938–951.

Benson, M., & O' Reilly, K. (2015). From lifestyle migration to lifestyle in migration: Categories, concept and way of thinking. *Migration Studies, 4*(1), 20–37.

Brandhagen, M., et al. (2012). Alcohol and macronutrient intake patterns are related to general and central adiposity. *European Journal of Clinical Nutrition, 66*, 305–313.

Casali, M. E., et al. (2015). Lifestyle and food habits changes after migration: a focus on immigrants women in Medona (Italy). *Ann Ig, 27*(1), 1–12.

Caspersen, C. J., et al. (1985). Physical activity, exercise, and physical fitness: definitions and distinctions for health-related research. *Public Health Reports, 100*(2), 126–131.

Chowdhury, T. K. (2017). Health and Lifestyle among Migrant Oraons: An anthropological Study, Ph.D. Thesis, Department of Anthropology, University of Culcutta.

Clarke, J. I. (1965). *Population geography* (p. 123). Pergamon Press.

Cockerham, W. C. (2000). Health lifestyles in Russia. *Social Science and Medicine, 51*, 1313–1324.

Dima-Cozma, C. et al. (2014). The importance of healthy lifestyle in modern society: a medical, social and spiritual perspective. *European Journal of Science and Theology, 10*(3), 111-120.

Fan et al. (2015). Marital status, lifestyle and dementia: A nationwide survey in Taiwan. *PLOS ONE*, pp. 1–8.

Featherstone, M. (1991). Consumer cultural and postmodernism, London, Sage.

Frankish, C. J., et al. (1998). A review of Relationships between active living and determinants of health. *Social Science & Medicine, 47*(3), 287–301.

Ginsberg, Y. (1975). joint leisure activities and social networks in two neighbourhoods in Tel Aviv. *Journal of Marriage and Family, 37*(3), 668–676.

Jagadeeswari, D. (2005) A study on changes in lifestyle under globalised economy of India, Ph.D. Thesis, Post Graduate and Research of Commerce, Kandaswami Kandar's College, Velur Periyar University, 2004-2005.

Jensen, M. (2007). Defining lifestyle. *Environmental Sciences, 4*(2), 63-73.

Jenum, et al. (2005). Ethnicity and sex are strong deteminants of diabetes in an urban Western society: Implications for prevention, *Diabetologia, 48*(3), 435–439.

Johnson, N. (2001). Tobacco use and oral cancer: a global perspective. *Journal of Dental Education, 65*(4), 328–339.

Katz-Gerro, T., & Shavit, Y. (1998). The stratification of leisure and taste: classes and lifestyles in Israel. *European Sociological Review, 14*(4), 369–386.

Khan, N. Pattern of Male Out-migration from Eastern Uttar Pradesh: A study of twelve villages. In Mandal, R.B., (1981). (Ed.), *Frontiers in Migration analysis*, p. 501. Concept Publishing Company, New Delhi.

Kim, S., et al. (2004). Contrasting socioeconomic profiles related to healthier lifestyles in China and the United States. *American Journal of Epidemiology, 159*(2), 186–191.

Kunsche, S., et al. (2009). Social roles and alcohol consumption: A study of 10 industrialized countries. *Social Science & Medicine, 68*, 1263–1270.

Kuper, H., et al. (2002). Tobacco use, cancer causation and public health impact. *Journal of Internal Medicine, 251*, 455–466.

Krueger, M., & Chang, V. W. (2008). Being poor and coping with stress: health behaviors and the risk of death, *American journal of public health, 98*(5), 889–896.

Lassetter, J. H., & Callister, L. C. (2008). The impact of migration on the health of voluntary migrants in western societies: a review of the literature. *Journal of Transcultural Nursing, SAGE*, 93-104.

Mazengo M. C. et al. (1997). Food consumption in rural and urban Tanzania. ActaTropica, vol.68, pp. 313–326. Migration: categories, concept and way of thinking. *Migration Studies*, 1-18

Mishra, A., & Ganda, O. P. (2007). Migration and its impact on adiposity and type 2 diabetes. *Nutrition, 23*, 696–708.
Nierkens, V. et al. (2015). Smoking in In-migrants: do socio-economic ingredients follow the pattern expected from the tobacco epidemic?, group.bmj.com. http://tobaccocontrol.bmj.com/.
Nowicka, P., & Flodmark, C. E. (2007). Physical activity-key issues in treatment of childhood obesity. *Acta Paediatrica, 96*(454), 39–45.
Ottesen, G. H., & Wandel, M. (2012). Changes in dietary habits after migration and consequences for health: a focus on South Asians in Europe. *Food and Nutrition Research, 56*(18891), 1–13.
Palosuo, H. (2000). Health-related Lifestyle and Alienation in Moscow and Helsinki. *Social Science and Medicine, 51*, 1325–1341.
Parrott, A. C. (1999). Does cigarettes smoking cause stress? *Am Psychol, 54*, 817–820.
Pulkkinen, L., & Kokko, K. (2000). Identity development in adulthood: A longitudinal study. *Journal Research in Personality, 34*, 445–470.
Rojek, C. (1985). Captalism and leisure theory, London, Tavistock.
Ross, J. A. (1982). *International encyclopedia of population* (p. 448). A Division of Macmillan Publishing Co., Inc, New York.
Salmon, P. (2001). Effects of physical exercise on anxiety, depression, and sensitivity to stress: a unifying theory. *ClinPsychol, 22*, 33–61.
Singh, M.S. (2013). Obesity and self reported morbidity in relation to lifestyle and diet among the Meitei of Manipur. Ph.D. Thesis, Department of Anthropology, North-Eastern Hill University, Shillong.
Sinha, B.R.K. (2005). Human migration: Concept and approaches, VITA.
Thrope et al. (2019). Education and lifestyle predict change in dietry patterns and diet quality of adults 55 years and over. *Nutrition Journal, 18*:67, 1-13
Utama, I. G. B. R. (2021). Review studies: lifestyle and social class in consumer behavior for services industries. *International Journal of Social Science and Education Research Studies, 1*, 19–24.
Van Dam, R. M., et al. (2003). Patterns of food consumption and risk factors for cardiovascular disease in the general Dutch population. *American Journal of Clinical Nutrition, 77*, 1156–1163.
Vari et al. (2016). Gender-related differences in lifestyle may affect health status. *Ann 1st Super Sanita, 52*(2), 158-166.
Veal, A.J. (2001). Leisure, Culture, and Lifestyle. *Leisure and Society, 24*(2), 359-376
Vincent, S., et al. (2002). Genetic polymorphisms and lipoprotein responses to diets. *ProcNutrSoc, 61*, 427–34.
Wandel, M. et al. (2008). Changes in Food Habits after Migration among South Asians settled in Oslo: The Effect of Demographic, Socio-economic and Integration Factors, Appetite, pp. 376–385.
Weeks, J. R. (1989). *Population: An introduction to concept and issues* (p. 186). Wordsworth Publishing Company, Belmont.
World Health Organization. (1998). *Health promotion glossary*. WHO.
World Health Organization. (2008). *2008–2013 action plan for the global strategy for the prevention and control of non-communicable diseases: prevent and control cardiovascular diseases, cancers, chronic respiratory diseases and diabetes*. WHO.
World Health Organization. (2011). *Global status report on non-communicable diseases 2010*. WHO.
World Health Organization, WHO. (2005a). *Preventing chronic diseases: A vital investment*. WHO.
Yadav, J. (2015). Migration dynamics, lifestyle and its effects on health status and expenditure pattern of male migrants in Delhi Slums. Ph.D. Thesis, International Institute for Population Sciences, Mumbai.
Yang, Z. et al. (2015). The influence of income, lifestyle and green spaces on interregional migration: policy implications for China, population, space and place, pp. 1–14.
Zelinsky, W. (1971). The hypothesis of the mobility transition. *Geographical Review, 61*. Worcester.

Zhang, S., & Xiang, W. (2019). Income gradient in Health-related quality of life- the role of social networking time. *International Journal for Equity in Health, 18*(44), 1–10.

Ziglio, E., et al. (2004). The WHO cross-national study of health behaviour in school-aged children from 35 countries: findings from 2001–2002. *Journal of School Health, 74*(6), 204–206.

Part IV
Urban Development, Planning and Policies

Chapter 26
Going Green in Urban Development: Conceptual Advancements and Applications in Selected Cities

Upuli Perera, Collins Adjei Mensah, and Lakshi Karunarathne

Abstract Going green in urban development focuses on adopting "environmental friendly" practices to enhance the health and well-being of urban dwellers and also protect the environment in which they live. Thus the idea of going green in urban development supports the sustainability and liveability of cities. This chapter traces the development of the going green concept, morals, and approaches within academia and how these have been put into practice to shape urban development in different regions of the world. The theoretical progression of going green concept was investigated by reviewing of 324 academic publications; looking at how it evolved from the nineteenth Century: commencement of modern planning; to the present day twenty-first Century. The urban practice in going green was looked at through four urban development case studies from England, Australia, Ghana, and Sri Lanka. These cases exemplify the way in which planners have been operationalizing the idea of going green to tackle different urban issues related to housing, transport, climatic change, public health, and many others to improve the living conditions of urban dwellers and achieve sustainable urban development. Understanding the academic (theoretical) and practical progressions in the implementation of "going green" development projects will help in shaping the future growth of urban areas (Sustainable Development Goals 11–Sustainable Cities and Communities) to make them healthy and liveable under post-COVID-19 and post-globalization era.

Keywords Going green · Urban development · Sustainable development goals · Urban areas · Concepts · Applications · Australia · England · Ghana · Sri Lanka

---

U. Perera (✉) · L. Karunarathne
Department of Estate Management and Valuation, University of Sri Jayewardenepura, Nugegoda, Sri Lanka
e-mail: tgup@sjp.ac.lk

L. Karunarathne
e-mail: lakshikarunarathne@sjp.ac.lk

C. A. Mensah
Department of Geography and Regional Planning, University of Cape Coast, Cape Coast, Ghana
e-mail: cmensah@ucc.edu.gh

B. R. K. Sinha (ed.), *Urban Dynamics, Environment and Health*,
https://doi.org/10.1007/978-981-99-5744-6_26

26.1 Introduction

Globally, urban development is progressive in nature. We now live in an urbanized world with more than half of world's population living in urban areas. This trend particularly started in 2008 when for the first time in the history of the world more people lived in urban areas than in rural setting (Population Reference Bureau, 2008). Currently about 55% of the world population are now living in urban areas. Future projections show that this situation will further increased by 2050 where about 68% of the world's population is estimated to live in urban areas with Africa and Asia projected to be greatly hit by this development (United Nations, 2019a).

Although this global high rate of urbanization is helping to accelerate the economic and social progress of many urban areas but at the same time it is having some environmental ramifications that cannot be overlooked. In terms of the economic benefits of urbanization, urban areas now account substantially to the global Gross Domestic Product, serve as where entrepreneurship and technological innovation thrive well due to the concentration of many businesses and diverse educated people, and hubs of development where economics of scale and government policies help to provide easy access to job opportunities and basic infrastructure. In the social front, easy access to basic infrastructure and social services in urban areas enhance opportunities for cultural and political participation and gender equality than in rural areas (UNDP, 2016; United Nations, 2019a).

These social-economic benefits are now being outpaced by various environmental challenges which are thwarting global efforts on sustainable development. Statistics show that over 70 per cent of the world's greenhouse gases are generated in urban areas due to their large consumption of the world's energy supply. Globally, it is estimated that urban dwellers generate about 1.2 kg of waste a day (amounting to 1.3 billion tonnes per year) with the resultant effects being excessive marine pollution especially from plastic waste, high air pollution, inadequate landfill sites to contain the waste generated, and many health hazards such as cholera and diarrhoea from informal dumps and untreated waste (UNDP, 2016). Excessive depletion of farmlands, wetlands, parks, forests, and shortage of water supply as a result of urbanization has also been emphasized in the literature.

The 2019 Progress Report on the attainment of the United Nation's Sustainable Development Goals showed abysmal performance of urban areas towards these goals especially Goal 11 which emphasizes on making cities and human settlements inclusive, safe, resilient, and sustainable (United Nations, 2019b). For example, the report found high urban air pollution with many urban areas not able to meet the WHOs air quality standard of 2.5 particulate matters or less. In addition, in most cities, there was poor access to public spaces due to high depletion of the green vegetation. The report further revealed a high rate of urban sprawl with urban areas expanding faster in their land area to cause several environmental problems and poor access to transport.

The foregoing is critical if sustainable cities are to be achieved since such cities require reversing many of the above environmental problems and at the same time promoting economic and social developments. This means most urban areas have to "go green" or encourage "green developments" which emphasize undertaken "environmental-friendly" projects that simultaneously preserve the natural environment and also stimulate the socio-economic well-being of urban dwellers. Although some urban areas both in the global north and south have heed to this call by pursing different green development plans and projects but information on the nature of those projects, and how such projects are contributing to the general welfare of urban dwellers are scarce. The available studies have remained silence on this and rather concentrate on urbanization trends, consequences, and future projections of urban development (Miller & Hutchins, 2017). It is as result of this backdrop that this chapter was written with the aim of tracing the development of the going green concept, morals, and approaches within academia and how these have been put into practice to shape urban development in both the global north and south (UK, Australia, Sri Lanka, and Ghana).

The conceptual progression discusses the reasons why the going green idea emerged and developed as a concept. The different case studies demonstrate how planners are pursing different going green projects as strategies to solve urban problems and challenges affecting the implementations of those projects.

This effort will help to show the progress made in some urban areas towards achieving sustainable urban development (Sustainable Development Goals (SDG) 11 –Sustainable Cities and Communities). It will unveil the nature of development projects that carries the values of going green in different and challenging urban context (global north and south) and the dividends they have started yielding to make those urban areas healthy and liveable. These will provide good lessons to shape urban areas of the world under post-COVID-19 and post-globalization context.

26.2 The Conceptual Advancements of Going Green

26.2.1 From Early 19th Century Up to Late 20th Century

The first milestone era that led to the development of going green (eco-centric) approaches was the 19th Century rapid industrialization which caused environmental degradation in terms of air and water pollution, public health, poverty, and degradation of working-class housing (Konteh, 2009; Wohl, 1983). In response to these, several visionary urban planners, sociologists, and architects promulgated several utopian forms of urban designs that could improve the living and working conditions of cities. Ebenezer Howard's *Garden City*, Frank Lloyd Wright's *Broadacre City, and* Le Corbusier's *Radiant City* forms are influential works to quote in this respect (Hobson, 1999). Despite the variety in urban designs, their general advocacy during the time was to surround the living and working environments with green spaces (greenbelt lands, agricultural lands etc.) for better air, calmness, orderly urban environment, spiritual renewal, ensure locally supplied food and locally serviced quality public housing projects (Hobson, 1999; Sharifi, 2016). The underlying economic model proposed to finance these urban forms were primarily through the businesses in which these workers were working for. By early twentieth Century (1914–1945 AD); the world war times brought the imperative that resources should be used thriftily. Consequently, the careful consumption of resources was emphasized through ideas such as recycling, reuse, and reduce (These later progressed as circular economic strategies[1]). Concerns were expressed that waste are not simply garbage but resources and wealth (Cooper, 2008). This started a stream of advocacy regarding the environmental concerns over uncontrolled consumption, municipal reclamation and waste handling of urban areas, and so on (e.g. UN Scientific Conference on the Conservation and Utilization of Resources—1949)(UN, 1951).

The most prominent second stage evolution was the post-World War II environmental movements from 1950s to 1970. They were lobbying for environmental care after the environment was largely degraded by the increased production exercises to foster economic revival and growth of cities after the war. Firstly, these movements drove urban planners in the 1980s to rediscover urban designs suitable for the time. Neo-traditional movement in the form of postmodern urbanism can be recognized as some form of response to these designs. Eco-city design by Richard Register (Ashique-Uz-Zaman, 2012); transit-oriented development by Peter Calthorpe (Ian, 2009); new urbanism using the Ahwahnee Principles signed by New Urbanist designers, including Andres Duany, Elizabeth Plater-Zyberk, Stefanos Polyzoides, Elizabeth Moule, Peter Calthorpe, Jane Jacobs and Michael Corbett (Fulton, 1996); and smart growth by American Planning Association are some urban design agendas which came about with such neo-traditional movements (Ashique-Uz-Zaman, 2012; Fulton, 1996). They valued walkability, less cars in cities, pleasant greenery surroundings, identifiable urban boundaries, and compactness of services to ensure self-sufficiency as solutions to bring environmental friendliness to development. Secondly, this period supported efforts to regain the importance of industrial economics, industrial ecological ideas that propagated closed loop circular economy—regenerative system in which resource input and waste emission, and energy leakages of institutions are minimized (Geissdoerfer et al., 2017). Consequently, resource production and consumption strategies such as reuse, recycle, reduce, and reengineering were reemphasized as methods that follow going green development agendas (Cooper, 2008). Parallel to this environmental movement, there was also a gradual upraise of social justice movement, particularly against the "less democratic" and "injustice" planning process for the urban renewal of the time–police brutality, inadequate public services, inequality etc. (Fainstein, 2014, pp. 1). Though there were no normative statement as to what the "justice" in cities is (Fainstein, 2000), the leftist scholars particularly had been critical on planning practices both theoretically and empirically (Fainstein, 2000, 2014). All such movements led the concerns to elevate the importance of natural environment and society to the same level as that of economic development. In 1987, Brutland Commission's

[1] The circular economic strategies was declined in the nineteenth Century in the face of hygiene issues of recycling and rise of incineration technology.

(initially known as World Commission on Environment and Development) report *Our Common Futures* established the term "sustainable development" as a global concern. The aim was particularly to influence the mindset of the masses (Robinson, 2004) that economic development cannot be over rated, but it needs to be in harmony with natural environmental systems and measures to ensure social well-being of people in order to sustain the systems in the long run. So this was a paradigm shift for the then going green concept which concerned– social acceptance, equity, society's well-being etc. to become more explicit ingredients in recognizing development as "environmentally friendly". All urban development projects of cities overtime were compelled to go through the lenses of different sustainability credentials. Some commonly accepted credentials were: (i) Urban development should revise land use priorities—include more green spaces (parks, playgrounds, agricultural lands, roadside trees, greenbelts etc.), (ii) Urban transportation should prioritize public transport and communities should be encouraged to live nearby transit nodes of the city to be less car driven and efficient commuting, (iii) Projects should ensure health, and spiritual well-being of humans, (iv) Urban areas should restore damaged environments, (v) Better housing should be available for people living in the city, (vi) Built environment should be live-able, (vii) Fair-share community– equity for all people and ecological systems, (viii) Prosperity in businesses in the city, (ix) Sufficient agriculture to ensure food supply, (x) Civil society rights should be ensured, (xi) Technology should be designed to be compatible with local setting, (xii) Community Economic Development (protect local businesses, worker cooperatives etc.) should be established, (xiii) Grassroot democracy should be encouraged, and (xiv) City should have no violence and crime. By the latter part of the twentieth Century, these were developed as the *criteria* to measure the "sustainability" credentials of urban development projects. They structured the different going green tenets, principles, and attributes recognized and implemented over the years into three broad categories known as economic, environmental, and social sustainability. Attracting funding from various financial institutes: development assistance, state budgetary allocations, and corporate funding required the projects to prove that they do meet such criteria, thus "sustainable".

26.2.2 Turn of the Twentieth Century to Twenty-First Century First Quarter

The turn of the twentieth Century to the first quarter of twenty-first Century was a period of intensified globalization. It was characterized with the blurring of geographical boundaries in which the economic, environmental, and social issues of urban environments were encountered (Jessop, 2000; Parker, 2015). Moreover, all such issues have become complex and dynamic in a way that single institution seems incapable of correctly characterizing and let alone addressing them (Healeyet al., 2017). Urban areas became often shaped by the influence of flexible capital and labour market drivers, limiting national government's control over these drivers. For instance, developing of buy-to-let markets with arm's length investments on housing (e.g. Rolnik, 2013), labour market being influenced by short-term immigrant workers, global surveillance issues, and pandemics (Gushulak et al., 2009). Economic, environment, and social urban settings have less stability and encounter increasing risks of change: climatic change risks, disruptions and economic uncertainty about downfalls or recessions, dynamic lifestyle changes of people, and the increasing risks on public with respect to their health and well-being (Marsden & Docherty, 2013; Gushulak et al., 2009).

Concurrently, the going green approaches in the millennium have progressed in several folds. The mainstream advocacy for sustainability—United Nations, advanced the three pillars (economic, environment, and social sustainability) into seventeen action areas; also known as Sustainable Development Goals (SDGs): (i) poverty; (ii) hunger; (iii) health and well-being; (iv) education; (v) gender equality; (vi) clean water and sanitation; (vii) affordable and clean energy; (viii) decent work and economic growth; (ix) industry innovation and infrastructure; (x) reduce equality; (xi) sustainable cities and communities; (xii) responsible consumption and production; (xiii) climate action; (xiv) life below water; (xv) life on land; (xvi) peace, justice and strong institutions; and (xvii) partnerships. Respective countries, regions, or local areas are expected to discover how each of the action areas can become relevant, respective of problem definitions, strategies, and methods which can be adopted to better-off the lives of the citizenry. Equitable access to clean water and energy is recognized to be one of the most valuable resources for urban area (UN Water, 2007). In a similar vein, other members of United Nations also called for different actions and initiatives to operationalize sustainability. For instance, United Nations Environmental Programme in 2009 called for a Global Green New Deal—responding to global crisis on food, fuel and finance with both developed and developing countries allocating 1% of their GDP for carbon reduction and, improving clean water, sanitation, health and education respectively (UNEP, 2009). Similarly, UNEP also upgraded the circular economic ideas of recycling, reuse, and reduce with different concepts (e.g. Cleaner Production) and impact assessment methods such as product life cycle assessments on cleaner production (Røyne et al., 2016).

With the world's pressing concern for climatic change (e.g. United Nations Framework Convention on Climate Change established in 1992, WHOs global strategy for health, environment, and climatic change 2019–2023), urban resilience was another burgeoning field of study that developed in parallel with the mainstream sustainability or going green literature. Urban resilience refers to the ability of the urban systems to withstand different forms of risks, shocks, or stresses (Marchese et al., 2018). The key goal is for urban systems to incorporate adaptation and mitigation strategies against these tensions. The urban resilience topics of inquiry developed the notion that increasing the resilience of a system makes that system more sustainable (Marchese et al., 2018).

Besides, the advancement of technology (e.g. Artifical Intelligence, Augmented Reality—AR, Virtual Reality—VR, Drones, Internet of Things–IoT, Blockchain technologies) was found to be a novel breakthrough to solve different urban environmental problems (Klaassen et al., 2013). These were sought as an opportunity to gain efficiency in reconciling different demand and supply conditions in markets (e.g. travel demand and supply) and, possibilities for collaborative consumption (sharing economy:- *the peer to-peer-based activity of obtaining, giving, or sharing the access to goods and services, coordinated through community-based online services*; Hamari et al., 2016, p. 2047). This thereby shifted the attitudes and options available with consumption i.e. optimize resource consumption (meeting effective demand) and minimize waste of resources (avoid excess, wasteful utilization of resources). This approach became the new moral of the first quarter of the twenty-first Century to meet the challenges of various urban development sectors: housing, transportation, business, disaster resilience, health, public administration, and so on to achieve sustainability through maximizing productivity of resource consumption without jeopardizing the people's quality of life.

The vast literature that grew after the 1990s inquiring the social vulnerability of cities (e.g. poverty, inequality, insecurities of their social capital, exploitation etc.) particularly against globalization and market powers (Fainstein, 2014; Carmon and Fainstein 2013), reinstated Utopia in the name of social justice—well-being, equitable development, and others as a significant proportion of going green (Fainstein, 2000). On top of that, urban issues amidst globalization found to be more complex in a way that a single institution cannot address them alone and people tend to live in groups that are multi-sided with scalar interests (Fainstein, 2014). In this background, collaborative governance as a progression of communicative rationality reasserted its hegemonic position in resolving going green or sustainability within urban development (Perera & Mensah, 2019; Feinstein, 2014). The argument here was that going green or sustainability was socially constructed bounded by time—it's the shared interests of different communities and stakeholders that reflect the different economic, environmental, and social perspectives to characterize and provide solutions to urban issues of time. Accordingly, there were a wave of literature in planning, housing, energy, transport etc. including SDGs which propagated the status quo that community and stakeholder participation is an unavoidable means to going green or sustainable urban development (Martin et al., 2014; Perera & Mensah, 2019).

26.3 Case Studies: Going Green Applications

26.3.1 Milton Keynes New Town Development, England

Introduction

The Milton Keynes (MK) New Town in Buckinghamshire-England established in 1967 is considered to be one of the most observant new towns in England with respect to its practices for sustainability in urban development (Alexander, 2009). With the initial designation area of 22,000 acres, the new town is located about 87.5 km away in North-West of London. With the targeted population of 250,000,

MK was primarily developed as a housing solution to the then over spilling conditions prevailing within London boroughs. Established under the New Town Act of 1946, the Milton Keynes Development Corporation (1967–1992) was to design, deliver, and manage the new town, which afterwards was handed over to MK Council. New Towns in England (including MK), are generally considered to be in Utopian traditions of modernist planning having strong links with the Garden city concept (Falk and Carley, 2012; TCPA, 2007). The MK local development plan 2016–2020 putting its aim at opportunity, affordability, and health reflects the retention of the Garden city concept values which it still takes forward. In tracing MK's traffic management, the road network, transport, green spaces, and housing arrangements with mixed use clearly show how going green has been applied there to tackle key urban development areas.

Traffic Management, Road Network, Transport

MK is well-connected to the public transportation system via rail and bus. The primary traffic and transport aim of MK was to have low Co_2 emissions and achieving mobility for all. The rail connection is available from MK Central Station, Bletchely, and Wolverton station. These rail services are complemented by the local bus services, cycle pathways, and shaded pedestrian ways. The internal road circulation is shaded in Boulevard styles and follows the American style grid system that tends to support better mobility by walking and cycling. The town has over 30 bus stops designed in an inclusive way—improved accessibility via elevated kerbs and shelter. These initial features enabled MK Development Corporation and MK local council to subsequently attract more national governments funding which could support the upgrading of "innovations of going green"-towards sophistication of transport. The MK local council received £420,000 from the Department for Transport's Sustainable Travel Transition Year (STTY) to implement the project that encouraged people to cycle, scoot, and walk (MK Council, 2018). The project branded cycling, scooting, and walking as a "smarter choice active travel and intelligent health", influenced the lifestyle choices of MK citizens. The city was also one of the four UK cities which could attract £9 million from £40 million project on "Go Ultra Cities Programme" led by UK's Department of Transport and Office for Low Emission Vehicles (OGL, 2016). This project aimed at encouraging MK people to choose electric cars over petroleum-driven cars. Owing to the grid road system, the local authority also attracted funding to start operating 40 self-driving car pods capable of traveling 15 km/hr with lasting for 60 miles in one-offcharge that could support the internal mobility of MK citizens—from Milton Keynes Central station to parks and boulevards that are primarily towards apartment housing and offices. A catalyst fund operated by the Higher Education Funding Council for England (HEFCE) was also secured to establish a smart central data hub (HEFCE, 2014). As a succession part of this project, MK's traffic management system was linked to Smart central data hub via 425 traffic sensors fixed across major junctions and 1800 parking sensors fixed at paid local authority parking spaces (Foreign and Commonwealth Office, 2018). This Smart data hub that receives real time data on traffic together with other data sets relevant to education, housing, energy, environment, business, and many others would enable stakeholders to manage the different demand and supply conditions related to occupants of MK.

Green Spaces

MK New Town—being a progression of Garden City, is designed with an extensive network of green and blue water spaces. MK has 6000 acres (1/5 of total land area) of parklands which falls into the best parks in the world (MK Council, 2018). These consist of (i) linear parks in pedestrian and cycle corridors, special dedicated wildlife corridors and spaces for informal outdoor recreation, (ii) different other parks open for public, (iii) playing fields and outdoor sports facilities, (iv) grid road curtilage, (v) amenity green spaces, (vi) church yards and cemeteries, (vii) agriculture allotments, and (viii) home gardens. The importance given to green spaces in MK local level development policies and the active participation of the civil society movements are found to be the primary success factors in achieving the said level of green space conditions. The sustainability appraisal for the MK local development plan (2017) recognizes the importance of conserving green landscape as an essential part to protect biodiversity and geo-diversity and for that matter recognized the need to establish action plans and management systems. The Buckinghamshire and MK Natural Environment Partnerships (a partnership arrangement with local authorities and institutions from state, private, health, and education for championing the environmental perspectives within urban development decision making) have recognized these green spaces as "green infrastructure"–different eco-services the green spaces provide to the living environment (MK Council, 2018). This has therefore resulted in the establishment of principles and priorities to improve the green spaces. MK particularly has community-supporting ownership of land and long-term stewardship and management of common assets in its localities. The 6000 acres of the parks and other green spaces are managed and looked after by the Parks Trust- the self-financing charity run by the locally lived residents. However, the continuous development pressure due to MKs proximity to London and Birmingham by train particularly has been the main challenge to these green spaces. This is similar to what happens at many other settlements around UK. It is noteworthy that urban areas having sufficient public green spaces near where people live became a vital matter during COVID-19 lockdown times, since many outside activities in populated places were halted. MK could provide approximately 47 m^2 of public parks and gardens per person in that respect- one of the highest in UK (MK Citizen, 2020).

Housing Arrangements

Inclusiveness and low carbon-energy efficiency are the two going green credentials that MK housing sector aims to achieve.

MK has approximately 115,222 houses (MK Council, 2020a, 2020b) which comprise both private and affordable housing units. MK planning attempted to achieve inclusivity in housing by offering both need and choice. The city's private housing is designed with a wider range of housing mix comprising apartments, detached, semi-detached, and terraced housing. The UK's national level emphasis for all housing developments to have affordable housing which is also inclusive (see, Sec. 106 of Town and Country Planning Act, 1990, Paragraph 56, National Planning Policies Framework (2012, 2019), Planning practice guidance etc.) is the practice that MK council's at the local level undertook following policies and initiatives. The supplementary planning document for MK development plan insists planning applications with 11 or more units should provide 31% affordable housing, out of which higher proportion (20%) is allocated for starter homes-tenant pay only 80% market value (MK Council, 2020b). Among the rest, 5% are to be social rent and 6% shared ownership having 25%-40% equity share (MK Council, 2020b). To further strengthen the inclusivity, the local policy stresses that the location of those housing should be pepper potted (clusters more than 12 should be avoided in one location), make the development flatter to encourage resident and community interaction among them and the design, and building materials and amenities provided should be tenure blind (MK Council, 2020b). In addition, particularly as a response to the local pressure groups, local council's housing strategy also insists "housing for older people and people with mobility difficulties".

In addition, planning for MK housing also aimed to follow low carbon emission and energy efficiency policies. Going by the national policy—Code for Sustainable Homes back in 2006 which aimed at zero carbon homes by 2016, the supplementary planning document of MK in 2007 requested all planning applications which have more than 5 dwellings to include a sustainability statement. That was to provide sufficient details on how new houses will have energy and water efficiency, sustainable building materials, and waste disposal aspects in a way that could lead to code 4 level low emission. When the said national policy however was repealed in 2015 and replaced with the new national technical standards which made carbon code an optional regulation, MK council yet set out its own 2019–2050 Sustainable Strategy, targeting the housing units to be carbon neutral by 2030 and carbon negative by 2050. Irrespective of such emphasis, the realization of energy efficiency housing number targets seemed challenging: by 2020 only 98 houses (less than 1%) in MK fell into the Grade A[2] and 29,243 (nearly 23%) could achieve at least up to Grade C level (MK Citizen, 2020).

However, arranging housing with mixed land use development (residential together with commercial and office), thus minimize commuting can be said to have generated successful results having looked at the MK's local job growth. During 2008, approximately 80% of MK residents had been working within the MK area itself. The local employment growth rate in Milton Keynes from 2004 to 2013 was 18.2% (MK Council, 2013)—one of the highest growth rates within UK but there was a drop after 2016 (ONS, 2018). In 2019, around 95% of economically active population gained employment (ONS, 2019). By January 2020, the number of unemployment benefit claimants have been dropped to 4,105 (MK Citizen, 2020) from 5400 in 2019 (ONS, 2019). The number of active business enterprises established within MK by 2019 was 14,105 with 75 large-scale businesses (ONS, 2019).

[2] Energy performance certificate provides a grade between A to G: Highest to the lowest.

26.3.2 Canberra City Climate Resilient-Australia

Introduction

Canberra in South East Australia is a planned capital just like Ottawa and Washington. The city accommodates a population of 397,397 (Australian Bureau of Statistics, 2016). Canberra is often threatened by climatic change risks–heat waves, urban heat, storms, bush fire, rain fall variations (ACT Government, 2019; Head et al., 2014). Those tend to affect the well-being and health of people, biodiversity losses, and also to the country's GDP growth. As a response to this, Canberra's going green trajectory (to Climate-Wise city as they call it) seemed to have focused on three aspects: green open spaces, compact mixed-use development, and energy efficient urban space.

Green Open Spaces

Green open spaces and water bodies are significant features of Canberra. This was an established fact since the initial Canberra plan (1912)—commonly known as Griffin's plan (by American Architects; Walter Burley Griffin and Marion Mahony Griffin) which was laid by the principles of Garden city and liveable city concepts. The city is affectionately called "Bush Capital". The green spaces include urban parks, open spaces, other landscape components such as pedestrian parklands which are for the movement of people within and between suburbs, semi-natural open spaces (grazing lands, ridges, hilltop areas, creek corridors, buffer areas between suburbs), native grasslands or woodlands, road verges or medians, non-irrigated open dry land grass areas, open spaces allocated for recreational activities or sporting events. Besides green spaces, the city has a number of natural and man-made water bodies-ponds, lakes, and rivers. The management of these water bodies is shared among different state organizations, volunteering businesses, and local government: Environment, Planning and Sustainable Development Directorate (EPSDD); Icon Water; Transport Canberra and City Services (TCCS); Environmental Protection Authority (ACT); National Capital Authority; Queanbeyan-Palerang Regional Council; and NSW Department of Planning, Industry and Environment. Under present context, these green spaces and water bodies are recognized as living infrastructure and government's commitment to provide and maintain such infrastructure is ensured via Canberra's Living Infrastructure Plan (2019). Moreover, the spatial planning for Canberra's growth since the Y-Plan[3] connected towns with arterial roads yet protected the natural setting of major mountains, grasslands, and ridges from developments. With a view to take forward this green space trajectory furthermore, Canberra's living infrastructure plan (2019) aims to increase the urban tree canopy up to 30% by 2045 from its 21% in 2019 (Environment, Planning and Sustainable Development Directorate, 2019a, 2019b). These green spaces have benefitted the city to become climatic resilient from temperature and rain fall variations—reduce urban

[3] National Capital Development Committee (NCDC)'s spatial plan (1967) - known as Y plan, connecting three satellite new towns; Woden, Belconnen and Tuggeranong to decentralize Canberra for accommodating growing population (Mees, 2014).

heat island effect, act as carbon sink, increase the access to shade, improve water penetration, retain ground water for sustainable vegetation, and preserve biodiversity, ultimately increasing well-being of people (Environment, Planning and Sustainable Development Directorate, 2019, 2019b; Rakhshandehroo et al., 2017). Besides, as reported in Canberra Times (2020), the planned green spaces in and around the city has also provided the capacity to ease social distancing tensions during COVID-19 pandemics—bring in wellness by allowing people to get outside while adhering to public health guidelines.

Compact Mixed Use Development

Canberra tends to be a car dependent city having 81% of Canberrans travelled to work by car (Environment and Sustainable Development Directorate, 2012). Keeping the city's vision to be carbon neutral by 2060, the city has applied compact mixed use development strategy (close proximity between housing, employment, shops, and transport)—though it doesn't tackle car ownership, yet handles the daily travel distances of people. The city targeted the residential intensification at specific locations that are in close proximity with employment. Accordingly, intensification took place in major urban renewal sites: Fyswick and Kingston, major employment centres: civic, town centres, and along the major avenues such as Northbourne and Constitution avenue (ACT Planning and Land Authority, 2004). However, as some community activist see, this intensification promoted wall-to-wall high-rise buildings, keeping the community farther away from access to quality green spaces. In addition, ACT government invested in public transport, walking and cycling infrastructure. Between the years 2010–2015, the ACT government invested $120 million on bus priority lanes of frequent networks, cycling, and walking infrastructure, new stops and stations, a real time passenger information system, frequent and rapid public transportation system, etc. (Environment & Sustainable Development Directorate, 2012). The government also invested on high-frequency light rail service between Gungahlin—highly densified residential area, and the city centre, Park and Ride or Bike and Ride facilities. The foundation laid by Griffin's plan with boulevard style and grid pattern developments with substantial green spaces complemented the ACT government's effort to motivate people to walk and cycle (active travel) between closer distances.

Energy Efficient Urban Space

Nationwide House Energy Rating Scheme (NatHERS) is a star rating system[4] that administers energy consumption levels of residential buildings in Australia which Canberra initiated in 1995. Canberra enforced this rule progressively under the

[4] Range of the star rating is from 0–10. 0-star means building shell dose nothing to reduce thermal discomfort, 6-star is minimum standard and indicates good but not outstanding. 10-star means house doesn't need artificial cooling or heating. Rating is calculated by considering house design, construction, climate and common pattern of house use (e.g.: size and function of rooms, type of construction, dwelling orientation, location, etc.).

Building Code of Australia (BCA). The NatHERS administer the thermal performance of housing by using both technology and guidelines to professional practice. Accredited software is used to calculate the thermal level of buildings during inspection. Assessors, building certifiers, and regulators are instructed and provided guidelines to monitor documentation related to buildings (Nationwide House Energy Rating Scheme, 2015). The Energy Efficiency Improvement Scheme introduced in 2013 under the Energy Efficiency (Cost of Living) Improvement Act 2012 and its amendments supplements NatHERS by regulating the energy consumption caused by equipment and appliances that sell in Australia. Rebates and subsidies are provided to business and households for the installation of solar, reversed cycle air conditioner, replacing old appliances, lighting upgrade for business etc. Besides progressively adhering to national policies, Canberra initiated its own applications to promote renewable energy. Places such as Ginninderry has developed into electricity only suburb replacing mandatory gas connection of housing in which electricity would be produced through solar power generated at household levels and renewable electricity (Green Energy News, 2020). Besides, the ACT Climate Change strategy 2019–2025 sets a pathway in which the community, business, and government achieving zero CO_2 emissions by 2045. It sets goals to reduce CO_2 emissions by reducing energy and water consumption and waste generation and treatment across all sectors of the city: housing, roads, transport, farming, health, and buildings. Community collaboration and supporting low-income residents and industry workers in order not to be left out in the process have been recognized as an essential element of the plan. From 2013–2019, the scheme shown to have reduced 500,000 t CO_2 –e and supported more than 74,000 households and businesses to save $400 million (approximately) on items installed (Environment, Planning, and Sustainable Development Directorate – Environment, 2019, p. 26).

26.3.3 Greater Accra Climate Resilient and Integrated Development Project

Introduction

Accra or Accra Metropolis which is the capital city of Ghana and serves as the nucleus of Greater Accra Metropolitan Area (GAMA) is vulnerable to the consequences of climate-related risks. This is due to the combination of factors such as the low-lying nature of the place below sea level, drainage network which is not keeping pace with the rapid expansion of the city, a lack of climate-sensitive urban planning, and a growing solid waste management crisis leading to siltation, environmental pollution and clogging of the drainage network. This problem is further exacerbated by high rise of informal settlements in flood prone areas fuelled largely by rural–urban migration from the Northern Ghana to Accra (World Bank, 2018). It is in view of these problems and others that Greater Accra Climate Resilient and Integrated

Development Project; a six-year development project (2019–2025) which is already underway was initiated.

The objectives of the project are to (i) strengthen flood and solid waste management, (ii) improve the living conditions of the most climate vulnerable communities in the Odaw basin, and (iii) foster adaptation of the flood management infrastructure to climate change in Greater Accra Region.

The Greater Accra Resilient and Integrated Development Project (GARIDP) is built around four components: (i) Drainage and flood management improvements within the Odaw Drainage Basin, (ii) Improvements in solid waste management capacity including minimizing solid waste in waterways, (iii) Support to most vulnerable communities within the Odaw Drainage Basin, and (iv) Strengthening capacity for planning, coordination, monitoring and evaluation (World Bank, 2018).

Sustainable Greening and Beautification of Accra

The Greening and Beautification Project which falls under the first component of (GARIDP) has two phases which will enable the Accra Metropolitan Assembly (AMA), the city authorities of Accra Metropolis to effectively overcome challenges associated with land acquisition and securing financing for implementation on larger plots of land (Accra Metropolitan Assembly, 2018). The emphasis of Phase one is on greening of interstitial spaces with special attention given to main avenues, triangles, circles, and open spaces along busy roads in the Central Business District (CBD). The earmarked areas will be provided with new landscaping, planting of trees and grasses, benches for relaxation, spaces for leisure, and shelter from heat. To collectively accomplish this, the city authorities have launched a campaign called "Adopt a Space" to invite the private sector and benevolent organizations to participate through this campaign to adopt a specific space of their choice within the project and provide the necessary funding to assist in acquiring the required facilities to make those spaces greener. In view of this, in October 2019 the Bank of Africa in Ghana signed a Memorandum of Understanding (MOU) with the Accra city authorities to adopt and transform the triangular open space located at the heart of Accra into a green park (Ghanaian Times, 2019). Other organizations that have also adopted open spaces in Accra are the Ghana Commercial Bank (now GCB Bank), Ecobank, and Church of Jesus Christ of Latter-day Saints. The Ecobank for instance has adopted the open space stretching from Ridge Roundabout through the median of African Union (AU) Circle to National Theatre Traffic light.

The Phase Two covers the expansion of twenty-one (21) green spaces within the Accra Metropolis. The basic challenge confronting this project is the problem of acquiring enough lands for the intended purpose. This is because traditional authorities (chiefs) who own many lands in the city are reluctant in making lands available and squatters who have encroached lands earmarked for green spaces proving difficult to vacate those lands.

Green and Energy Efficient Construction and Building in Accra

Energy efficiency measures which are keys in minimizing greenhouse effects and also the consequences of climate change have not received the needed attention in

Ghana's National Building Regulations. Due to this shortfall in the National Building Code of Ghana which is supposed to guide the constructions of all kinds of buildings in the country, contractors in Accra build without incorporating energy efficiency measures in their architectural designs and building aesthetics resulting in the generation of much heat in the city. Green building technologies that support energy efficiency measures with an appropriate rating system in Accra were to encourage environmental friendly and energy saving practices in newly constructed or retrofitted buildings. There is therefore currently an ongoing review of the National Building Regulations by the Ministry of Works and Housing with the assistance of the Energy Commission of Ghana to factor in energy efficiency measures and green building technologies to reduce the emissions of greenhouse gases especially in urban areas which harbours most the country's buildings.

To achieve energy efficiency in buildings in Accra which falls under the third component of the Greater Accra Climate Resilience and Integrated Project, the AMA is in partnership with the International Financial Corporation's (IFC's) existing Excellence in Design for Greater Efficiencies (EDGE) tool to make buildings in Accra environmental friendly. This partnership revolves around finding good software to determine the optimum combination of building design strategies for the best return on investment, achieving a global standard of a 20% reduction in energy and water use, comparing embodied energy in materials in a standard building, and offering certification that validates the developer's achievement (Accra Metropolitan Assembly, 2018). Furthermore, the AMA has started a process of developing an educational toolkit that lay emphasis on the building codes, energy savings, and measures to enable businesses to adhere to the codes.

The Accra city authorities are bent on building on earlier piloted project by the Accra Spatial Planning Committee to institute an incentive structure to promote the integration of renewable energy into new constructions. The piloted project by the Accra Spatial Planning Committee in July 2017 focused on offering a 10% reduction in the building permit fee for applications that have incorporated a 20 kW roof-mounted solar PV system in the building design (Accra Metropolitan Assembly, 2018). The Energy Commission of Ghana has therefore been tasked in this direction to develop clear guidelines for use by the City in the administration of incentives for building permit applications with renewable energy systems. Poor conformity to energy efficiency measures and green building technologies by the general public as against the high initial investment and lack of awareness have posed much challenge to this project.

Upgrading Informal Settlements and Living Environments

This project falls under the third component of GARIDP. It is estimated that about 60% of the residents of Accra live and work in informal settlements. In 2016, a study conducted by Ghana Federation of the Urban Poor (GHAFUP) on living conditions of Accra's informal settlements found a myriad of challenges facing these settlements (Accra Metropolitan Assembly, 2018). These challenges included dilapidated housing structures and poor water supply. The informal settlements were also negatively affected by fire outbreaks due to improper and illegal electricity connections, perennial flooding resulting from their poor locations on water ways and flood prone areas, and poor sanitation conditions causing outbreaks of diseases such as malaria, cholera, diarrhoea, and typhoid.

To upgrade the informal settlements to a befitting status and to also make them resilient in times of disasters, the Accra Metropolitan Assembly in collaboration with Greater Accra, Resilient and Inclusive Development Project (GARIDP) have started taking some initiatives which include the following:

- Assisting through a number of mechanisms to upgrade four main informal settlements: Alajo, Akweteyman, Nima, and Alogboshie. These mechanisms include providing engineering insights around acceptable infrastructure for these communities, brokering negotiations among key stakeholders, and reviewing environmental approvals to prepare a healthy environment for the selected communities.
- Providing basic infrastructural services for the residents of these informal settlements. This is done through collaboration with Slum Dwellers International, UN-Habitat, the World Bank and its partners, and GHAFUP.
- Providing appropriate capacity building and learning for all the stakeholders involved. A host of resilience benefits are expected to be produced after a successful completion of this project. Among these benefits are improved safety and security for residents, better public health conditions, and reduced risk from fires, floods, and disease outbreak. However, some residents of the affected neighbourhoods who see the upgrade as a means of temporal or permanent disruption of their livelihoods (perceive them as losers of the project) are mounting resistance campaigns against the implementation of the project.

Integrated Planning to Improve Resilience to Earthquake and Urban Sanitation

This is a major project under the second and fourth components of GARIDP. The AMA through partnership with the National Disaster Management Organisation (NADMO) has begun undertaking a vulnerability and capacity assessment to find out populations and systems at risk, and also the preparedness systems that can help to withstand and respond to earthquakes (Accra Metropolitan Assembly, 2018). As part of this, training workshops have been organized by the Engineering Council for artisans and technicians on the building regulations of Ghana. This is to enhance their skills and abilities to build houses that can withstand disasters such as earthquakes. In addition to this, existing building statutes and protocols are being reviewed to develop a solid regulatory framework to make Accra more resilient to earthquakes. Among the status being reviewed are National Disaster Management Act (ACT 927), Land Use and spatial Planning ACT (ACT 925), Engineering Council Act (ACT 819) and National Building Regulations (LI 1630).

In the area of urban sanitation, master plan for drainage, sanitation and solid waste management is being prepared. The master plan among other things is expected to provide a strategy for effective management of environmental sanitation including solid and liquid waste and storm water runoff from the Greater Accra Metropolitan Accra (GAMA). It is also expected to provide solutions to existing environmental sanitation and drainage problems, and also guide the implementation of future environmental sanitation and drainage improvements. Disruption of utility services and damage to public infrastructure which may be caused by laying of both surface and underground drainage pipes are potential challenges that need to be addressed to enhance the implementation of this problem. In addition, not-in-my-back-yard (NIMBY) syndrome which has affected the siting of land field sites to manage solid waste is a challenge that the project needs to tackle.

26.3.4 Makumbura Multimodal Transport Hub, Sri Lanka

Introduction

The Kottawa-Makumbura Multi Modal Transport Centre (KMMC), Sri Lanka is a ubiquitous public transit hub in West of Colombo established in 2019. Being the first of its kind in Sri Lanka, the project is part of the larger Township Development Component of Greater Colombo Urban Transport Development Project (3 K Project), undertaken by the Ministry of Mega Polis and Western Region Development Ministry. The pilot operation of the transit hub runs from 2019–2022 under a Project Management Unit while the full operations under a formally set up institution to be commenced subsequently. The project is a partial solution to the severe traffic and transport problem in Colombo – the most severe urban issue in the city. The project proposal is embedded with several going green credentials which help with low carbon emission in the city transport; integration of public transport modes, using of renewable energy to operate the transit hub, capacity building programs for bus operators and staff and, smart ticketing and turn around management system.

Integrated Public Transport Modes

The KMMC integrates public transport modes which include; the existing Kelani Valley commuting rail line (connecting Colombo West to East), long and short-distance buses that connect with A4-Colombo—Batticaloa highway and the EX01-Southern Expressway (toll way). By the end of first year's pilot operation, more than 50% of long and short-distance bus routes in A4-Colombo—Batticaloa highway and all bus routes in EX01-Southern Expressway have been integrated with the Kelani Valley commuting rail line. This integration facilitates public predominant journeys within Colombo: trips to manufacturing and service sector work places in Colombo, national universities, residential suburbs within Colombo Metro region, and affordable shopping markets situated within Colombo such as in Maharagama and Pettah. The passengers transit via KMMC also facilitates travel information, air-conditioned waiting areas, convenient shops, food outlets, resting places, differently abled access, park and ride facilities, schedule display of all rail and bus services, ATM machines, and luggage services. The aim is to attract more people to use public transport by enabling it to be accessible easily, convenient, and affordable.

Renewable Energy Use

Energy for KMMC operations is supplied through its in-situ produced solar energy. The KMMC's buildings has been designed with wider-spaced roof top and sufficient load bearing capacity in a way that it can mount solar photovoltaic and produce electricity. The consultants have estimated that KMMC at completion will have the capacity to produce 250 KWP. The estimations are that approximately 75% of such electricity production will be needed for KMMC's full daily operations: 24-h illumination of the centre, signage, and promotional panels, operation of escalators, elevators, functioning of office and shops and air conditioning, and so on. The KMMC's business plan proposes selling the surplus energy to the national electricity grid which was enabled by the government's "Sooryabala Sangramaya"[5] (Battle for Solar Energy program). Strategically, this will help KMMC to function with low carbon emitted renewable energy (during the first six month of 2020, it has saved 40,060.4 kg Co_2 emissions) while strengthening the financial viability of the facility and offering modern services to users. However, the extent to which KMMC can save energy as surplus to the national grid depends on behavioural aspect of KMMC users and the design of office buildings too. During the pilot stage of KMMC which mostly used temporarily partitioned office spaces that did not support energy saving to the level expected has resulted in a limited electricity supply to the national grid.

Smart Ticketing and Turn Around Management

KMMC at its full fledge stage is planned to be upgraded with smart operations: integrated ticketing for passengers and turn around management system for bus operations. The integrated ticketing aims to provide the opportunity for passengers to purchase a single ticket for their entire journey (i.e. bus, rail, taxi combined as required in a trip). The aim is to provide passenger convenience, and at the same time generate real time data about passengers' travel demand. This informs the management about the required supply levels of buses and rail services thus helps avoiding energy waste caused by inefficient service delivery. The turnaround of buses at KMMC aimed to be managed through an App provided to the bus operators' Smart phone-when they should start the engine for loading of passengers at the bay. The aim is to help reducing the fuel consumption of bus operations: if not for such a turnaround management system the practice is that the bus operators continue to keep the engine on till their trip turns. However, the KMMC's plans of upgrading of its operations to smart systems has been delayed owing to reasons such as: procurement interruptions caused by the economic shut downs due to COVID-19 lock downs of 2020 and Easter terrorist attack on Colombo in 2019, institutional bureaucracy and several other reasons that are of political in nature.

[5] The "Sooryabala Sangramaya" is a micro power producer scheme promoted by Ministry of Power and Renewable Energy in Sri Lanka that encourages institutional electricity consumers such as KMMC to produce their own electricity via solar energy and export the surpluses they generate to the national electricity grid.

Capacity Building of Bus Operators and Other Staff

The KMMC's operational and maintenance plan provided significant budgetary allocations and identified actions needed for capacity building of bus operators who would be attached to the hub and other service providing staff that include technical, ground handling, security services, janitorial etc. Even though private sector provides nearly 70% of bus service in the country, the private bus operators work for individual bus owners as casual labourers. They largely come from less educated backgrounds. The state agencies for transport such as National Transport Commission, Western Province Passenger Transport Authority etc. regulate only the route permits and the bus schedules. The KMMC carried out this capacity building programs in collaboration with University of Sri Jayewardenepura with the understanding that attracting people to public transport and adopting planned smart systems require not only the physical facility, but also the staff adherence to the system and quality service provided by the service providers. Training programs were held both prior and after the pilot commencement of the facility, and were designed, branded, and conducted by university academics in the field of transport, planning, real estate management, and human resource management. The capacity building program focused on adopting the service providers to KMMC's vision and aims, sophisticated equipment and systems, new operational procedures, communication skills, and positive work attitudes. However, the high turnover of casually employed bus operators which required the KMMC to train the numbers more than expected, together with funding constraints imposed a challenge to the conduct of the capacity building programs persistently.

26.4 Discussion and Conclusion

Going green has been understood, recognized and practised within urban development as a human-centric concept -It had mattered, because it affects human lives. In modern societies since nineteenth Century, going green has been a solution to issues and challenges of urban development that affects the livelihood of urban dwellers. As shown above, the key interwoven factors that formed and reformed such "going green" solutions have been the ideologies and values that people are built into and those becoming mainstream in the society time to time and, the knowledge and technological innovations that people are progressing towards.

From nineteenth Century up to latter parts of twentieth Century which were dominated by periods of industrialisation and urbanization, poverty, public health problems, and pollution were among the major urban issues. This period was enriched with substantial level of imaginary and dreamy ideologies and influential movements for social and environmental justice (*Garden City movement, Broadacre City, Radiant City, compact city, transit-oriented development, new urbanism movement, smart growth agenda, eco-city, sustainable development likewise*). These eventually influenced the thinking of policy makers, practitioners, academia and masses that a "good" urban development means a built-up area that is in harmony with the natural environment and which promotes equity in the society. Hence the idea of going green was interweavingly viewed as; all living in better "quality" environments, reduce, reuse, and recycle of resources at production, distribution, and consumption stagesand, balancing the environmental, economic, and societal interfaces.

Between the late twentieth Century and early part of twenty-first Century witnessed intensification of globalization, and complexity of the society and environment. The academic concerns for urban development during this time were particularly on climatic issues and disaster management, water and energy conservation, general well-being of urban dwellers. Accordingly, the going green concept at its early stage while being influenced by previous development ideologies, paid more attention to possible methods, approaches, and techniques to operationalize those development ideologies under the given complex context. In that respect, stakeholder participation collaborative governance, and smart systems were predominantly stressed as the means of going green.

The four case studies from England, Australia, Ghana, and Sri Lanka demonstrate going green practices in urban developments across global north and south. The experiences from the case studies show that going green projects have provided multiple benefits in urban areas such as providing better housing, transport, amenities, employment, minimizing risks from climatic changes, attracting funding for cities and, and increasing quality of life. In terms of practising going green in urban development, the specific measures put in place from the selected urban areas include: (i) Conservation of one-fifth to one-third of planned green open space land use within urban limits, (ii) Creating mixed land uses with housing, employment, and services well-connected through rail and buses, (iii) Developing inclusive housing mix which has equity access to sanitation, services, employment, and green spaces, (iii) Developing transport system that encourage active traveling (walking, cycling), public transport and any other mode that emits lower CO_2 levels, and (iv) Designing of building structures to be energy efficient and emit minimum levels of CO_2. Such physical development have been established with systems which include: (a) urban vulnerability, climatic risk and disaster risk identification and responsive planning, (b) national level planning, laws, strategies, and approaches together with local government's adherence and further reinforcement of these regulations, (c) Stakeholder and community participation in managing urban common property, and constructive criticism and communication about system or structural weaknesses, (d) Smart Technology to make urban systems and functions convenient and affordable, (e) Multiple types of funding modelling to attract state, corporate, individual capital or capacities to be invested in urban developments and management of building structures.

The case studies further show a variety of challenges hampering successful implementation of going green projects. Interweavingly, these challenges cover development pressure imposed on green open spaces, high cost associated with constructing energy efficient housing and other infrastructure; incompatibilities of technological advancement and diversity of people's behaviour that cannot operationalize smart systems as planned; societies living in scalar interests that are often subject to change and the dynamics of social, economic and physical environmental context as a whole.

The going green development practices in urban areas show the efforts being made in different urban context (global north and south) to create sustainable cities and communities (SDG 11). Even though going green is not an easy and definitive target to achieve, continuous learning from practice and sharing of ideas will help to overcome this herculean endeavour to protect the natural environment and at the same time enhance the well-being of urban dwellers in future eras to come.

Acknowledgements The first author thanks Professor Baratha Dodankotuwa and Prof Hettige Don Karunaratne for the opportunities given to speak about this topic at different platforms which eventually helped developing the ideas into a chapter. The Centre for Real Estate Studies, University of Sri Jayewardenepura, Sri Lanka is also acknowledged for providing various assistance in developing this chapter.

References

Accra Metropolitan Assembly. (2018). *Accra resilience strategy.* Accra Metropolitan Assembly.

ACT Government. (2019). https://www.environment.act.gov.au/__data/assets/pdf_file/0005/1413770/Canberras-Living-Infrastructure-Plan.pdf.

ACT Planning & Land Authority. (2004). *Canberra spatial plan.* http://apps.actpla.act.gov.au/plandev/sp-pdf/spatialplan.pdf.

Alexander, A. (2009). *Britain's new towns: Garden cities to sustainable communities.* Routledge.

Ashique-Uz-Zaman, Q. M. (2012). *Adaptation of the eco-city concept for cities in developing countries (Master Thesis).* Department of Architecture, National University of Singapore.

Australian Bureau of Statistics. (2016). https://www.abs.gov.au/.

Canberra Times. (2020). https://www.canberratimes.com.au/story/6726907/griffins-garden-city-plan-pays-off-during-pandemic/.

Carmon, N., & Fainstein, S.S. (Eds.).(2013). *Policy, planning, and people: Promoting justice in urban development.* University of Pennsylvania Press.

Cooper, T. (2008). Challenging the 'refuse revolution': War, waste and the rediscovery of recycling, 1900–1950. *Historical Research, 81*(214), 710–731.

Environment and Sustainable Development Directorate. (2012). *Transport for Canberra: Transport for a sustainable city 2012–203.* https://www.transport.act.gov.au/__data/assets/pdf_file/0017/1237013/TransportForCanberra_Policy.pdf.

Environment, Planning and Sustainable Development Directorate. (2019). *Canberra's living infrastructure plan: Cooling the city.* https://www.environment.act.gov.au/__data/assets/pdf_file/0005/1413770/Canberras-Living-Infrastructure-Plan.pdf.

Environment, Planning and Sustainable Development Directorate–Environment. (2019). https://www.environment.act.gov.au/__data/assets/pdf_file/0003/1414641/ACT-Climate-Change-Strategy-2019-2025.pdf/_recache.

Fainstein, S. S. (2000). New directions in planning theory. *Urban Affairs Review, 35*(4), 451–478.

Fainstein, S. S. (2014). The just city. *International Journal of Urban Sciences, 18*(1), 1–18.

Falk, N., & Carley, M. (2012). *Sustainable urban neighbourhoods: Building communities that last.* Joseph Rowntree Foundation.

Foreign and Commonwealth Office. (2018). https://smartsustainablecities.uk/vivamk-milton-keynes-smart-city/.

Fulton, W. (1996). *The new urbanism: Hope or hype for American communities?* Lincoln Institute of Land Policy.

Geissdoerfer, M., Savaget, P., Bocken, N. M., & Hultink, E. J. (2017). The circular economy–A new sustainability paradigm? *Journal of Cleaner Production, 143*, 757–768.

Ghanaian Times. (2019). *Ghana: Greening Accra—Bank of Africa adopts open space.* https://allafrica.com/stories/201910150395.html.

Green Energy News. (2020). https://greentechnewstoday.com/renewable_energy/canberra-households-reject-gas-as-act-moves-to-end-mandatory-connections/.

Gushulak, B.D., Weekers, J., & MacPherson, D. W. (2009). Migrants and emerging public health issues in a globalized world: Threats, risks and challenges, an evidence-based framework. *Emerging Health Threats Journal, 2*(1), 7091.

Hamari, J., Sjöklint, M., & Ukkonen, A. (2016). The sharing economy: Why people participate in collaborative consumption. *Journal of the Association for Information Science and Technology, 67*(9), 2047–2059.

Head, L., Adams, M., McGregor, H., & Toole, S. (2014). Climate change and Australia. *Wiley Interdisciplinary Reviews: Wires Climate Change, 5*(2), 175–197.

Healey, P., Cars, G., Madanipour, A., & De Magalhaes, C. (2017). Transforming governance, institutionalist analysis and institutional capacity. In *Urban governance, institutional capacity and social milieu* (pp. 6–28). Routledge.

HEFCE. (2014). https://assets.publishing.service.gov.uk/government/uploads/system/uploads/attachment_data/file/311717/HEFCE_An_Rep_and_Acc_2014.pdf.

Hobson, J. (1999). New towns, the modernist planning project and social justice. *The Cases of Milton Keynes.*

Ian, C. (2009). *Histories of transit-oriented development: Perspectives on the development of the TOD concept* (Working Paper, No. 2009,02). University of California, Institute of Urban and Regional Development (IURD), Berkeley, CA.

Jessop, B. (2000). Globalisation, entrepreneurial cities, and the social economy. *Urban movements in a globalising world*, 81–100.

Klaassen, E. A. M., Veldman, E., Slootweg, J. G., & Kling, W. L. (2013) Energy efficient residential areas through smart grids. In *2013 IEEE Power & Energy society general meeting* (pp. 1–5). IEEE.

Konteh, F. H. (2009). Urban sanitation and health in the developing world: Reminiscing the nineteenth century industrial nations. *Health & Place, 15*(1), 69–78.

Marchese, D., Reynolds, E., Bates, M. E., Morgan, H., Clark, S. S., & Linkov, I. (2018). Resilience and sustainability: Similarities and differences in environmental management applications. *Science of the Total Environment, 613*, 1275–1283.

Marsden, G., & Docherty, I. (2013). Insights on disruptions as opportunities for transport policy change. *Transportation Research Part a: Policy and Practice, 51*, 46–55.

Martin, N. J., Rice, J. L., & Lodhia, S. K. (2014). Sustainable development planning: A case of public participation using online forums. *Sustainable Development, 22*, 265–275.

Mees,. (2014). A centenary review of transport planning in Canberra, Australia. *Progress in Planning, 87*, 1–32.

Miller, J. D., & Hutchins, M. (2017). The impacts of urbanisation and climate change on urban flooding and urban water quality: A review of the evidence concerning the United Kingdom. *Journal of Hydrology: Regional Studies, 12*, 345–362. https://doi.org/10.1016/j.ejrh.2017.06.006.

MK Citizen. (2020). https://www.miltonkeynes.co.uk/lifestyle/outdoors/milton-keynes-ranks-top-green-space-2551060.

MK Council. (2013). file:///C:/Users/User/Downloads/Local_Economic_Assessment_Refresh_March_2013_Final.pdf.

MK Council. (2018). http://www.miltonkeynesforum.org/uploads/7/5/5/2/75526607/green_infrastructure_stategy_-_mk_forum_response.pdf.

MK Council. (2020a). https://www.milton-keynes.gov.uk/your-council-and-elections/statistics/housing-statistics.

MK Council. (2020b). Affordable housing supplementary planning document. https://www.milton-keynes.gov.uk/planning-and-building/planning-policy/adoption-affordable-housing-supplementary-planning-document.

Nationwide House Energy Rating Scheme. (2015). https://www.nathers.gov.au/sites/default/files/Admin%2520Governance%2520Arrangements-August%25202015.pdf.

OGL. (2016). https://www.gov.uk/government/news/40-million-to-drive-green-car-revolution-across-uk-cities.

ONS. (2018). https://www.ons.gov.uk/census/2011census.

ONS. (2019). https://www.ons.gov.uk/census/2011census.

Parker, S. (2015). *Urban theory and the urban experience: Encountering the city*. Routledge.

Perera, U., & Mensah, C. A. (2019). Conceptualizing sustainability in the real estate development process. *Sustinere. Journal of Environment and Sustainability, 3*(1), 1–14.

Population Reference Bureau. (2008). *2008 world population data sheet*. Population Reference Bureau.

Rakhshandehroo, M., Yusof, M. J. M., Arabi, R., Parva, M., & Nochian, A. (2017). The environmental benefits of urban open green spaces. *AlamCipta, 10*(1), 10-15.

Robinson, J. (2004). Squaring the circle? Some thoughts on the idea of sustainable development. *Ecological Economics, 48*, 369–384. https://doi.org/10.1016/j.ecolecon.2003.10.017

Rolnik, R. (2013). Late neoliberalism: The financialization of homeownership and housing rights. *International Journal of Urban and Regional Research, 37*(3), 1058–1066.

Røyne, F., Penaloza, D., Sandin, G., Berlin, J., & Svanstrom, M. (2016). Climate impact assessment in life cycle assessments of forest products: Implications of method choice for results and decision-making. *Journal of Cleaner Production, 116*, 90–99.

Sharifi, A. (2016). From Garden City to Eco-urbanism: The quest for sustainable neighborhood development. *Sustainable Cities and Society, 20*, 1–16.

TCPA. (2007). Best practices in urban extensions and new settlements. https://www.tcpa.org.uk/Handlers/Download.ashx?IDMF=55474980-8f1e-4298-a86f-5fec267dc530.

UN. (1951). Proceeding of the United Nations scientific conference on the conservation and utilization of resources. 3, 2–333.

UNDP. (2016). *Sustainable urbanization strategy*. UNDP.

UNEP. (2009). A global green new deal. https://sustainabledevelopment.un.org/index.php?page=view&type=400&nr=670&menu=1515. [last accessed on 09 Jun 2020].

United Nations. (2019a). *World Urbanization Prospects 2018: Highlights*. Department of Economic and Social Affairs, Population Division.

United Nations. (2019b). *Progress towards the Sustainable Development Goals: Report of the secretary general*. United Nations.

UN Water. (2007). Coping with water scarcity: Challenge of twenty-first century. http://www.fao.org/3/a-aq444e.pdf.

Wohl, A.S.(1983). *Endangered lives: Public health in Victorian Britain*. JM Dent and Sons Ltd.

World Bank. (2018). *Concept note: Greater Accra climate resilient and integrated development project*. World Bank.

Chapter 27
Internal Migration and Urban Development: Recent Experiences of China

Xingna Nina Zhang

Abstract China has seen an unprecedented urbanization and industrialization process since 1978, during which internal migration has played an important part in shaping the overall urban development across the country. Based on China's recent urban development experiences, this chapter explores the links between internal migration and socio-economic development, with a focus on the critical role of state policies. It first examines major trends and characteristics of internal migration in mainland China, and then elaborates links between internal migration and socio-economic development. China is a vast country with diverse geographical, demographic, cultural, and socio-economic conditions across different regions. It is therefore important to evaluate its urban development in a systematic manner regarding issues of regional inequalities, as well as the associated challenges and opportunities of internal migration in affecting urban employment and poverty. Some local governments strategically implement state policies of migration management and regional development in accordance with local needs, which have greatly facilitated harnessing the merits of internal labour relocation as a source of urban socio-economic development. Such practices have also helped to cultivate the sense of "localised urbanisation" as a nationwide state-led development manifesto. The chapter suggests that China's experiences of internal migration and urban development have important implications for policy making regarding migration management and regional development governance, and can be shared by other developing countries.

Keywords Migration · Urban development · China · Policies and governance · Migration management

X. N. Zhang (✉)
Department Public Health, Policy and Systems, Institute of Population Health, University of Liverpool, Liverpool, UK
e-mail: xingna.zhang@liverpool.ac.uk

B. R. K. Sinha (ed.), *Urban Dynamics, Environment and Health*,
https://doi.org/10.1007/978-981-99-5744-6_27

27.1 Introduction

Since the economic reform in 1978, China has witnessed large volumes of internal migrants. Based on the definition of Chinese censuses, internal migrants refer to people who currently reside in one place but have their household registration in another county (city or district) and have left their registered residence for more than six months (Fan, 2005a; Shen, 2012). Because of its huge population size, the scale of such population movement is astonishing and often surpasses that of other countries. For instance, there were 294 million internal migrants in China in 2015 (Source: National Bureau of Statistics of China), which is 20% larger than the total number of international migrant stock of the same year in the world (Source: UN DESA). This massive internal migration is reported as primarily labour redistribution and economically driven, which has been essential to China's rapid urbanization and industrialization (Fan, 1997; Shen & Yang, 2014).

Apart from the impressive scale, internal migration in China has other interesting features, including its unique state policies towards migration management and socio-economic development. China has achieved fast industrialization and rapid economic growth, which is accompanied by massive internal population movement within the country (Fan, 1997, 2005a; Lewis, 1954). Behind this massive internal migration, the major cause is economic growth but not vice versa (Zhang & Shunfeng, 2003). The rapid economic growth is achieved by combining market forces and state interventions such as the implementation of a wide range of regional development policies (Fan, 2005a; Lewis, 1954). Studying China's recent internal migration therefore offers a valuable opportunity to further the understanding of migration in the context of a socialist transitional economy.

Research on internal migration in China has grown steadily over the past few decades but further advances in theory and empirical study strategies are still arguably in need. This chapter therefore reviews relevant literature in order to identify research frontiers in understanding the relationship between internal migration and urban development in China. The first section explains the characteristics and recent changes of internal migration in China and how internal migration is related to urbanization. The second section then discusses major macro- and micro-level drivers of China's internal migration. The final section summarizes how perspectives afforded by the literature have allowed the development of a greater understanding of internal migration and urbanization in China.

27.2 Characteristics and Recent Changes of Internal Migration in China: How is It Distinctive?

27.2.1 Characteristics

Compared with other developing countries, China's recent internal migration and urbanization story is unique for two reasons: first, the state keeps relatively strict and effective management on internal migration through multiple policies; second, China's internal migration forms one of the largest "temporary" population movements in human history with a peak volume of 298 million in 20,141. The temporariness of China's internal migration lies in its extensive institutional interventions (Fan, 2007; Liang & White, 1997); so much so that some migrants have to adopt circular migratory strategies (Liang & White, 1997; Zhao, 2003). Also due to the temporariness of China's internal migration, measurements of internal migration have been notoriously inconsistent and difficult to detect to the extent of being "statistically invisible" (Chan, 2013; Fielding, 2010). This makes the internal migration in China a remarkable phenomenon but rather complicated to study.

For most countries around the world, nationwide internal-migraiton-related policies usually address three major issues—overpopulation and poverty in rural areas, rapid expansion of metropolitan areas, and unbalanced regional development (Findley, 1977). In the context of China, it has the one-child policy (recently abolished at the end of 2015) to cope with overpopulation (Greenhalgh, 2008), the infamous Hukou system (户口, namely the household registration system) to manage urbanization speed and to regulate metropolitan expansion (Bao et al., 2011; Chan, 2009; Chan & Zhang, 1999), and various regional development policies to coordinate regional development with the spatial distribution of labour (Bao et al., 2002; Chen & Groenewold, 2011; Fan, 2005a; Fielding, 2010).

China has maintained a long-standing unbalanced regional development policy until very recently. Since the establishment of the People's Republic of China in 1949, its regional discriminative development policies have gone through five stages in general (Sun, 2013). Stage one lasted from 1949 to 1978 and involved almost 30 years of central planning, resulting in the overly agriculture-based economic structure before the economic reform in 1978. Stage two spanned from 1979 to the early 1980s and featured a clear coastal region orientation with limited trade freedom in Guangdong and Fujian of the east region. The next stage saw the enhancement of the coast-favoured strategy by widening free trade rights and regions between 1985 and 1989. 1990s marked the arrival of stage four, extending free trade rights to several inland cities along the Yangtze River. Post-2000 has seen the emergence of stage five, when the whole of China was opened up with three major economic region policies—China Western Development, Revitalize Northeast China, and Rise of Central China.

Among all the migration management policies, the Hukou system is the most significant and widely accepted as a culprit of rural and urban divide (Liu, 2005). The current form of Hukou system came into effect in 1958, but with its origins dating back to ancient times (Young, 2013). It is designed to restrict and regulate population movements by assigning migration quota through managing the admission systems, regulations of entry, duration of stay, work permits, and access to social welfare support. The government has the residential registration information of "urban households" or "rural households" for the citizens and every status holder is either recognized as "agricultural" or "non-agricultural" accordingly. The labour force transfer from rural-to-urban areas has faced many constraints from the Hukou system (Vendryes, 2011; Wang, 2012), which allows urban residents to have privileges on employment and other public services.

China also has numerous regional and provincial migration policies tailored to more nuanced local needs. For intance, first-tier cities such as Beijing and Shanghai have the most strigent Hukou policies in China, which only allow internal migrants with specifically needed qualifications and skills to obtain the local Hukou to permanently settle down (Guo & Iredale, 2004; Wang & Zuo, 1999). As Goldstein and Goldstein (1987) summarized, China's state and regional policies had been primarily oriented towards internal migration management, but it had led to a striking contrast between the designed effects of such policies and their achieved results. They have also created tremendous pressure in simultaneously achieving development and migration control for many local governments. For instance, in the late 1970s China adopted a nationwide policy of promoting small towns and rural enterprises to reduce migration (Byrd et al., 1990; Liang & White, 1997); after a few years, many rural areas with successful rural enterprises did see an evident decrease in out-migration, whereas they also experienced more significant increase in in-migration (Liang & White, 1997; Ma & Fan, 1994). Despite being very effective in promoting socio-economic development (Ma & Fan, 1994), such a policy was greatly revised in 1993 due to its very limited success in reducing total migration (Liang & White, 1997).

As shown above, politics and the state are essential to understanding the characteristics of China's internal migration (Brettell & Hollifield, 2014). The following sub-section will be devoted to discussing major recent changes of China's internal migration to depict the research context.

27.2.2 *Recent Changes*

Despite the two long-standing, main characteristics of internal migration in China—namely, being temporary in nature and being strictly regulated by policies—it is experiencing important structural changes and new trends are emerging. According to *Reports on China's Migrant Population Development* and *China Statistical Yearbook*, China's internal migration has shown seven major recent changes since 2010, which will be explained in detail below.

First, even though the total volume of internal migration remains vast, it has seen a slow decline since 2015.[1] This is caused by a continued decelerating natural population growth with slackening fertility and declining mortality. Internal migration, however, continues to be the leading factor in shaping population changes in China, and is likely to remain so for a long time. If put into Zelinsky's mobility transition theoretical framework (Zelinsky, 1971), this new development signposts the onset of "Phase C- The Late Transitional Society" to "The Vital Transition" for the Chinese society (Zhu, 2018). Most importantly, this new development may have marked a significant structural and compositional change in China's internal migration: the continued growth of rural–urban migration during the past four decades has finally levelled off, while urban-urban migration is gaining momentum and becoming an increasingly important component (Zhu, 2018; Zhu et al., 2016). This may set China's internal migration apart from its peers in many modernizing societies such as India and Vietnam (Anh et al., 2012; Chen & Ravallion, 2004; Czaika, 2012; Deshingkar, 2006; Nath, 1988; Nguyen & Locke, 2014; Phuong & McPeak, 2010). Because these developing countries are still experiencing increasing rural–urban migration and large volumes of marriage-oriented rural-rural migration (Lucas, 1997; Parida & Raman, 2018). China's current internal migration therefore offers a valuable opportunity to advance migration research in the structure and component of internal migration at this critical transitional time point.

Second, there are signs of increasing stability, as the average settlement period for migrants grows from 4.8 years in 2014 to 5.7 years in 2016.[2] This new development reflects that China's internal migration has entered the stabilizing stage where migrants are increasingly seeking for settlement (Lucas, 1997). It may also be relevant to the continual growth of the Chinese economy, where capitalist development proceeds and regions become more integrated with widespread grass-root private entrepreneurships and state-owned entities gaining momentum simultaneously (Fielding, 2010; Yiu et al., 2007). Consequently, Chinese economy is becoming more capable of retaining "fluid" labour locally overall, while enjoying economies of scale as a whole (Appadurai, 1990; Fukuyama, 1995).

Third, there has been continued growth of new-generation migrants (born in 1980 and after), whose share has risen from 48.8% in 2013 to 64.7% in 20,163. Compared with the former generation, this new-generation migrants are not only younger, but tend to have little agricultural experience. They also have a lesser sense of rural identity (Wang, 2001). Many of them prefer to relocate to bigger cities for employment opportunities with better income prospects and career trajectories (Huang & Zhan, 2005; Wang, 2001). This new development is unsurprising due to increases in education and skills among migrants (T. Fielding, 2010; Huang & Zhan, 2005).

Fourth, internal migrants have experienced a faster ageing process than the national population. The average age of migrants has grown from 27.3 to 29.8 years old from 2011 to 20,163, while the national population aged about 2.0 years during

[1] According to China Statistical Yearbook 2017, the peak value was 298 million in 2014, which reduced to 294 and 292 million in 2015 and 2016 respectively.

[2] Data source is Reports on China's Migrant Population Development.

this period. This is widely recognized as one consequence of strong state intervention, i.e. the one-child policy (Chen & Powell, 2012). China is likely to get older before it gets richer, whilst an ageing population is typically seen as a burden rather than a blessing to socio-economic development (Chen & Powell, 2012; Fan, 2007). This fast-ageing migration population could cause systemic and structural changes to social support systems of both the origin and the destination (Chan & Zhang, 1999; Fan, 2007; Rozelle et al., 1999). It may also relate to the aforementioned signs of increasing stability—as migrants get older, they tend to have stronger intention and desire to settle down, such as starting a family or taking up care responsibilities (Rogers & Castro, 1981; Rogers & Willekens, 1986; Schanbacher, 2007; Wang & Fan, 2006).

Fifth, internal migrants have enjoyed a larger household income growth than the national average. There has been a 72.6% increase between 2013 and 2016 for migrants, in contrast to 30.1% of the national average.[3] This is in line with the neoclassical economic paradigm, which expects population movement originating from low-income places and settling down in (comparatively) high-income destinations with migrants seeking to maximize their utility (King, 2012; Massey et al., 1993; Sjaastad, 1962). Higher economic returns are indeed the fundamental booster powering China's urbanization process marked by long-standing massive nationwide population relocation (Fan, 2005b, 2007). Migration has important impacts upon migrant households beyond increasing family income, such as empowering female migrants and restructuring family relations (Fan, 2003; Fan & Wang, 2008; Fan et al., 2011; Wang & Fan, 2006, 2012). It has also created spill-over effects of consumption and entrepreneurship in local neighbourhoods and communities of the origin (Cao et al., 2017; Fan, 2007; Gao, 2012; Han & Chen, 2016; Rozelle et al., 1999; Snyder & Chern, 2009; Wahba & Zenou, 2012; Wenfei & Fan, 2006). Nevertheless, none of these would have happened without the onset of the famous "Reform and Open-up" policy in 1978, which opened the Chinese market to international capitals (Fujita & Hu, 2001; Nee et al., 2007; Prahalad & Lieberthal, 1998; Tsai, 2007).

Sixth, there has been a growing trend of familiarization among internal migrants with rising household size from 2.5 to 2.6 persons between 2013 and 2015 (see footnote 3). This stands as a contrast to the prevailing conjugal separations and family splits observed among earlier migrants for three decades (Cheng, 1991; Fan, 1995; Fan & Sun, 2008; Fan et al., 2011; Fan & Wang, 2008; Gordon, 1995; Ma & Fan, 1994; Wang & Fan, 2007; Ye et al., 2013). It, therefore, lends further support to the hypothesis that China is currently in the late transitional stage of Zelinsky's mobility transition framework (Zelinsky, 1971; Zhu, 2018). There are mainly three groups of left-behind family members, namely wives, the elderly, and children (Biao, 2007; Fan, 2007; Wen & Lin, 2012; Ye et al., 2013). The left-behind population encounter various challenges, such as agricultural production, family caring, and children's education and well-being (Chang et al., 2011; Lu et al., 2016; Wen & Lin, 2012; Ye et al., 2013). Compared with non-migrant households, however, the situation of the

[3] Data source is the National Bureau of Statistics of China.

left-behind population of migrant households in general is not much worse (Biao, 2007), which may be relevant to the remittance sent back by the migrants (Chang et al., 2011; Cong & Silverstein, 2011; Deshingkar, 2006). It is widely believed that the fundamental cause of conjugal separations and family splits is the long-standing unbalanced regional development policy, although on the surface family separations within migrant household may seem like voluntary decisions at the individual level (Biao, 2007; Fan, 1997, 2007). These institutional factors set fundamental constraints on household decisions of who migrates and who stays behind (Biao, 2007). Therefore, migration familiarization is relevant to the increasing stability of China's internal migration, as family migration is generally seen as a positive prelude to permanent settlement. Migration familiarization has also contributed to reducing the gender imbalance of migration within China (Fan, 2003; Fan & Huang, 1998; Raghuram, 2004). Family migration may also be a possible solution to the hardship encountered by the left-behind population (Wen & Lin, 2012; Ye et al., 2013).

Seventh, social welfare coverage has seen observable improvement for internal migrants. 91.1% of internal migrants have joined at least one type of health insurances in 2016, standing as a striking contrast with less than 30% coverage rate in 2011 (see footnote 3). This is in line with the overall recent trend in China (Liu, 2018; Saich, 2000), as part of the national development strategy is to reduce regional inequalities (Fan & Sun, 2008; Zhang & Kanbur, 2005). Such a positive development reduces migrants' living expense and probably has helped them achieve familiarization and stability (Wong et al., 2007; Wang & Fan, 2012).

These seven major recent trends reflect important new development and changing demographic features of China's internal migration. They have not only signposted the onset of the late transitional stage, but an ongoing critical transition of China's urbanization process and socio-economic development (Kissinger & Hormann, 2011; Shambaugh, 2013). Studying internal migration of the largest developing country, namely China, could thus provide valuable lessons for other developing countries. Researching on translating internal migration into a factor contributing to Third World development therefore remains as a prevailing and important task (Zhu, 2018).

27.3 Drivers of Internal Migration in China: Why is the Economic Inter-Provincial Perspective Important?

27.3.1 Macro-level Drivers: Rural and Urban Divide and Regional Inequality

Among macro-level drivers of China's internal migration, some may have played a predominant role and have been most effective in relocating population across different regions, such as economic factors and regional development policies (Fan, 1996, 1997, 2005a; Shen, 2012). Others could be peripheral in certain areas or periods by exerting varying impacts upon different age groups or some particular households

(Ye et al., 2013). This sub-section therefore serves as a detailed discussion of macro-level internal migration drivers in China to further clarify why regional inequality is the major driver.

Following the economic reform in 1978, large-scales internal migration started to emerge and spread all over China. The economic reform therefore has been commonly regarded as symbolizing the advent of initial large-scale internal migration in China's recent history. Neo-Marxist approaches have been widely used to explain the onset of the recent massive internal migration in China after the economic reform (Chan, 2013; Fan, 1995; Ye et al., 2013), attributing to the joint forces of the "Reform and Open-up" policy and the subsequent penetration of international capitals (Fujita & Hu, 2001; Nee et al., 2007; Prahalad & Lieberthal, 1998; Tsai, 2007).

Regional inequality is essential to understanding how mechanisms, patterns, and dynamics of internal migration play out in China. In empirical studies of international settings, a strong association between regional inequality and migration flows has been repeatedly identified (De Haas, 2010; Vignoli, 2008; Ye et al., 2013). The predominant role of regional inequality has also been stressed by a wide range of theories—Lewis's unlimited supply of labour (Lewis, 1954), Harris and Todaro's expected income (Harris & Todaro, 1970), and Stark's relative deprivation (Stark & Taylor, 1991). China has experienced severe regional inequality for many years (Fan & Sun, 2008; Fan et al., 2009; Jian et al., 1996; Wang et al., 2013), which has been well reflected by the scale and direction of internal migration (Fan & Sun, 2008; Ye et al., 2013). Regional inequality in China is recognized to have three major dimensions of regional inequality—rural and urban, province, and region (Fan & Sun, 2008). All these three dimensions of regional inequality are believed to be closely related to the unbalanced regional development policies adopted in China (Sheng, 2011; Tian, 2001; Wang & Piesse, 2010; Zhang & Song, 2003).

Regional inequality could also influence internal migration through conditioning macro-level demographic characteristics such as education, further impacting migrants' accessibility to employment opportunities and economic returns of human capital on the aggregated level. For instance, Yu et al. (2014) revealed that the economic return for migration varied greatly across different geographic regions, verifying that more developed regions tended to have more employment opportunities and higher returns to human capital and weaker institutional and cultural barriers for migrants. Such a finding is also related to the important concept of the "escalator region", which features strong both in- and out-migration flows but with distinctive age structures and socio-economic statuses (Fielding, 1992). Indeed, there is strong evidence for the existence of a positive relationship between social and geographical mobility in mainland China. Based on the 2000 and 2005 census data, inter-provincial migrants who moved to "escalator regions" such as Beijing and Shanghai were particularly more likely to search for occupational promotion and improved social status (Fielding, 2010). Other regularities between education and internal migration on the macro-level observed around the world also hold true for China's internal migration. For instance, Li (2009) found that inter-provincial migration and education had a relationship of inverted-U shape, with the number of inter-provincial migrants peaking at middle-school (or equivalent) education; such a relationship was particularly pronounced among rural–urban migrants. Liu et al. (2014) found that young people were more likely to migrate in response to economic inequalities, and that educated and skilled migrants showed more signs of clustering in socio-economically developed destinations than the less-educated.

Some other political structures other than the unbalanced regional development policies could nevertheless serve as migration drivers or barriers. These macro-level political structures could influence household or individual migration processes. For instance, Yan et al. (2014) found that China's unique land ownership and tenure system had played a significant role in farmer's migration decision-making process within the household. Other macro-level political structures, such as the Hukou system, have had complicated effects in shaping and conditioning internal migration flows (Bosker et al., 2012; Chan, 2010). As explained earlier, China's internal migration has two major characteristics of being temporary in nature and being regulated by heavy institutional interventions, which have very close links to macro-level migration drivers such as regional development and migration management policies.

The Hukou system has strong influences upon internal migration in terms of directions, volumes, and durations. The complicated effects of the Hukou system have led to mixed explanations and competing discourses among migration policy studies. Part of the reason is the very dynamic and always-changing migration management policies in China (Chan, 2009). Contending with the wide belief of the deterring effect of the Hukou system upon internal migrants, some scholars argued that the combined effects of both national and local migration management policies had led to mixed but overall positive outcomes for the socio-economic development of the country as a whole (Naybor, 2018; Zhan, 2011). Recently there has been a growing belief that the current Hukou system causes little deterring impact upon internal migrants (Naybor, 2018; Zhan, 2011). It has fallen in significance and abolishing it could help little in improving the life quality of most migrant workers (Zhan, 2011). As a case in point, one recent study showed that there were few rural-to-urban migrants willing to give up their rural Hukou in exchange for the urban Hukou (Naybor, 2018).

27.3.2 Micro-level Drivers

On the whole, findings on micro-level migration drivers of China's internal migration confirm those of existing empirical studies on internal migration in other countries. The following will provide evidence for it by reviewing relevant empirical studies of micro-level migration drivers in China, including micro-level demographic factors and individual's migration experience.

A large body of studies have elaborated upon demographic characteristics of migrants across and above different stages of migration decision-making process for China's internal migration. Regarding demographic characteristics of individuals, important factors are aspiration, education, skill, sex, age, and past migration experience (Chen et al., 2013; Cheng, 1991; Liu, 2005; Piotrowski & Tong, 2013;

Poncet, 2006; Yang, 2000; Ye et al., 2013). For example, Liu and Shen (2014) found that Chinese skilled workers valued career prospects over life qualities when making migration decisions. Qu and Zhao (2014) argued that rural migrants depended a lot upon former migration experiences to make migration decisions. Chiang et al. (2014) pointed out that the migration decisions of young migrants were profoundly influenced by gender as boy and girl migrants behaved substantially differently in the process. Similar to other countries, age selection is also evident in China's internal migration. For instance, Wan (1995) revealed that family migration was significantly negatively influenced by age, that marriage had a greater deterrent impact on individual migration, and that family network was a more significant determinant for individual rather than family migration. This also points to the argument by Yang and Guo (1999) that individuals often do not act alone and that migration could be a group decision within the household and community. Such a family strategy links closely with observations of migrants' life course, family norm and social expectations in terms of fertility, education, marriage, inheritance, and risk spreading (Findley, 1977).

Apart from the age selectivity, gender preference for men is also prevalent in China and male and female migrants tend to behave differently. For instance, Yang (2000) revealed that men were more inclined to conduct individual migration than women. Wan (1995) found that women were less active in migration, and more sensitive to migratory distance than men. This is in line with the general finding in migration research literature that men tend to have a predominant role in migration, especially for pioneering or long-distance migration (Findley, 1977), while many women migrants take part in marriage-oriented rural-rural migration with short migratory distances (Fan & Huang, 1998; Lucas, 1997). Sex could also co-function with other factors to profoundly influence migration activities. For example, there exists strong evidence that household structure exerts more impacts on women than men by limiting female migrants' participation in economic activities (Fan & Huang, 1998; Fan & Li, 2002; King & Skeldon, 2010; Kofman, 2004; Yang & Guo, 1999). The preference for men is also common in internal migration of many other countries (Lucas, 1978). Conversely, however, Zuo (2008) argued that within the household joint decision-making was favoured by adult men out-migrants after marriage, and that married women usually had more say in-migration decision-making as they started to take more household management responsibilities. Such findings of Zuo (2008) contrasted with those of other studies where women were found to be depowered one way or another when making migration decisions (Chiang et al., 2014; Liu, 2012; Nguyen & Locke, 2014; Zhang, 2013; Zhang et al., 2013).

Internal migrants in China not only suffer from aggregated-level exclusions from both local society and institutional systems in the destination as a social group, but also encounter within-group discrimination from peer migrants on the individual level. For instance, Zhan (2011) showed that both local residents and institutional systems held discrimination against migrants, and that social exclusion and discrimination existed among different migrant ethnic groups and even among migrants of the same ethnic group themselves (Ling, 2015; Lu & Zhou, 2013; Shin et al., 2013; Wang et al., 2015a, 2015b; Wang & Fan, 2012). As socio-economic, institutional, and cultural barriers to migrants persistently exist and influence their behaviour, it is therefore common to find migrants suffering from issues such as occupational injuries and identity crisis. Many studies have confirmed that migrants shoulder a disproportionate burden of occupational injuries and mortalities as well as higher risks of other non-occupational diseases in China (Fitzgerald et al., 2013; Kang et al., 2016; Sun & Liu, 2014; Wang et al., 2016; Yang et al., 2015). Gui et al. (2012) showed that similar to international immigrants, integration was also the best strategy for internal migrants to attain social well-being. Without sufficient supply of medical services in the host society (Dai, 2011; Ngok, 2012; Qiu et al., 2011; Sun & Liu, 2014; Xu et al., 2011; Zhuang, 2009), mental health is increasingly a problem for Chinese migrants (Chan & Qiu, 2011; Qiu et al., 2011; Wong et al., 2008; Zhong et al., 2015; Zhu et al., 2013). Indeed, individual migrants differ in demographic and socio-economic characteristics and expectations, which affects their degrees of satisfaction with current destinations and their awareness and strategies to adapt to destination societies. All of these contribute to their different migratory behaviours.

In summary, micro-level drivers of China's internal migration show consistencies and similarities to those of internal migration in other regions and countries. Two major characteristics of China's internal migration (temporariness and heavy institutional regulation) have very close linkages to macro-level migration drivers such as regional development and migration management policies. These macro-level drivers of China's internal migration are typically external to either individual migrants or their households or communities, and are very unique to China. This section highlights the overall important role of macro-level factors (more specifically, the regional inequality) in understanding the unique processes of China's internal migration.

27.4 Conclusions

As China is currently going through a structural mobility transition, it provides an excellent opportunity to further our understanding of the development trajectory of migration on the aggregated level (Zhu, 2018). This chapter therefore takes up this opportunity and endeavours to explain how such a mobility transition unfolds in a socialist transitional economy by primarily focusing on macro-level economic factors and regional differences (Fan, 1999). This is because these macro-level factors have the most profound impact on overall China's internal migration (Bao et al., 2005; Barro & Sala-i-Martin, 1990, 1992a, 1992b; Cai & Wang, 2003a, 2003b; Li, 2009; Shen, 2012, 2013; Song & Wang, 2005; Wang et al., 2005). Yan (2007) provided strong evidence for this argument by establishing a "flow chain" migration model based on the population mobility theories. Yan (2007) also stressed that the recent regional economic inequality and consequently the uneven distribution of wealth originated from the imbalanced implementation of "Reform and Open-up" policy across different regions. The macro-level economic factors along with regional development and migration management policies therefore consist of the fundamental context in studying the internal migration of China. This could also provide valuable lessons for other developing countries.

One underlying assumption of this chapter is that China is likely to arrive at the stabilization stage of migration soon based on the evidence of recent migration trends, as this structural transition of mobility is important and interesting to research on; unless China is going to make some severe mistakes in the near future to stop the current socio-economic development trend, such as crushing productivity with excessive state control, embargoing international trades entirely, or initiating and engaging in large-scale military confrontations and wars. Under these rare conditions, however, the stabilization of mobility and migration in the general population might not take place in the internal migration system of China due to such grave socio-economic interruptions and crises.

The implications of the findings presented in this chapter are that economic and institutional factors are major migration determinants that play significant roles in internal migration processes of China. China's internal flows are spatially imbalanced primarily due to macro-level regional inequalities and institutional policies, while these internal flows are powered by individual migrants substantially differing in migratory behaviours and experiences. The ultimate goal of any migrant management policies should always been made to be fit to the needs and wellbing of people. It is therefore vital that China must learn from other countries' experience in redistributing population with a much greater degree of efficiency. As the stabilization of mobility with improved movement freedom is likely to be the foreseeable future for China, it is crucial to understand how to harness positive impacts of migration while minimizing its negative side-effects.

References

Anh, N. T., Rigg, J., Luong, T. T. H., & Dieu, D. T. (2012). Becoming and being urban in Hanoi: Rural-urban migration and relations in Viet Nam. *Journal of Peasant Studies, 39*(5), 1103–1131. https://doi.org/10.1080/03066150.2011.652618

Appadurai, A. (1990). Disjuncture and difference in the global cultural economy. *Theory, Culture & Society, 7*(2–3), 295–310.

Bao, S., Shi, A., & Hou, W. (2005). Zhongguo renkou qianyi de kongjian xingtai bianhua fenxi. *Chinese Journal of Population Science, 2005*(5), 28–36.

Bao, S. M., Bodvarsson, O. B., Hou, J. W., & Zhao, Y. H. (2011). The Regulation of Migration in a Transition Economy: China's Hukou System. *Contemporary Economic Policy, 29*(4), 564–579. https://doi.org/10.1111/j.1465-7287.2010.00224.x

Bao, S. M., Chang, G. H., Sachs, J. D., & Woo, W. T. (2002). Geographic factors and China's regional development under market reforms, 1978–1998. *China Economic Review, 13*(1), 89–111.

Barro, R. T., & Sala-i-Martin, X. (1990). *Economic growth and convergence across the United States*. Retrieved from

Barro, R. T., & Sala-i-Martin, X. (1992a). Convergence. *Journal of political Economy*, 223–251.

Barro, R. T., & Sala-i-Martin, X. (1992b). Regional growth and migration: A Japan-United States comparison. *Journal of the japanese and International Economies, 6*(4), 312–346

Biao, X. (2007). How far are the left-behind left behind? A preliminary study in rural China. *Population, Space and Place, 13*(3), 179–191.

Bosker, M., Brakman, S., Garretsen, H., & Schramm, M. (2012). Relaxing Hukou: Increased labor mobility and China's economic geography. *Journal of Urban Economics, 72*(2–3), 252–266. https://doi.org/10.1016/j.jue.2012.06.002

Brettell, C. B., & Hollifield, J. F. (2014). The Politics of International Migration How Can "We Bring the State Back In"? In *Migration theory* (pp. 239–300): Routledge.

Byrd, W. A., Byrd, W., Lin, Q., & Lin, C.-s. (1990). *China's rural industry: Structure, development, and reform*: Oxford University Press.

Cai, F., & Wang, D. (2003a). Migration as marketization: What can we learn from China's 2000 census data? *China Review, 3*(2), 73–93.

Cai, F., & Wang, D. (2003b). Zuowei shichanghua de renkou liudong diwuci renkou pucha shuju fenxi. *Zhongguo Renkou Kexue, 2003*(5), 11–19.

Cao, G. Z., Li, K., Wang, R. M., & Liu, T. (2017). Consumption Structure of Migrant Worker Families in China. *China & World Economy, 25*(4), 1–21.

Chan, K. W. (2009). The Chinese hukou system at 50. *Eurasian Geography and Economics, 50*(2), 197–221.

Chan, K. W. (2010). The Household Registration System and Migrant Labor in China: Notes on a Debate. *Population and Development Review, 36*(2), 357–364.

Chan, K. W. (2013). China, internal migration. *The Encyclopedia of Global Human Migration, 2*, 980–995.

Chan, K. W., & Zhang, L. (1999). The hukou system and rural-urban migration in China: Processes and changes. *The China Quarterly, 160*, 818–855.

Chan, S. H., & Qiu, H. H. (2011). Loneliness, job satisfaction, and organizational commitment of migrant workers: Empirical evidence from China. *International Journal of Human Resource Management, 22*(5), 1109–1127.

Chang, H. Q., Dong, X. Y., & Macphail, F. (2011). Labor Migration and Time Use Patterns of the Left-behind Children and Elderly in Rural China. *World Development, 39*(12), 2199–2210. https://doi.org/10.1016/j.worlddev.2011.05.021

Chen, A. P., & Groenewold, N. (2011). Regional Equality and National Development in China: Is There a Trade-Off? *Growth and Change, 42*(4), 628–669. https://doi.org/10.1111/j.1468-2257.2011.00570.x

Chen, S., & Powell, J. L. (2012). *Aging in China: Implications to social policy of a changing economic state* (Vol. 2): Springer Science & Business Media.

Chen, S. H., & Ravallion, M. (2004). How have the world's poorest fared since the early 1980s? *World Bank Research Observer, 19*(2), 141–169.

Chen, Z., Liu, X. F., & Lu, M. (2013). Beyond Lewis: Rural-to-urban migration with endogenous policy change. *China Agricultural Economic Review, 5*(2), 213–230. https://doi.org/10.1108/17561371311331106

Cheng, C. (1991). Internal Migration in Mainland China - the Impact of Government Policies. *Issues & Studies, 27*(8), 47–70.

Chiang, Y. L., Hannum, E., & Kao, G. (2014). It's Not Just About the Money: Gender and Youth Migration from Rural China. *Chinese Sociological Review, 47*(2), 177–201.

Cong, Z., & Silverstein, M. (2011). Intergenerational Exchange Between Parents and Migrant and Nonmigrant Sons in Rural China. *Journal of Marriage and Family, 73*(1), 93–104.

Czaika, M. (2012). Internal Versus International Migration and the Role of Multiple Deprivation: Some Evidence from India. *Asian Population Studies, 8*(2), 125–149.

Dai, H. J. (2011). Surviving in "Localistic Communitas": Endogenous Multicultural Community Organizing among Migrant Workers in Post-Socialist China. *Journal of Social Service Research, 37*(2), 165–179.

De Haas, H. (2010). Migration and Development: A Theoretical Perspective. *International Migration Review, 44*(1), 227–264. https://doi.org/10.1111/j.1747-7379.2009.00804.x

Deshingkar, P. (2006). Internal migration, poverty and development in Asia: Including the excluded. *Ids Bulletin-Institute of Development Studies, 37*(3), 88-+.

Fan, C. C. (1995). Of Belts and Ladders - State Policy and Uneven Regional-Development in Post-Mao China. *Annals of the Association of American Geographers, 85*(3), 421–449.

Fan, C. C. (1996). Economic opportunities and internal migration: A case study of Guangdong Province. *China. Professional Geographer, 48*(1), 28–45. https://doi.org/10.1111/j.0033-0124.1996.00028.x

Fan, C. C. (1997). Uneven development and beyond: Regional development theory in post-Mao China. *International Journal of Urban and Regional Research, 21*(4), 620-+. https://doi.org/10.1111/1468-2427.00105

Fan, C. C. (1999). Migration in a socialist transitional economy: Heterogeneity, socioeconomic and spatial characteristics of migrants in China and Guangdong province. *International Migration Review, 33*(4), 954–987. https://doi.org/10.2307/2547359

Fan, C. C. (2003). Rural-urban migration and gender division of labor in transitional China. *International Journal of Urban and Regional Research, 27*(1), 24–47.

Fan, C. C. (2005a). Interprovincial migration, population redistribution, and regional development in China: 1990 and 2000 census comparisons. *The Professional Geographer, 57*(2), 295–311.

Fan, C. C. (2005b). Modeling interprovincial migration in China, 1985–2000. *Eurasian Geography and Economics, 46*(3), 165–184. https://doi.org/10.2747/1538-7216.46.3.165

Fan, C. C. (2007). *China on the Move: Migration, the State, and the Household*: Routledge.

Fan, C. C., & Huang, Y. Q. (1998). Waves of rural brides: Female marriage migration in China. *Annals of the Association of American Geographers, 88*(2), 227–251. https://doi.org/10.1111/1467-8306.00092

Fan, C. C., & Li, L. (2002). Marriage and migration in transitional China: A field study of Gaozhou, western Guangdong. *Environment and Planning A, 34*(4), 619–638. https://doi.org/10.1068/A34116

Fan, C. C., & Sun, M. (2008). Regional inequality in China, 1978–2006. *Eurasian Geography and Economics, 49*(1), 1–18.

Fan, C. C., Sun, M., & Zheng, S. (2011). Migration and split households: A comparison of sole, couple, and family migrants in Beijing China. *Environment and Planning-Part A, 43*(9), 2164.

Fan, C. C., & Wang, W. W. (2008). The household as security: Strategies of rural–urban migrants in China. *Migration and Social Protection in China, 14*, 205.

Fan, S., Kanbur, R., & Zhang, X. (2009). *Regional inequality in China: trends, explanations and policy responses*: Routledge.

Fielding, A. J. (1992). Migration and social mobility: South East England as an escalator region. *Regional Studies, 26*(1), 1–15.

Fielding, T. (2010). *Inter-regional migration in a transition economy: The case of China.* Retrieved from

Findley, S. E. (1977). *Planning for internal migration: a review of issues and policies in developing countries*: US Dept. of Commerce, Bureau of the Census, International Statistical Programs Center: for sale by the Supt. of Docs., US Govt. Print. Off.

Fitzgerald, S., Chen, X., Qu, H., & Sheff, M. G. (2013). Occupational injury among migrant workers in China: A systematic review. *Injury Prevention, 19*(5), 348–354.

Fujita, M., & Hu, D. (2001). Regional disparity in China 1985–1994: The effects of globalization and economic liberalization. *The Annals of Regional Science, 35*(1), 3–37.

Fukuyama, F. (1995). Social capital and the global economy. *Foreign Affairs, 74*, 89.

Gao, S. (2012). Commodification of place, consumption of identity: The sociolinguistic construction of a 'global village' in rural China. *Journal of Sociolinguistics, 16*(3), 336–357. https://doi.org/10.1111/j.1467-9841.2012.00534.x

Goldstein, A., & Goldstein, S. (1987). Migration in China - Methodological and Policy Challenges. *Social Science History, 11*(1), 85–104.

Gordon, I. (1995). Migration in a Segmented Labour Market. *Transactions of the Institute of British Geographers, 20*(2), 139–155. https://doi.org/10.2307/622428

Greenhalgh, S. (2008). *Just one child: Science and policy in Deng's China*: Univ of California Press.

Gui, Y. X., Berry, J. W., & Zheng, Y. (2012). Migrant worker acculturation in China. *International Journal of Intercultural Relations, 36*(4), 598–610. https://doi.org/10.1016/j.ijintrel.2011.11.007

Guo, F., & Iredale, R. (2004). The Impact of Hukou Status on Migrants' Employment: Findings from the 1997 Beijing Migrant Census 1. *International Migration Review, 38*(2), 709–731.

Han, X. R., & Chen, Y. F. (2016). Food consumption of outgoing rural migrant workers in urban area of China A QUAIDS approach. *China Agricultural Economic Review, 8*(2), 230–249.

Harris, J. R., & Todaro, M. P. (1970). Migration, unemployment and development: A two-sector analysis. *The American Economic Review, 60*(1), 126–142.

Huang, P., & Zhan, S. (2005). Internal migration in China: Linking it to development. In: International Organization for Migration.

Jian, T., Sachs, J. D., & Warner, A. M. (1996). Trends in regional inequality in China. *China Economic Review, 7*(1), 1–21.

Kang, G. Y., Xiao, J. M., Wang, J. Y., Chen, J. H., Li, W., Wang, Y. T., & Yang, F. (2016). Congenital Heart Disease in Local and Migrant Elementary Schoolchildren in Dongguan China. *American Journal of Cardiology, 117*(3), 461–464.

Keung Wong, D. F., Li, C. Y., & Song, H. X. (2007). Rural migrant workers in urban China: Living a marginalised life. *International Journal of Social Welfare, 16*(1), 32–40.

King, R. (2012). Theories and typologies of migration: an overview and a primer.

King, R., & Skeldon, R. (2010). 'Mind the gap!'Integrating approaches to internal and international migration. *Journal of Ethnic and Migration Studies, 36*(10), 1619–1646.

Kissinger, H., & Hormann, N. (2011). *On China: Penguin Press.*

Kofman, E. (2004). Family-related migration: A critial review of European studies. *Journal of Ethnic and Migration Studies, 30*(2), 243–262.

Lewis, W. A. (1954). Economic development with unlimited supplies of labour. *The Manchester School, 22*(2), 139–191.

Li, P. (2009). Zhongguo chengxiang renkou qianyi de shikong tezheng jiqi yingxiang yinsu. *Economist*(1), 50–57.

Liang, Z., & White, M. J. (1997). Market transition, government policies, and interprovincial migration in China: 1983–1988. *Economic Development and Cultural Change, 45*(2), 321–339.

Ling, M. H. (2015). Bad Students Go to Vocational Schools!": Education, Social Reproduction and Migrant Youth in Urban China. *China Journal, 73*, 108–131.

Liu, G. G. (2018). *Urbanization and Social Welfare in China*: Routledge.

Liu, Y., & Shen, J. (2014). Jobs or amenities? Location choices of interprovincial skilled migrants in China, 2000–2005. *Population, Space and Place, 20*(7), 592–605.

Liu, Y., Stillwell, J., Shen, J., & Daras, K. (2014). Interprovincial Migration, Regional Development and State Policy in China, 1985–2010. *Applied Spatial Analysis and Policy, 7*(1), 47–70. https://doi.org/10.1007/s12061-014-9102-6

Liu, Y. Z. (2012). Decisions to leave home: An examination of rural married women's labour migration in contemporary China. *Womens Studies International Forum, 35*(5), 305–313.

Liu, Z. Q. (2005). Institution and inequality: The hukou system in China. *Journal of Comparative Economics, 33*(1), 133–157.

Lu, S., Lin, Y. T., Vikse, J. H., & Huang, C. C. (2016). Well-being of migrant and left-behind children in China: Education, health, parenting, and personal values. *International Journal of Social Welfare, 25*(1), 58–68.

Lu, Y., & Zhou, H. (2013). Academic Achievement and Loneliness of Migrant Children in China: School Segregation and Segmented Assimilation. *Comparative Education Review, 57*(1), 85–116.

Lucas, R. E. (1997). Internal migration in developing countries. *Handbook of Population and Family Economics, 1*, 721–798.

Lucas, R. E. B. (1978). Internal Migration in Developing-Countries - Review of Theory, Evidence, Methodology and Research Priorities - Todaro Mp. *Journal of Development Economics, 5*(1), 101–103.

Ma, L. J., & Fan, M. (1994). Urbanisation from below: The growth of towns in Jiangsu China. *Urban Studies, 31*(10), 1625–1645.
Massey, D. S., Arango, J., Hugo, G., Kouaouci, A., Pellegrino, A., & Taylor, J. E. (1993). Theories of international migration: A review and appraisal. *Population and development review*, 431–466.
Nath, V. (1988). Studies in Migration - Internal and International Migration in India - Rao Msa. *Economic and Political Weekly, 23*(21), 1061–1066.
Naybor, D. (2018). The Impact of China's Changing Hukou on Women's Rural Land Rights. In *Environment, Politics, and Society* (pp. 119–132): Emerald Publishing Limited.
Nee, V., Opper, S., & Wong, S. (2007). Developmental state and corporate governance in China. *Management and Organization Review, 3*(1), 19–53.
Ngok, K. (2012). Serving Migrant Workers: A Challenging Public Service Issue in China. *Australian Journal of Public Administration, 71*(2), 178–190.
Nguyen, M. T. N., & Locke, C. (2014). Rural-urban migration in Vietnam and China: Gendered householding, production of space and the state. *Journal of Peasant Studies, 41*(5), 855–876.
Parida, J. K., & Raman, K. R. (2018). 14 India: rising trends of international and internal migration. *Handbook of Migration and Globalisation*, 226.
Phuong, N. H., & McPeak, J. (2010). Leaving or Staying: Inter-Provincial Migration in Vietnam. *Asian and Pacific Migration Journal, 19*(4), 473–500.
Piotrowski, M., & Tong, Y. Y. (2013). Straddling Two Geographic Regions: The Impact of Place of Origin and Destination on Return Migration Intentions in China. *Population Space and Place, 19*(3), 329–349. https://doi.org/10.1002/psp.1715
Poncet, S. (2006). Provincial migration dynamics in China: Borders, costs and economic motivations. *Regional Science and Urban Economics, 36*(3), 385–398. https://doi.org/10.1016/j.regsciurbeco.2005.11.003
Prahalad, C. K., & Lieberthal, K. (1998). The end of corporate imperialism. *Harvard Business Review, 76*, 68–79.
Qiu, P. Y., Caine, E., Yang, Y., Chen, Q., Li, J., & Ma, X. (2011). Depression and associated factors in internal migrant workers in China. *Journal of Affective Disorders, 134*(1–3), 198–207.
Qu, Z. F., & Zhao, Z. (2014). Evolution of the Chinese rural-urban migrant labor market from 2002 to 2007. *China Agricultural Economic Review, 6*(2), 316–334.
Raghuram, P. (2004). The difference that skills make: Gender, family migration strategies and regulated labour markets. *Journal of Ethnic and Migration Studies, 30*(2), 303–321.
Rogers, A., & Castro, L. J. (1981). Model migration schedules.
Rogers, A., & Willekens, F. (1986). Migration and settlement: A multiregional comparative study.
Rozelle, S., Taylor, J. E., & DeBrauw, A. (1999). Migration, remittances, and agricultural productivity in China. *American Economic Review, 89*(2), 287–291.
Saich, T. (2000). Negotiating the state: The development of social organizations in China. *The China Quarterly, 161*, 124–141.
Schanbacher, G. L. (2007). *Migration patterns : stories*. Golden, Colo.: Fulcrum Pub.
Shambaugh, D. L. (2013). *China goes global: The partial power* (Vol. 111): Oxford University Press Oxford.
Shen, J., & Yang, X. (2014). Analyzing Urban Competitiveness Changes in Major Chinese Cities 1995–2008. *Applied Spatial Analysis and Policy, 7*(4), 361–379.
Shen, J. F. (2012). Changing Patterns and Determinants of Interprovincial Migration in China 1985–2000. *Population Space and Place, 18*(3), 384–402. https://doi.org/10.1002/Psp.668
Shen, J. F. (2013). Increasing internal migration in China from 1985 to 2005: Institutional versus economic drivers. *Habitat International, 39*, 1–7.
Sheng, Z. (2011). Towards China's urban-rural integration: Issues and options. *International Journal of China Studies, 2*(2), 345–367.
Shin, S. S., Wan, X., Wang, Q., Raymond, H. F., Liu, H. L., Ding, D., & Novotny, T. E. (2013). Perceived Discrimination and Smoking Among Rural-to-Urban Migrant Women in China. *Journal of Immigrant and Minority Health, 15*(1), 132–140. https://doi.org/10.1007/s10903-012-9599-1

Sjaastad, L. A. (1962). The costs and returns of human migration. *Journal of political Economy, 70*(5, Part 2), 80–93.
Snyder, S., & Chern, W. S. (2009). The impact of remittance income on rural households in China. *China Agricultural Economic Review, 1*(1), 38–57.
Song, X., & Wang, Y. (2005). Zhongguo renkou kongjian fenbu bianhua de shoulianxing fenxi. *Zhongguo renkou kexue*(4), 54–58.
Stark, O., & Taylor, J. E. (1991). Migration incentives, migration types: The role of relative deprivation. *The Economic Journal, 101*(408), 1163–1178.
Sun, L., & Liu, T. (2014). Injured but not Entitled to Legal Insurance Compensation - Ornamental Institutions and Migrant Workers' Informal Channels in China. *Social Policy & Administration, 48*(7), 905–922.
Sun, Z. (2013). *Explaining Regional Disparities of China's Economic Growth: Geography, Policy and Infrastructure.* Universtiy of California, Berkeley.
Tian, Q. J. (2001). China's new urban-rural divide and pitfalls for the Chinese economy. *Revue Canadienne D Etudes Du Developpement-Canadian Journal of Development Studies, 22*(1), 165–190.
Tsai, K. S. (2007). *Capitalism without democracy: The private sector in contemporary China*: Cornell University Press.
Vendryes, T. (2011). Migration constraints and development: Hukou and capital accumulation in China. *China Economic Review, 22*(4), 669–692. https://doi.org/10.1016/j.chieco.2011.08.006
Vignoli, J. R. (2008). Spatial distribution, internal migration and development in Latin America and the Caribbean. *Cepal Review*(96), 137–157.
Wahba, J., & Zenou, Y. (2012). Out of sight, out of mind: Migration, entrepreneurship and social capital. *Regional Science and Urban Economics, 42*(5), 890–903.
Wan, G. H. (1995). Peasant Flood in China - Internal Migration and Its Policy Determinants. *Third World Quarterly, 16*(2), 173–196.
Wang, C. (2001). Social Identity of the New Generation of Rural Hobo and Merger of Urban and Rural [J]. *Sociological Research, 3*, 63–76.
Wang, F., & Zuo, X. (1999). Inside China's cities: Institutional barriers and opportunities for urban migrants. *American Economic Review, 89*(2), 276–280.
Wang, G., Wei, X., & Shen, J. (2005). Study on Inter-Provincial Migration's Influence on Regional Economy Development in China. *FUDAN JOURNAL (social Sciences), 3*, 148–161.
Wang, H. N., Guo, F., & Cheng, Z. M. (2015a). Discrimination in Migrant Workers' Welfare Entitlements and Benefits in Urban Labour Market: Findings from a Four-City Study in China. *Population Space and Place, 21*(2), 124–139.
Wang, H. N., Guo, F., & Cheng, Z. M. (2015b). A distributional analysis of wage discrimination against migrant workers in China's urban labour market. *Urban Studies, 52*(13), 2383–2403.
Wang, L. (2012). Social exclusion and education inequality: Towards an integrated analytical framework for the urban-rural divide in China. *British Journal of Sociology of Education, 33*(3), 409–430. https://doi.org/10.1080/01425692.2012.659455
Wang, W. F. W. N., & Fan, C. C. (2012a). Migrant Workers' Integration in Urban China: Experiences in Employment, Social Adaptation, and Self-Identity. *Eurasian Geography and Economics, 53*(6), 731–749. https://doi.org/10.2747/1539-7216.53.6.731
Wang, W. W., & Fan, C. C. (2006a). Success or failure: Selectivity and reasons of return migration in Sichuan and Anhui. *China. Environment and Planning A, 38*(5), 939–958.
Wang, W. W., & Fan, C. C. (2006b). Success or failure: selectivity and reasons of return migration in Sichuan and Anhui, China. *California Center for Population Research.*
Wang, W. W., & Fan, C. C. (2012). Migrant workers' integration in urban China: Experiences in employment, social adaptation, and self-identity. *Eurasian Geography and Economics, 53*(6), 731-749.
Wang, X. B., & Piesse, J. (2010). Inequality and the Urban-rural Divide in China: Effects of Regressive Taxation. *China & World Economy, 18*(6), 36–55. https://doi.org/10.1111/j.1749-124X.2010.01219.x

Wang, X. B., Piesse, J., & Weaver, N. (2013). Mind the gaps: A political economy of the multiple dimensions of China's rural-urban divide. *Asian-Pacific Economic Literature, 27*(2), 52–67. https://doi.org/10.1111/Apel.12026

Wang, Y. L., Hu, J. X., Lin, W., Wang, N., Li, C., Luo, P., & Shen, C. F. (2016). Health risk assessment of migrant workers' exposure to polychlorinated biphenyls in air and dust in an e-waste recycling area in China: Indication for a new wealth gap in environmental rights. *Environment International, 87*, 33–41.

Wen, M., & Lin, D. H. (2012). Child Development in Rural China: Children Left Behind by Their Migrant Parents and Children of Nonmigrant Families. *Child Development, 83*(1), 120–136.

Wong, D. F. K., He, X. S., Leung, G., Lau, Y., & Chang, Y. L. (2008). Mental health of migrant workers in China: Prevalence and correlates. *Social Psychiatry and Psychiatric Epidemiology, 43*(6), 483–489.

Xu, Q. W., Guan, X. P., & Yao, F. F. (2011). Welfare program participation among rural-to-urban migrant workers in China. *International Journal of Social Welfare, 20*(1), 10–21.

Yan, S. (2007). Inter-provincial migration and its determinants in the 1990's China. *Chinese Journal of Population Science, 1*, 71–77.

Yan, X. H., Bauer, S., & Huo, X. X. (2014). Farm Size, Land Reallocation, and Labour Migration in Rural China. *Population Space and Place, 20*(4), 303–315.

Yang, H., He, F., Wang, T. H., Liu, Y., Shen, Y., Gong, J., . . . Zhu, S. Z. (2015). Health-Related Lifestyle Behaviors among Male and Female Rural-to-Urban Migrant Workers in Shanghai, China. *Plos One, 10*(2).

Yang, X., & Guo, F. (1999). Gender differences in determinants of temporary labor migration in China: A multilevel analysis. *International Migration Review*, 929–953.

Yang, X. S. (2000). Determinants of migration intentions in Hubei province, China: Individual versus family migration. *Environment and Planning A, 32*(5), 769–787.

Ye, J. Z., Wang, C. Y., Wu, H. F., He, C. Z., & Liu, J. (2013). Internal migration and left-behind populations in China. *Journal of Peasant Studies, 40*(6), 1119–1146. https://doi.org/10.1080/03066150.2013.861421

Yiu, D. W., Lau, C., & Bruton, G. D. (2007). International venturing by emerging economy firms: The effects of firm capabilities, home country networks, and corporate entrepreneurship. *Journal of International Business Studies, 38*(4), 519–540.

Young, J. (2013). *China's hukou system: Markets, migrants and institutional change*: Springer.

Yu, L., Xu, W., Zhu, Y., & Lin, L. Y. (2014). Geographic structure of the migrant labour market in China: A case study of four city-regions in Fujian. *Canadian Studies in Population, 41*(3–4), 22–48.

Zelinsky, W. (1971). The hypothesis of the mobility transition. *Geographical review*, 219–249.

Zhan, S. H. (2011). What Determines Migrant Workers' Life Chances in Contemporary China? Hukou, Social Exclusion, and the Market. *Modern China, 37*(3), 243–285.

Zhang, C. H., Gao, Q. J., & Li, X. Y. (2013). The Impact of Rural-urban Migration on Gender Relations in Chinese Households. *Asian Journal of Womens Studies, 19*(1), 41–66.

Zhang, K. H., & Shunfeng, S. (2003). Rural–urban migration and urbanization in China: Evidence from time-series and cross-section analyses. *China Economic Review, 14*(4), 386–400.

Zhang, K. H., & Song, S. (2003). Rural–urban migration and urbanization in China: Evidence from time-series and cross-section analyses. *China Economic Review, 14*(4), 386–400. https://doi.org/10.1016/j.chieco.2003.09.018

Zhang, N. N. (2013). Rural women migrant returnees in contemporary China. *Journal of Peasant Studies, 40*(1), 171–188. https://doi.org/10.1080/03066150.2012.749867

Zhang, X., & Kanbur, R. (2005). Spatial inequality in education and health care in China. *China Economic Review, 16*(2), 189–204.

Zhao, Y. (2003). The role of migrant networks in labor migration: The case of China. *Contemporary Economic Policy, 21*(4), 500–511.

Zhong, B. L., Liu, T. B., Chan, S. S. M., Jin, D., Hu, C. Y., Dai, J., & Chiu, H. F. K. (2015). Prevalence and correlates of major depressive disorder among rural-to-urban migrant workers in Shenzhen, China. *Journal of Affective Disorders, 183*, 1–9.

Zhu, C. Y., Geng, Q. S., Yang, H. L., Chen, L., Fu, X. H., & Jiang, W. (2013). Quality of life in China rural-to-urban female migrant factory workers: a before-and-after study. *Health and Quality of Life Outcomes, 11*.

Zhu, Y. (2018). Advancing research on internal migration in Asia: The mobility transition hypothesis revisited. *Asian Population Studies, 14*(1), 1–4.

Zhu, Y., Lin, L., & Ke, W. (2016). Trends in internal migration and mobility: International experiences and their implications for China. *Renkou Yanjiu (population Research), 40*(5), 50–60.

Zhuang, M. (2009). The Social Support Network for Rural Migrant Workers in Chengdu, China: Local Governance and Civil Society in the Fight Against Poverty and Exclusion. *Ids Bulletin-Institute of Development Studies, 40*(6), 41–49.

Zuo, J. (2008). Marital construction of family power among male-out-migrant couples in a Chinese village - A relation-oriented exchange model. *Journal of Family Issues, 29*(5), 663–691.

Chapter 28
Countering Counter-Urbanization: A Comparative Analysis of Resiliency and Redevelopment in Postindustrial Cities of the United States and Poland

Bartosz Bartosiewicz and Christopher Cusack

Abstract While the world has hosted a majority urban population for more than a decade, the ebbs and flows of urban dynamics are far from uniform. Many urban areas currently experiencing substantive declines in population and associated economic well-being are former centres of industry and production. The challenge for such cities, then, is to determine solutions that best address shrinking populations and a changing economic base. This challenge is not place-specific; rather it is commonplace to an abundance of cities around the globe. Two of the manufacturing belts wherein multiple cities face this plight are that of Great Lakes region of the United States and the central industrial region of Poland. These regions are highlighted by the cities of Detroit and Lodz, respectively, both of which are noted for declining populations and shifting economic bases. Within these urban areas, notable variation exists between the fortunes of the urban core and urban fringe and is reflected in changes to population density and housing. In response, both Detroit and Lodz have implemented redevelopment strategies focusing on resiliency and innovation in their attempts to transition to a postindustrial economy. The ultimate objective of this chapter is to counter the processes of counter-urbanization that each city experiences.

Keywords Postindustrial cities · Economic well-being · Shrinking populations · Shifting economic bases · Resiliency · Innovation

B. Bartosiewicz (✉)
Institute of the Built Environment and Spatial Policy, Faculty of Geographical Sciences, University of Lodz, Lodz, Poland
e-mail: bartosz.bartosiewicz@geo.uni.lodz.pl

C. Cusack
Department of Geography, Keene State College, Keene, NH 03435-2001, USA
e-mail: ccusack@keene.edu

B. R. K. Sinha (ed.), *Urban Dynamics, Environment and Health*,
https://doi.org/10.1007/978-981-99-5744-6_28

28.1 Introduction: The Shrinking Cities Phenomenon

More than a decade has passed since the world became host to a majority urban population and fully 55 per cent of the global population now resides in urban areas. This trend of global urbanization continues virtually unabated across the globe and projections reveal that nearly all future population growth in the world will occur in cities. Yet this past and projected growth is far from uniform. Explosive growth of the urban population is projected for such countries as India, China, and Nigeria. There is, however, another extreme and pockets of cities wherein population decline and economic malaise have long been set. Although these disparate trajectories are well noted and all are worthy of analysis, the focus of this chapter is to examine the causes and consequences of urban decline. Focus is placed on "shrinking cities" because of the sheer magnitude of this phenomenon. In fact, one analysis has determined that already there are more than 5,000 shrinking cities around the world (Meng, et al., 2021). These cities are scattered across the globe but primarily are concentrated in Europe and the United States, as well as eastern Asia. With the number and location of shrinking cities established, attention can be turned to the genesis of the shrinking cities phenomenon.

The term "shrinking cities" originated in 2002 and as yet has no uniformly accepted definition. In fact, the concept of urban shrinkage has even been labelled as "conceptually incoherent" and lacking in consensus amongst scholars, planners, politicians, and the public (Pallagst et al., 2017). The literature abounds with a myriad of posited symptoms, causes, and solutions to shrinkage. Symptoms may include population loss, economic decline, unemployment, blight, disinvestment, and more. Among the listed causes of shrinkage are globalization, natural disasters, low birth rate, deindustrialization, and suburbanization, among others. A partial listing of identified solutions includes greening/urban agriculture, planned shrinkage/rightsizing, demolition, and perhaps somewhat surprisingly, gentrification (Berglund, 2020). Still other veins of research identify shrinking cities not by what they are, but rather, by what they are not. Bontje (2016) observes that shrinking cities are usually not "attractive places" or "motors of growth" or "powerful regional centers" or beacons of "social progress". Of course, even this collection of variables does not account for all the complexities of a shrinking city. For instance, a shrinking city may contain pockets of growth while experiencing an overall decline in population. Likewise, a shrinking city may be embedded within a larger growth region.

One commonality among the research is that population loss is identified as the core characteristic of a shrinking city. Yet even the acceptance of population decline as the defining characteristic poses some difficulty in assessment (Hartt, 2019). Two broad approaches highlighted in associating shrinking cities with population loss are the binary method and the threshold method. An example of the binary method is simply to record population loss between two periods in time, whereas the threshold method employs a predetermined value, such as 10 or 25%, for population loss over a given time period (Weaver et al., 2017). Neither is perfect, though both are widely used. Hartt (2021) expands upon these approaches by underscoring how the physical manifestation of population loss, as revealed through housing vacancies and housing unit loss, is yet another indicator of a shrinking city. Thus, the reality endured by a shrinking city extends beyond absolute population loss. Likewise, the reality is that the extent of urban shrinkage on a global scale and the complexities of its causes are far less established. Rather than being a deterrent, this lack of clarity regarding shrinking cities makes the need for further investigation paramount and this is why shrinking cities have been labelled as an emerging topic in spatial planning since 2005 (Pallagst et al., 2017).

28.2 Background to Case Study Analysis

In addition to population loss, and in light of the complexities regarding shrinking cities, the literature does at least agree that the most pervasive underpinnings of shrinking cities are deindustrialization and economic restructuring, as well as the collapse of the Soviet Union and the opening of the Iron Curtain. The waning of cities in industrial regions across the globe establishes opportunity for substantive cross-country analysis. Specifically, a comparative analysis of a shrinking city in the West and a shrinking city in a country formerly under Communist control can provide for rich insight into the causes and consequences of shrinkage. For this purpose, the city of Detroit in the United States is evaluated in concert with the city of Łódź in Poland. Both cities have experienced sustained population loss. Likewise, both are former centres of industry and production and the process of deindustrialization experienced by both cities enables their comparison (Gebalksa, 2017). Without question, each is a prototypical shrinking city. Yet Detroit and Łódź also possess important distinctions that enhance their value for analysis by comparison. Among the foremost of their differences is that Detroit sits amidst an entire region of shrinking cities while Łódź is more of a standalone in central Poland. A second distinction is that of political ecology and the fact that, unlike Detroit, Łódź endured an abrupt transition from Communist rule. These dissimilarities, along with their common industrial heritage and population loss, contribute to the opportunity for rich comparative analysis.

Perhaps the most well-known and hardest hit region of urban decline in the entire world is that of the Rust Belt of the United States. Spanning roughly 500 miles, the Rust Belt contains a majority of shrinking cities in the United States (Thompson & de Beurs, 2018). Situated in the middle of the Rust Belt, and serving as the poster child for the postindustrial decline, is the city of Detroit (Hannigan et al., 2015). The US Rust Belt, of course, is hardly alone and similar urban hardship and postindustrial regional decline can be found in the Ruhr Valley in Germany, the Midlands in England, the Canadian industrial belt of the Great Lakes, and beyond (Carter, 2012; Wilson, 2014). As such, these regions have received attention in scholarly literature both individually and collectively. Other cities and regions have experienced similar trends of deindustrialization and concomitant population loss but have not yet received commensurate attention. One such place is, of course, the industrial city of Łódź in central Poland. In comparison to the US Rust Belt, the city of

Łódź exhibits a similar industrial backbone and today faces comparable hardships of economic adversity and population decline. One important geographic distinction is that Łódź has no regional industrial counterparts.

Łódź also stands as a good comparator because of its overall population size and the fact that its rate of population decline over the past decade has nearly perfectly matched that of Detroit in the United States (Fig. 28.1). Of all postindustrial US cities, no other city can lay claim to such a sizable and sustained population loss as Detroit. From its 1950 high population of nearly 1.85 million residents, the city has lost nearly two-thirds of its population. The hemorrhaging of population is a point that bears repeating, as Detroit has lost nearly two out of every three residents from its peak population in 1950. In comparison, the population trend for Łódź at first looks markedly different. In fact, from 1950 to 1970, the rate of population growth for Łódź nearly matched the rate of decline for Detroit. Łódź continued to grow, albeit more slowly until the end of the Communist era and the change in political structure at which point the textile industry collapsed and the city's economic structure changed markedly. Since then, Łódź has lost nearly 20% of its population and presently represents the fastest depopulating large city in Poland (Szafranska, 2019). Ultimately, the rates of decline for the two cities have nearly coincided over the past ten years. Only time will tell if the plans and policies being implemented and executed today in Detroit and Łódź will have an impact on the future outlook of these cities.

Due to sustained population loss and concomitant economic malaise, Detroit and Łódź have each earned the aforementioned moniker of a shrinking city and warrant case study analysis. While Detroit and Łódź are representatives of shrinking cities, they also somewhat incongruously reveal patterns of land consumption and spatial growth. In the Rust Belt, Detroit and other cities have long experienced a twofold pattern of migration. The first is a movement to the south and west and the second is a movement from central city to suburb. Long a hallmark of cities in the United States, sprawl is now extending its tentacles from the city to the hinterland of Łódź. That both cities now grapple with disparate fortunes of the urban core and urban fringe has created yet another similarity between Detroit and Łódź. Beyond these urban parallels, a significant difference between the two cities that makes them ripe for an

Fig. 28.1 Population decline in the cities of Detroit and Łódź

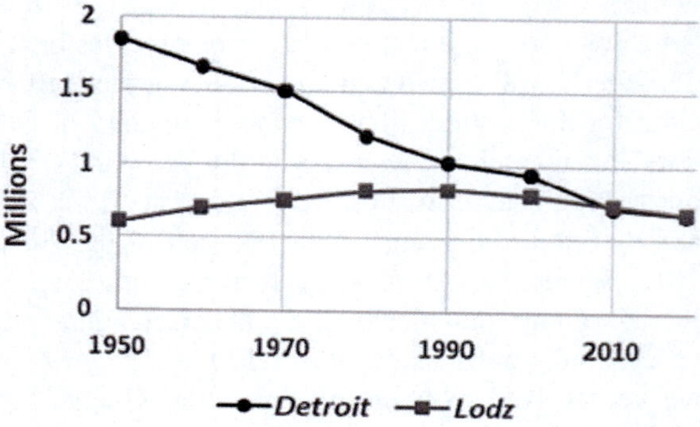

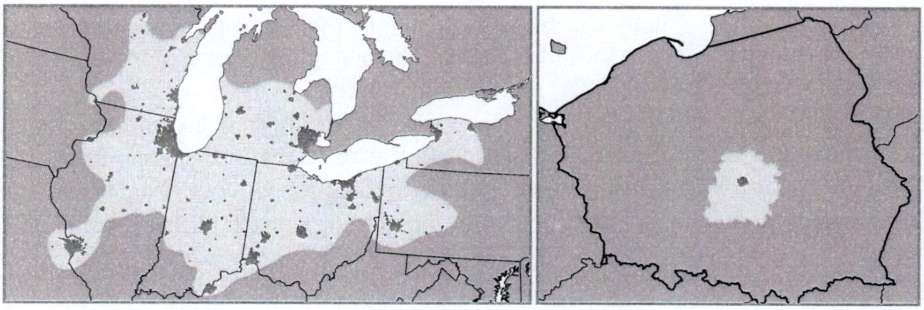

Fig. 28.2 Cities of the Rust Belt region in the United *States and the standalone postindustrial city of Łódź within the central Łódź province of Poland*

explorative assessment is that Detroit sits almost centrally located amidst an entire region of declining cities while Łódź stands alone with no regional comparators (Fig. 28.2). To execute this comparative analysis, both cities are afforded individual attention to explicate strategies aimed at stopping the exodus of people, jobs, and dollars.

In comparison to Detroit, the historical experience of Łódź in terms of centralized planning and socialist government structure affords still another important distinction that ripens these cities for research by comparison. The noted declines of Detroit and Łódź have parallels in terms of deindustrialization. However, the shrinkage of Łódź was also highly influenced by decades of centralist government planning. Having set economic priorities through a hierarchical top-down government for decades, the ultimate collapse of the Soviet Union resulted in tremendous difficulty for Łódź and many other cities facing a difficult transition to a market economy. The resulting economic hardship following the collapse of socialism was marked by low wages for workers and significant emigration to Western Europe (Wichowska, 2021). This political change of the late 1980s and early 1990s contributed to the collapse of the city's textile industry and contributed to high rates of unemployment from which the city has never recovered (Szafranska et al., 2019).

Contrasting the political experiences of Łódź, the city of Detroit has never operated under a centralized government or planning structure. To the contrary, and as commonplace to US cities, Detroit has operated independently. While affording autonomy, this local control also means that Detroit is separated from its hinterland and operates in something of an isolated state within a larger region. This means that Detroit competes not only with cities across the United States and around the globe, but also with the communities that lie just beyond its own city borders. A principal consequence is that a number of jurisdictions even within the greater Detroit metropolitan area have been enjoying economic and population gains at the same time that Detroit has been enduring decades of difficult shrinkage. While Detroit can tap into Federal programmes and grant money in order to address issues of housing, the environment, and more, there is no national government planning structure. Even the state of Michigan takes a largely hands-off approach to individual city planning.

Michigan does not require communities to adopt a comprehensive plan, nor does it even provide guidelines to locales interested in conducting comprehensive planning (Jun, 2017). The upshot is that, unlike Łódź, the city of Detroit has always been responsible for its own destiny. Yet while experiencing different historic government structures, the cities of Łódź and Detroit find themselves in the same unenviable position as shrinking cities. This then leads to need for individual analysis of the current trends, planning strategies, and future outlook of each of these comparative cities.

28.3 Detroit, Michigan, USA

In a point that bears repeating, of all the cities in the world it is Detroit that now has the unenviable distinction of being the poster child of the shrinking cities phenomenon (Safransky, 2017). Although not a desirable defining characteristic, Detroit can at the very least take solace from the fact that it is far from alone. The near universal factor underlying such decline is a historic dependence on a singular economic activity. Indeed, as Pijawka (2015) notes, one need only look at Detroit to understand the detrimental effects brought about by an absence of economic diversification. This lesson has been a hard one to learn and today's Detroit is only a shell of its former self. After having enjoyed decades of economic and population growth based on a booming auto industry, Detroit reached its apex at the midpoint of the twentieth century. As the birthplace of the automobile, and where the standardized form of mass production known as "Fordism" began, Detroit was an industrial and economic powerhouse for much of the first half of the twentieth century. These heady economic days brought a population boom that lasted until the beginnings of disinvestment following World War II. This early period of disinvestment snowballed into a series of plant closures that ultimately resulted in decades of ruination for the city (Apel, 2015).

As was the case with most of its Rust Belt counterparts, the 1950 Census marked the peak population for Detroit. Yet the decade of the 1950s was to bring about unprecedented urban change as, for the first time, suburbanites outnumbered central city residents. Highlighting a commonality amongst Rust Belt cities, the role of geographic scale was becoming an important distinction. In a notable trend, the city of Detroit was undergoing population decline while the population of the overall metropolitan area continued to grow. Consequently, it was becoming more and more clear that "central cities were becoming less and less central" (Mallach, 2018, 24). These countervailing trends of central city decline and overall metropolitan growth continue to this day. Since 2010, the city of Detroit has lost 43,746 residents while the overall metropolitan area has increased in population by a closely comparable 49,728 residents. While Detroit is the vanguard it is hardly alone in this spatial pattern of growth and decline as, New York City aside, every one of the 25 biggest cities in 1950 situated on or east of the Mississippi River has experienced population decline. At the same time, however, the metropolitan regions of all these shrinking cities have grown.

The deindustrialization of Detroit and Rust Belt cities has been so pronounced as to be almost fanciful. Beyond declines in population, more than 60 census tracts in Detroit have lost at least 50 per cent of their housing stock since 1970 (Hackworth, 2018). As noted by Schramm (2020), to even begin to recreate its economy from more than a half-century earlier would require Detroit to acquire fully 800,000 new residents. The city's decline has become so extreme that more than 20 square miles, roughly the size of Manhattan, just owned by the city now sits vacant. Likewise, the 28,000 vacant lots in Cleveland and 25,000 vacant lots in Gary, Indiana, highlight these cities as accompanying prototypes of the postindustrial city (Mallach, 2018, 177). Deindustrialization has undeniably been the driving force of decline throughout the Rust Belt, as evidenced by the city of East Cleveland, Ohio, which had 45 times fewer manufacturing jobs in 2012 than it did a mere four decades earlier (Race and the production of extreme land abandonment... 53). Urban decline is nothing new and the statistics roll readily off the pages of history; however, viable solutions are far more difficult to ascertain.

28.3.1 Revitalization: Politics and Planning

Part of the problem in attempting to develop a planning vision for shrinking cities in the United States is that the US planning discourse remains, for the most part, entrenched in the ideology of growth. Critics of this worldview note that the planning approach to urban shrinkage has long been to employ a variety of responses rather than devise a specific strategy, such as rightsizing. However, optimism that a new shift in this archetypal planning response to urban decline from descriptive to prescriptive is taking place. Within this shift, planners are moving from "lamenting urban decline to celebrating it as an opportunity to reinvent planning" (Safransky, 2017, 1085). A fresh approach to planning that is more proactive than reactive is taking shape in many Rust Belt communities and being integrated into local comprehensive plans. Yet these plans remain largely place-specific and do not account for regional variation among Rust Belt communities. Furthermore, cities in Michigan are not required to adopt a comprehensive plan nor is there any state-based comprehensive planning programme (Jun, 2017). While cities of the Rust Belt have a common history and share in the experience of deindustrialization, they are generally going solo in terms of planning.

In Detroit, it is the 2010 *Detroit Future City* (DFC) land use plan that is providing the foundation for rightsizing and generating movement on multiple fronts. Beyond specific objectives and goals is the notable fact that the DFC espouses the need for a fundamental reinvention of Detroit. As reviving the city's manufacturing base is not a viable option, innovation must be at the forefront of a new Detroit. With an emphasis on community development and quality of life issues, the DFC also signals a willingness to experiment and rejects the single-minded pursuit of economic growth (Schindler, 2016, 831). These are interesting times in the history of Detroit and its approach to planning. Although the city's manufacturing base is not likely to return in full force, there is certainly a manufacturing rebound currently underway across the United States as manufacturing companies added nearly 1.5 million jobs between 2010 and 2019 (Stettner & Yudken, 2019, 88). With these jobs comes opportunity for innovation and a way forward in Detroit, throughout the Rust Belt, and beyond. For a city such as Detroit, it is not the number of manufacturing jobs that is critical; it is that the manufacturing sector is providing the basis for innovation.

With the *Detroit Future City* plan and its historical base of auto manufacturing, the city is now reinventing itself as a hub of technology (Deslatte, 2020). While its legacy stands as the automotive capital of the world, the future of Detroit is most closely connected to R&D, engineering, and other innovation-driven forces (Foggin, 2018). Optimistic assessments of the city are looking beyond the abandoned factories and vacant lots and positing that the decline of the manufacturing base does not mark the end of the auto industry in Detroit, but only a transition. Reinforcing these claims are findings of an analysis of Detroit that show the level of continued innovation within the city frequently exceeds the level of productivity of the United States as a whole (Hannigan et al., 2015). Thus, the foundation for future economic success is connected to, but not dependent on, the city's industrial past. Detroit in this way may emulate other Rust Belt cities, such as Pittsburgh and Milwaukee, which have received recognition for their ability to recreate and reinvent themselves. These are not mere pipe dreams but are evident through such developments as Detroit's Innovation District, which encourages continued diversification of the city's economy and helps establish density thresholds critical for entrepreneurial collaboration. Indeed, much has been done, and the groundwork has been laid for a more economically sustainable city.

28.3.2 Resiliency and a Sustainable Future

The question facing Detroit, and Łódź, and countless other shrinking cities, is how to turn sustained decline into a sustainable future. To that end, what opportunities exist for Detroit to capitalize on its strengths, re-foster a sense of place, and become a more efficient and more resilient city? The answer is that despite enduring decades of hardship and decline, Detroit does possess a handful of strengths that collectively afford it a lifeline and afford it some reason for optimism. Whether these strengths are enough to overcome a seemingly intractable drift towards decay and decline remains to be seen. However, there are indeed opportunities for rehabilitation and development. Foremost among them is the city's historic development and industrial heritage. Though an ageing housing stock certainly provides an intractable set of difficulties, the fabric of the city's urban heritage is of true value. Furthermore, the removal of aged housing can provide opportunity for urban greening and agriculture. Greening the city does not have to happen in isolated fashion but rather can be used to make intraurban connections through creation of green ribbons. Thus, there remain reasons for optimism and opportunities for Detroit to chart a path towards a more resilient and sustainable future.

In order to stem the decline, Detroit appropriately understands the need to embrace its legacy and heritage. In fact, the term "legacy cities" was coined by the American Assembly in 2011 in an effort to put a positive spin on the shrinkage of Detroit and other former centres of industry. The term "legacy" acknowledges the challenges facing these cities while evoking their collective heritage and assets (Tighe & Ryberg-Webster, 2019). Rather than accepting a ruinous fate, many view legacy cities as places of opportunity that are now "ground zero for individuals, nonprofits, the public sector, and corporations to try new ideas and policies to remake urbanism in these postindustrial cities" (Clouse et al., 2019, 38). The value of, and appreciation for, historic buildings and walkable streetscapes is only gaining in appreciation and increasing in demand. The history of Detroit's buildings reflects its culture and industrial past, and presents a real opportunity for the city to capitalize on this new appreciation and consumer demand. In fact, Detroit has been identified as the city in which challenges of heritage and urbanism are most clearly on display. The attitude regarding these places and spaces in Detroit has changed with changing consumer tastes and with growing appreciation for their myriad cultural, economic, and ecological benefits (Locke et al., 2018).

Such an approach to urban regeneration is far from fanciful and already there are numerous projects completed or underway that involves renovation of Detroit's historic buildings (Korkmaz, 2018). These projects are providing opportunity for the revitalization of Detroit and may well contribute to helping to stem the seemingly unending population decline. Moreover, they appeal to the type of young urbanites of the digital and creative industries. Detroit is right to play on its rather limited strengths and in doing so through these renovation projects, the city is striving to appeal to a mix of employment clusters. Rather surprisingly, Detroit today actually ranks among the strongest cities in the United States for diversity of employment. Employment clusters include innovative industries, medical and educational institutions, and, yes, automotive production. These clusters and the workers therein offer opportunity for the city to create a more resilient future (Hannigan et al., 2015; Kinkead, 2016). Young urbanites are attracted to Detroit not just because of its status as a legacy city or its economic opportunity, but also because the city has taken impressive steps at regenerating its urban environment.

In addition to celebrating its historic buildings, Detroit is likewise reinventing itself as a greener and more environmentally friendly city as it strives to transition from "Motown to 'Growtown'" (Safransky, 2017, 1083). Like so many other shrinking cities, Detroit is taking an urban greening approach as a means of "rightsizing". The Dequindre Cut Greenway (DCG) is an exemplar of this approach and was one of the first segments of greenway to be completed in Detroit (Fig. 28.3). Formerly a railway, the DCG provides a two-mile green ribbon connecting city neighbourhood with Milliken State Park and the riverfront. Milliken State Park was itself created from city parkland and riverfront brownfields in 2004. Without question, urban greening is being embraced by city and state leaders as a means of revitalization (Cialdella, 2020). However, establishing a structural approach to greening the city's 120,000 vacant lots is a point of concern. Some contend that scarce resources need to be directed towards downtown development at the expense of vulnerable neighbourhoods. Detroit is an

Fig. 28.3 The Dequindre cut greenway in Detroit. **Photo** Authors

example where a form of urban triage is taking place through the withdrawal of public funds from the poorest areas of the city in favour of the central business district (Kirkpatrick et al., 2018).

The same urban triage approach is being implemented through the city's greening efforts, as areas with the highest vacancy rates are being targeted for cost-saving green projects while the gentrifying downtown area receives public investment. As services are withdrawn, the affected neighbourhoods are converted to landscape infrastructure in the form of parks, greenspace, and urban agriculture. While the reconceptualization of landscape as infrastructure is a means to make the city more sustainable, there are noted concerns of categorizing the city into zones of investment and divestment as being a form of modern day redlining (Safransky, 2017). Yet the benefits of greening cannot be discounted. Aside from the obvious environmental and human health benefits, the expansion of greenspace in Detroit will also improve the city's resiliency by increasing water absorption and management of biowaste (Berglund, 2020). Furthermore, the addition of greenspace to historically distressed neighbourhoods can help Detroit address social equity goals. Without question, Detroit and other shrinking cities are incorporating urban greening as a significant feature of sustainability planning and in an effort to make strides towards equity (Schwarza et al., 2018). The greening of Detroit does afford opportunity to become a more equitable, economically viable, and environmentally friendly city. In such a way, Detroit can tout its comparative advantages, attract talent, and perhaps even stem its economic malaise and continuing population exodus.

28.4 Łódź, Poland

Located in central Poland, Łódź is currently the country's third largest city and the metropolitan area of Łódź has a population of more than one million people. The number of people in the city is rapidly decreasing as a result of persistent negative natural growth. The city is characterized by one of the oldest societies in Poland, and is an example of a rapidly crumbling city as evidenced by the fact that as recently as 1989 (at the beginning of the political transformation) the population of the city was 851,000. Natural growth is of course crucial for this process, but since the beginning of the twenty-first century, it has been further stimulated by the migration of people from the city to neighbouring municipalities. The enrichment of residents and the development of transport infrastructure have activated the suburbanization processes that persist to this day. The demographic situation of the city, which is one of the main challenges, is the result of a history in which Łódź began as a small settlement with a dominant agricultural function on a secondary trade route.

Everything changed in 1820 when the political decision was made to include Łódź in the team of industrial settlements in the Kalisz-Mazovian industrial district. By the threshold of World War I in 1914, the city's population had reached 500,000 inhabitants. The success of the city's industry is largely the result of its location. Before 1918, Łódź was within the borders of the Russian Empire, which at that time was subject to industrialization and urbanization process. In independent Poland (1918) Łódź next to Warsaw had become the largest city and the largest industrial centre. During this period, the first housing estates were built, railway connections were developed, and one of the first museums of contemporary art in Europe was created. In 1939 (outbreak of World War II) the population of Łódź had reached 672,000. After the war, the city did not lose its industrial character although the entire industry was nationalized. Łódź from a multinational city has become a purely Polish city. The majority of the Jewish population was killed (in the Łódź ghetto), the German population fled with the arrival of the Red Army. During the socialist period, new factories were also built in Łódź, mainly in the fields of clothing, textiles, and wool.

Unfortunately, the city's factories were becoming less and less viable in relation to foreign competition. This ultimately led to the collapse of industry in Łódź in the early 1990s. Practically overnight, all factories dissolved and tens of thousands of people lost their jobs, especially women. The city received no assistance from the government and a most difficult time began. The lack of assistance and favour from the central government, the lack of an idea for the city by successive local authorities, and the lack of external investment caused a major socio-economic crisis in the city. The result has been an ageing population and deteriorating housing conditions in the city center where a large part of the population lives. Łódź entered the twenty-first century with a neglected, destroyed (by lack of renovations), and underfunded downtown. The municipality owns most of the tenement houses (after 1990 state property was transferred to local governments) but could ill-afford renovations. Residents did not care about apartments, which they did not own. Public services also stood at a lower level than in large modern housing estates. The emergence of investors (including foreign ones) in recent years and the accession of Poland to the European Union in 2004 is at long last resulting in positive change. Whether these changes are enough to counter population loss and economic stagnation remains to be seen, and undoubtedly require substantive political and planning support.

28.4.1 Revitalization: Politics and Planning

Local government and planning leaders have a vision of development for Łódź that amounts to a rebirth of the city centre, particularly around the new railway station. The motto of change is "Return to the City". The vision of this "new center" also concerns the use of land for investment areas, and includes construction of office space, new housing, and new service spaces. Within a dozen years, more than a dozen residential investments are being established in the city centre, the same number as office investments. These actions also translate into the decision of other entities operating in the area, such as the hotel industry. The ultimate goal of the physical regeneration of the city is to create a denser urban form and thereby augment economic performance. While this approach is not without its critics who deem the city's approach to the physical structure as being overly focused on attraction of business and tourism, it maximizes both public and private investment. Perhaps a larger issue is the fact that Łódź is not capable of maximizing the financial returns on these investments as property taxes in Poland are based on land and building size, as opposed to value. Thus, the potential financial windfall associated with these regenerative efforts is not captured through local property taxes (Krawchenko & Hainc, 2017).

An even bigger problem for Łódź than unrealized tax revenue is the virtual absence of citywide spatial planning. Without comprehensive planning, the ability of the city to address current needs and direct future development is curtailed by the piecemeal approach to planning by which it is hamstrung. Poland is notorious for the length of time it takes for land use planning to take effect, and municipalities frequently lack the proper policy and regulatory instruments to manage local land use practices. Indeed, the city of Łódź is ill-equipped to address either the depopulation of its urban core or the development of its outer fringe. Although Poland has spatial planning at every level of government, from national to regional to local, it is at the local level that land use and the built environment are regulated and spatial development plans are produced. However, plan coverage is particularly low in Łódź, even in comparison to most other municipalities of Poland. As a result, planning decisions frequently take on a piecemeal approach and can lead to fragmented forms of development. The ultimate development pattern may then become one of unplanned and undesired sprawl (OECD, 2016).

As a municipality that owns a significant portion of the land and buildings in its city centre, Lodz is certainly interested in both preserving and revitalizing its inner core. The spatial strategy of the city is therefore oriented around promotion of infill and brownfield development. However, any possible tax-based or other financial inducements remain essentially non-existent. Revitalization of costly brownfield sites is therefore frequently less attractive than developing on open greenfields located beyond the city centre. An additional planning problem facing Lodz is one shared by Detroit and other cities, and that is the difficulty of implementing any regional spatial plans. The upshot is that municipalities compete with one another for economically beneficial development. The result is that while the population of Lodz (as with Detroit) is in decline, the adjacent smaller communities have been enjoying a relatively robust inflow of residents. This leaves Lodz responsible for the entirety of the costs of providing services and maintaining the infrastructure for the benefit of both its own residents and the increasing numbers of commuters (Krawchenko & Hainc, 2017). In recent years, the highest growth of new housing development is clearly being concentrated on the outskirts of Lodz and spreading even further afield from there. Urban sprawl has undoubtedly come to Lodz, and while the city remains under-equipped to adequately plan for this recent and ongoing pattern of development, it is taking steps towards brighter and more sustainable future.

28.4.2 Resiliency and a Sustainable Future

As is the case in Detroit, much of the revitalization efforts in Łódź are oriented around improving resiliency and enhancing sustainability. Conception of the so-called new city centre in the areas around the railway station and adjacent post-rail and postindustrial areas is largely oriented around public transit. The main element is the construction of an underground station and a railway tunnel that leads trains from the centre towards Warsaw, though this is also accompanied by a broader reorganization of public transport in the city. The principal focus is the replacement of rolling stock with electric rolling stock in the case of buses, and the expansion of the tram system with its continued separation from automobile traffic. In addition, the implementation of a new parking policy is intended to reduce traffic in the centre. The road infrastructure around the railway station is also being restructured to reduce congestion and facilitate connectivity with other parts of the city. The city's activities, however, are not limited to its centre and include the first pilot programmes aimed at improving the housing stock. These programmes include the highly praised "City of Townhouses" programme, which is striving to modernize more than 150 city townhouses in the city centre.

However, real efforts at sustainability only started after the 2015 launching of the so-called area revitalization programme. Activities commenced in eight key areas designed to bring about comprehensive change. These include modernization of residential buildings, reconstruction of road infrastructure (including priority is given to bicycle roads), revalorization of green spaces or their reconstruction (parks, squares, greenery), and improvement of the quality of public services. At the same time, smaller actions resulting from social initiatives (participatory budget) are being carried out. The so-called woonerf, a street with limited car traffic, where pedestrian

Fig. 28.4 Sign for a woonerf in Łódź. **Photo**: Authors

lines and greenery (such pedestrian streets) have a large share, is a showcase of change in Łódź. Thus far, more than 20 have been created in the city (see Fig. 28.4). These actions are part of the city's efforts to develop blue-green infrastructure. All these activities, by improving living conditions, are expected to bring an increase in interest in the city centre in the near future, especially among young people who have so far chosen to move out of town and build their own home.

Łódź is indeed making sustainability and resiliency a hallmark of its revitalization efforts and has joined 44 other cities in Poland by declaring that it will achieve environmental resilience by the end of the decade. In 2020, the city initiated the *Eco-pact for Łódź* programme and began taking steps towards greening the city by reducing waste, increasing recycling, adding greenspace, and other environmental initiatives such as supporting rain gardens, xeriscaping, and alternative transportation (City of Łódź Office, 2019). Without question, these efforts have important environmental, economic, and equity consequences, but greening the city is not the only revitalization strategy being undertaken by Łódź. While the city had a nearly singular economic base during its time as a textile centre during the mid-nineteenth century, today the city is utilizing its wealth of historic mill structures as a means for revitalization. That numerous historic structures have been renovated and adapted for reuse fits well with the city's spatial strategy of promoting infill development and the reuse of brownfields over development of outer greenfield areas. Beyond the clear connection to sustainability, the reuse of these buildings also plays well with and reinforces the city's cultural heritage and identity.

Albeit having received the label "Textilopolis" (see Haase et al., 2020), the city's history is not limited to the textile industry and due to its historic heritage in cinematography, in 2017 the city joined the UNESCO Creative Cities Network as a city of film. Yet the film industry is not just something of the past, as the city hosts numerous film festivals and is home to upwards of 500 companies associated with the sector. The city both supports and is supported by the world-famous Łódź Film School, which produces actors, directors, photographers, and many others, and was attended by Roman Polanski. The heritage of Łódź provides many great opportunities. Original monuments of the industrialization of the city, including monumental factory palaces and great film traditions, are both reminders of the past and harbingers of the future. Łódź is refocusing its image as based on culture and artistic heritage, and is doing so in tandem with its efforts at enhancing the quality of life in the city through its sustainability initiatives and by revitalizing its urban core (Szapakowska-Loranc & Matusik, 2020).

28.5 Detroit and Łódź: Comparable Pasts, Indeterminate Futures

Shared histories that include painful deindustrialization, population loss, urban sprawl, and social distress, are the ties that bind the shrinking cities of Detroit and Łódź. Yet it is not only their similarities but also their differences that make Łódź and Detroit worth comparing (Zysiak, 2021). Socialism is proffered as one primary difference between the cities of Łódź and Detroit. Without question, the cities operated under disparate political structures in the decades following the Second World War. Łódź then experienced a painful transition to a market economy upon the eventual collapse of socialism. While difficult for Łódź to endure, this transition marks another important distinction between the two cities as these types of structural shocks can also generate constructive political creativity. In comparison, Detroit and many of its Rust Belt counterparts are noted to suffer from a political monoculture that robbed them of any potentially beneficial ideological diversity (Schramm, 2020).

In addition to a disparate political history, the cities of Łódź and Detroit also differ in the urban planning process. In Łódź, the limitations of planning are particularly apparent in regard to housing. Although local comprehensive plans are deemed critical to ensuring a sufficient housing supply (see Jun, 2017), only 20% of Łódź has spatial plans in place (Szapakowska-Loranc & Matusik, 2020). Despite a similarly aged housing stock, the city of Łódź suffers from a structural housing shortage inherited from the communist period. As a result, demolition of residential dwellings is quite rare and limited only to residences exhibiting only the very worst of conditions. In comparison, the problem in Detroit is an abundance of housing vacancies resulting from prolonged population decline. Subsequent clearance programmes have razed and removed tens of thousands of derelict houses throughout the city. In terms of the planning process, the cities do coincide when it comes to the issues of planning and urban sprawl. In Łódź, sprawl is creating a "doughnut effect" and is recognized as one of the most pressing planning issues (Gebalksa, 2017). As a result, the city has embarked on a strategy of increasing the density of its urban core, which has the opportunity to improve the city's economy while helping preserve its greenspace.

While regulatory limitations in Łódź hamper the process, greening the city is a planning strategy shared by both Łódź and Detroit alike. To reach its proclaimed goal of achieving environmental resilience by 2030, Łódź is taking measures to provide greener, reduce waste, and promote overall environmental quality. Through its blue-green infrastructure, the city is linking its complementary amenities of open space and water. Łódź is also taking great strides in revitalizing former brownfields into spectacular new uses such as a planetarium and a festival marketplace. This redevelopment adds economic prosperity while improving the city's image and preventing development of greenfields. Preserving brownfields also helps to preserve the city's heritage and built environment. Heritage urbanism provides opportunity for Łódź and Detroit to both stand alone as independent cities. Detroit's Brownfield Redevelopment Authority receives funding from 17 different sources and has taken on numerous rehabilitation projects of historically and culturally important buildings. Preserving and giving new life to these buildings provides benefits that may well spread beyond the central business district. The heritage of these legacy cities is a comparative advantage that helps foster an unrivaled sense of place that remains absent in surrounding suburbs and hinterlands.

The cities of Detroit and Łódź stand as testament to their legacies of production and pollution. After a dark period of decline that has lasted decades, both cities are now using the edifices of their past production as a means of promoting their cultural heritage as an opportunity for a brighter future. They are also reversing course on the environment. Whereas during the industrial era, both Detroit and Łódź became associated with dreadful air and water pollution, the cities are now turning towards the environment for regenerative purposes. Cleaner fleets of public transit, added greenspace, new bike lanes, and so much more, are serving to collectively enhance the vitality and resiliency of these legacy cities. While the future remains far from certain, for the cities of Detroit and Łódź, it is shining brighter than in the past.

28.6 Conclusion

Cities are economic engines that serve as centres of employment and production, but also of history, culture, entertainment, and more. They are also magnets of talent and innovation. These assets can provide cities with a roadmap towards a more prosperous and sustainable future. Such opportunities are particularly valuable to those formerly industrial cities that have experienced years and even decades of economic hardship. Notable examples of this type of city include Łódź in Poland and Detroit in the United States. These cities share a common history of an economic base that was dependent on manufacturing and mass production. The transition of these cities to a postindustrial economy proved fraught with difficulty and marked by population decline, economic decay, and social disorder. Efforts at urban improvement were largely unsuccessful and Detroit and Łódź became associated with the shrinking cities phenomenon. However, far from a death knell, this distinction provided a clarion call and new approaches towards urban revitalization were put forth. Most prominent have been the efforts based around urban resiliency and sustainability that have focused around greenspace, alternative forms of transportation, and cultural heritage. The old paradigm of growth above all else has been replaced with a focus on quality of life considerations in both Detroit and Łódź as they transition towards a more sustainable future.

References

Apel, D. (2015). *Beautiful Terrible Ruins: Detroit and the Anxiety of Decline.* Rutgers University Press.

Berglund, Lisa. 2020. Critiques of the Shrinking Cities Literature from an Urban Political Economy Framework. *Journal of Planning Literature* XX(X): 1–17.

Bontje, M. (2016). Creative cities and shrinking cities: False opposites? In V. Mamadouh & A. van Wageningen (Eds.), *Urban Europe: 50 Tales of the City* (pp. 153–159). Amsterdam University Press.

Bratton, William, Eide, Stephen, Goldsmith, Stephen, Hendrix, Michael, Husock, Howard, Miller, Judith, Murad, Jon, Renn, Aaron, and Peter Salins. 2018. *Urban Policy 2018.* Manhattan Institute for Policy Research (Online). Available at: https://media4.manhattan-institute.org/sites/default/files/MI_Urban_Policy_2018.pdf#page=107 [accessed 3/12/2020].

Carter, Donald K. 2012. Hope for the Future of the Postindustrial City. In, *SynergiCity: Reinventing the Postindustrial City*, Paul Hardin Kapp and Paul J. Armstrong (eds.). pp. 3–15. Champaign: University of Illinois Press.

Cialdella, J. H. (2020). *Motor City Green: A Century of Landscapes and Environmentalism in Detroit.* University of Pittsburgh Press.

City of Łódź Office. 2019. Eco-pact for the City of Łódź (Online). accessed 6/30/2020.

Clouse, Candi, Jinhee Yun, and Merissa C. Piazza. 2019. The Economics of Legacy Cities. In, *Legacy Cities: Continuity and Change Amid Decline and Revival.* J. Rosie Tighe and Stephanie Ryberg-Webster (eds.). pp. 37–50. Pittsburgh: University of Pittsburgh Press.

Deslatte, Aaron. 2020. Revisiting Bureaucratic Entrepreneurialism in an Age of Urban Austerity: Framing Issues, Taking Risks, and Building Collaborative Capacity. In, *Urban Studies and Entrepreneurship.*Iftikhar, M., Justice, J., and D. Audretsch (eds.), pp. 143–162. Switzerland: Springer International Publishing.

Foggin, M. (2018). Shaping a Future for Detroit: Small-Batch Production. *Challenge, 61*(3), 215–230. https://doi.org/10.1080/05775132.2018.1469296

Gebalksa, B. A. (2017). Using global trends as catalysts for city transition. *Procedia Engineering, 198*, 600–611.

Hackworth, J. (2018). Race and the Production of Extreme Land Abandonment in the American Rust Belt. *International Journal of Urban and Regional Research, 42*, 51–73. https://doi.org/10.1111/1468-2427.12588

Hannigan, T. J., Cano-Kollmann, M., & Mudambi, R. (2015). Thriving innovation amidst manufacturing decline: The Detroit auto cluster and the resilience of local knowledge production. *Industrial and Corporate Change, 24*(3), 613–634.

Hartt, M. (2019). The Prevalence of Prosperous Shrinking Cities. *Annals of the American Association of Geographers, 109*(5), 1651–1670.

Hartt, M. (2021). The Elasticity of Shrinking Cities: An Analysis of Indicators. *The Professional Geographer., 73*(2), 230–239.

Jun, H.-J. (2017). The Link between Local Comprehensive Plans and Housing Affordability. *Journal of the American Planning Association, 83*(3), 249–261.

Kinkead, D. (2016). Detroit Case Study. In D. K. Carter (Ed.), *Remaking Post-Industrial Cities: Lessons from North America and Europe* (pp. 46–65). Routledge.

Korkmaz, K. A. (2017). Analytical Approach for Detroit Urban Redevelopment and Rehabilitation. *IOP Conference Series: Materials Science and Engineering, 245*, 1–7.

Korkmaz, K. A. (2018). Urban Redevelopment and Rehabilitation for Detroit. *Journal of Civil Engineering and Architecture, 12*, 502–506.

Locke, R., Mehaffy, M., Haas, T., & Olsson, K. (2018). Urban Heritage as a Generator of Landscapes: Building New Geographies from Post-Urban Decline in Detroit. *Urban Science, 2*(3), 1–16.

Mallach, A. (2018). *The Divided City: Poverty and Prosperity in Urban America*. Island Press.

Meng, X., Jiang, Z., Wang, X., & Long, Y. 2021. Shrinking cities on thee: Evidence from LandScan 2000–2019. *Environment and Planning A: Economy and Space*. 0308518X2110061. https://doi.org/10.1177/0308518X211006118.

OECD. (2016). *Governance of Land Use in Poland: The Case of Łódź*. OECD Publishing. https://doi.org/10.1787/9789264260597-en

Pallagst, K., Mulligan, H., Cunningham-Sabot, E., & Fol, S. (2017). The shrinking city awakens: Perceptions and strategies on the way to revitalization? *The Town Planning Review, 88*(1), 9–13.

Pijawka, K. D. (2015). *Sustainability for the 21st Century*. Kendall Hunt Publishing Company.

Safransky, S. (2017). Rethinking Land Struggle in the Postindustrial City. *Antipode, 49*(4), 1079–1100.

Schindler, S. (2016). Detroit after bankruptcy: A case of degrowth machine politics. *Urban Studies., 53*(4), 818–836.

Schramm, C. J. (2020). Save America's Dying Cities. *Issues in Science and Technology, 36*(4), 62–70.

Stettner, A., & Yudken, J. S. (2019). Policies and Strategies for Revitalizing America's Manufacturing Communities. *Journal of Management Policy and Practice., 20*(4), 85–100.

Szafranska, Ewa, de Lille, Lydia Coudroy, and JaroslawKazimierczak. 2109. Urban Shrinkage and housing in a post-socialist city: relationship between the demographic evolution and housing development in Łódź, Poland. *Journal of Housing and the Built Environment* 34: 441–464.

Szapakowska-Loranc, E., & AngieszkaMatusik. (2020). Łódź – Towards a resilient city. *Cities, 107*, 1–14. https://doi.org/10.1016/j.cities.2020.102936

Thompson, E., & de Beurs, K. M. (2018). Tracking the removal of buildings in rust belt cities with open-source geospatial data. *International Journal of Applied Earth Observation and Geoinformation, 73*, 471–481.

Tighe, J. R., & Ryberg-Webster, S. (2019). *Legacy Cities: Continuity and Change Amid Decline and Revival*. University of Pittsburgh Press.

United Nations, Department of Economic and Social Affairs, Population Division (2019). *World Urbanization Prospects: The 2018 Revision* (ST/ESA/SER.A/420). New York: United Nations.

Weaver, R., Bagchi-Sen, S., Knight, J., & Frazier, A. E. (2017). *Shrinking cities: Understanding urban decline in the United States*. Routledge.

Wilson, D. (2014). The global trope and urban redevelopment: The American experience. *Geographica Helvetica, 69*, 79–88. https://doi.org/10.5194/gh-69-79-204

Zysiak, A. (2021). "Good" and "Bad" Workers and the Collapse of the Expected Life Course: The Postwar Working Class in Detroit (USA) and Łódź (Poland), 1940s–1980s. *East European Politics and Societies and Cultures, 35*(1), 3–25.

Chapter 29
Polycentric Megacity Regions and Regional Planning and Governance: Insights from New York and the Rhine-Ruhr?

Linda McCarthy

Abstract This paper reviews the academic and public policy literatures on megacity regions and regional planning and governance in an effort to establish the state of the theoretical and empirical research for the United States and Europe and to develop a conceptual framework for examining polycentric megacity regions and regional planning and governance in the United States and Europe from a comparative perspective. New York and the Rhine-Ruhr were selected as case studies because they are the largest polycentric megacity regions (containing multiple large urban centres) in the United States and Europe respectively. Having established a conceptual framework, this paper examines the regional planning and governance efforts in New York and the Rhine-Ruhr. Not so much despite but rather because neither has an especially stellar record for having achieved regional planning and governance, these cases offer theoretical insights and policy implications concerning the processes, dynamics, and issues associated with the challenges as well as the opportunities at a polycentric megacity regional scale. As such, notwithstanding differences such as location and political context, demographic characteristics, economic development, and environmental quality, the lessons learned from New York and the Rhine-Ruhr may be of interest for policy and planning in other large polycentric megacity regions.

Keywords Megacity regions · Polycentric megacity regions · Regional planning and governance · Cooperation · Competition

29.1 Introduction

This paper reviewed the academic and public policy literatures on polycentric megacity regions and regional planning and governance in an effort to establish the state of the theoretical and empirical research for the United States and Europe

L. McCarthy (✉)
Department of Geography, University of Wisconsin-Milwaukee, Milwaukee, USA
e-mail: lmccarth@uwm.edu

B. R. K. Sinha (ed.), *Urban Dynamics, Environment and Health*,
https://doi.org/10.1007/978-981-99-5744-6_29

and to develop a conceptual framework for examining polycentric megacity regions and regional planning and governance from a comparative perspective. According to Hall and Pain (2006, 3), a polycentric megacity region can be defined as between ten and fifty towns and cities that are separated spatially but networked functionally based on a new functional division of labour in a cluster around a number of central cities connected by freeways, high-speed railways, and telecommunications lines that facilitate the flow of people and information.

New York and the Rhine-Ruhr were selected as case studies because they are the largest polycentric megacity regions in the United States and Europe respectively. The New York and Rhine-Ruhr polycentric megacity regions contain multiple large urban centres. In addition to New York City, other large centres in the New York megacity region include Newark and Jersey City. Large urban centres in the Rhine-Ruhr megacity region of Germany include Cologne, Düsseldorf, Dortmund, and Essen.

Polycentricity as a concept and as a policy are addressed as part of the regional context for planning and governance for megacity regions. A critical take on two aspects of regional planning and governance is useful for developing a conceptual framework for understanding polycentric megacity regions from a comparative perspective: the "region" as the (obvious) scale of focus and the tensions between cooperation and competition.

Having established a conceptual framework, this paper examines the regional planning and governance efforts in the New York and the Rhine-Ruhr megacity regions. Not so much despite but rather because neither has an especially stellar record for having achieved regional planning and governance. These two cases offer theoretical insights and policy implications concerning the processes, dynamics, and issues associated with the challenges as well as the opportunities at a polycentric megacity regional scale. As such, notwithstanding differences such as location and political context, demographic characteristics, economic development, and environmental quality, the lessons learned from New York and the Rhine-Ruhr may be of interest for policy and planning in other large polycentric megacity regions.

29.2 Megacity Regions and Regional Planning and Governance

Polycentricity has recently received attention in the academic literature related to its spatial pattern and spatial relations, as well as to polycentricity as policy and as a planning approach. A distinction has been made between a more monocentric and a more multicentric spatial pattern and spatial relations: the more monocentric pattern of clustering of people and economic activities involves interaction and commuting between surrounding smaller urban centres and a dominant core, such as London and Paris; the more multicentric pattern involves interaction and commuting among historically more independent centres, such as the Randstad in the Netherlands (Burger & Meijers, 2012; Kloosterman & Musterd, 2001).

The more multicentric pattern of polycentricity applies well to the Rhine-Ruhr. The New York megacity region can also be viewed as comprising a multicentric pattern of polycentricity despite the importance of New York City. There is a sharp political divide separating New York on one side of the New York–New Jersey state border from other large urban centres such as Newark and Jersey City on the other. There is an economic divide captured, for example, in the pattern of employment distribution, with jobs in Manhattan comprising less than 20 per cent of the megacity regional total (Cox, 2020). Some scholars identify a cultural divide, captured, for example, by something as simple as the generalization that in New York, people walk, and in New Jersey, people drive (Rockland, 2008).

Polycentricity has also received recent attention in the public policy literature. As a planning approach, polycentricity has been used to promote more balanced urban development among and within city regions. At the inter-urban or inter-city regional scale, the European Union's (EU) European Spatial Development Perspective (ESDP) incorporates polycentricity to achieve a more balanced distribution of urban centres across its entire territory (European Commission, 1999; Hall & Pain, 2006; Vandermotten et al., 2008). At the intra-urban or intra-city regional scale, polycentricity has been used as a policy to promote more balanced development within individual city regions: either by making a city region even more polycentric, such as by establishing new centres as nodal points for development or by maintaining and better utilizing existing polycentricity to achieve positive synergies. The underlying assumption is that the individual urban centres connect in a synergistic fashion, making a polycentric city region more than the sum of its parts due to complementary relations and positive externalities (Knapp et al., 2006, 157). Polycentricity as a policy is usually conceptualized as an entirely strategic one, and so is related frequently to other planning goals such as establishing a decentralized regional transportation system (Schmitt et al., 2015, 7). Although there is no consensus in the literature, some studies have found that polycentricity can be a sustainable development policy, for example, if it helps to decentralize both housing and jobs and leads to less commuting (Knaap et al., 2015; Lin et al., 2013).

Based on a review of the academic and public policy literature, a critical take on two important aspects of regional planning and governance is particularly useful for developing a conceptual framework for understanding polycentric megacity regions from a comparative perspective: the "region" as the (obvious) scale of focus and the tensions between cooperation and competition.

29.2.1 *The "Region" as the (Obvious) Scale of Planning and Governance*

Not surprisingly for research and policy related to city regions, the "region" is the (obvious) scale of focus in the planning and governance literature. Despite political fragmentation and urban sprawl making coordinated regional planning and development difficult to achieve at the city regional scale, and even more challenging to tackle at the megacity regional scale, many scholars and planners accept both the region as a notion, and regional planning and governance as essential, up to and including for some the need for a regional tier of government (Knaap & Lewis, 2011; Ziegler, 2009). In this connection, Seltzer and Carbonell (2011, 1), for example, argue that, because the regions that people inhabit are usually defined by function rather than local government boundaries, regional planning is necessary because local planning efforts cannot properly address local issues without concern for the regional context.

At the same time, geographers and other academic and policy analysts and practitioners have long been challenged by definitional issues associated with identifying regions; consequently, identifying city regions is no less fraught. Moreover, given political fragmentation and urban development reaching far beyond municipal boundaries, some researchers are sceptical about the very essence of city regions (Healey, 2009). For most urban inhabitants, a region is merely an abstraction. People do not tend to think regionally given that they encounter community and trust at a human and cognitive scale closer to themselves as individuals, such as at the scale of their street or neighbourhood (Seltzer & Carbonell, 2011, 2–3; Wheeler, 2009, 867). Residents or businesspeople can lack any perception of a region as the space with which they identify, on the one hand, or operate in, on the other; most people are fairly localized within their larger region in their travel to work, shopping, recreation, and other trips (Gordon, 2012, 50). Concerns have been raised that this lack of legitimacy for the region may undermine its possibility or value as a planning and governance concept for promoting an integrated policy approach (Healey, 2009).

Certainly, no one would dispute the difficulty of the concept of their region becoming part of peoples' imagination and action space, and the resultant challenges for tackling the region, and especially the megacity region, as a scale of planning and governance. But the regional scale does not have to be the sole scale at which to imagine or consider planning and governance. Regional and local actions do not necessarily need to be seen as either/or scales. Taking a pragmatic view, it is evident that bottom-up localism within polycentric city regions is already occurring in conjunction with explicit top-down regional planning (or not). And so the conceptual framework for polycentric megacity regional planning and governance in this paper incorporates the reality that bottom-up local inter-jurisdictional efforts operate at the same time as top-down or regional inter-local governmental planning and governance. This is the case even if the bottom-up local efforts are sub-regional, and perhaps involve only a few local governments within the larger polycentric city region, and the efforts may not be tied explicitly into regional policy, that is not completely regional in scale but feed into regional planning and governance nonetheless. As such, it is necessary to consider and incorporate bottom-up local efforts as well as top-down regional policy and practice in order to understand regional planning and governance.

So the conceptual framework in this paper questions using (only) polycentricity as a policy and planning approach in the many situations where it is difficult or impossible to achieve a regional focus or cooperation across an entire region and instead accepts reality and includes bottom-up localism in terms of it contributing results that benefit the entire region nonetheless. This relates to the distinction between functional versus territorial cooperation. There may be times when it is important to cooperate as a territory (an entire region) where the members represent their region on regional scale activities, but then there may also be (many) times when functional cooperation (on sub-regional projects and activities) makes more sense among members expert in particular issues and where delineating the region as the scale for planning is not relevant or useful (Fürst, 2009; Kunzmann, 2015).

This conceptualization of localism as part of regionalism incorporates what is actually occurring on the ground in a number of ways. For example, sporadic or piecemeal bottom-up efforts need to be conceptualized in addition to, or at least rather than solely in response to, top-down regional efforts. In fact, some analysts have found that the impetus for local governments to cooperate with each other is determined by imperatives other than regionalism, for example, due to the pressing need for change prompting local communities to become more self-reliant. These local community efforts may not even be driven by proximity relations within a polycentric city region, and so regional planning and/or aspirations of polycentricity are not the driving force (Groth & Smidt-Jensen, 2007, 82). While Turok's (2009, 845) case study was not about cooperation, his finding is relevant in terms of the tension between how a regional agenda for housing provision conflicted with the local needs of individual communities for jobs, training, and better public services; a focus on accomplishing particular regional policies can be at odds with achieving the needs of individual local communities within the larger region.

29.2.2 Tensions Between Regional Cooperation and Competition

Not surprisingly for planning and governance related to city regions, coordination of development and government efforts is a major focus in the literature. At the same time, the tensions between cooperation and competition within the context of regional planning and governance need to be taken into consideration in practice and in developing a conceptual framework for understanding polycentric megacity regions from a comparative perspective. Certainly, cooperation and competition are not mutually exclusive. Given that localities cooperate to compete, from small groups of localities to the entire city region, cooperation and competition are not either/or propositions.

In terms of how the tensions between cooperation and competition play out, four bodies of literature provide insights into regional planning and governance: political rescaling and the city regional scale; economic competitiveness of city regions in a global economy; social issues related to democratic processes associated with uneven development and inequalities; and environmental quality and sustainable development.

Politically, research on the rescaling of the national state has focused increasingly on territorial competitiveness at the city regional scale (Brenner, 2003; Neuman & Hull, 2009). This literature facilitates an understanding of the difficulties of cooperation both within city regions and between different levels of government. Some of the arguments about the changing relations between different levels of government in conjunction with the "hollowing out" of the national state, however, neglect the continued, albeit reconfigured, role of the national level of government. In certain contexts, some scholars even argue that the spatial reach of regional strategies can in fact be constrained by, rather than transcend, existing formal structures of governance and political decision-making (Walsh, 2010).

Economically, localities often cooperate within city regions in quite straightforward ways to provide more cost-effective intra-regional services such as emergency response or water infrastructure. Localities may also attempt to cooperate within city regions to be more competitive globally. Allen Scott (2001: 4–5) has argued that global city regions operate as the motors of the world economy and so globalization and city region development are two parts of an integrated process. But as Jonas et al., (2005, 240) have argued, any attempt to reduce the impulse to engage in city-regionalism to a rational and similar set of reactions to economic competitiveness imposed from above neglects the varied and contested competing strategies or conceptualizations of the city region as a functional urban region.

Moreover, in contrast to competition among companies that leads to innovation, competition among localities may merely, if at all, affect the location of economic development within a particular city region. Studies have found that competition among localities within a city region usually leads to the duplication of ideas and policies rather than, for example, complementarity that maximizes the strengths of each locality, which is a goal of cooperation for polycentric urban regions in the ESDP policy framework (European Commission, 1999; Maier et al., 2006; McCarthy, 2015).

Socially, the tensions between cooperation and competition in polycentric city regions relate to issues of uneven development and spatial equity. Jonas and Ward (2007, 174) have argued that "age-old contradictions of spatial growth in capitalism come to rest in and through the territorial structures of city-regions". Jones and Etherington (2009) have contended that city regional policies may reinforce and promote rather than reduce uneven development and socio-economic disparities. This relates to concerns about the operation of democratic processes if city regional policies transcend or circumvent the formal structures of government and political accountability (Jonas & Ward, 2007; Walsh, 2010).

Environmental quality as an issue crosses political boundaries within a city region and so cooperative efforts at the city regional scale make sense. Certainly, the environment cross-cuts the political, economic, and social: environmental degradation in city regions comes with negative social effects, involving economic costs and requiring political intervention. Moreover, environmental and economic goals overlap because the environment cannot be separated from the economic context; increasingly, environmental quality is seen as an aspect of the competitiveness of city regions.

So, for polycentricity as a policy and planning approach, it is necessary to consider the extent to which cooperation is competition; that is, to take into account that even if cooperation is occurring, it may also be competition when considered from the next scale up, and in some cases, where the cooperation appears constructive and productive at a sub-regional or even regional scale, it may be wasteful zero-sum competition at the national scale.

Meijers (2006) has contended that given a long history of rivalry, localities need a basis for cooperation that minimizes the incentive for unproductive opportunism and wasteful competition. Healey (2009, 840) has argued against any imposed formal arrangement and instead for allowing a "soft institutional form" albeit supported possibly by enabling legislation. Megacity regions' huge size means that they are economically important in their own right as well as a national priority as part of a country's urban system (Gordon, 2012). To a greater or lesser extent, there is some acceptance of the role of national government to support and help manage with funding the onward momentum of the diverse efforts within a region as part of the "new regionalism". The new regionalism involves a variety of diverse kinds of coordination and cooperation that is not based on a predetermined plan or policy but accepts instead incremental iterative processes and change over the long term (Gordon, 2012; Kübler, 2012). Some researchers, however, fear that localities will merely work together to the minimum extent possible in order to secure available national government funding; they also raise concerns about the emphasis on economic or administrative efficiencies at the expense of citizens' concerns about quality of life issues (Healey, 2009; Jonas et al., 2005).

29.3 The New York and Rhine-Ruhr Megacity Regions: Regional Planning and Governance

Informed by the conceptual framework in the previous section, this section of the paper examines the regional planning and governance efforts in the New York and the Rhine-Ruhr megacity regions respectively. For each megacity region, after a short description, representative examples of regional cooperation and sub-regional cooperation are discussed.

29.3.1 The New York Megacity Region

The New York megacity region, with a population of about 23 million people, closely approximates the U.S. Census Bureau's combined statistical area of New York-Newark, NY-NJ-CT-PA. This megacity region is politically fragmented, comprising 35 counties containing multiple municipalities and numerous single-purpose agencies stretching across four US states (New York, New Jersey, Connecticut, and Pennsylvania).

The perceived need for a regional scale of planning and governance goes back to at least 1929 and the first long-range regional plan devised by the Regional Plan Association (RPA) (Table 29.1). The fourth regional plan in 2017 follows the third plan in 1996 and the second plan in 1968. In terms of polycentricity used as policy, while not explicitly applied as in the EU, the 2017 regional plan's goal of "growth and prosperity" involves a concern with balanced development in terms of the "where" and the "how" of growth based on investments in infrastructure, economic development, housing, and environmental sustainability; so, for example, more growth and jobs are designated to reinforce the urban core to promote the resurgence of small to large urban centres surrounding the core, and more population growth is envisaged in existing or new walkable communities associated with increased transit ridership and reduced greenhouse gasses to achieve more sustainable development with less sprawl (RPA 2016).

In terms of the spatial extent of this region, it should be noted that while the regional plans include most of the megacity region—the tri-state area—the part in Pennsylvania is not included (Table 29.1). While the RPA is focused from the top-down at the regional scale, in developing the 2017 plan, it continues efforts at inclusion since the first plan in actively attempting to engage civic groups, community members, policy experts, etc. in the planning process. As such, it constantly encounters the tensions between cooperation and competition, particularly given that a major goal of the 2017 plan is to cooperate regionally to be more competitive globally. In fact,

Table 29.1 Regional planning and governance: examples from New York and the Rhine-Ruhr

New York	The Rhine-Ruhr
Regional (top-down) cooperation	*Regional (top-down) cooperation*
Regional Plan Association plans:	EU and national government:
• Region excludes Pennsylvania	• "European metropolitan region"
• Municipal tax base competition	• Ruhr and Lower Rhine in competition
Sub-regional (bottom-up) cooperation	*Sub-regional (bottom-up) cooperation*
• Hudson Valley Econ. Development Corp	• RVR and Koln/Bonn und Nachbarn
• Sterling Forest State Park	• WiN Emscher-Lippe GmbH
• MAP Forum (5 MPOs)	• 6 city planning association and regional land use plan

the RPA has acknowledged that mobilizing the public, private, and nonprofit sectors to implement its regional plan is difficult. Politicians and planners tend not to be focused at a regional scale but are concerned with the more limited area of their own jurisdictions.

A perhaps more major impediment to implementation that has been an obstacle to implementing the previous three plans is the fact that the RPA is an independent nonprofit body; it is not a regional tier of government and has no regulatory or other powers to implement the plans. At the same time, the elite status of the RPA and its association with big business has also been raised as an issue in general, and in terms of a criticism of the third plan's focus on the urban core which primarily benefitted businesses because it fuelled real estate profits (Angotti, 2008; Henwood, 2013; Juhasz, 2014).

On the one hand, the overall assessment of the regional plans has been mixed (Gillham, 2002); on the other, the plans are seen to have served as a stimulus for investment "in transit infrastructure and open space protection and creative thinking about the shape and function of urban and suburban centers" (Yaro, 2000, 68). Moreover, the four regional plans are at least comprehensive in their focus, and apply to the majority of the New York megacity region's spatial extent. The regional transportation governance recommendations, however, have not been implemented (Gillham, 2002, 234) and the New York megacity region's transportation system is not coordinated across state lines.

The two other major entities in the megacity region—the Port Authority of New York and New Jersey, and the Metropolitan Transportation Authority (MTA)—have a limited spatial extent of responsibility and operation and focus only on a single policy issue, transportation. In the 1990s, there was renewed interest in the concept and process of regional planning in the United States. This shift may have been due in large part to federal statutes which mandated that state and local governments develop regional strategies or work with regional planning agencies in order to qualify for federal funding, including for transportation (Salkin, 1993). This brought to the fore metropolitan planning organizations (MPOs). MPOs are councils of government (COGs) comprising elected officials from each municipality, or regional planning councils (RPCs) staffed by planning professionals. MPOs are responsible for devising metropolitan transportation plans and determining which local transportation projects receive federal funding.

The New York Metropolitan Transportation Council (NYMTC), for example, was designated as the MPO for New York City, Long Island, and the lower Hudson River Valley. Despite having metropolitan in its title, the MTA was established by the New York state legislature to oversee transportation (only) from New York City to Long Island.

The NYMTC lobbies for federal transportation funding, but, despite the fact that MPOs are directly associated with federal laws that require the coordination of transportation policies within metropolitan regions, coordination is not always achieved; although at least the MPOs as a group cover the entire megacity region, active cooperation is not region-wide, being only ongoing among small groups of MPOs within the region. For example, the Metropolitan Area Planning (MAP) Forum comprises five MPOs across three US states (New York, New Jersey, and Connecticut) that agreed to cooperate to reduce duplication and promote consistency in their efforts (Table 29.1).

By far the most common cooperative efforts within metropolitan regions like New York are arrangements such as joint powers agreements, or special districts in which two or more municipalities agree to co-finance services like law enforcement or emergency response. A larger version of the special district is a regional district or authority. A long-established, since 1921, example is the Port Authority of New York and New Jersey. The Port Authority was an initiative by New York and New Jersey to move beyond competition over the use of a port spanning both states.

Although not having jurisdiction for the entire New York megacity region, the Port Authority could do more to promote greater coordination in its capacity as a regional planning agency; instead, it focuses on infrastructure projects that generate revenue. The Port Authority has considerable revenue-raising capacity based on, for example, the tolls on the bridges between the two states (Ross & Levine, 2001). It should be acknowledged, however, that the Port Authority has attempted to encourage cooperation in the past, such as devising a regional marketing strategy that localities refused to sign onto. The Port Authority did convince state and local governments to sign a regional "non-competition pact" in 1991 in which they promised not to offer public subsidies to poach companies away from each other. The pact fell apart when tested.

As a result, wasteful competition continues apace (Table 29.1). An example of zero-sum competition between New York and New Jersey is the more than $200 million in public subsidies for Citigroup over two decades. A 2007 report questioned the value of the state and local subsidies of $113.9 from New York and $101.1 from New Jersey that essentially poached (the same) jobs from one side of the Hudson River to the other over time. These subsidies would have been better spent cooperatively by the states and localities of New York and New Jersey to make the megacity region more competitive globally by investing in improving the basic requirements of all businesses, including the quality of the labour force, schools, and infrastructure (Stecker & Steinberg, 2007).

Competition also occurs within individual states. Many jurisdictions operate their own economic development agencies. The New York City Economic Development Corporation (NYCEDC), for example, is a public–private economic development agency that competes with other government entities for investment and jobs. The NYCEDC administers the New York City Industrial Development Agency (NYCIDA) which seeks to enhance New York City's ability to attract corporate investment by issuing revenue bonds to support economic development projects.

There is also competition within individual jurisdictions. The State of New York passed a bill in 1996, for example, that gave the five boroughs of New York City more opportunity to compete with each other for new investment. The bill allocated a quarter of the NYCIDA's budget among the five boroughs to fund their own economic development projects and provide tax subsidies for businesses. An editorial in the New York Times (1996) argued that given how it was already problematical that New York City felt compelled to compete with Stamford or Jersey City for investment and businesses, it made no sense to allow the Bronx, Brooklyn, Manhattan, Queens, and Staten Island to join the competition.

At the same time, there are ample examples of cooperation to be more competitive by localities within individual states. A top-down initiated example comes from the State of New Jersey, which created a regional body in 1968, the New Jersey Meadowlands Commission (which is now part of the New Jersey Sports and Exposition Authority (NJSEA)). Cooperation among the 14 localities across two counties along the Hackensack River comprises public and private participants who operate as a zoning board to address economic development, environmental protection, and waste management. Importantly, the 14 localities agreed to tax-base sharing; this prevents individual municipalities from suffering financially because they are not allowed to develop protected land for commerce or industry. These localities cooperated with environmentalists, businesses, mayors, and state and federal agencies to draw up a master plan that emphasizes brownfield redevelopment, job creation, and wetland and open space preservation.

In contrast, New York's Hudson Valley Economic Development Corporation (HVEDC) was established in a bottom-up initiative by seven counties, their cities, and a nonprofit environmental group; the environmental and economic aspects involve marketing sites including brownfields along the river to businesses for cleanup and redevelopment (Davis, 2003) (Table 29.1).

Another bottom-up effort came about after a three-decade struggle to preserve from development a 20,000-acre watershed greenbelt privately owned by the Sterling Forest Corporation in the New York–New Jersey Hudson Highlands. A coalition of contributors that included the federal government, New York State, the State of New Jersey, private foundations, environmental groups, and local individuals established the first regional park in the New York area since 1940, the Sterling Forest State Park (Collins, 2006; Trust for Public Land, 2006) (Table 29.1).

29.3.2 The Rhine-Ruhr Megacity Region

The Rhine-Ruhr megacity region, with a population of about 13 million people, stretches from Mönchengladbach in the west, and Bonn in the south, to Hamm in the north-east (Dieleman & Faludi, 1998; Schön, 2010). The Rhine-Ruhr contains multiple municipalities and districts across three administrative units in the German state of North Rhine-Westphalia.

The EU's ESDP and Germany's strategic spatial planning have the goal of a polycentric regional structure with balanced development comprising urban centres that are well-connected via an efficient transportation network in conjunction with enhanced European and German integration (Botequilha-Leitão, 2012, 3; European Commission, 1999; Knapp et al., 2006, 158; Schmidt & Buehler, 2007, 55) (Table 29.1). In the Rhine-Ruhr megacity region in particular, however, although polycentricity as a policy is viewed as beneficial, the multiple stakeholders and conflicting interests and agendas are seen as significant obstacles to regional planning and cooperation. The Rhine-Ruhr is politically fragmented in general, as well as particularly in terms of the different historical development of the Ruhr area in the north-east (Ruhrgebiet, with Bochum and Essen) given the legacy of its industrial past and continued efforts to restructure the economy; and the Rhine corridor in the south-west (Rheinschiene, with Düsseldorf, Cologne, and Bonn) that is more prosperous with an economy already based more on advanced producer services (Knapp et al., 2006).

Yet the Rhine-Ruhr Metropolitan Region (Metropolregion Rhein-Ruhr) has been designated as one of eleven "European Metropolitan Regions" in Germany (Table 29.1). German national strategic spatial planning identifies these metropolitan regions as motors of economic, social, and cultural development to maintain and enhance German and EU competitiveness and integration (Jurczek, 2008, 2). Despite the idea of a Rhine-Ruhr metropolitan region being raised first in 1995 in the North Rhine-Westphalia Regional Development Plan (Grier, 2002, 4), a number of analysts see the state government as compounding the internal division between the Ruhr and the Rhine areas. The state government created the Rhine-Ruhr metropolitan region despite the political, economic, and social differences between the Ruhr and the Rhine areas. Politically, there was limited interest by local stakeholders, an important obstacle in the German system where, compared to the national government, local and state governments are strong participants with self-governing ability enshrined in the basic German national legal framework. Designating the metropolitan region from the top down did not erase the political and administrative fragmentation, strong rivalry among historically independent urban centres in a multicentric urban system, and lack of regional identity (Knapp et al., 2006, 157) (Table 29.1).

Moreover, the top-down designation of the Rhine-Ruhr metropolitan region by the state government actually worked against the goal of regional cooperation because it evoked apprehension among local governments that they would be forced to relinquish their self-governing authority (Knapp et al., 2006, 160). Partly in response, the state government was seen to be indecisive and hesitant to take an active role and develop concrete regional policy proposals for implementation (Yamai, 2015, 56). Instead, the state government concentrated its efforts on supporting existing cooperation within the more well-defined Ruhr area, and allowed the urban centres along the Rhine to work on their own to promote polycentricity by improving their individual roles as growth poles.

So instead of cooperation across the entire metropolitan region as is more common in the other ten German metropolitan regions, cooperation occurs separately within the Ruhr area and within the Rhine area (Kunzmann, 2015). Examples of cooperation to become more competitive include mergers of public transit companies. In the Rhine area, the port operations for Neuss and Düsseldorf have been combined. In the Ruhr area, all of the district governments cooperated to develop a regional air quality plan.

Instead of one Rhine-Ruhr agency (proposed but not realized), there are two separate independent networking and lobbying organizations (for strategic spatial planning and to secure state, national, and EU funding): the Ruhr Regional Association (RVR), and for the Lower Rhine, Regio Köln/Bonn and Nachbarn (Cologne, Bonn, and neighbours, including Leverkusen) (Table 29.1).

Regio Köln/Bonn and Nachbarn is a cooperative effort to achieve regional marketing and regional planning and projects partly through securing funding from higher levels of government (Knapp et al., 2004, 338). Examples of cooperation to be more competitive include the Cologne-Bonn Airport and the Rhine-Sieg Transport Association (Verkehrsverbund Rhein-Sieg), a public transport association for Cologne and Bonn. It should be noted that Regio Köln/Bonn and Nachbarn does not include all the urban centres along the Lower Rhine. Discussion and studies are ongoing about including more "neighbours", including Düsseldorf, so that membership covers the Lower Rhine more completely, especially compared to the more established Ruhr Regional Association (RVR).

The Ruhr's RVR is a cooperative effort among 53 municipalities (including 11 cities: Bochum, Bottrop, Dortmund, Duisburg, Essen, Gelsenkirchen, Hagen, Hamm, Herne, Mülheim, and Oberhausen) across four districts. The RVR seeks to achieve regional networking and marketing, regional projects (including the Industrial Heritage Trail and the Emscher Landscape Park) with funding secured from higher levels of government, and regional planning including a regional plan to be coordinated with the municipal and district plans and be legally-binding. All the municipalities and districts are represented at the RVR parliament which is a regional assembly elected for 5 year terms with voting rights on joint measures for the Ruhr area including projects and a regional plan; it is composed of about 70 municipal and district delegates with the 15 Lord Mayors and district administrators.

An example of bottom-up cooperation is WiN Emscher-Lippe GmBH established in 1990 as a public–private corporation for improving the economy of the Emscher-Lippe region including supporting jobs, marketing, and innovative projects (WiN Emscher-Lippe GmBH, 2021) (Table 29.1). This regional development company founded by 36 companies cooperates with other entities in the northern part of the Ruhr area including municipalities and chambers of commerce and trade as well as the Westdeutsche Landesbank and its savings banks (Eckart et al., 2003, 138–139).

At the same time, competition also occurs. In the Ruhr area, for example, efforts to cooperate on the visioning exercise associated with the regional plan were impeded by the lack of communication between the municipalities and districts (RVR, 2010). Likewise, the RVR has bemoaned the fact that the redevelopment of brownfields has rarely been done in conjunction with a regional planning perspective. Competition among localities for developers to invest in brownfield redevelopment has resulted in piecemeal efforts.

29.4 Discussion and Conclusions

As Labbé and Sorensen (2020, 2) argue, very large urban regions confront "distinctive governance and environmental challenges, present significant questions for research, and occupy a major place in urban imaginaries and understandings". The New York and the Rhine-Ruhr megacity regions offer theoretical insights and policy implications concerning the processes, dynamics, and issues associated with the challenges as well as the opportunities at a polycentric megacity regional scale. With global population growth and urbanization projected to slow during the next 50 years, "the particular timing, governance institutions, and contingent choices of planning and infrastructure building will … have profound long-term consequences" for megacity regions around the world (Sorensen, 2019, 35). As such, the historical lessons learned from longer-established, already slower-growth polycentric megacity regions such as New York and the Rhine-Ruhr may be of interest for policy and planning associated with the ongoing transformations taking place in other large polycentric megacity regions.

In terms of spatial patterns and spatial relations, although polycentricity as a policy of higher levels of government with the goal of balanced development of urban centres in a city region is viewed generally as beneficial, it may be easier to accomplish in city regions with a more monocentric pattern. The multicentric Rhine-Ruhr metropolitan region, for example, lacks the regional planning and cooperation across the entire megacity region found more commonly in more monocentric German metropolitan regions such as Stuttgart, Munich, Hamburg, Bremen, and Nuremberg (Kunzmann, 2015).

In terms of a more multicentric pattern of polycentricity then, there can be a greater possibility not only of a lack of regional identity as the basis for regional cooperation but also strong rivalry among politically fragmented and historically independent urban centres that can contribute to a lack of buy-in by localities for regional cooperation. So given the difficulties in achieving regional planning from the top-down in the Rhine-Ruhr megacity region, and the mixed results of the RPA plans for the New York megacity region, it needs to be considered that regional planning cannot be the only focus. Certainly, a top-down regional planning focus to manage interdependencies within a megacity region is necessary given the need for a regional perspective that cannot be achieved by individual municipalities acting alone. Despite in the New York megacity region, for example, where the "RPA has developed an aspirational vision for the metropolitan region" (2016, 5) based on regional cooperation, the reality is that regional planning and cooperation across the entire megacity region is not going to happen in the short- or even long-term.

Given that reality, a variety of sub-regional bottom-up cooperative efforts could also be viewed as beneficial even if they are not at the scale of the entire region as long as they benefit the region as a whole. Functional cooperation with a focus on a particular issue or issues could also be acceptable, especially if ongoing efforts to achieve territorial cooperation across the entire megacity region are challenging. This relates to Gordon (2012) highlighting the need to develop an appreciation of interdependencies within a region and how to manage them better, on the one hand, and the costs of ignoring these interdependencies, on the other.

This focus could open up opportunities and acceptance of cooperation among localities from the bottom-up of the kind that met criticism in the Rhine-Ruhr, for example. Cooperation by Bochum, Essen, Gelsenkirchen, Herne, Mülheim, and Oberhausen that led to their establishment of a planning association to develop a regional land use plan (Table 29.1) met disapproval in a report commissioned by the state government; the criticism was that by focusing the plan on only part of the Rhine-Ruhr metropolitan region, it limited the opportunities for future cooperation across the entire region (Wegener, 2012). Likewise, Yamai (2015, 59) concluded that, for the Rhine-Ruhr, "the growth of sub-regional cooperation has hindered the development of a metropolitan region for the entire area". Yet sub-regional cooperation such as this could also be considered as part of the path towards regional cooperation if it helps establish the basis for wider cooperation including engendering relations of trust. So instead of looking at it so much from the perspective of the entire region from the top down in terms of the whole being greater than the sum of the parts; instead also look at it from the perspective of sub-regional efforts from the bottom-up in terms of the sum of the parts making a greater whole.

In this connection, Wegener (2012) has argued for what he terms more government, less governance, that is, more top-down government to achieve regional planning. But why not have government *and* governance? If regional planning is nowhere in sight, real cooperative efforts by a sub-set of cities may yet benefit their region. This is the kind of sub-regional cooperation that is not just done for funding capture from higher tiers of government. Moreover, in terms of the tensions between cooperation and competition, for example, because of their different economic characteristics, the Ruhr area and the Rhine area are not in direct competition with each other for the same corporate investment anyway.

Recognizing the value of bottom-up sub-regional cooperation at the same time as top-down regional planning may also come with unanticipated positive results including from opening space for innovative collaborations that are not considered in the regional plans and regional planning process, such as the Sterling Forest State Park (Table 29.1). This can also help address the rigidity of regional plans by not just allowing but deliberately building into these plans the possibility and mechanisms of adjustment and revision that facilitate flexibility and incremental action; it could also allow greater openness and transparency and so partly address some criticisms of regional plans that they benefit vested interests such as urban elites and real estate partners.

Many academics and public policy analysts have researched and published on regional planning and cooperation amid competition, and they invariably constantly refer to the challenges to regional planning and cooperation. They also usually identify the factors that can encourage regional cooperation, including the involvement of a higher level government offering funding and other incentives with the goal of engendering inter-locality trust and habits of cooperation (Gordon, 2012).

But, in thinking about it, how are these efforts to promote cooperation often promoted? Top-down initiatives with higher-tier government funds including for regional conferences to which a representative of each locality is invited. But if the underlying impetus for attending a regional conference is a competitive urge, to discover if there is anything happening with other jurisdictions that an individual locality needs to be concerned about in terms of its competitiveness, or if it is just to find out whether there is an opportunity for funding capture to benefit one locality as much as or more than another, then the top-down organized conferences may in fact merely be reinforcing existing rivalries instead of promoting sincere cooperative efforts!

So in terms of a research agenda, if researchers were to focus more on sub-regional cooperation, as well as regional cooperation, not by using individual case studies of best practice, but by using groups of cases within a megacity region to determine the genuine sub-regional impulses for cooperation (as well as competition), including if and how they individually and collectively contribute to positive regional outcomes, adding this kind of "scaling up" may offer a more realistic picture of how regional cooperation (as competition) is actually occurring.

References

Angotti, T. (2008). *New York for Sale: Community Planning Confronts Global Real Estate*. MIT Press.

Botequilha-Leitão, A. (2012). Eco-polycentric urban systems: An ecological region perspective for network cities. *Challenges, 3*, 1–42.

Brenner, N. (2003). Metropolitan institutional reform and the rescaling of state space in contemporary Western Europe. *European Urban and Regional Studies, 10*(4), 297–324.

Burger, M., & Meijers, E. (2012). Form follows function? Linking morphological and functional polycentricity. *Urban Studies, 49*(5), 1127–1149.

Collins, G. (2006). Ending years of dispute, New York buys the final piece of Sterling Forest. *New York Times*, November 28.

Cox, W. (2020). Escape from New York? *New Geography*. October 25.

Davis, L. (2003). Hudson River communities breathe new life into former industrial sites. *Nation's Cities Weekly, 26*, 4–6.

Dieleman, F., & Faludi, A. (1998). Randstad, Rhine-Ruhr and Flemish Diamond as one polynucleated macro-region? *Tijdschrift Voor Economische En Sociale Geografie, 89*, 320–327.

Eckart, K., Kowalke, H., & Mazeland, J. (2003). *Social, Economic and Cultural Aspects in the Dynamic Changing Process of Old Industrial Regions: Ruhr District (Germany), Upper Silesia (Poland), Ostrava Region (Czech Republic)*. Lit Verlag Münster.

European Commission. (1999). *ESDP: European Spatial Development Perspective: Towards Balanced and Sustainable Development of the Territory of the European Union*. Luxembourg: Office for Official Publications of the European Communities. Available at http://ec.europa.eu/regional_policy/sources/docoffic/official/reports/pdf/sum_en.pdf

Fürst, D. (2009). Metropolitan governance in Germany. In H. Heinelt & D. Kübler (Eds.), *Metropolitan Governance in the 21st Century: Capacity, Democracy and the Dynamics of Place* (pp. 151–168). Routledge.

Gillham, O. (2002). *The Limitless City: A Primer on Urban Sprawl*. Island Press.

Gordon, I. (2012). London—Beyond the city state. In M. Ward & S. Hardy (Eds.), *Changing Gear—Is Localism the New Regionalism?* (pp. 45–54). The Smith Institute.

Grier, C. H. (2002). *Comparative Analysis of the Rhine-Ruhr Metropolitan Region*. Düsseldorf Regional Government.

Groth, N. B., & Smidt-Jensen, S. (2007). The ideo-centricity of urban poly-centricity. In: Cattan, N. (ed.) *Cities and Networks in Europe: A Critical Approach of Polycentricism.* Esher, Surrey: John Libbey Eurotext, 81–91.

Hall, P., & Pain, K. (Eds.). (2006). *The Polycentric Metropolis: Learning from Mega-City Regions in Europe.* Earthscan.

Healey, P. (2009). City regions and place development. *Regional Studies, 43*(6), 831–843.

Henwood, D. (2013). How the 1 percent rules. *The Nation* April 16.

Jonas, A. E. G., Gibbs, D. C., & While, A. (2005). Uneven development, sustainability, and city-regionalism contested: English city-regions in the European context. In H. Halkier & I. Sagan (Eds.), *Regionalism Contested: Institution, Society and Territorial Governance* (pp. 223–246). Ashgate.

Jonas, A. E. G., & Ward, K. (2007). Introduction to a debate on city-regions: New geographies of governance, democracy and social reproduction. *International Journal of Urban and Regional Research, 31*(1), 169–178.

Jones, M., & Etherington, D. (2009). City-regions: New geographies of uneven development and inequality. *Regional Studies, 43*(2), 247–265.

Juhasz, T. (2014). Pareto optimum efficiency between chaos and order when seeking consensus in urban planning. In: Ambrose, D., Sriraman, B. and Pierce, K. M. (eds.) *A Critique of Creativity and Complexity: Deconstructing Clichés.* Rotterdam: Sense Publishers.

Jurczek, P. (2008). *European Metropolitan Regions in Germany: A New Spatial Planning Strategy in Europe.* Chemnitz University of Technology.

Kloosterman, R. C., & Musterd, S. (2001). The polycentric urban region: Towards a research agenda. *Urban Studies, 38*(4), 623–633.

Knaap, E., Ding, C., Niu, Y., & Mishra, S. (2015). Polycentricism as a sustainable development strategy: Empirical analysis from the State of Maryland. *Journal of Urbanism, 9*(1), 73–92.

Knaap, G.-J., & Lewis, R. (2011). Regional planning for sustainability and hegemony of metropolitan regionalism. In E. Seltzer & A. Carbonell (Eds.), *Regional Planning in America: Practice and Prospect* (pp. 176–208). Lincoln Institute of Land Policy.

Knapp, W., Kunzmann, K. R., & Schmitt, P. (2004). A cooperative spatial future for RheinRuhr. *European Planning Studies, 12*(3), 323–349.

Knapp, W., Scherhag, D., & Schmitt, P. (2006). RhineRuhr: 'Polycentricity at its best'? In P. Hall & K. Pain (Eds.), *The Polycentric Metropolis: Learning from Mega-City Regions in Europe* (pp. 154–162). Earthscan.

Kübler, D. (2012). Governing the metropolis: Towards kinder, gentler democracies. *European Political Science, 11*(3), 430–445.

Kunzmann, K. R. (2015). *Creative Governance in City Regions: A German Perspective.* Paper presented at symposium on Planning Nordic City Regions: Experience and Agendas, Oslo, May. Available at http://www.nordregio.se/Global/News/Creative%20Governance%20in%20City%20Regions.pdf

Labbé, D., & Sorensen, A. (2020). *Handbook of Megacities and Megacity-Regions.* Edward Elgar.

Lin, D., Allan, A., & Cui, J. (2013). *Does Polycentric Urban Spatial Development Lead to Less Commuting: A Perspective of Jobs-Housing Balance.* Paper presented at 49th ISOCARP Congress, Brisbane, Australia, October.

Maier, K., Mourato, J., & Ryser, J. (2006). *How can Polycentricity of Territorial Development Improve Functional Integration?* Discussion Paper, 10th UNECE Conference on Urban and Regional Research, Bratislava, May.

McCarthy, L. (2015). Something new or more of the same in the bidding war for big business? *Growth and Change, 46*(2), 153–171.

Meijers, E. (2006). *The Notion of Complementarity in Urban Networks: Definition, Value, Measurement and Development.* Response paper presented at the 10th UNECE Conference on Urban and Regional Research, Bratislava, May.

Neuman, M., & Hull, A. (2009). The futures of the city region. *Regional Studies, 43*(6), 777–787.

New York Times. (1996). The wrong way to lure business. *New York times Editorial, September, Section, 4*, 16.

Rockland, M. A. (2008). *The George Washington Bridge: Poetry in Steel*. Rutgers University Press.

Ross, B., & Levine, M. (2001). *Urban Politics* (6th ed.). F. E. Peacock.

RPA (Regional Plan Association). (2016). *Charting a New Course: A Vision for a Successful Region.* New York, NY: RPA.

RVR. (2010). *Regionalkunde Ruhrgebiet: Main Topic: Perspectives for the Future*. Available at http://www.ruhrgebiet-regionalkunde.de/vertiefungsseiten/perspectives_for_the_future.php

Salkin, P. E. (1993). Regional planning in New York State: A State rich in national models, yet weak in overall statewide planning coordination. *Pace Law Review, 13*(2), 505–558.

Schmidt, S., & Buehler, R. (2007). The planning process in the U.S. and Germany: A comparative analysis. *International Planning Studies*, 12(1), 55–75.

Schmitt, P., Volgmann, K., & Münter, & Reardon, M. (2015). Unpacking polycentricity at the city-regional scale: Insights from Düsseldorf and Stockholm. *European Journal of Spatial Development, 59*, 1–26.

Schön, K. P. (2010). *Metropolitan Areas in Europe: Abstract of a New BBSR Study*, Bonn: Federal Institute for Research on Building, Urban Affairs and Spatial Development (BBSR) within the Federal Office for Building and Regional Planning (BBR).

Scott, A. (Ed.). (2001). *Global City-Regions.* Oxford University Press.

Seltzer, E., & Carbonell, A. (Eds.). (2011). *Regional Planning in America: Practice and Prospect.* Lincoln Institute of Land Policy.

Sorensen, A. (2019). Tokaido megalopolis: Lessons from a shrinking mega-conurbation. *International Planning Studies, 24*(1), 23–39.

Stecker, S., & Steinberg, D. (2007). *Pay, Or We (Might) Go: How Citigroup games the states and Cities.* Trenton, NJ and New York, NY: NJ Policy Perspective/Good Jobs NY.\

Trust for Public Land. (2006). *Sterling Forest capstone protected (NY)*. Available at https://www.tpl.org/media-room/sterling-forest-capstone-protected-ny

Turok, I. (2009). Limits to the mega-city region: Conflicting local and regional needs. *Regional Studies, 43*(6), 845–862.

Vandermotten, C., Roelandts, M. G., Cornut, P., & Halbert, L. (2008). European planning and the polycentric consensus: Wishful thinking? *Regional Studies, 42*(8), 1205–1217.

Walsh, C. 2010. *Territorial Politics and Formal Structures of Governance in Spatial Planning: Insights from the Dublin City-Region.* National Institute for Regional and Spatial Analysis (NIRSA) Working Paper, 64, November, Maynooth: National University of Ireland (NUI).

Wegener, M. (2012). Government or governance? The challenge of planning for sustainability in the Ruhr. In Hartmann, T. and Needham, B. *Planning by Law and Property Rights Reconsidered.* London and New York: Routledge, 157–170.

Wheeler, S. (2009). Regions, megaregions, and sustainability. *Regional Studies, 43*(6), 863–876.

WiN Emscher-Lippe GmBH. (2021). *Welcome to WiN Emscher-Lippe GmBH*. Available at https://www.emscher-lippe.de/

Yamai, T. (2015). The "metropolitan region:" A farewell to a decentralized regional structure? Regional policy in Germany since the 1990s. *Konan Economics Papers, 55*(3–4), 47–52.

Yaro, R. D. (2000). Growing and governing smart: A case study of the New York Region. In: Katz, B. (ed.) *Reflections on Regionalism.* Washington, DC: Brookings Institution Press.

Ziegler, E. H. (2009). The Case for megapolitan growth management in the 21stcentury: Regional urban planning and sustainable development in the United States. *International Journal of Law in the Built Environment, 1*(2), 105–129.

Chapter 30
Medium-Sized Spanish Cities: Urban Planning and Urban Policies

Carmen Vázquez-Varela and José M. Martínez-Navarro

Abstract This chapter addresses the debate on the restatement of the dynamics of urbanization in medium-sized Spanish cities in the last forty years. Since 1979 Spanish cities have undergone a remarkable transformation and show a rapid expansion of urban areas in fragmented and dispersed form as well as a notable restructuring of their inner features. These shifts must be viewed in the context, on the one hand, of a general framework of far-reaching political, social, and economic changes within the country and, on the other hand, of the changes arising from its links with European and global circuits, especially since 1986, when Spain joined the EU. Thus, in recent decades in Spain there has been a redefinition of the traditional city pattern, moving from a compact urban structure to another defined by the growing role of sprawl, which hinders the precise identification of its limits and features. The analysis of urban policies (land policy, planning, and urban projects) will be studied after selecting a sample of cities between 50,000 and 300,000 inhabitants, whose surrounding urban area does not exceed 400,000 inhabitants. From this initial sample, we will focus on a few case studies as examples. The two main hypotheses that the authors will try to address and validate as expected findings in this contribution are: (1) The planning and urban policies deployed at the local level meet the combination of the traditions and approaches existing at the time (visions of the urban and the city) and the socio-political and economic context and (2) Local urbanism in the last forty years has gone from an "urban" and regulatory urbanism, between 1979 and 1992, to an expansive and corporate urbanism, 1993–2007, to arrive, at present, at a series of "dual urbanism" in which the previous tendency coexists and, is rehearsed in parallel, with a new more participative, transversal, and less technocratic urbanism.

Keywords Urbanization · Urban policies · Medium-sized Spanish cities · Financial and real estate boom · Urban planning · Neoliberal urbanism

C. Vázquez-Varela (✉)
Department of Geography and Spatial Planning, Universidad de Castilla-La Mancha, Cuenca, Spain
e-mail: carmen.vazquez@uclm.es

J. M. Martínez-Navarro
Department of Geography, Universidad Autónoma de Madrid, Madrid, Spain

B. R. K. Sinha (ed.), *Urban Dynamics, Environment and Health*,
https://doi.org/10.1007/978-981-99-5744-6_30

30.1 Introduction

In the last 40 years, Spanish cities have experienced a great deal of change within a more general framework of far-reaching political, social, and economic transformation of the country. Different surveys have shown the great variability of urban policies that change depending on the geographical context, institutional framework, political and social situation, and characteristics of local actors (Iglesias et al., 2014). In fact, cities are the ones that most intensely reflect the economic, political, and social changes that have taken place in recent decades. Among the many related factors of change that help us to read the recent evolution of urban policies, we would highlight those associated with contemporary globalization, which have turned cities into leading economic, political, and cultural actors. Consequently, the role of the State and, in parallel, of local governments has been repositioned (Martí-Costa & Tomàs, 2016). The municipal authority is now deploying development strategies aimed at defining the competitive profile of cities and their promotion, an aspect which is very prominent in the context of widespread neo-liberal policies since the 1980s. Within this framework, economic and urban policies through different planning and intervention procedures (such as strategic plans, urban marketing, and big urban projects) will gradually become the main actors in urban public policies. Our contribution is deployed from a series of basic approaches, developed within the framework of the research project, which we can specify in the following (Bellet, 2020, pp. 3–4):

1. The urban planning policies implemented by local public authorities have played—and still play—a leading role in the dynamics and characteristics of urbanization, as well as in the patterns and visions of the urban and the city that are adopted and prevailed at each moment (Burriel, 2008; Capel, 2013a). As Cochrane points out: "Urban policy is both an expression of what is understood at any given time to be the urban, which makes cities what they are, and contributes to shaping those understandings" (Cochrane, 2007, p. 13).
2. Urban planning and policies displayed at a local level meet the combination of traditions and approaches (visions of the urban and the city) and the socio-political and economic contexts existing at different times (Díez & Monclús, 2017; López de Lucio, 1993a). We would like to point out here that despite the differences in the urban planning policies found in the different Autonomous Communities, there are common lines that allow for the definition of a general explanatory framework for municipal urban planning deployed in the country since the establishment of the first democratic town councils in 1979 (Calderón & García Cuesta, 2017).
3. Neo-liberalism and the financialization of cities explain why city government is increasingly conceived as a purely managerial function, depriving it of its own political dimension (Harvey, 2013). The new forms of city governance would be based on the inclusion of private market actors, civil society groups, and parts of the traditional state apparatus and would generally be described as "participatory". However, in many cases, the participatory processes do not avoid being "autocratic", since the objectives or the final image cannot be questioned because

in many of the participation processes, they are previously defined (Sennet, 2018; Swyngedouw, 2005). Furthermore, as Jordi Borja has already highlighted, "although urbanization is expressed in the local sphere, the political and economic framework that makes it possible is state and global" (Borja, 2011, p. 85).

Through bibliographical and documentary analysis, in which other specialists have also made great contributions (Bellet, 2020), the chapter attempts, first, to contextualize and characterize the evolution of urban policies in Spanish medium-sized cities, especially spatial planning, from the first democratic town councils to the present day and to describe the change in the model of the city on which these are based throughout the different periods. Thus, it questions: (1) the origin and characteristics of current urban policies; (2) how these and the urban model have changed since 1979; and (3) how their implementation has led to the transformation of cities. To this end, three different study periods are established (1979–1992; 1993–2007, and 2008–2019), which are delimited according to the characteristics of urban planning deployed on a municipal scale, the dates of municipal elections and the evolution of real estate dynamics. The planning documents in force today in the sample of cities analysed, in many cases with countless revisions since they were first drawn up, are a good example of the turnaround of urban planning in medium-sized Spanish cities.

The public policies of the late seventies and eighties faced, on the one hand, problems and scenarios that were still typical of a classical industrial society (Subirats, 2016) and, on the other hand, cities with strong urban deficits inherited from the decades of Franco's dictatorship (Ferrer & Sabaté, 1999; López de Lucio, 1993a).

Later, and clearly in the 1990s, urban policies, and with them urban planning policies, would change with the implementation of economic liberalization policies, the full deployment of urban competitiveness strategies, and city marketing, through which the authorities of a growing number of cities would explicitly seek to attract external resources and capital. This increased the commoditization of cities through the promotion of private real estate investment (and privatization of the management of urban services) in the context of the growing mobility of capital produced by financial globalization (Méndez, 2018).

The chapter looks at the last of the periods (2008–2019), to explore the characteristics and composition of the new urban agendas that some local governments have been pursuing since the crisis, and especially since the local elections of 2015 (Díaz Orueta et al., 2018). The new policies and actions carried out, especially by the so-called New Left councils, tried to innovate in a context of limited public resources, very diverse and politically weak governments, and a demand for greater citizen participation in an increasingly complex and unequal society (Nel-lo, 2015; Subirats, 2016). Thus, the new initiatives and policies, some of them innovative and experimental, have led to a significant change in the type of interventions conducted. Traditional methodologies and governance approaches assumed alternative and participatory logics that, however, did not question the functioning of the land market within the capitalist system or the logic of neoliberal urbanism (Sennet, 2018; Swyngedouw, 2018).

The bibliographic and documentary review of planning documents, as well as information on the main plans and projects reviewed in various papers and the main urban planning and urban studies journals in the country, are the main sources of work. To study the dynamics of urbanization, we have used the data on municipal planning presented in the Spanish Land Occupancy Information System (SIOSE) with information from the Ministry of Public Works in the version available online in 2018.

Regarding the sample of selected cities, we will work with those whose population, as of 2019, is between 50,000 and 300,000 inhabitants. This is one of the groups of urban settlements that has witnessed the fastest and most dynamic growth of the Spanish urban network since the middle of the last century (Vázquez-Varela & Martínez, 2020) (see Table 30.1). Nevertheless, for the purposes of our research, we will disregard all those whose urban area exceeds 400,000 inhabitants.

We removed 73 of the 136 average cities identified according to the chosen statistical criterion (50,000 to 300,000 inhabitants), either because their urban area exceeds 400,000 inhabitants or because they are included in another urban area that exceeds that same figure. To the remaining 63, we then add the cases of Soria (39,398 inhabitants) and Teruel (35,890 inhabitants) which, despite not reaching the minimum threshold chosen in 2019, are in both cases provincial capitals and, as such, have the functions of an intermediate city. Overall, then, our sample totals 65 medium-sized cities.

30.2 Municipal Spatial Planning, Urban Planning, and Real Estate Dynamics (1979–2019)

Many scholars have stressed the strong link between economic growth and the expansion of the real estate sector in Spain since the middle of the twentieth century (Rodríguez López, 2006), which has led to an expansive, speculative, and unsustainable urban model (Naredo, 2010). Thus, the different periods of economic growth in the country have been based on booms in real estate activity with the intense production of infrastructures, urban land, and real estate assets. These developments have already been sequenced in other surveys in three moments of real estate expansion (Rodríguez López, 2006; Górgolas, 2017).

Urban (and territorial) planning conveys the socio-economic and cultural conditions in which they are conceived and deployed and, therefore, their foundations, strategic orientations, and guidelines are adapted to these (Calderón & García Cuesta, 2017). But, in addition, within the framework of the Spanish model of real estate capitalism, which has prevailed since the middle of the twentieth century, urban planning and management from the public sector has become the basic instrument for catalysing expectations of economic growth, based almost exclusively on the consumption of land (Górgolas, 2017). Thus, the city, and the production of the urban "becomes a mechanism for absorbing surplus capital and integrating itself into the circuit of accumulation" (Lois et al., 2016, p. 7). In this way, both regulation and urban development activity seem to have been feeding the real estate dynamic and the production of the urban, turning the city and the territory into the business itself (Calderón & García Cuesta, 2017).

The first of these periods (1979–1991) began in the midst of the severe crisis that erupted in the Western economies between 1975 and 1985. Added to this, in Spain, was the uncertainty associated with the end of the dictatorship. The country's integration into the European Economic Community reinforces the internationalization of the economy and the centrality that the financial economy would acquire, becoming a destination for international investment which, in Spain, was mainly directed at the financial and real estate markets (López & Rodríguez, 2010). This concludes in a speculative bubble caused by the "wealth effect" created in families by the apparent increase in the value of their assets (Naredo, 1996). The economic recovery of the mid-1980s (1985–1992), a period during which the GDP grew at an annual rate of 6%, was also fed by the dynamization of tourism and real estate that had been introduced during the dictatorship in the 1960s.

Based on the contributions of some scholars, it can be concluded that during this period of economic expansion, which was strongly based on financial and real estate income, two spatial processes were feeding off each other in parallel (Górgolas, 2017; Rullán, 2012):

- Investment in transport infrastructure, financed by the Structural Funds, became the trigger for the real estate market. In Spain, this results in important business for the sector as there is a coincidence between the business groups that operate infrastructures and those that are dedicated to real estate production. Furthermore, investment in transport infrastructure helps to eliminate the friction of distance.
- The expansion of the real estate market thanks to the easy access to financing resulting from market deregulation. However, in a context of low population growth and low migratory movement, the effect of the bubble mainly affected prices, not so much the volume of construction or excessive demand for land (Naredo, 1996).

The upward cycle that began in 1985 ended due to financial problems after 1992 (Rodríguez & López, 2006). But the foundations of economic and land policy, based on market deregulation, remained hidden during the subsequent crisis and became hegemonic between 1997 and 2007 during the last real estate boom. The introduction of the euro, low interest rates and a favourable political framework led the Spanish economy to flee forward, triggering a new upward cycle that was much more speculative and intense than the previous one (Burriel, 2014; Naredo, 2010).

Specifically, mortgages have increasingly become standardized, securitized, packaged, and commercialized in new and specialized financial markets, transforming one of their most important financial characteristics: their relatively low level of liquidity. Financialization has broadly affected the expansionary dynamics of the secondary circuit in Spain. But there is no doubt that, in order to be able to increase its size for more than a decade, the real estate bubble had to have significant public support, both direct and indirect, which has allowed interpretations of the process that incorporate an explicit political component (García, 2010; Romero et al., 2012).

In this new context, urban and territorial planning came to set limits and were restrictive, so they were subdued by the project, strategic planning, and urban marketing (Fernández Durán, 1993). These new instruments were perceived by the real estate market as an opportunity to erode planning regulations in pursuit of land liberalization. With no urban and territorial reference model, it was easier to convey the individualized demands of private stakeholders and, for this purpose, the concrete scale of the project represented the ideal instrumental supporting tool. Together with this, the liberalization of the land market and the favourable financial and political climate acted as a stimulus for a process of urbanization and building which has already been characterized as intensive, extensive, and private (Bellet, 2007; Calderón, 2004).

At this point, if we look at the analysis of the Municipal Master Plans currently in force for the sample of selected cities (see Table 30.2), we realize that most of them were approved during the second period, coinciding with the financial-real estate boom, which we have previously described. Only ten cities keep in force Master Plans approved between 1979 and 1992, of course successively altered until the present day by more or less far-reaching modifications. It is noteworthy that except for two cases, these are coastal/tourist municipalities (Marbella, Torrevieja, Orihuela, Benidorm…), mostly focused in the south-eastern province of Alicante (see Fig. 30.1), which have experienced significant growth and urban expansion processes being driven by the activity of private stakeholders. Neither these stakeholders nor the various municipal governments have felt the need or had the capacity to approve up-to-date and truly binding urban planning.

The second period, especially between 1996 and 2007, has the highest number of approved Master Plans in force, 34. These plans are mostly designed to accommodate and encourage the speculative bubble, classifying land for development for a volume of population and activity that had never been reached, yet which was modified as many times as necessary to comply with the development and location interests of private agents. In other words, the patterns, direction, and layout of spatial growth have been altered when the proposals and demands of private developers must be considered.

The third and last period analysed has witnessed the approval of 16 plans which, from the second decade of this century, generally incorporate concepts such as regeneration and strengthening, sustainability, and resilience. They are committed to moderate growth or even sometimes to the envisaged declassification of land for development, but without overlooking the inclusion of strategic operations in central areas where the vested interests of capital are still represented.

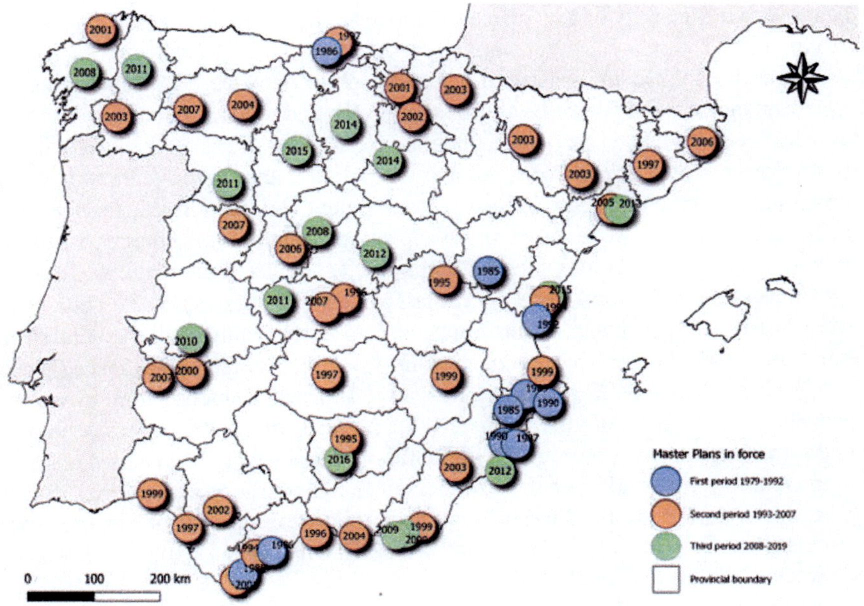

Fig. 30.1 Medium-sized cities according to the approval date of the Master Plan still in force in 2019. *Source* Urban Information System—SIU, 2018. (http://siu.fomento.gob.es/)

30.3 The Evolution of Spatial Planning Policies at Municipal Scale (1979–2019)

30.3.1 First Period (1979–1992). Urban and Reformist Planning: Valladolid as an Example

The new democratic town councils, recently constituted, undertook the construction of a municipal political agenda in which town planning would occupy, in this first period, a leading role. However, unlike what happened in Central and Northern Europe, where this was already done through the development of urban projects, in Spain the Italian tendency to do so through the Structural Plan was followed (Hall, 1996; Sainz, 2006). Urbanism, and especially urban planning, was consolidated as the great political project of the first democratic town councils, the banner of the new municipal policy that was inspired by a certain "spatial Keynesianism" (Subirats, 2016).

Thus, the new town councils began by reviewing the master municipal planning, in a sense that went beyond the mere adaptation of the new Land Act of 1975. The revision introduced new visions, according to the moments of change, which prioritized: restructuring, reform and improvement of the existing city. In fact, the eighties have been deemed the "golden age" of Spanish urban planning not only because of the quantity of new plans but, above all, because of their quality (MOPU, 1990).

The cities looked inwards with a general philosophy based on the values of urbanism that we could describe as urban, reformist and social. At first, it is described as urban because this type of urban planning focused its efforts on the improvement of the structure and already consolidated urban fabric and on the redressing of the urban deficits inherited from Franco's development (infrastructures, services, and facilities), especially in the most disadvantaged peripheral neighbourhoods. In this respect, medium-sized cities such as Tarragona, A Coruña, Santa Cruz de Tenerife, Salamanca, and Valladolid stood out (Solà-Morales & Parcerisa, 1987). Thus, plans were drawn up which, under Italian influence, paid great attention to the general structure, form, and relationship between the built-up fabric and the public space (Bellet, 1995). Secondly, there was a need for urban reform and transformation in view of growth (Gaja, 2008). And all of this gives a notable role to the collective interest and sometimes to the processes of collaboration of the new and expanded teams of municipal technical staff with the still dynamic neighbourhood associations. Examples of this period are the Master Plans for Urban Development of medium-sized cities such as Salamanca (1984) Valladolid (1984), Gijón (1985), Toledo (1986), and Orihuela (1990) (see Fig. 30.2).

Plans which, on the whole, can be described as conservative in terms of growth forecasts, focusing on the restructuring of the whole, densification, and the improvement of the centres and dysfunctional neighbourhoods on the periphery. It was also intended to regulate the uses and intensities of buildings, through more concise regulations, both for the existing city and for new developments (Font, 2000). In fact, the most disadvantaged and deprived neighbourhoods of the first urban periphery were the object of preferential intervention, acquiring in the first years of the reinstated democracy the character of a political issue, largely due to the impulse of neighbourhood movements during the late Franco era. This contributed to the definition of an urban agenda with a remarkable social character during the first years (Rivas et al., 2017).

Valladolid is a particularly telling case, a city that in 1981 had 320,293 inhabitants, reaching its peak in 1991 with 330,700 residents. Since then, it has been losing population in favour of the surrounding municipalities, reaching a total of 298,412 inhabitants in 2019. The 1984 Valladolid Master Plan provided one of the best urban diagnoses ever made of the city (Fernández & Santos, 2020). Its drafting took place during the mandate of a socialist government in the mayor's office, which lasted from the first democratic municipal elections (1979) until 1995. Among its goals were to halt the destruction process of the inherited city, to lay the foundations for a redistribution of uses that would allow for a better quality of life, to rationalize the process of city development, to optimally implement legal management formulas, and to provide the city with accurate and up-to-date urban planning documents (Ruiz, 2013). A radio-centric planning scheme was chosen that sought to combine three realities: a historic centre to be protected, peripheral neighbourhoods to be upgraded, and outer rings for controlled and isotropic urban growth (Rivas, 2008). The plan involved

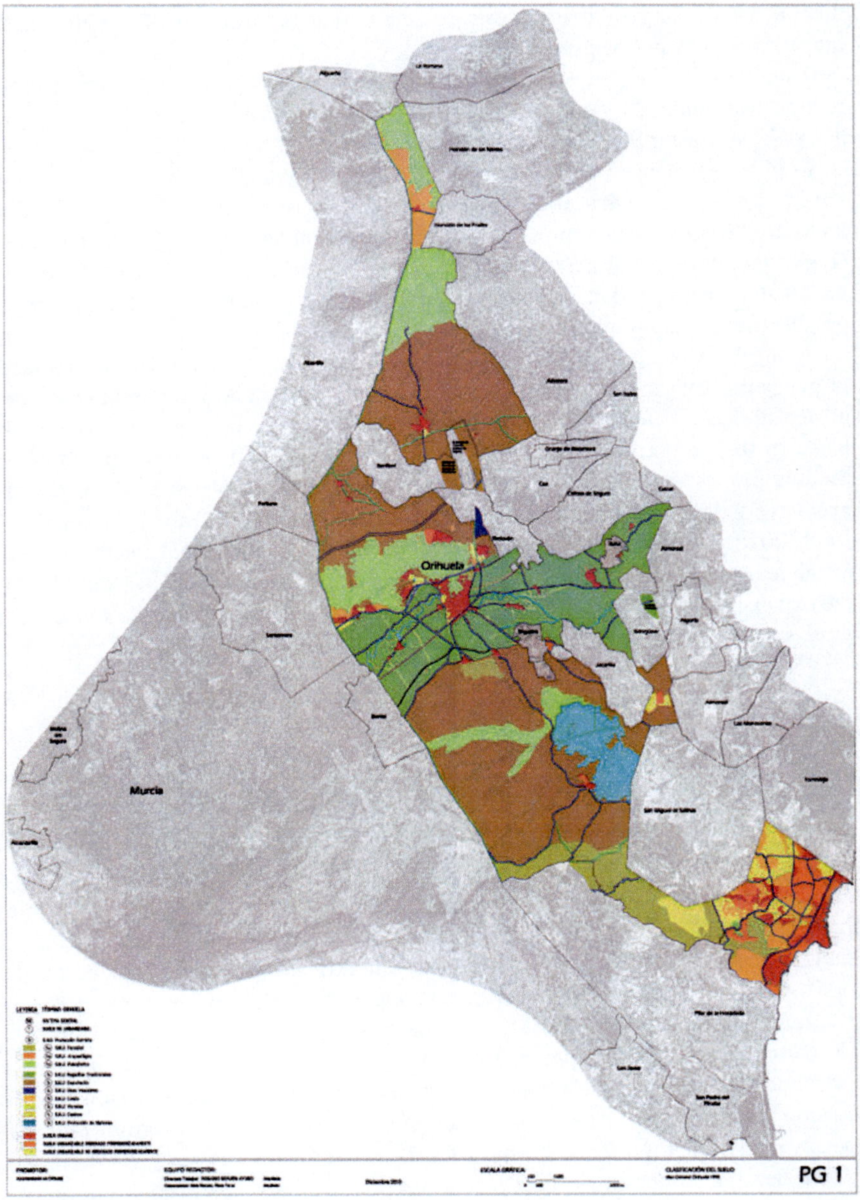

Fig. 30.2 Orihuela Master Plan for Urban Development 1990. *Source* Orihuela City Council

the renewal of marginal housing estates, small urban redevelopment operations, and the study of new peripheries (see Fig. 30.3).

This 1984 Valladolid Master Plan for Urban Development was subsequently revised and updated by a new Master Plan passed in 1997, and in 2020, the latest and current Master Plan has been approved.

In plans so focused on the completion of the city and in interventions mainly aimed at modifying what already exists, territorial issues, processes of population, and activities de-concentration in the territory were, at least, insufficiently addressed (López de Lucio, 1993b). Suburbanization and peri-urbanization exceeded the scale on which planning had been concentrated in these years, not without bringing about notable internal contradictions as well.

Soon, the values of urban, reformist, and social urbanism would be replaced in our context by the new principles of competitiveness and growth that the entry of neo-liberal urbanism (López de Lucio, 1993a) and the changes that would take place in the property market brought about. The problem of urban planning, of the building of the city, gradually yielded to the problem of land, which became exclusively the management of the land market (Calderón & García Cuesta, 2017; Roch, 2015). The alliance forged between land market management and the interests of the real estate and financial sectors (Méndez, 2014; Naredo, 2010) would lead to the expansive urbanization cycle of the mid-1990s (first boom) and the subsequent, more speculative cycle of the 2000s (Lois et al., 2016; Rullán, 2012).

30.3.2 Second Period (1993–2007). Embracing Neoliberal Policies, Managerialism, and Sprawling Urbanism: Toledo and Cuenca as Examples

The second period is characterized by the wide application of the neoliberal principles that were tested, especially in the management and governance of the city. The use of management forms and business methods was already being deployed at the end of the 1980s. During this period, public–private cooperation schemes were already being tried out in some cities, especially through the private management of public services. The use of urban marketing proved to be a multi-purpose instrument since it not only contributed to the sale of the city-commodity by preparing it for consumption but also became a useful tool for "the construction of consensus and social legitimization of the city model and the different projects for the city" (Alcalá-Santaella, 2004, p. 310).

The new guidelines would aim at ensuring the growth of cities, promoting competitiveness, and implementing new forms of local government to meet the challenges of economic globalization (Mendez, 2014; Brenner et al., 2015). Instead of controlling urban growth, urbanism would now be devoted to promoting it.

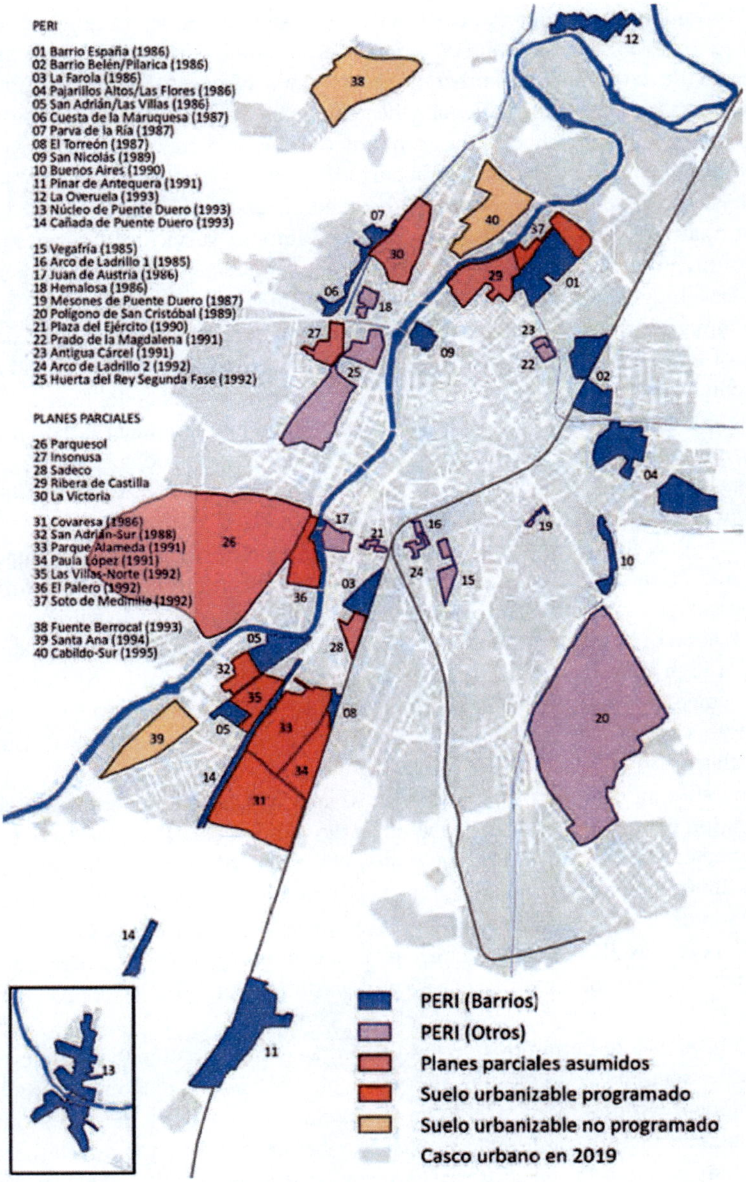

Fig. 30.3 Valladolid Master Plan for Urban Development 1984. In blue and violet Special Plans for Inner Redevelopment, in red and yellow land to be developed. In grey built-up area in 2019. *Source* Fernández and Santos (2020): 86

Conventional planning and the use of plans and guidelines to regulate land use seemed to have fallen into total disrepute (Ferrer, 2005). Conversely, strategic planning and the development of urban projects were considered the only possible alternatives for cities to successfully face the challenges of global competitiveness (Hall, 1996). Once the basic shortcomings inherited from the dictatorship were overcome, some local governments began to think in terms of urban strategy, defining objectives, and trying to generate complicities with economic and social stakeholders. In doing so, normative/regulatory urbanism was turned into the service of the mass production of urban land, housing, and urban artefacts, imposing the vision of a so-called Neo-Developmental Urbanism (Capel, 2013b; Romero et al., 2015).

Although it might initially be assumed that strategic planning and the development of large urban projects belong to large cities, the fact is that it spread rapidly to medium-sized cities, with such well-known cases as Alicante, Elche, Cuenca, Tarragona, Donostia/San Sebastián, etc. Among the most frequent urban initiatives in this period are those involving image and branding and spectacle architecture, as well as those related to improving habitability, materialized in residential quality of life and environment, and sustainability (Precedo et al., 2010).

The key strategies of the neoliberal city (deregulation, flexibility, competitiveness, and privatization) were transferred to urban planning through the following processes: reclassification and rezoning of land, development of large urban projects, privatization of basic urban services and public facilities, and the commodification of areas of high expected value (Bellet, 2017; García-Herrera, 2017). Agreements and management bodies (private, semi-public, public–private cooperation, etc.) became the perfect vehicle for making rules and planning frameworks more flexible and deregulated.

The building of large projects, facilities, and infrastructure, combined with the celebration of major events, contributed to the public debt. Thus, oversized spaces were created (the City of Culture in Santiago, the City of Light in Alicante, the City of the Environment in Soria, the Circus City in Alcorcón); large facilities with little use, underused infrastructures, some closed or sold to the highest tender (Ciudad Real airport, Castelló de la Plana airport, Lleida airport, the Jaén tramway, etc.), all of which are proof of an era of excess and waste (Bellet, 2017; Romero González, 2019).

The laws passed during this period therefore seem to assume that making a city consists essentially of exploiting the value of the land by making it suitable for development (Burriel, 2009; Calderón & García Cuesta, 2018), thereby promoting new land use and urbanization as opposed to the preservation and refurbishment of the already built fabric and heritage (Fariña & Naredo, 2010). When the government turned over in 1996—with the triumph of the neoliberal right—economic growth accelerated. This scenario, together with the adoption of the single European currency and the strict monetary policy, inflation and banking rates containment carried out from Brussels, contributed to the emergence of the new real estate boom (1997–2008) (López & Rodríguez, 2010). The so-called "urbanizing tsunami" and by others "the prodigious decade of Spanish urbanism" (1997–2008) were underway, sponsored by the public authorities through urban planning laws, favourable economic and fiscal policy, and depleted local finances. Many of the local governments also encouraged these processes as the returns associated with urbanization and land development were used to finance themselves in the context of municipal financing difficulties.

The expanding urban and developable land (Olazabal, 2018) has more to do with speculative interests than with a proven need arising from the increase in the resident population. In other words, most of Spain's medium-sized cities have exhibited oversized, messy urban growth that is increasingly driven by business rather than by the general interest. From the perspective of urban form, all this has led to the appearance of large low-density expansions in close contact with the built-up area and scattered and strung-out urban fabrics in suburban areas, with a tendency for discontinuity.

In this context, we find Master Plans that, first, give up attempts at comprehensive planning and relax their long-term approaches, claiming to seek greater flexibility and efficiency. Secondly, the Plans envisage very high growth expectations and provide for significant amounts of land for development.

If we look at the data recently published by Bellet (2020), 64% of the municipalities with more than 10,000 inhabitants in the Urban Information System (SIU) database have approved the Master Plan in force during the period 1992–2008, the time frame during which the urbanizing tsunami occurred. Overall, the indicator of urban expansion (the ratio of land for development to built-up land), which is key to the growth model outlined in the Master Plan, shows a ratio of 1.3. This is a good example of the growth expectations fostered by urban planning in many of the country's municipalities in the 1992–2008 period. In 19.6% of the municipalities (of the 449 with approved plans in the period) this figure is exceeded, with a diverse profile and location. However, the following medium-sized cities stand out for their growth expectations:

- Municipalities with a clear tourist character on the first or second waterfront.
- Municipalities located on the periphery of different metropolitan areas or in the urban belt.
- The indicators for a fair number of provincial capitals are also noteworthy: Toledo (with 2.1); Badajoz (with 1.6); Cuenca (with 1.6); Ciudad Real (1.3) and Segovia (1.1). The data for the municipality of Toledo refers to the Master Plan, approved in 2007 (see Fig. 30.4), but nullified in 2018. Nowadays, the Master Plan of 1986 remains in force (Escudero, 2018).

It can be noticed how the same gravitational and centrifugal effect that in previous decades had characterized the Spanish urban system, concentrating the largest urban expansion projects on the periphery of the large cities and the main tourist destinations, has come to characterize the medium-sized cities during the boom period. However, the greatest growth continues to be concentrated in the municipalities of the metropolitan and coastal medium-sized city areas. Therefore, urban sprawl processes become a trend independent of the size of the city, medium or large.

As we have pointed out earlier, Toledo is a paradigmatic case, both for the indicators of land classified for urban expansion and for the failure of its planning and the proliferation of large urban projects that several specialists dismiss as nonsense (Escudero & García, 2022).

Toledo, a city of 78,618 inhabitants in 2007 and capital of the Castilla-La Mancha region, more than doubled its urbanized area between 1990 and 2006. The municipality approved a Master Plan in 2007 that classified more developable land than built-up urban areas in the city (Fig. 30.4) and which envisaged the development of more housing than the total number of dwellings registered at the time. The Master Plan was basically an instrument at the service of speculative profit and the private growth management model (Escudero, 2018; Vázquez-Varela & Martínez, 2013). However, its approval took place a year before the 2008 crisis and this fact prevented practically nothing of what had been planned from being implemented. Furthermore, the Master Plan underwent a long judicial journey, analysed by Escudero (2018), until its definitive annulment by the Supreme Court in 2018. The city council then reinstated the 1986 Master Plan by adapting it in order to legalize what was built by the annulled Master Plan and to accommodate the urban developments designed by the 2007 Master Plan when a private developer dares to develop them (Escudero & García, 2022). Toledo's urban planning blunder can also be seen in the big urban projects and schemes developed in recent years (high-speed train, cultural facilities, University Hospital, Puy du Fou España theme park, etc.).

Cuenca is another excellent case study of urban planning trends and processes during the period under consideration. It was an intermediate city of 52,980 inhabitants in 2007, set in a territory of deep rurality, which reached its population peak in 2012 (57,032 inhabitants) and has been slowly losing population ever since. The

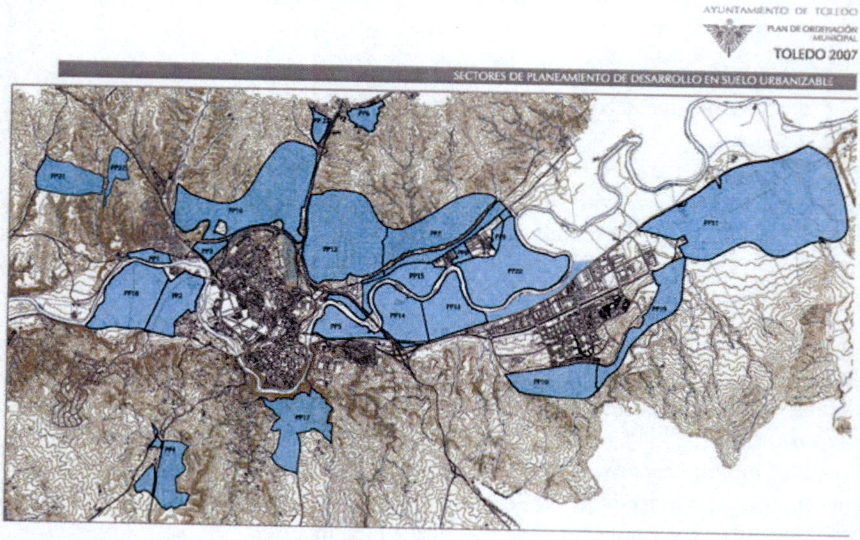

Fig. 30.4 Classification of land for new urban developments (in light blue), Toledo Master Plan, 2007. *Source* Toledo City Council

context of euphoria and financial and real estate speculation in the first decade of the twenty-first century is coinciding in Cuenca with local circumstances which explain the ordering of a new Master Plan, and which resulted in the desired improvements in accessibility (motorway connection with Madrid and a new station on the high-speed rail network). In January 2003, the municipality commissioned the drafting of a new Master Plan, which was submitted for public information in the summer of 2006 and which, had it been approved, would have meant the largest classification of land for development in its history (Fig. 30.5), with a planned expansion to accommodate more than 230,000 inhabitants in the municipality. The problems revealed and the process of discussion and debate that ensued explain why successive city councils were unable to push through the definitive approval of the plan so that the 1996 Master Plan remains in force.

On the other hand, in both the two cases analysed and in many other cases, it is clear that the municipal scale is not always appropriate for dealing with the physical planning of the territory, and other supra-municipal scales are needed to deal with the problem and the existence of very ruralized urban structures in the neighbourhood of some medium-sized cities (e.g. León, Burgos, Palencia, Cuenca, Lugo). Spatial planning shows blatant failures and has not been accepted as an adequate management tool to order the extension and growth towards the area of influence of these medium-sized cities (Andrés & González, 2020).

Whatever the case, it seems clear that the dynamics and stakeholders that allow the process in Spain's medium-sized cities to be understood are complex and diverse. There are the medium-sized cities located in metropolitan areas, where transformation is closely associated with a process of metropolitan expansion in a territory that merges with its new uses and functions as its external periphery, as well as with its

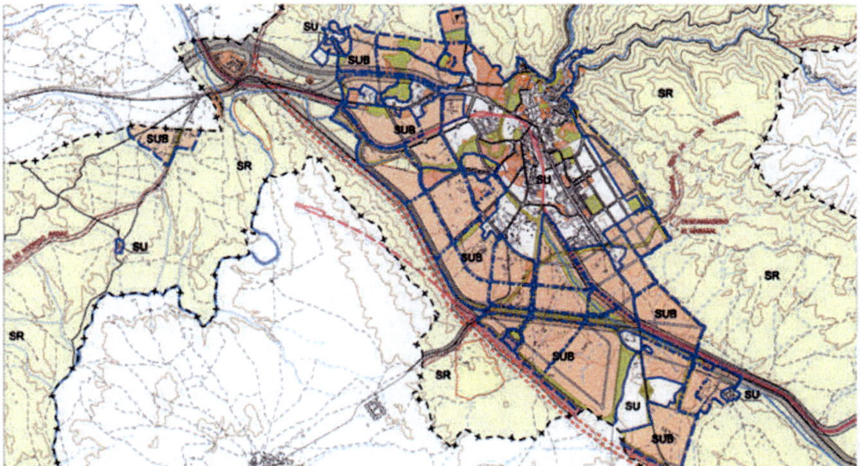

Fig. 30.5 Classification of land for new urban developments (in light orange), Cuenca Master Plan Proposal, 2006. *Source* Cuenca City Council

own dynamic of suburbanization which would add to the former. On the other hand, the coastal territories, where the dynamics of housing, infrastructure, and facilities that cater for both seasonal tourism demand and residential tourism have encouraged and fed back into the market. And finally, isolated medium-sized inland cities, where the dynamics and processes have local drivers that fit and adapt to the global framework (Vázquez-Varela & Martínez, 2018).

The legacy of the period bequeathed plans with disproportionate growth forecasts, out-of-scale urban projects and, above all, large amounts of urbanized land or unfinished developments that were scattered throughout the country. And so, landscapes still remain everywhere that are the footprint of the collapse of a real estate bubble that generated a strong economic, social, and political crisis (Gaja, 2015; Méndez & Plaza, 2016; Méndez et al., 2015), the consequences of which are still felt today.

Another of the outstanding features of the urban planning of the period is the protection of the environment, the landscape, and the urban heritage which, however, in many cases too often become another business possibility through their own commoditization (Fernández Salinas, 2005; Tomé, 2007; Boira, 1995). The shift in pattern and rationale led to a greater rupture and spatial fragmentation, especially in the small and medium-sized cities of the hinterland, which were based on less complex structures, more tied to local logics and with more everyday-related space-time (Bellet, 2017). Furthermore, all this deepened the urban socio-spatial segregation and ended up burdening the already dwindling municipal finances (Burriel, 2014; Nel-lo, 2015; Subirats, 2016; Górgolas, 2017).

30.3.3 Third Period (2008–2019). The Crisis, the Reassessment of Urban Policies and New Local Agendas: Sustainability, Circularity, and Public Participation—Just Rhetoric or Reality?

In this period, the process of neo-liberal evolution of the State is taking a further step forward, in which the management of the social sphere is clearly subordinated to financial interests. High unemployment and increasing social inequalities were growing because of the crisis, coinciding with the reform agenda carried out by the state and the European Union under the austerity discourse. The reforms entailed budget cuts, the deepening of the application of neoliberal policies, and the recentralization of policies (especially those related to welfare) to the detriment of regional and local governments (Martí-Costa & Tomàs, 2016).

At the local level, a new city project aimed at promoting competitiveness is being consolidated in which an increasingly polarized class structure and an overarching governance model are being reinforced (Díaz-Orueta et al., 2018; Iglesias et al., 2014). Since the crisis, the new real estate strategies seem to have shifted from large-scale operations in the periphery to the development of more limited projects in the city's core, which, through rehabilitation or renewal, will bring about major changes in the functionality and socio-economic profile of the population, generating displacement and physical and symbolic gentrification (García-Herrera, 2018).

These operations coexist with the resurgence of a good number of interventions which question the model and appear as an alternative to the market and to institutional operations carried out through self-management: squatting, cooperative management of housing and public facilities, etc. These are proposals by urban stakeholders that challenge the neo-liberal model of the city in the absence and withdrawal of public policies. Paradoxically, these challenges will inspire some of the new municipal policies.

Many of the changes brought about by the "new urbanism" are based on the cultural and socio-political changes that have been taking place in the country prior to the crisis, and which have clearly crystallized in the strong social mobilization of May 2011 and the new political movements that began to be involved in local government in the municipal elections of 2015, the so-called New Left. Since May 2015, alternative platforms led the government in four of Spain's five main cities, but they also managed to win and lead the government in such important medium-sized cities as Santiago de Compostela, A Coruña, Pamplona, and Cádiz.

Municipalism—the democratic autonomy of municipalities (from town parishes to metropolitan boroughs to city regions) over political and economic life vis-'a-vis the nation-state—is renascent (Thompson, 2020). In all the Spanish cities concerned by the change in municipal government and the coming to power of new political parties or citizens' platforms, the new political scenario implies a break with governments that were deeply rooted in previous cycles, regardless of their political background. The new political proposals have succeeded in capturing the disenchantment that other traditional parties have ignored, in captivating a wide-ranging alternative left, and in generating an unusual enthusiasm among the citizens. The agendas "for change" generally emphasize social policies and the issue of housing and social urbanism, as well as democratic regeneration, transparency, austerity of representatives and local citizen participation, the (re)municipalization of previously privatized public assets or services, and a commitment to the design, appropriation, and control methods of public space. In most cases, however, these governments face difficulties with regard to governance, whether due to strong pressure from global capital in the big and medium-sized cities or due to the widespread economic crisis and austerity of local treasuries. Moreover, in many cases, the challenges are assumed from a position of relative weakness (Janoschka & Mota, 2018).

These movements were demanding, among others, new forms of local governance—less technocratic and more based on the direct participation of citizens in the production and management of the public and of the city—and new policies closer to the everyday problems of citizens. So how could one characterize these new municipal policies (including urban policies) launched especially from what were called "New Left councils" but also in many other councils? In general, and based on the comprehensive contributions of Paisaje Transversal (2018), LaCol (2018), and Bellet (2020), the following should be noted:

- Urban policies become more plural and complex by addressing social, cultural, and environmental aspects, all without neglecting the urban economic side. Many of the initiatives or interventions will be more comprehensive, embracing very diverse disciplines that are difficult to fit into a single sectoral policy.
- The new projects were drawn up based on spatial/territorial criteria (revitalization of vulnerable neighbourhoods, regeneration of historic centres, etc.), thematic criteria (gender, exclusion, immigration, etc.), age groups (young people, the elderly, etc.), or were aimed at specific groups (single-parent families, the disabled, etc.).
- The involvement of citizens in the management of public affairs becomes a priority and, on many occasions, entails new forms of vertical (between institutions) and horizontal (citizenship) coordination.

An example of the difficulties of harmonising the interests of global capital with urban policies more attentive to social, cultural, and environmental problems can be traced in the case of the city of A Coruña. This municipality holds a Master Plan commissioned by the socialist local government to the urbanist Joan Busquets and approved in 2013 (Fig. 30.6). However, the amendments to the Master Plan introduced by the conservative municipal government (2011–2015) and the misguided decoupling of the urban planning model from mobility models by the New Left government (2015–2019) have undermined the original radical mobility proposals.

Another issue addressed in the plan was the new uses of port areas on the city's traditional waterfront. The relocation of port industrial activity to new facilities left an underused space in the centre of A Coruña. A revalued area incorporated into urban planning as the backbone of the change in the productive model imposed by the global advance of flexible capitalism. In this changing context, the "official" discourse of the area's transformation process, supported mainly by the Port Authority, the body with direct control over the area (García Morales, 2010; Zambonino, 2015), champions speculation and privatization as a method for obtaining economic liquidity to cover investments committed to in other territories. On the other side, an "alternative" discourse is emerging, supported by the municipal authorities and some citizens' platforms, which advocates giving more prominence to the collective and cultural uses of publicly owned spaces. Whatever the case, the public status of the land, its functional and economic importance, and the historical, heritage, and landscape value of the industrial port area would require the use of participatory instruments so that the residents of A Coruña could give their opinion on how the historic centre of their city is to be configured and to whom it will belong (Rodríguez-Barcón et al., 2018).

Fig. 30.6 A Coruña master pan for urban development 2013. *Source* A Coruña City Council

30.4 Conclusion

With the unfolding of neoliberal policies in the 1990s, the old paradigms based on planning and normative urbanism fell into crisis and were overcome by strategic planning and the logic of large-scale projects, both in the case of large cities and in the medium-sized cities. The onset of the crisis and the new approaches that emerged from the May 2011 movements helped to question the city model and the type of urbanism implemented during the real estate boom, which was accused of being technocratic and lacking in transparency. Thus, some initiatives arose that came to modify the contents, forms, and methods of traditional urbanism and urban planning, adopting new government forms that have coexisted, since then, with the old and market-oriented models of the neoliberal city.

Detailed studies and comparative evaluation of the legacy of this last period are still required. Anyway, although the policies pursued so far have not succeeded in presenting an alternative model, at least they have managed to change the focus of some municipal urban planning agendas towards the problems of everyday life and to try out new formulas of more participatory government.

Contemporary to all these processes, Spain's medium-sized cities have undergone functional and morphological changes derived from their new role in the context of the regionalisation process at the beginning of the 1980s. The establishment of quasi-federal regions in Spain helps to understand the emergence of new infrastructures and institutions bound up with the new powers transferred from the central government. Medium-sized cities then adapt, competing and cooperating, within the national urban network and the regional urban networks themselves. All medium-sized cities exhibit dynamics and processes with local keys that are embedded and adapted to the regional (emergence of new regional capitals and hierarchies) and global framework, displaying a range of situations that follow the dominant trends regarding demographic and urbanization dynamics. The role of large-scale projects and new urban centres (high-speed train stations, university campuses, new hospitals, big shopping malls, waterway corridors, cultural and sports facilities, science and technology parks, etc.) has become strategic, although they have not always proved to enhance growth. What has become clear is its involvement in territorial branding and urban marketing strategies.

The logic of property development and urban planning *à la carte* is still deeply rooted in our cities, despite being portrayed, time and again, as an exhausted model. However, and unlike the second period, the new initiatives are mostly central and highly strategic, guided by the principles of accumulation of global financial capitalism. An extractive urbanism is thus accomplished that multiplies the logic of the splintering city.

According to the same rationale, different scholars have already proved that in the twenty-first century political emancipation is highly unlikely to be accomplished at the scale of the city (Brenner, 2018; Brenner & Schmid, 2015). Progressive regimes might be viable in felicitous circumstances, but whether they can mobilize the resources required for a comprehensive break with neoliberalism, at what scale, remains to be seen. Nevertheless, a variety of surveys shows cities to be critical actors in setting the tone for regional, national, and international politics (Davies & Blanco, 2017). For all these reasons, further studies of austerity and neoliberal urbanism and its antagonists—welfarism and contentious politics—among the medium-sized Spanish cities will prove very fruitful as we have just overcome the 10th anniversary of the crash while being plunged into a new global crisis resulting from the COVID-19 pandemic.

Table 30.1 Spanish urban system according to population size, 1960–2019

Category of urban settlement	1960			2019		
	Number of cities/towns	Total population	% of the Spanish population	Number of cities/towns	Total population	% of the Spanish population
Cities with more than 1,000,000 inhabitants	2	3,817,794	12.48	2	4,902,888	10.44
Cities between 500,001 and 1,000,000 inhabitants	1	505,066	1.65	4	2,732,531	5.82
Cities between 300,001 and 500,000 inhabitants	3	1,069,664	3.50	6	2,256,679	4.80
Medium-sized cities between 50,001 and 300,000 inhabitants	55	5,532,850	18.09	136	15,015,546	31.96
Small cities/towns between 10,001 and 50,000 inhabitants	362	6,438,416	21.05	713	12,553,110	26.72
Small towns between 5,000 and 10,000 inhabitants	618	4,346,078	14.21	545	3,849,133	8.19
Rural municipalities below 5,000 inhabitants	8,161	8,873,068	29.01	6,813	5,674,403	12.08
Total		30,582,936	100.00		46,984,290	100.00

Source Spanish National Bureau of Statistics

Table 30.2 Approval date of master plans still in force for the sample of selected cities

Medium-sized cities (in brackets date of approval of the Master Plan currently in force)	Municipal Master Plans approved in the first period 1979–1992	Municipal Master Plans approved in the second period 1993–2007	Municipal Master Plans approved in the third period 2008–2019
La Línea de la Concepción (1985); Elda (1985); Teruel (1985); Marbella (1986); Torrelavega (1986); Torrevieja (1987); Alcoy (1989); Orihuela (1990); Benidorm (1990); Sagunto (1992)	**10**		
Estepona (1994); Vila-Real (1994); Linares (1995); Cuenca (1995); Vélez-Málaga (1996); Aranjuez (1996); Santander (1997); Manresa (1997); Ciudad Real (1997); Sanlúcar de Barrameda (1997); Almería (1999); Albacete (1999); Huelva (1999); Gandía (1999); Mérida (2000); Vitoria-Gasteiz (2001); Algeciras (2001); Ferrol (2001); Logroño (2002); Utrera (2002); Pamplona/Iruña (2003); Lleida (2003); Ourense (2003); Lorca (2003); Huesca (2003); León (2004); Motril (2004); Reus (2005); Girona (2006); Ávila (2006); Badajoz (2007); Salamanca (2007); Ponferrada (2007); Toledo (2007)		**34**	
Santiago de Compostela (2008); Segovia (2008); Palencia (2008); Roquetas de Mar (2009); El Ejido (2009); Cáceres (2010); Lugo (2011); Talavera de la Reina (2011); Zamora (2011); Cartagena (2012); Guadalajara (2012); Tarragona (2013); Burgos (2014); Soria (2014); Castellón de la Plana (2015); Jaén (2016)			**16**

Source Urban Information System—SIU, 2018. (http://siu.fomento.gob.es/)

Acknowledgements This contribution was supported as part of the project "Evolution of Urban Planning Policies in Medium-sized Spanish Cities and their Urban Areas: 1979–2019" (RTI2018-096435-B-C21), funded by the Ministry of Science and Innovation, within the call for R+D+i Projects "Research Challenges in 2018 (Spanish Government).

References

Alcalá-Santaella, F., Díaz de Orueta, F; Ginés, X. & Lourés, Mª L. (2011). Una nueva agenda urbana para las grandes ciudades: crecimiento y competitividad. In M. Iglesias, M. Martí-Costa, M. Tomàs, & J. Subirats (Ed.). *Políticas urbanas en España. Grandes ciudades, actores y gobiernos locales* (pp. 307–334). Barcelona: Icària.

Andrés, G. & González, M.J. (2020). Crecimiento y extensión reciente en ciudades medias: una aproximación a los cambios en los usos del suelo en las áreas urbanas de Castilla y León. In Cebrián Abellán, F. (Coord.) *Dinámicas de urbanización en ciudades medias interiores: ¿hacia un urbanismo más urbano?* (pp. 147–174). Valencia: Tirant lo Blanch, Tirant Humanidades.

Bellet, C. (1995). *Activitat immobiliària i polítiques urbanes dels anys vuitanta a les ciutats mitjanes catalanes. El cas de la ciutat de Lleida* (Doctoral dissertation, Universitat de Lleida, Spain). Retrieved from https://www.tdx.cat/handle/10803/8189

Bellet, C. (2007). Los espacios residenciales de tipo privativo y la construcción de la nueva ciudad: visiones de privatopía. *Scripta Nova*, XI, 245(08). Retrieved from http://www.ub.edu/geocrit/sn/sn-24508.htm

Bellet, C. (2014). La activación de solares urbanos: de práctica alternativa a objeto de programas municipales. *Biblio 3W*, XIX, 1058. Retrieved from http://www.ub.es/geocrit/b3w-1058.htm

Bellet, C. (2017). Proyectos y grandes operaciones urbanas. *Naturaleza, territorio y ciudad en un mundo global* (pp. 1249–1265). Universidad Autónoma de Madrid & AGE.

Bellet, C. (2020). Las políticas urbanísticas municipales en España: 40 años de ayuntamientos democráticos (1979–2019). *Boletín de la Asociación de Geógrafos Españoles*, 85, 2877, 1–38. https://doi.org/10.21138/bage.287

Boira, J. V. (1995). La rehabilitación urbana en los centros históricos valencianos. *Cuadernos de Geografía*, 58, 241–258. Retrieved from https://ojs.uv.es/index.php/CGUV/article/view/14723

Borja, J. (2011). ¿Un cambio de ciclo o un cambio de época? Siete líneas para la reflexión y la acción. *Urban* (NS01), 83–88. Retrieved from http://polired.upm.es/index.php/urban/article/view/412/1760

Brenner, N. (2004). *New State Spaces: Urban Governance and the Rescaling of Statehood*. Oxford University Press.

Brenner, N. (2018). Debating planetary urbanization: For an engaged pluralism. *Environment and Planning D: Society and Space*, 36(3), 570–590. https://doi.org/10.1177/0263775818757510

Brenner, N., Peck, J., & Theodore, N. (2015). Urbanismo neoliberal. La ciudad y el imperio de los mercados. In Observatorio Metropolitano de Madrid (Ed.). *El mercado contra la ciudad. Sobre globalización, gentrificación y políticas urbanas* (pp. 211–244). Madrid: Traficantes de Sueños.

Brenner, N., & Schmid, C. (2015). Towards a new epistemology of the urban? *City*, *19*(2–3), 151–182. https://doi.org/10.1080/13604813.2015.1014712

Burriel, E. (2008). La 'década prodigiosa' del urbanismo español (1997–2006). In *Diez años de cambios en el Mundo, en la Geografía y en las Ciencias Sociales, 1999–2008. Actas del X Coloquio Internacional de Geocrítica*. Barcelona: Universidad de Barcelona. Retrieved from http://www.ub.es/geocrit/-xcol/383.htm

Burriel, E. (2009). Los límites del planeamiento urbanístico municipal. El ejemplo valenciano. *Documents d'Anàlisi Geogràfica*, 54, 33–54. Retrieved from https://ddd.uab.cat/pub/dag/02121573n54/02121573n54p33.pdf_error_estat_espanyol

Burriel, E. (2014). El estallido de la burbuja inmobiliaria y sus efectos en el territorio. In J. M. Alberto & J. L. Sánchez (Coord.). *Geografía de la crisis económica en España* (pp. 101–140). València: Universitat de València- Servei de Publicacions.

Calderón Calderón, B. (2004). La ciudad del todo urbanizable: estrategias del sector inmobiliario y nuevas e insostenibles formas de urbanización. *Revista Ciudades*, 8, 135–155. Retrieved from http://bit.ly/2g4NiOV

Calderón Calderón, B., & García Cuesta, J. L. (2017). Legislación urbanística y planeamiento urbano en España, 1998–2015. Del despilfarro a la sostenibilidad. *Scripta Nova*, 21, 570. Retrieved from https://doi.org/10.1344/sn2017.21.19429

Calderón Calderón, B., & García Cuesta, J. L. (2018). The structure of Spanish cities: a complex interweaving of continuity and change, uses and forms. *Boletín de la Asociación de Geografía Española*, 77, 283–314. http://dx.doi.org/https://doi.org/10.21138/bage.2542

Capel, H. (2013a). *La morfología de las ciudades. III Agentes urbanos y mercado inmobiliario*. Serbal.

Capel, H. (2013b). Crisis de los modelos urbanos. Una mirada hacia el futuro. *Mercator. Revista de Geografía da Universidade Federal do Ceará*, 12(2), 7–27. https://doi.org/10.4215/RM2013.1202.0001

Cochrane, A. (2007). *Understanding Urban Policy. A critical approach*. Blackwell.

Davies, J. S., & Blanco, I. (2017). Austerity urbanism: Patterns of neo-liberalisation and resistance in six cities of Spain and the UK. *Environment and Planning A, 49*(7), 1517–1536. https://doi.org/10.1177/0308518X17701729

Díaz-Orueta, F., Lourés, M. L., & Pradel-Miquel, M. (2018). Transformando los modelos de crecimiento y cohesión: cambios en la gobernanza de Barcelona y Madrid. *Eure*, 44(131), 173–191. Retrieved from https://www.eure.cl/index.php/eure/article/view/2295/1064

Díez Medina, C., & Monclús, J. (2017). *Visiones urbanas. De la cultura del plan al urbanismo paisajístico*. Abada editores.

Escudero Gómez, L. A. (2018). El fracaso del planeamiento urbanístico actual de la ciudad de Toledo: El Plan de Ordenación Municipal de 2007. *Anales De Geografía De La Universidad Complutense, 38*(2), 213–339. https://doi.org/10.5209/AGUC.62482

Fariña, J., & Naredo, J. M. (2010). *Libro Blanco de la sostenibilidad en el planeamiento urbanístico español*. Ministerio de la Vivienda.

Fernández Durán, R. (1993). *La Explosión del Desorden. La metrópoli como espacio de la crisis global*. Editorial Fundamentos.

Fernández Maroto, M., & Santos Ganges, L. (2020). *Ayuntamiento de Valladolid, transición democrática y transformación urbana. Cambios de rumbo y nuevas estrategias urbanísticas (1979–1995)*. Universidad de Valladolid. Instituto Universitario de Urbanística

Fernández Salinas, V. (2005). De la protección a la legitimación social del patrimonio urbano en España. *Scripta Nova*, XX (539). https://doi.org/10.1344/sn2005.9.957

Ferrer, A., & Sabaté, J. (1999). L'urbanisme municipal. In O. Nel·lo (Ed.). *20 anys d'Ajuntaments democràtics* (pp. 117–160). Barcelona: Federació de Municipis de Catalunya.

Ferrer, A. (2005). Introducción: la última deriva del urbanismo municipal. *Papers. Regió Metropolitana de Barcelona*, 43, 11–13. Retrieved from https://ddd.uab.cat/record/23956?ln=ca

Font, A. (2000). La experiencia reciente de Cataluña Planeamiento urbanístico para el siglo XXI. *Urban*, 5, 60–82. Retrieved from http://polired.upm.es/index.php/urban/article/view/512

Gaja Díaz, F. (2008). El "tsunami urbanizador" en el litoral mediterráneo. El ciclo de hiperproducción inmobiliaria 1996–2006. *Scripta Nova*, XII, 270.Retrieved from http://www.ub.es/geocrit/sn/sn-270/sn-270-66.htm

Gaja Díaz, F. (2015). Urbanismo concesional. Modernización, privatización y cambio de hegemonía en la acción urbana. *Ciudades, Instituto Universitario de Urbanística*, 18(1), 103–126. https://doi.org/10.24197/ciudades.18.2015

García, M. (2010). The Breakdown of the Spanish urban model: Social and territorial effects of the Global Crisis. *International Journal of Urban and Regional Research, 34*(4), 967–980.

García Morales, V. Y. (2010). Políticas sectoriales que condicionan la competencia de ordenación del territorio y urbanismo: Especial referencia a puertos. *Revista Digital Facultad De Derecho, 2*, 1–36.

García-Herrera, L. M. (2018). Mercantilización del espacio urbano bajo la lógica neoliberal: Gentrificación y redefinición de los espacios públicos en España. *Naturaleza, territorio y ciudad en un mundo global* (pp. 858–877). Universidad Autónoma de Madrid & AGE.

Górgolas Martín, P. (2017). Burbujas inmobiliarias y planeamiento urbano en España: una amistad peligrosa. *Cuadernos de Investigación urbanística*, 111, 3–65. https://doi.org/10.20868/ciur.2017.111.3536

Hall, P. (1996). *Cities of Tomorrow: An Intellectual History of Urban Planning and Design in the Twentieth Century*. Oxford, U.K.: Blackwell Publishers; 2nd Revised edition.

Harvey, D. (2013). *Rebel Cities: From the Right to the City to the Urban Revolution*. Verso Books.

Iglesias, M., Martí-Costa, M., Tomàs, M., & Subirats, J. (Eds.). (2014). *Políticas urbanas en España. Grandes ciudades, actores y gobiernos locales*. Icaria.

Janoschka, M. & Mota, F. (2018). De la protesta a la propuesta. Un diagnóstico de los Ayuntamientos 'del cambio'. *Working Paper Series CONTESTED_CITIES*, 1–9. Retrieved from http://contested-cities.net/working-papers/wp-content/uploads/sites/8/2018/01/WPCC_180001_JanoschkaMota-Intro.pdf

LaCol, Arquitectura cooperativa (2018). *Construir en col·lectiu. Participació en arquitectura i urbanisme*. Barcelona: Pol·len.

Lois, R., Piñeira, M. J., & Vives, S. (2016). The urban bubble process in Spain: An interpretation from the theory of the circuits of capital. *Journal of Urban and Regional Analysis*, 8 (1), 5–20. Retrieved from https://www.jurareview.ro/resources/pdf/volume_8_issue_1_2016_file.pdf

López de Lucio, R. (1993a). *Ciudad y urbanismo a finales del siglo XX*. Universitat de València-Servei de Publicacions.

López de Lucio, R. (1993b). El ensimismamiento en el 'Urbanismo Urbano" como respuesta a las limitaciones del planeamiento territorial. *Ciudades*, 1, 61–66. https://doi.org/10.24197/ciudades.01.1993.61-66

López, I., & Rodríguez, E. (2010). *Fin de ciclo. Financiarización, territorio y sociedad de propietarios en la onda larga del capitalismo hispano (1959–2010)*. Madrid: Traficantes de sueños.

Martí-Costa, M., & Tomàs, M. (2016). Crisis y evolución de la gobernanza urbana en España. *Ciudad y territorio-Estudios Territoriales*, XLVIII (188),187–200. Retrieved from https://apps.fomento.gob.es/CVP/handlers/pdfhandler.ashx?idpub=BP1016

Méndez Gutiérrez del Valle, R. (2014). Expansión y crisis del modelo neoliberal en Madrid. In R. Hidalgo & M. Janoscka (Ed.). *La ciudad neoliberal. Gentrificación y exclusión en Santiago de Chile, Buenos Aires, Ciudad de México y Madrid* (pp. 217–232). Santiago de Chile: Pontificia Universidad Católica de Chile.

Méndez Gutiérrez del Valle, R., Abad, L.D., & Echaves, C. (2015). *Atlas de la crisis. Impactos socioeconómicos y territorios vulnerables en España*. València: Tirant Humanidades

Méndez Gutiérrez del Valle, R., & Plaza Tabasco, J. (2016). Crisis inmobiliaria y desahucios hipotecarios en España: una perspectiva geográfica. *Boletín de la Asociación de Geógrafos Españoles*, 71, 99–127. https://doi.org/10.21138/bage.2276

Méndez Gutiérrez del Valle, R. (2018). *La telaraña financiera. Una Geografía de la financiarización y su crisis*. Santiago de Chile: RiL editores, Instituto de Geografía de la Pontificia Universidad Católica de Chile.

Ministerio de Fomento (2018). *Sistema de Información Urbana*. Madrid: Ministerio de Fomento; Dirección General de Arquitectura, Vivienda y Suelo. Retrieved from https://www.mitma.gob.es/portal-del-suelo-y-politicas-urbanas/sistema-de-informacion-urbana/sistema-de-informacion-urbana-siu

Naredo, J. M. (1996). *La burbuja económico-financiera en la coyuntura económica reciente (1985–1995)*. Siglo XXI de España.

Naredo, J. M. (2010). El modelo inmobiliario español y sus consecuencias. *Boletín CF+S*, 44, 13–27. Retrieved from http://habitat.aq.upm.es/boletin/n44/ajnar.html

Nel·lo, O. (2015). *La ciudad en movimiento. Crisis social y respuesta ciudadana*. Madrid: Díaz Pons.

Olazabal, E. (2018). *Dinámicas y caracterización del proceso de urbanización en España (1987–2011). Las áreas urbanas articuladas por ciudades medias* (Doctoral dissertation, Universitat de Lleida). Retrieved from https://www.tdx.cat/handle/10803/664412#page=1

Paisaje Transversal (2018). *Escuchar y transformar la ciudad. Urbanismo colaborativo y participación ciudadana*. Madrid: Catarata.

Precedo Ledo, A., Orosa González, J. J., & Míguez Iglesias, A. (2010). De la planificación estratégica al marketing urbano: Hacia la ciudad inmaterial. *EURE (santiago), 36*(108), 5–27. https://doi.org/10.4067/S0250-71612010000200001

Rivas, J. L. de las (2008). Ayuntamiento y urbanismo. In Merino Estrada, V. & Orduña Rebollo, E. (coord.) *La ciudad de Valladolid y su Ayuntamiento: 100 años de historia común* (pp. 271–294). Valladolid, Ayuntamiento de Valladolid.

Rivas, J. L. de las, Fernández Maroto, M., Rodrigo González, E., & Martínez Sierra, M. (2017). Recuperando el concepto urbanístico de barrio: unidades urbanas y regeneración urbana en Castilla y León. *Ciudad y Territorio. Estudios Territoriales CYTET*, XLIX (191), 45–62. Retrieved from https://apps.fomento.gob.es/CVP/handlers/pdfhandler.ashx?idpub=BP1019

Roch, F. (2015). Revolución neoliberal y utopía ciudadana, una batalla inaplazable. *Ciudades, Instituto Universitario de Urbanística*, 18(1), 49–68. Retrieved from https://revistas.uva.es/index.php/ciudades/article/view/1116/961

Rodríguez Barcón, A., Calo, E., & Otero-Enríquez, R. (2018). Reconversión de espacios portuarios y privatización de la fachada litoral de A Coruña: una lectura crítica. *Ciudades: Revista del Instituto Universitario de Urbanística de la Universidad de Valladolid*, (21), 99–120. https://doi.org/10.24197/ciudades.21.2018.99-120

Rodríguez López, J. (2006). Los booms inmobiliarios en España. Un análisis de tres períodos, *Papeles de Economía Española*, 109, 76–90. Retrieved from https://dialnet.unirioja.es/servlet/articulo?codigo=2142662

Romero, J., Jiménez, F., & Villoria, M. (2012). (Un)sustainable territories: Causes of the speculative bubble in Spain (1996–2010) and its territorial, environmental and sociopolitical consequences. *Environment and Planning C: Government and Policy, 30*(3), 467–486. https://doi.org/10.1068/c11193r

Romero, J., Brandis, D., & Melo, C. (2015). El giro neoliberal de las políticas para la Ciudad en España. Balance a partir de los casos de Madrid y Valencia. *Boletín de la Asociación de Geógrafos Españoles*, 69, 369–386. http://dx.doi.org/https://doi.org/10.21138/bage.1901

Romero González, J., & (Coord.),. (2019). *Geografía del despilfarro en España*. Ediciones Universitat de València.

Ruiz Varona, A. (2013). Valladolid. Del crecimiento sobre sí misma a la articulación interurbana de fin de siglo. *Biblio 3W: Revista Bibliográfica de Geografía y Ciencias Sociales*, (18), 27. http://www.ub.es/geocrit/b3w-1036.htm

Rullán, O. (2012). Urbanismo expansivo en el estado español: de la utopía a la realidad. In AA.VV. *Geografía y desafíos territoriales en el siglo XXI* (pp. 165–209). Alicante: AGE/Universidad de Alicante.

Sainz Gutiérrez, V. (2006). *El proyecto urbano en España. Génesis y desarrollo de un urbanismo de los arquitectos*. Publicaciones de la Universidad de Sevilla.

Sennett, R. (2018). *Building and Dwelling: Ethics for the City*. Allen Lane.

Solà Morales, M., & Parcerisa, J. (1987). El urbanismo urbano. Forma urbana y planeamiento urbanístico en siete capitales españolas. *Estudios Territoriales*, 24, 33–51. Retrieved from https://dialnet.unirioja.es/servlet/articulo?codigo=81659

Subirats, J. (2016). *El poder de lo próximo. Las virtudes del municipalismo*. Los Libros de la Catarata.

Swyngedouw, E. (2005). Governance Innovation and the Citizen: The Janus Face of Governance-beyond-the-state. *Urban Studies, 42*(11), 1991–2006. https://doi.org/10.1080/00420980500279869

Swyngedouw, E. (2018). The perverse lure of Autocratic Democracy. *South Atlantic Quarterly, 112*(2), 267–286. https://doi.org/10.1215/00382876-7381134

Swyngedouw, E. (2019). The perverse lure of Autocratic Postdemocracy. *South Atlantic Quarterly, 112*(2), 267–286. https://doi.org/10.1215/00382876-7381134

Thompson, M. (2020). What's so new about New Municipalism?. *Progress in Human Geography*, 0309132520909480. 1–26. https://journals.sagepub.com/doi/pdf/https://doi.org/10.1177/0309132520909480

Tomé Fernández, S. (2007). Los Centros Históricos de las Ciudades Españoles. *Eria*, 72, 75–88. Retrieved from https://www.unioviedo.es/reunido/index.php/RCG/article/view/1571
Vázquez-Varela, C. y Martínez, J. M. (2013). Los criterios de sostenibilidad en el urbanismo y en las políticas urbanas durante la primera década del siglo XXI. Las ciudades de Castilla-La Mancha como observatorio. In Valenzuela, M. (coord.). *Las ciudades españolas en la encrucijada: Entre el 'boom' inmobiliario y la crisis económica* (pp. 111–166). Madrid: Real Sociedad Geográfica.
Vázquez-Varela, C. & Martínez, J.M. (2018). Dinámicas de expansión urbana en ciudades medias interiores. Análisis comparado de ciudades aisladas y ciudades situadas en entornos metropolitanos y corredores de transporte: los casos de Cuenca y Guadalajara. In Cebrián Abellán, F. (coord.): *Ciudades medias y áreas metropolitanas. De la dispersión a la regeneración* (pp. 175–199). Cuenca: Ediciones de la Universidad de Castilla-La Mancha.
Vázquez-Varela, C., & Martínez, J. M. (2020). Medium-sized Spanish inland cities and regional development in the Iberian Peninsula: Challenges and opportunities. In J. Banski (Ed.), *Dilemmas of Regional and Local Development* (pp. 147–168). Routledge.
Zambonino Pulito, M. (2015). El dominio público portuario. In González García, Julio V. (coord.) *Derecho de los bienes públicos* (pp. 1519–1576). Valencia, Tirant lo Blanch.

Chapter 31
Unfolding Binh Duong's Formula to Success: A Breakthrough Strategy Towards Industry 4.0, Knowledge Economy, and Sustainable City

Nguyen Viet Long and Hana Nguyen

Abstract In spite of a historical establishment as an agricultural province, Binh Duong carried out a provincial industrialization revolution about 25 years ago and became one of the most impressive regions of economic and societal development in Vietnam. However, such a labour-intensive local economic model was based on traditional processing production, which was no longer sustainable with the rapid gross regional domestic product (GRDP) growth of the province, particularly in the context of the inevitable Industry 4.0 era today. In fact, the multitude of challenges faced by developing countries and their manufacturing cities provided significant motivation for Binh Duong to embark on a radical transformation. As a result, in 2016, the provincial leaders strived for a crucial breakthrough to develop a uniquely comprehensive socio-economic strategy called the Smart City Program, with supporting partnership from Eindhoven city of the Netherlands. The goal is to build solid foundations, creating an innovative community and favourable business environment, increasing economic competitiveness, improving living standards, and gravitating towards knowledge and digital economy. Binh Duong succeeded in achieving exponential growth over the past years. The success was so influential that the province aspired to reimplement and reinforce the strategy for the next period. Through a combination of documentary research, qualitative methods, participant observation, and semi-structured interviews with experts and leaders in Vietnam, this paper unfolds Binh Duong's unique breakthrough strategy and its effectiveness in embracing the challenges presented by the Fourth Industrial Revolution.

Keywords Binh Duong · Smart city · Sustainable development · Innovation 4.0 · Digital economy · Knowledge economy

N. V. Long (✉)
International University, National University of Ho Chi Minh City, Ho Chi Minh City, Vietnam
e-mail: nguyenvietlong.com@gmail.com

H. Nguyen
Georgia State University, Atlanta, USA
e-mail: hnguyen200@gsu.edu

B. R. K. Sinha (ed.), *Urban Dynamics, Environment and Health*,
https://doi.org/10.1007/978-981-99-5744-6_31

31.1 Introduction

For developing countries with limited resources like Vietnam, the challenges of integrating with the Fourth Industrial Revolution are multiplied. Far from developed countries whose robust infrastructure facilitates a higher baseline for such an integration, Vietnam is at a disadvantage in the race to Industry 4.0, knowledge economy, and smart sustainable city. On top of that, the global COVID-19 outbreak gave rise to destructive post-pandemic impacts, exposing the country to manifold layers of hurdles. Then, how could Binh Duong, a province in Vietnam, thrive in this race? In order to explicate the answer to this question, we first elucidate the three key notions that make up the context for this race: Industry 4.0, knowledge economy, and smart sustainable city.

Industry 4.0 denotes an era of smart and connected machines and systems—the virtual and physical systems of manufacturing that flexibly cooperate with each other, also known as "smart factories". The term "Industry 4.0", originated in 2011 at the world's largest trade fair Hannover Messe, and was used to describe how "smart factories" will revolutionize the organization of global value chains. This revolution was then formalized by Schwab (2016) as the Fourth Industrial Revolution—an umbrella that encompasses not only Industry 4.0, but also the simultaneous breakthroughs beyond manufacturing due to the fusion of technologies and their interactions. In other words, Industry 4.0 indicates interactive interconnectedness between analogue production and the digital world, while the Fourth Industrial Revolution comprehensively addresses production, businesses, economies, and societies (Wolter et al., 2015). As both terms "Industry 4.0" and the "Fourth Industrial Revolution" characterize a new era of humankind, we forgo their nuanced differences and use the two concepts interchangeably in this study. Pfohl et al. (2015) and Mittal et al. (2017) enumerated the enablers and drivers of this 4.0 era, including virtual reality, big data analytics, cloud computing, 3D printing, smart sensors, smart data, smart factory, autonomous robots and systems, cyber security, etc.

Distinctly, Industry 4.0 is based on intellectual capabilities and knowledge-intensive activities. As a result, Industry 4.0 accelerates the transition to "knowledge economy", which is the most advanced practice of economic system and production to date. In "knowledge economy", goods and services production is based on knowledge-intensive activities (Powell & Snellman, 2004; Unger, 2019). Knowledge economy goes hand in hand with "digital economy"—an economy in which digital technologies facilitate trade and business (Bukht and Heeks, 2019). In this human-capital-based economy, the labour force is required to be competent in managing data, developing models, and innovating systems (OECD, 2001). The requirement of an upskilling workforce is a massive constraint that hinders developing nations from adopting a knowledge economy.

In turn, the creation of knowledge economy is promoted by "smart sustainable cities" (Ivaldi et al., 2020). These are smart cities—urban development entailing a network of human capital, social capital, and information and communications technologies (ICTs)—that integrate different dimensions of sustainability to optimize the cities' functionality, protect the environment, stimulate economic growth, and improve the quality of life through infrastructures, processes, and services supported by ICTs (Bibri & Krogstie, 2017; Cocchia, 2014; Hollands, 2008). The three key definitions above lay out the current global trend and contextualize the challenges that many nations, especially developing ones, have been facing. Within this race, Binh Duong basically overcame the bottlenecks in growth prevalent in developing nations and made its breakthrough towards Industry 4.0, knowledge economy, and smart sustainable city.

This article unfolds the strategy that has shaped Binh Duong's success. Binh Duong is a sizeable industrial province of more than 2.5 million in population and 2,695 km^2 of land. Its GRDP predominantly comprises foreign investments and exports. Remarkably, the visionary leaders of Binh Duong profoundly understand that they must turn the province's weaknesses into its strategic niche to get ahead. In this light, as the entire province is fighting against the potential crisis posed by the global pandemic, the government leaders of Binh Duong unanimously decided to reimplement and reinforce its 2016–2021 Smart City Program for the next period. This critical decision was based on the successful outcomes of the Smart City Program during the 2016–2021 period, guiding Binh Duong towards Industry 4.0. This time, the Program focuses its resources on the most critical aspects, setting a new level of momentum for the economy.

This study applies two research methodologies to analyse Binh Duong's strategy. The first is the method of semi-structured interviews by Edwards and Holland (2013), which is applied in the context of Vietnam in this paper (Nguyen Viet Long, 2015; Le Thi Hoa and Nguyen Viet Long, 2011; 2012). 64 professionals and leaders in Binh Duong participated in these interviews, of which 39 come from businesses, 7 work in research institutions, and 18 are authorities in the local government. The second is the method of participant observation by Geertz (1973). This method integrates the author's participation in planning and implementing Binh Duong's Smart City Program, as well as Binh Duong's strategic directions. These research methods provide the means to valuable documents, government publications, and insights imperative to fulfil the process of documentary research—one of the three major types of social research most widely used throughout the history of social sciences (Scott, 2006).

This study comprises the following sections: Sect. 31.1 describes the Smart City Program of Binh Duong throughout the period of 2016–2021. Section 2 elaborates the essential project clusters within the Smart City Program.

31.2 The Period of 2016–2021: The Start of the Smart City Program in Binh Duong—A Foundational Program for a New Development

Binh Duong is known to carry the tradition of innovativeness. Even though Binh Duong had a humble start as an agricultural province, its successful industrialization in 1997 has made it one of the most essential industrial provinces in Vietnam. By 2015, alongside the beginning of the Industry 4.0 era, Binh Duong was facing a booming scientific and technological development, which created new challenges and opportunities. In such a critical time, during which the traditional manufacturing-based growth model had also reached its peak, Binh Duong strategically strived for a breakthrough by building a new development platform to address the emerging needs of the locals and businesses for the 2016–2021 period. This new platform is the Smart City Program. Learning from and collaborating with the Netherlands' city of Eindhoven, which is one of the leading technology hubs in Europe, the Smart City Program is created not only to transform Binh Duong into a services-based and high-tech production economy but also to create an innovative ecosystem within sustainable development and green urban cities (Nguyen Viet Long, 2020a). The Program gradually formed a platform for Binh Duong to move towards a knowledge economy and a digital one (Nguyen Viet Long, 2019a).

Accordingly, unlike any conventional approach that only focuses on the application of technology in urban cities, Binh Duong's Smart City Program uniquely made impacts in both technological and non-technological fronts. The Program focused on increasing the innovation content in economic activities. In this approach, a smart city is an innovative and well-connected ecosystem, within which the members/actors were constantly optimized and innovated. Pushing forward, Binh Duong has delivered an action plan that is based on the Triple Helix model: the "Binh Duong Navigator 2021" (Nguyen Viet Long, 2020b).

31.2.1 The Triple Helix Model

Etzkowitz (2008) theorized the Triple Helix as a model that promotes and formalizes collaboration among three entities: (1) government, (2) academia, and (3) industry. Hereafter, each of the three components in the model is called the "partner". By definition, the first partner—government—includes not only the local government but also the central government and the regional ones. The second partner—academia—is composed of universities, research institutions, and other training institutes. The third partner—industry—contains startups, regional businesses, as well as large multinational corporations. The partnership of the three partners aims for a united goal of fostering socio-economic development for the province.

By the end of the 90s, many parts of the world had already adopted the Triple Helix model. A case in point is the city of Eindhoven with its typical success story in applying this model (Van Gijzel, 2016). In 2016, Binh Duong learned from Eindhoven and adopted a similar model (Long & Ha, 2020). Specifically, the Triple Helix places the people and the intellects at the centre. It encourages the three partners to cooperate in a flexible and voluntary manner via joint discussions. Such an open policy allows for the sharing of visions, challenges, opportunities, and resources among the partners to develop mutually agreed directions for the province, while motivating their commitments to carry out the initiatives. Each partner has the flexibility to work on any preferred projects, as such can be beneficial for its own organization to gain a competitive advantage. In fact, the Triple Helix model not only benefits all partners jointly, but also supports each partner separately. In Binh Duong, the Triple Helix has become the driving force for any trailblazing projects.

31.2.2 Line of Actions

The "Binh Duong Navigator 2021" action plan outlines definitive goals and guiding directions. The plan indicates a united initiative to transform Binh Duong into a top world-class technological and economic hub. The Navigator is attainable thanks to the joint efforts of multiple Triple Helix partners, following the guidance of the provincial leaders. Under the umbrella of an overall shared vision, the Binh Duong Navigator can be adaptably adjusted to suit particular situations as deemed fit. There are 46 actions that are organized under 14 programmes and are quartered into the following domains: "People", "Technology", "Business", and "Basics" (Long, 2019a).

i. People: At the heart of the Smart City Program are the people. Within this domain, the labour force is the centre, along with the labour's education, capacity, collaboration skills, and suitability for the job. On its journey to be a competitive global player, Binh Duong endeavours to engage a talented and efficient labour force that plays pivotal roles in fostering innovation and driving economic growth. TechLab, FabLab, and Maker Space would be developed in Binh Duong. These lab centres provide facilities to share among businesses and academic institutions, enabling creativity, new ideas, and entrepreneurship. These objectives are feasible thanks to high-quality educational infrastructure and dynamically cooperative working environment founded on a solid financial system.

ii. Technology: This domain is imperative to the action plan. Correspondingly, Binh Duong provincial government specially offers support to unite educational institutions and businesses at the local and regional levels to motivate R&D, knowledge transfer, collaboration, up-to-date technology application in business and daily lives. Innovations are timely tested at living lab facilities for real-life application to promote research activities and enhance the living standards of the residents. In building up the know-how database, Binh Duong is aiming for a long-term future that accommodates many more advanced manufacturers and R&D facilities to serve much further beyond the province of Binh Duong, i.e. the Southern Key Economic Zone (Fig. 31.1). Every partner will work together to enhance its competitiveness.

By using advanced technology as a social-economic driver, Binh Duong can and will create a better future for its people and businesses.

iii. Business: In the business domain, the action programmes are designed to strengthen established businesses, attract investments, and promote entrepreneurship. One of Binh Duong's niche market advantages is its ability to bring in the abundant number of advanced manufacturing companies, many of which form large-scale and modern industrial parks in the region. Taking this opportunity of growth, Binh Duong is working on forming a regional supply-chain portal and a business database, while maintaining the development of supporting industries, tech-focused industrial parks, and business incubators. Within a future not so distant, Binh Duong targets to form an entrepreneurial and innovative ecosystem, in which a strong supply-chain network of service and industry can constantly attract and actively integrate high-tech investors.

iv. Basics: This domain focuses on the preconditions required in building a strong region. Examples of preconditions include sustainable cities with appealing residential setting and distinctive culture, favourable business environment with high-quality infrastructure and reputable brands, modern amenities such as broadband, ICTs, sport facilities, and recreation centres. Binh Duong prioritizes both transportation infrastructure (logistic centres, container depots, etc.) and broadband ICT infrastructure to guarantee power supply to high-tech manufacturers. In the long run, as Binh Duong is becoming a part of globally accredited organizations, it aims to be influential globally, to not only be a great place for commercial and residential activities but also a smart-dynamic-innovative region.

This detailed road map to a smart city is designed to attract locally and internationally dynamic people, intellectuals, and investors, including foreign and technological ones. In learning from the Triple Helix model adopted by Eindhoven, which specifically focuses on the people instead of technology, Binh Duong is realizing its vision of a dynamic environment, within which local entities and international communities join forces in developing smart ideas, finding solutions, and developing business models for the betterment of all partners. This comprehensive road map is crafted in a step-by-step level of detail to avoid resource fragmentation. The step starts with operations that strengthen the economy, focuses next on strategically located urban areas, and finally extends its implementation to the whole province.

31.3 Contribution to the Development of Binh Duong

The direct and strong supervision of Binh Duong's leaders has put forth drastic implementation, resource mobilization, and active participation across the community. Only after a few years of implementation, the Smart City Program has delivered impressive results, greatly contributing to the expeditious growth of the province and opening up many new potential opportunities for the years to come (Long, 2020b; Binh Duong People's Committee—Executive Committee of Smart City Program, 2020).

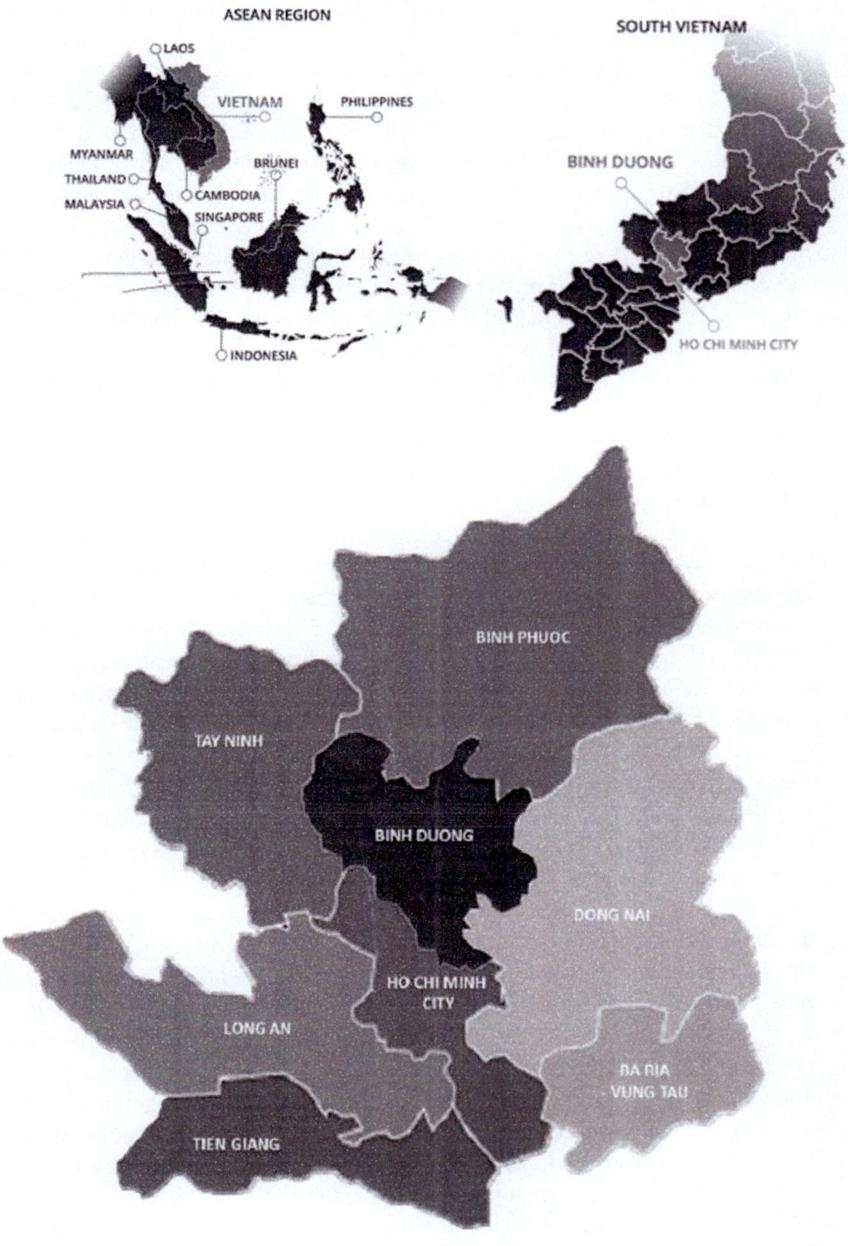

Fig. 31.1 Binh Duong, a part of the Southern Key Economic Zone, VietnamNguyen Viet Long (2020b)

Over the years, thanks to the Smart City Program giving rise to many innovation initiatives, Binh Duong has exceeded its targets, excelling far ahead of its original ones in terms of international cooperation, attraction of foreign investment, trade surplus, etc. (Long, 2020b). The number of enterprises in the province from early 2016 to the end of 2021 has multiplied by 2.6 times, from over 20,000 to about 52,000. Within Vietnam, in 2021, Binh Duong was consecutively ranked first in infrastructure development, third in attraction of foreign investment, third in domestic production, etc. Binh Duong is not only the first in the country to build affordable housing that supports low-income households but also the first to completely eradicate poverty. Furthermore, the province has promoted activities to improve the quality of life, organized cultural festivals, and enacted modern urban planning. For instance, Thu Dau Mot, the capital city of Binh Duong, has built hundreds of parks to greenify the city and allocated recreational areas that promote active lifestyles among the community. In the northern districts, particularly the new district of Bau Bang, industrial production has developed exponentially. In the south, Thu Dau Mot city has been upgraded to a class-I urban area, while Di An, Thuan An, and Tan Uyen have officially been upgraded to cities.

Programs to encourage innovation, science, and technology research have been implemented by the local government, training providers, and businesses. Binh Duong has been at the forefront of introducing numerous programmes in Vietnam that promote the application of science and technology into daily activities. By 2021, these programmes include, but not limited to, (1) the first modern centralized administrative centre of Vietnam, (2) geographic information system (GIS) digitized maps for Binh Duong, (3) automated environmental monitor and industrial waste management, (4) land and environmental databases for close and active management, (5) a business database to facilitate the analysis/management/and promotion of investment, (6) fast-speed emergency services via GIS optimization, (7) artistic lighting system for the Provincial Administration centre building to promote cultural activities of the people in the New City via using state of the art smart technology to save energy from Philips, (8) call centre 1022 providing social channels (Zalo, Facebook, e-mail, etc.) to connect the public and the government 24/7, and particularly (9) the Intelligent Operational centre for the entire province, etc. Additionally, Binh Duong province promoted and supported enterprises to modernize and apply technology 4.0 to improve production. Becamex Corporation is a shining embodiment whose digitization and application of BIM technology have achieved a lot of success.

Following the Smart City strategy, Binh Duong has been attracting many leading corporations in science and technology such as Schneider Group, Philips, Bosch, NXP, Intel, Amazon Web Service, Viettel, VNPT, Lego, etc. The province has also welcomed prestigious training institutions, such as the National University of Singapore (NUS) from Singapore, Portland State University from the United States, Industrial Technology Research Institute (ITRI) from Taiwan, Chungnam National University from South Korea, Eindhoven University of Technology from the Netherlands, Vietnam National University from Ho Chi Minh City, etc. The province has been collaborating with these institutions in building research, innovation, and training centres at scale. Binh Duong leaders' aspiration for innovation and strategic vision towards the 4.0 era has helped the province expand connectivity, and enhance cooperation with many prosperous provinces around the world, such as Eindhoven and Emmen in the Netherlands, Emilia Romagna in Italy, Daejeon in Korea, Yamaguchi in Japan, etc.

Keenly integrating with the world, Binh Duong successively hosted global economic, science, and technology events such as the Global Innovation Forum with UNESCO and the World Technopolis Association (WTA), the Asian Economic Cooperation Forum with Horasis, the 15th International Asian Urbanization Conference with Asian Urban Research Association, etc. These events facilitate the avid participation of leaders of thousands of enterprises, multinational companies, universities, political parties, provinces, and cities around the world, accompanying significant benefits through joint investment, experience sharing, and local brand promotion opportunities. As of 2019 and 2020, the per capita real income of Binh Duong was the highest in Vietnam, while the gap between the rich and the poor was impressively minuscule. Binh Duong and its regional locales have been granted memberships by many international organizations such as the Intelligent Community Forum (ICF), the World Trade centre Association (WTCA), etc. Since 2019, Binh Duong's Smart Region was honoured many times by the ICF as the top 21 and then the top 7 international exemplars in smart city development strategy of the years.

31.4 Resource Mobilization, Revision, and Adoption of New Strategies, Planning for the Innovation Region, and Development of the Smart City Program for the Next Period Until 2030

The Vietnamese State has recently issued abundant guidance, directions, and policies in a timely manner in response to swift changes brought about by Industry 4.0. These policies are an important foundation upon which the local governments develop their strategies to be in line with the national vision and global trends. Guidance includes Resolution No. 52-NQ/TW by the Vietnam Politburo, dated 27th September in 2019, providing a number of guidelines and policies to prepare for active participation and transformation in light of Industry 4.0; Decision No. 950/QD-TTg by the Vietnamese Prime Minister, dated 1st August in 2018, approving the development scheme for smart sustainable cities in Vietnam in the 2018–2025 period, with a vision towards 2030; Decision No. 749/QD-TTg, dated 3rd June in 2020, was the approval of the digital transformation programme of Vietnam by the Prime Minister, etc. These guidelines are the concrete groundwork on which Binh Duong builds, revises, and updates its Smart City Program.

At the onset of 2022, major disruptions caused by the COVID-19 pandemic gave rise to extensive changes in the way we live and do business (McKinsey&Company, 2022). At the same time, conflicts and upheavals in the world are also creating unprecedented problems. In fact, the international production chain is changing unpredictably, bringing a lot of challenges but also opening up opportunities for Vietnam and developing countries. Meanwhile, the people and businesses are longing for innovation, given the flourishing progress in science and technology development in the era of Industry 4.0.

Taking into account the factors above, this section focuses on three key aspects of the Binh Duong Smart City Program in the 2022–2030 period: (a) Overarching vision for the master plan, (b) Specific key directions, and (c) Development of the Innovation Region at the heart of the master plan.

31.4.1 Overarching Vision for the Master Plan

Given the context above, Binh Duong province planned to especially invest more in the Smart City Program, strategically using it as the key to excel in the international arena. The interval between now and 2030 is a crucial period for Binh Duong's breakthrough as the trailblazer in overcoming the middle-income-trap. This objective is scheduled for discussion at the Provincial Government for the next term. The Program keeps its ambitious original goals while focusing more intensively on promoting innovation and a digital society and economy. With its strengths and potentials, Binh Duong is now better positioned to take bolder, stronger, and more specific actions to spread motivation and influence across the region and Vietnam (Nguyen Viet Long, 2020b).

In terms of strategic directions, Binh Duong continues to promote its industrial strength, which is the heart of its actions, especially advancing the application of high technology, progressive production, and innovation. At the same time, the province puts more focus on restructuring the economy towards high-quality services, namely e-commerce, large-scale commerce, higher education, modern urban living, etc. For the forthcoming period, achieving a larger share of services is a precondition for fostering rapid and sustainable development while concurrently bolstering industrial growth in a more effective manner. In addition, the Binh Duong government aims to create enabling conditions for business growth and expansion to other provinces and cities in the country in order to not only contribute to their economic development but also expand Binh Duong's economic development resources. Regarding planning, the province maintains its commitments to advance industrial development towards the North, provide favourable conditions for the establishment of new and high-quality infrastructure, and upgrade service-oriented and livable urban areas in the South where the New City is planned to be the modern and inclusive centre of the province.

Grounded by the vision above, the leaders of Binh Duong are engaging its provincial departments, relevant businesses, and academia in research, revision, and update of the master plan, and in the development of the Smart City Program, to come together to solve the needs of the people and businesses in the next period. The two keywords will be "innovation" and "digital technology"—the enablers for a breakthrough. The upcoming strategy will leverage the solid foundation established in the previous period, thereby maintaining the four Navigator domains from the past Binh Duong Navigator 2016–2021. Furthermore, the strategy will place significant emphasis on achieving a harmonious blend of Social, Green, and Economic Growth.

To further promote focused investments, Becamex Corporation has been commissioned by the province to deliver the fresh initiative of Binh Duong's Innovation Region, which is the heart of the Smart City Program for the period of 2022–2030. This effort is a joint collaboration with provincial departments and international partners from Eindhoven—Netherlands, Daejeon—South Korea, and Singaporean experts. The Innovation Region, by definition, contains the following cities: Thu Dau Mot, Di An, Thuan An, Bau Bang, Tan Uyen, and Ben Cat. Thus, Binh Duong's Innovation Region will border Thu Duc City—the eastern innovative urban area of Ho Chi Minh City to establish an innovation sub-region in the Southern Key Economic Zone, which runs all the way through Binh Duong up to Binh Phuoc province Fig. 31.2.

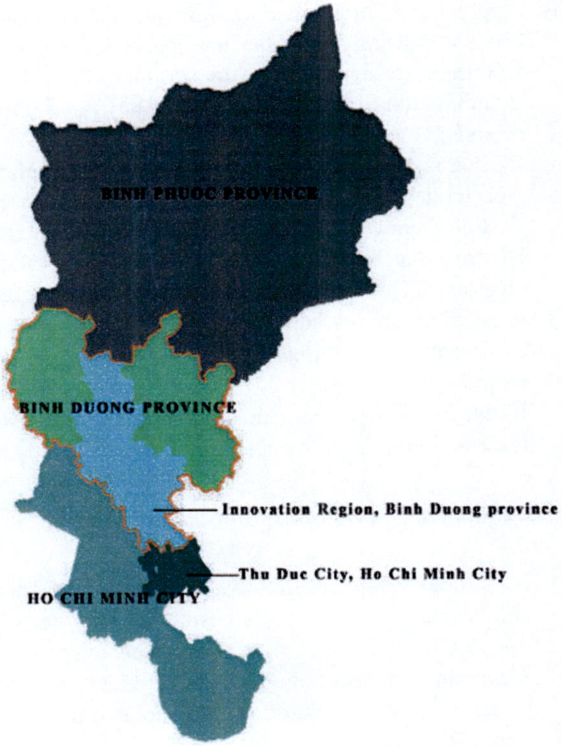

Fig. 31.2 Innovation Region, Binh Duong province. *Source Organized by This Research*

31.4.1.1 Some Major Directions for the Binh Duong Smart City Program in the Next Period

To fulfil the vision above, Binh Duong, especially the Innovation Region, focuses on the following five major directions, or layers (Binh Duong People's Committee, 2021). Each direction includes multiple important projects to directly solve the province's main challenges. These five key directions are also highlighted and integrated into the four domains of the Navigator of Smart City.

i. Urban Planning, Technical and Transportation Infrastructure

- Facilitate the implementation of the master plan, with special attention to the development of the Innovation Region, where Binh Duong's New City—Thu Dau Mot City is to be the centre of development before the expansion of science parks and industrial parks in the future.
- Apply transit-oriented development (TOD), which is a form of urban development that maximizes leisure space, residential living, and businesses within walking distances of public transportation. That is, TOD develops the urban areas in adjunction to the construction of bus rapid transit (BRT) routes that connect regions.
- Further vitalize the roadway system and solving the problem of traffic congestion. Accelerate the construction of north–south/east–west arterial roads and ring roads in Binh Duong to improve transportation connectivity.
- Devise a plan to develop new forms of transportation, such as waterways and port clusters on Dong Nai and Saigon rivers, railways, and smart logistics. The goal is to alleviate the pressure on the road system, reduce the transportation lead time and product costs for businesses, and meet urgent needs for the next development period.
- Invest in the development of high-quality infrastructure, telecommunications, information technology, and digital—broadband technology.

ii. Building a Culture of Innovation and Startup

- Focus on developing supportive mechanisms and policies, administrative reforms, and provision that supports businesses and motivates startup movement. Additionally, Binh Duong strives to enhance awareness of and education on Industry 4.0 models and technologies to government personnel, bringing their knowledge to the next level so that these officers can make more creative, responsive, and impactful decisions.
- Further apply Eindhoven's Triple Helix model, encouraging closer cooperation between the Government, Industry, and Universities to serve the public interests.
- Develop a strategy to promote a culture of innovation in society, expanding the depth and width of this movement by spreading the spirit of innovation to help Binh Duong deliver a breakthrough in socio-economic development in the next period.

iii. Comprehensive and Balanced Economic Development

- Promote the industrial development that goes hand in hand with science and technology advancement, the production progress linking with high-quality research and education, the quality of life, green space, and renewable energy for sustainable development.
- Promote the development of trade and services, e-commerce, international cooperation, further expansion of international partnerships (WTC, Horasis, ICF, etc.) to attract investment resources. Enhance collaboration in science and human resources. Develop the sectors of finance, banking, and industrial services. These tasks aim to support the current ecosystem, narrowing down the difference in percentage share between the industry sector and the service sector in the current economic structure of Binh Duong.
- Enhance agricultural development via improving rural areas through programmes such as smart village, smart agriculture, ecotourism, etc.

iv. Digital Transformation and Industry 4.0, Digital Government, E-commerce, Digital Social Economy

Develop a province-specific strategy for digital transformation that focuses on expediting e-government development, digitalization in the industry sector, smart manufacturing, smart factories, etc. Apply core Industry 4.0 technologies such as artificial intelligence, big data, blockchain, etc. Via these changes, develop new industrial park models in the digital age, and contribute to upgrading existing traditional industrial parks. The implementation of digital transformation may not demand much infrastructure but requires a shift in mindset and approaches to implementation.

In order to create the foundation for this digital transformation, Binh Duong focuses on leveraging provincial resources such as:

- Broadband infrastructure—the groundwork for digital economic and social development.
- Shared, integrated, and opened databases as a platform for all analytical applications and as a driving force for innovation and growth.
- Administrative reforms towards e-government and technology application in management to address the needs of the people and businesses.

v. Enhancing and Attracting Human Resources

Reinforce education and training on digital skills at all levels of education. Introduce STEM and STEAM into educational programmes to encourage creativity among the younger generations, preparing them for future digital transformation and innovation.

Universities and colleges will retain their important roles in fulfilling the requirements above, especially as distance learning has widely become the new normal post-COVID-19, thereby ushering in a new form of education. Under the direction of the provincial leaders, the government will further invest in and provide enabling conditions for these academic institutions to strengthen their collaboration with businesses and other institutions around the world, to engage in experience-sharing opportunities, to develop advanced research and training programmes, and to enhance online education—smart training courses to reflect the new global trends.

Develop a selective number of key vertical centres for excellence, with initial priority given to the centres for digital technology and smart manufacturing in order to support the digital transformation strategy and the digital economy development, as guided by the Politburo's Resolution No. 52. The centres for excellence will take a role in creating projects, research, and high-skilled jobs in novel fields, thereby attracting high-quality human resources to Binh Duong. Via collaborative research projects, these centres will facilitate the cooperation among researchers, scientists, and experts around the world.

31.5 The Innovation region—A New Focus of the Smart City Program in the Next Period

Binh Duong's Innovation Region is the focus of the Binh Duong Smart City Program in the next period (Binh Duong People's Committee, 2021). Covering Di An, Thuan An, Thu Dau Mot, Tan Uyen, Ben Cat, and Bau Bang, the Region will be expanded starting from the Smart Region planned under the 2016–2021 Smart City Program. Along with the current potentials and strengths of the province, the Innovation Region is expected to leverage and further promote the smart development strategy, with multi-layered and inter-sectoral solutions that are based on the framework of an innovative ecosystem and on a culture of innovation built-up over generations. This Region will create an attractive investment environment that step-by-step turns Binh Duong into a smart and innovative city. Promotion of the Innovation Region—the expanse of the Smart Region—continues to bring successes, empowering Binh Duong to join international organizations and to win worldwide recognition.

In order to meet the expectations above, the Innovation Region will follow the overall development directions of the Smart City Program, particularly the five major layers above, and will give an increased focus on the implementation of innovative projects. Among these projects, this paper presents the following two as the key elements for disruptive growth: one in the service sector and the other in the industry sector (Nguyen Viet Long, 2020b) Figs. 31.3 and 31.4.

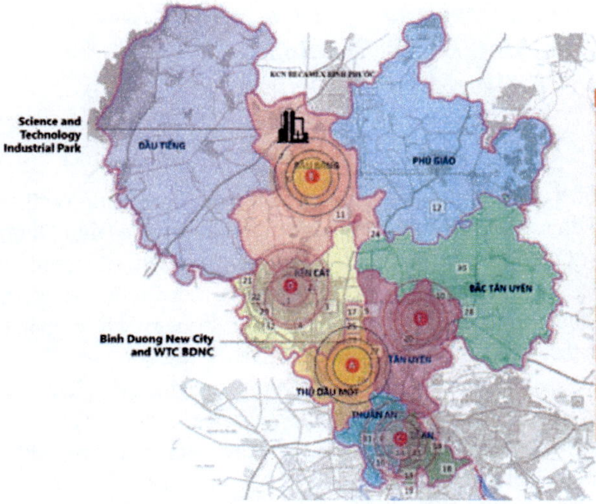

No.	Industrial Park	No.	Industrial Park
1	MY PHUOC 1	18	BINH AN
2	MY PHUOC 2	19	BINH DUONG
3	MY PHUOC 3	20	NAM TAN UYEN
4	THOI HOA	21	RACH BAP
5	BAU BANG	22	AN TAY
6	BAU BANG MR	23	MAPLETREE
7	CAY TRUONG	24	TAN BINH
8	VSIP 1	25	PHU TAN
9	VSIP 2	26	DAI DANG
10	VSIP 3	27	KIM HUY
11	LAI HUNG	28	DAT CUOC
12	VINH LAP	29	BD RIVERSIDE
13	SONG THAN 1	30	TAN LAP 1
14	SONG THAN 2	31	VIET HUONG 1
15	SONG THAN 3	32	VIET HUONG 2
16	DONG AN	33	TAN DONG HIEP A
17	DONG AN 2	34	TAN DONG HIEP B

Fig. 31.3 Master plan for the Innovation Region, which includes 2 breakthrough projects: Science and Technology Industrial Park and World Trade Center Binh Duong New City. *Source Becamex Corporation, 2021*

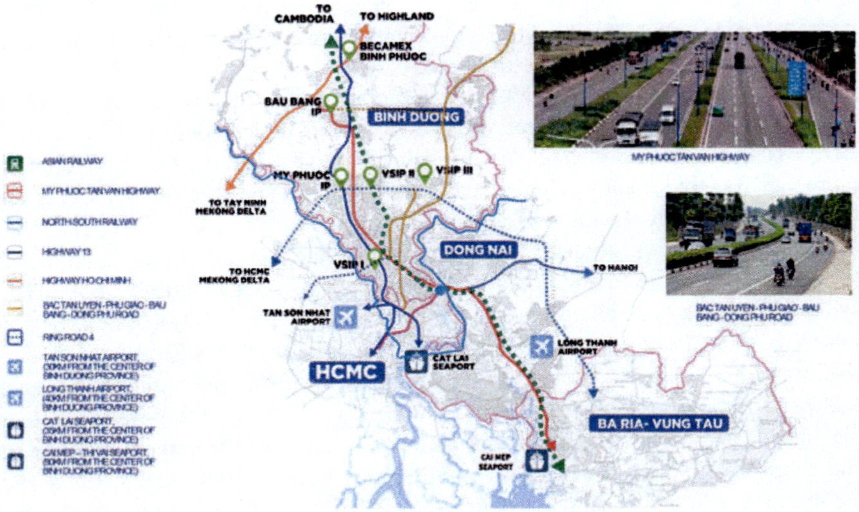

Fig. 31.4 Transportation system for regional connectivity and new forms of transportation. *Source Becamex corporation, 2021*

31.5.1 Boosting Services, Developing the Cohort of Three Cities (Thu Dau Mot, Di An, Thuan An), and Founding the World Trade Center Binh Duong New City

Binh Duong service sector has been growing, in terms of services serving both the industry, urban areas, and education. Building on such momentum, the Binh Duong Smart City Program must take more drastic actions to drive growth, especially in knowledge-based services, if Binh Duong was to become a more prosperous and sustainable society. After the period of industrialization, this strategy will be one of the most critical decisions that propel the province to thrive.

Currently, the three cities within the south of Binh Duong province (i.e. Thu Dau Mot, Di An, and Thuan An) display substantial development potentials and opportunities for disruptive growth. As a result, these cities will receive increased investment in order to attract abundant resources, setting forth to become civilized, modern, livable, green cities that provide high-quality services for the entire region. The three cities must quickly update the urban development master plan to enhance services and connectivity in the region and the Southern Key Economic Zone. This updated master plan must be based on the new directions that target environmental aspects, social and cultural issues, business environment, prevention of traffic congestion, upgrades of electrical systems, water supply and drainage, infrastructure, information technology, and broadband infrastructure.

In 2019, in Binh Duong New City, the combination of projects in Areas of complexity was recognized as the World Trade Center Binh Duong New City (WTC BDNC) and granted membership to the World Trade Centers Association (including more than 315 trade centres from 92 countries/regions). These achievements are the foundation for Binh Duong's disruptive development of services and international trade. WTC BDNC was strategically planned under the Smart City Program, expected to strongly contribute to an impressive and prestigious image of Binh Duong. The WTC BDNC is a symbolic project that generates value for the region, becomes an integrated global trade and service centre, attracts partners, investors, and skilled workforce, facilitates an ecosystem for international trade, and connects regional transportation system. Via such roles, the Center reinforces the development of the Science and Technology Industrial Park, innovates new disruptive technologies, and opens up innumerable business opportunities for industrial sectors, small and medium-sized enterprises. In particular, the WTC BDNC will be an important driving force that set in motion the momentum for Binh Duong's e-commerce development, which has been thriving due to the COVID-19 pandemic.

Besides, Becamex corporation and the private equity firm Warburg Pincus (USA) have also cooperated to develop a research area for cross-border e-commerce, in close proximity to the WTC BDNC. This area aims to promote trade and service development that focuses on e-commerce logistics in Binh Duong, and at the same time attract human resources and brainpower in this field. Along with the construction of bonded warehouses is the application of a technology platform with the participation of all related stakeholders. Cross-border e-commerce is one of the important pillars to promote and develop trade and digital economic development.

In the near future, the Binh Duong Smart City Program and the Innovation Region will continue to launch projects that promote the service sector. Concurrently, the key WTC BDNC project requires ample support and participation from the Triple Helix partners, especially the "government" partners. Specifically, the direct guidance of the local and central government is a prerequisite to implement many large-scale plans, such as design completion, development of services for international trade, admission of memberships, construction of commercial real estate, etc. It is also Binh Duong's central transit-oriented development (TOD) to facilitate the development of BRT routes and other TODs along My Phuoc Tan Van roadway, while connecting the Northern part of the province to the Science and Technology Industrial Park.

31.5.2 Science and Technology Industrial Park and Advanced Manufacturing Development

Science and Technology Industrial Park will be a long-term and groundbreaking project designed to assist in the formation of an innovative ecosystem and to attract skilled human resources along with technology investment in the strategic areas of production in the region. This industrial park will create a linkage between industries and areas along the main transport axis of My Phuoc Tan Van—Bau Bang—New City.

To fulfil the vision above, under the direction of the provincial government, Binh Duong's Becamex corporation, in collaboration with Dutch, South Korean, and Singaporean partners, proposed that the Science and Technology Industrial Park includes the following: (1) an academic zone for human resource training and technology transfer; (2) a zone for technology industrial parks hosting innovative and startup companies; (3) a business zone for offices and research institutions—technology laboratories; (4) a high-end urban-service zone that provides a high-quality environment for scientists and experts; and (5) a cultural and recreational zone for the community.

Besides synchronous and modern infrastructure, the success of the Science and Technology Industrial Park stems from incentive policies, modern governance systems, and enhanced innovative capabilities facilitated by collaborative efforts among key regional stakeholders, with the Triple Helix framework serving as a pivotal foundation. Additionally, Binh Duong will form a system that fosters the region's strategic industries and provides support to businesses, forming incubators and attracting technology companies.

Capitalizing on its abundance of local production facilities, which serves as a competitive advantage, Binh Duong and the Science & Technology Industrial Park are actively researching to develop smart and advanced manufacturing. That is, to cooperate with major corporations that can offer modern and advanced designs to facilitate transformation, enhance production capacities, and tailor human resources sufficiently competent for Industry 4.0. The smart manufacturing includes digitization, automation, artificial intelligence, the Internet of Things (IoT), training 4.0, etc. The visionary effort above will open up potential markets and give rise to new investments in high-tech production for Binh Duong. Recently, in the New City, the smart manufacturing centre was formed. It is attracting active participation from provincial partners that are experts in the fields. These experts include, but are not limited to, Amazon Web Services, Philips, Schneider, Intel, Bosch, TMA, ITRI, National University of Singapore (NUS), Vietnam National University-Ho Chi Minh, etc.

Together with the current WTC BDNC, the long-term project Science and Technology Industrial Park is the groundbreaking and visionary strategy that elevates Binh Duong's identity, sufficing the provincial requirements for the next development stages in the future.

31.6 Conclusion

This paper unfolds the breakthrough strategy that is transforming a developing country's province in the Industry 4.0 era, highlighting the untraditional successful story like no other. After more than 20 years of successful industrialization, amid the science and technology boom by Industry 4.0 in 2016, Binh Duong teamed up with the city of Eindhoven-Netherlands to implement the groundbreaking Smart City Program that targets better services, high-tech production, middle-income-trap solution, and development of livable cities. Using the Triple Helix model that places people and knowledge at its core and positions cooperation as its guiding principle to promote innovation, the Smart City Program is the main contribution to the breakthrough progress of Binh Duong. It laid the foundation that has enabled the province to reach a new height in the international arena.

The Fourth Industrial Revolution, the aftermath of COVID-19, as well as conflicts and geopolitical changes in the world, have ushered in a new era, especially in industrial development and digital applications. Against this backdrop, the 2022–2030 Smart City Program is being finalized to provide strategic actions for disruptive growth. The Program especially focuses on developing the Innovation Region containing many innovative projects, such as the Science and Technology Industrial Park and the World Trade Center Binh Duong New City. The Smart City Program is expected to further contribute and initiate a new driving force for Binh Duong's socio-economic development. The Program is set to enhance the province's added value and international competitiveness while reaching for a knowledge-based economy, digital economy, and innovation, turning Binh Duong into a smart city in the Industry 4.0 era.

References

Bibri, S. E., & Krogstie, J. (2017). Smart Sustainable Cities of the Future: An Extensive Interdisciplinary Literature Review. *Sustainable Cities and Society, 31*, 183–212.

Binh Duong People's Committee - Executive Committee of Smart City Program. (2020). *Plan 09/ KH-BĐH, 22/01/2020: Plan 2020 of Binh Duong Smart City Program.*

Binh Duong People's Committee. (2021). *Announcement 01/TB-UBND on Conclusion of Binh Duong Provincial People's Committee Chairman on the Binh Duong Innovation Region.*

Bukht, R., Heeks, R. (2017).Defining, Conceptualising and Measuring the Digital Economy.*Centre for Development Informatics, Global Development Institute, SEED,* Working paper No. 68. SSRN 3431732.

Cocchia, A. (2014). Smart and Digital City a Systematic Literature Review. In R. P. Dameri & C. Rosenthal-Sabroux (Eds.), *Smart city how to create public and economic value with high technology in urban space* (pp. 13–43). Springer.

Edwards, Rosalind & Holland, Janet. (2013). What is Qualitative Interviewing?.*London: Bloomsbury.*

Etzkowitz, H. (2008). The Triple Helix: University-Industry-Government Innovation in Action. *Routledge.*

Geertz, C. (1973). *The interpretation of cultures: Selected essays.* Basic Books.

Hollands, R. G. (2008). Will the Real Smart City Please Stand Up? Intelligent, Progressive or Entrepreneurial? *City, 12*(3), 303–320.

Ivaldi, E., Penco, L., & Isola, G. (2020). Smart Sustainable Cities and the Urban Knowledge-Based Economy: A NUTS3 Level Analysis. *Social Indicators Research, 150*, 45–72.

Le Thi Hoa., Nguyen Viet Long. (2011). Psychology and Communication. *Vietnam National University Ho Chi Minh City Publishing House.*

Le Thi Hoa., Nguyen Viet Long. (2012). Psychology and Management. *Vietnam National University Ho Chi Minh City Publishing House.*

McKinsey & Company. (2020). Coronavirus' business impact: Evolving perspective. *McKinsey.*

Mittal et al. (2017): Mittal, S., Kahn, M. A., Romero, D., & Wuest, T. (2017). Smart Manufacturing: Characteristics, Technologies and Enabling Factors. *Proceedings of the Institution of Mechanical Engineers, Part B: Journal of Engineering Manufacture.*

Nguyen Viet Long., Doan Thi Anh Ngoc. (2020). Development of the center for public administrative services in Binh Duong province, towards the building of a smart city. *State Management Review - National Academy of Public Administration.*

Nguyen Viet Long., Nguyen Thi Thu Ha. (2020). The Triple Helix: The key to innovation, science and technology, and development of knowledge-based society and economy. *Journal of Political Science - Ho Chi Minh National Academy of Politics - Academy of Politics Region II, 01/2020.*

Nguyen Viet Long. (2015). Management Challenges of International Firms in the Vietnamese Cultural Context. *University Paris Dauphine.*

Nguyen Viet Long. (2019a). The groundbreaking strategic socio-economic development—Smart city planning: A case study of Binh Duong. *Journal of Architecture and Planning* (Taiwan), 19(2).

Nguyen Viet Long. (2019b). Vision and criteria for platform development to move forwards digital economy: Approach by Intelligent Community Forum. *Journal of Political Science - Ho Chi Minh National Academy of Politics - Academy of Politics Region II*, 06/2019b.

Nguyen Viet Long. (2020a). Development of a Smart Society: Case Study of Eindhoven—The Netherlands. *Journal of Political Science - Ho Chi Minh National Academy of Politics - Academy of Politics Region II*, 03/2020a.

Nguyen Viet Long. (2020b). Strategic planning vision and some key directions for economic breakthroughs during 2021–2025 period: A study of Binh Duong-Vietnam and Smart City Program. *Journal of Architecture and Planning* (Taiwan), 20(1).

OECD (2001). Competencies for the Knowledge Economy. *Organization for Economic Co-operation and Development.*

Pfohl HC, Yahsi B, Kuznaz T (2015). The impact of Industry 4.0 on the supply chain. *C.M.R.Wolfgang Kersten, Thorsten Blecker (ed) Proceedings of the Hamburg international conference of logistic (HICL)-20.* Hamburgo: innovations and strategies for logistics an supply chain.

Powell, W., & Snellman, K. (2004). The Knowledge Economy. *Annual Review of Sociology, 30*(1), 199–220.

Schwab, K. (2016). The fourth industrial revolution. *The World Economic Forum.*

Scott, J. (2006). *Documentary Research.* Sage Publications Ltd.

Unger, R. (2019). *The Knowledge Economy.* Verso.

Van Gijzel, R. (2016). A city creating the future. *Lecturis.*

Wolter, M. I., A. Mönnig, M. Hummel, C. Schneemann, E. Weber, G. Zika, R. Helmrich, T. Maier and C Neuber-Pohl (2015). Industry 4.0 and the consequences for labour market and economy Scenario calculations in line with the BIBB-IAB qualifications and occupational field projections. *IAB Forschungsbericht.*

Chapter 32
A Future Design for the Sustainable Urban Renewal of Manila Megacity of the Philippines

Ria Liza Centeno-Canlas, Joseph D. Retumban, and Custer C. Deocaris

Abstract Manila Megacity is also known as the Metro Manila in Asia and was once described during the pre-war era as the "Paris of Asia". The historical narrative of its unprecedented post-war urbanization led to the proliferation of informal settlers having inadequate or no basic services. An increase in urban population aggravated by in-migration from the rural areas created a myriad of complex socio-economic, health, and environmental problems. This chapter first traces the evolution of the city, from its colonizer up until the independence of the nation and discusses the initiatives by the local government for urban regeneration under the Mega Manila Dream Plan. Second, we developed a framework for a renewed and sustainable urban environment that incorporates the hard and soft infrastructures (physical and organizational structure). Through Delphi methodology, stakeholders have identified 12 Key Elements (KEs) contextualized to the urban renewal of Metro Manila: *Kaunlaran* (Progress and Development), *Kabuhayan* (Livelihood), *Kapayapaan* (Peace), *Kaayusan* (Order), *Kalinisan* (Cleanliness), *Kapaligiran* (Environment), *Kalusugan* (Health), *Karunungan* (Education), *Kultura* (Culture), *Kabataan* (Youth), *Kalinga* (Elderly) and *Kaisahan* (Unity). Sub-indicators of the renewed and sustainable urban environment were crafted based on these 12 KEs. Lastly, based on the KEs and their sub-indicators, we propose a future framework by integrating other methodological approaches, including data analytics, simulation, and meta-analysis. Some comments are provided on the application of a future framework to assess the urbanization and revival of the megacity.

R. L. Centeno-Canlas (✉)
Center for Innovation, National University, Manila, Philippines
e-mail: rlccanlas@national-u.edu.ph

J. D. Retumban
Center for Resilient Philippines, National University, Manila, Philippines
e-mail: jdretumban@national-u.edu.ph

C. C. Deocaris
Department of Science and Technology, Philippine Nuclear Research Institute, Diliman, Quezon City, Philippines
e-mail: ccdeocaris@pnri.dost.gov.ph

B. R. K. Sinha (ed.), *Urban Dynamics, Environment and Health*,
https://doi.org/10.1007/978-981-99-5744-6_32

Keywords Delphi methodology · Development policy research · Informal settlers · Megacities · Meta-analysis · Paris of Asia · Simulation · Sustainable urban environment · Urban planning

32.1 Introduction

After the end of the Spanish-American War, Spain ceded the Philippines and several other territories to the United States for $20 million as part of the Treaty of Paris on 10 December 1898 (Schurman et al., 1900). With its new "bounty", the City of Manila was transformed from the political-religious-economic nucleus of the Spanish Empire in South East Asia into a colonial capital by the United States. From 1898 to 1946, the City of Manila, formerly Spain's "*Insigne y Siempre Leal Ciudad*" (Distinguished and Ever Loyal City) (Gomà, 2012), was morphed into a monument of the US' benevolent assimilation of the entire nation.

The United States regarded Manila as its "grand city from another era fallen on hard times" (McCoy, 1985). American rule officially started with the hoisting of the US flag on 13 August 1898. This symbolic event was at Fort Santiago, a famous 1593 citadel built by Spanish navigator and governor Miguel López de Legazpi, located at the heart of the 0.67-square kilometre historic walled area of Intramuros (Latin = "within the walls"). Intramuros was then one of the most celebrated colonial structures of the seventeenth and eighteenth centuries, with a grid plan configuration, arranged about a large-sized urban space (*plaza mayor*) and grand ecclesiastical edifices, e.g. the San Augustin complex of church and monastery. Fort Santiago became the headquarters of the Philippine Division of the United States Army in 1901 (Verdejo et al., 2014).

As with its urban plan, a significant influence in Manila's re-design was the "City Beautiful Movement" of Daniel Hudson Burnham. Burnham launched the movement during the World's Columbian Exposition held in Chicago in 1893 to celebrate the 400th anniversary of Christopher Columbus' discovery of the New World in 1492 (Burnham, Bennett, & Bennett, 1909). The City Beautiful Movement no longer views a city as a mere symbol of economic development and industrialization. Instead, the urban planners during this era prioritize enhancing the aesthetic environment of the city inhabitants (Freestone, 2019). Burnham recommended during his visit to the Philippines in 1904–1905 that development should revolve around the shoreline of Manila Bay which he thought was the most aesthetically pleasing feature of the city. Taking notice of the Pasig River running through the city, Burnham aligned his plan for the new Manila to be the next Paris or Italy. Quoting Burnham, "while possessing the bay of Naples, the winding river of Paris, and the canals of Venice, Manila has before it an opportunity unique in the history of modern times, the opportunity to create a unified city equal to the greatest of the Western world with the unparalleled and priceless addition of a tropical setting". Not only were these bodies of water seen as romanticized transportation routes and elements of a beautiful city but Burnham also suggested that a city plan should revolve around these bodies of water to help mitigate the heat of the tropical climate (Hines, 1972).

Burnham's plans encouraged establishing more expansive parklands because these areas are salient for recreation and for connecting the city inhabitants to nature. Burnham proposed neoclassical public buildings within the city, radial, and broad boulevards as part of the Manila Plan of 1905 (Gamboa et al., 2019). From 1906 to 1914, significant edifices were erected, such as the Philippine General Hospital, Manila City Aquarium, Paco Train Station, Tondo Intermediate School, Normal School, University Hall of the University of the Philippines, and YMCA Building. Along the seafront of Manila Bay, as recommended by Burnham in 1905, the Elk's Club, Army and Navy Club, and Manila Hotel were among the urban amenities erected to support American expatriates. These infrastructures were also notable from Burnham's urban plans for Cleveland, San Francisco, and Washington, D.C. (Hines, 2009).

Following the Second World War, more than 90% of the world's architectural jewel, Intramuros was reduced to rubble together with the rest of Manila. The "Battle for Manila" from 3 February to 3 March 1945, emerged as one the deadliest, most atrocity-stained, and physically destructive battles in the world. Filipinos were first-hand witnesses to the Japanese's "*senko-seisaku*" (kill all, burn all, destroy all) atrocities (Mathias, 1996; Steinberg, 2008). After the war and upon the declaration of Philippine independence on 4 July 1946, the rehabilitation efforts of Manila became a top priority of the newly instituted Philippine government. The National Urban Planning Commission was established in 1946 and tasked to rehabilitate and propose a new plan for the city. In 1949, the Revised Charter of the City of Manila was enacted (Republic Act No. 409) giving more autonomy to Manila's city government to implement the reconstruction plan. Under the Charter, Manila was divided into 896 administrative divisions called *barangays*, the native Filipino term for villages, districts, or wards. The *barangay* chairperson and seven councillors that head each *barangay* are elected posts. Manila was also subdivided into six legislative districts, which correspond to the City's political representation in the Philippines' House of Representatives, the country's national legislative body. For administrative convenience, all the *barangays* in Manila are grouped into 100 zones and further grouped into 16 administrative districts.

During the past decades, Metro Manila (also called the National Capital Region, NCR) faced immense sustainability challenges due to rapid urbanization. Based on the 2020 census (Philippine Statistics Authority, 2020), Metro Manila has a total population of 13,484,462, which accounts for about 12.37% of the Philippine total population. Metro Manila is composed of 16 cities and one municipality. Quezon City (2,960,048 persons), the City of Manila (1,846,513 persons), and the City of Caloocan (1,661,584 persons) are the top three highly urbanized cities in NCR with the largest population, respectively. Moreover, Metro Manila was the most densely populated region in the Philippines with a population density of 21,765 persons per square kilometre. Among the 16 cities in Metro Manila, the City of Manila has the highest population density of 73,920 persons per square kilometre of land. With an official population of more than 1.85 million, it represents 13.7% of the total urban population and about 1.69% of the Philippines' national population. The City of Mandaluyong (45,830 persons per square kilometre) and Pasay City (31, 543 persons per square kilometre) are ranked 2nd and 3rd respectively, as the cities with the highest population densities in NCR. According to The Rockefeller Foundation's Informal City Dialogues, about 20 to 35% of residents were informal settlers. "Informal settlers" refer to families living in slums with no legal tenure over the land they occupy. Informal settlers densely inhabit private and public properties for decades, many at locations vulnerable to natural and artificial disasters, e.g. along the esteros (estuarial creeks), under the bridges, and adjacent to railways (Tomeldan et al., 2014). Now, the absence of adequate and formalized housing, sanitation, and clean water, aggravated by a lack of access to education, public health provision, and skilled employment shape the day-to-day lives of millions of Manileños (Shatkin, 2004). Metro Manila also faces a solid waste problem, water, and air pollution. Based on the 2020 projected waste generation, the city produced 3,466,469 metric tonnes of garbage, equivalently almost twice the volume of garbage produced by the residents in other areas. This can be accounted for by the estimated weighted average per capita in Metro Manila of almost 0.7 kilos per day (Mayuga, 2021).

Calling to mind the modern day state of Intramuros, the walled city was reconstructed gradually in the decades that followed and formed the theatre for the 1998 centennial celebrations of Philippine independence. Intramuros has become one of Manila's most popular attractions, with visitors walking along its restored walls (Mason & Istvandity, 2018). However, outside of its elegant fortitude, Manila's present picture is truly far-fetched from its "City Beautiful" image as its local government faces multifaceted and complex challenges like in other developing economies of South East Asia. Aside from managing environmental health and basic services for urban dwellers, the local government struggles to incorporate environmental issues into urban planning to reduce the risks of further ecological degradation and to uphold peace and order. To mitigate the impact of rapid urban expansion, both the municipal and national governments have financed infrastructure, transportation, and social welfare services with approval from the National Economic and Development Authority (NEDA). The Roadmap for Transport Infrastructure Development for Metro Manila and Surrounding Areas was known as the "Mega Manila Dream Plan" (Estoque, 2017).

To establish a holistic framework for the sustainable urban renewal of Metro Manila, the authors adopt an interdisciplinary lens considering the socio-economic and cultural perspectives and sustainability science. Through a modified Delphi methodology, 12 Key Elements (KEs) were developed, namely, *Kaunlaran* (Progress & Development), *Kabuhayan* (Livelihood), *Kapayapaan* (Peace), *Kaayusan* (Order), *Kalinisan* (Cleanliness), *Kapaligiran* (Environment), *Kalusugan* (Health), *Karunungan* (Education), *Kalinangan* (Culture), *Kabataan* (Youth), *Kalinga* (Elderly), and *Kaisahan* (Unity) as indexes of an urban-renewed and sustainable environment. With a "dying Manila" as a backdrop, the people-centred Future Design Framework proposes an "old renewal" pathway by asking the crucial questions: *How do we describe the present urban environment? What indicators can be used to help envision a sustainable and urban-renewed environment for the future?*

32.2 Methodology

32.2.1 Study Design and Sampling Strategy

The study area was Intramuros which is geographically situated in the City of Manila with coordinates 14.5896° N, 120.9747° E (see Fig. 32.1). To accurately and objectively evaluate the condition of Intramuros, it is essential to measure the physical and organizational structure in terms of the present practices and patterns of various social aspects. Using a modified Delphi approach, this paper established the 12 KEs for an urban-renewed and sustainable environment.

Phase 1 consists of micro-studies for each identified KE in Intramuros. Table 32.1 presents the profiling of the pre-test study areas that applied the 12 KEs for urban renewal. A sample micro-study on *Kalinangan* is shown in Fig. 32.2.

Phase 2 of the study is a macro-study. In this phase, the 12 KEs were defined and sub-indices (or indicators) of urban renewal were determined. The general framework of the indices together with its measures applies to any parts and parcel of any city under study. The result of the study is intended for possible policy implications for local government units and other government units/agencies, as well as the development of intervention programmes. Different stakeholders will benefit from any outcome derived from a city analysis utilizing the said indexes of a sustainable urban-renewed environment. Thus, the list of indexes was the research instrument whenever employed in an actual survey.

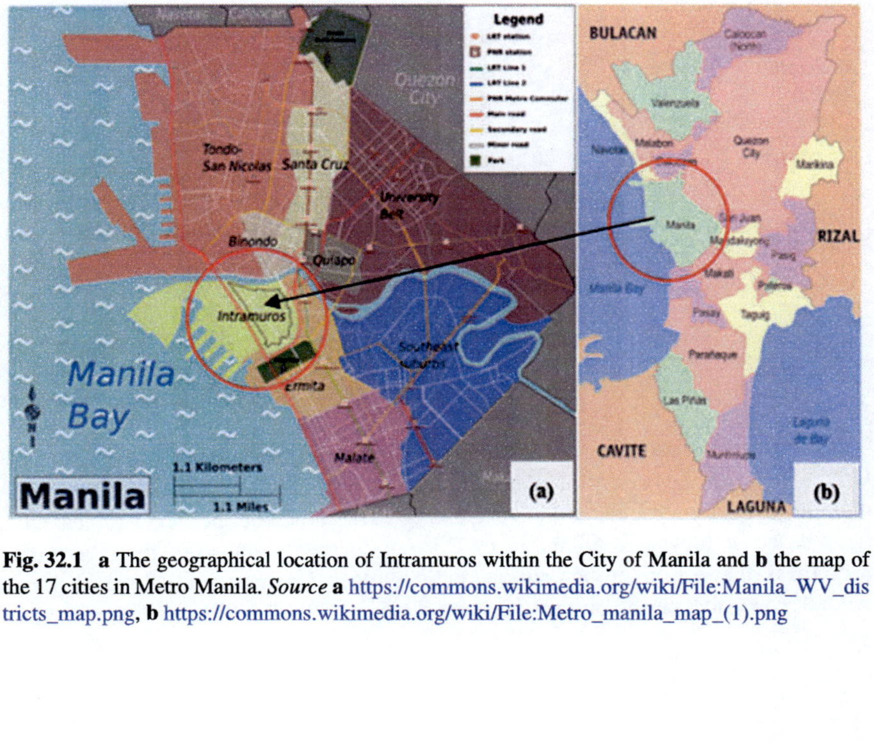

Fig. 32.1 **a** The geographical location of Intramuros within the City of Manila and **b** the map of the 17 cities in Metro Manila. *Source* **a** https://commons.wikimedia.org/wiki/File:Manila_WV_districts_map.png, **b** https://commons.wikimedia.org/wiki/File:Metro_manila_map_(1).png

Table 32.1 Profiling of the socio-demographic, economic, and institutional Aspects of the 12 KEs

Key elements	Pre-Test Area	Respondents/Subject	Proposed Indicators
Kaunlaran (Progress & Development)	Binondo, Manila	10 business establishments	Corporate social responsibility programme
Kabuhayan (Livelihood)	Luneta Park, Manila	30 individuals from different livelihood opportunities	Stability of business and livelihood opportunities
Kapayapaan (Peace)	Paco Park, Manila	31 security members from four security organizations	Community policing
Kalinga (Elderly)	Sampaloc, Manila	107 senior citizens	Satisfaction patterns of benefits and privileges
Kalinisan (Cleanliness)	Luneta Park, Manila	21 foodservice establishments	Food safety policies and practices
Kapaligiran (Environment)	Navotas, Manila	10 establishments comprising offices/firms, food chains, government agencies, schools, banks, and households	Solid waste disposal practices
Kalinangan (Culture)	Paco, Manila	30 institutions collecting and housing cultural artefacts including universities, museums, churches, and libraries	Cultural heritage preservation programmes
Kabataan (Youth)	Malabon, Manila	30 youth respondents from different *barangays* from four distinct groups such as *Sangguniang Kabataan*, fraternity groups, religious groups, and those not affiliated with any youth organizations	The contemporary societal interest of youth
Kaayusan (Order)	Binondo, Manila	10 building owners	Road infrastructure and transportation and traffic management
Kalusugan (Health)	PacoPark, Manila	30 health infrastructure	Health infrastructure status
Karunungan (Education)	Recto, Manila	5 universities	Program evaluation and employability of students
Kaisahan (Unity)	Malate, Manila	30 individuals from *barangays* and establishments	Governance and unity in project management of *barangays*

The research paradigm is illustrated in Fig. 32.3. The Research for a Sustainable Urban-Renewed Environment (ReASURE) was anchored to five pillars: political, cultural, economic, social, and environmental. The indicators per KE were categorized based on these five pillars. If a KE is not deemed sustainable, then there is a need for further action through policy modification and programmatic interventions. After a reasonable period of implementing remedial activities, the city is expected to

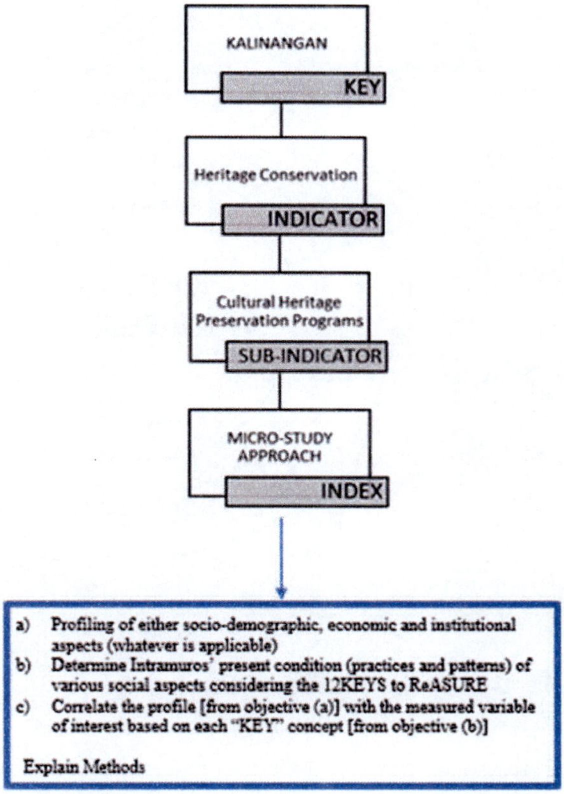

Fig. 32.2 Sample micro-study on the *Kalinangan*

be in a condition termed as "*a sustainable urban-renewed environment*". Thus, the outcomes can be measured.

The concept of identifying indicators and sub-indicators is similar to a work breakdown structure (WBS). A complex project is made manageable by breaking it down into individual components. Such a structure defines tasks that can be completed independently of other tasks and facilitates resource allocation, assignment of responsibilities, and measurement. In this chapter, the authors term it an indicator-breakdown structure (IBS). The whole concept of urban renewal will be measured by the 12 KEs each associated with an indicator of a sustainable urban-renewed environment. Each indicator also had sub-indicators or a measurable index. A typical 4-level IBS is illustrated in Fig. 32.4 using sample indicators for the KE *Kalinangan.*

Some indices can be measured directly from available secondary data or publications or through direct measures (i.e. per cent of population, area in hectares, number of buildings, the volume of wastes, etc.). In most cases, the indices need a more complex or longer process to measure (i.e. the effectiveness of a conservation programme, patterns of behaviour, food safety practices, etc.) and correlate with socio-demographic-economic profiles.

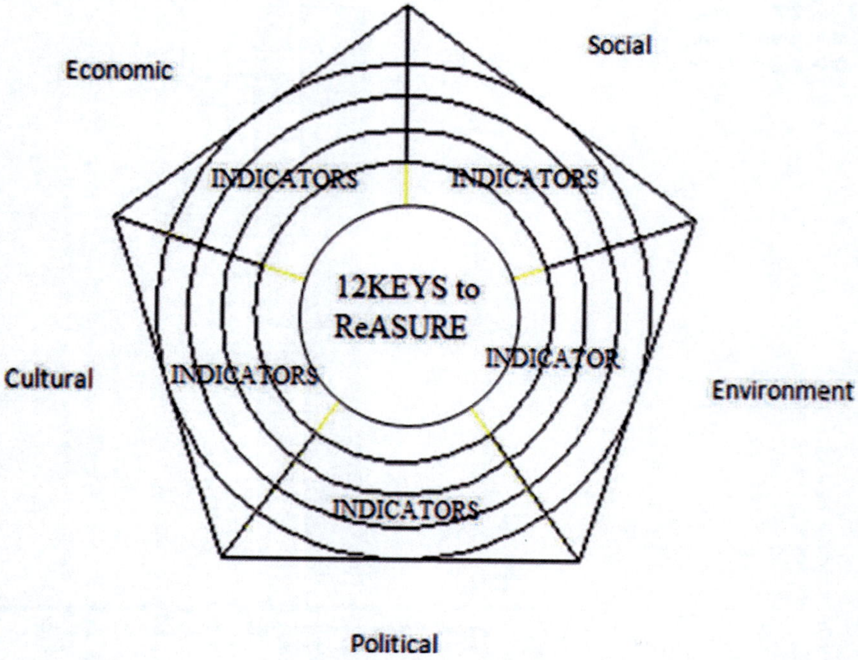

Fig. 32.3 The research framework of ReASURE

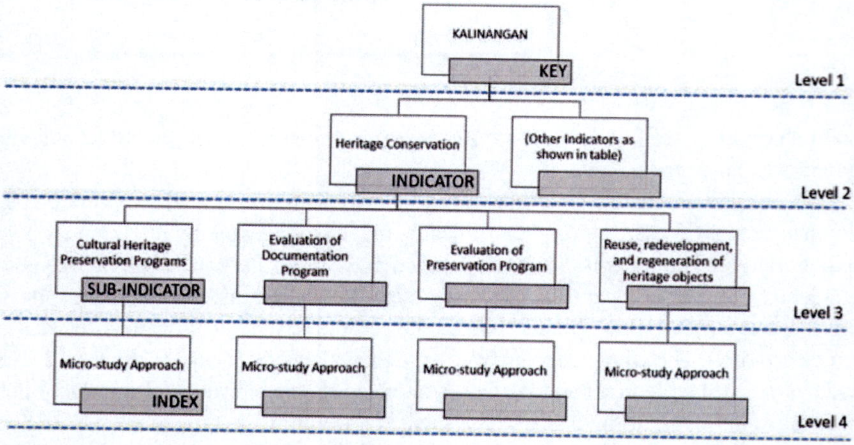

Fig. 32.4 An example of a four-level indicator-breakdown structure (IBS) for *Kalinangan*

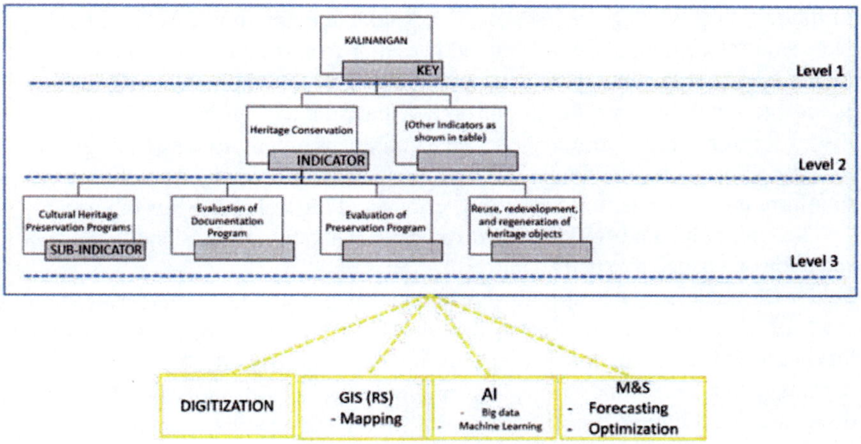

Fig. 32.5 A future design framework

Phase 3 of the study is the future design framework (see Fig. 32.5). The framework aims to shorten the assessment and decision-making process through appropriate technology adoption. Various technologies are available which can singly, or in combination, be utilized to measure an indicator. The data from which an indicator can be measured are of various types. Some are numerical and can be extracted from secondary publications and other government documents (i.e. census, etc.). Others are visual, specifically the hard infrastructure (i.e. buildings, flora, fauna, etc.) and particle type (i.e. pollution, particulate matter, bacteria, viruses, etc.). Some of the technology that may be employed are wireless technology: radio frequency (RF) and infrared (IR) waves for SMS and Wifi; geospatial information system (GIS), including remote sensing and mapping of land use; artificial intelligence (AI), including deep learning, internet of things (IoT), decision support system, and big data analytics; and modelling and simulation including forecasting/prediction and optimization.

32.2.2 A Modified Delphi Approach

The Delphi method is an iterative process that employs a structured progression of repeated rounds of voting. The method is a commonly used and accepted approach for reaching a consensus regarding practical knowledge from experts in their respective fields (Eubank et al., 2016). The Delphi method was originally developed in the 1950s at the RAND Corporation to apply expert opinion in choosing the best industrial target system in the United States (Dalkey & Helmer, 1963). The methodology is a group communication process for in-depth analysis of specific issues to set goals (Singg & Webbe, 1979), examine policies (Sandrey & Bulger, 2008), and forecast occurrences of future events (Gallego & Bueno, 2014; Kauko & Palmaroos, 2014). According to Delbecq et al. (1975), the Delphi method can help to accomplish the following goals: (1) to identify or develop a variety of potential programme alternatives; (2) to examine or uncover underlying presumptions or data that produce various conclusions; (3) to discover information that may generate a consensus among the response group; (4) to correlate well-informed opinions on a subject covering many disciplines; and (5) to inform the respondent group of the various and interrelated facets of a topic.

The Delphi method typically involves three evaluation rounds and begins with a selection of a panel of experts. During the first round, an open-ended questionnaire is given to the panel to gather specific information about a subject or content area (Custer et al., 1999). The panel is then asked to review the items summarized by the investigators based on the information gathered during the second round. The rank order of the items is usually determined to establish preliminary priorities and areas of agreement/disagreement (Ludwig, 1994). In the third round, the panellists receive a questionnaire containing the summarized items and ratings from the previous round. They are requested to recalibrate their positions on specific topics and explain their rationale for remaining outside the consensus (Pfeiffer, 1968). A fourth and final round may be necessary for the panellists to revise judgments after given the summarized items and ratings from the previous rounds (Delbecq et al., 1975; Ludwig, 1994). It should be noted that the number of iterations of the procedure may vary depending on the degree of consensus that the study demands. According to Dalkey (1972), the method employs three features: (1) anonymous response—members of the panel utilize a formal questionnaire to relay their opinion; (2) iteration and controlled feedback—interaction is conducted in several iterations, and feedback is regulated in between rounds; and (3) statistical group response—the panel's opinion on the last round is taken as the collective sum of the individual opinions of the members of the panel.

A modified version of the Delphi method employs the same procedure. However, the difference is that the methodology starts with a structured questionnaire based on an extensive literature review (Kerlinger, 1973). The questionnaire contains preselected items from diverse sources, including associated competency profiles, interviews with subject-matter experts, and published literature. This modified Delphi method is more advantageous than the standard one because it also ensures panel anonymity and thereby allowing unhindered expression of opinions. The approach also reduces bias by lowering coercion incidents attributed to the influence of dominant personalities and panel members' status (Judd, 1972; Jones & Hunter, 1995; Adams, 2001; Hsu & Sanford, 2007). Studies on the modified Delphi method (Gustafon et al., 1973; Graefe & Armstrong, 2001) demonstrated its efficiency.

The study utilizes a modified Delphi approach (Canlas et al., 2013) which includes: (1) descriptive research describing the economic, demographic, institutional, and industrial profile, including patterns and practices, (2) explanatory/exploratory research analysing the effects of the profile and patterns of current practices, and (3) triangulation method describing the combination of several qualitative methods or the combination of a qualitative study following a qualitative one which provides validation for the qualitative findings.

A structured questionnaire was developed. The validity test was performed qualitatively and quantitatively. Four field experts were assigned per KE as panel members. The qualitative review allowed feedback on the instrument. The panel independently assessed each item in the instrument as (1) essential; (2) useful but not essential; and (3) not necessary (Cooper & Schindler, 2003). Content validity was employed to test the validity by using the content validity ratio (CVR) formula given in (1), where n is the number of essentials and N is the total number of the panel. A minimum CVR value of 0.75 was considered statistically significant. The objective was to determine the ability of the research instrument to measure what it purported to measure. The test was conducted after the initial draft of the instrument was edited following the results of the qualitative review.

$$CVR = \left(\frac{n_e - \frac{N}{2}}{\frac{N}{2}} \right) \quad (1)$$

The reliability test of the instrument was through pilot testing on selected institutions. A pilot test was conducted to detect any weaknesses in the instrument and to provide proxy data for the probable samples. It should, therefore, draw subjects from the target population and simulate the procedures and protocols designated for data collection. The size of the pilot group ranged from 25 to 100 subjects, depending on the method tested. The respondents do not have to be statistically selected (Cooper & Schindler, 2003).

32.3 Data Analysis

The success of the measurement relies on the criteria for the selection of respondents/subjects. The number of samples must be more than 400. The sampling technique was purposive. The acquired data were analysed statistically:

Descriptive statistics—The data collected were processed using the necessary statistical tool and with the aid of SPSS. The data were presented in graphs and tables.

Inferential statistics—The results of the analysis were interpreted and inferred based on related readings. The explanatory powers of each determinant were explained.

Triangulation—To have an in-depth analysis, certain variables were treated qualitatively and combining it with quantitative analysis, thereby making a quasi-treatment of data.

Correlational Analysis—The correlational method, as one of the many types of explanatory methods, defined the relationship between two or more variables which was used in determining the relationships between profiles of selected respondents/subjects and patterns. Spearman's Rho Correlation Coefficient was used to determine the relationship between the profile of selected business establishments in Intramuros, Manila, and the CSR patterns.

32.4 Results and Discussion

32.4.1 On the 12 KEs of Urban Renewal

Hard infrastructures are said to be tangible and readily observable. In contrast, soft infrastructures are regarded as intangible and difficult to define and measure. The latter cannot always be described by quantitative indicators and are often assessed subjectively. Examples of hard infrastructures are the provision of basic utilities such as water, gas, and electricity; waste collection facilities; transport provision (e.g. roads, rail, air); community facilities and public buildings, such as libraries, town halls, recreation and cultural facilities, meeting rooms, childcare facilities, office spaces for community groups, community and neighbourhood centres, and spaces for young and older people. For soft infrastructures, examples are the provision of human services such as equitable, accessible, and appropriate community services; individual skills, knowledge, and abilities; and local networks, relationships, and collaborative responses.

The need to operationalize our definitions of the sustainable urban-renewed environment is integral in the context of the study. Urbanization is generally defined as “growth” in the population of heterogeneous people engaging in complex activities impacting on social, economic, cultural, political, and environmental conditions of a district or city. Sustainability implies the development, implementation, and evaluation of programmes and policies ensuring the quality of life of current and future generations. Merging these two (words) concepts, thus, equates to a sustainable urban-renewed environment.

Since the indicators provide a vehicle for understanding and addressing community issues from a holistic and outcomes-oriented perspective, we then derived the Key Elements (KEs) to quantify their contribution to a certain sustainability theme. KEs have become a widely used tool to measure the status and progress of the variable intended to be improved (Swain & Hollar, 2003). As in this chapter, the identified objectives of measurement indicators (or basis in the identification of indexes of each key) are in monitoring the progress, establishing global and regional norms, and measuring impacts. The indices in each KE were crafted based on the results of the micro-studies.

After implementing a modified Delphi technique in the context of Metro Manila, 12 KEs were identified as measures of a sustainable and urban-renewed environment. Using the concepts of hard and soft infrastructures, objectives of index measurements, and sustainable urban renewal, the following KEs were defined as shown in Table 1:

(1) *Kaunlaran* (Progress and Development), which refers to progress and development brought about by hard infrastructure (transport, energy, utilities, telecommunication, and social infrastructures); (2) *Kabuhayan* (Livelihood) which refers to the financial, physical, human, social and natural capital, as well as economic, food, and health security; (3) *Kapayapaan* (Peace) which refers to the level of local security; (4) *Kaayusan* (Order) which refers to infrastructural services and urban planning, policies and regulations, stewardship, aesthetic value, historicity, financing measures, road infrastructure, traffic management, disturbance, imageability, habitat quality, urban structure, air quality, waste generation management, landscape character, resource use, coherence, visual scale, and urban design; (5) *Kalinisan* (Cleanliness) which refers to the global climate indicator, acidification indicator, ecosystem-toxification indicator, urban mobility, energy consumption, water consumption, nuisance, social justice, housing quality, urban safety, and green public space; (6) *Kapaligiran* (Environment) which refers to the natural resources; (7) *Kalusugan* (Health) which refers to health infrastructure status and overall human well-being; (8) *Karunungan* (Education) which refers to globalization and liberalization, future employability, value and norm level, quality of teachers, programme design, enrolees, collaboration with local agencies, sustaining involvement with the community, support for training to academic staff, supporting climate, physical level, university environmental/sustainability policy, school strategic plan, school teaching and learning strategy, students' needs, programme evaluation, accountability, educational system as a whole, shortages, schooling process, and community assessment; (9) *Kalinangan* (Culture) which refers to heritage conservation, community involvement and participation, research projects in cultural organizations, cultural tourism, cultural labour market, public–private cooperation, entrepreneurship, accessibility of arts facilities, number of cultural sites and monuments, cultural vitality, indigenous knowledge and traditional practices, community cultural development, arts, education and youth, inclusion of heritage assets in plans by other agencies, cultural policy, and local government; (10) *Kabataan* (Youth) which refers to contextual indicators such as education and training, employment and entrepreneurship, health and well-being, social inclusion, participation in political community, voluntary activities of the population aged 25 years or younger; (11) *Kalinga* (Elderly) which refers to poverty rate at old age, welfare, life expectancy, healthy life expectancy, safety, relative psychological well-being, presence of hospital, primary care, etc., employment, education, social connections, transportation, walkable neighbourhood for senior citizens (or the sector of the population aged 60 years or older); and (12) *Kaisahan* (Unity) which refers to gender equality, cultural development interventions, female labour force, consensus, unifying concept, rural urban infrastructure, decentralization, government, urban development policies, protection, governance and redevelopment.

Sub-indicators for each KE were also derived according to pillars of urbanization namely, social, economic, cultural, political, and environmental (Table 2). The social aspect refers to relationships, interaction with the community, linkage, involvement, and a sense of belongingness. Economic refers to financial resources, benefits, and growth. Environmental refers to natural settings. Culture refers to arts, language, religion, beliefs, values, tradition, and way of life. The political aspect refers to government, laws, policies, and regulations.

32.4.2 *Micro-Case Studies on Selected KEs*

In this section, the authors present three micro-case studies of selected KEs to demonstrate the practical application of assessing the attempts of both the communities and local governments, albeit to limited coverage, in attaining a sustainable urban-renewed environment for Metro Manila.

32.4.2.1 Kalinisan and Food Safety Practices

The pre-test area for the *Kalinisan* (Cleanliness) was Luneta Park, a historic urban park located in Ermita, Manila, that is adjacent to Intramuros. The park plays a significant role, especially during national events, presidential inaugurations, and political rallies. During regular days, the park is visited by locals and tourists making it an ideal study area to investigate *Kalinisan*. The respondents were the managers from 21 food service establishments that are located within and along the vicinity of the park. As an inclusion criterion, the business entities that employed five or more staff, excluding the manager, were selected.

This micro-case study focused on the current standard operating procedures (SOPs) among the food service establishments on food safety policies and practices such as proper personal hygiene, time–temperature abuse, cross-contamination, employee training, and facilities and equipment attempted to find out if these current SOPs were related to some company profiles such as capitalization, net income or net loss, number of employees, number of years in business, and type of business ownership. Data were collected from personal interviews with food service establishments' managers. Of the 21 food service establishments legally registered in the area, only 11 or 52% responded to the study. The responding food service establishments had a mean capitalization of Php 1.7 million, the mean score was a net loss of less than one million, which means the majority (54.6%) of them had net income above the mean. In terms of the number of employees, the mean score was 20.7273, the majority of them managed to be in the business for less than ten years and most of them were a stock corporation type of business ownership.

Our results reported that food safety practices are acceptable based on food industry standards set by the local government. Significant relationships are (a) capitalization and employee training; (b) the number of years of business operation and proper personal hygiene; (c) the number of employees and cross-contamination, time–temperature abuse; and (d) type of business ownership and employee training, cross-contamination, facilities, and equipment. The results of this micro-study were forwarded to the Intramuros Administration to assist in the monitoring and improvement of the food service establishments.

32.4.2.2 Kapaligiran and Solid Waste Management

For *Kapaligiran* (Environment), the researchers chose Navotas as a pre-test area. Navotas is known as the Commercial Fishing Hub of the Philippines. It is classified as a highly urbanized city in Metro Manila, which makes it a suitable study site for investigating the implementation of proper solid waste management. To evaluate the indicators of the *Kapaligiran*, the researchers selected ten establishments comprising offices, food chains, government agencies, schools, banks, and households. The profiling covered capitalization, net income or net loss, number of employees, years in business, and type of business ownership. The solid waste practice indicators were the collection, segregation, transport, and disposal of waste practices in conformance to a standard set by RA 9003, otherwise known as the Ecological Solid Waste Management Act of 2000, and by LGU units. The nature, volume, type, and composition of solid wastes were assessed.

The stakeholders reported good practices for solid waste disposal. Biodegradable wastes accounted for 379 kg/day or 5.05%, while non-biodegradable waste accounted for 7114 kg/day or 94.96%. The result showed no significant relationship between the practices with the economic profile.

Amidst the bleak scenario of Metro Manila's worsening garbage problem are the environmental management programmes of the City of Manila regarding *Kapaligiran* in the following areas: air pollution, solid waste pollution levels in Manila Bay, Pasig River, esteros within the City of Manila and other tributaries, monitoring of solid waste disposal of industries, commercial establishment, private and public institutions, citywide implementation of waste management programme, and citizens participation in garbage segregation. It should also be noted that Intramuros is one of the centres of tourism in the "Wow Philippines!" tourism branding programme. Environmental programmes must be instituted to sustain the goal of a cleaner Philippines from the perspective of global tourism.

32.4.2.3 Kabataan and the Contemporary Societal Interest Among the Youth

The youth play a vital role in building a strong and upright community. Malabon, Manila served as a pre-test area for *Kabataan (Youth)*. Malabon is a first-class and highly urbanized city in Metro Manila. Malabon is one of the most densely populated cities and is primarily both a residential and an industrial area. In 2015, the youth comprised 20.19% of the population of Malabon. These are individuals in the 15–24 age group as defined by the World Health Organization (WHO). The respondents of the survey were 30 respondents from the youth sector living in different *barangays* from the *Sangguniang Kabataan* (an elected youth council), fraternity groups, religious groups, and others who were not affiliated with any organizations. The profiling covered interest, frequency of time, costs, and duration of time spent on an activity. The indicators for the contemporary societal interest of youth were physical, religious, social, community involvement, and leisure.

Being aware of the present condition of Filipino youth need not go very far. In the city of Manila, there are places known for their history and era—the Intramuros. This micro-case study was exploratory in design and aimed to describe the patterns of youth activities in Intramuros. The researchers categorized the types of youth activities as physical or recreational, religious, social, and community involvement. Physical or recreational can be in the form of sports or cyber games. Social activity may also involve interest in a cyber social network. Partially, the study attempted to gauge how many percentages of youth in Intramuros were interested in a computer-related type of activity. The study reported that the youth was not inclined towards different societal interests. Significant relationships were (a) religious activities and gender and (b) the fathers' educational background and parents' income relation to their social activities.

32.5 A Proposal for a Future Design Framework for Metro Manila

The future design framework involves the use of technology to readily assess the condition of an area towards a sustainable and urban-renewed environment. Digitization, mapping using GIS, the use of AI for big data analytics and machine learning, and modelling and simulation are some of the current technologies that can be adopted for this design framework.

Table 32.2 shows the measures per indicator per category of the KE *Kalinangan*. Taking the sub-indicator "Cultural heritage preservation programmes" under Heritage Conservation (see Table 32.3), a web-based mapping system for data collection can be developed for each cultural organization by inputting the data from the sub-indicators. These data are, however, intractably scattered in the knowledge management offices of various agencies. We view that by consolidating their profiles, programmes, and assessment data in this proposed framework and by tapping advanced analytics, such as AI, the local and national governments can employ data-driven, efficient, and more cost-effective programmes for the communities. Further, mathematical and statistical algorithms can be embedded in the system to determine the relationship among variables. The data can further be mapped. Visualizations can be shown as heat maps.

32.6 Concluding Remarks

This chapter expounds on a future design framework for a sustainable urban-renewed environment incorporating hard and soft infrastructures as indicators. The 12 KEs were identified using a modified Delphi approach to prioritize specific measures or indices that can assist in guiding the urban renewal of Metro Manila. These KEs

Table 32.2 Sample Indicators and Measures for the Future Design Framework per Category of the KE *Kalinangan*

Category	Indicator	Measure/Sub-indicator
Cultural Economic	Heritage conservation	• Cultural heritage preservation programmes • Evaluation of documentation programmes • Evaluation of preservation programmes • Reuse, redevelopment, and regeneration of heritage objects
	Community involvement and participation	• Engagement of the general population in ethnic culture • Attendance in ethnic and cultural events • Percentage of the population who support cultural diversity in arts • Number of events representing diverse cultural tradition • Prevalence of outreach policies among arts organizations • Audience composition in terms of diversity • The cultural diversity of festivals • Cultural organizations with diverse cultural traditions
	Research projects in cultural organizations	• Number of research projects in cultural organizations • Percentage of arts organizations creating new works • Community satisfaction with a range of city events and festivals
	Cultural tourism	• Number of visits to arts and culture facilities • Spending on arts and culture by tourists • Contribution of cultural tourism to the economy • Improvement in business performance due to engagement with arts and culture • Value of cultural tourism • Value-added contributions by cultural industries
	Cultural labour market	• Employment opportunities in arts and heritage • Presence of diverse labour market (artists) • Number of creative jobs • Workers in creative jobs by age, sex, ethnic background, qualification, and region • Wage levels for artistic work
	Public–private cooperation	• Financing innovation • Availability and level of funding for community capacity building by a cultural organization • Capital projects for the regeneration of urban areas

(continued)

Table 32.2 (continued)

Category	Indicator	Measure/Sub-indicator
	Entrepreneurship	• Number of business establishments
Environmental	Accessibility of art facilities	• Appropriate public transit for cultural activities • Percentage of artists who think the cultural zone is a good place to live
	Number of cultural sites and monuments	• Number of cultural facilities in compliance with standards • Number of listed heritage buildings • Number of heritage buildings demolished • Number of new buildings meeting architectural standards • Number of newly built projects
	Impact of the natural and built environment in inspiring art and culture	• Community perceptions regarding local places of significance
Social	Indigenous knowledge and traditional practices	• Recovery and protection of cultural health and the culture of indigenous knowledge in society
	Community cultural development	• Community engagement through arts and cultural activities • Ownership over own community resources and identity
	Arts, education, and youth	• Number of students who visit arts and culture facilities • Amount of cultural content in schools • Label of cultural organization involved in education • Cultural programmes contribute to health, education, and disease prevention • Percentage of organizations offering participation opportunities for adult • Percentage of arts organizations offering art education and youth outreach
	Quality of life	• Area of natural green space available for community use • Percentage living below the poverty line • Well-being of locals

(continued)

Table 32.2 (continued)

Category	Indicator	Measure/Sub-indicator
Political	Inclusion of heritage assets in plans by other agencies	• Integration of existing community building and development plans
	Cultural policy and local government	• Existence of policy and plans • Evaluation of the success of plans implemented • Cultural life community is considered part of the cultural strategy • Availability of leadership support and development • Local authority investment in arts and culture

Table 32.3 The utilization of different technologies in assessing the sub-indicator "Cultural heritage preservation programmes"

Objective	Proposed Auxiliary Method
1. Profiles of cultural organizations	Digitization and GIS
2. Cultural preservation programmes implemented by the organizations	Digitization and GIS
3. Level of the effectiveness of the preservation programmes in terms of (a) Level of visibility and awareness (b) Degree of deterioration and weathering	AI
4. A significant relationship between the level of effectiveness of the programmes and the profile of the organization	AI

are (1) *Kaunlaran* (Progress and Development), (2) *Kabuhayan* (Livelihood), (3) *Kapayapaan* (Peace), (4) *Kaayusan* (Order), (5) *Kalinisan* (Cleanliness), (6) *Kapaligiran* (Environment), (7) *Kalusugan* (Health), (8) *Karunungan* (Education), (9) *Kultura* (Culture), (10) *Kabataan* (Youth), (11) *Kalinga* (Elderly), and (12) *Kaisahan* (Unity). To help improve the granularity of describing the current state of the city, sub-indicators were further established per KE grounded on the five pillars: political, cultural, economic, social, and environmental. These indexes were able to comprehensively evaluate the present status of a certain city in terms of the defined measures of a sustainable and urban-renewed environment. The KE and their sub-indicators served as bases for utilizing technologies for "future design framework" to enhance the assessment and decision-making process for urban planning. The study however was limited to quantitative and case study approaches limited to specific areas of Manila, it is recommended that the proposed future design framework in the paper be empirically tested and applied to urban renewal assessment of Metro Manila as well as other megacities in the world.

Acknowledgements Dr. Ria Liza Canlas has been the Project Leader of the Research for an Urban-renewed Environment. Phase 1 and Phase 2 were possible because of the contributions from

Devy Galang, Maricel Abellera, Roque Gajo, Jesusa Manaloto, Marlon Fortunato, Sylvia Abracosa, Lilibeth Aragon, Arlene Caballero, Rovena Dellova, Dolores Montesines, Zenaida Jesalva, Sheila Dawang, Margaret Moreno, Dahlia Tanquezon, Christopher Cunanan, and Lyceum of the Philippines University, Intramuros, Manila.

Phase 3 has been a contribution from a Grant-in-aid Project funded by the Commission of Higher Education (CHED) of the Philippines with Project Leaders Dr. Ria Liza Canlas of National University - Philippines and Dr. Nuna Almanzor of Adamson University. The co-investigators are Leonila Valdez of First City Providential College, Ma. Artes Deceña Cabiles of Polytechnic University of the Philippines, Rita Aranda of National University-Philippines, Delia Myrna Fainsan of Lyceum of the Philippines University, Mark Lawrence Gale of Adamson University, Irene Cruz of Lyceum of the Philippines University, Judith Principe, and Gladys Lorenzo.

We also acknowledge the support given by John Gabriel Decena and Frederick Carlo Diaz in revising the paper.

The authors would like to express their gratitude to all individuals and institutions abovementioned.

References

Adams, S. J. (2001). Projecting the next decade in safety management. computer science, 3(4.00/4.00), 0–8303.

Burnham, D. H., Bennett, E. H., & Bennett, E. H. (1909). Plan of Chicago: Prepared under the direction of the Commercial Club during the years MCMVI, MCMVII, and MCMVIII: Commercial Club.

Canlas, R. L., Borja, M. R., Macapagal, L. D., Hinojales, E. (2013). Determinants of the best practices on Practicum Programs in engineering courses of selected Higher Education Institutes in the Philippines. Proceedings of 2013 IEEE Inter-national Conference on Teaching, Assessment and Learning for Engineering (TALE). https://doi.org/10.1109/tale.2013.6654513

Cooper, D. R., & Schindler, P. S. (2003). *Business Research Methods* (8th ed., p. 19). McGraw-Hill Irwin.

Custer, R. L., Scarcella, J. A., & Stewart, B. R. (1999). The modified Delphi technique: A rotational modification. *Journal of Vocational and Technical Education, 15*(2), 1–10.

Dalkey, N. C., & Helmer, O. (1963). An experimental application of the Delphi method to the use of experts. *Management Science, 9*(3), 458–467.

Dalkey, N. C. (1972). The Delphi method: an experimental application of group opinion. In N. C. Dalkey, D. L. Rourke, R. Lewis, & D. Snyder (Eds.) Studies in the quality of life. Lexington, MA: Lexington Books

Delbecq, A. L., Van de Ven, A. H., & Gustafson, D. H. (1975). Group techniques for program planning. Glenview, IL: Scott, Foresman, and Co.

Deosaran, R. (2002) Community Policing in the Caribbean Context: Community and Police Capability. Policing: An Inter-national Journal of Police Strategies and Management 25 (1)

Estoque, R. C. (2017). Manila metropolitan area Urban Development in Asia and Africa (pp. 85–110): Springer.

Eubank, B. H., Mohtadi, N. G., Lafave, M. R., Wiley, J. P., Bois, A. J., Boorman, R. S., & Sheps, D. M. (2016). Using the modified Delphi method to establish clinical consensus for the diagnosis and treatment of patients with rotator cuff patholo-gy. *BMC Medical Research Methodology, 16*(1), 1–15.

Freestone, R. (2019). City Beautiful Movement. The Wiley Blackwell Encyclopedia of Urban and Regional Studies, 1–6.

Gallego, D., & Bueno, S. (2014). Exploring the application of the Delphi method as a forecasting tool in Information Sys-tems and Technologies research. *Technology Analysis & Strategic Management, 26*(9), 987–999.

Gamboa, M. A. M., Rivera, R. R. B., & Reyes, M. R. D. (2019). City Profile: Manila. *Philippines. Environment and Urbani-Zation ASIA, 10*(2), 331–358. https://doi.org/10.1177/0975425319859149

Gomà, D. (2012). Control, espaciourbano e identidaden la Filipinas colonial española: elcaso de intramuros, Manila (siglos XVI–XVII). Scripta Nova. RevistaElectrónica de Geografía e CienciasSociales, 16(418), 19.

Graefe, A., & Armstrong, J. S. (2011). Comparing face-to-face meetings, nominal groups, Delphi, and prediction markets on an estimation task. *International Journal of Forecasting, 27*(1), 183–195.

Gustafson, D. H., Shukla, R. K., Delbecq, A., & Walster, G. W. (1973). A comparative study of differences in subjective likelihood estimates made by individuals, interacting groups, Delphi groups, and nominal groups. *Organizational Behavior and Human Performance, 9*(2), 280–291.

Hines, T. S. (1972). The imperial façade: Daniel H. Burnham and American architectural planning in the Philippines. The Pacific Historical Review, 33–53.

Hines, T. S. (2009). *Burnham of Chicago: Architect and Planner*. The University of Chicago Press.

Hsu, C. C., & Sandford, B. A. (2007). The Delphi technique: making sense of consensus. Pract assess res. Evaluation, 12(10).

Jones, J., & Hunter, D. (1995). Consensus methods for medical and health services research. BMJ: British Medical Journal, 311(7001), 376.

Judd, R. C. (1972). *Forecasting to Consensus Gathering*. College and University Busi-ness.

Kauko, K., & Palmroos, P. (2014). The Delphi method in forecasting financial markets—An experimental study. *Internation-Al Journal of Forecasting, 30*(2), 313–327.

Kerlinger, F. N. (1973). *Foundations of behavioral research* (p. 20). Holt, Rinehart, and Winston, Inc.

Ludwig, B. G. (1994). Internationalizing Extension: An exploration of the characteristics evident in a state university Exten-sion system that achieves internationalization. Unpublished doctoral dissertation, The Ohio State University, Columbus.

Mason, R., & Istvandity, L. (2018). Intramuros: Memory, violence and national becoming in Manila. *International Journal of Heritage Studies, 24*(10), 1053–1067. https://doi.org/10.1080/13527258.2018.1475405

Mathias, F. F. (1996). The Battle for Manila: The Most Devastating Untold Story of World War II: JSTOR.

Mayuga, J. L. (2021, January 10). The garbage conundrum. Retrieved May 5, 2022, from https://www.denr.gov.ph/images/DENR_News_Alerts/DENR_News_Alerts_10_January_2021_Sunday.pdf

McCoy, A. W. (1985). Philippine cartoons: Political caricature of the American era, 1900–1941: Vera-Reyes.

Philippine Statistics Authority. (2021, August 23). Retrieved January 23, 2022, from https://psa.gov.ph/content/highlights-national-capital-region-ncr-population-2020-census-population-and-housing-2020

Sandrey, M. A., & Bulger, S. M. (2008). The Delphi method: an approach for facilitating evidence-based practice in athletic training. Athletic Training Education Journal, 3(4), 135–142.

Schurman, J. G., Dewey, G., Otis, E. S., Denby, C., & Worcester, D. C. (1900). Report of the Philippine Commission to the President January 31, 1900 [-December 20, 1900]: US Government Printing Office.

Shatkin, G. (2004). Planning to forget: Informal settlements as' forgotten places' in globalizing Metro Manila. *Urban Studies, 41*(12), 2469–2484.

Singg, R. N., & Webbe, B. R. (1979). Use of Delphi Methodology to Assess Goals and Social Impacts of a Watershed Project. *Water Resources Bulletin, American Water Resources Association, 15*(2), 136–143.

Steinberg, F. (2008). Revitalization of historic inner-city areas in Asia: The potential for urban renewal in Ha Noi, Jakarta, and Manila: Asian Development Bank.
Swain, D., & Hollar, D. (2003). Measuring progress: Community indicators and the quality of life. *International Journal of Public Administration, 26*(7), 789–814. https://doi.org/10.1081/pad-120019247
The Rockefeller Foundation's Informal City Dialogues. (n.d.). Manila Philippines. Next City. Retrieved May 2, 2022, from https://nextcity.org/informalcity/city/manila
Tomeldan, M. V., Antonio, M., Arcenas, J., Beltran, K. M., & Cacalda, P. A. (2014). "Shared Growth" Urban Renewal Ini-tiatives in Makati City, Metro Manila. *Philippines. Journal of Urban Management, 3*(1), 45–65. https://doi.org/10.1016/S2226-5856(18)30083-9
Turalba, M. (2008). Revitalization of HIstoric Inner-City Areas in Asia: Manila Case Study. Asian Development Bank. Retrieved from http://www.adb.org/Documents/Reports/revitalization-inner-city/appendix3.pdf
Verdejo, J. J., Cabeza-Lainez, J., Pulido-Arcas, J., & Rubio-Bellido, C. (2014). Spanish fortifications in Asia: A case study of Intramuros district in Manila–current situation and future prospects. *WIT Transactions on the Built Environment, 143*, 387–398.

Index

B. R. K. Sinha (ed.), *Urban Dynamics, Environment and Health*,
https://doi.org/10.1007/978-981-99-5744-6

C

D

F

G

J

K

L

M

T

V